SpringerWienNewYork

Yuriy A. Knirel • Miguel A. Valvano
Editors

Bacterial Lipopolysaccharides

Structure, Chemical Synthesis, Biogenesis and Interaction with Host Cells

SpringerWienNewYork

Yuriy A. Knirel
N.D. Zelinsky Institute of
Organic Chemistry
Russian Academy of Sciences
Leninsky Prospekt 47
119991 Moscow, V-334
Russia
yknirel@gmail.com

Miguel A. Valvano
Centre for Human Immunology and
Department of Microbiology and Immunology
University of Western Ontario
London, ON N6A 5C1
Canada
mvalvano@uwo.ca

SpringerWienNewYork is a part of Springer Science+Business Media
springer.at

Cover design: WMXDesign GmbH, Heidelberg, Germany
Typesetting: SPi, Pondicherry, India

Printed on acid-free and chlorine-free bleached paper
SPIN: 12599509

With 65 Figures

Library of Congress Control Number: 2011932724

ISBN 978-3-7091-0732-4 e-ISBN 978-3-7091-0733-1
DOI 10.1007/978-3-7091-0733-1
SpringerWienNewYork

Preface

The lipopolysaccharide (LPS) is the major component of the outer leaflet of the outer membrane of Gram-negative bacteria. It contributes essentially to the integrity and stability of the outer membrane, represents an effective permeability barrier towards external stress factors, and is thus indispensable for the viability of bacteria in various niches, including animal and plant environment. On the other hand, the presence of the LPS on the cell surface is beneficial for the host as it serves as a pathogen-associated molecular pattern recognized by, and thus activates, the host immune system resulting normally in elimination of the pathogen. Being unable to get rid of the LPS, bacteria evolved various mechanisms for LPS structure modification to make them invisible for the immune system and resistant to defense factors such as complement and antibiotics. This highlights the LPS as the most variable cell wall constituent.

Since its discovery in the late 19th century the LPS, then named endotoxin, has attracted the curiosity of many researchers virtually in all fields of life science such as medicine, microbiology, pharmacology, chemistry, biochemistry, biophysics, immunology, cell biology, and genetics. Attesting this in part, more than 71,000 and 79,000 publications are cited in PubMed at the beginning of 2011 using LPS and endotoxin as queries, respectively. LPS has also attracted interest in biotechnology and the pharmacological industry for the development of diagnostic and therapeutic methods and reagents.

Early in the history of endotoxin, it was appreciated by Peter L. Panum in 1874 that putrid fluids contained a water-soluble, alcohol-insoluble, heat-resistant, non-volatile substance, which was lethal to dogs. Later, Richard Pfeiffer, a disciple of Robert Koch, showed that *Vibrio cholerae*, the cause of cholera, produced a heat-stable toxic substance that was associated with the insoluble part of the bacterial cell, coining the name "endotoxin" (from the Greek 'endo' meaning 'within'). Through pioneer discoveries by Otto Westphal, Otto Lüderitz, Hiroshi Nikaido and Mary J. Osborn in the mid 1950s, we learned that the endotoxin corresponds to the LPS. Efficient purification protocols of the LPS were elaborated and principles of its structural organization, genetics and biochemistry were then established. These early studies propelled a long and productive road of chemical and biochemical research to reveal the details of structure and biosynthesis of each of the components of the LPS molecule. In parallel a large body of work resulted in the biological characterization of the LPS in terms of its function as a potent elicitor of innate immune responses. This work culminated with the discovery by Bruce Beutler of the mouse gene encoding the TLR4 receptor molecule and the subsequent elucidation of the structural basis of the activation of the immune system by the LPS.

The purpose of this book is not to provide a comprehensive examination of all aspects related to the LPS but rather to give an up do date overview of research that applies to its chemistry, biosynthesis, genetics, and activities toward eukaryotic cells from structural and mechanistic perspectives.

Yuriy A. Knirel
Miguel A. Valvano

Contents

1 **Lipid A Structure** 1
Alba Silipo and Antonio Molinaro

2 **Structure of the Lipopolysaccharide Core Region** 21
Otto Holst

3 **Structure of O-Antigens** 41
Yuriy A. Knirel

4 **Chemical Synthesis of Lipid A and Analogues** 117
Shoichi Kusumoto

5 **Chemical Synthesis of Lipopolysaccharide Core** 131
Paul Kosma and Alla Zamyatina

6 **Genetics and Biosynthesis of Lipid A** 163
Christopher M. Stead, Aaron C. Pride, and M. Stephen Trent

7 **Pathways for the Biosynthesis of NDP Sugars** 195
Youai Hao and Joseph S. Lam

8 **Lipopolysaccharide Core Oligosaccharide Biosynthesis and Assembly** 237
Uwe Mamat, Mikael Skurnik, and José Antonio Bengoechea

9 **Genetics, Biosynthesis and Assembly of O-Antigen** 275
Miguel A. Valvano, Sarah E. Furlong, and Kinnari B. Patel

10 **Lipopolysaccharide Export to the Outer Membrane** 311
Paola Sperandeo, Gianni Dehò, and Alessandra Polissi

11 **Evolution of Lipopolysaccharide Biosynthesis Genes** 339
Monica M. Cunneen and Peter R. Reeves

12 **The Molecular Basis of Lipid A and Toll-Like Receptor 4 Interactions** 371
Georgina L. Hold and Clare E. Bryant

13 Modulation of Lipopolysaccharide Signalling Through TLR4 Agonists and Antagonists 389
Francesco Peri, Matteo Piazza, Valentina Calabrese, and Roberto Cighetti

14 Lipopolysaccharide and Its Interactions with Plants 417
Gitte Erbs and Mari-Anne Newman

Index 435

Contributors

José Antonio Bengoechea Laboratory Microbial Pathogenesis, Consejo Superior Investigaciones Científicas, Fundación de Investigación Sanitaria Illes Balears, Recinto Hospital Joan March, Carretera Sóller Km12; 07110, Bunyola, Spain, bengoechea@caubet-cimera.es

Clare E. Bryant Department of Veterinary Medicine, University of Cambridge, Madingley Road, Cambridge, UK CB3 0ES, ceb27@cam.ac.uk

Valentina Calabrese Dipartimento di Biotecnologie e Bioscienze, Università di Milano-Bicocca, Piazza della Scienza 2, 20126 Milan, Italy, valentina.calabrese@unimib.it

Roberto Cighetti Dipartimento di Biotecnologie e Bioscienze, Università di Milano-Bicocca, Piazza della Scienza 2, 20126 Milan, Italy, cighetti.roberto@hotmail.it

Monica M. Cunneen Division of Microbiology, School of Molecular and Microbial Biosciences, University of Sydney, Sydney, NSW 2006, Australia, monica.cunneen@sydney.edu.au

Gianni Dehò Dipartimento di Scienze biomolecolari e Biotecnologie, Università di Milano, Via Celoria 26, 20133 Milan, Italy, gianni.deho@unimi.it

Gitte Erbs Department of Plant Biology and Biotechnology, University of Copenhagen, Thorvaldsensvej 40, 1871 Frederiksberg, Denmark, ger@life.ku.dk

Sarah E. Furlong Centre for Human Immunology and Department of Microbiology and Immunology, University of Western Ontario, London, Ontario, Canada, N6A 5C1, sfurlon@uwo.ca

Youai Hao Department of Molecular and Cellular Biology, University of Guelph, 50 Stone Road E., Guelph, Canada, ON, N1G 2W1, haoy@uoguelph.ca

Georgina L. Hold Division of Applied Medicine, University of Aberdeen, Institute of Medical Sciences, Foresterhill, Aberdeen, UK AB25 2ZD, g.l.hold@abdn.ac.uk

Otto Holst Division of Structural Biochemistry, Research Center Borstel, Leibniz-Center for Medicine and Biosciences, Parkallee 4a/c, D-23845 Borstel, Germany, oholst@fz-borstel.de

Yuriy A. Knirel N.D. Zelinsky Institute of Organic Chemistry, Russian Academy of Sciences, Leninsky Prospekt 47, 119991 Moscow, V-334, Russia, yknirel@gmail.com

Paul Kosma Department of Chemistry, University of Natural Resources and Life Sciences, Muthgasse 18, A-1190 Vienna, Austria, paul.kosma@boku.ac.at

Shoichi Kusumoto Suntory Institute for Bioorganic Research, Wakayamadai 1–1–1, Shimamoto-cho, Mishima-gun, Osaka 618–8503, Japan, skus@sunbor.or.jp

Joseph S. Lam Department of Molecular and Cellular Biology, University of Guelph, 50 Stone Road E., Guelph, Canada ON, N1G 2W1, jlam@uoguelph.ca

Uwe Mamat Division of Structural Biochemistry, Research Center Borstel, Leibniz-Center for Medicine and Biosciences, Parkallee 4a/4c, D-23845 Borstel, Germany, umamat@fz-borstel.de

Antonio Molinaro Dipartimento di Chimica Organica e Biochimica, Università di Napoli Federico II, Via Cintia 4, 80126 Napoli, Italy, molinaro@unina.it

Mari-Anne Newman Department of Plant Biology and Biotechnology, University of Copenhagen, Thorvaldsensvej 40, 1871 Frederiksberg, Denmark, mari@life.ku.dk

Kinnari B. Patel Centre for Human Immunology and Department of Microbiology and Immunology, University of Western Ontario, London, Ontario, Canada, N6A 5C1, kpatel59@uwo.ca

Francesco Peri Dipartimento di Biotecnologie e Bioscienze, Università di Milano-Bicocca, Piazza della Scienza 2, 20126 Milan, Italy, francesco.peri@unimib.it

Matteo Piazza Dipartimento di Biotecnologie e Bioscienze, Università di Milano-Bicocca, Piazza della Scienza 2, 20126 Milan, Italy, matteo.piazza1@unimib.it

Alessandra Polissi Dipartimento di Biotecnologie e Bioscienze, Università di Milano-Bicocca, Piazza della Scienza 2, 20126 Milan, Italy, alessandra.polissi@unimib.it

Aaron C. Pride Institute of Cellular and Molecular Biology, University of Texas at Austin, Austin, TX 78712, USA, acpride@mail.utexas.edu

Peter R. Reeves Department of Microbiology, School of Molecular and Microbial Biosciences, University of Sydney, Sydney, New South Wales 2006, Australia, peter.reeves@sydney.edu.au

Alba Silipo Dipartimento di Chimica Organica e Biochimica, Universitá di Napoli Federico II, Via Cintia 4, 80126 Napoli, Italy, silipo@unina.it

Mikael Skurnik Department of Bacteriology and Immunology, Haartman Institute, University of Helsinki, P.O. Box 21, Haartmaninkatu 3, FIN-00014 Helsinki, Finland, mikael.skurnik@helsinki.fi

Paola Sperandeo Dipartimento di Biotecnologie e Bioscienze, Università di Milano-Bicocca, Piazza della Scienza 2, 20126 Milan, Italy, paola.sperandeo@unimib.it

Christopher M. Stead Georgia Health Sciences University, Department of Biochemistry and Molecular Biology, Augusta, GA 30912, USA, cstead@georgiahealth.edu

M. Stephen Trent Section of Molecular Genetics and Microbiology and Institute of Cellular and Molecular Biology, University of Texas at Austin, Austin, TX 78712, USA, strent@mail.utexas.edu

Miguel A. Valvano Centre for Human Immunology and Department of Microbiology and Immunology, University of Western Ontario, London, Canada ON N6A 5C1, mvalvano@uwo.ca

Alla Zamyatina Department of Chemistry, University of Natural Resources and Life Sciences, Muthgasse 18, A-1190 Vienna, Austria, alla.zamyatina@boku.ac.at

Lipid A Structure

1

Alba Silipo and Antonio Molinaro

1.1 Introduction

Bacteria and Archaea account for the largest amount of biomass on earth and are major reservoirs of essential nutrients and energy. They have a simpler internal cell structure than eukaryotic cells, and in most cases they lack membrane-enclosed organelles. Archaea includes extremophilic prokaryotic organisms living in habitats that are unusual for most other organisms, such as high salinity or pressure, extreme temperatures and critic pH. Bacteria include saprophytic and pathogenic species. They are divided into Gram-negative and Gram-positive bacteria based on the Gram stain, which reflects differences in the cell envelope architecture.

Gram-positive and -negative bacteria possess a cytoplasmic membrane made of a phospholipid bilayer, which surrounds the cytosol and provides a physical, semi-permeable barrier regulating the movement of molecules in and out the cell. The cell wall peptidoglycan or murein, a rigid layer that confers shape and osmotic strength to the bacterial cell, encloses the cytoplasmic membrane. Peptidoglycan is a polymeric mesh formed by carbohydrate backbone chains of *N*-acetylglucosamine and *N*-acetylmuramic acid that are cross-linked by penta-peptide chains. In Gram-positive bacteria, the peptidoglycan layer is thick and constitutes the external portion of the cell wall. In Gram-negative bacteria, there is a thin layer of peptidoglycan surrounded by the outer membrane (OM).

The OM is a unique asymmetric phospholipid bilayer. The inner leaflet consists of glycerophospholipids while the external leaflet is rich in lipopolysaccharide (LPS), which covers up to 75% of the cell surface. Embedded in the OM there are also integral membrane proteins like porins, which serve as channels for the passage of small hydrophilic molecules, and lipoproteins [1, 2].

A. Silipo • A. Molinaro (✉)
Dipartimento di Chimica Organica e Biochimica, Università di Napoli Federico II, Via Cintia 4, 80126 Napoli, Italy
e-mail: silipo@unina.it; molinaro@unina.it

Y.A. Knirel and M.A. Valvano (eds.), *Bacterial Lipopolysaccharides*,
DOI 10.1007/978-3-7091-0733-1_1, © Springer-Verlag/Wien 2011

The LPS is indispensable for viability and survival of Gram-negative bacteria, as it contributes to the correct assembly of the OM. LPS is a heat-stable complex of amphiphilic macromolecules that provides an extraordinary permeability barrier to many different classes of molecules including detergents, antibiotics, and toxic dyes and metals. The barrier properties of the OM depend on its low fluidity, which is due to the highly ordered structure of the LPS monolayer. Owing to their external location, LPS molecules interact with other biological systems by participating in host-bacterium interactions like adhesion, colonization, virulence, and symbiosis. LPS, also called endotoxin, is a potent elicitor of innate immune responses and plays a key role in the pathogenesis of Gram-negative infections in both plant and animal hosts [3].

In most bacteria, LPS displays a common structural architecture that includes three domains: a lipophilic moiety termed lipid A, a hydrophilic glycan called the O-specific polysaccharide (also known as O-chain or O-antigen), and a joining core oligosaccharide (OS) (Fig. 1.1). The core OS can be further separated into two regions, one proximal to lipid A (inner core OS), and another one distal from lipid A but proximal to the O-antigen (outer core OS). The inner core OS contains at least one residue of 3-deoxy-D-*manno*-oct-2-ulosonic acid (Kdo), and several heptoses. Kdo is rarely found in other glycans and therefore can be considered as a marker for the presence of LPS. The inner core OS can often be decorated with other substituents, usually present in non-stoichiometric amounts. These are phosphate (P), diphosphate, 2-aminoethyl phosphate (PEtN) or 2-aminoethyl diphosphate, uronic acids as D-GalA, and 4-amino-4-deoxy-L-arabinose (L-Ara4N) (for core and O-antigen structures see Chaps. 2 and 3). Whereas the carbohydrate chain is oriented outwards, lipid A is embedded in the outer leaflet of the OM and anchors the LPS molecules through electrostatic and hydrophobic interactions [2, 4]. The complete LPS comprising all three regions is termed "smooth" (S) LPS, while LPS lacking the O-chain and/or portions of core OS the LPS is called "rough" (R) LPS.

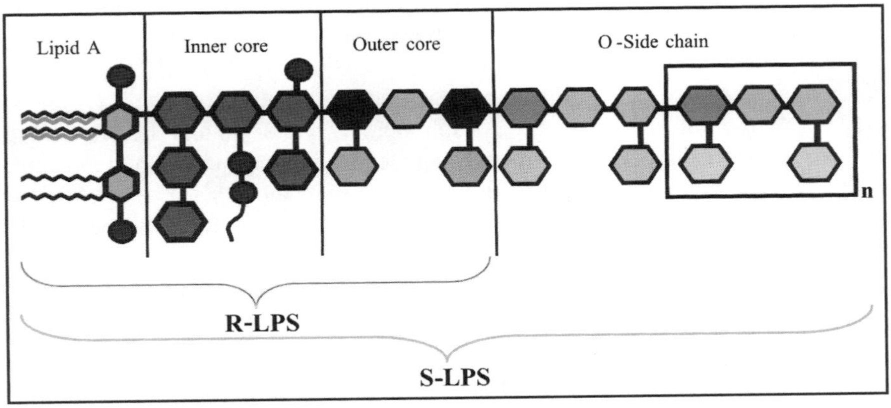

Fig. 1.1 General structure of the lipopolysaccharide of Gram-negative bacteria

Lipid A is essential for bacterial viability and carries the endotoxic properties of the LPS [4–6]. As such, lipid A acts as a potent stimulator of the innate immune system via recognition by the toll-like receptor TLR4 (see Chap. 12), often causing a wide variety of biological effects that range from a significant enhancement of the resistance to the infection to an uncontrolled and massive immune response resulting in sepsis and septic shock. The bioactivity of lipid A, including the capacity to interact and activate receptors of the immune system, is strongly influenced by its primary structure.

The first complete lipid A structure was elucidated in 1954 by Westphal and Lüderitz and finally established in detail in 1983 [7, 8]. Since that time, an increasing number of novel lipid A variants have been isolated and structurally elucidated in many bacteria, mainly due to major improvements in the procedures for their extraction and purification and dramatic advances in methods and instrumentation for structural analysis. This chapter discusses structures of the most biologically significant lipid A species published recently, including all structures reported after 2009. Distinct sections are devoted to structural analysis and supramolecular structure of lipid A. Additional data on lipid A structures are available or cited in Refs. [4–6, 9].

1.2 General Aspects of Lipid A Structure

Lipid A is the most conserved portion of the LPS. In most bacteria studied to date lipid A has a β-(1 → 6)-linked D-glucosamine disaccharide backbone. The backbone is phosphorylated at positions 1 of the proximal α-GlcN (GlcN I) and 4′ of the distal β-GlcN (GlcN II), and acylated with 3-hydroxy fatty acids at positions 2 and 3 of both GlcN residues by amide and ester linkages. The acyl groups that are directly linked to the sugar backbone are defined as primary. Some of the primary fatty acids are further acylated at the hydroxy groups by secondary acyl chains. One or both GlcN residues may be replaced with 2,3-diamino-2,3-dideoxy-D-glucopyranose (GlcN3N) residues. The phosphate groups can be substituted by polar groups or replaced with another acid, or one of the phosphate groups may be absent. The first monosaccharide of the core, a Kdo residue or its 3-hydroxylated derivative, is linked at position 6′ of GlcN II [4].

Despite its general structural conservation, lipid A also has considerable structural microheterogeneity. Therefore, it is more appropriate to consider lipid A as a family of structurally related molecular species with different acylation and phosphorylation patterns rather than as an individual, homogeneous molecule. Microheterogeneity depends on various factors including bacterial adaptation to changing environment and external stimuli, incomplete biosynthesis, and breakdown products and/or chemical modifications resulting from the procedures used for lipid A isolation.

The first structurally elucidated lipid A was from *Escherichia coli*, and consists of a P-4-β-D-Glc*p*N-(1 → 6)-α-D-Glc*p*N-1 → P backbone N-acylated at positions 2 and 2′ and O-acylated at positions 3 and 3′ of both GlcN residues, with (*R*)-3-hydroxymyristoyl groups (3-OH-14:0) as primary fatty acids

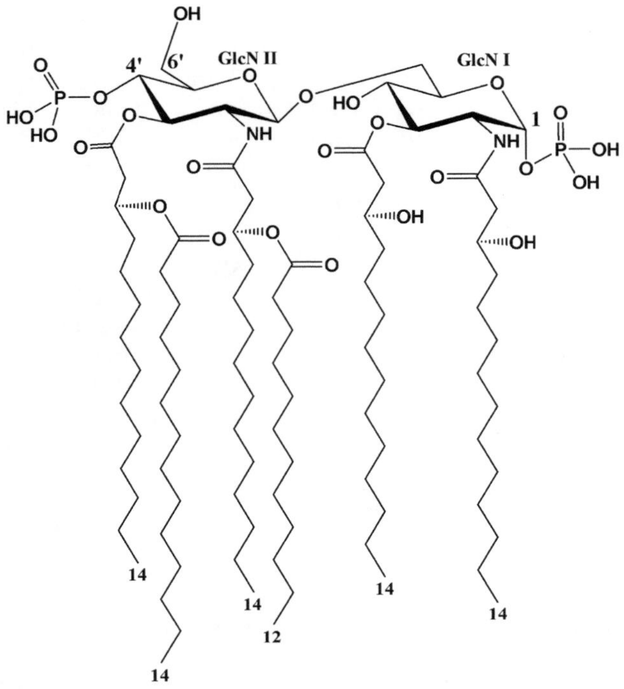

Fig. 1.2 Structure of lipid A of *E. coli*

(Fig. 1.2). Both primary acyl groups attached to GlcN II are esterified at their 3-hydroxy group with two secondary fatty acids: the amide-linked acid bears a lauroyl group (12:0) and the ester-linked acid a myristoyl group (14:0).

Concerning biological activity, the bisphosphorylated hexaacylated disaccharide lipid A with an asymmetric (4 + 2) distribution of the acyl groups represents the most active agonistic structure for LPS-responsive human cells [1–6]. Lipid A with a moderate agonistic activity, such as hexaacylated monophosphorylated lipid A from *Salmonella*, has adjuvant properties [10–12]. In contrast, the classical antagonistic structure for the human immune system is a tetraacyl biosynthetic precursor from *E. coli* named lipid IV_A. E5531, a synthetic compound with ether-bound acyl residues at positions 3 and 3′ and a methyl group at position 6′, exhibits strong antagonistic activity in vivo and in vitro [13] (see Chap. 13).

1.3 Diversity of Lipid A Structures

Lipid A structures are diverse among different genera, and sometimes also within species of the same genus, with respect to the sugar backbone, phosphate substitutions, as well as the number, type, and distribution of fatty acids (Table 1.1).

Table 1.1 Substitution patterns of the lipid A disaccharide backbone in recently studied bacteria

Bacteria	GlcN II (GlcN3N II)			GlcN I (GlcN3N I)			References
	O-4′	O-3′	N-2′	O-3	N-2	O-1	
Acinetobacter radioresistens	P	12:0 [3-O(12:0)]	12:0 [3-O(12:0)]	12:0 (3-OH)	12:0/14:0 [3-O(12:0)]	P	[73]
Acinetobacter baumannii	P	12:0 [3-O (12:0)]	14:0 [3-O(12:0(2-OH)]	12:0 (3-OH)	14:0 [3-O (12:0]	P	[37]
Aeromonas salmonicida	P	14:0 [3-O(16:1)]	14:0 [3-O(12:0)]	14:0 (3-OH)	14:0 (3-OH)	-	[74]
Agrobacterium tumefaciens	P	14:0 (3-OH)	16:0 [3-O(28:0 (27-O(4:0 3-OH)[a]]	14:0 (3-OH)	16:0 (3-OH)	P	[43]
Alteromonas macleodi	P	12:0 (3-OH)	12:0 [3-O (12:0)]	12:0 (3-OH)	12:0 (3-OH)	P	[75]
A. pyrophilus[b]	GalA	14:0 [3-O (18:0)]	16:0 (3-OH)	14:0 (3-OH)	14:0 (3-OH)	GalA	[14]
A. lipoferum	–	14:0 (3-OH)	16:0 [3-O (18:1/18:0)]	14:0 (3-OH)	16:0 (3-OH)	GalA	[30]
B. stolpii	EtNPP	14:0 [3-O (i13:0)][a]	14:0 [3-O (i13:0)][a]	14:0/i15:0 (3-OH)	14:0/i15:0 (3-OH)	P	[16]
Bartonella henseale[b]	P	12:0 (3-OH)	16:0 [3-O(28:0(27-OH)]	12:0 (3-OH)	16:0 (3-OH)	P	[20]
Bdellovibrio bacteriovorus[b]	Man	13:0 [3-O (13:0 (3-OH)]	13:0 [3-O(13:0 (3-OH)]	13:0 (3-OH)	13:0 (3,4-OH)	Man	[76]
Bordetella parapertussis	P	10:0 (3-OH)	14:0 [3-O(14:0)]	16:0	14:0 (3-OH)	P	[25]
B. pertussis	GalN-P	14:0 (3-OH)	14:0 [3-O(14:0)]	10:0 (3-OH)	14:0 (3-OH)	P-GalN	[25]
B. cepacia complex	Ara4N-P[a]	14:0 (3-OH)[a]	16:0 [3-O(14:0)]	14:0 (3-OH)[a]	16:0 (3-OH)	P-Ara4N[a]	[59]
B. abortus[b]	P[a]	16:0 [3-O(28:0(27-OH)]	14:0 (3-OH)	14:0 [3-O (18:0)]	12:0 (3-OH)	P[a]	[15]
Chlamydia trachomatis	P	14:0/16:0	20:0 [3-O(18:0–21:0)]	14:0/15:0	20:0 (3-OH)	P	[77]
Coxiella burnetii	P	16:0/15:0	16:0 (3-OH)	16:0	16:0 (3-OH)	P	[78]
Francisella victoria	Man-P	–	18:0 [3-O (16:0)]	18:0 (3-OH)	18:0 (3-OH)	P-GalN	[79]
Fusobacterium nucleatum	P	14:0/16:0 [3-O(14:0)]	16:0 [3-O(14:0)]	14:0 (3-OH)	14:0/16:0 (3-OH)	P	[80]
H. alvei	P	14:0 [3-O (14:0)]	12:0 [3-O (12:0)]	14:0 (3-OH)	14:0 [3-O (16:0)]	P	[39]
H. magadiensis	P	12:0[a] [3-O (18:1/16:0[a])]	12:0[a] [3-O (14:0)[a]]	12:0 (3-OH)	12:0 [3-O (10:0)]	P[a]	[38]
L. interrogans[b]	–	12:0 [3-O (12:1/14:1)]	16:0 [3-O (12:1/14:1)]	12:0 (3-OH)	16:0 (3-OH)	P-Me	[29]
M. vaga	–	–	12:0 [3-O (12:0 (3-OH)]	12:0 (3-OH)	12:0 [3-O (10:0/12:0)]	P	[81]
Porphyromonas gingivalis	P[a]	i15:0 (3-OH)	i17:0 [3-O (16:0)]	16:0 (3-OH)	i17:0 (3-OH)	P/PEtN	[82]
Pseudoalteromonas nigrifaciens	P	10:0 (3-OH)	12:0 (3-OH)	10:0 (3-OH)	12:0 [3-O (12:0)]	P	[83]
Rhodospirillum fulvum	heptose	14:0 [3-O (12:0)]	14:0 [3-O (16:0)]	14:0 (3-OH)	14:0 (3-OH)	GalA	[84]
Shewanella pacifica	P	13:0 [3-O (13:0)]	12:0 [3-O (13:0)]	13:0 (3-OH)	13:0 (3-OH)	P	[85]
Xanthomonas campestris	EtNPP	10-13:0 [3-O (10:0/11:0)]	12:0 (3-OH)	10-13:0 [3-O (10:0/11:0)	12:0 (3-OH)	PPEtN[a]	[40]

[a]The substituent is present in a non-stoichiometric amount
[b]The GlcN3N backbone

Whereas a β-(1 → 6)-linked GlcN disaccharide backbone is most common, a similarly linked GlcN3N disaccharide backbone has been identified in a number of bacterial species, such as *Aquifex pyrophilus* [14], *Brucella abortus* [15], *Bacteriovorax stolpii* [16], *Caulobacter crescentus* [17], *Mesorhizobium huakuii* [18], *Bradyrhizobium elkanii* [19], *Bartonella henselae* [20], *Legionella pneumophila* [21]. The major lipid A structural variant of *Campylobacter jejuni*, a bacterium responsible for gastrointestinal diseases, contains a hybrid backbone of a β-GlcN3N-(1 → 6)-GlcN disaccharide (Fig. 1.3a) [22, 23]; two other lipid A variants present have GlcN disaccharide and GlcN3N disaccharide backbones. No occurrence of an alternative hybrid GlcN-(1 → 6)-GlcN3N backbone has been documented so far in any lipid A.

The most common polar substituents of the phosphate groups, which are usually present non-stoichiometric quantities, are a second phosphate (giving rise to a diphosphate group), EtN, PEtN, and Ara4N. Other charged and non-charged substituents are listed in Table 1.1. Charged groups allow the bacterium to modulate the net surface charge and may vary significantly depending on growth conditions.

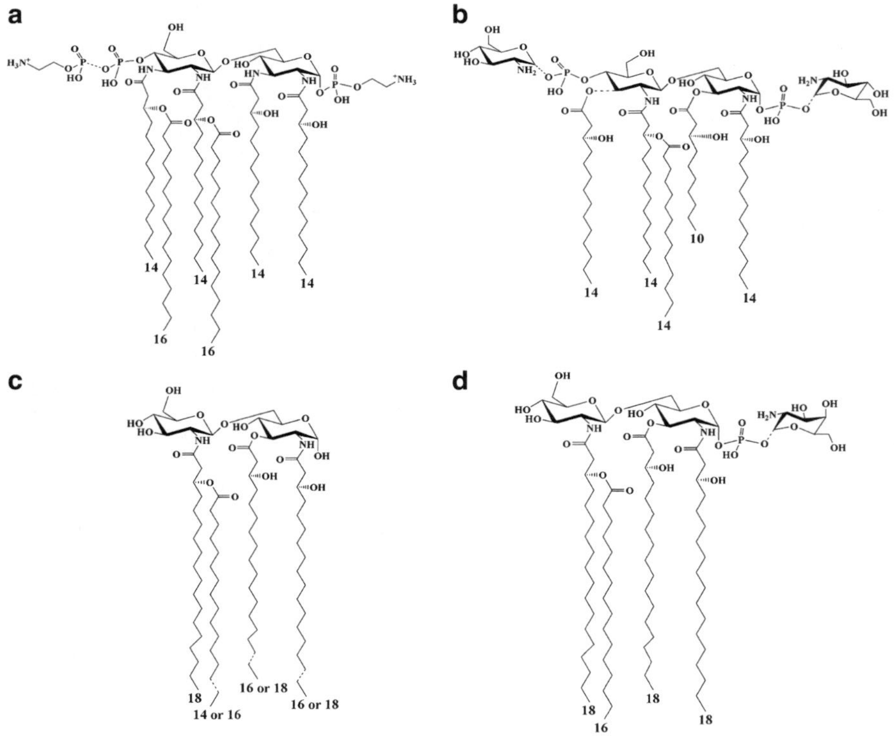

Fig. 1.3 Structures of lipid A of *C. jejuni* (**a**), *B. pertussis* Tohama I and *B. bronchiseptica* 4650 (**b**), *F. tularensis* ATCC 29684 (**c**), *F. tularensis* ssp. *holarctica* 1547-57 and ssp. *novicida* (**d**)

The abundant anionic groups in the lipid A-core OS region are tightly associated by electrostatic interactions with divalent cations (Mg^{2+} and Ca^{2+}), which help connecting LPS molecules to each other. This phenomenon contributes to the remarkable stability of the OM and to a significant reduction in membrane permeability, resulting in an efficient protective barrier.

The negatively charged groups are selectively targeted by cationic antimicrobial peptides. However, many bacteria can decorate their lipid A with Ara4N (see Sect. 1.4). The presence of Ara4N shields the negative charges of the lipid A and confers resistance to antimicrobial peptides. Lipid A of *Neisseria meningitidis*, a bacterium responsible for meningococcal infections, carries two PPEtN groups at positions 1 and 4′ of the disaccharide backbone [24], and the lipid A of *C. jejuni* contains a PEtN at position 1 and P or PPEtN at position 4′ (Fig. 1.3a) [22, 23]. *Bordetella pertussis* and *Bordetella bronchiseptica*, important pathogens that cause a range of pathologies in different hosts, have lipid A with free non-acylated GlcN substituents on both phosphate groups (Fig. 1.3b) [25]. It has been speculated that these lipid A modifications provide bacteria with the ability to modulate host immune responses, which represents an evolutionary advantage to adapt and survive during infection [1–4].

Lipid A of the tularemia bacterium *Francisella tularensis* lacks one of the phosphate groups at position 4′ or both phosphate groups [26–28] (Fig. 1.3c). These features appear to be responsible for a low bioactivity of the *Francisella* LPS [28]. Moreover, lipid A of *F. tularensis* ssp. *holarctica* and ssp. *novicida* has a galactosamine 1-phosphate group at position 1 (Fig. 1.3d) [27, 28]. Underphosphorylation of lipid A has been reported also in some other bacteria, including the Weil's disease pathogen *Leptospira interrogans* whose lipid A is not phosphorylated at position 4′ and the 1-phosphate group is O-methylated (Table 1.1) [29]. The LPS of this bacterium activates TLR2 rather TLR4, which seems to be due to the unusual lipid A structure [1–4, 29].

The phosphate groups can be replaced by other substituents. For example, two GalA residues replace both phosphate groups at the GlcN3N disaccharide backbone of the lipid A of the hyperthermophilic bacterium *A. pyrophilus* [14]. Similarly, lipid A of *C. crescentus* has an α-D-Gal*p*A-(1 → 4)-β-D-Glc*p*N3N-(1 → 6)-α-D-Glc*p*N3N-(1 → 1)-α-D-Gal*p*A tetrasaccharide backbone [17], whereas rhizobacteria *M. huakuii* [18] (Fig. 1.4a) and *Azospirillum lipoferum* [30] possess a β-D-Glc*p*N3N-(1 → 6)-α-D-Glc*p*N3N-(1 → 1)-α-D-Gal*p*A trisaccharide backbone. As the glycosidic linkage of uronic acids is more resistant to cleavage than the ester phosphate linkage, this modification may contribute to the membrane stability under non-canonical physico-chemical environmental conditions.

The lipid A of the marine bacterium *Loktanella rosea* has a very unusual structure as the molecule is non-phosphorylated, both GlcN residues are β-linked, and the proximal β-GlcN I forms with α-GalA a unique mixed trehalose-like structure [31]. Lipid A of a slow-growing rhizobacterium *Bradyrhizobium elkanii* lacks negatively charged groups. Instead, it has a single D-mannose residue at the reducing end and a D-mannose disaccharide attached to the non-reducing GlcN3N residue (Fig. 1.4b) [19]. In lipid A of *Rhizobium etli*, the phosphate group at

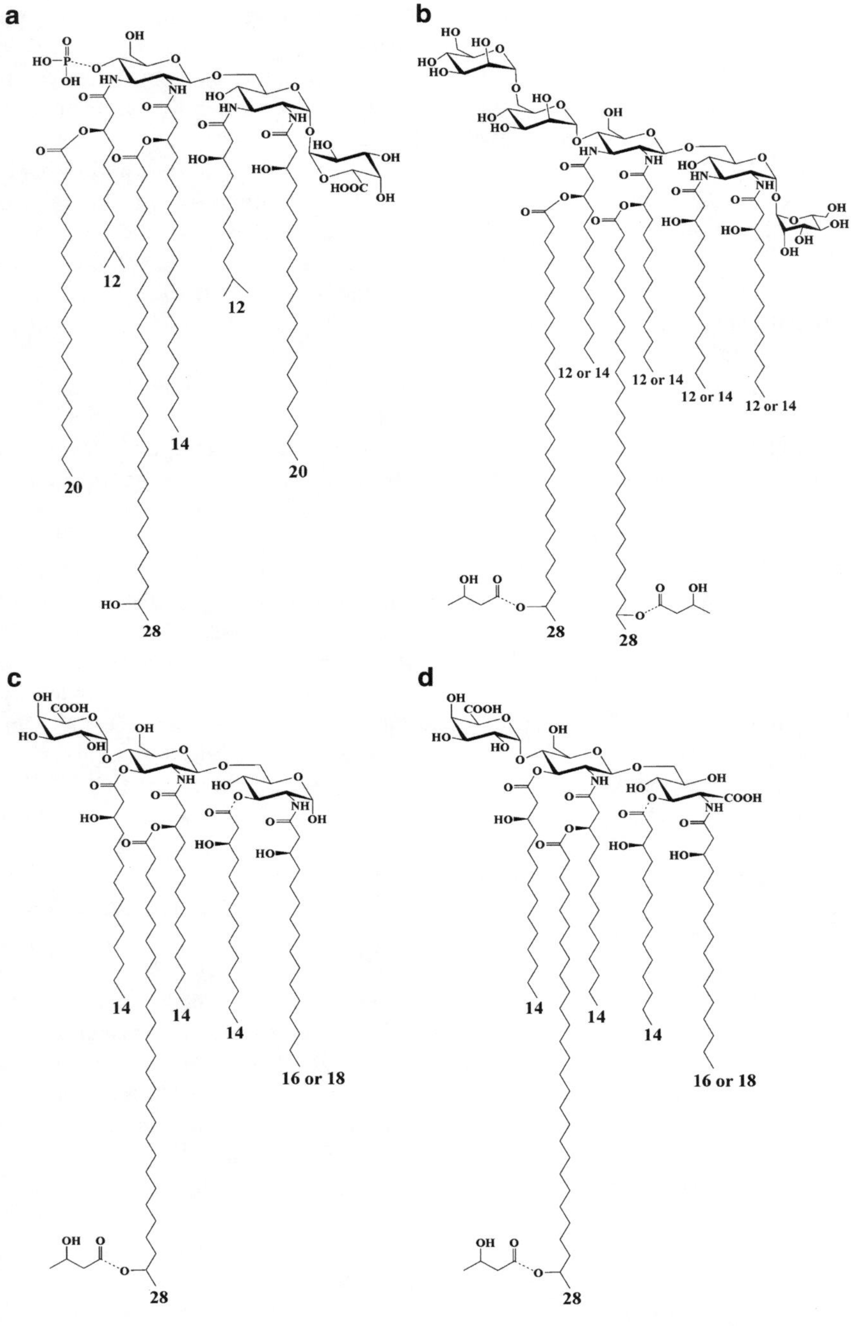

Fig. 1.4 Structures of lipid A of *M. huakuii* IFO 15243 (**a**), *B. elkanii* (**b**), and *Rhizobium etli* CE3 (**c** and **d**). Dotted lines indicate non-stoichiometric substitution

position 4′ of the distal GlcN II is replaced with GalA and the proximal GlcN I is either devoid of the 1-phosphate group (Fig. 1.4c) or is oxidized into 2-amino-2-deoxygluconic acid (Fig. 1.4d) [32]. The last component is present also in lipid A of *Rhizobium leguminosarum* bv. *phaseoli* [33, 34] and *Rhizobium* sp. Sin-1 [35] but has not been reported in any non-rhizobial LPS.

Lipid A moieties of various bacteria differ also in number, type and distribution of fatty acids (Table 1.1). Their number varies from four, as in *F. tularensis* [2, 4, 6], *Helicobacter pylori* [2, 4, 6], and *Pseudoalteromonas issachenkonii* KMM 3549^T [36] (Table 1.1), to seven, as in *Erwinia carotovora*, *Acinetobacter* [2, 6, 37], *Halomonas magadiensis* [38], *Klebsiella*, *Salmonella enterica*, and *Hafnia alvei* [2, 4, 6, 39]. In *Burkholderia* and some other bacteria, the highest acylated form is a pentaacyl lipid A (Table 1.1). Triacyl species often accompany higher acylated lipid A variants. In hexaacyl lipid A of most bacteria, as in *E. coli*, the acyl groups are asymmetrically distributed (4 + 2), whereas in some other species, as *P. aeruginosa*, the distribution of fatty acids is symmetric (3 + 3).

Fatty acids in lipid A are generally saturated and usually possess an even number (10–22) of carbon atoms (Table 1.1). *Iso* and *ante-iso* fatty acids have been identified in some bacteria like for example in *Bacteroides* [4, 9] and *Xanthomonas* [40]. The primary acyl groups are 3-hydroxylated with (*R*)-configuration, the most frequently found being 3-OH-10:0, 3-OH-12:0, 3-OH-14:0, 3-OH-16:0, and 3-OH-18:0. Less common are odd numbered, branched, and unsaturated fatty acids. A branched 2,3-dihydroxy fatty acid, 2,3-di-OH-i-14:0 has been found in *L. pneumophila* [21], while 3-keto fatty acids are present in *Rhodobacter sphaeroides* [41] and *R. capsulatus* [4].

In lipid A of *C. jejuni* with a hybrid β-GlcN3N-(1 → 6)-GlcN backbone, three primary acids are amide-linked and one ester-linked (Fig. 1.3a). Some other bacterial species, including *L. interrogans* [29], synthesize lipid A with only amide linked acyl chains. It has been suggested that lipid A with more ester-bound acyl groups is more biological active towards TLR4 than lipid A with more amide-bound acyl groups [2], and, therefore, an increased *N*-acylation, as in *C. jejuni*, may help the bacterium to evade activation of the innate host defense.

In many bacteria, lipid A possesses only one type of 3-hydroxy fatty acids, whereas in others, ester- and amide-linked primary acyl chains have different length (Table 1.1). For example, lipid A of *Bordetella* has the same amide-linked fatty acids (two 3-OH-14:0) and different ester-linked acids (3-OH-10:0 and 3-OH-14:0) (Fig. 1.3b) [25]. From four amide-linked 3-hydroxy acyl residues in lipid A of *M. huakuii*, two relatively short-chain acids are the same (3-OH-i-13:0) and two longer-chain acids are different (3-OH-14:0 and 3-OH-20:0) (Fig. 1.4a) [18].

The secondary fatty acids are more variable, comprising saturated and unsaturated acyl chains of different length (Table 1.1). In some cases [4], they are further modified, e.g. by non-stoichiometric (*S*)-2-hydroxylation, as in *Salmonella*, *Pseudomonas*, *Acinetobacter*, and *Bordetella* (Table 1.1). (ω-1)-Hydroxylated or differently functionalized long-chain fatty acids occur in some bacterial species. The secondary 27-OH-28:0 group present in Rhizobiaceae [18, 19, 42] (Fig. 1.4) and *Agrobacterium* [43] (Table 1.1) is partially O-acylated itself with 3-hydroxybutanoic acid. In *B. elkanii*, two long-chain acyl groups are present, which is unusual, and either of them may be non-stoichiometrically O-acylated [19] (Fig. 1.4b). An increased chain-length of fatty acids may be another reason of a low endotoxic activity of lipid A, e.g. as in *L. pneumophila* [21] lipid A, which contains four primary amide-linked acyl groups, two of which have 18–22 carbon atoms, and two secondary acyl groups, one having a long chain (27-oxo-28:0 or 26-carboxy-26:0).

1.4 Additional Lipid A Modifications and Their Biological Implications

Strain-specific modifications of lipid A, although not required for growth, modulate virulence of some Gram-negative pathogens. Structure of lipid A can vary also depending on the growth and environmental conditions. These adaptive and dynamic changes affect the polar heads and the acyl groups. Expression of different lipid A variants by the same strain has an impact on the vulnerability to antibiotics and the immunostimulatory power. A recent comprehensive review [44] has been dedicated to these aspects of lipid A; hence, they are discussed below only briefly.

When cultures are grown at low temperature (10–15°C), *E. coli* and *S. enterica* incorporates unsaturated fatty acids into lipid A [45]. Modification of lipid A by palmitoylation catalyzed by PagP (Fig. 1.5) has been demonstrated in such bacterial species as *S. enterica*, *E. coli*, *L. pneumophila*, *B. bronchiseptica*, and *Y. pseudotuberculosis*. Introduction of palmitate 16:0 is under control of the PhoP/PhoQ signal transduction system, which responds to the presence of antimicrobial peptides and is activated by low concentration of Mg^{2+} [44]. *S. enterica* Typhimurium mutants that are unable to add palmitate to lipid A, are sensitive to certain cationic antimicrobial peptides, including representatives of amphipatic α-helical (C18G) and β-sheet (protegrin) structural classes but excluding polymyxin. In *B. bronchiseptica*, palmitoylation of lipid A at O-3′ [46] is required for persistent colonization of the respiratory tract and for resistance to antibody-mediated complement lysis [47].

In contrast to *Y. pseudotuberculosis*, in which palmitoylated lipid A species predominate at the body temperature of the infected warm-blooded host, the plague pathogen *Y. pestis* cannot incorporate the 16:0 group into lipid A but can remodel the acylation pattern in a temperature-dependent manner [48]. When grown at 26°C, *Y. pestis* expresses a hexaacyl lipid A containing an unsaturated secondary fatty acid 16:1 (Fig. 1.6a), whereas at 37°C mainly a tetraacyl lipid A (lipid IV_A) is synthesized (Fig. 1.6b). The latter is poorly recognized by TLR4 on the immune cells of the mammalian host, and thus the systemic infection is allowed, demonstrating that the evasion of the LPS-induced inflammatory response is critical for *Y. pestis* virulence. By the modification of the lipid A structure *Y. pestis* achieves also a high bacterial load in mammalian blood required for efficient flea infection before the induction of the lethal shock [47, 48].

In *S. enterica*, OM lipases, PagL and LpxR, are responsible for removing an acyl group from position 3 and an acyloxyacyl group from position 3′ of the lipid A backbone, respectively (Fig. 1.5). As an underacylated lipid A possesses a low

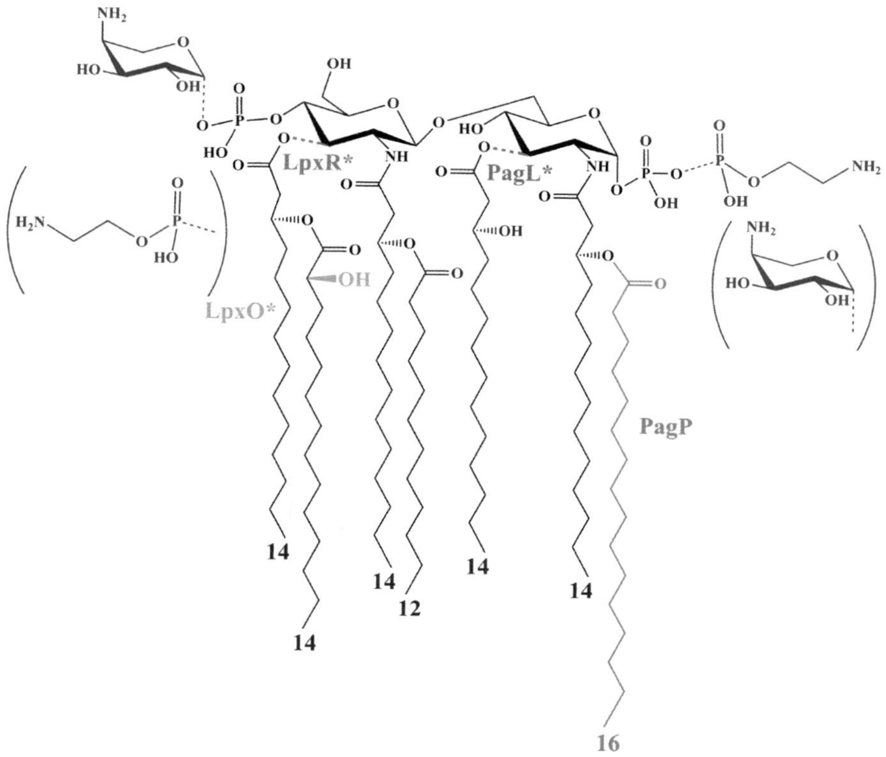

Fig. 1.5 Lipid A modifications in *E. coli* and *S. enterica* (Adapted from Raetz et al. [44]). Also shown are enzymes involved in the modifications (see Chap. 6). *Asterisk* refers to modifications absent from *E. coli*

immunostimulant potential, this modification results in attenuation in vivo of the cytokine-inducing ability of the LPS [45]. PagL homologues that may change the signal information by detoxifying the LPS in vivo, are widely distributed in bacteria. When not cleaved, the secondary acyl group at position 3 may be 2-hydroxylated by the inner membrane enzyme LpxO in the presence of O_2 (Fig. 1.5). The biological significance of such modification remains unknown.

Lipid A may be modified also by a human neutrophil enzyme, an acyloxyacyl hydrolase, which cleaves the secondary acyl residues from the acyloxyacyl groups and is active toward the LPS of *S. enterica* Typhimurium, *E. coli*, *P. aeruginosa*, *Haemophilus influenzae*, *N. meningitidis*, and *N. gonorrhoeae* [49].

A number of bacteria (*E. coli*, *S. enterica*, *Yersinia*, *Pseudomonas*, *Burkholderia*, *Francisella*, *Neisseria*) are able to add PEtN or Ara4N to the phosphate groups of lipid A [44, 50–52]. Temperature-dependent glycosylation of the phosphate groups with Ara4N regulated by the PhoP/PhoQ system [53] makes *Y. pestis* resistant to the cyclic polycationic antibiotic peptide polymyxin B [52, 53]. Mutants of *S. enterica* Typhimurium and *E. coli* that are able to synthesize

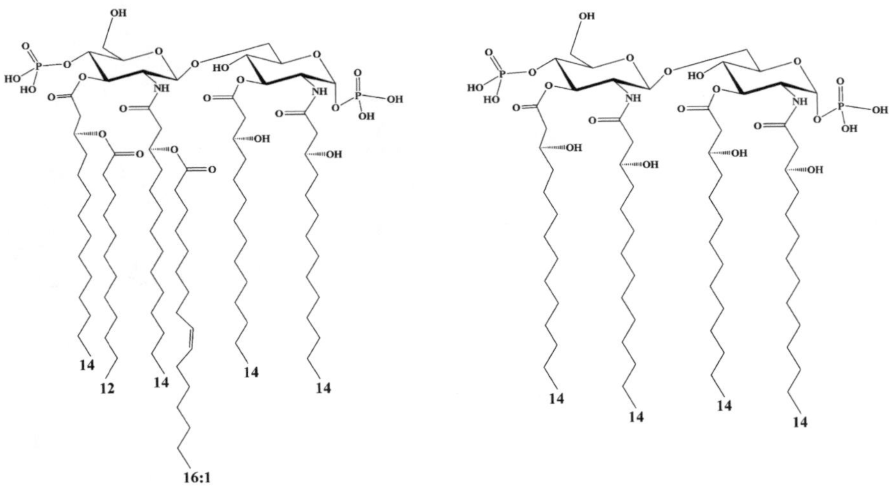

Fig. 1.6 Structures of the main lipid A variants isolated from *Yersinia pestis* KIM5 grown in liquid culture at 26°C (**a**) and 37°C (**b**). Polar substituents of the phosphate groups are not shown

significant amounts of lipid A species bearing palmitate, Ara4N, and/or PEtN (Fig. 1.5), are polymyxin-resistant [54]. In *N. gonorrhoeae*, the PEtN substitution of the 4′-phosphate in lipid A confers resistance to both polymyxin B and normal human serum [55].

Lipid A modifications play a special role in *P. aeruginosa* and a group of related species known as the *Burkholderia cepacia* complex, which are the dominant Gram-negative opportunistic pathogens that infect the respiratory tract in cystic fibrosis patients. Variations in *P. aeruginosa* lipid A structure can influence the pathogenesis of the chronic lung disease [56, 57]. Chemical structure and biological activities of *P. aeruginosa* lipid A in serial isolates of mucoid and non-mucoid strains collected from a cystic fibrosis patient over a period of up to 7.5 years have been examined [58]. Early and late mucoid *P. aeruginosa* isolates synthesized mainly tetra-, penta-, and hexaacyl lipid A lacking the primary 3-OH-10:0 fatty acid (Fig. 1.7a). However, lipid A of the late non-mucoid isolate carried an additional 3-OH-10:0 residue, e.g. it consists of hexa- and heptaacyl species (Fig. 1.7b). These structural differences in the lipid A in *P. aeruginosa* isolates that are clonal may explain the reduced ability of late, non-mucoid strains to promote reduced recruitment of leukocytes in bronchoalveolar lavage and reduced cytokine levels in lung homogenates upon experimental infections. It is possible that non-mucoid isolates, which are more susceptible to phagocytosis, can escape detection by the immune system by reducing their inflammatory potential.

The *Burkholderia* lipid A consists of a glucosamine disaccharide backbone, two phosphate groups non-stoichiometrically substituted with Ara4N, and four

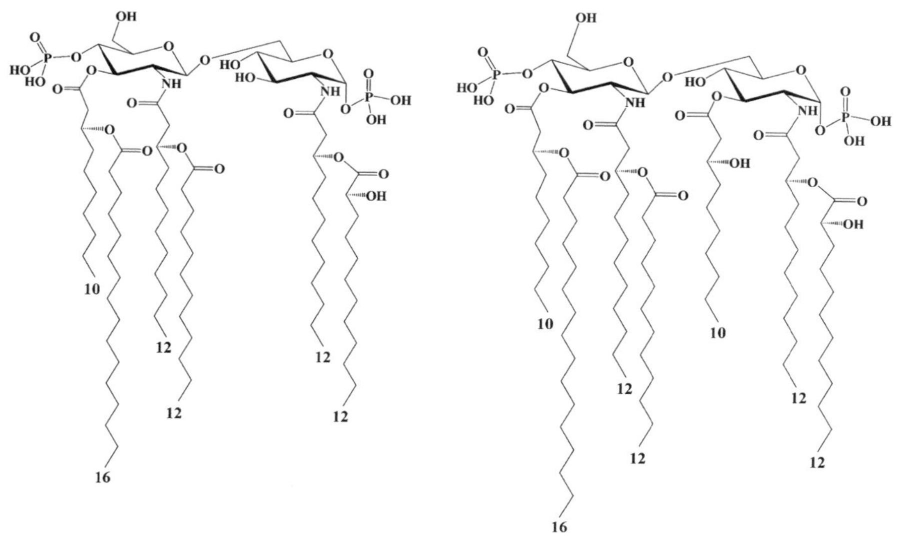

Fig. 1.7 Structures of lipid A variants in *Pseudomonas aeruginosa* clinical isolates from a CF patient: in mucoid strains after 6 months and 7.5 year of colonization (**a**) and in a non-mucoid strain after 7.5 year of colonization (**b**). Polar substituents of the phosphate groups are not shown

or five acyl groups: 3-OH-14:0 and 3-OH-16:0 as primary and 14:0 as secondary [59]. Unlike other bacteria, the presence of Ara4N is absolutely essential for bacterial viability, making its synthesis an optimal target to develop new antibiotics [60].

1.5 Structural Elucidation of Lipid A

The structural elucidation of the lipid A is of pivotal importance to better understand the lipid A/LPS three-dimensional structure and biological properties, including the agonist/antagonist action in the elicitation of the host innate immune response. The lipid A fraction is typically obtained as a sediment after mild acid hydrolysis of the LPS, most commonly by acetic acid (100°C, 2 h). This mild hydrolysis exploits the lability of the bond between Kdo and GlcN II of lipid A and therefore releases lipid A from the rest of the LPS molecule. As discussed above, the lipid A fraction consists of a mixture of intrinsically heterogeneous lipid A molecules, which differ in the number, distribution, and stoichiometry of fatty acids and polar heads. The state-of-art structural approach implies a careful use of mass spectrometry (MS) analysis by matrix-assisted laser desorption/ionization time-of-flight (MALDI-TOF) MS, and electrospray ionization (ESI) MS techniques both on intact and selectively degraded lipid A preparations. The MS allows gaining insights into the number of lipid A species present in the fraction, the number and nature of acyl residues and polar heads, and their distribution on each GlcN unit of the disaccharide backbone.

In our laboratory, we have developed a novel approach to deduce the location and distribution of lipid A components. The methodology to establish the primary and secondary fatty acid distribution combines MALDI MS or ESI MS and selective chemical degradation of lipid A by ammonium hydroxide hydrolysis [61–63]. The selective degradation must be supported with classical chemical analyses, which involve either a removal of all ester-linked fatty acids by mild alkaline hydrolysis or complete decollation by strong alkaline treatment, followed by GLC-MS analysis of derived fatty acids [64].

Nuclear magnetic resonance (NMR) investigation of lipid A is a less usable approach but equally important. It is typically employed to establish the carbohydrate backbone structure. In some cases, it has been possible to examine a partially deacylated lipid A by NMR spectroscopy, whereas analysis of the unmodified lipid A is usually precluded by its amphiphilic nature, which results in poor solubility of the sample in any solvent system (see Chap. 2 for more details on core OS structure and lipid A-core OS structure elucidation).

1.6 Supramolecular Structure of Lipid A

The variations of the primary structure of lipid A influence its physicochemical and biological behaviour. Indeed, the lipid A intrinsic conformation is responsible for its agonistic and antagonistic activity [65–69]. Due to their amphiphilic nature, LPS molecules tend to form aggregates whose structure changes according to the lipid A primary structure. Only LPS molecules that adopt a nonlamellar (cubic or hexagonal) aggregate are biologically active whereas those possessing a lamellar structure are not. The shape of individual lipid A molecules has been determined from these superstructures based on the assumption of a fixed angle between the two GlcN residues due to the tendency of acyl chains to pack parallel to each other and to stabilize the aggregates via hydrophobic interactions.

The inclination angle of the glycosidic linkage depends on the number of acyl chains. Also, the overall inclination angle of the sugar backbone with respect to the packed acyl chains is pivotal in determining the endotoxic activity. Hexaacylated bisphosphorylated lipid A possessing an asymmetric (4 + 2) distribution of acyl groups, such as the *E. coli* lipid A, exhibit a tilt angle >50° and the highest cytokine-inducing capacity. These biologically active molecules have a conical molecular shape and their hydrophobic portion occupies a larger cross-section than the hydrophilic part. Species with a tilt angle <25°, such as lipid IV_A, and the pentaacylated and symmetrically (3 + 3) hexaacylated species of *Rhodobacter sphaeroides* and *R. capsulatus* are endotoxically inactive but can be antagonist. The molecular shape of these species is cylindrical. Species with an angle between 25° and 50°, as monophosphorylated lipid A, have a lower bioactivity and are located between these two values.

An absolute prerequisite for both agonistic and antagonistic lipid A is a sufficiently negative charge density of the polar heads. Two parameters influence the interaction with the immune receptors: the tilt angle of the disaccharide backbone with the acyl chains and the molecular shape. The inclination of the sugar linkage relative to the hydrophobic region places the anomeric phosphate on the GlcN I outside of the OM surface. The specificity of lipid A binding to the target receptors is mediated by its hydrophilic backbone, but the hydrophobic region largely determines the subsequent activation of the immune response. Only species whose primary structure confers a conical molecular shape are endotoxically active. Their hydrophobic portion has a large cross-section and can interact with the hydrophobic binding sites of the TLR4/MD-2 complex to activate transmembrane signalling, likely due to mechanical stress-induced conformational change of the proteins. Species with a cylindrical shape are not able to induce the appropriate mechanical stress to activate the receptors. NMR conformational studies [69] in aqueous media of the biosynthetic precursor lipid IV_A agree with other studies [68–71]. Recently, the structure and dynamics of complex LPS molecules in aqueous solution have been analyzed by solubilising the LPS in perdeuterated dihexanoylphosphatidylcholine micelles, which mimic the bacterial membrane [70].

Structural and functional studies on liposomes formed by the LPS to better understand the stability, low-permeability and resistance of the OM have been performed by a variety of physico-chemical techniques (e.g., differential calorimetry, fluorescence polarization and fluorescence resonance energy transfer spectroscopy, Fourier transform infrared spectroscopy, small- and wide-angle X-ray scattering, microscopy, electron microscopy). Different liposomes formulations included LPS from different sources, but mostly from *E. coli* and *S. enterica*. To evaluate the relationship between the lipooligosaccharide molecular structure and the liposome functional behaviour, large unilamellar liposomes formed by the Re LPS (Kdo_2-lipid A) from *S. enterica* Minnesota strain 595 (Re mutant) have been prepared by the extrusion techniques [71] and characterized at different observation scales, from the morphological to the microstructural level.

Temperature also causes substantial changes in the bilayer microstructure and functionality. Although the liposome structure does not change with temperature, the bilayer fluidity increases as a function of the temperature. At 30–35°C, a progressive transition of the acyl chain self-organization from a gel to a liquid crystalline phase is detected. At 20°C, the fluidity of the LPS bilayer is quite low, and exhibits reduced water permeability. Above this temperature, the bilayer becomes more fluid and its water permeability increases correspondingly.

The LPS structure determines the local self-organization of the individual molecules, which in turn influences the architecture and dynamics of the aggregates. Investigation of the microstructure of liposomes formed by LPS from *Burkholderia* and *Agrobacterium* reveal that the LPS molecular structure determines the morphology of the aggregates in aqueous medium through a complex interplay of hydrophobic, steric, and electrostatic interactions [72]. In all of these cases LPS-derived liposomes mainly present a multilamellar arrangement. The thickness of the hydrophobic domain of each bilayer and the local ordering of the acyl chains are determined not only by lipid A but also, indirectly, by the bulkiness of the saccharide portion. Biologically, these results suggest that the rich biodiversity of the LPS molecular structure could be instrumental for the fine-tuning of the structure and functional properties of the Gram-negative bacterial OM.

References

1. Silipo A, De Castro C, Lanzetta R, Parrilli M, Molinaro A (2010) Lipopolysaccharides. In: König H, Claus H, Varma A (eds) Prokaryotic cell wall compounds – structure and biochemistry. Springer, Heidelberg, pp 133–154
2. Raetz CR, Whitfield C (2002) Lipopolysaccharide endotoxins. Annu Rev Biochem 71:635–700
3. Takeuchi O, Akira S (2010) Pattern recognition receptors and inflammation. Cell 140:805–820
4. Holst O, Molinaro A (2009) Core oligosaccharide and lipid A components of lipopolysaccharides. In: Moran A, Brennan P, Holst O, von Itszstein M (eds) Microbial glycobiology: structures relevance and applications. Elsevier, San Diego, pp 29–56
5. Silipo A, Molinaro A (2010) The diversity of the core oligosaccaride in lipopolysaccharides. Subcell Biochem 53:69–99
6. Zähringer U, Lindner B, Rietschel ET (1994) Molecular structure of lipid A, the endotoxic center of bacterial lipopolysaccharides. Adv Carbohydr Chem Biochem 50:211–276
7. Westphal O, Lüderitz O (1954) Chemische Erforschung von Lipopolysacchariden Gram-Negativer Bakterien. Angew Chem 66:407–417
8. Takayama K, Qureshi N, Mascagni P (1983) Complete structures of lipid A obtained from the lipopolysaccharide of the heptoseless mutant of *Salmonella typhimurium*. J Biol Chem 258:12801–12803
9. Kabanov DS, Prokhorenko IR (2010) Structural analysis of lipopolysaccharides from Gram-negative bacteria. Biochemistry (Moscow) 75:383–404
10. Vernacchio L, Bernstein H, Pelton S, Allen C, MacDonald K, Dunn J, Duncan DD, Tsao G, LaPosta V, Eldridge J, Laussucq S, Ambrosino DM, Molrine DC (2002) Effect of monophosphoryl lipid A (MPL) on T-helper cells when administered as an adjuvant with pneumocococcal-CRM197 conjugate vaccine in healthy toddlers. Vaccine 20:3658–3667
11. Cluff CW (2009) Monophosphoryl lipid A (MPL) as an adjuvant for anti-cancer vaccines: clinical results. Adv Exp Med Biol 667:111–123
12. Mata-Haro V, Cekic C, Martin M, Chilton PM, Casella CR, Mitchell TC (2007) The vaccine adjuvant monophosphoryl lipid A as a TRIF-biased agonist of TLR4. Science 316:1628–1632
13. Christ WJ, Asano O, Robidoux AL, Perez M, Wang Y, Dubuc GR, Gavin WE, Hawkins LD, McGuinness PD, Mullarkey MA, Lewis MD, Kishi Y, Kawata T, Brisson JR, Rose JR, Rossignol DP, Kobayashi S, Hishinuma I, Kimura A, Asakawa N, Katayama K, Yamatsu I (1995) E5531, a pure endotoxin antagonist of high potency. Science 268:80–83
14. Plötz BM, Lindner B, Stetter KO, Holst O (2000) Characterization of a novel lipid A containing D-galacturonic acid that replaces phosphate residues. The structure of the lipid A of the lipopolysaccharide from the hyperthermophilic bacterium *Aquifex pyrophilus*. J Biol Chem 275:11222–11228
15. Qureshi N, Takayama K, Seydel U, Wang R, Cotter RJ, Agrawal PK, Bush CA, Kurtz R, Berman DT (1994) Structural analysis of the lipid A derived from the lipopolysaccharide of *Brucella abortus*. J Endotoxin Res 1:137–148
16. Beck S, Müller FD, Strauch E, Brecker L, Linscheid MW (2010) Chemical structure of *Bacteriovorax stolpii* lipid A. Lipids 45:189–198

17. Smit J, Kaltashov IA, Cotter RJ, Vinogradov E, Perry MB, Haider H, Qureshi N (2008) Structure of a novel lipid A obtained from the lipopolysaccharide of *Caulobacter crescentus*. Innate Immun 14:25–37
18. Choma A, Sowinski P (2004) Characterization of *Mesorhizobium huakuii* lipid A containing both D-galacturonic acid and phosphate residues. Eur J Biochem 271:1310–1322
19. Komaniecka I, Choma A, Lindner B, Holst O (2010) The structure of a novel neutral lipid A from the lipopolysaccharide of *Bradyrhizobium elkanii* containing three mannose units in the backbone. Chem Eur J 16:2922–2929
20. Zähringer U, Lindner B, Knirel YA, van den Akker WM, Hiestand R, Heine H, Dehio C (2004) Structure and biological activity of the short-chain lipopolysaccharide from *Bartonella henselae* ATCC 49882T. J Biol Chem 279:21046–21054
21. Zähringer U, Knirel YA, Lindner B, Helbig JH, Sonesson A, Marre R, Rietschel ET (1995) The lipopolysaccharide of *Legionella pneumophila* serogroup 1 (strain Philadelphia 1): chemical structure and biological significance. Prog Clin Biol Res 392:113–139
22. van Mourik A, Steeghs L, van Laar J, Meiring HD, Hamstra HJ, van Putten JP, Wösten MM (2010) Altered linkage of hydroxyacyl chains in lipid A of *Campylobacter jejuni* reduces TLR4 activation and antimicrobial resistance. J Biol Chem 285:15828–15836
23. Moran AP, Zähringer U, Seydel U, Scholz D, Stütz P, Rietschel ET (1991) Structural analysis of the lipid A component of *Campylobacter jejuni* CCUG 10936 (serotype O:2) lipopolysaccharide. Description of a lipid A containing a hybrid backbone of 2-amino-2-deoxy-D-glucose and 2,3-diamino-2,3-dideoxy-D-glucose. Eur J Biochem 198:459–469
24. Kulshin VA, Zähringer U, Lindner B, Frasch CE, Tsai CM, Dmitriev BA, Rietschel ET (1992) Structural characterization of the lipid A component of pathogenic *Neisseria meningitides*. J Bacteriol 174:1793–1800
25. Marr N, Tirsoaga A, Blanot D, Fernandez R, Caroff M (2008) Glucosamine found as a substituent of both phosphate groups in *Bordetella* lipid A backbones: role of a BvgAS-activated ArnT ortholog. J Bacteriol 190:4281–4290
26. Vinogradov E, Perry MB, Conlan JW (2002) Structural analysis of *Francisella tularensis* lipopolysaccharide. Eur J Biochem 269:6112–6118
27. Wang X, Ribeiro AA, Guan Z, McGrath SC, Cotter RJ, Raetz CR (2006) Structure and biosynthesis of free lipid A molecules that replace lipopolysaccharide in *Francisella tularensis* subsp. *novicida*. Biochemistry 45:14427–14440
28. Phillips NJ, Schilling B, McLendon MK, Apicella MA, Gibson BW (2004) Novel modification of lipid A of *Francisella tularensis*. Infect Immun 72:5340–5348
29. Que-Gewirth NL, Ribeiro AA, Kalb SR, Cotter RJ, Bulach DM, Adler B, Girons IS, Werts C, Raetz CR (2004) A methylated phosphate group and four amide-linked acyl chains in *Leptospira interrogans* lipid A. The membrane anchor of an unusual lipopolysaccharide that activates TLR2. J Biol Chem 279:25420–25429
30. Choma A, Komaniecka I (2008) Characterization of a novel lipid A structure isolated from *Azospirillum lipoferum* lipopolysaccharide. Carbohydr Res 343:799–804
31. Ieranò T, Silipo A, Nazarenko EL, Gorshkova RP, Ivanova EP, Garozzo D, Sturiale L, Lanzetta R, Parrilli M, Molinaro A (2010) Against the rules: a marine bacterium, *Loktanella rosea*, possesses a unique lipopolysaccharide. Glycobiology 20:586–593
32. Que NL, Lin S, Cotter RJ, Raetz CR (2000) Purification and mass spectrometry of six lipid A species from the bacterial endosymbiont *Rhizobium etli*. Demonstration of a conserved distal unit and a variable proximal portion. J Biol Chem 275:28006–28016
33. Muszyński A, Laus M, Kijne JW, Carlson RW (2011) Structures of the lipopolysaccharides from *Rhizobium leguminosarum* RBL5523 and its UDP-glucose dehydrogenase mutant (exo5). Glycobiology 21:55–68
34. Bhat UR, Forsberg LS, Carlson RW (1994) Structure of lipid A component of *Rhizobium leguminosarum* bv. *phaseoli* lipopolysaccharide. Unique nonphosphorylated lipid A containing 2-amino-2-deoxygluconate, galacturonate, and glucosamine. J Biol Chem 269: 14402–14410

35. Jeyaretnam B, Glushka J, Kolli VS, Carlson RW (2002) Characterization of a novel lipid-A from *Rhizobium* species Sin-1. A unique lipid-A structure that is devoid of phosphate and has a glycosyl backbone consisting of glucosamine and 2-aminogluconic acid. J Biol Chem 277:41802–41810
36. Leone S, Silipo A, Nazarenko E, Lanzetta R, Parrilli M, Molinaro A (2007) Molecular structure of endotoxins from Gram-negative marine bacteria: an update. Mar Drugs 5:85–112
37. Fregolino E, Fugazza G, Galano E, Gargiulo V, Landini P, Lanzetta R, Lindner B, Pagani L, Parrilli M, Holst O, De Castro C (2010) Complete lipooligosaccharide structure of the clinical isolate Acinetobacter baumannii, strain SMAL. Eur J Org Chem 1345–1352
38. Silipo A, Sturiale L, Garozzo D, De Castro C, Lanzetta R, Parrilli M, Grant W, Molinaro A (2004) Structure elucidation of the highly heterogeneous lipid A from the lipopolysaccharide of the Gram-negative extremophile bacterium *Halomonas magadiensis* strain 21 M1. Eur J Org Chem 2263–2271
39. Lukasiewicz J, Jachymek W, Niedziela T, Kenne L, Lugowski C (2010) Structural analysis of the lipid A isolated from *Hafnia alvei* 32 and PCM 1192 lipopolysaccharides. J Lipid Res 51:564–574
40. Silipo A, Molinaro A, Sturiale L, Dow JM, Erbs G, Lanzetta R, Newman MA, Parrilli M (2005) The elicitation of plant innate immunity by lipooligosaccharide of *Xanthomonas campestris*. J Biol Chem 280:33660–33668
41. Salimath PV, Weckesser J, Strittmatter W, Mayer H (1983) Structural studies on the non-toxic lipid A from *Rhodopseudomonas sphaeroides* ATCC 17023. Eur J Biochem 136:195–200
42. De Castro C, Molinaro A, Lanzetta R, Silipo A, Parrilli M (2008) Lipopolysaccharide structures from *Agrobacterium* and other *Rhizobiaceae* species. Carbohydr Res 343: 1924–1933
43. Silipo A, De Castro C, Lanzetta R, Molinaro A, Parrilli M (2004) Full structural characterization of the lipid A components from the *Agrobacterium tumefaciens* strain C58 lipopolysaccharide fraction. Glycobiology 14:805–815
44. Raetz CH, Reynolds CM, Trent MS, Bishop RE (2007) Lipid A modification systems in Gram-negative bacteria. Annu Rev Biochem 76:295–329
45. Wollenweber HW, Schlecht S, Lüderitz O, Rietschel ET (1983) Fatty acid in lipopolysaccharides of *Salmonella* species grown at low temperature. Identification and position. Eur J Biochem 130:167–171
46. Zarrouk H, Karibian D, Bodie S, Perry MB, Richards JC, Caroff M (1997) Structural characterization of the lipids A of three *Bordetella bronchiseptica* strains: variability of fatty acid substitution. J Bacteriol 179:3756–3760
47. Diacovich L, Gorvel JP (2010) Bacterial manipulation of innate immunity to promote infection. Nat Rev Microbiol 8:117–128
48. Montminy SW, Khan N, McGrath S, Walkowicz MJ, Sharp F, Conlon JE, Fukase K, Kusumoto S, Sweet C, Miyake K, Akira S, Cotter RJ, Goguen JD, Lien E (2006) Virulence factors of *Yersinia pestis* are overcome by a strong lipopolysaccharide response. Nat Immunol 7:1066–1073
49. Erwin AL, Munford RS (1990) Deacylation of structurally diverse lipopolysaccharides by human acyloxyacyl hydrolase. J Biol Chem 265:16444–16449
50. Zhou Z, Lin S, Cotter RJ, Raetz CR (1999) Lipid A modifications characteristic of *Salmonella typhimurium* are induced by NH_4VO_3 in *Escherichia coli* K12. Detection of 4-amino-4-deoxy-L-arabinose, phosphoethanolamine and palmitate. J Biol Chem 274:18503–18514
51. Zhou Z, Ribeiro AA, Raetz CR (2000) High-resolution NMR spectroscopy of lipid A molecules containing 4-amino-4-deoxy-L-arabinose and phosphoethanolamine substituents. Different attachment sites on lipid A molecules from NH_4VO_3-treated *Escherichia coli* versus *kdsA* mutants of *Salmonella typhimurium*. J Biol Chem 275:13542–13551
52. Knirel YA, Lindner B, Vinogradov EV, Kocharova NA, Senchenkova SN, Shaikhutdinova RZ, Dentovskaya SV, Fursova NK, Bakhteeva IV, Titareva GM, Balakhonov SV, Holst O, Gremyakova TA, Pier GB, Anisimov AP (2005) Temperature-dependent variations and

intraspecies diversity of the structure of the lipopolysaccharide of *Yersinia pestis*. Biochemistry 44:1731–1743

53. Rebeil R, Ernst RK, Gowen BB, Miller SI, Hinnebusch BJ (2004) Variation in lipid A structure in the pathogenic Yersiniae. Mol Microbiol 52:1363–1373
54. Zhou Z, Ribeiro AA, Lin S, Cotter RJ, Miller SI, Raetz CR (2001) Lipid A modifications in polymyxin-resistant *Salmonella typhimurium*: PMRA-dependent 4-amino-4-deoxy-L-arabinose, and phosphoethanolamine incorporation. J Biol Chem 276:43111–43121
55. Lewis LA, Choudhury B, Balthazar JT, Martin LE, Ram S, Rice PA, Stephens DS, Carlson R, Shafer WM (2009) Phosphoethanolamine substitution of lipid A and resistance of *Neisseria gonorrhoeae* to cationic antimicrobial peptides and complement-mediated killing by normal human serum. Infect Immun 77:1112–1120
56. Ernst RK, Yi EC, Guo L, Lim KB, Burns JL, Hackett M, Miller SI (1999) Specific lipopolysaccharide found in cystic fibrosis airway *Pseudomonas aeruginosa*. Science 286:1561–1565
57. Ernst RK, Moskowitz SM, Emerson JC, Kraig GM, Adams KN, Harvey MD, Ramsey B, Speert DP, Burns JL, Miller SI (2007) Unique lipid A modifications in *Pseudomonas aeruginosa* isolated from the airways of patients with cystic fibrosis. J Infect Dis 196: 1088–1092
58. Cigana C, Curcuru L, Leone ML, Ieranò T, Lore NI, Bianconi I, Silipo A, Cozzolino F, Lanzetta R, Molinaro A, Bernardini ML, Bragonzi A (2010) *Pseudomonas aeruginosa* exploits lipid A and muropeptides modification as a strategy to lower innate immunity during cystic fibrosis lung infection. PLoS ONE 4:e8439
59. De Soyza A, Silipo A, Lanzetta R, Govan JR, Molinaro A (2008) Chemical and biological features of *Burkholderia cepacia* complex lipopolysaccharides. Innate Immun 14:127–144
60. Ortega XP, Cardona ST, Brown AR, Loutet SA, Flannagan RS, Campopiano DJ, Govan JR, Valvano MA (2007) A putative gene cluster for aminoarabinose biosynthesis is essential for *Burkholderia cenocepacia* viability. J Bacteriol 189:3639–3644
61. Silipo A, Lanzetta R, Amoresano A, Parrilli M, Molinaro A (2002) Ammonium hydroxide hydrolysis: a valuable support in the MALDI-TOF mass spectrometry analysis of lipid A fatty acid distribution. J Lipid Res 43:2188–2195
62. Sforza S, Silipo A, Molinaro A, Marchelli R, Parrilli M, Lanzetta R (2004) Determination of fatty acid positions in native lipid A by positive and negative electrospray ionization mass spectrometry. J Mass Spectrom 39:378–383
63. Sturiale L, Garozzo D, Silipo A, Lanzetta R, Parrilli M, Molinaro A (2005) MALDI mass spectrometry of native bacterial lipooligosaccharides. Rapid Commun Mass Spectrom 19:1829–1834
64. De Castro C, Parrilli M, Holst O, Molinaro A (2010) Microbe-associated molecular patterns in innate immunity: extraction and chemical analysis of Gram-negative bacterial lipopolysaccharides. Methods Enzymol 480:89–115
65. Brandenburg K, Mayer H, Koch MH, Weckesser J, Rietschel ET, Seydel U (1993) Influence of the supramolecular structure of free lipid A on its biological activity. Eur J Biochem 218:555–563
66. Rietschel ET, Kirikae T, Schade FU, Mamat U, Schmidt G, Loppnow H, Ulmer AJ, Zähringer U, Seydel U, Di Padova F, Schreier M, Brade H (1994) Bacterial endotoxin: molecular relationships of structure to activity and function. FASEB J 8:217–225
67. Seydel U, Oikawa M, Fukase K, Kusumoto S, Brandenburg K (2000) Intrinsic conformation of lipid A is responsible for agonistic and antagonistic activity. Eur J Biochem 267:3032–3039
68. Fukuoka S, Brandenburg K, Müller M, Lindner B, Koch MH, Seydel U (2001) Physico-chemical analysis of lipid A fractions of lipopolysaccharide from *Erwinia carotovora* in relation to bioactivity. Biochim Biophys Acta 1510:185–197
69. Oikawa M, Shintaku T, Fukuda N, Sekljic H, Fukase Y, Yoshizaki H, Fukase K, Kusumoto S (2004) NMR conformational analysis of biosynthetic precursor-type lipid A: monomolecular state and supramolecular assembly. Org Biomol Chem 2:3557–3565

70. Wang W, Sass HJ, Zähringer U, Grzesiek S (2008) Structure and dynamics of ^{13}C,^{15}N-labeled lipopolysaccharides in a membrane mimetic. Angew Chem Int Ed Engl 47:9870–9874
71. D'Errico G, Silipo A, Mangiapia G, Molinaro A, Paduano L, Lanzetta R (2009) Mesoscopic and microstructural characterization of liposomes formed by the lipooligosaccharide from *Salmonella minnesota* strain 595 (Re mutant). Phys Chem Chem Phys 11:2314–2322
72. D'Errico G, Silipo A, Mangiapia G, Vitiello G, Radulescu A, Molinaro A, Lanzetta R, Paduano L (2010) Characterization of liposomes formed by lipopolysaccharides from *Burkholderia cenocepacia*, *Burkholderia multivorans* and *Agrobacterium tumefaciens*: from the molecular structure to the aggregate architecture. Phys Chem Chem Phys 12:13574–13585
73. Leone S, Molinaro A, Pessione E, Mazzoli R, Giunta C, Sturiale L, Garozzo D, Lanzetta R, Parrilli M (2006) Structural elucidation of the core-lipid A backbone from the lipopolysaccharide of *Acinetobacter radioresistens* S13, an organic solvent tolerating Gram-negative bacterium. Carbohydr Res 341:582–590
74. Wang Z, Li J, Altman E (2006) Structural characterization of the lipid A region of *Aeromonas salmonicida* subsp. *salmonicida* lipopolysaccharide. Carbohydr Res 341:2816–2825
75. Liparoti V, Molinaro A, Sturiale L, Garozzo D, Nazarenko EL, Gorshkova RP, Ivanova EP, Shevcenko LS, Lanzetta R, Parrilli M (2006) Structural analysis of the deep rough lipopolysaccharide from Gram-negative bacterium *Alteromonas macleodii* ATCC 27126: the first finding of β-Kdo in the inner core of lipopolysaccharides. Eur J Org Chem 4710–4716
76. Schwudke D, Linscheid M, Strauch E, Appel B, Zähringer U, Moll H, Müller M, Brecker L, Gronow S, Lindner B (2003) The obligate predatory *Bdellovibrio bacteriovorus* possesses a neutral lipid A containing α-D-mannoses that replace phosphate residues: similarities and differences between the lipid As and the lipopolysaccharides of the wild type strain *B. bacteriovorus* HD100 and its host-independent derivative HI100. J Biol Chem 278: 27502–27512
77. Rund S, Lindner B, Brade H, Holst O (1999) Structural analysis of the lipopolysaccharide from *Chlamydia trachomatis* serotype L2. J Biol Chem 274:16819–16824
78. Toman R, Garidel P, Jorg A, Slaba K, Hussein A, Koch MHJ, Brandenburg K (2004) Physicochemical characterization of the endotoxins from *Coxiella burnetii* strain Priscilla in relation to their bioactivities. BMC Biochem 5:1
79. Kay W, Petersen BO, Duus JØ, Perry MB, Vinogradov E (2006) Characterization of the lipopolysaccharide and β-glucan of the fish pathogen *Francisella victoria*. FEBS J 273: 3002–3013
80. Asai Y, Makimura Y, Kawabata A, Ogawa T (2007) Soluble CD14 discriminates slight structural differences between lipid as that lead to distinct host cell activation. J Immunol 179:7674–7683
81. Krasikova IN, Kapustina NV, Isakov VV, Dmitrenok AS, Dmitrenok PS, Gorshkova NM, Solov'eva TF (2004) Detailed structure of lipid A isolated from lipopolysaccharide from the marine proteobacterium *Marinomonas vaga* ATCC 27119. Eur J Biochem 271:2895–2904
82. Kumada H, Haishima Y, Umemoto T, Tanamoto K-I (1995) Structural study on the free lipid A isolated from lipopolysaccharide of *Porphyromonas gingivalis*. J Bacteriol 177:2098–2106
83. Krasikova IN, Kapustina NV, Svetashev VI, Gorshkova RP, Tomshich SV, Nazarenko EL, Komandrova NA, Ivanova EP, Gorshkova NM, Romanenko LA, Mikhailov VV, Solov'eva TF (2001) Chemical characterization of lipid A from some marine proteobacteria. Biochemistry (Moscow) 66:1047–1054
84. Schromm AB, Brandenburg K, Loppnow H, Moran AP, Koch MH, Rietschel ET, Seydel U (2000) Biological activities of lipopolysaccharides are determined by the shape of their lipid A portion. Eur J Biochem 267:2008–2013
85. Leone S, Molinaro A, De Castro C, Baier A, Nazarenko EL, Lanzetta R, Parrilli M (2007) Absolute configuration of 8-amino-3,8-dideoxyoct-2-ulosonic acid, the chemical hallmark of lipopolysaccharides of the genus *Shewanella*. J Nat Prod 70:1624–1627

Structure of the Lipopolysaccharide Core Region

2

Otto Holst

2.1 Introduction

Most Gram-negative bacteria (exceptions are *Treponema pallidum*, *Borrelia burgdorferi* and *B. hispanica*, *Sphingomonas capsulata* and *S. paucimobili*, *Thermus thermophilus*, and *Meiothermus taiwanensis* [1–7]) contain lipopolysaccharide (LPS) in their outer membrane [8]. In general, the LPS may be present either in the smooth form (S, possessing the polysaccharide region) or in the rough form (R, lacking the polysaccharide, also called lipooligosaccharide, LOS). Both forms contain lipid A [9] and a core oligosaccharide (OS) that comprises up to 15 sugar residues [8–17]. In the S-form LPS, the core OS region is substituted by a polysaccharide, which most often is an O-specific polysaccharide (O-antigen, for structures see Chap. 3), and in other cases is the enterobacterial common antigen (only in *Enterobacteriaceae* [18]) or a capsular polysaccharide [19]. For a long time a dogma existed claiming that mutants of LPS-containing Gram-negative bacteria are not viable without a minimal core OS structure. However, viable mutants synthesizing only lipid A or a precursor thereof have been recently isolated [20, 21].

2.2 Historic Outline

Richard Pfeiffer identified endotoxin in 1892 [22]. However, it took some 60 years to establish the appropriate extraction protocol (the hot phenol–water procedure [23, 24]) by which the isolation of rather pure endotoxin was possible. It became

O. Holst (✉)
Division of Structural Biochemistry, Research Center Borstel, Leibniz-Center for Medicine and Biosciences, Parkallee 4a/c, D-23845 Borstel, Germany
e-mail: oholst@fz-borstel.de

Y.A. Knirel and M.A. Valvano (eds.), *Bacterial Lipopolysaccharides*,
DOI 10.1007/978-3-7091-0733-1_2,

clear soon after that endotoxin consists of sugars, phosphates, and fatty acids, and is a lipoglycan termed "lipopolysaccharide" [25]. However, it was also shown that LPS purity depends on the method of extraction. Purified LPS from *Salmonella* displayed endotoxic and antigenic activities. The former could be assigned to the lipid moiety and the latter to the polysaccharide, which was also called O-antigen. At that time, researchers were convinced that LPS was simply built up from only a lipid and a polysaccharide (the lipid was later named lipid A since also a second lipid, lipid B [26], and a third one were identified; but it turned out that the latter two were not part of LPS and thus of no importance here). In the early 1960s, it had become clear from analyses of samples extracted from *Salmonella* that LPS possessed two classes of sugars: common ones and those that occurred only in particular LPS [27]. Consequently, a working hypothesis was generated claiming that all *Salmonella* LPS should possess a common carbohydrate core substituted by the O-antigen.

The structure of a complete LPS molecule was thought to be much more complex than it is known today [28, 29]. The lipid A was believed to consist of a poly-(*N*-β-hydroxymyristoyl-D-glucosamine phosphate), which is substituted by ester-linked long-chain fatty acids and, via 3-deoxy-D-*manno*-oct-2-ulosonic acid (Kdo) [30] at a then unknown position, by heptose phosphate chains via phosphodiester bridges (Fig. 2.1). These chains in turn are substituted by short OSs of hexoses and GlcN as well as by the O-antigen consisting of repeating units. Thus, the LPS was proposed to possess a highly polymeric branched comb-like architecture. This overall structure was revised at the turn of the decade, and in 1971, a general LPS architecture was reported, which is still valid for most LPS today [31]. At that time, newer data had indicated that lipid A is a phosphorylated and acylated GlcN disaccharide. However, phosphodiester bridges between these units could still not be excluded. This was also the case concerning the core region but its general structure had become that of an OS.

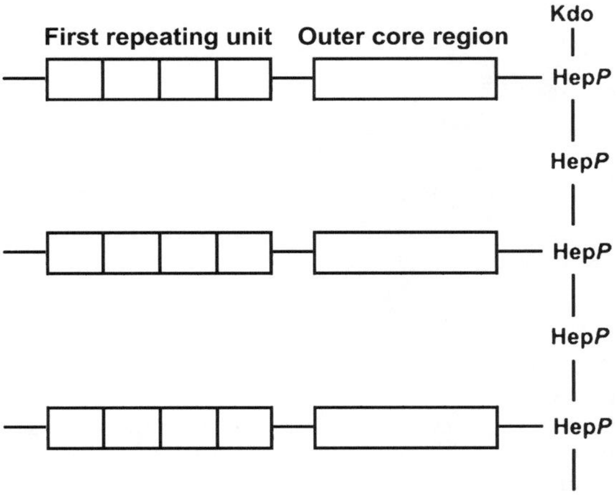

Fig. 2.1 The architecture of the core-O-antigen regions proposed in 1966 [28, 29]

The basis for this progress was the availability of a variety of rough (R)-type mutants, first obtained from *Salmonella* [29]. Also, a new method for the extraction of particularly R-type LPS was invented [32]. All R-mutants synthesized only short-chain LPS (LOS) comprising lipid A and the core region or a part thereof. They are very helpful for the detailed structural analysis of the enterobacterial LPS core region, which still holds true to date. Investigations on such LPS had been performed already in the early 1960s and led to the identification of an RI and an RII core, which were later re-named Rb and Ra (the complete core), respectively. The Ra-mutant was defective in O-antigen biosynthesis, and the Rb mutant additionally in core biosynthesis. Other mutants with shorter core regions were also identified and termed Rc through Re [28]. Later, additional *Salmonella* mutants (named Rd_1 and Rd_2) and mutants obtained from other bacteria, like *Escherichia coli* and *Citrobacter*, were obtained and their LPS investigated (summarized in Ref. [30]). Also, so-called SR-mutants were available which possessed in their LPS a complete core region plus one repeating unit; thus, the linkage position of the O-antigen at the core region could be identified.

In the following years, significant progress concerning the structural analysis of the core regions of LPS from *Salmonella* (one core type) and *E. coli* (in which five core types were identified, R1–R4 and K-12) could be achieved (summarized in Ref. [33]). It was proposed that the core OS could be subdivided into inner (containing Kdo and heptose) and outer (consisting of hexoses and hexosamines) regions. However, no complete structures were available due to difficulties in chemical analyses of the core region at that time (see below). Also, it had been proposed that there were LPS that contained no Kdo, like that of *Vibrio cholerae*. This was revised later when the presence of Kdo-4-phosphate was identified in this LPS, due to which the photometric determination of Kdo had failed in earlier studies.

From the mid-1980s on, owing to improved analytical tools and methods, including modern nuclear magnetic resonance (NMR) spectroscopy and mass spectrometry (MS), and to development of protocols, which e.g. brought about the isolation of a complete lipid A-core backbone, a lot of core structures from a range of bacterial species have been reported (summarized in Ref. [8–17]; see also below). A comparison of these structures clearly verifies the hypothesis of the 1960s, which proposed a broad variety of core OS structures. It is clear now that all LPS molecules contain at least one Kdo residue (which in some LPS may to a certain extent be replaced by D-*glycero*-D-*talo*-oct-2-ulosonic acid, Ko) and that not all core regions contain heptose or phosphate residues and/or not all may be subdivided into inner and outer core.

2.3 Structural Analysis of the Core OS Region

Owing to (1) the difficult chemistry of Kdo and the phosphate substitution, (2) the lack of procedures for the isolation of highly purified OSs, and (3) limitations in the application of NMR spectroscopy and MS, the structural analysis of the core region could not be completely elucidated for quite some time. This was improved since the late 1980s when an appropriate protocol for the determination of Kdo was developed (reviewed in Ref. [34]). However, a full analysis of the phosphate substitution still remains problematic.

The current analytical strategy includes the investigation of the isolated and purified LPS by MS prior to the application of any degradation protocol. This provides valuable information on the LPS composition. Apart from matrix-assisted laser desorption/ionization time-of-flight (MALDI-TOF) MS [35], highly resolved spectra have been obtained from electrospray ionization (ESI) MS [36], also in combination with capillary electrophoresis of either O-deacylated LPS [37] or native LPS or lipid A [38]. A degradation pathway to obtain phosphorylated OSs from LPS [39] was developed, which utilizes successive mild hydrazinolysis and 4 M KOH treatment followed by isolation of pure compounds by high-performance anion-exchange chromatography. In the recent years, this protocol has been applied by several investigators and has helped characterizing a variety of core OS regions from various bacteria.

This approach has some limitations in the presence of phosphodiester, diphosphate, diphosphodiester, acetyl and carbamoyl groups, which are components of many LPS and are readily cleaved under strong alkaline conditions, making it impossible to identify their positions in the core OS. Also, the alkaline treatment leads to phosphate migration in the case of substitution by 2-aminoethyl phosphate, and to β-elimination of 4-substituted uronic acid residues [39]. In some cases, the O-deacylated product obtained after mild hydrazinolysis can be analyzed by NMR spectroscopy particularly, using $^{31}P,^{1}H$-correlated experiments, which helps to demonstrate the presence and location of phosphodiester, diphosphate, and diphosphodiester groups. However, in other cases spectra may be poorly resolved and NMR spectroscopy is not applicable. MS may help to identify the sugar residue that is substituted but the identification of the linkage position may still hardly be possible.

It is recommended to employ (in addition) the "traditional" protocol to isolate the core region, namely mild acetic acid (buffer) hydrolysis in order to cleave the linkage between the core region and lipid A [31, 40]. This alternative approach is applicable in any case and not only if a deacylated product cannot be obtained (e.g., when the anomeric position of the lipid A backbone is not substituted) or the deacylation procedure results only in a partial structure (e.g., due to β-elimination of a 4-substituted uronic acid). In a variant, the addition of 1% sodium dodecyl sulfate improves the cleavage [41]. Two other protocols have been described, using either a mixture of isobutanoic acid and 1 M ammonium hydroxide (5:3, v:v) [42] or triethylamine citrate [43].

From S-form LPS, usually a mixture of an O-antigen-core polysaccharide and a core OS is obtained, and sometimes investigators have the good luck to yield in addition an SR-OS fraction. Such mixture can be routinely separated by gel-permeation chromatography, e.g. on a column of Sephadex G-50 eluted with a pyridine-acetic acid buffer.

From R-form LPS (LOS), mild acid hydrolysis yields (an) incomplete core OS(s), since if branched Kdo-saccharide is present, the branching Kdo residue(s) is/are also cleaved. Thus, the structure of this moiety cannot be determined. Also, Kdo residues that may be present at the nonreducing terminus of the core (like in *Klebsiella pneumoniae* [44]) are cleaved and structural information is lost. Still, products can be isolated that represent the major part of the core region. With regard to the substitution pattern, phosphodiester, diphosphate, diphosphodiester, acetyl and carbamoyl groups are at least in part retained (see for example Refs. [45, 46]). The isolation of pure phosphorylated compounds can be achieved by high-performance anion-exchange chromatography utilizing a gradient of sodium acetate in water at pH 6 [39].

Prior to any degrading chemical analysis, pure OSs are extensively investigated by NMR spectroscopy, applying in particular homonuclear and heteronuclear two-dimensional experiments. Also, small amounts (10–20 μg) are used for MS studies. In chemical analysis of LPS as well as the O-antigen and the core region (the "wet chemistry" approach [47]), investigators have to deal with different sugars possessing different stability under acidic hydrolysis conditions. Routine hydrolysis protocols comprise e.g. the use of 0.1 M HCl (100°C, 48 h) or trifluoroacetic acid (2 M, 120°C, 2 h, or 4 M, 100°C, 4 h). After such hydrolyses followed by reduction and acetylation, hexoses and pentoses are identified by GLC. However, more stable linked amino sugars are identified only in small amounts and acid-labile compounds like Kdo and dideoxy sugars are largely destroyed.

To get an idea about which sugars are present in the core region, it is recommended to begin with two different methanolysis protocols, e.g. analysis of acetylated methyl glycosides after mild (0.5 M methanolic HCl, 85°C, 45 min) and strong (2 M methanolic HCl, 85°C, 16 h) methanolysis conditions. Here, it is possible to detect deoxyhexoses, hexoses, pentoses, uronic acids, amino sugars, and Kdo as well as fatty acids of lipid A. After that, appropriate conditions for the production of alditol acetates can be chosen for the identification and quantification of different monosaccharides. To determine the substitution pattern of the monosaccharide residues, methylation analysis is performed, utilizing either the Hakomori methylation protocol [48] or that developed by Ciucanu and Kerek [49]. Particular protocols on methylation analysis of the Kdo region have been published [50]. If uronic acids are present, these are esterified and the methyl ester function is reduced prior to the hydrolysis step. Samples are finally analyzed by GLC-MS.

2.4 General Structural Features of the Core Region

The chemical structures of the core regions of bacterial LPS have been regularly and extensively summarized since 1992 [8–17]. Therefore, mostly general features of core structures are described here.

All core region chemical structures identified so far are less varied than those of O-antigens. Still, only one structural element is present in all core regions, namely that Kdo residue which links the core to the lipid A. In many bacteria, the core region contains L-*glycero*-D-*manno*-heptose (L,D-Hep) and an L-α-D-Hep-(1 → 3)-L-α-D-Hep-(1 → 5)-[α-Kdo-(2 → 4)]-α-Kdo tetrasaccharide (Hep II, Hep I, Kdo II, and Kdo I, respectively), which may be further substituted by other sugars or phosphate residues, or sometimes by acetyl groups or amino acids. In addition to L,D-Hep, several LPS contain its biosynthetic precursor, D-*glycero*-D-*manno*-heptose (D,D-Hep). There are other LPS that contain only D,D-Hep or even lack any heptose. Either Kdo I (in *Acinetobacter*) or Kdo II (in Burkholderiaceae, *Yersinia pestis*, and *Serratia marcescens*) may be replaced by the stereochemically similar sugar acid Ko, the biosynthesis of which and regulation of the exchange between Kdo and Ko have not been elucidated so far.

2.5 Core Structures of Some Important Pathogenic Bacteria

2.5.1 Enterobacteriaceae

Two types of enterobacterial core regions have been recognized, namely the *Salmonella* type core and the core different from the *Salmonella* type. The first is characterized by the common structural element L-α-D-Hep-(1 → 7)-L-α-D-Hep-(1 → 3)-L-α-D-Hep-(1 → 5)-Kdo substituted at O-3 of Hep II by glucopyranose. Hep I and II are phosphorylated and O-4 of Hep I is not substituted by a saccharide. In the second core type, the same common partial structure is present but lacks Glc*p* at O-3 of Hep II, the heptose residues are not generally phosphorylated, and Hep I is substituted by a hexose residue or an OS at O-4. The core regions of LPS from *E. coli*, *Providencia* and *Yersinia* species are shown as examples. Those of *Providencia* and *Yersinia* occur as two glycoforms differing in sugar residues in the outer regions (D-Glc*p* vs. D-Glc*p*NAc or β-D-Gal*p* vs. D-α-D-Hep, respectively).

Escherichia coli R4 (*Salmonella* type) [51]

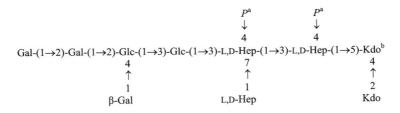

```
                                                        P^a               P^a
                                                        ↓                 ↓
                                                        4                 4
Gal-(1→2)-Gal-(1→2)-Glc-(1→3)-Glc-(1→3)-L,D-Hep-(1→3)-L,D-Hep-(1→5)-Kdo^b
                    4                              7                             4
                    ↑                              ↑                             ↑
                    1                              1                             2
                  β-Gal                         L,D-Hep                         Kdo
```

Providencia rustigianii O-34 (*Salmonella* type) [52]

```
                                                        PEtn           PPEtn^c
                                                         ↓               ↓
                                                         6               4
Glc or GlcNAc-(1→2)-Glc-(1→2)-Gal-(1→3)-Glc-(1→3)-L,D-Hep-(1→3)-L,D-Hep-(1→5)-Kdo^b
                                                         7                      8
                                                         ↑                      ↑
                                                         1                      1
                                       β-GalA-(1→7)-L,D-Hep               β-L-Ara4N
```

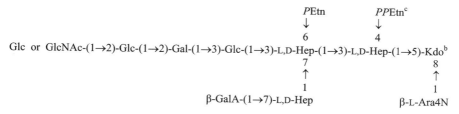

Yersinia pestis (other than *Salmonella* type), various strains [53]

```
      β-GlcNAc^d-(1→3)-L,D-Hep-(1→3)-L,D-Hep-(1→5)-Kdo^b
                           7             4         4
                           ↑             ↑         ↑
                           1             1         2
β-Gal or D,D-Hep-(1→7)-L,D-Hep         β-Glc    Kdo or Ko
```

Here and below where not stated otherwise, sugars are α-D-pyranosides. Abbreviations: *P*Etn, 2-aminoethyl phosphate; *PP*Etn, 2-aminoethyl diphosphate; L-Ara4N, 4-deoxy-L-arabinose. [a]Product obtained after *O*-deacylation. Further substituents are not known. [b]This Kdo residue links the core region to lipid A. [c]Either *PP*EtN or *P* is present at this position. [d]Non-stoichiometric substitution.

2.5.2 Pasteurellaceae

A high number of core structures have been reported for LPS of the genera *Haemophilus*, *Histophilus* and *Pasteurella*.

The core region of LPS from *Haemophilus influenzae* possesses as common partial structure the L-α-D-Hep-(1 → 2)-[*P*EtN → 6]-L-α-D-Hep-(1 → 3)-[β-D-Glc*p*-(1 → 4)]-L-α-D-Hep-(1 → 5)-[*PP*EtN → 4]-α-Kdo saccharide. Only this one Kdo residue is present which links the core region to lipid A. Other substitutions occur mainly at β-D-Glc*p* and Hep III. In particular, core regions of non-typeable *H. influenzae* strains have been investigated in the past years [9, 17].

Haemophilus influenzae strains 1200, 1268 [54]

```
                                       β-Glc-(1→4)-L,D-Hep-(1→5)-Kdo-4←PPEtn
                                          PCho           3
                                           ↓             ↑
                                           6             1
β-GalNAc-(1→3)-Gal-(1→4)-β-Gal-(1→4)-β-Glc-(1→3)-L,D-Hep-6←PEtn
                                                         2
                                                         ↑
                                                         1
β-GalNAc-(1→3)-Gal-(1→4)-β-Gal-(1→4)-β-Glc-(1→2)-L,D-Hep
```

Abbreviation: *P*Cho, 2-trimethylammonioethyl phosphate (phosphocholine).

Haemophilus influenzae [55–59]

General structure

R^1→6)-[R^2→4)]-β-Glc[a]-(1→4)-L,D-Hep[b]-(1→5)-Kdo[c]-4←R^3
3
↑
1
R^4→2)-L,D-Hep[d]-(1→2)-L,D-Hep-6←*P*Etn
3
↑
R^5

Strain	R^1	R^2	R^3	R^4	R^5
1003	*P*Cho	OAc	*PP*Etn	Neu5Ac-(2→3)-β-Gal-(1→4)-β-GlcOAc-(1	OAc
RM118	*P*Cho	Neu5Ac-(2→3)-β-Gal-(1→4)-β-GlcNAc-(1→3)-β-Gal-(1	*PP*Etn	Gal-(1→4)-β-Gal-(1→4)-β-Glc-(1	H
162	*P*Cho	H	n.d.	β-GalNAc-(1→3)-β-Gal-(1→4)-β-Gal-(1→4)-β-Glc-(1	H
981 Form 1	*P*Cho	β-GalNAc-(1→3)-Gal-(1→4)-β-Gal-(1→4)-β-Glc-(1	*PP*Etn	H	*P*Etn
981 Form 2	β-Gal-(1→4)-D,D-Hep-(1	Gal-(1→4)-β-Gal-(1→4)-β-Glc-(1	*PP*Etn	H	*P*Etn
723	*P*Cho	OAc	n.d.	H	*P*Etn

[a]O-Acetylated at position 3 in strain 723. [b]O-Acetylated at position 2 in strain 723. [c]This Kdo residue links the core region to lipid A. [d]O-Acetylated at unknown position in strain 723.

In LPS of *Histophilus* (*Haemophilus*) *somnus*, the core region comprises the α-Kdo-(2 → 4)-α-Kdo disaccharide and the common L-α-D-Hep-(1 → 3)-[β-D-Glc*p*-(1 → 4)]-L-α-D-Hep-(1 → 5)-α-Kdo structure. The β-D-Glc*p* residue is further substituted and Hep II carries either α-D-Glc*p*NAc or β-D-Gal*p* at O-2. Hep II may contain one or two *P*EtN residues, and *N*-acetylneuraminic acid may be incorporated which is important for serum resistance and reduction of antibody binding. Two examples are shown below.

Histophilus somnus [60, 61]

```
738      β-Gal-(1→4)-β-Glc-(1→4)-L,D-Hep-(1→5)-Kdo
                                  3          4
                                  ↑          ↑
                                  1          2
                  GlcNAc-(1→2)-L,D-Hep      Kdo

129Pt    Gal-(1→4)-β-Gal-(1→4)-β-Glc-(1→4)-L,D-Hep-(1→5)-Kdo
                                            3          4
                                            ↑          ↑
                                            1          2
                            GlcNAc-(1→2)-L,D-Hep      Kdo
                                            6
                                            ↑
                                           PEtn
```

The core region of the LPS from *Pasteurella multocida* has as a common feature of the L-α-D-Hep-(1→2)-L-α-D-Hep-(1 → 3)-[β-D-Glc*p*-(1 → 4)]-[α-D-Glc*p*-(1 → 6)]-L-α-D-Hep-(1 → 5)-α-Kdo hexasaccharide [9]. In strain VP161, both α-Kdo-(2 → 4)-α-Kdo disaccharide and α-Kdo-4 ← *PP*EtN are present [62, 63].

2.5.3 Pseudomonadaceae

In the core of LPS of *Pseudomonas aeruginosa*, a α-D-Gal*p*N residue is present which in most strains studied is N-acylated by L-alanine. It substitutes O-3 of Hep II of the L-α-D-Hep-(1 → 3)-L-α-D-Hep-(1 → 5)-[α-Kdo-(2 → 4)]-α-Kdo tetrasaccharide, in which Hep II is further substituted at O-7 by a carbamoyl group. In all strains studied so far, the core region is highly phosphorylated. The outer region occurs as two glycoforms and is randomly O-acetylated.

Pseudomonas aeruginosa [45]

```
Glycoform 1
                      Ala              CONH2          P
                       ↓                 ↓            ↓
                       2                 7            4
Glc-(1→6)-β-Glc-(1→3)-GalN-(1→3)-L,D-Hep-(1→3)-L,D-Hep-(1→5)-Kdo
                       4                 6            2          4
                       ↑                 ↑            ↑          ↑
                       1                 P          PPEtn^b      2
Glc^a-(1→2)-L-Rha-(1→6)-Glc                                     Kdo
```

Glycoform 2

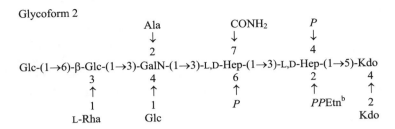

Ala CONH₂ *P*
↓ ↓ ↓
2 7 4
Glc-(1→6)-β-Glc-(1→3)-GalN-(1→3)-L,D-Hep-(1→3)-L,D-Hep-(1→5)-Kdo
3 4 6 2 4
↑ ↑ ↑ ↑ ↑
1 1 *P* *PP*Etn[b] 2
L-Rha Glc Kdo

O-Acetylation is not shown. [a]This glucose residue is a non-oligatory component. [b]Either *PP*EtN or *P* is present at this position.

In a *wbjE* mutant of *P. aeruginosa* PA103 (serogroup O-11), the full glycoform 1 core contains Gal*p*NAc in place of Gal*p*NAla and two or three phosphate groups as mono- or di-phosphates but no *P*EtN [64].

The inner core region of *Pseudomonas syringae* is similar to that of *P. aeruginosa*, whereas the outer region is different but also occurs as two glycoforms, one of which contains Kdo.

Pseudomonas syringae pv. *phaseolicola* [65]

Ala CONH₂ *P*
↓ ↓ ↓
2 7 4
β-GlcNAc-(1→2)-β-Glc-(1→3)-GalN-(1→3)-L,D-Hep-(1→3)-L,D-Hep-(1→5)-Kdo
4 6 2 4
↑ ↑ ↑ ↑
1 *P* *PP*Etn 2
L-Rha-(1 or Kdo-(2→6)-Glc Kdo

2.5.4 Moraxellaceae

Various core structures have been identified in the past in this family, either possessing Ko or not. Only one novel core region was identified recently, namely that of the LPS from the allergy-protective bacterium *Acinetobacter lwoffii* F78 [66]. Unexpectedly, it possesses the α-Kdo-(2 → 8)-α-Kdo disaccharide which was found earlier only in the core regions of LPS from the genera *Chlamydia*/*Chlamydophila* [12].

Acinetobacter haemolyticus ATCC 17906 [67]

Dha-(1→6)-β-Glc-(1→4)-Dha-(2→6)-Glc-(1→6)-Glc-(1→6)-β-Glc-(1→4)-Glc-(1→5)-Sug[a,b]
6 4
↑ ↑
P 2
Kdo

Acinetobacter baumannii NCTC 10303 [68]

```
GlcNAc-(1→4)-GlcNA^d-(1→4)-Kdo-(2→5)-Kdo^b
                              7         4
                              ↑         ↑
                              2         2
L-Rha-(1[→3)-L-Rha-(1→]n8)-Kdo        Kdo          n = 1-4, mainly 2
```

Acinetobacter lwoffii F78 [66]

```
GalNAc-(1→3)-β-GalNAc-(1→6)-Glc-(1→5)-Kdo
                                       2
                                       ↓
                                       8
                β-Glc-(1→6)-Glc-(1→5)-Kdo
                                       4
                                       ↑
                                       2
                                      Kdo
```

Abbreviations: Dha, 3-deoxy-D-*lyxo*-heptulosaric acid; GlcNA, 2-amino-2-deoxy-D-glucuronic acid. [a]Sug stands for Kdo (minor) or Ko (major). [b]This Kdo residue links the core region to lipid A.

2.5.5 Vibrionaceae

Several structures of LPS core regions from *Vibrio cholerae* have been established [8–13]. They have as common feature one Kdo residue phosphorylated at O-4 and substituted at O-5 by the α-D-Glc*p*NAc-(1 → 7)-L-α-D-Hep-(1 → 2)-L-α-D-Hep-(1 → 3)-L-α-D-Hep tetrasaccharide, in which Hep I is substituted by β-Glc*p* at O-4 and α-Glc*p* at O-6. Variations in core structures occur by different substituents at O-6 of both Glc*p* residues and Hep II as well as at O-2 or O-4 of Hep III. In all characterized structures, a β-linked fructose or sedoheptulose (D-*altro*-hept-2-ulose) residue is present at O-6 of the β-linked Glc*p*.

Vibrio cholerae [8–13]

```
General structure              R1-Glc
                                  1        P
                                  ↓        ↓
                                  6        4
           R2-β-Glc-(1→4)-L,D-Hep-(1→5)-Kdo^a
                                  3        ↑
                                  ↑        R3
                                  1
GlcNAc-(1→7)-L,D-Hep-(1→2)-L,D-Hep-R4
                                  ↑
                                  R5
```

[a]This Kdo residue links the core region to lipid A. R^1 may be H or Glc-(1→6), R^2 D-fructofuranose or D-sedoheptulofuranose (Sed*f*), R^3 H or *P*Etn→7, R^4 H or L,D-Hep-(1→6), R^5 H or O-specific polysaccharide (OPS); in strain H11, R^5 is OPS-(1→4)-β-Sed*f*-(2→3)-β-Gal-(1→3).

Several LPS core regions of the genus *Aeromonas* have also been investigated [69–71]. For example, *A. salmonicida* ssp. *salmonicida* comprises four L-α-D-Hep residues, of which Hep I is α-(1 → 5)-linked to Kdo-4-phosphate and in turn carries the L-α-D-Hep-(1 → 6)-α-D-Glc*p* disaccharide at O-4 [69]. The heptose residue of this unit is differently substituted at O-4, resulting in heterogeneity. In addition to these four L-α-D-Hep residues, the core region of *A. hydrophila* AH-3 contains a D-α-D-Hep disaccharide substituting O-6 of β-D-Glc*p* at O-4 of Hep I [70].

2.5.6 Burkholderiaceae

Bacteria of the genus *Burkholderia* possess LPS core structural features of chemotaxonomic value [72, 73]. The core regions that have been identified so far include in various amounts both α-Kdo-(2 → 4)-α-Kdo and α-Ko-(2 → 4)-α-Kdo disaccharides. The common partial structure is L-α-D-Hep-(1 → 7)-L-α-D-Hep-(1 → 3)-[β-D-Glc*p*-(1 → 4)]-L-α-D-Hep-(1 → 5)-α-Kdo.

In the LPS core of *Burkholderia caryophylli*, O-3 of Hep II is substituted by a branched glycan to which one of the O-specific polysaccharides (caryophyllan or caryan) is linked [73]. This core region possesses as a unique feature two L-α-D-Hep-(1 → 5)-α-Kdo moieties.

The core region of the LPS of *Burkholderia pyrrocinia* contains as a peculiar feature a linear heptose pentasaccharide consisting of four L,D-Hep and one terminal D,D-Hep residues.

Burkholderia pyrrocinia BTS7 [74]

```
                                                                         β-Glc
                                                                           1
                                                                           ↓
                                                                           4
D-α-D-Hep-(1→7)-L-α-D-Hep-(1→3)-L-α-D-Hep-(1→7)-L-α-D-Hep-(1→3)-L-α-D-Hep-(1→5)-Kdo
                                                   2                              4
                                                   ↑                              ↑
                                                   1                              2
                                                 L-Rha           β-L-Ara4N-(1→8)-Ko
```

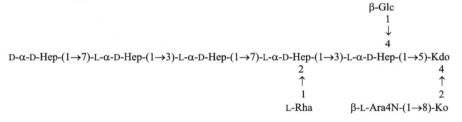

B. pyrrocinia shares the α-L-Rha-(1 → 2)-L-α-D-Hep-(1 → 3)-[β-D-Glc*p*-(1 → 4)]-L-α-D-Hep-(1 → 5)-[-Ko-(2 → 4)]-α-Kdo hexasaccharide with *Burkholderia cepacia* [71] and *Ralstonia solanacearum* [75]. The latter differs from representatives of the genus *Burkholderia* in the attachment of β-L-Ara*p*4N to Hep I rather than Ko.

Ralstonia solanacearum Toudk-2 [75]

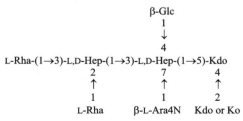

```
                             β-Glc
                               1
                               ↓
                               4
L-Rha-(1→3)-L,D-Hep-(1→3)-L,D-Hep-(1→5)-Kdo
              2                7         4
              ↑                ↑         ↑
              1                1         2
            L-Rha         β-L-Ara4N   Kdo or Ko
```

2.5.7 Neisseriaceae

In general, neisserial LPS are of the rough type. Thus, LOS and core regions consist of the L-α-D-Hep-(1 → 3)-L-α-D-Hep-(1 → 5)-Kdo trisaccharide, which is substituted by β-D-Glc*p* at O-4 of Hep I and by α-D-Glc*p*NAc at O-2 of Hep II. Structural variations occur by different substituents at O-4 of β-D-Glc*p* and at Hep II.

Neisseria [10]

```
General structure    R¹-(1→4)-β-Glc-(1→4)-L,D-Hep[a]-(1→5)-Kdo[b]
                                               3              4
                                               ↑              ↑
                                               1              R²
                              GlcNAc[c]-(1→2)-L,D-Hep[a]-(3←R³
                                               6/7
                                               ↑
                                               R⁴
```

Neisseria meningitidis M986 [76]

R^1	R^2	R^3	R^4
β-Gal-(1→4)-β-GlcNAc-(1→3)-β-Gal	Kdo-(2	*P*Etn	*P*Etn

[a]The absolute configuration was not determined in *N. gonorrhoeae*. [b]This Kdo residue links the core region to lipid A. [c]Non-stoichiometric O-acetylation (~0.5) occurs at O-3 of this residue.

Six LPS serotypes have been identified in the causative agent of gonorrhoea, *N. gonorrhoeae* [77]. The resistance of *N. gonorrhoeae* in patients and when grown in serum-containing media against the bactericidal effects of complement may be caused by sialylation of the LPS which occurs at a terminal β-D-Gal*p* residue (1 → 4)-linked to α-D-Glc*p*NAc. A further interesting observation is that phase variation of LPS caused by a different degree of sialylation controls both entry of *N. gonorrhoeae* into human mucosa cells (low degree of sialylation) and the resistance to complement-mediated killing (high degree of sialylation).

N. meningitidis is a causative agent of severe human diseases, as sepsis and meningitis. It is differentiated into 12 LPS serotypes [77], of which 8 possess lacto-N-neotetraose β-Gal*p*-(1 → 4)-β-Glc*p*NAc-(1 → 3)-β-Gal*p*-(1 → 4)-β-Glc*p*. This common partial structure is identical to the glycosyl moiety of lactoneotetraglycosylceramide, which is present on the membrane of human erythrocytes and thought to represent a mimicry antigen that helps to evade the host defense. Sialylation occurs also in some meningococcal LPS where α-Neu5Ac is (2 → 3)-linked to the terminal Gal*p* residue of lacto-N-neotetraose.

Other substituents of the core structures have been reported, e.g. two *P*EtN residues at Hep I in *N. meningitidis* strain BZ157 *galE* [78], glycine on O-7 of Hep II in various immunotypes [78, 79], and an *O*-acetyl group may be present at O-3 of Glc*p*NAc [80]. It has been proven that *P*EtN is located at O-6 of Hep II [81].

The considerable structural heterogeneity has been profiled by MALDI MS and related to cytokine induction [82].

2.6 Conclusions

To date, after more than 60 years of intensive structural research, a rather high number of core structures from LPS of various bacterial species have been elucidated. The general principle consists of a negatively charged core region (provided by phosphoryl substituents and/or sugar acids like Kdo and uronic acids), which strengthens the rigidity of the Gram-negative cell wall through intermolecular cationic cross-links.

The linkage of the core region to lipid A occurs always via a Kdo residue. However, in *Acinetobacter* Ko may replace Kdo non-stoichiometrically at this position. Also, there are core structures containing and lacking heptoses. In the first type, L,D-Hep or D,D-Hep alone, or both may be present. When present, D,D-Hep either decorates the inner core region (e.g. in *Yersinia*) or is attached to more remote parts of the carbohydrate chain. Furthermore, many core regions possess as partial structure the L-α-D-Hep-(1 → 3)-L-α-D-Hep-(1 → 5)-α-Kdo trisaccharide, and a genus often contains in its LPS a common core structural theme that varies by certain (non-)carbohydrate substituents.

Although mutants of *E. coli* K-12 were isolated that are viable possessing only lipid IV_A in their cell envelope, the common structural principle of LPS may still be described as core OS plus lipid A. Any (O-specific) polysaccharide expression in LPS is not a prerequisite for bacterial survival; still, the finding that the polysaccharide portion in S-form LPS may be furnished either by the O-chain or a capsular structure or the ECA clearly indicates that this LPS form is highly advantageous for many species.

Acknowledgements Current research of O. H. is supported by the Deutsche Forschungsgemeinschaft (grants SFB-TR 22, project A2, and FOR 585, project 8).

References

1. Takayama K, Rothenberg RJ, Barbour AG (1987) Absence of lipopolysaccharide in the Lyme disease spirochete, *Borrelia burgdorferi*. Infect Immun 55:2311–2313
2. Hardy PH Jr, Levin J (1983) Lack of endotoxin in *Borrelia hispanica* and *Treponema pallidum*. Proc Soc Exp Biol Med 174:47–52
3. Kawahara K, Seydel U, Matsuura M, Danbara H, Rietschel ET, Zähringer U (1991) Chemical structure of glycosphingolipids isolated from *Sphingomonas paucimobilis*. FEBS Lett 292:107–110
4. Kawasaki S, Moriguchi R, Sekiya K, Nakai T, Ono E, Kume K, Kawahara K (1994) The cell envelope structure of the lipopolysaccharide-lacking Gram-negative bacterium *Sphingomonas paucimobilis*. J Bacteriol 176:284–290

5. Kawahara K, Moll H, Knirel YA, Seydel U, Zähringer U (2000) Structural analysis of two glycosphingolipids from the lipopolysaccharide-lacking bacterium *Sphingomonas capsulata*. Eur J Biochem 267:1837–1846
6. Leone S, Molinaro A, Lindner B, Romano I, Nicolaus B, Parrilli M, Lanzetta R, Holst O (2006) The structures of glycolipids isolated from the highly thermophilic bacterium *Thermus thermophilus* Samu-SA1. Glycobiology 16:766–775
7. Yang YL, Yang FL, Huang ZY, Tsai YH, Zou W, Wu SH (2010) Structural variation of glycolipids from *Meiothermus taiwanensis* ATCC BAA-400 under different growth temperatures. Org Biomol Chem 8:4252–4254
8. Holst O, Müller-Loennies S (2007) Microbial polysaccharide structures. In: Kamerling JP, Boons G-J, Lee YC, Suzuki A, Taniguchi N, Voragen AGJ (eds) Comprehensive glycoscience. From chemistry to systems biology, vol 1. Elsevier, Oxford, pp 123–179
9. Holst O, Molinaro A (2009) Core region and lipid A components of lipopolysaccharides. In: Moran AP, Holst O, Brennan PJ, von Itzstein M (eds) Microbial glycobiology. Structures, relevance and applications. Elsevier, Amsterdam, pp 29–55
10. Holst O, Brade H (1992) Chemical structure of the core region of lipopolysaccharides. In: Morrison DC, Ryan JL (eds) Bacterial endotoxic lipopolysaccharides, vol I. CRC Press, Boca Raton, pp 135–154
11. Knirel YA, Kochetkov NK (1993) The structure of lipopolysaccharides of Gram-negative bacteria. II. The structure of the core region: a review. Biokhimya 58:182–201
12. Holst O (1999) Chemical structure of the core region of lipopolysaccharides. In: Brade H, Opal SM, Vogel SN, Morrison DC (eds) Endotoxin in health and disease. Marcel Dekker, New York, pp 115–154
13. Holst O (2002) Chemical structure of the core region of lipopolysaccharides – an update. Trends Glycosci Glyc 14:87–103
14. Vinogradov E, Sodorczyk Z, Knirel YA (2002) Structure of the lipopolysaccharide core region of the bacteria of the genus *Proteus*. Aust J Chem 55:61–67
15. Frirdich E, Whitfield C (2005) Lipopolysaccharide inner core oligosaccharide structure and outer membrane stability in human pathogens belonging to the *Enterobacteriaceae*. J Endotoxin Res 11:133–144
16. Holst O (2007) The structures of core regions from enterobacterial lipopolysaccharides – an update. FEMS Microbiol Lett 271:3–11
17. Schweda EKH, Twelkmeyer B, Li J (2008) Profiling structural elements of short-chain lipopolysaccharide of non-typeable *Haemophilus influenzae*. Innate Immun 14:199–211
18. Kuhn HM, Meier-Dieter U, Mayer H (1988) ECA, the enterobacterial common antigen. FEMS Microbiol Rev 4:195–222
19. Cescutti P (2009) Bacterial capsular polysaccharides and exopolysaccharides. In: Moran AP, Holst O, Brennan PJ, von Itzstein M (eds) Microbial glycobiology. Structures, relevance and applications. Elsevier, Amsterdam, pp 93–115
20. Meredith TC, Aggarwal P, Mamat U, Lindner B, Woodard RW (2006) Redefining the requisite lipopolysaccharide structure in *Escherichia coli*. ACS Chem Biol 1:33–42
21. Mamat U, Meredith TC, Aggarwal P, Kühl A, Kirchhoff P, Lindner B, Hanuszkiewicz A, Sun J, Holst O, Woodard RW (2008) Single amino acid substitutions in either YhjD or MsbA confer viability to 3-deoxy-D-*manno*-oct-2-ulosonic acid-depleted *Escherichia coli*. Mol Microbiol 67:633–648
22. Beutler B, Rietschel ET (2003) Innate immune sensing and its roots: the story of endotoxin. Nat Rev Immunol 3:169–176
23. Lüderitz O, Westphal O, Bister F (1952) Über die Extraktion von Bakterien mit Phenol/Wasser. Z Naturforsch 7b:148–155
24. Westphal O, Jann K (1965) Bacterial lipopolysaccharides extraction with phenol water and further applications of the procedure. Meth Carbohydr Chem 5:83–91
25. Shear MJ, Turner FC (1943) Chemical treatment of tumors. V. Isolation of the hemorrhage-producing fraction from *Serratia marcescens* (*Bacillus prodigiosus*) culture filtrate. J Natl Cancer Res 4:81–97

26. Lüderitz O, Jann K, Wheat R (1968) Somatic and capsular antigens of Gram-negative bacteria. In: Florkin M, Sotz H (eds) Comprehensive biochemistry, vol 26A. Elsevier, Amsterdam, pp 105–228
27. Kauffmann F, Braun OH, Lüderitz O, Stierlin H, Westphal O (1961) Zur Immunchemie der O-Antigene von *Enterobacteriaceae*. Zentralbl Bakteriol 182:57–66
28. Lüderitz O, Staub AM, Westphal O (1966) Immunochemistry of O and R antigens of *Salmonella* and related *Enterobacteriaceae*. Bacteriol Rev 30:192–255
29. Lüderitz O, Westphal O (1966) The significance of enterobacterial mutants for the chemical investigation of their cell-wall polysaccharides. Angew Chem Int Ed Eng 5:198–210
30. Unger F (1983) The chemistry and biological significance of 3-deoxy-D-*manno*-2-octulosonic acid (Kdo). Adv Carbohydr Chem Biochem 348:323–387
31. Lüderitz O, Westphal O, Staub AM, Nikaido H (1971) Isolation and chemical and immunological characterization of bacterial lipopolysaccharides. In: Weinbaum G, Kadis S, Ajl SJ (eds) Microbial toxins. IV. Bacterial endotoxins. Academic, New York, pp 145–233
32. Galanos C, Lüderitz O, Westphal O (1969) A new method for the extraction of R-lipopolysaccharides. Eur J Biochem 9:245–249
33. Galanos C, Lüderitz O, Rietschel ET, Westphal O (1977) Newer aspects of the chemistry and biology of bacterial lipopolysaccharides, with special reference to their lipid A component. In: Goodwin TW (ed) International review of biochemistry, biochemistry of lipids II, vol 14. University Park Press, Baltimore, pp 239–335
34. Brade H, Brade L, Rietschel ET (1988) Structure-activity relationships of bacterial lipopolysaccharides (endotoxins). Current and future aspects. Zentralbl Bakteriol Hyg A268:151–179
35. Lindner B (2000) Matrix-assisted laser desorption/ionization time-of-flight mass spectrometry of lipopolysaccharides. In: Holst O (ed) Bacterial toxins. Methods and protocols, vol 145. Humana Press, Totowa, pp 311–325
36. Kondakova A, Lindner B (2005) Structural characterization of complex bacterial glycolipids by Fourier-transform mass spectrometry. Eur J Mass Spectrom 11:535–546
37. Thibault P, Richards JC (2000) Applications of combined capillary electrophoresis-electrospray mass spectrometry in the characterization of short-chain lipopolysaccharides. In: Holst O (ed) Bacterial toxins. Methods and protocols, vol 145. Humana Press, Totowa, pp 327–344
38. Hübner G, Lindner B (2009) Separation of R-form lipopolysaccharide and lipid A by Fourier-transform ion cyclotron resonance MS. Electrophoresis 30:1808–1816
39. Holst O (2000) Deacylation of lipopolysaccharides and isolation of oligosaccharide phosphates. In: Holst O (ed) Bacterial toxins. Methods and protocols, vol 145. Humana Press, Totowa, pp 345–353
40. Rosner MR, Tang J-Y, Barzilay I, Khorana HG (1978) Structure of the lipopolysaccharide from an *Escherichia coli* heptose-less mutant. I. Chemical degradations and identification of products. J Biol Chem 254:5906–5916
41. Caroff M, Tacken A, Szabó L (1988) Detergent-accelerated hydrolysis of bacterial endotoxins and determination of the anomeric configuration of the glycosyl phosphate present in the "isolated lipid A" fragment of the *Bordetella pertussis* endotoxin. Carbohydr Res 175:273–282
42. Caroff MG, Karibian D (1990) Several uses for isobutyric acid-ammonium hydroxide solvent in endotoxin analysis. Appl Environ Microbiol 56:1957–1959
43. Caroff M, Chafchaouni-Mossaaou I, Basheer S, Opota O, Lemaitre B, Perry M, Novikov A (2010) A new mild and easy micro-hydrolysis of lipopolysaccharides using triethylamine citrate. Paper presented at the annual meeting of the society for leukocyte biology and the international endotoxin and innate immunity society, Vancouver, Canada, 7–9 October 2010
44. Vinogradov E, Frirdich E, MacLean LL, Perry MB, Petersen BO, Duus JØ, Whitfield C (2002) Structures of lipopolysaccharides from *Klebsiella pneumoniae*. Eluicidation of the structure of the linkage region between core and polysaccharide O chain and identification of the residues at the non-reducing termini of the O chains. J Biol Chem 277:25070–25081

45. Bystrova OV, Knirel YA, Lindner B, Kocharova NA, Kondakova AN, Zähringer U, Pier GB (2006) Structures of the core oligosaccharides and O-units in the R- and SR-type lipopolysaccharides of reference strains of *Pseudomonas aeruginosa* O-serogroups. FEMS Immunol Med Microbiol 46:85–99
46. Olsthoorn MM, Petersen BO, Schlecht S, Haverkamp J, Bock K, Thomas-Oates JE, Holst O (1998) Identification of a novel core type in *Salmonella* lipopolysaccharide. Complete structural analysis of the core region of the lipopolysaccharide from *Salmonella enterica* sv. Arizonae O62. J Biol Chem 273:3817–3829
47. De Castro C, Parrilli M, Holst O, Molinaro A (2010) Microbe-associated molecular patterns in innate immunity: extraction and chemical analysis of Gram-negative bacterial lipopolysaccharides. Meth Enzymol 480:89–115
48. Hakomori S (1964) A rapid permethylation of glycolipid, and polysaccharide catalyzed by methylsulfinyl carbanion in dimethyl sulfoxide. J Biochem 55:205–208
49. Ciucanu I, Kerek F (1984) A simple and rapid method for the permethylation of carbohydrates. Carbohydr Res 131:209–217
50. Tacken A, Rietschel ET, Brade H (1986) Methylation analysis of the heptose/3-deoxy-D-*manno*-2-octulosonic acid region (inner core) of the lipopolysaccharide from *Salmonella minnesota* rough mutants. Carbohydr Res 149:279–291
51. Müller-Loennies S, Lindner B, Brade H (2002) Structural analysis of deacylated lipopolysaccharide of *Escherichia coli* strains 2513 (R4 core-type) and F653 (R3 core-type). Eur J Biochem 269:5982–5991
52. Kocharova NA, Kondakova AN, Vinogradov E, Ovchinnikova OG, Lindner B, Shashkov AS, Rozalski A, Knirel YA (2008) Full structure of the carbohydrate chain of the lipopolysaccharide of *Providencia rustigianii* O34. Chem Eur J 14:6184–6191
53. Knirel YA, Lindner B, Vinogradov E, Kocharova NA, Senchenkova SN, Shaikhutdinova RZ, Dentovskaya SV, Fursova NK, Bakhteeva IV, Titareva GM, Balakhonov SV (2005) Temperature-dependent variations and intraspecies diversity of the structure of the lipopolysaccharide of *Yersinia pestis*. Biochemistry 44:1731–1743
54. Lundström SL, Twelkmeyer B, Sagemark MK, Li J, Richards JC, Hood DW, Moxon ER, Schweda EK (2007) Novel globoside-like oligosaccharide expression patterns in non-typeable *Haemophilus influenzae* lipopolysaccharide. FEBS J 274:4886–4903
55. Månsson M, Hood DW, Li J, Richards JC, Moxon ER, Schweda EKH (2002) Structural analysis of the lipopolysaccharide from nontypeable *Haemophilus influenzae* strain 1003. Eur J Biochem 269:808–818
56. Cox AD, Hodd DW, Martin A, Makepeace KM, Deadman ME, Li J, Brisson J-R, Moxon ER, Richards JC (2002) Identification and structural characterization of a sialylated lacto-N-neotetraose structure in the lipopolysaccharide of *Haemophilus influenzae*. Eur J Biochem 269:4009–4019
57. Schweda EKH, Landerholm MK, Li J, Moxon ER, Richards JC (2003) Structural profiling of lipopolysaccharide glycoforms expressed by non-typeable *Haemophilus influenzae*: phenotypic similarities between NTHi strain 162 and the genome strain Rd. Carbohydr Res 338:2731–2744
58. Tinnert A-S, Månsson M, Yildrim HH, Hood DW, Schweda EKH (2005) Structural investigation of lipooligosaccharides from non-typeable *Haemophilus influenzae*: investigation of inner-core phosphoethanolamine addition in NTHi strain 981. Carbohydr Res 340:1900–1907
59. Månsson M, Hood DW, Li J, Moxon ER, Schweda EKH (2002) Structural characterization of a novel branching pattern in the lipopolysaccharide from nontypeable *Haemophilus influenzae*. Eur J Biochem 270:2979–2991
60. Cox AD, Howard MD, Inzana TJ (2003) Structural analysis of the lipooligosaccharide from the commensal *Haemophilus somnus* strain 1P. Carbohydr Res 338:1223–1228
61. St. Michael F, Howard MD, Li J, Duncan AJ, Inzana TJ, Cox AD (2004) Structural analysis of the lipooligosaccharide from the commensal *Haemophilus somnus* genome strain 129Pt. Carbohydr Res 339:529–535

62. St. Michael F, Li J, Vinogradov E, Larocque S, Harper M, Cox AD (2005) Structural analysis of the lipooligosaccharide of *Pasteurella multocida* strain VP161 identification of both Kdo-P and Kdo-Kdo species in the lipopolysaccharide. Carbohydr Res 340:59–68
63. Harper M, Cox AD, St. Michael F, Wilkie IW, Boyce JD, Adler B (2007) A heptosyltransferase mutant of *Pasteurella multocida* produces a truncated lipopolysaccharide structure and is attenuated in virulence. Infect Immun 72:3436–3443
64. Choudhury B, Carlson RW, Goldberg JB (2008) Characterization of the lipopolysaccharide from a *wbjE* mutant of the serogroup O11 *Pseudomonas aeruginosa* strain, PA103. Carbohydr Res 343:238–248
65. Zdorovenko EL, Vinogradov EV, Zdorovenko GM, Lindner B, Bystrova OV, Shashkov AS, Rudolph K, Zähringer U, Knirel YA (2004) Structure of the core oligosaccharide of a rough-type lipopolysaccharide of *Pseudomonas syringae* pv. *phaseolicola*. Eur J Biochem 271:4968–4977, Corrigendum in: FEBS J 272: 1815 (2005)
66. Hanuszkiewicz A, Hübner G, Vinogradov E, Lindner B, Brade L, Brade H, Debarry J, Heine H, Holst O (2008) Structural and immunochemical analysis of the lipopolysaccharide from *Acinetobacter lwoffii* F78 located outside Chlamydiaceae with a *Chlamydia*-specific lipopolysaccharide epitope. Chem Eur J 14:10251–10258
67. Vinogradov EV, Müller-Loennies S, Petersen BO, Meshkov S, Thomas-Oates JE, Holst O, Brade H (1997) Structural investigation of the lipopolysaccharide from *Acinetobacter haemolyticus* strain NCTC 10305 (ATCC 17906, DNA group 4). Eur J Biochem 247:82–909
68. Vinogradov EV, Petersen BO, Thomas-Oates JE, Duus JØ, Brade H, Holst O (1998) Characterization of a novel branched tetrasaccharide of 3-deoxy-D-*manno*-oct-2-ulopyranosonic acid. The structure of the carbohydrate backbone of the lipopolysaccharide from *Acinetobacter baumannii* strain NCTC 10303 (ATCC 17904). J Biol Chem 273:28122–28131
69. Wang Z, Li J, Vinogradov E, Altman E (2006) Structural studies of the core region of *Aeromonas salmonicida* subsp. *salmonicida* lipopolysaccharide. Carbohydr Res 341:109–117
70. Knirel YA, Vinogradov E, Jiminez N, Merino S, Tomás JM (2004) Structural analysis on the R-type lipopolysaccharide of *Aeromonas hydrophila*. Carbohydr Res 339:787–793
71. Jimenez N, Canals R, Lacasta A, Kondakova AN, Lindner B, Knirel YA, Merino S, Regué M, Tomás JM (2008) Molecular analysis of three *Aeromonas hydrophila* AH-3 (serotype O34) lipopolysaccharide core biosynthesis gene clusters. J Bacteriol 190:3176–3184
72. Isshiki Y, Zähringer U, Kawahara K (2003) Structure of the core oligosaccharide with a characteristic D-*glycero*-α-D-*talo*-oct-2-ulosonate-(2 → 4)-3-deoxy-D-*manno*-oct-2-ulosonate [α-Ko-(2 → 4)-Kdo] disaccharide in the lipopolysaccharide of *Burkholderia cepacia*. Carbohydr Res 338:2659–2666
73. De Castro C, Molinaro A, Lanzetta R, Holst O, Parrilli M (2005) The linkage between O-specific caryan and core region in the lipopolysaccharide of *Burkholderia caryophylli* is furnished by a primer monosaccharide. Carbohydr Res 340:1802–1807
74. Silipo A, Molinaro A, Comegna D, Sturiale L, Cescutti P, Garozzo P, Lanzetta R, Parrilli M (2006) Full structural characterisation of the lipooligosaccharide of a *Burkholderia pyrrocinia* clinical isolate. Eur J Org Chem 4874–4883
75. Zdorovenko EL, Vinogradov E, Wydra K, Lindner B, Knirel YA (2008) Structure of the oligosaccharide chain of the SR-type lipopolysaccharide of *Ralstonia solanacearum* Toudk-2. Biomacromolecules 9:2215–2220
76. Tsai C-M, Jankowska-Stephens E, Mizanur RM, Cipollo JF (2009) The fine structure of *Neisseria meningitidis* lipooligosaccharide from the M986 strain and three of its variants. J Biol Chem 284:4616–4625
77. Giardina PC, Preston A, Gibson B, Apicella MA (1992) Antigenic mimicry in *Neisseria* species. In: Morrison DC, Ryan JL (eds) Bacterial endotoxic lipopolysaccharides, vol I. CRC Press, Boca Raton, pp 55–65
78. Cox AD, Li J, Brisson J-R, Moxon ER, Richards JC (2002) Structural analysis of the lipopolysaccharide from *Neisseria meningitidis* strain BZ157 *galE*: localization of two phosphoethanolamine residues in the inner core oligosaccharide. Carbohydr Res 337:1435–1444

79. Cox AD, Li J, Richards JC (2002) Identification and localization of glycine in the inner core lipopolysaccharide of *Neisseria meningitidis*. Eur J Biochem 269:4169–4175
80. Kahler CM, Lyons-Schindler S, Choudhury B, Glushka J, Carlson RW, Stephens DS (2006) O-Acetylation of the terminal *N*-acetylglucosamine of the lipopolysaccharide inner core in *Neisseria meningitidis*. Influence on the inner core structure and assembly. J Biol Chem 281:19939–19948
81. St. Michael F, Vinogradov E, Wenzel CQ, McIntosh B, Li J, Hoe JC, Richards JC, Cox AD (2009) Phosphoethanolamine is located at the 6-position and not at the 7-position of the distal heptose residue in the lipopolysaccharide from *Neisseria meningitides*. Glycobiology 19:1436–1445
82. John CM, Liu M, Jarvis GA (2009) Profiles of structural heterogeneity in native lipooligosaccharides of *Neisseria* and cytokine induction. J Lipid Res 50:424–438

Structure of O-Antigens

3

Yuriy A. Knirel

3.1 Introduction

The lipopolysaccharide (LPS) is the major constituent of the outer leaflet of the outer membrane of Gram-negative bacteria. Its lipid A moiety is embedded in the membrane and serves as an anchor for the rest of the LPS molecule. The outermost repetitive glycan region of the LPS is linked to the lipid A through a core oligosaccharide (OS), and is designated as the O-specific polysaccharide (O-polysaccharide, OPS) or O-antigen. The O-antigen is the most variable portion of the LPS and provides serological specificity, which is used for bacterial serotyping. The OPS also provides protection to the microorganisms from host defenses such as complement mediated killing and phagocytosis, and is involved in interactions of bacteria with plants and bacteriophages. Studies of the OPSs ranging from the elucidation of their chemical structures and conformations to their biological and physicochemical properties help improving classification schemes of Gram-negative bacteria. Furthermore, these studies contributed to a better understanding of the mechanisms of pathogenesis of infectious diseases, as well as provided information to develop novel vaccines and diagnostic reagents.

Composition and structures of O-antigens have been surveyed repeatedly [1–7]. The number of OPSs with complete structural elucidation is rapidly growing and an annually updated Bacterial Carbohydrate Structure Database (BCSDB) is available online at http://www.glyco.ac.ru/bcsdb3/. The present chapter provides an updated collection of data on composition and structures of the OPSs published until the end of 2010. To avoid extensive citation of structures already reported, only earlier reviews are referenced. Whenever known OPS structures are presented in an earlier review or, in the case of *Escherichia coli*, in a permanently updated database, they

Y.A. Knirel (✉)
N.D. Zelinsky Institute of Organic Chemistry, Russian Academy of Sciences, Leninsky Prospekt 47, 119991 Moscow, V-334, Russia
e-mail: yknirel@gmail.com

Y.A. Knirel and M.A. Valvano (eds.), *Bacterial Lipopolysaccharides*,
DOI 10.1007/978-3-7091-0733-1_3, © Springer-Verlag/Wien 2011

are only briefly discussed in this chapter. Various OPS structures were established by older methods and required reinvestigation using new techniques. For structures already revised, only the publication reporting the final structure is cited.

Classification of Gram-negative bacteria is subject to change. In this review, the current names for bacterial classes, families, genera and species are used according to the NCBI Taxonomy Browser (http://www.ncbi.nlm.nih.gov/Taxonomy/). When an OPS structure was reported under a different bacterial name, the old name is indicated in parentheses.

3.2 Composition of O-Antigens

Typical components of the OPSs are both monosaccharides widely distributed in nature and uncommon sugars (Table 3.1), including those that have not been found elsewhere (here and below, the descriptor D in abbreviations of monosaccharides of the D series is omitted).

Most monosaccharides exist in the pyranose form (in the OPS structures below, the descriptor *p* for this form is omitted) but several are present as furanosides (Ara, Rib, L6dAlt, xylulose) or may occur in both forms (Gal, Fuc, paratose); in a few OPSs, Rib and L6dAlt are present as pyranosides and GalNAc as a furanoside.

From non-carbohydrate constituents (Table 3.2), commonly occurring are *N*-acetyl and *O*-acetyl groups. Less common is a methyl group, which is linked to hydroxyl or amino groups or esterifies a hexuronic acid. In various OPSs, hexuronic acids exist as a primary amide (this is indicated below by letter N, e.g. GalAN) or an amide with an amino compound like 2-amino-2-deoxyglycerol (GroN) or amino acids (in case of L-lysine and its N^{ε}-(1-carboxyethyl) derivatives hexuronic acids are linked to their α-amino group). Phosphate has been found only as diesters, including a cyclic phosphate.

3.3 Structures of O-Antigens

3.3.1 General Aspects

The OPS is the most variable LPS component in terms of composition and structure. The high diversity of O-antigens results mainly from genetic variations in the O-antigen gene clusters, and is further expanded by various prophage genes, which cause additional modifications such as lateral glycosylation or/and O-acetylation (see Chap. 11). The OPS is made of oligosaccharide repeats (O-units) consisting of two to eight different monosaccharide residues (heteroglycans) or, in some bacteria, of identical sugars (homoglycans). The O-unit is first assembled on a lipid carrier and then polymerized, whereas homoglycans and part of the heteroglycans with disaccharide O-units are synthesized by an alternative pathway including a sequential transfer of single monosaccharides to the growing chain (see Chap. 9). Lateral

Table 3.1 Monosaccharide components of OPSs

Pentoses, hexoses, heptoses and their deoxy derivatives	
D-arabinose (Ara)	D-glucose (Glc)
D-, L-xylose (Xyl, LXyl)	D-mannose (Man)
D-ribose (Rib)	D-galactose (Gal)
4-deoxy-D-*arabino*-hexose (4daraHex)	6-deoxy-D-gulose (6dGul)
6-deoxy-L-glucose (L-quinovose, LQui)	3,6-dideoxy-D-*arabino*-hexose (tyvelose, Tyv)
6-deoxy- D-, L-galactose (D-, L-fucose; Fuc, LFuc)	3,6-dideoxy-L-*arabino*-hexose (ascarylose, Asc)
6-deoxy-D-, L-mannose (D-, L-rhamnose; Rha, LRha)	3,6-dideoxy-D-*ribo*-hexose (paratose, Par)
6-deoxy-L-altrose (L6dAlt)	3,6-dideoxy-D-*xylo*-hexose (abequose, Abe)
6-deoxy-D-, L-talose (6dTal, L6dTal)	3,6-dideoxy-L-*xylo*-hexose (colitose, Col)
D-*glycero*-D-*manno*-heptose (DDmanHep)	L-*glycero*-D-*manno*-heptose (LDmanHep)
D-*glycero*-D-*galacto*-heptose (DDgalHep)	6-deoxy-D-*manno*-heptose (6dmanHep)
2-Amino-2-deoxyhexoses, amino and diamino 6-deoxyhexoses	
D-glucosamine (GlcN)	3-amino-3-deoxy-D-fucose (Fuc3N)
D-galactosamine (GalN)	4-amino-4-deoxy-D-quinovose (Qui4N)
D-mannosamine (ManN)	4-amino-4-deoxy-D-, L-rhamnose (Rha4N, LRha4N)
D-, L-quinovosamine (QuiN, LQuiN)	4-amino-4-deoxy-D-fucose (Fuc4N)
L-rhamnosamine (LRhaN)	2,3-diamino-2,3-dideoxy-L-rhamnose (LRhaN3N)
D-, L-fucosamine (FucN, LFucN)	2,4-diamino-2,4-dideoxy-D-quinovose (QuiN4N)
6-deoxy-L-talosamine (L6dTalN)	2,4-diamino-2,4-dideoxy-D-fucose (FucN4N)
3-amino-3-deoxy-D-, L-quinovose (Qui3N, LQui3N)	
Hexuronic acids, amino and diamino hexuronic acids	
D-glucuronic (GlcA)	D-glucosaminuronic (GlcNA)
D-mannuronic (ManA)	D-mannosaminuronic (ManNA)
D-galacturonic (GalA)	D-, L-galactosaminuronic (GalNA, LGalNA)
L-altruronic (LAltA)	L-altrosaminuronic (LAltNA)
L-iduronic (LIdoA)	L-gulosaminuronic (LGulNA)
3-amino-3-deoxy-D-glucuronic (Glc3NA)	2,3-diamino-2,3-dideoxy-D-glucuronic (GlcN3NA)
2,3-diamino-2,3-dideoxy-D-mannuronic (ManN3NA)	2,3-diamino-2,3-dideoxy-D-galacturonic (GalN3NA)
2,3-diamino-2,3-dideoxy-L-guluronic (LGulN3NA)	2,4-diamino-2,4-dideoxyglucuronic (GlcN4NA)
Keto sugars	
D-, L-*threo*-pent-2-ulose (D-, L-xylulose; Xlu, LXlu)	
2-amino-2,6-dideoxy-D-*xylo*-hexos-4-ulose	
3-deoxy-D-*manno*-oct-2-ulosonic acid (ketodeoxyoctonic acid, Kdo)	
5-amino-3,5-dideoxy-D-*glycero*-D-*galacto*-non-2-ulosonic acid (neuraminic acid, Neu)	

(*continued*)

Table 3.1 (continued)

5,7-diamino-5,7,9-trideoxynon-2-ulosonic acid[a]
5,7-diamino-3,5,7,9-tetradeoxy-L-*glycero*-L-*manno*-non-2-ulosonic (pseudaminic) acid (Pse)
5,7-diamino-3,5,7,9-tetradeoxy-D-*glycero*-D-*galacto*-non-2-ulosonic (legionaminic) acid (Leg)
5,7-diamino-3,5,7,9-tetradeoxy-D-*glycero*-D-*talo*-non-2-ulosonic (4-epilegionaminic) acid (4eLeg)
5,7-diamino-3,5,7,9-tetradeoxy-L-*glycero*-D-*galacto*-non-2-ulosonic (8-epilegionaminic) acid (8eLeg)
5,7,8-triamino-3,5,7,8,9-pentadeoxynon-2-ulosonic acid[b]
3-deoxy-D-*lyxo*-hept-2-ulosaric acid
Branched sugars[c]
3-*C*-methyl-D-mannose (Man3*C*Me)
3-*C*-methylrhamnose (Rha3*C*Me)[a]
3,6-dideoxy-4-*C*-[(*R*)-, (*S*)-1-hydroxyethyl]-D-*xylo*-hexose (yersiniose A, yersiniose B)
3,6,8-trideoxy-4-*C*-[(*R*)-1-hydroxyethyl]-D-*gulo*-octose (erwiniose)
3,6,10-trideoxy-4-*C*-[(*R*)-hydroxyethyl]-D-*erythro*-D-*gulo*-decose (caryophillose)
2-amino-4-*C*-(2-carbamoyl-2,2-dihydroxyethyl)-2,6-dideoxy-D-galactose (shewanellose)
4,8-cyclo-3,9-dideoxy-L-*erythro*-D-*ido*-nonose (caryose)

[a]The configuration of the monosaccharide remains unknown.
[b]The monosaccharide has the L-*glycero*-L-*manno* or D-*glycero*-L-*manno* configuration.
[c]For structures of branched monosaccharides see also review [7].

glycosyl groups and *O*-acetyl groups may be added to the growing OPS chain or after polymerization, and their content is often non-stoichiometric.

Some bacteria have LPS lacking OPS due to the absence or inactivation of the O-antigen gene cluster. When bacteria are able to assemble but unable to polymerize the O-unit, they elaborate LPS containing a single O-unit linked to the core OS. Several LPS forms may coexist in one strain. In some cases, LPS forms lacking O-antigen are designated as lipooligosaccharide. The length of the OPS chain varies considerably from one O-unit to more than 50 O-units. The chain length distribution is modal (except for bacteria which possess an S-layer) and is specific to each bacterial strain. It appears to be fine-tuned to give bacteria advantages in particular niches.

Most chemical data reported on OPSs are limited to the structure of the so-called chemical repeating unit, which may or may not agree with the structure of the biological O-unit that is based on the order of synthesis and that is the substrate for the O-antigen polymerization. Therefore, the monosaccharide sequence of the chemical repeating unit may be any cyclic permutation of the biological unit. Recently, it has been shown that in many heteroglycans, the first monosaccharide of the O-unit whose transfer to a lipid carrier initiates biosynthesis of the O-antigen, is a derivative of a 2-amino-2-deoxy-D-hexose (GlcN, GalN) or a 2-amino-2,6-dideoxy-D-hexose (QuiN, FucN, QuiN4N, FucN4N), all having the D-*gluco* or D-*galacto* configuration. One can assume that, when present, such an amino sugar is the first in other OPSs too. In several bacteria, e.g. *Salmonella enterica*, the first monosaccharide of the O-unit is Gal, whereas in many other species, the biological O-unit structure remains unknown.

Table 3.2 Non-carbohydrate components of OPSs

O-Linked (O-alkyl groups and acetals)	
(*R*)-, (*S*)-1-carboxyethyl (lactic acid ethers; *R*lac, *S*lac)	
(*1R,3R*)-, (*1S,3R*)-1-carboxy-3-hydroxybutyl (2,4-dihydroxypentanoic acid 2-ethers)	
(*R*)-, (*S*)-1-carboxyethylidene (pyruvic acid acetals; *R*pyr, *S*pyr)	
N-Linked (N-acyl groups)	
formyl (Fo)	acetimidoyl (Am)
(*R*)-, (*S*)-2-hydroxypropanoyl (*R*2Hp, *S*2Hp)	3-hydroxypropanoyl (3Hp)
(*R*)-, (*S*)-3-hydroxybutanoyl (*R*3Hb, *S*3Hb)	4-hydroxybutanoyl (4Hb)
L-glyceroyl (LGroA)	(*S*)-2,4-dihydroxybutanoyl
(*3S,5S*)-3,5-dihydroxyhexanoyl	malonyl
succinyl	(*R*)-, (*S*)-2-hydroxy-4-succinyl (4-D-malyl, 4-L-malyl)
(*S*)-2-hydroxy-5-glutaryl	glycyl (Gly)
D-, L-alanyl (DAla, LAla)	L-seryl (LSer)
D-homoseryl (DHse)	L-allothreonyl (LaThr)
D-, L-4-aspartyl (4DAsp, 4LAsp)	*N*-(1-carboxyethyl)alanyl[a]
(*2R,3R*)-3-hydroxy-3-methyl-5-oxoprolyl	3-hydroxy-2,3-dimethyl-5-oxoprolyl[a]
2,4-dihydroxy-3,3,4-trimethyl-5-oxoprolyl[a]	(*2R,3R,4S*)-3,4-dihydroxy-1,3-dimethyl-5-oxoprolyl
Carboxyl-linked (amides)	
2-amino-2-deoxyglycerol (GroN)	L-serine (LSer)
glycine (Gly)	L-threonine (LThr)
D-, L-alanine (DAla, LAla)	D-allothreonine (DaThr)
L-lysine (LLys)	
N^{ε}-[(*R*)-, (*S*)-1-carboxyethyl]-L-lysine ('alaninolysine'; *R*alaLys, *S*alaLys)	
Phosphate-linked (phosphodiesters)	
glycerol (Gro)	D-glyceric acid (DGroA)
ribitol (Rib-ol)	L-arabinitol (LAra-ol)
2-aminoethanol (ethanolamine, EtN)	2-[(*R*)-1-carboxyethylamino]ethanol
2-(trimethylammonio)ethanol (choline)	2-amino-2-deoxy-2-*C*-methylpentonic acid[a]

[a]The configuration of the amino acid remains unknown.

The core OS may carry a polysaccharide that is structurally different from the O-antigen and is encoded by a locus different from the O-antigen gene cluster. Examples of this are the enterobacterial common antigen produced by the Enterobacteriaceae [8] and the A-band O-antigen in *Pseudomonas aeruginosa* [9]. On the other hand, a repeat of the same structure as the O-unit may be employed as a building block for another surface polymer, e.g. a capsular polysaccharide [5] or a glycoprotein [10]. More than one structurally related or sometimes unrelated OPSs, may occur in one strain. In the latter case, one of the glycans may not be a part of the LPS but for example a capsular polysaccharide that is coextracted with the LPS [11].

The repetitive OPS structure is often masked by one or more non-stoichiometric modifications, including glycosylation, O-acetylation, methylation, phosphorylation or amidation (in the structures shown below, non-stoichiometric substituents are indicated in italics). Less common are epimerization at C-5 of hexuronic acids and alternative N-acylation of an amino group by different acyl groups. A rare reason for the lack of the strict regularity is a random or in another manner irregular distribution of α- and β-linked monosaccharide residues along the polymer chain.

Many LPSs, especially with homopolysaccharide O-chains, have additional nonrepetitive domains, which result from specific initiation and termination steps of the OPS biosynthesis. For instance, incorporation of an O-methylated sugar or a different monosaccharide to the non-reducing end is thought to be a signal for cessation of the OPS chain synthesis, which allows termination of the O-chain at a specific sugar residue rather than at any residue. Another non-repetitive domain may occur between the OPS and the core OS, such as a primer of a 2-*N*-acetylamino sugar whose transfer to a lipid carrier initiates the O-antigen synthesis. More complex reducing-end domains have been found in a few OPSs but they may be much more common than anticipated. Further information on OPS-associated non-repetitive structures is given in a recent review [7], whereas the present review focuses on the O-unit structures.

3.3.2 γ-Proteobacteria

3.3.2.1 Enterobacteriaceae

A majority of the bacteria, whose O-antigen structures have been elucidated, belong to the family Enterobacteriaceae.

Salmonella

Salmonella species, the agents of salmonellosis, are a leading cause of food-borne infections in many countries; several serovars are responsible for more severe diseases, such as typhoid fever. Currently, strains of *S. enterica* are combined into 46 O-serogroups, including former serogroups A–Z. Serovar names are used for strains of ssp. *enterica*, whereas Latin numbers are used to designate other subspecies: II for ssp. s*alamae,* IIIa for ssp. *arizonae*, IIIb for ssp. *diarizonae*, etc. The structures of the OPSs of *S. enterica* established by that time have been reviewed in 2006 [12], and more structures are shown below (Table 3.3).

Strains of serogroups A, B, D and E were the first bacteria whose O-antigen structures were elucidated in detail. They possess similar Man-LRha-Gal- main chains, in which the position of substitution of Man and the configuration of the linkages of Man and Gal vary both between and within O-serogroups. In serogroup D_3, α-Man- and β-Man-containing O-units coexist. In serogroups A, B and D, Man bears a 3,6-dideoxyhexose having D-*ribo* (paratose), D-*xylo* (abequose) or D-*arabino* (tyvelose) configuration, respectively, whereas in serogroup E, no 3,6-dideoxyhexose is present. Outside these serogroups, the OPSs display a variety of structures. Neutral sugars (Man, Glc, Gal, LRha, LFuc), GlcNAc and GalNAc

Table 3.3 Structures of *Salmonella* OPSs

O2 (A) Paratyphi [13,14]	2)Man(α1-4)LRha*2Ac*(α1-3)Gal(α1- Par(α1-3)┘ Glc(α1-4)┘
O4 (B) Typhimurium, Agona,[a] Abortusequi[a] [13,15-18]	2)Man(α1-4)LRha(α1-3)Gal(α1- Abe*2Ac*(α1-3)┘ *Glc(α1-4)*┘
O4 (B) Bredeney, Typhimurium SL3622[a] [13,16,19]	2)Man(α1-4)LRha(α1-3)Gal(α1- Abe*2Ac*(α1-3)┘ *Glc(α1-6)*┘
O6,7 (C_1) Livingstone [20]	2)Man(β1-2)Man(α1-2)Man(α1-2)Man(β1-3)GlcNAc(β1- Glc(α1-3)┘
O6,7 (C_1) Thompson [21]	2)Man(β1-2)Man(α1-2)Man(α1-2)Man(β1-3)GlcNAc(β1- and 2)Man(β1-2)Man(α1-2)Man(α1-2)Man(β1-3)GlcNAc(β1- Glc(α1-3)┘
O6,7 (C_1) Ohio [22]	2)Man(β1-2)Man(α1-2)Man(α1-2)Man(β1-3)GlcNAc(β1- Glc(α1-3)┘
O6,7 (C_4) Livingstone var. 14^{+} (*S. eimsbuttel*) [23]	2)Man(β1-2)Man(α1-2)Man(α1-2)Man(β1-3)GlcNAc(β1- Glc(α1-3)┘
O8 (C_2) Newport [13,24]	4)LRha*2Ac*(β1-2)Man(α1-2)Man(α1-3)Gal(β1- Abe(α1-3)┘ *Glc2Ac(α1-3)*┘
O8 (C_3) Kentucky I.S. 98 [13]	4)LRha(β1-2)Man(α1-2)Man(α1-3)Gal(β1- Abe(α1-3)┘ *Glc2Ac(α1-4)*┘
O8 (C_3) Kentucky 98/39 [25]	4)LRha(β1-2)Man(α1-2)Man(α1-3)Gal(β1- Abe(α1-3)┘ *Glc(α1-2)*┘
O9 (D_1) Typhi, Enteritidis SE6[a], Gallinarum bv. Pullorum 77[a] [26-28]	2)Man(α1-4)LRha(α1-3)Gal(α1- Tyv(α1-3)┘ *Glc2Ac(α1-4)*┘
O9 (D_1) Enteritidis I.S. 64, Gallinarum bv. Pullorum 11 [28,29]	2)Man(α1-4)LRha(α1-3)Gal(α1- Tyv(α1-3)┘
O9,46 (D_2) Strasbourg [13]	6)Man(β1-4)LRha(α1-3)Gal(α1- Tyv(α1-3)┘ *Glc(α1-4)*┘
O9,46 (D_2) II (*S. haarlem*) [30]	6)Man(β1-4)LRha(α1-3)Gal(α1- Tyv(α1-3)┘
O9,46,27 (D_3) II (*S. zuerich*) [31]	6)Man(α/β1-4)LRha(α1-3)Gal(α1- Tyv(α1-3)┘ *Glc(α1-6)*┘
O3,10 (E_1) Anatum [26,32]	6)Man(β1-4)LRha(α1-3)Gal*6Ac*(α1-
O3,10 (E_1) Muenster [13]	6)Man(β1-4)LRha(α1-3)Gal(α1- *Glc(α1-4)*┘
O3,10 (E_2) Anatum var. 15^{+} (*S. newington*) [26]	6)Man(β1-4)LRha(α1-3)Gal(β1-
O3,10 (E_3) Lexington var. 15^{+},34^{+} (*S. illinois*) [26]	6)Man(β1-4)LRha(α1-3)Gal(β1- *Glc(α1-4)*┘
O1,3,19 (E_4) Senftenberg [13,26]	6)Man(β1-4)LRha(α1-3)Gal(α1- *Glc(α1-6)*┘
O11 (F) Aberdeen [33]	3)Gal(α1-4)LRha(α1-3)GlcNAc(β1- Man(β1-4)┘
O13 (G) [34]	2)LFuc(α1-2)Gal(β1-3)GalNAc(α1-3)GlcNAc(α1-

(*continued*)

Table 3.3 (continued)

O6,14 (H) Boecker, Carrau [35,36]	6)Man(α1-2)Man(α1-2)Man(β1-3)GlcNAc(α1- and 6)Man(α1-2)Man(α1-2)Man(β1-3)GlcNAc(α1- Glc(α1-3)┘
O6,14 (H) Madelia [37]	6)Man(α1-2)Man(α1-2)Man(β1-3)GlcNAc(α1- and 6)Man(α1-2)Man(α1-2)Man(β1-3)GlcNAc(α1- and Glc(α1-3)┘ 6)Man(α1-2)Man(α1-2)Man(β1-3)GlcNAc(α1- Glc(α1-4)┘
O16 (I) [38]	4)GalNAc(α1-6)Man*3Ac*(α1-3)LFuc(α1-3)GalNAc(β1- └(3-1α)LFuc *Glc(β1-4)*┘
O17 (J) [39]	2)Gal(α1-3)ManNAc(β1-6)Gal*f2Ac*(β1-3)GlcNAc(β1- └(4-1α)Gal*f*
O18 (K) Cerro [40]	4)Man(α1-2)Man(α1-2)Man(β1-3)GalNAc(α1-
O21 (L)[b] [41]	4)GalNAc(β1-3)Gal(α1-4)Gal(β1-3)GalNAc(β1- └(3-1α)GlcNAc
O28 (M, $O28_1,28_2$) Telaviv [42]	4)Qui3NAc(β1-3)Rib*f*(β1-4)Gal(β1-3)GalNAc(α1- *Gal(α1-3)*Gal(α1-3)┘ *Glc(α1-4)*┘
O28 (M, $O28_1,28_3$) Dakar [43]	4)Qui3NAc(α1-3)LRha(α1-4)Gal(β1-3)GalNAc(α1- Glc(β1-4)┘
O30 (N) Landau [44]	2)Rha4NAc(α1-3)LFuc(α1-4)Glc*6Ac*(β1-3)GalNAc(α1-
O30 (N) Urbana, Godesberg [45]	2)Rha4NAc(α1-3)LFuc(α1-4)Glc(β1-3)GalNAc(α1 Glc(β1-4)┘
O35 (O) Adelaide [46]	4)Glc(α1-4)Gal(α1-3)GlcNAc(β1- Col(α1-3)┘ └(6-1α)Col
O38 (P) [38]	3)Gal(β1-4)Glc(β1-3)GalNAc(β1- Gal(β1-4)┘ └(2-1β)GlcNAc
O39 (Q) Mara[c] [47]	2)Qui3NAc(α1-3)Man(α1-3)LFuc(α1-3)GalNAc(α1-
O40 (R) Riogrande [48]	4)GalNAc(α1-3)Man(β1-4)Glc(β1-3)GalNAc(α1- GlcNAc(β1-2)┘
O41 (S) [49]	2)Man(β1-4)Glc(α1-3)LQuiNAc(α1-3)GlcNAc(α1-
O42 (T) [50]	3)LRha(α1-2)LRha(α1-2)Gal(α1-3)GlcNAc(β1- └(2-1β)ManNAc
O43 (U) Milwaukee [51]	4)LFuc(α1-2)Gal(β1-3)GalNAc(α1-3)GlcNAc(β1- Gal(α1-3)┘
O44 (V) [52]	2)Glc(α1-6)Glc(α1-4)Gal(α1-3)GlcNAc(β1- GlcNAc(β1-3)┘
O45 (W) IIIa (*S. arizonae*) [53]	4)GlcA(β1-4)LFuc*3Ac*(α1-3)Rib*f*(β1-4)Gal(β1-3)GlcNAc(β1- LFuc(α1-2)┘
O47 (X) [54]	2)Rib-ol(5-*P*-6)Gal*4Ac*(α1-3)LFucNAm(α1-3)GlcNAc(α1-
O48 (Y) Toucra [55,56]	4)Neu5Ac*7,9Ac*(α2-3)LFucNAm(α1-3)GlcNAc(β1-
O50 (Z) II (*S. greenside*) [1,46]	6)GlcNAc(β1-3)Gal(α1-3)GalNAc(β1- └(3-1β)Gal(2-1α)Col
O50 IV (*S. arizonae*) [57]	6)GlcNAc(β1-3)Gal(α1-3)GlcNAc(β1- └(3-1β)Gal(2-1α)Col

(*continued*)

Table 3.3 (continued)

O51 [58]	6)Glc(α1-4)Gal(β1-3)GalNAc(α1-3)GlcNAc(β1- GlcNAc(β1-3)┘
O52 [50]	2)Rib*f*(β1-4)Gal(β1-4)GlcNAc(α1-4)Gal(β1-3)GlcNAc(α1-
O53 [59]	2)Gal*f*(α1-4)GalNAc(β1-4)LRha*2,3Ac*(α1-3)GlcNAc(β1-
O54 Borreze [60]	4)ManNAc(β1-3)ManNAc(β1-
O55 [61]	2)Glc(β1-2)Fuc3NAc(β1-6)Glc(α1-4)GalNAc(α1-3)GlcNAc(β1-
O56 [62]	3)Qui4N(LSerAc)(β1-3)Rib*f*(β1-4)GalNAc(α1-3)GlcNAc(α1-
O57 [63]	3)LRha(α1-2)LRha(α1-4)Glc(α1-3)GalNAc(β1- └(2-1β)GlcNAc
O58 [64]	3)Qui4N(DAla*S*3Hb)(β1-6)GlcNAc(α1-3)LQuiNAc(α1-3)GlcNAc(α1-
O59[d] [65]	2)Gal(β1-3)GlcNAc(α1-4)LRha(α1-3)GlcNAc(β1-
O60 [66]	2)Man(β1-3)Glc(β1-3)GlcNAc(β1- Fuc3NFo(α1-3)┘
O61 IIIb (*S. arizonae*) [67]	8)8eLeg5(*R*3Hb)7Ac(α2-3)LFucNAm(α1-3)GlcNAc(α1-
O62 IIIa (*S. arizonae*)[e] [68]	3)LRha(α1-2)LRha(α1-3)LRha(α1-2)LRha(α1-3)GlcNAc(β1- └(2-1α)GalNAcAN
O63 IIIa (*S. arizonae*) [69]	3)Gal(β1-4)Glc(α1-4)GalNAc(α1-3)GalNAc(β1- └(4-1α)Fuc3NAc
O65 [50]	4)GlcNAc(β1-4)Man(β1-4)Man(α1-3)GlcNAc(β1-
O66 [70]	2)Gal(α1-6)Gal(α1-4)GalNAc(α1-3)GalNAc*6Ac*(β1- Glc(β1-3)┘

[a]The OPS lacks O-acetylation.
[b]This structure has been published erroneously as that of *S. enterica* ssp. *arizonae* O64 (*Arizona* 29) and *Citrobacter* O32 [71]. Earlier, another structure has been established for *S. enterica* ssp. *arizonae* O21 [72], which, in fact, may belong to *Citrobacter braakii* O37 [73].
[c]The absolute configuration of Qui3NAc has been revised from L to D [74].
[d]Earlier, another structure has been reported for *S. enterica* ssp. *arizonae* O59 [75], which, in fact, may belong to *Citrobacter braakii* O35 [76] or *E. coli* O15 [65].
[e]Amidation of GalNAcA has not been originally reported [68] but demonstrated later [50].

are common constituents, and ManNAc is present in three OPSs, including the O54 antigen, which is a homopolymer of ManNAc. There are present also 6-deoxyamino sugars, such as LQuiN, Qui3N, Qui4N, LFucN, Fuc3N and Rha4N, which often bear uncommon *N*-acyl groups, such as formyl, acetimidoyl, (*R*)-3-hydroxybutanoyl, *N*-[(*S*)-3-hydroxybutanoyl]-D-alanyl and *N*-acetyl-L-seryl. A few OPSs are acidic, from which the O48 and O61 antigens contain derivatives of higher acidic sugars: neuraminic acid (Neu) and 8-epilegionaminic acid (8eLeg), respectively. The O47 antigen is phosphorylated and has a ribitol teichoic acid-like structure. The O62 antigen contains GalNAcA but is neutral as the acid occurs in the amide form. Additional modifications by glucosylation or/and O-acetylation further extend the diversity of the O-antigen forms within several O-serogroups, including serogroups A-E. In serogroups B, C_1, D_3 and H, the glucosylated and non-glucosylated forms are discrete polymer chains. The O-polysaccharides of serovars

Telaviv ($O28_1,28_2$) and Dakar ($O28_1,28_3$) are significantly different in composition and structure of both main and side chains that is unusual for strains belonging to the same *Salmonella* serogroup.

A polysaccharide different from the O-antigen may be a part of the LPS of *Salmonella*. For instance, the T1-specificity of a transient form of *S. enterica* is defined by 6)Gal*f*(β1-3)Gal*f*(β1-3)Gal*f*(β1- and 2)Rib*f*(β1- homopolymers [1], whose synthesis is determined by the *rft* locus. The T1-antigen as well as the O54 antigen, which is encoded by genes located on a plasmid [60], can be co-expressed with various *S. enterica* O-antigens. Infection of a serovar Typhimurium strain with the ColIb drd2 plasmid suppressed the normal O-antigen synthesis and induced synthesis of an altered LPS O-chain, probably by activation of a chromosomal operon inactive in the wild strain [77]:

```
3)LRha(α1-6)Glc(α1-2)Man(α1-3)GlcNAc(β1-
   └(2-1α)Galf
```

Citrobacter, Edwardsiella

Bacteria of the genus *Citrobacter* are normal inhabitants of human and animal intestine but may cause gastrointestinal diseases, urinary tract infections and bacteremia. The OPS structures have been established for the majority of the existing 43 O-serogroups and several nontypable strains [78]. Many from them consist only of neutral monosaccharides, such as common hexoses, pentoses (Xyl, Rib) and deoxy sugars: both enantiomers of Rha and Fuc, a unique monosaccharide 4-deoxy-D-*arabino*-hexose (4daraHex) and abequose. A minority of the OPSs are acidic due to the occurrence of an acidic sugar (GlcA, Neu5Ac), glycerol phosphate or ethanolamine phosphate as a substituent or a glycosyl phosphate group in the main chain. Remarkably, in the O32 antigen, L-glyceric acid (LGroA) interlinks the Fuc3N residues being in each pair N-linked to one residue and glycosylated by the other. Another uncommon amino sugar, Rha4NAc, builds up various homopolysaccharides of serogroup O9 strains and is present also in the heteropolysaccharide of two nontypable strains (Table 3.4).

In the O12 and O41 antigens, GlcN and Fuc3N bear a (*R*)-3-hydroxybutanoyl group. The same OPS may be characteristic for more than one O-serogroup. For instance, a 4dAraHex homopolymer is present in serogroups O4, O36 and O27, and variations in the LPS core OS are the reason for classification of the corresponding strains in three different O-serogroups [78]. The O-antigens of serogroups O1-O3 and O7 possess similar 4)Sug(α1-3)Sug(β1-4)Sug(β1- main chains, where Sug indicates either Man or Rha. Two pairs of strains of serogroups O7 and O12 have quite different structures, and their classification to one O-serogroup is thus questioned.

Various *Citrobacter* O-antigens are identical with, or structurally related to, the O-antigens of other bacteria, including *S. enterica* (serogroups O21, O22, O24, O38), *E. coli* (O23, O35, *C. rodentium* ATCC 51459), *Klebsiella pneumoniae* (O28, O39), *Hafnia alvei* (O16, O41) and *Eubacterium sabbureum* (O32) [78]. The main

Table 3.4 Structures of *Citrobacter* OPSs

C. youngae O1 [79]	4)Rha(α1-3)Man(β1-4)Man(β1- Rib*f*(α1-4)┘
C. youngae O2, O25, *C. werkmanii* O20 [80]	4)Rha(α1-3)Man(β1-4)Rha(β1- Xyl*f*(α1-4)┘
C. youngae O3 [78]	4)Man(α1-3)Rha(β1-4)Rha(β1-
C. youngae O4, O36, *C. werkmanii* O27 [78]	2)4daraHex(β1-
C. braakii O5, *Citrobacter* sp. PCM 1487 [78]	6)GlcNAc(α1-4)GalNAc(α1- 4daraHex(β1-3)┘
C. braakii O6 [81]	3)Fuc(α1-3)LRha2Ac(β1-3)Fuc(α1 4daraHex(α1-4)┘
C. braakii O7 (PCM 1503) [82]	4)Man(α1-3)Rha(β1-4)Rha(β1- Glc(α1-2)┘
C. braakii O7 (PCM 1532) [78]	3)Man(α1-3)Man(α1-2)Man(α1-2)Man(α1-2)Man(α1- Glc(α1-3)┘
C. braakii O8 [78]	3)Rha(α1-3)Rha(α1-2)Rha(β1- Xyl*f*(α1-2)┘
C. gillenii O9 (PCM 1537) [78]	3)Rha4NAc(α1-2)Rha4NAc(α1-2)Rha4NAc(α1-3)Rha4NAc2Ac(α1- and 2)Rha4NAc(α1-
C. youngae O9 (PCM 1538) [83]	2)Rha4NAc(α1- and 3)Rha4NAc(α1-3)Rha4NAc(β1-
C. gillenii O11(PCM 1540) [84]	3)Man(β1-4)Glc*p*(β1-3)FucNAc4Ac(α1-4)GalNAc(α1- └(2-1β)GlcNAc Glc(α1-6)┘
C. gillenii O12 (PCM 1542) [78]	6)GlcN(*R*3Hb)(β1-3)GalNAc(α1-3)GalNAc(β1- Glc(α1-6)┘ └(4-1α)GlcNAc
C. gillenii O12 (PCM 1544) [78]	3)LRha2Ac(β1-4)GlcNAc(β1-6)Gal(α1- *GlcNAc(β1-3)*┘
C. werkmanii O14 [85]	4)Glc6(*P*1Gro)(β1-3)GlcNAc-(β1- GlcNAc(β1-2)┘ └(6-1α)Glc
C. youngae O16 [78]	6)Gal(β1-4)GalNAc3(*P*1Gro)(β1-4)Glc(β1-3)Gal*p*NAc(β1- Glc(α1-2)┘ └(6-1α)Gal
C. werkmanii O21 [78]	6)Man*3Ac*(α1-2)Man(α1-2)Man(α1-3)GlcNAc(α1- Glc(α1-3)┘
C. freundii O22 [86]	2)Man(α1-4)LRha(α1-3)Gal(α1- └(3-1α)Abe
C. freundii O23 [78]	4)Man(α1-2)Man(α1-2)Man(β1-3)GalNAc(α1-
C. werkmanii O24 [78]	4)GlcA(β1-4)LFuc*3Ac*(α1-3)Rib*f*(β1-4)Gal(β1-3)GlcNAc(β1- LFuc(α1-2)┘
C. werkmanii O26 [78]	3)ManNAc(β1-4)Glc(β1- Glc(α1-2)┘
C. braakii O28 [78]	2)Rib*f*(β1-3)LRha(α1-3)LRha(α1-
C. braakii O29, O30 [78]	3)ManNAc(β1-4)Glc(β1-
C. youngae O32 [78]	2)LGroA(1-3)Fuc3N*2Ac*(α1-
C. braakii O35 [78]	2)Gal(β1-3)LFucNAc(α1-3)GlcNAc(β1-
C. braakii O37 [73]	7)Neu5Ac(α2-3)LFucNAm(α1-3)GlcNAc*6Ac*(β1-

(continued)

Table 3.4 (continued)

C. werkmanii O38 [78]	4)LRha(β1-2)Man(α1-2)Man(α1-3)Gal(β1- └(3-1α)Abe4Ac Glc(α1-2)┘
C. freundii O39 [87]	3)Gal*6(PEtN)*(β1-3)Gal(α1- and 3)Gal*f*(β1-3)Gal(α1-
C. freundii O41 [78]	2)Glc(β1-2)Fuc3N(R3Hb)(β1-6)GlcNAc(α1-4)Gal(β1-3)GalNAc(β1- Glc(α1-2)┘
Citrobacter sp. 396[a] [78]	2)Man(β1-2)Man(β1-2)Man(β1-2)Man(β1-3)GlcNAc(α1- Abe2Ac(α1-3)┘ └(3-1α)Glc
C. sedlakii NRCC 6070, *C. freundii* OCU 158 [78]	2)Rha4NAc(α1-3)LFuc(α1-4)Glc(β1-3)GalNAc(α1-
C. freundii NRCC 6052 [78]	2)Rha(α1-3)Rha(β1-4)Glc(β1-
C. rodentium ATCC 51459 [78]	3)GlcNAc(α1-P-6)Glc(α1-2)Glc(β1-3)GlcNAc(β1- └(4-1β)LRha

[a]The structure was established by older methods and requires reinvestigation.

Table 3.5 Structures of *Edwardsiella* OPSs

E. ictaluri MT 104 [88]	4)Gal(β1-4)Glc(α1-4)GalNAc(α1-3)GalNAc(β1-
E. tarda MT 108 [89]	4)GalNAc(β1-3)Gal(α1-4)LRha(α1-3)GlcNAc(β1- └(3-1α)GalA6LThr
E. tarda 1145, 1151 [90]	2)Man(α1-4)LRha(α1-3)Gal(α1- └(3-1α)Abe*2Ac*
E. tarda 1153 [90]	4)GalA6(GroN)(α1-4)Gal(α1-3)GalA(α1-3)GlcNAc(β-

chain of *C. braakii* O7 (PCM 1532) has the same structure as the linear mannan of *E. coli* O9, *K. pneumoniae* O3, and *H. alvei* PCM 1223. *C. sedlakii* NRCC 6070 and *C. freundii* OCU 158 share the OPS with *S. enterica* O30 and *E. coli* O157, and are serologically related also to some other bacteria whose OPSs contain various *N*-acyl derivatives of Rha4N.

Edwardsiella are occasional pathogens of humans; *E. tarda* can cause gastroenteritis and extraintestinal infections. The acidic OPS of *E. tarda* MT 108 includes an amide of GalA with L-threonine, and that of strain 1153 contains both GalA and its amide with 2-amino-2-deoxyglycerol (GroN) (Table 3.5). The OPS of strains 1145 and 1151 has the same carbohydrate structure as those of *S. enterica* O4 and *C. freundii* O22.

Escherichia, Shigella

Escherichia coli is a common component of the normal gut flora but certain strains also cause diarrhea, gastroenteritis, urinary tract infections and neonatal meningitis. *E. coli* O157 and several other virulent strains cause hemorrhagic colitis and hemolytic uremic syndrome. Strains of *Shigella*, mainly *S. dysenteriae*, *S. flexneri*, and *S. sonnei*, are causative agents of shigellosis (bacillar dysentery). The two genera are closely related, and genetically most *Shigella* strains are clones of *E. coli*. The complete O-antigen structures have been determined for all 46 *Shigella* serotypes and a majority of about 180 *E. coli* O-serogroups. Those of *S. dysenteriae, S. boydii* and *S. sonnei* [91] as well as most known *E. coli* OPS structures [92] have been summarized recently. The latter are also periodically updated in the *E. coli* O-antigens database (ECODAB) at http://www.casper.organ.su.se/ECODAB/. Therefore, the OPS structures of *E. coli* and *Shigella* species mentioned above are not shown here.

The OPSs of most *E. coli* and *Shigella* have linear or branched tri- to hexasaccharide O-units; less common are disaccharide O-units and homopolysaccharides. Almost all *Shigella* OPSs (except for most *S. flexneri* types, *S. boydii* type 18 and *S. dysenteriae* type 1) and many *E. coli* OPSs are acidic due to the presence of hexuronic acids, including such uncommon as LIdoA (*E. coli* O112ab), LAltNAcA (*S. sonnei*) and ManNAc3NAcA (*E. coli* O180), nonulosonic acids (Neu5Ac, *N*-acyl derivatives of 5,7-diamino-3,5,7,9-tetradeoxynon-2-ulosonic acids) and acidic non-sugar components, such as lactic, glyceric, pyruvic acids, amino acids or phosphate. Several OPSs possess glycerol or ribitol teichoic acid-like structures. Other constituent sugars rarely occurring in nature are colitose (*E. coli* O55 and O111), 6-deoxy-D-*manno*-heptose in *E. coli* O52, D-*threo*-pentulose (xylulose) in *E. coli* O97, *N*-acyl derivatives of various 6-deoxyamino and 6-deoxydiamino sugars, including LRhaN3N (*E. coli* O109 and O119) and FucN4N (*S. sonnei*). In *S. sonnei* and all other OPSs where FucN4N is present, it is 2-N-acetylated and has a free amino group at position 4. About half of *Shigella* serotypes have identical or almost identical OPS structures with *E. coli* [91]. Many other *E. coli* strains share OPSs with various bacteria, such as *Salmonella, Citrobacter, Klebsiella, Serratia, Hafnia, Yersinia* (see published review [92] and the corresponding sections in this chapter).

The OPSs structures of two other *Escherichia* species, *E. hermannii* and *E. albertii*, have been established. A group of *E. hermannii* strains produce homopolymers of Rha4NAc differing in the position of substitution of one of the sugar residues in the pentasaccharide O-units (Table 3.6).

The neutral OPSs of *S. flexneri* types 1–5, X and Y as well as newly proposed types 7a and 7b possess a common basic structure, and a diversity of the O-antigen forms depends on prophage-encoded glucosylation or/and O-acetylation at different positions of the basic glycan (Table 3.7). These serotype-converting modifications add new and may mask existing antigenic determinants, and strains with

Table 3.6 Structures of *E. hermanii* and *E. albertii* OPSs

E. hermannii ATCC 33650, 33652 [93]	2)Rha(α1-3)Rha(β1-4)Glc(β1- └(3-1α)Gal
E. hermannii ATCC 33651 [94]	3)Rha*2Ac*(β1-
E. hermannii NRCC 4262 [95]	3)Rha4NAc(α1-2)Rha4NAc(α1-2)Rha4NAc(α1- 3)Rha4NAc(α1-2)Rha4NAc(α1-
E. hermannii NRCC 4297-4300 [95]	3)Rha4NAc(α1-2)Rha4NAc(α1-3)Rha4NAc(α1- 3)Rha4NAc(α1-2)Rha4NAc(α1-
E. albertii (former *Hafnia alvei* 10457) [96]	3)Gal(β1-6)Gal*f*(β1-3)GalNAc(β1- └(6-2α)Neu5Ac

Table 3.7 Structures of *S. flexneri* OPSs

1a [99]	2)LRha*3,4Ac*(α1-2)LRha(α1-3)LRha(α1-3)GlcNAc(β1- Glc(α1-4)┘
1b [99]	2)LRha*3,4Ac*(α1-2)LRha(α1-3)LRha*2Ac*(α1-3)GlcNAc(β1- Glc(α1-4)┘
2a [99]	2)LRha*3,4Ac*(α1-2)LRha(α1-3)LRha(α1-3)GlcNAc*6Ac*(β1- Glc(α1-4)┘
2b [100]	2)LRha(α1-2)LRha(α1-3)LRha(α1-3)GlcNAc(β1- └(3-1α)Glc Glc(α1-4)┘
3a [74]	2)LRha(α1-2)LRha(α1-3)LRha2Ac(α1-3)GlcNAc*6Ac*(β1- └(3-1α)Glc
3b [100]	2)LRha(α1-2)LRha(α1-3)LRha2Ac(α1-3)GlcNAc(β1-
4a[a] [101]	2)LRha3(*P*EtN)(α1-2)LRha(α1-3)LRha(α1-3)GlcNAc(β1- Glc(α1-6)┘
4b [100]	2)LRha(α1-2)LRha(α1-3)LRha2Ac(α1-3)GlcNAc(β1- Glc(α1-6)┘
5a [98]	2)LRha*3,4Ac*(α1-2)LRha(α1-3)LRha(α1-3)GlcNAc(β1- └(3-1α)Glc
5b [102]	2)LRha(α1-2)LRha(α1-3)LRha(α1-3)GlcNAc(β1- └(3-1α)Glc └(3-1α)Glc
X [102]	2)LRha(α1-2)LRha(α1-3)LRha(α1-3)GlcNAc(β1- └(3-1α)Glc
Y [74]	2)LRha*3,4Ac*(α1-2)LRha(α1-3)LRha(α1-3)GlcNAc*6Ac*(β1-
6, 6a[b] [74]	2)LRha*3,4Ac*(α1-2)LRha(α1-4)Gal*p*A(β1-3)GalNAc(β1-
7a [103]	2)LRha(α1-2)LRha(α1-3)LRha(α1-3)GlcNAc(β1- Glc(α1-2)Glc(α1-4)┘
7b [103]	2)LRha(α1-2)LRha(α1-3)LRha*2Ac*(α1-3)GlcNAc(β1- Glc(α1-2)Glc(α1-4)┘

[a]Type 4a strains may lack phosphorylation.
[b]Types 6 and 6a differ only in the degree of O-acetylation.

glycosylated O-antigens are increased in virulence [97]. *S. flexneri* types 6 and 6a have a distinct acidic OPSs but share a 2)LRha(α1-2)LRha(α1- disaccharide fragment with the other serotypes. Recently, a phosphorylated variant of the type 4a OPS has been found. The OPSs of *S. flexneri* types 4b and 5a are shared by *E. coli* O129 and O135, respectively [98].

Klebsiella, Raoultella, Serratia

Klebsiella pneumoniae is a common cause of nosocomial infections. Outside the hospital, these bacteria are often responsible of pneumonia and urinary tract

Table 3.8 Structures of *K. pneumoniae* OPSs

O1, O6 [1,104]	3)Gal(α1-3)Gal*f*(β1- and 3)Gal(β1-3)Gal(α1-
O2a, 2a,b [104,106]	3)Gal(α1-3)Gal*f*(β1-
O2a,c [104,106]	3)Gal(α1-3)Gal*f*(β1- and 5)Gal*f*(β1-3)GlcNAc(β1-
O2a,e, O2a,e,h, O9[a] [107,108]	3)Gal(α1-3)Gal*f*(β1- *Gal(α1-2)*┘
O2a,f,g [108]	3)Gal(α1-3)Gal*f*(β1- Gal(α1-4)┘
O3 [1,104]	2)Man(1-2)Man(1-2)Man(1-3)Man(1-3)Man(1-
O4, O11 [1,104]	4)Gal(α1-2)Rib*f*(β1-
O5 [1,104]	3)Man(β1-2)Man(α1-2)Man(α1-
O7 [1]	2)LRha(α1-2)Rib*f*(β1-3)LRha(α1-3)LRha(α1-
O8 [109]	3)Gal(α1-3)Gal*f2,6Ac*(β1- and 3)Gal(β1-3)Gal(α1-
O12 [1,104]	3)GlcNAc(β1-4)LRha(α1-
22535 [110]	3)LRha(1-3)LRha(1-2)LRha(1-2)LRha(1-2)LRha(1-
i28/94 [111]	4)Glc(α1-3)LRha(α1-

[a]Serotypes O2a,e, O2a,e,h and O9 differ in the degree of galactosylation and O-acetylation at unknown position.

infections. Their O-antigens are all neutral and many are linear. The OPSs of serogroups O1, O2 and O8 share a 3)Gal(α1-3)Gal*f*(β- chain called galactan I, and are serologically related (Table 3.8). The distal end of this chain may bear another homoglycan (galactan II in case of O1 and O8). The OPSs of some other serogroups are homopolysaccharides (mannans or an L-rhamnan) too. The O4 and O12 antigens are terminated with an α- or β-linked residue of 3-deoxy-D-*manno*-oct-2-ulosonic acid (Kdo) [104], and the O5-mannan with 3-O-methylated Man [1]. The terminating group in the O3-mannan is a methyl group too but it is linked presumably via a phosphate group rather than directly to a mannose residue [105]. The OPSs are linked to the core OS through a β-GlcNAc primer. In serogroups O3 and O5, a 3)Man(α1-3)Man(α1-3)- disaccharide bridge (so called adaptor) is located between the OPS and the primer [104]. The O-antigens of *K. pneumoniae* O3, O4 and O5 are shared by *E. coli* O9, O20a,b and O8, respectively [92]. The O5 antigen is shared by *Burkholderia cepacia* O2 and E (see below) and *S. marcescens* O28. *K. pneumoniae* O10 has been reclassified as *Enterobacter* sp.

Raoultella (former *Klebsiella*) are isolated from plants, soil and water. *R. terrigena* ATCC 33257 has the same OPS structure as *K. pneumoniae* O12 [112], and the OPS of another *R. terrigena* strain is acidic due to the presence of a pyruvic acid acetal and has a unique structure [113]:

2)Man4,6*S*pyr(β1-3)ManNAc(α1-3)LRha(β1-4)GlcNAc(α1-

Serratia marcescens is a widely distributed environmental bacterium, which can causes outbreaks of infection, and occasionally death, in hospitalized patients. Their OPSs are neutral and many of them are similar to each other (for structures see review [114]). Rather common are disaccharide O-units containing usual sugars (Glc, Gal, LRha, GlcNAc, GalNAc), which are occasionally O-acetylated [114]. The O14 antigen has the same structure as that of *P. aeruginosa* O15 and *B. cepacia* O3, whereas the O2 antigen is shared by *H. alvei* 38. The O4 antigen is an O-acetylated variant of the OPS of *K. pneumoniae* i28/94. *S. marcescens* O19 antigen is composed of two separate blocks of disaccharide O-units; the shorter chain is proximal to the core OS and shares the O-unit with *K. pneumoniae* O12, and the longer distal chain differs in substitution of LRha (at position 3 rather than 4) and is terminated with β-Kdo [115]. The OPS of *S. plymuthica* S90/4625 consists of the same two galactan blocks as *K. pneumoniae* O1 but is O-acetylated at unknown position [116].

Hafnia

Strains of *H. alvei* are isolated from natural environments and also hospital specimens. A serotyping scheme including 39 O-groups has been proposed for *H. alvei* strains but not correlated with known O-antigen structures [117]. In addition to common monosaccharides, Rib, LFucN, Qui3N, Qui4N are components of several *H. alvei* OPSs, and single OPSs include LFuc, 6dTal, ManN, Fuc3N and Neu. Amino sugars are usually N-acetylated but several bear an (*R*)-3-hydroxybutanoyl group; in strain 1204, Qui3NFo is present. Most OPSs are acidic, and many are phosphorylated. Several of the latter possess teichoic acid-like structures with glycerol or, in strain 1191, a unique L-arabinitol component; the others have a phosphate bridge between the O-units or are decorated with glycerol 1-phosphate or ethanolamine phosphate. The OPS of strain 1206 is the only known glycan that contains D-allotreonine amide-linked to GalA. The O-antigen of strain 2 has the largest octasaccharide O-unit, and that of strain 1189 consists of hexa-, hepta- and octasaccharde O-units owing to non-stoichiometric glucosylation at two sites.

There are two groups of strains with the O-antigens that are structurally and serologically related to strains 1187 and 1199 (Table 3.9). The OPSs of each group have the same main chain but differ in the patterns of glucosylation or/and O-acetylation. It has been suggested to combine these strains in two serogroups and to place the remaining strains having the strain-specific O-antigens to a separate serogroup each [117]. Several O-antigens of *H. alvei* are shared by other bacteria: the hexosaminoglycan of strain 38 by *S. marcescens* O2, the mannan of strain 1223 by *E. coli* O9 and *K. pneumoniae* O3, and two galactans of strain Y166/91 by *K. pneumoniae* O1.

Cronobacter, Enterobacter, Pantoea

Cronobacter species (former *Enterobacter sakazakii*) are food-borne pathogens causing bacteremia, necrotizing enterocolitis and neonatal meningitis. Most OPSs of the genus are acidic due to the presence of hexuronic acids or, in *C. malonaticus,* Kdo (Table 3.10). The latter is a common constituent of the LPS core OS and occur in other non-repetitive LPS domains but is uncommon in O-units. The only neutral OPS is that of *C. sakazakii* ZORB A 741, which contains a tyvelose side chain. The O-antigens of *C. sakazakii* O1 and HPB 3290 have the same composition, including

Table 3.9 Structures of *H. alvei* OPSs

1187 [117]	2)Glc(α1-*P*-6)GlcN(*R*3Hb)(α1-4)GalNAc(α1-3)GalNAc(β1-
744, 1194, 1219, 1221, 114/60 [117,118]	2)Glc(α1-*P*-6)GlcN(*R*3Hb)(α1-4)GalNAc(α1-3)GalNAc(β1- Glc(α1-6)┘
537 (ATCC 13337) [117]	2)Glc(α1-*P*-6)GlcN(*R*3Hb)*3Ac*(α1-4)GalNAc(α1-3)GalNAc(β1- Glc(α1-6)┘
1199 [117]	3)Qui4NAc(β1-3)Gro(1-*P*-3)Gal(β1-3)GlcNAc*6Ac*(α1- GlcNAc*6Ac*(β1-2)┘
1200, 1203, 1205[a] [117,119]	3)Qui4NAc(β1-3)Gro(1-*P*-3)Gal(β1-3)GlcNAc*6Ac*(α1- GlcNAc*3,6Ac*(β1-2)┘ └(4-1α)Glc
2 [117]	4)Neu5Ac(α2-6)Glc(α1-6)Gal(β1-3)GalNAc(β1- Glc(α1-4)Gal(β1-6)Glc(β1-3)┘ └(6-1α)Glc
23 [117]	3)Qui4NAc(β1-3)6dTal4Ac(α1-3)LFuc(α1-6)Glc(α1-*P*-3)GlcNAc(α1-
32 [120]	4)GalA*2,3Ac*(α1-2)LRha(α1-4)Gal(β1-3)GalNAc(β1-4)GlcNAc(α1-
38 [117]	4)ManNAc(β1-4)GlcNAc(α1-
39 [117]	3)Gal(β1-4)Glc(β1-3)GalNAc(β1- Gal(β1-4)┘ └(2-1β)GlcNAc
1185[b] [121]	2)Qui3N(*R*3Hb)(β1-6)Glc(α1-4)GlcA*2Ac*(β1-3)GlcNAc(α1- Glc(α1-4)┘
1188 [117]	4)GlcA(β1-2)Man(α1-4)Gal(β1-3)GlcNAc(β1- LRha*2,3,4Ac*(α1-3)┘
1189 [122]	6)Glc(α1-4)GlcA(β1-4)GalNAc(β1-3)Gal(α1-3)GalNAc(β1- └*(4-1α)Glc* └(6-1α)Glc*(2-1α)Glc*
1190 [117]	3)LRha(α1-2)Rib*f*(β1-4)GalA(α1-3)GlcNAc(β1- Gal*f*(α1-2)LRha(α1-2)┘ └(5-1α)Glc
1191[c] [123]	4)Glc(β1-1)LAra-ol*2Ac*(5-*P*-3)Gal(β1-3)GalNAc(β1- GlcNAc(β1-2)┘ └(4-1α)Glc
1192[b] [124]	3)LRha(α1-3)LRha(β1-4)LRha(α1-3)GlcNAc(β1- └(2-1α)GlcA*2Ac*(4-1β)Rib*f*
1195 [125]	3)LFucNAc(α1-4)Glc(α1-*P*-4)Glc(α1-3)LFucNAc(α1-3)GlcNAc(α1- GlcNAc(α1-4)┘
1196 [126]	2)Gal(β1-6)Glc(α1-6)GlcNAc(α1-4)GalA(α1-3)GlcNAc(β1-
1204[b] [127]	2)Qui3NFo(β1-3)GalNAc(α1-4)GlcA*3Ac*(α1-3)Man(α1-2)Man(α1-3)GlcNAc(β1-
1206 [117]	4)GalA6DaThr(α1-2)LRha(α1-2)Rib*f*(β1-4)Gal(β1-3)GalNAc(β1-
1207[b] [128]	4)GalNAc3(*P*1Gro)(β1-3)Gal(α1-4)Gal(β1-3)GalNAc(β1- Glc(α1-6)┘
1209 [117]	3)Gal(β1-4)Glc(α1-4)GlcA(β1-3)GalNAc(β1- └(4-1α)LRha
1210 [117]	3)GlcNAc(α1-*P*-6)Gal(α1-4)Gal(β1-3)GlcNAc(β1- └(4-1β)LRha
1211[d] [129]	2)Glc(β1-2)Fuc3N(*R*3Hb)4Ac(β1-6)GlcNAc(α1-4)GalNAc(α1-3)GlcNAc(β1- Glc(β1-3)┘
1216 [117]	4)Qui3N(*R*3Hb)(α1-4)Gal6Ac(β1-4)GlcNAc(β1-4)GlcA(β1-3)GlcNAc(β1-

(*continued*)

Table 3.9 (continued)

1220 [117]	3)Gro(1-*P*-6)Glc(β1-4)LFucNAc(α1-3)GlcNAc(β1- Glc(α1-6)Gal(α1-3)┘ Glc(α1-6)┘
1222 [130]	2)LRha(α1-2)LRha*3(PEtN)4Ac*(α1-2)Ri*f*(β1-4)Gal(α1-3)GlcNAc(α1- └(3-1β)Gal*f*
1223 [131]	2)Man(α1-2)Man(α1-2)Man(α1-3)Man(α1-3)Man(α1-
1529 [132]	2)LRha(α1-3)LRha(α1-4)GalA(α1-3)GlcNAc*6Ac*(β1- └(3-1α)LRha
1546 [133]	6)Glc*3Ac*(α1-4)GlcA(β1-4)GalNAc3Ac(β1-3)Gal(α1-3)GalNAc(β1-
Y166/91 [134]	3)Gal(β1-3)Gal(α1- and 3)Gal(α1-3)Gal*f*(β1-
481-L [135]	4)GalNAc(α1-*P*-6)Gal(β1-3)GalNAc(β1-4)GlcNAc(α1- └(3-1β)Glc Glc(α1-4)┘

[a]The OPS lacks *O*-acetyl groups at position 6 of α-GlcNAc in strain 1205, position 6 of β-GlcNAc in strain 1203 or at both positions in strain 1200.
[b]The OPS is non-stoichiometrically O-acetylated at unknown position.
[c]Arabinitol may be partially replaced by xylitol (~3:1).
[d]In ~10% α-GlcN, the *N*-acetyl group is replaced by a 3-hydroxybutanoyl group.

Table 3.10 Structures of *Chronobacter* OPSs

C. malonaticus [136]	4)Kdo(β2-6)Glc(β1-6)Gal(β1-3)GalNAc(β1- GlcNAc(β1-2)┘
C. muytjensii [137]	4)Qui3NAc(α1-3)LRha(α1-6)GlcNAc(α1-4)GlcA(β1-3)GalNAc(α1-
C. sakazakii O1 [138]	2)Qui3N(LAlaAc)(β1-6)Glc(β1-3)GalNAc(α1- Glc(α1-4)GlcA(α1-4)┘
C. sakazakii HPB 3290 [139]	2)Qui3N(LAlaAc)(β1-6)Glc(α1-3)GlcA(β1-3)GalNAc(α1- Glc(α1-2)┘
C. sakazakii O2[a] [140], *C. sakazakii* HPB 2855 [141]	3)LRha4Ac(α1-4)Glc(α1-2)LRha(α1-3)GlcNAc(β1- └(2-1α)GalA(4-1α)LRha*2,3,4Ac*
C. sakazakii 767 [142]	3)LRha*4Ac*(α1-4)Glc(α1-2)LRha(α1-3)GlcNAc(β1- └(2-1α)GalA(4-1α)LRha └(4-1α)Glc
C. sakazakii ZORB A 741 [143]	3)LRha(α1-3)Gal6Ac(α1-3)Gal(α1- Tyv(α1-2)┘

[a]In the O2 antigen, LRha in the main chain is not acetylated.

an *N*-acetyl-L-alanyl derivative of Qui3N, but a different O-unit topology and sugar sequence. *C. sakazakii* O2 and two more strains possess the same main chain and a disaccharide side-chain but differ in the pattern of O-acetylation and the presence of a lateral Glc in strain 767.

Enterobacter cloacae is sometimes associated with urinary tract and respiratory tract infections. The structure has been established for the O10 antigen [144]:

6)Man(α1-2)Man(α1-2)Man(β1-3)FucNAc(α1-
Glc(α1-4)┘

Table 3.11 Structures of *P. agglomerans* OPSs

FL1 [145]	2)Rha(α1-2)Rha(β1-3)Rha(α1-2)Rha(α1-
62D$_1$[a] [146]	2)Qui3NAc(β1-3)LRha(α1-3)Gal(β1-3)FucNAc(α1- Gal(α1-6)┘
CIP 55.49 [147]	3)LFucNAc(α1-3)LFucNAc(α1-3)GlcNAc(β1- Glc(α1-2)LRha(α1-6)┘

[a]Strain was originally classified as *E. coli*, then as *Erwinia herbicola*.

The OPS of an *Enterobacter* sp. strain, formerly classified as *K. pneumoniae* O10, is a linear riborhamnan terminated with 3-O-methylated LRha [1]:

3)LRha(α1-3)Rib*f*(β1-4)LRha(α1-3)Rib*f*(β1-4)LRha(α1-

Pantoea (former *Enterobacter*) *agglomerans* is commonly isolated from plant surfaces, seeds, fruits, animal or human feces, and is known to causing wound, blood, and urinary tract infections. The OPSs of this species studied are neutral and enriched in 6-deoxyhexoses (Table 3.11).

Proteus, Providencia, Morganella

O-antigen structures have been established for all 76 known *Proteus* O-serogroups and more than half of 61 *Providencia* O-serogroups. The former have been summarized in a recent review [148], and the OPS structures of *Providencia* are shown below. The O-antigens of both genera possess some peculiar features in common. Most of them are acidic due to the presence of hexuronic acids, including a rare isomer LAltA, nonulosonic acids: Kdo, pseudaminic acid (Pse) and 8-epilegionaminic acid (8eLeg), and non-sugar acids, such as carboxyl-linked amino acids, including stereoisomers of N^{ε}-(1-carboxyethyl)-L-lysine, N-linked dicarboxylic acids [malonic, succinic, aspartic acids, *N*-(1-carboxyethyl)alanine], ether-linked hydroxy acids (lactic and 2,4-dihydroxypentanoic acids) and a pyruvic acid acetal. Phosphate-linked non-sugar groups are both occurring in other bacterial OPS: ethanolamine, glycerol and ribitol, which are found mainly in *Proteus*, and unique: *N*-(1-carboxyethyl)ethanolamine, choline and D-glyceramide in *Proteus mirabilis* O14, O18 and *Providencia alcalifaciens* O22, respectively. Man and LFuc have been detected only in *Providencia* but some other monosaccharides (LRha, L6dTal, various 6-deoxyamino sugars) are common for both genera. LQui present in the OPS of *P. stuartii* O44 is a rare component of O-antigens. From diamino sugars, LRhaN3N has been found in *Proteus penneri* O66, whereas FucN4N in both *Proteus* and *Providencia*. The main chain of the OPS of *P. alcalifaciens* O6 has the same structure as hyaluronic acid. The O-unit of *P. alcalifaciens* O38 and O45 contains D-alanine linked to the carboxyl group of *N*-acetylmuramic acid and thus represents a fragment of the bacterial cell-wall peptidoglycan (Table 3.12).

Table 3.12 Structures of *Providencia* OPSs

P. stuartii O4 [149]	3)Gal(β1-6)GlcNAc(β1-6)Gal(β1-3)GlcNAc(β1- └(6-1β)Qui4N(4LAspAc)
P. alcalifaciens O5 [150]	4)Qui3NAc(β1-3)Gal(α1-3)Gal(β1-3)GlcNAc(β1-
P. alcalifaciens O6 [151]	4)GlcA(β1-3)GlcNAc(β1- Col(α1-2)Gal(β1-3)GlcNA(β1-6)┘
P. alcalifaciens O7 [152]	3)LRha2Ac(β1-4)GlcNAc(β1-3)GlcA(α1-4)GlcNAc(α1-
P. alcalifaciens O8[a] [153]	3)GlcNAc4R(β1-3)Gal(β1-2)Gro(1-*P*-3)FucNAc4N(β1-
P. alcalifaciens O9 [154]	2)Glc(β1-6)Gal(α1-6)GalNAc(α1-4)GalNAc(α1-3)GalNAc(α1- Glc(β1-3)┘
P. alcalifaciens O12 [155]	4)Gal(β1-3)GalNAc(α1-4)Gal(β1-3)GalNAc(β1- GlcNAc(β1-3)┘ └(2-1β)Glc(2-1β)GlcNAc
P. rustigianii O14 [156,157]	3)GalA6(2*S*alaLys)(α1-4)GalNAc(α1-3)GlcNAc(α1-
P. rustigianii O16 [158]	6)GlcNAc3(*R*lac)(α1-3)LRha(β1-4)GlcNAc(β-
P. stuartii O18 [159]	4)Qui3NAc(β1-6)GlcNAc(α1-4)GlcA(β1-3)GalNAc(α1-
P. alcalifaciens O19 [160, 161]	2)Fuc3NAc4Ac(β1-3)GlcNAc4,6(*S*pyr)(α1-4)Gal(α1-4)Gal(β-3)GlcNAc(β1-
P. stuartii O20 [162]	8)8eLeg5Ac7Ac(α2-4)GlcA(β1-4)GlcA(β1-3)GlcNAc(α1-
P. alcalifaciens O21 [163]	3)GalA(α1-4)GalNAc(α1-4)GalNAc(α1-3)GalNAc(β1- └(4-1α)Fuc3NFo
P. alcalifaciens O22 [164]	4)GalNAc3(*P*2DGroAN) (β1-4)Gal(β1-3)FucNAc4N(β1-
P. alcalifaciens O23 [165]	4)GlcA6(2*R*alaLys)(β1-6)Gal(β1-6)Glc(β1-3)GalNAc(β1-
P. alcalifaciens O25 [166]	6)GalNAc(β1-4)GlcA(β1-3)GlcNAc(β1- └(4-1α)GalA(2*R*alaLys)
P. alcalifaciens O27 [167]	2)Qui4NFo(α1-4)GlcA(α1-4)Glc(β1-3)GalNAc6*Ac*(β1-
P. alcalifaciens O28 [168]	3)GlcNAc(β1-3)LFuc(α1-3)GlcNAc(β1- └(4-1α)LFuc(3-1α)GlcA
P. alcalifaciens O29 [169]	6)GlcNAc(α1-3)LFucNAc(α1-3)GlcNAc(α1- Glc(β1-4)┘
P. alcalifaciens O30 [170]	2)Qui4NFo(β1-2)Rib(β1-4)GlcA(β1-4)GlcA(β1-3)FucNAc4N(α1-
P. alcalifaciens O31[b] [171]	3)Gal(α1-4)GalNAc(β1-3)GalNAc(β1- Man4R(β1-4)┘
P. alcalifaciens O32 [172]	6)GlcNAc3(*S*lac)(α1-3)LFucNAc(α1-3)GlcNAc(α1- Glc(β1-4)┘
P. stuartii O33 [173]	3)Qui4N(4DAspAc)(β1-6)GlcNAc(α1-4)GalA(α1-3)GlcNAc(α1-
P. rustigianii O34 [174]	4)GlcA(β1-4)LFuc(α1-2)Man(α1-2)LFuc(α1-2)Glc(β1-3)GlcNAc(β1- GalNAc(α1-3)┘
P. alcalifaciens O35[c] [164]	4)GalNAc(α1-6)Glc(α1-4)GlcA(β1-3)GalNAc(β1 └(6-1β)Qui4NR
P. alcalifaciens O36 [175]	7)Kdo(β2-3)L6dTal2*Ac*(α1-3)GlcNAc(α1-
P. alcalifaciens O38, O45 [164]	4)GlcNAc3(*R*lac-DAla)(α1-4)GlcNAc(β1-
P. alcalifaciens O40 [164]	4)Qui3NFo(β1-3)Gal(α1-3)GlcA(β1-3)GalNAc(β1-
P. stuartii O43 [167]	2)Qui4NAc(α1-4)GlcA(β1-3)GalA6LSer(β1-3)GlcNAc(β1-

(*continued*)

Table 3.12 (continued)

P. stuartii O44 [176]	4)GalNAc(α1-3)LFuc(α1-3)Glc(α1-4)LQui(α1-3)GlcNAc(α1- GlcA(β1-4)┘
P. alcalifaciens O46 [177]	3)GlcA(β1-4)LFuc(α1-4)LFuc(α1-2)Glc(β1-3)GlcNAc*6Ac*(α1- Glc(α1-3)┘
P. stuartii O47 [178]	2)Gal(β1-4)Man*6Ac*(β1-3)Man(β1-4)GlcA(β1-3)GlcNAc(α1- LRha(α1-3)┘
P. alcalifaciens O48 [179]	3)Man(α1-2)LFuc(α1-2)GlcA4Ac(β1-3)GalNAc(α1-
P. stuartii O49 [180]	4)Gal(α1-6)Gal(β1-3)GalNAc(β1-
P. stuartii O57 [181]	2)Gal(α1-3)LRha*2Ac*(α1-4)Glc(α1-4)GalA6LAla(α1-3)GlcNAc(β1-
P. alcalifaciens O60 [182]	4)GlcA6LSer(β1-3)GalNAc(β1-4)Glc(β1-3)Gal(α1-4)GalNAc(β1-

[a]R indicates (*1S,3R*)-3-hydroxy-1-carboxybutyl. In the original publication [153], Gro(3-*P* has been shown in the structure erroneously.
[b]R indicates (*1R,3R*)-3-hydroxy-1-carboxybutyl.
[c]R indicates *N*-(1-carboxyethyl)alanine of unknown configuration.

Morganella morganii is commonly found in the environment and in the intestinal tract of humans, mammals and reptiles as normal flora. A remarkable feature of the OPS of *M. morganii* is the presence of two rare sugars: a 5-*N*-acetimidoyl-7-*N*-acetyl derivative of 8-epilegionaminic acid and a higher branched ketouronamide called shewanellose, which occurs in the pyranose form in some O-units or in the furanose form in the others [183] (Fig. 3.1).

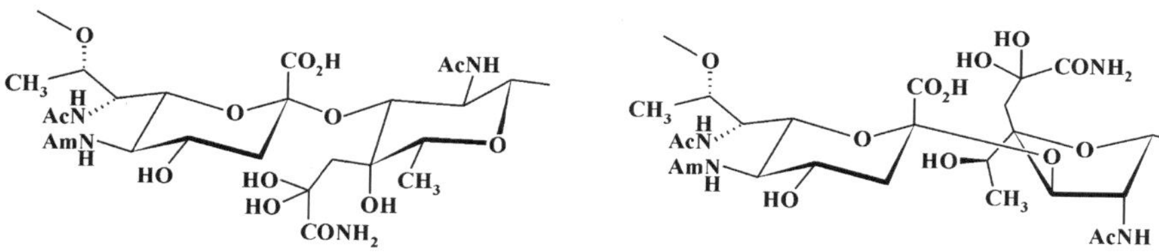

Fig. 3.1 Structures of the O-units of *Morganella morganii* [183]

A similar structure but with shewanellose exclusively in the pyranose form has been reported for a polysaccharide of *Shewanella putrefaciens* A6 [184].

Yersinia

Most important *Yersinia* species are *Yersinia pestis*, the cause of bubonic and pneumonic plague, *Yersinia pseudotuberculosis* and *Yersinia enterocolitica*, which cause less severe diseases usually restricted to gastrointestinal tract. *Y. pestis* has a cryptic O-antigen gene cluster and does not express any O-antigen [186]. Minireviews on the OPS structures of other *Yersinia* have been published [185–188].

Yersinia pseudotuberculosis is the only bacterium that produces all known natural 3,6-dideoxyhexose, and most of its OPSs have a side chain of one of the isomers. Paratose occurs as either pyranose (serogroup O3) or furanose (serogroup

O1); other 3,6-dideoxyhexoses are always pyranosidic. Two OPSs have an L6dAlt*f* side chain (Table 3.13). The 6-deoxy- and 3,6-dideoxy-hexoses are linked either directly to the main chain or through another uncommon monosaccharide: 6-deoxy-D-*manno*-heptose (6dmanHep) or, in serogroup O6, a branched sugar 3,6-dideoxy-4-*C*-[(*S*)-1-hydroxyethyl]-D-*xylo*-hexose (yersiniose A). When synthesis of 6dmanHep is impaired, its biosynthetic precursor, D-*glycero*-D-*manno*-heptose, is incorporated into the O-unit in place of 6dmanHep [189]. Between O-serogroups, the OPSs differ in the side chain or the main chain or both. Within complex O-serogroups, division to subgroups is based either on different side chains linked to the same main chain as in serogroup O5, or different main chains bearing the same side chain as in the other serogroups. The OPS of *Y. pseudotuberculosis* O10 is remarkably similar to that of *E. coli* O111 and *S. enterica* O35.

Many linear OPSs and main chains of branched OPSs of *Y. enterocolitica* and several other *Yersinia* species are homopolymers of Rha, LRha or L6dAlt (Table 3.14). The lateral monosaccharides are enantiomers of xylose and xylulose (Xlu), yersiniose A and its (*R*)-stereoisomer yersiniose B. The O-antigens of *Y. enterocolitica* O6,31 and O8 are the only known polysaccharides that contain 6dGul. The O5,27 and O10 antigens have comb-like structures with each rhamnose residue of the main chain substituted with a xylulose residue. The OPSs of two *Y. kristensenii* strains resemble glycerol teichoic acids. The *Y. ruckerii* OPSs are acidic due to the presence of *N*-acetylmuramic acid or a derivative of 8eLeg with a 4-hydroxybutanoyl group at N-5. An α1-2-linked homopolymer of Rha4NFo is shared by *Y. enterocolitica* O9 and *Brucella abortus* [203]. The OPS of *Y. ruckerii* O1 is remarkably similar to that of *Salmonella arizonae* O61, and those of *Y. enterocolitica* O5,27 and *Y. kristensenii* O11,23 are identical with the OPSs of *E. coli* O97 and O98, respectively.

Other Genera

Plesiomonas shigelloides, the only species in the genus, is a ubiquitous micro-organism, which may cause water- and food-born gastrointestinal infections and illnesses in immunocompromised hosts and neonates. Its OPSs contains various unusual components, including D-*glycero*-D-*manno*-heptose (DDmanHep), 6dmanHep, L6dTalN, QuiN4N and GlcN3NA as well as *N*-acyl groups: acetimidoyl, (*S*)-3-hydroxybutanoyl or 3-hydroxy-2,3-dimethyl-5-oxoprolyl (Table 3.15). The O17 antigen possesses a disaccharide O-unit composed of two uncommon sugars: one acidic, LAltNAcA, and one basic, FucNAc4N. It has the same structure as the plasmid-encoded OPS of *Shigella sonnei* [91].

Yokenella regensburgei is recovered from wounds and knee fluid, respiratory tract, urine, sputum and stool. It is an opportunistic pathogen, especially under immunocompromised conditions. The OPSs of four strains studied have the same trisaccharide O-unit containing LDmanHep and 2-O-acetylated or, in one strain, 2,4-di-O-acetylated L6dTal [227]:

2)LDmanHep(α1-3)L6dTal2(4)Ac(α1-3)FucNAc(α1-

Table 3.13 Structures of *Y. pseudotuberculosis* OPSs

O1a [190]	3)Gal(α1-3)GlcNAc(β1- Par*f*(α1-3)6dmanHep(β1-4)┘
O1b [191]	2)Man(β1-4)Man(α1-3)LFuc(α1-3)GlcNAc(α1- Par*f*(β1-3)┘
O1c [192]	2)Man(α1-3)LFuc(α1-3)GalNAc(β1- Par*f*(β1-3)┘
O2a [189,193]	3)Gal(α1-3)GlcNAc(β1- Abe(α1-3)6dmanHep(β1-4)┘
O2b [194]	2)Man(α1-3)LFuc(α1-3)GalNAc(β1- Abe(α1-3)┘
O2c [195]	6)Man(α1-2)Man(α1-2)Man(β1-3)GalNAc(α1- Abe(α1-3)┘
O3 [186, 195]	2)Man(α1-3)LFuc(α1-3)GalNAc(α1- Par(β1-4)┘
O4a [196]	6)Man(α1-2)Man(α1-2)Man(β1-3)GalNAc(α1- Tyv(α1-3)┘
O4b [197]	3)Gal(α1-3)GlcNAc(β1- Tyv(α1-3)6dmanHep(β1-4)┘
O5a [185,186]	2)LFuc(α1-3)Man(α1-4)LFuc(α1-3)GalNAc(α1- Asc(α1-3)┘
O5b [185,186]	2)LFuc(α1-3)Man(α1-4)LFuc(α1-3)GalNAc(α1- L6dAlt*f*(α1-3)┘
O6[a] [185,186,198]	3)GlcNAc(β1-6)GalNAc(α1-3)GalNAc(β1- Col(α1-2)Sug(β1-3)┘
O7 [187]	6)Glc(β1-3)GalNAc(α1-3)GalNAc(β1- Col(α1-2)┘ Glc(α1-6)┘
O9 [199]	4)GlcNAc3Ac(β1-4)LFucNAm(α1-3)GlcNAc(α1- Gal(α1-3)┘
O10 [200]	4)Glc(α1-4)Gal(α1-3)GalNAc(β1- Col(α1-3)┘ └(6-1α)Col
O11 [201]	2)Man(β1-4)Man(α1-3)LFuc(α1-3)GlcNAc(α1- L6dAlt*f*(α1-3)┘
O15 [202]	2)LFuc(α1-3)Man(α1-4)LFuc(α1-3)GalNAc(α1- Par*f*(β1-3)┘

[a]Sug indicates yersiniose A.

Budvicia aquatica, Pragia fontium, Rahnella aquatilis are the only species in each of the three new genera of Enterobacteriaceae. They are isolated mainly from fresh water, water pipes and sometimes from clinical specimens but the

Table 3.14 Structures of other *Yersinia* sp. OPSs

Y. enterocolitica O1,2a,3[a], O2a,2b,3 [185,204]	2)L6dAlt*f3Ac*(β1-2)L6dAlt*f3Ac*(β1-3)L6dAlt*f*(β1-
Y. enterocolitica O2,3, O3 [185,204]	2)L6dAlt(β1-
Y. enterocolitica O4,32, *Y. intermedia* O4,33[a,b] [185,198]	3)GalNAc(α1-3)GalNAc(β1- Sug1'Ac(α1-4)┘
Y. enterocolitica O5,27[c] [185]	3)LRha(α1-3)LRha(β1- Xlu*f*(β2-2)┘ └(2-2β)Xlu*f*
Y. enterocolitica O6,31 [185]	2)Gal(β1-3)6dGul(α1-
Y. enterocolitica O8[d] [185]	4)Man(1-3)Gal(1-3)GalNAc(α1- 6dGul(1-3)┘ └(2-1)LFuc
Y. enterocolitica O9 [185]	2)Rha4NFo(α1-
Y. enterocolitica O10 [205]	3)Rha(α1- LXlu*f*(β2-2)┘
Y. kristensenii O11,23, O11,24[a] [206]	3)LQuiNAc(α1-4)GalNAcA3Ac(α1-3)LQuiNAc(α1-3)GlcNAc(β1-
Y. kristensenii O12,25 [207]	2)Gro(1-P-6)Glc(β1-4)LFucNAc(α1-3)GlcNAc(β1- Glc(α1-6)GalNAc(α1-3)┘ GlcNAc(β1-4)┘
Y. kristensenii O12,26 [208]	2)Gro(1-P-6)Glc(β1-6)GalNAc(α1-3)LFucNAc(α1-3)GlcNAc(β1- Glc(α1-2)┘ Glc(α1-4)┘
Y. frideriksenii O16,29[e] [209]	2)Rha(α1-3)Rha(β1-3)Rha(α1- *Sug(β1-2)*┘
Y. kristensenii O25,35 [210]	2)Gro(1-P-6)Glc(β1-4)LFucNAc(α1-3)GlcNAc(β1- Glc(α1-6)Gal(α1-3)┘ Glc(α1-4)┘
Y. kristensenii O28 [211]	3)LRha(α1-3)LRha(α1-3)LRha(α1-3)GlcNAc(β1- └(2-1α)GalNAcA(4-1α)LRha
Y. aldovae 6005 [212]	2)Glc(β1-2)Fuc3N(*R*3Hb)(β1-6)GlcNAc(α1-4)GalNAc(α1-3)GlcNAc(β1- Glc(β1-3)┘
Y. bercovieri O10[e] [213]	3)Rha(α1-3)Rha(α1- Sug(α1-2)┘
Y. mollarettii [214]	2)Gal(β1-3)6dGul(α1-
Y. rohdei WA 339 [215]	3)LRha(α1-3)LRha(α1-3)LRha(β1-
Y. ruckerii O1 [67, 216]	8)8eLegp5(4Hb)7Ac(α2-3)LFucNAm(α1-3)GlcNAc(α1- GlcNAc(β1-4)┘
Y. ruckerii O2[f] [217]	4)GlcNAc*6Ac*3(*R*lac)(α1-3)LQuiNAc(α1-3)GlcNAc(β1-

[a]The OPS lacks O-acetylation.
[b]Sug indicates yersiniose B.
[c]An alternative structure with one more LRha residue in the O-unit has been reported for the O5 and O5,27 antigens [218].
[d]The configurations of most glycosidic linkages have not been determined.
[e]Sug indicates yersiniose A.
[f]Details of the structure elucidation have not been reported.

Table 3.15 Structures of *P. shigelloides* OPSs

O1 [219,220]	3)L6dTalNAc*4Ac*(β1-4)LFucNAc(α1-4)LFucNAc(α1- 4) LFucNAc(α1-3)QuiNAc4N(S3Hb)(β1-
O17 [221]	-4)LAltNAcA(α1-3)FucNAc4N(β1-
O51 [222]	4)GlcNAc3N(S3Hb)A(β1-4)LFucNAm3Ac(α1-3)QuiNAc(α1-
O54 [223,224]	4)DDmanHep(β1-3)6dmanHep2Ac(β1-4)LRha(α1-3)GlcNAc(β1- └(3-1α)LRha(4-1β)Gal*f*
O74[a] [225]	2)Qui3NR(β1-3)LRha2Ac(α1-3)FucNAc(α1-
22074, 12254 [226]	3)LRha(α1-2)LRha(α1-2)LRha(α1-4)GalA(α1-3)GlcNAc(α1-

[a]R indicates 3-hydroxy-2,3-dimethyl-5-oxoprolyl of unknown configuration.

Table 3.16 Structures of *R. aquatilis* OPSs

33071^{T} [231]	3)Man(α1-2)Man(α1-3)Gal(β1- and 4)Rha(α1-3)Rha(α1-3)Man(β1- └(2-1α)GlcA(4-1α)Gal(3-1β)Glc
1-95 [233]	3)Gal*f*(β1-3)Fuc(α1- Gal(α1-2)┘
3-95 [234]	2)Man(α1-3)Man(α1-6)Man(α1- and 6)Glc(α1-

medical significance of the three genera remains uncertain. The OPS of *B. aquatica* has a 1,3-poly(glycerol phosphate) main chain decorated with β1-2-linked Glc residues [228].

The OPS of *P. fontium* 27480 is acidic due to the presence of ManNAc3NAcA [229]:

4)ManNAc3NAcA(β1-2)LRha(α1-3)LRha(β1-4)GlcNAc(α1-

and that of *P. fontium* 97 U116 is neutral [230]:

2)Gal*f*(α1-3)LRha2Ac(α1-4)GlcNAc(α1-2)LRha(α1-3)GlcNAc(β1-

Both acidic and neutral OPSs have been found in *R. aquatilis* 33071^{T} [231], the former being shared by strain 95 U003 [232]. In *R. aquatilis* 3–95, two neutral homoglycans, a mannan and a glycan, are present (Table 3.16).

Erwinia and *Pectobacterium* are pathogens of plants. The former causes wilts or blight diseases and the latter soft rot. The OPS of *E. amylovora* T is structurally similar to that of *R. aquatilis* 1–95 [233] but galactofuranose is replaced by glucofuranose [235]. The latter sugar has not been reported elsewhere in natural carbohydrates, and the structure may need revision [1]. The OPS of *P. atrosepticum* ssp. *carotovora* (formerly *E. carotovora*) is enriched in deoxy sugars [236]:

3)LRha(β1-4)LRha(α1-3)Fuc(α1-
  Glc(α1-3)┘

and a higher branched monosaccharide erwiniose has been identified in the OPS of *P. atrosepticum* ssp. *atroseptica* [237] (Fig. 3.2).

3)Man(α1-4)LRha(α1-3)Gal(α1-
  └(2-1α)Gal2Ac(3-1β)Sug

Sug =

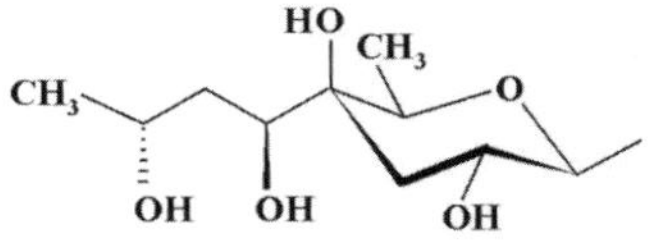

erwiniose

Fig. 3.2 Structure of the OPS of *Pectobacterium atrosepticum* ssp. *atroseptica* [237]

3.3.2.2 Aeromonadaceae

Aeromonas species are ubiquitous water-borne bacteria responsible for a wide spectrum of diseases in aquatic and terrestrial animals as well as in humans. *A. hydrophila* and *A. caviae* are often associated with gastrointestinal diseases in adults and acute gastroenteritis in children. Most OPSs of the genus studied so far are neutral. The O-unit of *A. hydrophila* O34 contains two L6dTal residues, one of which is randomly O-acetylated. The OPSs of various *A. salmonicida* types possess a main chain of 4)LRha(α1-3)ManNAc(β1- and differ in the modes of O-acetylation and glucosylation (Table 3.17). Under *in vivo* growth conditions, *A. salmonicida* type A strain A449 produces a different OPS with a side chain elongated by four more Glc residues and more sites of O-acetylation [238]. In encapsulated type A strain 80204-1, the OPS includes a partially amidated GalNAcA residue and an *N*-acetyl-L-alanyl derivative of Qui3N [239]. The OPSs of *A. caviae* are acidic due to the presence of GlcA or glycerol 1-phosphate. The O-antigen of *A. bestiarum* with an L-rhamnan backbone is shared by *Pseudomonas syringe* pv. *atrofaciens* [240, 241]. *A. trota*, *Vibrio cholerae* O22 and O139 and *Pseudoalteromonas tetraodonis* have a branched tetrasaccharide fragment in common, which represents a colitose (3-deoxy-L-fucose) analogue of the Le^b antigenic determinant.

3.3.2.3 Pseudoalteromonadaceae, Shewanellaceae, Idiomarinaceae

These families combine microorganisms of the marine origin, whose O-antigen structures have been summarized recently [251, 252]. The OPSs of obligatory marine bacteria *Pseudoalteromonas* (formerly *Alteromonas*) are neutral or acidic and contain various unusual components, such as LIdoA, amino and diamino hexuronic acids, their primary amides and amides with amino acids, keto sugars, including Kdo and Pse, an ether of Glc with (*R*)-lactic acid (glucolactilic acid) and glycerol phosphate; constituent amino sugars bear various N-linked hydroxy and amino acids (Table 3.18). An agarolytic strain *P. agarivorans* KMM 232 (former

Table 3.17 Structures of *Aeromonas* OPSs

A. bestiarum [242]	3)LRhap(α1-3)LRhap(α1-2)LRhap(α1-2)LRhap(α1- └(2-1β)GlcNAc
A. caviae 11212 [243]	6)ManNAc(β1-4)GlcA(β1-3)GalNAc(β1- LRha(α1-3)┘ └(4-1β)Gal
A. caviae ATCC 15468 [244]	4)GalNAc3(*P*1Gro)(β1-4)GlcNAc(β1-4)LRhap(α1-3)GalNAc(β1-
A. hydrophila SJ-44[a] [245]	4)LRha*2Ac*(α1-3)GlcNAc(β1-
A. hydrophila O34[b] [246]	4)Man(α1-3)L6dTal2Ac(α1-3)GalNAc(β1- └(3-1α)L6dTal*2,3,4Ac*
A. salmonicida type A [247,248]	4)LRha*2Ac*(α1-3)ManNAc(β1- Glc(α1-3)┘
A. salmonicida type B [248]	4)LRha(α1-3)ManNAc(β1-
A. salmonicida type C [248]	4)LRha*Ac*(α1-3)ManNAc(β1-
A. salmonicida SJ-15[c] [249]	4)LRha(α1-3)ManNAc*4Ac*(β1- Glc(α1-4)Glc(α1-3)┘
A. salmonicida 80204-1 [239]	4)Qui3N(LAlaAc)(β1-3)GalNAcA*N*(1-3)QuiNAc(β1-
A. trota [250]	3)Gal(β1-3)GlcNAc(β1-4)LRha(α1-3)GalNAc(α1- Col(α1-2)┘ └(4-1α)Col

[a]*A. hydrophila* O11 antigen has the same structure but, in addition to LRha2Ac, includes minor LRha3Ac [74].

[b]Lateral L6dTal carries no, one or two *O*-acetyl groups at any positions.

[c]The structure seems to need reinvestigation [248].

P. marinoglutinosa) synthesizes different polysaccharides in the S- and R-form colonies: a linear sulfated glycan, which is highly uncommon for O-antigens, or a branched OPS enriched in amino sugars, including an *N*-acetyl-L-threonyl derivative of FucN, respectively. The OPS of *P. rubra* has a similar structure to that of *Vibrio vulnificus* CECT 5198 [253] but the latter incorporates QuiNAc into the O-unit in place of its biosynthetic precursor 2-acetamido-2,6-dideoxy-D-*xylo*-hexos-4-ulose in *P. rubra*.

Bacteria of the genus *Shewanella* are responsible for spoilage of protein-rich foods and are opportunistic pathogens of marine animals and humans. All OPSs of *Shewanella* studied are acidic and many contain GlcA, GalA or amides of GalA with 2-amino-1,3-propanediol (GroN) or N^{ε}-[(*S*)-1-carboxyethyl]-L-lysine (*S*alaLys) (Table 3.19). The OPS of *S. japonica* KMM 3601 is one of a few O-antigens that contain a derivative of 4-epilegionaminic acid (4eLeg). In *S. algae* BrY, an LRha residue is linked to a neighbouring LFucN through O2 of an L-malyl group, which is the *N*-acyl substituent of the latter.

The OPS of *Idiomarina zobellii* is unique in the presence of two amino sugars, Qui4N and LGulNA, with free amino groups [258]:

3)Qui4N(α1-4)GlcA(α1-6)GlcNAc(α1-4)LGulNA(α1-3)FucNAc(β1-

Table 3.18 Structures of *Pseudoalteromonas* OPSs

Pseudoalteromonas sp. KMM 634 [251]	4)ManNAc3NAcA6LAla(β1-4)GlcNAc3NAcA(β1- 4)GlcA(β1-3)QuiNAc4N(*S*3Hb)(α1-
Pseudoalteromonas sp. KMM 637 [251]	4)Glc(β1-4)GalA(β1-4)Man(β1-
Pseudoalteromonas sp. KMM 639 [251]	3)LRha(α1-3)Gal6(*P*2Gro)(α1-
P. agarivorans (R-from) [254]	3)LRha(α1-3)FucN(LThrAc)(α1-3)GalNAc(α1- ManNAcA(β1-4)┘
P. agarivorans (S-from)[a] [251]	4)LRha2R(α1-3)Man(β1-
P. aliena [252]	3)GlcA6LSer(β1-4)GlcNAc(α1-4)ManNAcA6LSer(β1-4)GlcNAc(β1- └(4-1α)Qui4NAc
P. atlantica [255]	3)Gal(α1-6)GlcNAc(α1-4)GalA(α1-3)QuiNAc(β1- └(6-2β)Pse5Ac7Ac
P. distincta [251]	4)Pse5Ac7Fo(α2-4)QuiNAc(β1- GlcA(α1-4)GalNAc(β1-4)GalNAcA*3Ac*(α1-3) ┘
P. elyakovii [251]	6)Glc(α1-2)Glc(α1-4)GalNAc(β1-3)Gal(α1-3)GalNAc(β1-
P. flavipulchra [251]	7)Kdo(α2-3)L6dTal4Ac(α1-3)Gal(1β-
P. haloplanktis ATCC 14393[T] [251]	2)Qui3N(DAlaAc)(β1-4)GalNAcA(α1-4)Gal*2,6Ac*(α1- 4)LGalNAcA(α1-3)QuiNAc4NAc(β1-
P. haloplanktis KMM 156 [251]	2)LRha(α1-3)LRha(β1-4)GlcNAc(β1- └(3-1α)Glc3*R*lac
P. haloplanktis KMM 223 [251]	2)LIdoA(α1-4)GlcA(β1-4)GlcA(β1-3)QuiNAc4N(*S*3Hb)(β1- └(4-1α)QuiNAc4N(*S*3Hb)
P. mariniglutinosa (*Alteromonas marinoglutinosa*) [256]	3)Gal(α1-3)GlcNAc(β1 └(4-1β)ManNAc
P. nigrifaciens [251]	3)Gal(α1-4)LGulNAcA(α1-4)GlcNAc3Ac(β1- └(4-1α)Fuc3N(4Hb)
P. rubra[b] [253]	4)GlcNAc3NRAN(β1-4)LGalNAmA3Ac(α1-3)Sug(α1-
P. tetraodonis [251], *P. carrageenovora* [252]	2)Col(α1-4)GlcNAc(β1-4)GlcA(β1-3)GalNAc(1β- └(3-1β)Gal(2-1α)Col

[a]R indicates sulfate.
[b]R indicates 4-L-malyl, and Sug indicates 2-acetamido-2,6-dideoxy-D-*xylo*-hexos-4-ulose.

Table 3.19 Structures of *Shewanella* OPSs

S. algae 48055 [251]	3)GalA6(GroN)(α1-4)Neu5Ac(α2-3)GalA6(GroN)(β1-3)GlcNAc(β1-
S. algae BrY[a] [251]	3)LRha(α1-2)LRha(α1-2)R(4-2)LFucN(α1-3)QuipNAc4N(*R*3Hb)(α1-
S. fidelis KMM 3582[T] [252]	2)GalA6(2*S*alaLys)(α1-3)GalNAc(β1-4)GlcA(β1-3)GalNAc(β1-
S. japonica KMM 3299[T] [252]	3)Fuc4NAc(α1-4)GalA(α1-3)LFucNAc(α1-3)QuiNAc4NAc(β1-
S. japonica KMM 3601 [257]	4)4eLeg5Ac7Ac(α2-4)GlcA3Ac-(β1-3)GalNAc(β1-

[a]R indicates 4-L-malyl.

3.3.2.4 Pasteurellaceae

Bacteria *Aggregatibacter* (former *Actinobacillus*) *actinomycetemcomitans* are associated with aggressively progressing periodontitis and also cause serious infections, such as endocarditis. The O-antigens of serotypes a-f are neutral polysaccharides with di- or tri-saccharide O-units enriched in 6-deoxy sugars (Table 3.20). In serotypes a and c, they are distinctly O-acetylated homopolymers of enantiomers of 6-deoxytalose.

Actinobacillus (*Haemophilus*) *pleuropneumoniae* is a primary swine pathogen that causes hemorrhagic necrotizing pneumonia. *A. pleuropneumoniae* O-antigens are neutral polysaccharides, including galactans and glucogalactans present in many serogroups (Table 3.21).

Actinobacillus suis is a pathogen of pigs too. The O1 antigen of *A. suis* is a β1-6-linked glucan [268]. The O2 antigen that occurs in the majority of isolates in sick animals is a heteropolysaccharide [269]:

3)Gal(β1-4)Glc(β1-6)GlcNAc(β1-
   Gal(α1-6)┘

Mannheimia (*Pasteurella*) *haemolytica* is associated with several diseases of cattle and sheep. The OPSs of both biotypes A and T are neutral and as simple as the other O-antigens in the family Pasteurellaceae (Table 3.22). The OPS of serotypes T4 and T10 has the same structure as galactan I of *Klebsiella pneumoniae* present also in *S. marcescens* O20 and some other bacteria. Serotype T3 shares the OPS with *S. marcescens* O19.

Although *Haemophilus influenzae* is perceived to lack any O-antigen, it has been found that when grown on a solid medium enriched in sialic acid, a group of *H. influenzae* strains synthesize LPSs, in which a tetrasaccharide is attached *en bloc* to the core OS and may be considered thus as an O-unit in an SR-type LPS [273]. As in *S. enterica* serogroups A-E, the first sugar of the O-unit is Gal. Two glycofroms are coexpressed, which differ only in the terminal non-reducing sugar, which is either Neu5Ac or phosphoethanolamine-bearing GalNAc:

Neu5Ac(α2-3)Gal(β1-4)GlcNAc(β1-3)Gal(β1- and
GalNAc6*P*EtN(α1-6)Gal(β1-4)GlcNAc(β1-3)Gal(β1-

Table 3.20 Structures of *A. actinomycetemcomitans* OPSs

a [259]	3)6dTal2Ac(α1-2)6dTal(α1-	d [259]	3)Glc(β1-4)Man(β1-4)Man(α1- LRha(α1-3)┘
b [260]	3)Fuc(α1-2)LRha(α1- GalNAc(β1-3)┘	e [259]	4)GlcNAc(α1-3)LRha(α1-
c [259]	3)L6dTal4Ac(α1-2)L6dTal(α1-	f [261]	2)LRha(α1-3)LRha(α1- GalNAc(β1-2)┘

Table 3.21 Structures of *A. pleuropneumoniae* OPSs

1, 9,[a] 11 [262,263]	2)LRha(α1-2)LRha(α1-6)Glc(α1- GlcNAc(β1-3)┘
2 [262]	2)LRha(α1-2)Gal(α1-3)Glc(β1-4)Glc*6Ac*(α1-4)GalNAc(β1-
3, 8, 15 [262,264]	3)Glc(α1-2)Gal*f*(β1-6)Gal(α1-6)Glc(β1-3)Gal*f*(β1-
4 [262]	4)LRha(α1-3)Gal(β1-4)GalNAc(β1- Glc(β1-3)┘
5[b] [262]	6)Gal(β1-
6 [262]	3)Glc(α1-2)Gal*f*(β1-6)Glc(α1-6)Glc(β1-3)Gal*f*(β1-
7, 13 [262,265]	4)LRha(α1-3)Gal(β1-4)GalNAc(β1- Gal(β1-3)┘
10 [262]	2)Gal*f*(β1-
12 [266]	5)Gal*f*(β1-6)Gal*f*(β1- Gal(α1-6)┘
14 [267]	5)Gal*f*(β1- Gal(α1-2)┘

[a]In serotype 9, GlcNAc is present in a non-stoichiometric amount.
[b]In several strains, the polysaccharide is randomly O-acetylated.

Table 3.22 Structures of *M. haemolytica* OPSs

A1, A6, A9 [270]	4)Gal(β1-3)Gal(β1-3)GalNAc(β1-
T3 [271]	4)LRha(α1-3)GlcNAc(β1-
T4, T10 [272]	3)Gal(α1-3)Gal*f*(β1-

3.3.2.5 Pseudomonadaceae

Pseudomonas aeruginosa is an important opportunistic pathogen causing human infections, primarily in immunocompromized hosts and cystic fibrosis patients. O-antigen structures of this bacterium have been studied in detail and surveyed repeatedly [274–276]. In serogroups O1-O13, the OPSs have linear acidic tri- or tetra-saccharide O-units typically containing LRha, 6-deoxyamino sugars (QuiN, FucN, LFucN, QuiN4N) and acidic amino sugars, including GalNA, LGalNA, GlcN3NA, ManN3NA, LGulN3NA, Pse and 8eLeg. 2,3-Diamino-2,3-dideoxyhexuronic acids and both 5,7-diamino-3,5,7,9-tetradeoxynon-2-ulosonic acids have been found in *P. aeruginosa* for the first time in nature. Most amino sugars are N-acetylated but formyl, acetimidoyl, (*R*)- and (*S*)-3-hydroxybutanoyl occur as *N*-acyl groups too. Similar OPSs within complex O-serogroups differ in: (1) the pattern of O-acetylation, (2) an *N*-acyl group (acetyl *versus* 3-hydroxybutanoyl), (3) a monosaccharide (QuiN *versus* FucN, ManN3NA *versus* LGulN3NA, the presence of lateral Glc), and (4) a linkage (α1-3 *versus* α1-2 or β1-3, α1-4 *versus* β1-4).

Another bacterium well studied in respect to the O-antigen structure is *Pseudomonas syringae*, an important phytopathogen that infects a wide range of plants. The OPSs of *P. syringae* and related species are linear D- or L-rhamnan, a mixed D/L-rhamnan or branched polysaccharides with a rhamnan backbone and side chains of Rha, Fuc, GlcNAc or Fuc3NAc [240, 241, 277, 278]. In several D-rhamnan-based OPSs, Rha may be O-methylated. Characteristic features of the OPSs of this group are (1) irregularity owing to a non-stoichiometric glycosylation or O-methylation, (2) the presence of O-units of different types in one strain, (3) O-antigen diversity within one pathovar, and (4) sharing an OPS by different pathovars.

Structures of the OPSs have been determined also in an ubiquitous microorganism *P. fluorescens*, a phytopathogen *P. cichorii*, a mushroom pathogen *P. tolaasii*, a mushroom-associated bacterium *P. reactans*, a rhizosphere colonizer *P. putida* and several other *Pseudomonas* species. They are diverse in composition and include various 6-deoxyamino sugars (QuiN, LQuiN, FucN, Fuc3N, Fuc4N, QuiN4N), which may bear uncommon *N*-acyl groups, such as (*S*)-3-hydroxybutanoyl, *N*-acetyl-L-alanyl and 3-hydroxy-2,3-dimethyl-5-oxoprolyl. The last substituent resides on Qui3N in the OPSs of both *P. fluorescens* IMV 2366 and 361, which differ only in one monosaccharide (LRha *versus* L6dTal4Ac) (Table 3.23). The OPS of the type strain *P. fluorescens* ATCC 13525 is structurally related to that of several *P. syringae* strains [240, 241]. The OPS of *P. fluorescens* ATCC 49271 is a homoglycan composed of a partially 8-O-acetylated 5-*N*-acetimidoyl-7-*N*-acetyl derivative of Leg. Essentially the same homopolymer is the O-antigen of *Legionella pneumophila* serogroup 1 [67, 279]. The OPS of *P. corrugate* contains a derivative of a unique higher sugar 5,7-diamino-5,7,9-trideoxynon-2-ulosonic acid [280]; both OPS structure and configuration of the acid remain to be determined. *Pseudomonas* sp. (former *P. stutzeri*) OX1 has an OPS consisting of two 4-amino-4,6-dideoxyhexose derivatives, Rha4NAc and Fuc4NFo, but in the presence of the azo dye Orange II, it produces another, acidic OPS with such rarely occurring constituents as LGulNAcA and an amide of GalNAcA with L-serine. LGulNAcA in the amide form is present also in the OPS of *P. tolaasii*.

3.3.2.6 Moraxellaceae

Bacteria of the genus *Acinetobacter* are soil organisms, which participate in mineralization of various organic compounds. Several species are a key source of hospital infections in debilitated patients and are responsible for cases of community-acquired meningitis and pneumonia. The OPS structures have been studied in *A. baumanni* as well as several other species and unnamed DNA groups. A sugar pyruvic acid acetal is a component of the only known OPS of *A. calcoaceticus* (DNA group 1), whereas other strains of this species produce R-type LPSs. The OPSs of *A. haemolyticus* (DNA group 4) are similar in the presence of various 2-amino-2-deoxyhexuronic acids and derivatives of QuiN4N. The OPSs of *Acinetobacter* (DNA group 2) are either neutral or acidic due to the presence of hexuronic acids (GlcA, GalNAcA, GlcNAc3NAcA) or a derivative of Leg. The other OPSs studied, including those of *A. junii* and *A. lwoffii* (DNA groups 5 and 8,

Table 3.23 Structures of *Pseudomonas* OPSs

P. fluorescens A (ATCC 13525[T]) [281]	3)LRha(α1-3)LRha(α1-2)LRha(α1- Fuc3NAc(α1-2)┘ └*(2-1α)Fuc3NAc*
P. fluorescens A (IMV 472) [282]	3)LRha2Ac(β1-4)LRha(α1-3)Fuc(α1- GlcNAc(β1-2)┘
P. fluorescens A (IMV 1152) [283]	3)Fuc4NAc(α1-4)LQuiNAc(α1-3)QuiNAc(β1-
P. fluorescens B (IMV 247) [284]	2)Qui3N(*S*3Hb)(β1-2)LRha(α1-4)GalNAcA(α1-3)QuiNAc4N(*S*3Hb)(α1-
P. fluorescens C[a] (IMV 2366) [285]	2)Qui3NR(β1-3)LRha(α1-3)FucNAc(α1-
P. fluorescens 361[a] [286,287]	4)Qui3NR(β1-3)L6dTal4Ac(α1-3)FucNAc(β1-
P. fluorescens G (IMV 2763)[b] [288]	4)Man(α1-2)Man(α1-3)GalNAc(βα1- L6dTal2Ac(α1-3)┘
P. fluorescens ATCC 49271 [67,289]	4)Leg5Am7Ac*8Ac*(α2-
P. chlororaphis ssp. aurantiaca (*P. aurantiaca*) [290]	3)LFucNAc(α1-3)LFucNAc(α1-3)QuiNAc4NAc(β1-
P. cichorii [291]	3)LFucNAc(α1-2)Qui3NAc(β1-3)LFucNAc(α1-3)QuiNAc(α1-
P. putida [292]	2)Rha(α1-3)Rha(α1-3)Man(β1-
P. reactans [293]	3)QuiN(LAlaAc)4N(LAlaAc)(β1-3)GlcNAm(α1-3)QuiNAc4NAc(α1-
P. tolaasii [294]	4)LGulNAcAN*3Ac*(α1-3)QuiNAc(β1-
Pseudomonas sp. OX1 [295]	2)Rha4NAc(α1- *Fuc4NFo(α1-3)*┘
Pseudomonas sp. OX1[c] [296]	4)GalNAcA6Ser(α1-4)ManNAcA(β1-4)LGulNAcA(α1-3)QuiNAc4N(*S*3Hb)(β1- └(3-1β)Glc

[a]R indicates 3-hydroxy-2,3-dimethyl-5-oxoprolyl of unknown configuration.
[b]Later, classification of this strain as *P. fluorescens* was questioned.
[c]Configuration of serine has not been determined.

respectively), are all neutral. In *A. lwoffii* EK30 and *Acinetobacter* sp. 4 (DNA group 11), Qui4N and Fuc3N bear uncommon *N*-acyl groups: D-homoseryl (DHse) and (*S*)-2-hydroxypropanoyl, respectively (Table 3.24). A peculiar feature of three *Acinetobacter* OPSs is alternating *N*-acetyl and *N*-[(*S*)-3-hydroxybutanoyl] groups on Leg, QuiN4N or DHse. The OPSs of *A. baumanni* O7 and O10 have the same main chain, and those of *A. haemolyticus* 57 and 61 differ only in the configuration of the linkage between the O-units.

3.3.2.7 Vibrionaceae

From about 200 *V. cholerae* O-serogroups, O1 and O139 strains cause Asiatic cholera, whereas others are opportunistic pathogens responsible for travel diarrhea and other enteric diseases. The OPS structures of both pathogenic and several non-O1, non-O139 serogroups have been established and most of them reviewed recently [322]. Homopolymers of (*R*)- and (*S*)-2-hydroxypropanoyl derivatives of LRha4N have been found in the O144 and O76 antigens, respectively, and the O1 antigen consists of an (*S*)-2,4-dihydroxybutanoyl derivative of Rha4N.

Table 3.24 Structures of *Acinetobacter* OPSs

A. calcoaceticus 7 [297]	2)Gal4,6*R*pyr*3Ac*(β1-3)GlcNAc(β1-4)GlcA(β1-3)GalNAc(β1-
A. baumanni O1 [298]	3)GlcNAc(α1-3)GalNAc(β1- Gal(α1-6)┘
A. baumanni O2 [299]	4)Gal(α1-6)Gal(β1-3)GalNAc(β1- └(3-1β)GalNAc(3-1α)GalNAc(3-1β)Fuc3N(*R*3Hb)
A. baumanni O5 [300,301]	3)GalNAcA(α1-3)LFucNAc(α1-3)GlcNAc(β1- └(4-1α)LFucNAc
A. baumanni O7 [302]	2)LRha(α1-2)LRha(α1-3)LRha(α1-3)GlcNAc(α1- └(3-1β)GlcNAc(4-1β)LRha
A. baumanni O10 [303]	2)LRha(α1-2)LRha(α1-3)LRha(α1-3)GlcNAc(α1- └(3-1α)ManNAc
A. baumanni O11[a] [304,305]	4)GalNAc(β1-3)Gal(α1-6)Gal(β1-3)GalNAc(α1- └(6-1β)Glc
A. baumanni O12[a] O23 [306]	3)GalNAc(β1-3)Gal(α1-3)GlcNAc(β1- └(4-1α)GlcNAc(6-1β)Qui3N(*R*3Hb)
A. baumanni O16 [305]	6)GlcNAc(α1-4)GalNAc(α1-3)GlcNAc(α1- Glc(β1-3)┘
A. baumanni O18 [307]	3)Gal(β1-3)GalNAc(β1- ManNAc(β1-4)Gal(α1-4)┘
A. baumanni O22 [308]	3)Glc(β1-3)GalNAc(β1- Gal(α1-6)┘
A. baumanni O24[b] [67,309]	4)Leg5R7Ac(β2-6)GlcNAc(α1-3)LFucNAc(α1-3)GlcNAc(α1-
A. baumanni ATCC 17961 [310]	3)Gal(α1-6)Glc(β1-3)GalNAc(β1- GlcNAc3NAcA(β1-4)┘ └(6-1β)GlcNAc
A. baumanni [311]	3)Qui4NAc(β1-3)GalNAc(α1-4)GalNAc(α1-3)GalNAc(α1- Gal(α1-6)┘
A. baumanni 24[b] [312]	4)GlcNAc*6Ac*(α1-4)GalNAcA(α1-3)QuiNAc4NR(β1-
Acinetobacter sp. 44 (DNA group 3 [313]	3)LRha(α1-3)LRha(α1-2)LRha(α1-3)GlcNAc(β1- └(2-1α)LRha(2-1β)GlcA(4-1α)LRha
A. haemolyticus ATCC 17906 [314]	4)GalNAcA6DAla(α1-4)GalNAcA(α1-3)QuiNAc4NAc(β1-
A. haemolyticus 57 [315]	4)ManNAcA(β1-4)LGulNAcA3Ac(α1-3)QuiNAc4N(*S*3Hb)(α1-
A. haemolyticus 61 [315]	4)ManNAcA(β1-4)LGulNAcA3Ac(α1-3)QuiNAc4N(*S*3Hb)(β1-
A. junii 65 [316]	2)LRha(α1-3)LRha(α1-2)LRha(α1-3)LRha(α1-3)Gal(β1-
A. lwoffii EK30[b] [317]	3)Qui4N(DHseR)(β1-6)Gal(α1-4)GalNAc(α1-3)FucNAc(α1-
A. lwoffii EK67, *Acinetobacter* sp. VS-15 [318]	2)LRha(1-6)Gal(1-4)GalNAc(1-3)QuiNAc(1- GlcNAc(β1-3)┘
Acinetobacter sp. 90 (DNA group 10) [319]	3)Gal(α1-4)GalNAc(β1-3)Gal(α1-3)GlcNAc(β1- └(4-1α)Fuc4N(*R*3Hb)
Acinetobacter sp. 94 (DNA group 11) [320]	3)Gal(α1-3)GalNAc(β1- └(4-1β)GalNAc(4-1α)Fuc3N(*S*2Hp*Ac*)
Acinetobacter sp. 96 (DNA group 11) [321]	4)Man(β1-3)Man(α1-3)LFuc(α1-3)GlcNAc(β1- └(3-1α)LFuc
Acinetobacter sp. 108 (DNA group 13) [301]	4)Gal(α1-6)Gal(β1-3)GalNAc(β1- └(3-1β)GalNAc(3-1α)GalNAc(3-1β)Fuc3N(*R*3Hb)

[a]Another OPS having the same structure as the *A. baumanni* O16 antigen is also present.
[b]R indicates acetyl or (*S*)-3-hydroxybutanoyl.

2-O-Methylation of the terminal non-reducing Rha4N residue in the O1 antigen results in seroconversion from variant Inaba to Ogawa. There are present also other unusual monosaccharide components, such as ascarylose, DDmanHep and a 5-*N*-acetimidoyl-7-*N*-acetyl derivative of Pse. Several other unusual *N*-acyl groups present on amino sugars are 3,5-dihydroxyhexanoyl, (*2R,3R*)-3-hydroxy-3-methyl-5-oxoprolyl and *N*-acetyl-L-allothreonyl (Table 3.25). The O139 and O155 antigens, as well as that of *Vibrio mimicus* N-1990, include a cyclic phosphate group on Gal. The O22 and O139 antigens consist of only one O-unit with two colitose residues in each strain. The OPSs of *V. cholerae* O8, O10 and an unclassified strain H11 are similar to those of *Listonella anguillarum* O2a, *E. coli* O64 and *Shewanella algae* 48055, respectively.

Table 3.25 Structures of *V. cholerae* OPSs

O1[a] [322]	2)Rha4NR(α1-
O2 [323]	4)Qui*p*NAc(β1-4)Pse5Am7Ac(β2-4)Gal(β1-
O3[b] [324]	2)DDmanHep(α1-4)LFucNAc(α1-3)QuiNAc4NR(β1- └(3-1α)Asc
O5[c] [325]	4)ManNAcA(β1-3)QuiNAc4NAc(β1- Fuc3NR(α1-3)┘
O6 [326]	6)GlcNAc3Ac(α1-3)LRha2Ac(β1-4)GlcNAc(β1- └(4-1α)GlcA
O8 [327]	4)GlcNAc3N(LAlaFo)AN(β1-4)ManNAc3NAcAN(β1- 4)LGulNAc3NAcA(α1-3)QuiNAc4NAc(β1-
O9 [328]	4)Glc(α1-4)GalNAcA(α1-3)GalNAcA(α1-3)GlcNAc(α1- Glc(α1-4)┘
O10 [322]	3)ManNAc(α1-4)GlcA(β1-3)Gal(β1-3)GlcNAc(β1-
O21 [329]	7) DDmanHep(β1-3)GlcNAc(β1- LRha(α1-3)┘ └(4-1β)GalNAc
O22 [322]	GalA*3,4Ac*(β1-3)GlcNAc(α1-4)GalA(α1-3)QuiNAc(β1- └(2-1α)Col └(4-1α)Col
O43 [330]	3)Qui4N(LaThrAc)(β1-3)Gal*p*NAcA(α1-4)GalNAc(α1-3)QuiNAc(α1-
O76 [331]	2)LRha4N(*S*2Hp)(α1-
O139 [322]	Gal4,6*P*(β1-3)GlcNAc(β1-4)GalA(α1-3)QuiNAc(β1- Col(α1-2)┘ └(4-1α)Col
O140 (bioserogroup Hakata [332]	2)Rha4NAc(α1-
O144 [333]	2)LRha4N(*R*2Hp)(α1-
O155 [334]	4) LFuc(α1-3)FucNAc(β1- └(3-1α)GlcNAc(4-1α)LFuc(3-1α)Gal4,6*P*
H11 [335]	4)GalA6(GroN)(α1-4)NeuAc(α2-3)GalA6(GroN)(β1-3)QuiNAc(β1-

[a]R indicates (*S*)-2,4-dihydroxybutanoyl.
[b]R indicates 3,5-dihydroxyhexanoyl of unknown configuration.
[c]R indicates (*2R,3R*)-3-hydroxy-3-methyl-5-oxoprolyl.

Among non-cholerae vibrios, there are marine bacteria, including fish pathogens *V. vulnificus* and *V. ordalii*, as well as opportunistic pathogens of humans, such as *V. fluvialis* and *V. mimicus*. Their OPSs contain various unusual components too, e.g. a (*S*)-3-hydroxybutanoyl derivative of LRhaN3N, 2-acetamido-2,6-dideoxy-D-*xylo*-hexos-4-ulose, a 2-*N*-acetimidoyl derivative of LGalNA, a partially O-acetylated 4-D-malyl derivative of GlcN3N and 3-*O*-[(*R*)-1-hydroxyethyl]-L-rhamnose (rhamnolactilic acid). The OPS of *V. fluvialis* O19 and *Vibrio* bioserogroup 1875 is a homopolymer of a 3-hydroxypropanoyl derivative of Rha4N; in the latter bacterium, the monosaccharide at the non-reducing end is 2-O-methylated [336]. The SR-type LPS of *V. fluvialis* M-940 has a single heptasaccharide O-unit (Table 3.26). The OPS of *V. alginoluticus* includes di-*N*-acetyllegionaminic acid [67, 337] but the O-unit structure remains unknown.

In the OPSs of a fish pathogen *Listonella* (former *Vibrio*) *anguillarum,* derivatives of amino and diamino sugars and hexuronic acids are abundant (Table 3.27). In strain 1282, an *N*-formyl-L-alanyl derivative of GlcN3NAN at the non-reducing end of the OPS is 4-O-acetylated, and in an unnamed strain, the terminal LQui3NAc residue is 4-O-methylated. The discrimination of strains

Table 3.26 Structures of other *Vibrio* sp. OPSs

V. fluvialis sv. 3 [338]	4)LRha(α1-3)ManNAc(β1-
V. fluvialis OKA-82-708 [339]	2)LRha(α1-3)LRha(α1-3)LRha(α1-3)LRha(α1- GlcNAc(β1-2)┘
V. fluvialis AQ-0002B [340]	2)Man(β1-4)GalNAc(α1-4)GalA(α1-3)GlcNAc(α1- └(3-1α)LRha3*R*lac
V. fluvialis M-940 [341]	LRha(α1-2)LFuc(α1-2)Gal(α1-2)LFuc(α1-3)GlcA(β1- 4)LRha(α1-3)GlcNAc(β1-
V. fluvialis O19, *Vibrio* bioserogroup 1875 [342,343]	2)Rha4N(3Hp)(α1-
V. fluvialis AA-18239 [344]	4)GalNAc(α1-2)Rib*f*(β1-
V. mimicus N-1990 [345]	4)GalNAc(α1-3)GalNAc(β1-2)Gal4,6*P*(β1-3)GalNAc(α1-
V. mimicus W-26768 [346]	3)Qui3N(*R*3Hb)(β1- *GalNAc(α1-2)*┘
V. ordalii O2[a] [347,348]	4)GlcNAc3N(LAlaFo)AN(β1-4)GlcNAc3NAmA(β1- 4)LGulNAc3NAcA(α1-3)Sug-(β1-
V. vulnificus CECT 4602[b] [349]	4)GlcNAc(α1-3)LRha(α1-3)GlcNAc(β1- └(3-1β)LRhaNAc3N(*S*3Hb)
V. vulnificus YJ016 [350]	3)LGalNAmA(α1-3)QuiNAc4NAc(β1-3)LFuc(α1- 3)GlcNAc(α1- └(4-1β)GlcNAc*6Ac*
V. vulnificus CECT 5198[c] [253]	4)GlcNAc3NRAN(β1-4)LGalNAmA(α1-3)QuiNAc(α1-

[a]Sug indicates 2-acetamido-2,6-dideoxy-D-*xylo*-hexos-4-ulose.
[b]The presence of ~20% (*R*)-3-hydroxybutanoyl group reported [349] could be due to a partial racemization in the course of acid hydrolysis.
[c]R indicates 4-D-malyl or 2-*O*-acetyl-4-D-malyl.

Table 3.27 Structures of *L. anguillarum* OPSs

L. anguillarum O2a; O2b[a] [347,351,352]	4)GlcNAc3N(LAlaR)AN(β1-4)ManNAc3NAmA(β1- 4)LGulNAc3NAcA(α1-3)QuiNAc4NAc(β1-
L. anguillarum 1282 [352]	4)GlcNAc3N(LAlaFo)AN(β1-4)ManNAc3NAmA(β1- 4)Qui3NAc(β1-3)FucNAc4NAc-(α1-
L. anguillarum V-123[b] [353]	3)GalNAcAN(α1-4)GalNFoA(α1-3)QuiNAc(α1-3)Qui4NR(β1-
L. anguillarum[c] [354]	4)LQui3NAc(β1-4)LQui3NAc(β1- QuiNAc(α1-2)┘

[a]R indicates Fo in serotype O2a or Ac in serotype O2b [351].
[b]R indicates 2,4-dihydroxy-3,3,4-trimethyl-5-oxoprolyl of unknown configuration.
[c]Presumably, an *O*-propanoyl group is present at position 3 or 4 of QuiNAc.

of O2a and O2b serotypes is based on the nature of a 3-*N*-acyl group on GlcN3NAN, which is either *N*-formyl-L-alanyl or *N*-acetyl-L-alanyl, respectively.

3.3.2.8 Xanthomonadaceae

Xanthomonas campestris and related species cause several plant diseases. Their OPS structures have been examined [240, 278]. With a few exceptions, the OPSs have a D- or L-rhamnan backbone and many from them have Xyl or LXyl side chains. In *X. campestris* pv. *pruni,* there are three sites of non-stoichiometric xylosylation of the main chain, and totally 0 to 2 LXyl residues per O-unit are present (Table 3.28). The OPSs of *X. campestris* pv. *vitians* and *X. fragariae* have main chains of α1-3- and β1-3-linked LRha residues, which lack strict regularity.

Table 3.28 Structures of *Xanthomonas* OPSs

X. campestris pv. *begoniae* [240]	2)LRha(α1-3)LRha(α1-3)LRha(α1- LXyl(β1-2)┘ └*(4-1β)LXyl*
X. campestris pv. *campestris* 8004 [355]	3)Rha(α1-3)Rha(β1- Fuc3NAc(α1-2)┘
X. campestris pv. *malvacearum* [356]	2)Rha*3Me*(α1-3)Rha(α1-3)Rha(α1- Fuc*f*(α1-4)┘
X. campestris pv. *manihotis* [240]	2)LRha(α1-2)LRha(α1-3)LRha(β1- Xyl(β1-2)┘
X. campestris pv. *phaseoli* var. *fuscans* [356]	2)Rha(α1-3)Rha(α1-3)Rha(α1-
X. campestris pv. *pruni* [357]	2)LRha(α1-2)Glc(α1-3)LRha(α1- *LXyl(β1-4)*┘ *LXyl(β1-3)*┘ *LXyl(β1-4)*┘
X. campestris pv. *vignicola* [240]	2)Rha(α1-2)Rha(α1-3)Rha(β1- Rha(α1-3)┘ └*(3-1α)Rha*
X. campestris NCPPB 45 [240]	3)GalA(α1-2)LRha(α1-2)LRha(α1-3)LRha(α1-3)Gal(β1- └(4-1α)LRha
X. campestris 642 [240]	2)LRha(α1-3)LRha(α1-2)LRha(α1-3)LRha(α1-3)LRha(α1- Xyl(β1-2)┘ └(4-1β)Xyl
X. cassavae [278]	3)Rha(β1-3)Rha4NAc(α1- Xyl(β1-2)┘

In the former, parts of the polysaccharide chains are linear and the others bear α1-2-linked Fuc3NAc residues [240, 278], and in the latter, the rhamnan is decorated with α1-2-linked Fuc residues [240]. The OPS of *X. campestris* NCPPB 45 is exceptionally acidic due to the presence of GalA.

Stenotrophomonas (Xanthomonas or *Pseudomonas) maltophilia* is an emerging opportunist human pathogen, which can causes blood-stream infections and pneumonia with considerable morbidity in immunosuppressed patients. The OPSs of these bacteria are neutral, and most O-units are branched tri- and tetra-saccharides (Table 3.29). As in *X. campestris*, Xyl and Rha in both enantiomeric forms occur in many O-serogroups, and several xylorhamnans are structurally related in the two

Table 3.29 Structures of *S. maltophilia* OPSs

O1[a] [358]	3)L6dTal2Ac(α1-3)GlcNAc(β1- Ara*f*(α1-6)┘
O2 [359]	4)Man(α1-3)LRha(α1- LXyl(β1-2)┘
O3 [360]	3)Fuc(α1-3)GlcNAc(β1- └(4-1α)Fuc4NAc
O4 [361]	2)Rha(α1-3)Rha(α1-3)Rha(α1- Xyl(β1-2)┘ └(4-1β)Xyl
O6 [362]	3)LRha(α1-3)GlcNAc(β1- Xyl(β1-4)┘
O7 [363]	2)Rha(α1-3)Rha(α1-3)Rha(α1-
O8 [364]	2)LRha(α1-3)LRha(α1-4)LRha(α1- LXyl3Me(β1-4)┘
O10 [365]	2)LRha(β1-2)LRha(α1-2)LRha(α1- LXyl(β1-3)┘
O12/O27 [366]	3)Rha(β1-3)Rha4NAc(α1-3)Rha4NAc(α1-3)Rha4NAc(α1- └(2-1α)Fuc3NAc
O16[b] [367]	3)ManNAc(β1-4)GlcNAc(β1- Rib*f*(α1-4)┘
O18 [361]	2)LRha(α1-3)LRha(α1-3)LRha(α1- LXyl(β1-2)┘ *└(4-1β)LXyl*
O19 [368]	3)LRha(β1-4)LRha(α1-3)Fuc(α1- Glc(α1-3)┘
O20 [369]	2)Man(α1-3)Rha(β1-2)Rha(α1-2)Rha(α1-
O21 [370]	6)GlcNAc(α1-4)GalNAc(α1- Ara*f*(α1-3)┘
O25 [370]	6)GlcNAc(α1-4)GalNAc(α1-

[a]The location of the *O*-acetyl group is tentative.
[b]The OPS is non-stoichiometrically O-acetylated at unknown position.

species. The O4 and O18 antigens have the same structure but the constituent monosaccharides, Xyl and Rha, are either D or L, respectively. The O8 antigen contains 3-*O*-methyl-L-xylose as a component of each O-unit, and the O1 antigen is presumably terminated with 3-*O*-methyl-6-deoxytalose. Whereas Xyl is always pyranosidic, two other constituent pentoses, Ara and Rib, are present in the furanose form. Other uncommon monosaccharides, including L6dTal, Fuc3NAc, Fuc4NAc and Rha4NAc, are components of the OPSs. A linear D-rhamnan of serogroup O7 has the same structure as the common polysaccharide antigen of *P. aeruginosa* [9] and the O-antigen of several strains of *P. syringae* [240, 241, 277, 278] and *X. campestris* pv. *phaseoli*. A 6)GlcNAc(α1-4)GalNAc(α1- backbone of the O21 and O25 antigens is shared by several *Citrobacter* strains [78].

3.3.2.9 Other Families

Francisella tularensis from the family Francisellaceae is the etiologic agent of tularemia and one of the most virulent Gram-negative bacteria considered as a biological weapon or bioterrorist agent. From four subspecies, ssp. *tularensis* is the most infective and fatal for humans, whereas ssp. *novicida* is virulent for mice but not humans. The OPS common for *F. tularensis* ssp. *tularensis* and *holarctica* (types A and B) has a tetrasaccharide O-unit with two residues of GalNAcA, both in the amide form, and one residue each of QuiNAc and Qui4NFo [371]:

2)Qui4NFo(β1-4)GalNAcAN(α1-4)GalNAcAN(α1-3)QuiNAc(β1-

The 4)GalNAcAN(α1-4)GalNAcAN(α1- disaccharide fragment of this O-antigen is shared by *F. tularensis* ssp. *novicida,* in which QuiNAc is replaced by QuiNAc4NAc and Qui4NFo by the third GalNAcAN residue [371]:

4)GalNAcAN(α1-4)GalNAcAN(α1-4)GalNAcAN(α1-3)QuiNAc4NAc(α1-

A fish pathogen *Francisella victoria* possesses a non-repetitive polysaccharide part of the LPS containing 20 monosaccharides as well as alanyl, 3-aminobutanoyl and 4-acetamido-3-hydroxy-3-methyl-5-oxoprolyl on Qui3N, Qui4N and Fuc4N [372].

Legionella pneumophila from the family Legionellaceae is a facultative intracellular bacterium and the cause of legionellosis, pneumonia with sometimes-fatal progression. From 15 existing O-serogroups, strains of serogroup 1 are most often isolated from environmental samples and clinical specimens. Their O-antigen is polylegionaminic acid 4)Leg5Am7Ac(α2-, which is 8-O-acetylated in part of the strains and mostly nonacetylated in the others [67, 279]. Accordingly, serogroup 1 strains are divided into the Pontiac and non-Pontiac groups. The O-antigen of *L. pneumophila* serogroup 2 and most other non-1 serogroups, except for serogroups 7 and 13, is a homopolymer of a similar derivative of 4-epilegionaminic acid 4)4eLeg5Am7Ac(α2-, which is also 8-O-acetylated to a different extent (10–90%)

Table 3.30 Structures of *Halomonas* OPSs

H. alkaliantarctica [375]	3)LRha(β1-4)LRha(α1-3)LRha(α1-
H. magadiensis [376,377]	4)Glc(β1-3)Gal(β1- and Glc(α1-4)⌋ 4)LGulNAcA(α1-4)LGulNAcA(α1-6)Glc(α1-
H. pantelleriensis [374]	2)GlcA4Slac(β1-4)GlcA(β1-4)GalNAcA(α1-3)LQuiNAc(β1-
H. stevensii [378]	4)Glc(β1-3)Gal(β1- Glc(α1-4)⌋

[67, 373]. Both Leg and 4-eLeg have been found in *L. pneumophila* for the first time in nature.

The O-antigens have been studied in four species of halophilic bacteria of the genus *Halomonas* (family Halomonadaceae) (Table 3.30). The OPS of *H. alkaliantarctica* is an L-rhamnan, and that of *H. pantelleriensis* is highly acidic due to the presence of GlcA, GalNAcA and an ether of GlcA with (*S*)-lactic acid. The latter OPS is unusual in that an L-configurated monosaccharide, LQuiNAc, is the first sugar of the O-unit [374]. *H. magadiensis* (former *H. magadii*) produces two OPSs, one neutral (major) and one acidic enriched in LGulNAcA. The neutral OPS of *H. magadiensis* is shared by *H. stevensii*.

The OPS of the marine bacterium *Marinomonas communis* classified to the family Ocenospirillaceae is a 2)LRha(α1-3)LRha(α1-3)LRha(α1- rhamnan [379], which is shared by several *P. syringae* strains [241, 278].

The OPS of a mesophilic chemolithotroph *Acidithiobacillus* (*Thiobacillus*) *ferrooxidans* from the family Acidithiobacillaceae includes both rhamnose enantiomers and 3-*O*-methyl-L-rhamnose as a component of the O-unit [380]:

3)Glc(α1-3)Rha(α1-3)LRha(α1-3)Glc(β1-
LRha3Me(α1-4)⌋

3.3.3 α-Proteobacteria

3.3.3.1 Rhizobiaceae, Xanthobacteraceae

Rhizobacteria are unique in their ability to interact with roots of legumes and to form nitrogen-fixing nodules. The OPSs of *Rhizobium, Mesorhizobium* and *Sinorhizobium* (both former *Rhizobium* too) from the family Rhizobiaceae have a lipophilic character due to the abundance of 6-deoxyhexoses (Rha, LRha, LFuc, 6dTal, L6dTal), *O*-methyl and *O*-acetyl groups [381, 382] (Table 3.31).

A short-chain OPS of *R. etli* consisting of five O-units is enriched in O-methylated sugars, including methyl ester of GlcA present in the majority of the O-units. It is increased in the content of 2-*O*-methyl-L-fucose in bacteroids isolated from root nodules of the host plant *Phaseolus vulgaris* or in bacterial cultures grown in the presence of anthocyanin as compared with cultures grown

Table 3.31 Structures of rhizobial OPSs

R. etli[a] [383,384]	4)GlcA6Me(β1-4)LFuc*2Me*(α1- 6dTal3Me(α1-3)┘
R. leguminosarum bv. *viciae* 3841 [387]	4)Glc3NAmA(β1-4)LFuc*2Ac*(α1-3)LQuiNAc(α1- 6dTal2Ac3Me*4Me*(α1-3)┘
R. leguminosarum bv. *viciae* 5523[a] [388]	4)Glc(α1-3)QuiNAc(α1-
R. leguminosarum bv. *trifolii* 4s [382]	3)LRha(α1-3)LRha(α1-3)LRha(α1-4)GlcNAc(β1- ManNAc(α1-2)┘
R. leguminosarium bv. *trifolii* 24[b] [389,390]	3)L6dTal(α1-2)LRha(α1-5)Sug(2-
R. leguminosarum bv. *viciae* [382]	3)LRha(α1-3)LFuc(α1-3)LFuc(α1- Man(α1-2)┘
R. tropici [382]	3)6dTal2Ac(α1-3)LFuc(α1-4)Glc(β1-
M. amorphae ATCC 19655, *M. loti* HAMBI 1148 [391]	3)Rha(α1-3)Rha(α1-3)Rha(α1-3)Rha(α1-2)Rha*3Me*(α1- └*(2-1β)GlcNAc4Me*
M. loti NZP2213 [392]	3)L6dTal2Ac(α1-
M. loti 2213.1[c] [385]	3)L6dTal*2R*(α1-
M. loti Mlo-13 [386]	2)L6dTal(α1-3)L6dTal4Ac(α1-2)LRha*3Me*(α1-
M. huakuii [382]	2)L6dTal(α1-3)L6dTal(α1-2)LRha(α1-
S. fredii[c] [393]	4)GalA(α1-3)LRha2Ac(α1-3)Man2Ac*6R*(α1-
Sinorhizobium sp. NGR234 [394]	3)LRha(α1-3)LRha(α1-2)LRha*3Me*(α1-

[a]The OPS is O-acetylated at unknown position.
[b]Sug indicates 3-deoxy-D-*lyxo*-hept-2-ulosaric acid. The configuration of its linkage remains unknown.
[c]R indicates Ac or Me.

under standard laboratory conditions [383]. 2,3,4-Tri-*O*-methylfucose or, in a minority of molecules, 2-*O*-methyl- and 2,3-di-*O*-methylfucose terminates the nonreducing end of the OPS, and a non-repetitive tetrasaccharide with a Kdo residue at the reducing end is located between the O-antigen and the core OS [384].

The OPS of *R. leguminosarium* 3841 is also short and is built up of three to four O-units. It is the only known O-antigen that contains a derivative of 3-amino-3-deoxyhexuronic acid (Glc3NAmA). Another unique components, a dicarboxylic 3-deoxyhept-2-ulosaric acid, is present in the OPS of *R. leguminosarium* bv. *trifolii* (*R. trifolii*) 24. A Fix$^-$ mutant of this bacterium has a totally different OPS that lacks L6dTal but is rich in heptose and *O*-methylheptose [384]. The OPS of *M. loti* NZP2213 is a homopolymer of 2-*O*-acetyl-6-deoxy-L-talose with a small content of 2-*O*-methyl-6-deoxy-L-talose, which is significantly higher in a Tn5 mutant 2213.1 with impaired effectiveness of symbiosis with the host plant *Lotus corniculatus* [385]. In contrast, another Tn5 mutant of the same *M. loti* strain, Mlo-13, is symbiotically enhanced [386]. It has structurally different OPS that makes it resistant to bacteriophage A1, which requires the 6-deoxytalan of the parent strain as receptor.

6-Deoxyhexoses are abundant also in the OPSs of the genus *Agrobacterium* from the same family Rhizobiaceae but non-sugar groups are less common (for the known structures of six strains of *A. tumefaciens* and *A. radiobacter* see review [382]). Three O-antigens are homoglycans: (1) a 6-deoxy-L-talan in *A. tumefaciens* C58, which shares the carbohydrate structure with *M. loti* NZP2213 but differs in the pattern of O-acetylation, (2) an L-rhamnan in a *A. radiobacter* strain having the same structure as the main chain in several *P. syringae* strains [240, 241], and (3) a unique α1-3-linked L-*glycero*-D-*manno*-heptan in *A. radiobacter* M2/1. Two OPSs are elaborated by *A. tumefaciens* TT9, one of which is a homopolymer of a 3-O-methylated derivative of Fuc4N, in which the monomers are linked through a 4-N-linked 3,4-dihydroxy-1,3-dimethyl-5-oxoprolyl group [382].

The OPS of *Azorhizobium caulinodans* from the family Xanthobacteraceae is composed of a rarely occurring branched monosaccharide 3-*C*-methylrhamnose, together with rhamnose and 2-*O*-methylrhamnose, whose absolute configurations are either all D or all L [395]:

3)Rha2Me(α1-2)Rha3*C*Me(β1-3)Rha(α1-2)Rha3*C*Me(β1-3)Rha(α1-

3.3.3.2 Other Families

Bacteria of the genus *Brucella* (family Brucellaceae) are facultative intracellular pathogens that cause brucellosis, a group of closely related zoonotic diseases. The bacteria are rather homogeneous in terms of the O-antigens, which are homopolymers of α1-2-linked Rha4NFo in A-dominant smooth *Brucella* strains but every fifth residue is α1,3-linked in M-dominant strains [203]. Biotype 1 *B. abortus* and *B. melitensis* carry exclusively A or M epitopes, respectively. The existence of various intermediate AM biotypes in these species and *B. suis* with a reduced proportion of the α-1,3 linkage suggests that the two OPSs are coexpressed. The A-type OPS is characteristic also for *Y. enterocolitica* O9 (Hy 128) [185] that accounts for false-positive serological reactions in the serodiagnostics of the diseases caused by the two bacteria.

Bacteria of the genus *Ohcrobactrum* are taxonomically related to *Brucella* but have no medical importance. The only known OPS structure of *O. anthropi*, 3)GlcNAc(α1-2)LRha(α1- [396], resembles those of several *S. marcescens* serogroups [114].

The OPS of *Pseudaminobacter defluvii* THI 051^T (former *Thiobacillus* sp. IFO 14570), the only representative of the family Phyllobacteriaceae studied, consists of three diamino sugars, one of which, 2,4-diamino-2,4-dideoxyglucuronic acid, has not been found elsewhere in nature (the absolute configurations of the monosaccharides have not been proven) [397]:

4)GlcNAc3NAcA(β1-3)GlcNAc4NAcA(β1-3)QuiNAm4NAc(α1-

The O-antigens of several strains of *Acidomonas methanolica* (former *Acetobacter methanolicus*) from the family Acetobacteraceae are homopolysacharides

Table 3.32 Structures of *Azospirillum* OPSs

A. brasilense S17 [278]	4)LRha2Me(α1-3)ManN(*S*3Hb)(α1- and GlcNAc(β1-4)┘ 3)LRha(α1-3)LRha(α1-2)LRha(α1- Glc(β1-3)┘
A. lipoferum SpBr17, SR65[a] [278,399]	3)LRha(α1-3)LRha2Ac(α1-2)LRha(α1- Glc(β1-3)┘
A. brasilense SR15 [400]	2)Rha(α1-2)Rha(β1-3)Rha(α1-2)Rha(α1-
A. brasilense Sp245, S27, *A. lipoferum* RG20a [278,400]	2)Rha(α1-2)Rha(β1-3)Rha(α1-3)Rha(α1-2)Rha(α1-
A. brasilense Sp245.5 [401]	6)GalNAc(α1-4)ManNAcA(β1-
A. irakense KBC1 [278]	4)LRha(α1-3)Gal(β1- └(3-1α)LRha(3-1β)Man(3-1α)LRha(2-1α)Gal*f*
A. lipoferum Sp59b [278]	3)Gal(α1-3)Gal(β1- └(4-1β)Man(3-1α)LRha(2-1α)LRha(3-1α)LRha

[a]The OPS of strain SR65 lacks O-acetylation.

of common hexoses (for the structures see review [4]). The OPS of another representative of the family, *Gluconacetobacter* (former *Acetobacter*) *diazotrophicus*, has the following structure [398]:

2)Rib*f*(β1-3)LRha(α1-3)LRha(α1-2)LRha(α1-
Glc(β1-2)┘

In the family Rhodospirillaceae, studied are nitrogen-fixing soil bacteria of the genus *Azospirillum*, which colonize roots and promote growth of a broad range of plants. In most strains, the OPSs are D-rhamnans or heteroglycans enriched in LRha [278] (Table 3.32). In *A. brasilense* S17, two OPSs have been observed, one of which includes 2-*O*-methyl-L-rhamnose and a (*S*)-3-hydroxybutanoyl derivative of ManN. The OPSs of *A. irakense* KBC1 and *A. lipoferum* Sp59b are built up of hexasaccharide O-units having the same composition but different structures. A spontaneous mutant Sp245.5 of *A. brasilense* with a changed plasmid switched from the production of a D-rhamnan to an acidic hexosaminoglycan.

The OPS of *Brevundimonas* (*Pseudomonas*) *diminuta* from the family Caulobacteraceae is a partially O-acetylated 4)Man6*Ac*(α1-2)Man(α1- mannan [402].

3.3.4 β-Proteobacteria

3.3.4.1 Burkholderiaceae

Bacteria classified as *Burkholderia* and *Ralstonia* were known formerly as *Pseudomonas* species. Emergent pathogens *B. mallei* and *B. pseudomallei* are the etiologic agents of glanders and melioidosis, respectively, whereas a closely related bacterium *B. thailandensis* is avirulent. All these bacteria possess similar OPSs having a 3)L6dTal(α1-3)Glc(β1- backbone, where L6dTal may be non-modified or 2-O-acetylated (in all species), 2-O-methylated (in *B. mallei*) or 2-O-methylated and 4-O-acetylated (in *B. thailandensis* and *B. pseudomallei*) [403–406].

Microorganisms of the so-called *B. cepacia* complex (currently 17 species) including *B. cepacia, B. cenocepacia, B. vietnamiensis* and others are opportunistic pathogens in immunocompromised patients, especially in those with cystic fibrosis and chronic granulomatous disease. There are several O-serotyping schemes of these bacteria based on the O-antigens, whose structures have been reviewed earlier [407, 408] and are updated below. They are rather simple with linear di- or trisaccharide O-units consisting mainly of hexoses, 6-deoxyhexoses and *N*-acetylhexosamines (Table 3.33). In various strains, two structurally different OPSs coexist. The OPS of *B. cepacia* L is one of a few known O-antigens that contain L-*glycero*-D-*manno*-heptose, a common component of the LPS core OS of many bacteria (see Chap. 2). The OPS of *B. cepacia* O3 (CIP 8237) is shared by *P. aeruginosa* O15, *S. marcescens* O14 and *Vibrio fluvialis* AA-18239; that of *B. cepacia* O5 by *P. aeruginosa* O14, *Burkholderia plantarii* and *V. fluvialis* sv. 3.

Other representatives of *Burkholderia* with known OPS structures are phytopathogens, such as *B. gladioli* and *B. plantarii* [240], and plant growth-promoting bacteria (*B. phytofirmans, B. brasiliensis*) (Table 3.33). One of the OPS components of *B. brasiliensis* is yersiniose A, a branched monosaccharide found also in *Yersinia*.

Another phytopathogen, *Burkholderia caryophylli,* possesses two OPSs, which are homopolymers of unique higher monosaccharides caryophyllose and caryose (reviewed in ref. [240]). Caryophyllan is irregular owing to the presence of both α- (major) and β-linked monosaccharide units, and caryan is built up of blocks of O-acetylated and non-acetylated units. Caryan is linked to the core OS through a QuiNAc primer [416].

Phytopathogenic bacteria *Ralstonia solanacearum* cause wilt in tobacco and other plants. A large group of strains of this species have linear or branched OPSs with similar LRha-LRha-LRha-GlcNAc- main chains that differ in the configuration of the GlcNAc linkage, the position of substitution of a Rha residue and a lateral monosaccharide (L-xylose or L-rhamnose) (reviewed in ref. [240]). In many strains, more than one OPS of the sort occur [417]. The OPS of *Ralstonia pickettii* NCTC 11149 has a main chain of the same type [418]:

2)LRha(α1-2)LRha(β1-3)LRha2Ac(α1-3)GlcNAc(β1-

whereas that of another *R. pickettii* strain [419] resembles several OPSs of *P. aeruginosa* [276]:

4)Rha(α1-4)LGalNAcA(α1-3)QuiNAc4NAc(β1-

Table 3.33 Structures of *Burkholderia* OPSs

B. cepacia O1 [408]	4)Glc(α1-3)LGlcNAc(α1- and 4)Glc(α1-3)LRha(α1-
B. cepacia O2, E (McKevitt) [408]	2)Man(α1-2)Man(α1-4)Gal(β1- and 2)Man(α1-2)Man(α1-3)Man(β1-
B. cepacia O2, G (IMV 4137) [408]	2)LRha(α1-4)Gal(α1-
B. cepacia O2, G (IMV 598/2) [408]	2)LRha(α1-4)Gal(α1- and 4)Glc(β1-3)Man2Ac(β1-
B. cepacia O3 (CIP 8237) [408]	2)Rib*f*(β1-4)GalNAc(α1-
B. cepacia O3 (IMV 4176) [408]	4)GalNAc(α1-4)GalNAc(β1- and 2)Rib*f*(β1-4)GalNAc(α1-
B. cepacia O4, C, *B. vietnamiensis* LMG 6999 [408]	3)Gal(α1-3)Gal(β1-3)GalNAc(β1- and 4)LRha(α1-3)GalNAc(α1-3)GalNAc(β1-
B. cenocepacia K56-2 [409]	4)LRha(α1-3)GalNAc(α1-3)GalNAc(β1-
B. cepacia O5 [408]	4)LRha(α1-3)ManNAc(β1-
B.cepacia O6 [408]	3)Gal*f6Ac*(β1-3)Man(β1-
B. cepacia O7, A [408]	4)Glc(β1-3)Man2Ac(β1-
B. cepacia O9 [408]	4)Glc(α1-3)LRha(α1-
B. cepacia B [408]	3)Gal*f*(β1-3)Fuc(α1-
B. cepacia E [408]	3)Fuc(α1-3)GlcNAc(β1-
B. cepacia I [408]	3)Fuc(α1-4)GalNAc(β1- and 3)Fuc(α1-2)LRha(α1-
B. cepacia J [407]	3)LRha(α1-3)Man(β1-4)Man3Ac(α1-
B. vietnamiensis LMG 6998 [408]	3)LRha(α1-3)Man(β1-4)Man(α1-
B. cepacia K [408]	3)Rha(α1-3)Rha(α1-2)Rha(β1-
B. cepacia L [408]	3)Rha(α1-3)Rha(α1-2)LDmanHep(α1-
B. cepacia A (McKevitt) [408]	4)LRha(α1-3)GalNAc(α1-3)GalNAc(β1-
B. cepacia PVFi-5A [408]	3)Gal(α1-6)GlcNAc(α1-4)GalNAc(β1-
B. cepacia [410]	3)Rha(α1-3)Rha(α1-4)Gal(α1- and 3)Rha(α1-3)Rha(α1-2)Rha(α1-
B. cepacia ASP B 2D [278]	2)Rib*f*(β1-6)Glc(α1-
B. multivorans C1576 [411]	2)Man(α1-2)Rha(α1-3)Man(α1- and 2)Man(α1-2)Rha3Me(α1-3)Rha(α1-
B. vietnamiensis LMG 10926 [412]	4)LRha(α1-2)LRha(α1-3)LRha(β1- and 3)Fuc(α1-3)Fuc(α1-3)LRha(α1- LRha(α1-2)⌋
B. anthina LMG 20983 [413]	3)LRha(α1-2)LRha(α1-2)Gal(α1-
B. gladioli pv. gladioli [240]	3)Man2Ac(β1-4)LRha(α1-3)Gal(α1-
B. gladioli pv. *agaricicola* [414]	3)Man2Ac(α1-2)Rha(α1-4)Gal(β1-
B. gladioli pv. alliicola [240]	4)LRha(α1-3)Man2Ac(β1- └(2-1α)Fuc(3-1α)LRha
B. plantarii [240]	4)LRha(α1-3)ManNAc(β1-
B. phytofirmans [278]	3)L6dTal(α1-3)GalNAc(β1- *Xyl(β1-2)*⌋ └*(4-1β)Xyl*
B. brasiliensis[a] [415]	3)Rha(α1-3)Rha(α1-2)Rha(1- └(2-1α)Sug

[a]Sug indicates yersiniose A.

3.3.4.2 Alcaligenaceae

The genus *Bordetella* includes respiratory pathogens causing a variety of diseases in warm-blooded animals (*B. bronchiseptica, B. hinzii, B. avium*) and whooping cough in humans (*B. pertussis* and *B. parapertussis*). *B. trematum* has been found in human ear and blood infections. Except for *B. pertussis* having no long-chain O-antigen, the OPSs of *Bordetella* are homo- or hetero-glycans containing derivatives of various 2,3-diamino-2,3-dideoxyhexuronic acids (Table 3.34). These are fully amidated in *B. hinzii* or partially amidated in *B. bronchiseptica* and *B. parapertussis*. The OPSs of *B. hinzii* and *B. bronchiseptica* MO149 are rather short having not more than six O-units and that of *B. trematum* not more than two O-units.

The OPSs of *B. bronchiseptica* and *B. parapertussis* are terminated with various *N*-acyl derivatives of 2,3,4-triamino-2,3,4-trideoxygalacturonamide, which, together with variations in the amidation pattern of the uronic acids, confer clear serological distinctions between strains sharing the same LGalNAc3NAcA*N* homopolysaccharide [421]. The OPSs of *B. hinzii* and *B. bronchiseptica* MO149 are terminated with a 4-O-methylated GalNAc3NAcAN residue. In *B. bronchiseptica, B. parapertussis* and *B. hinzii,* the O-chain is linked to the core OS through a specific non-repetitive pentasaccharide domain enriched in 2,3-diamino-2,3-dideoxyhexuronic acid derivatives too [421, 423]. A portion of this domain proximal to the core OS, called A-band trisaccharide, is also present in the short-chain LPS of *B. pertussis* and synthesized by a pathway similar to that of an O-unit [425].

Taylorella equigenitalis is the cause of contagious equine metritis, a venereal disease of horses, whereas *Taylorella asinigenitali* is not pathogenic. They elaborate quite different acidic OPSs. That of *T. equigenitalis* consists of two partially amidated derivatives of 2,3-diamino-2,3-dideoxyhexuronic acids and is terminated with a 4-O-methylated LGulNAc3NAcA residue [426]:

4)LGulNAc3NAcA*N*(α1-4)ManNAc3NAcA*N*(β1-

The OPS of *T. asinigenitali* also has a disaccharide O-unit containing a unique *N*-acetimidoyl derivative of GlcNA [427]:

3)GlcNAmA(β1-3)QuiNAc4NAc(β1-

Alcaligenes faecalis shares the OPS structure with *S. maltophilia* O4 [428].

Table 3.34 Structures of *Bordetella* OPSs

B. avium[a] [420]	4)GlcNAm3N(3Hb)A(β1-
B. bronchiseptica, B. parapertussis [421]	4)LGalNAc3NAcA*N*(α1-
B. bronchiseptica MO149 [422]	4)GlcNAc3NAcA*N*(β1-4)LGalNAc3NAcA*N*(α1-
B. hinzii [422,423]	4)GlcNAc3NAcAN(β1-4)GlcNAc3NAcAN(β1-4)LGalNAc3NAcAN(α1-
B. trematum [424]	4)ManNAc3NAmA(β1-4)ManNAc3NAmA(β1-3)FucNAc(α1-

[a]The absolute configuration of the 3-hydroxybutanoyl group has not been determined.

3.3.4.3 Other Families

The OPS structures have been established for several soil- or/and water-inhabiting β-proteobacteria, including *Naxibacter alkalitolerans* from the family Oxalobacteraceae, *Sphaerotilus natans,* a non-classified bacterium of the order Burkholderiales, and *Chromobacterium violaceum* from the family Neisseriaceae (Table 3.35). The last bacterium has the only known OPS that contains D-*glycero*-D-*galacto*-heptose (DDgalHep).

3.3.5 ε-Proteobacteria

3.3.5.1 Campylobacteraceae

Campylobacter jejuni is a common cause of human gastroenteritis and is associated with postinfection autoimmune arthritis and neuropathy (Guillain-Barré syndrome). Molecular mimicry between the R-type LPS of *C. jejuni* and gangliosides in peripheral nerves plays a crucial role in the pathogenesis. Structures of LPS-associated polysaccharides have been established in various *C. jejuni* serotypes but later found to be capsular polysaccharides not related to LPS [432], whereas LPS is of R-type. The only documented exception is *C. jejuni* 81116, which produces a neutral OPS of the following structure [433]:

6)Glc(α1-4)Gal(α1-3)GlcNAc(β1-
GlcNAc(β1-3)┘

Polysaccharides characterized in several *Campylobacter lari* and *Campylobacter coli* strains do not seem to be O-antigens too. *Campylobacter fetus,* a causative agent of abortion in cattle and sheep, can cause bacteremia and thrombophlebitis in humans. The OPS of serotype A is an α1-2-linked homopolymer of partially (80–90%) 2-O-acetylated Man [434] and that of serotype B is a 3)Rha(β1-2)Rha (α1- rhamnan terminated with 3-O-methylated Rha [435].

3.3.5.2 Helicobacteraceae

Helicobacter pylori is a prevalent gastroduodenal pathogen of humans, which colonizes gastric mucosa. Once established, infection may persist in the stomach for life and is associated with active inflammation of gastric mucosa leading to gastritis, gastric and duodenal ulcer and increasing risk of gastric cancer. The LPSs of *H. pylori* have generally a poly(*N*-acetyl-β-lactosamine) chain, which in most strains is L-fucosylated to various degrees (see reviews [436, 437]). In several

Table 3.35 Structures of OPSs from other families of β-proteobacteria

N. alkalitolerans [429]	3)FucNAc(α1-2)Qui3N(*S*3Hb)(β1-2)Rha(α1-4)Gal(β1-
S. natans[a] [430]	4)Glc(α1-3)Rha(α1-3)Rha(α1-3)Rha(α1-3)Rha(α1-
C. violaceum [431]	4)DDgalHep(α1-2)LRha(α1-4)DDgalHep(β1-3)GlcNAc(α1-

[a]The absolute configurations of the monosaccharides have not been determined.

strains, an additional non-stoichiometric decoration of the main chain with Glc or Gal (Sug) has been reported [436, 438]:

3)Gal(β1-4)GlcNAc(β1- or 3)Gal(β1-4)GlcNAc(β1- or 3)Gal(β1-4)GlcNAc(β1-
LFuc(α1-3)┘ *LFuc(α1-3)*┘ └*(6-1α)Sug*

The terminal non-reducing unit usually carries one or two LFuc residues giving rise to Le^x trisaccharide or Le^y tetrasaccharide, respectively, which are interconvertible upon phase variation [438]. Less often, the OPS chain is terminated with another Lewis or related blood group antigenic determinant. In polylactosamine-lacking strains of *H. pylori* and several less studied non-human *Helicobacter* species, like *H. mustelae* from ferrets [436], the antigenic determinants may be expressed on the LPS core OS. These features have multiple biological effects on pathogenesis and disease outcome, including gastric adaptation due to molecular mimicry of Lewis antigens [437].

In *H. pylori* LPSs, there are also other core OS-linked polymers, such as heptans and glucans [436, 437]. Atypically of *H. pylori*, the O-antigen of strains D1, D3 and D6 is a 2)Man3*C*Me(α1-3)LRha(α1-3)Rha(α1- heteropolysaccharide composed of 3-*C*-methyl-D-mannose and both D- and L-rhamnose [439].

3.3.6 Flavobacteria

Flavobacteriaceae is the only family studied in the class Flavobacteria. Marine bacteria of the genus *Flavobacterium* are fish pathogens and are also associated with infectious diseases in humans. The OPSs of *F. columnare* A contains a keto amino sugar, namely 2-acetamido-2,6-dideoxy-D-*xylo*-hexos-4-ulose (Sug) [440] and is structurally related to the OPS of *Pseudoalteromonas rubra* [253]:

4)GlcNAcA3Ac(β1-4)LFucNAm3Ac(α1-3)Sug(α1-

An unusual 4-*N*-[(*3S,5S*)-3,5-dihydroxyhexanoyl] derivative of QuiN4N (QuiNAc4NR) is a component of the trisaccharide O-unit of *F. psychrophilum* [441]:

2)LRha(α1-4)LFucNAcA(α1-3)QuiNAc4NR(α1-

The OPS of another fish pathogen *Tenacibaculum maritimum* (former *Flexibacter maritimus*) includes a unique higher sugar 5-acetamido-8-amino-3,5,7,8,9-pentadeoxy-7-[(*S*)-3-hydroxybutanoylamino]non-2-ulosonic acid. The C-4–C-7 fragment of the acid has the β-L-*manno* configuration, whereas the configuration at C-8 is unknown. It is linked to the neighbouring QuiN4N residue through O-2 of a (*S*)-2-hydroxy-5-glutaryl group at the N-4 of the latter [442] (Fig. 3.3).

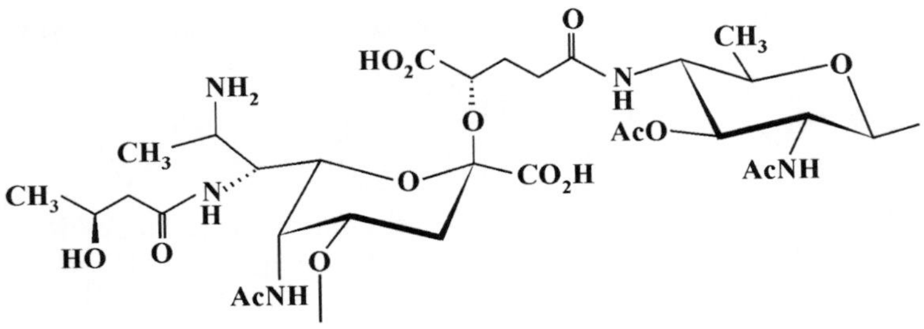

Fig. 3.3 Structure of the OPS of *Tenacibaculum maritimum* (former *Flexibacter maritimus*) [442]

The structures of the OPSs of two marine bacteria of the genus *Cellulophaga* have been established. That of *C. fucicola* contains a di-*N*-acetyl derivative of Pse [443]:

4)Pse5Ac7Ac(β2-4)Gal(β1-4)Glc(β1-

The OPS of *C. fucicola* is acidic too due to the presence of GlcA [444]:

2)Man(β1-3)Man*2Ac*(β1-4)GlcA(β1-3)GlcNAc(α1-

3.3.7 Other Classes

Fusobacterium necrophorum (class Fusobacteria, family Fusobacteriaceae) is an anaerobic bacterium associated with pyogenic infections in animals and humans. It has a teichoic acid-like O-antigen with a highly unusual polyalcohol, 2-amino-2-deoxy-2-*C*-methylpentonic acid (R), whose configuration remains unknown [445]:

4)R(5-*P*-4)Glc(α1-3)LFucNAm(α1-3)QuiNAc4N(*S*3Hb)(β1-

The genus *Pectinatus* from the family Veillonellaceae (class Clostridia) includes strictly anaerobic beer spoilage bacteria. The OPS of *P. frisingensis* consists of α- and β-linked L6dAlt, both in the furanose form [446]:

2)L6dAlt*f*(β1-3)L6dAlt*f*(β1-2)L6dAlt*f*(α1-
L6dAlt*f*(α1-2)┘

The OPS of *P. cerevisiiphilus* contains a fucofuranose residue as a component of the 2)Fuc*f*(β1-2)Glc(α1- discaccharide O-unit [446].

The genus *Porphyromonas* (class Bacteroidia, family Bacteroidaceae) includes etiologic agents for periodontal disease in adults (*P. gingivalis*) and animals: cats and dogs (*P. circumdentaria*). The OPS of *P. gingivalis* is distinguished by a non-stoichiometric phosphorylation of a rhamnose residue with phosphoethanolamine [447]:

3)Gal(α1-6)Glc(α1-4)LRha*2PEtN*(α1-3)GalNAc(β1-

The LPS of this bacterium has another phosphorylated branched α-mannan chain [448]. The OPS of *P. circumdentaria* consists of hexoses and *N*-acetylhexosamines only [449]:

6)Glc(β1-6)Gal(β1-6)Gal(β1-3)GlcNAc(β1-3)GalNAc(β1-

Bacteroides vulgatus from the same family is involved in the aggravation of colitis. It has a linear OPS with the 4)LRha(α1-3)Man(β1- disaccharide O-unit and a rhamnose residue at the non-reducing end [450].

Acknowledgements Y. A. K. is supported by the Russian Foundation for Basic Research (grants 10-04-00598 and 10-04-90047).

References

1. Kenne L, Lindberg B (1983) Bacterial polysaccharides. In: Aspinall GO (ed) The polysaccharides. Academic Press, New York, pp 287–363
2. Jann K, Jann B (1984) Structure and biosynthesis of O-antigens. In: Rietschel ET (ed) Chemistry of endotoxin (Handbook of endotoxin, vol. 1). Elsevier, Amsterdam, pp 138–186
3. Lindberg B (1998) Bacterial polysaccharides: components. In: Dumitriu S (ed) Polysaccharides: structural diversity and functional versatility. Marcel Dekker, New York, pp 237–273
4. Knirel YA, Kochetkov NK (1994) The structure of lipopolysaccharides of Gram-negative bacteria. III. The structure of O-antigens. Biochemistry (Moscow) 59:1325–1383
5. Wilkinson SG (1996) Bacterial lipopolysaccharides: themes and variations. Prog Lipid Res 35:283–343
6. Jansson P-E (1999) The chemistry of O-polysaccharide chains in bacterial lipopolysaccharides. In: Brade H, Opal SM, Vogel SN, Morrison DC (eds) Endotoxin in health and disease. Marcel Dekker, New York, pp 155–178
7. Knirel YA (2009) O-Specific polysaccharides of Gram-negative bacteria. In: Moran A, Brennan P, Holst O, von Itzstein M (eds) Microbial glycobiology: structures, relevance and applications. Elsevier, Amsterdam, pp 57–73
8. Kuhn H-M, Meier-Dieter U, Mayer H (1988) ECA, the enterobacterial common antigen. FEMS Microbiol Lett 54:195–222
9. Knirel YA, Kocharova NA (1995) Structure and properties of the common polysaccharide antigen of *Pseudomonas aeruginosa*. Biochemistry (Moscow) 60:1499–1507
10. Castric P, Cassels FJ, Carlson RW (2001) Structural characterization of the *Pseudomonas aeruginosa* 1244 pilin glycan. J Biol Chem 276:26479–26485
11. Isshiki Y, Matsuura M, Dejsirilert S, Ezaki T, Kawahara K (2001) Separation of 6-deoxy-heptane from a smooth-type lipopolysaccharide preparation of *Burkholderia pseudomallei*. FEMS Microbiol Lett 199:21–25
12. Gajdus J, Glosnicka R, Szafranek J (2006) Primary structure of *Salmonella* spp. O antigens. Wiad Chemiczne 60:621–653
13. Wilkinson SG (1977) Composition and structure of bacterial lipopolysaccharides. In: Sutherland IW (ed) Surface carbohydrates of the prokaryotic cell. Academic Press, London, pp 97–175
14. Hellerqvist CG, Lindberg B, Samuelsson K, Lindberg AA (1971) Structural studies on the O-specific side-chains of the cell-wall lipopolysaccharide from *Salmonella paratyphi* A var. *durazzo*. Acta Chem Scand 25:955–961

15. Hellerqvist CG, Lindberg B, Svensson S, Holme T, Lindberg AA (1969) Structural studies on the O-specific side-chains of the cell-wall lipopolysaccharide from *Salmonella typhimurium* LT2. Carbohydr Res 9:237–241
16. Svenson SB, Lönngren J, Carlin N, Lindberg AA (1979) *Salmonella* bacteriophage glycanases: endorhamnosidases of *Salmonella typhimurium* bacteriophages. J Virol 32:583–592
17. Szafranek J, Kumirska J, Czerwicka M, Kunikowska D, Dziadziuszko H, Glosnicka R (2006) Structure and heterogeneity of the O-antigen chain of *Salmonella* Agona lipopolysaccharide. FEMS Immunol Med Microbiol 48:223–236
18. Kaczynski Z, Gajdus J, Dziadziuszko H, Stepnowski P (2009) Chemical structure of the somatic antigen isolated from *Salmonella* Abortusequi (O4). J Pharm Biomed Anal 50:679–682
19. Hellerqvist CG, Larm O, Lindberg B, Holme T, Lindberg AA (1969) Structural studies on the O-specific side chains of the cell wall lipopolysaccharide from *Salmonella bredeney*. Acta Chem Scand 23:2217–2222
20. Di Fabio JL, Brisson J-R, Perry MB (1989) Structure of the lipopolysaccharide antigenic O-chain produced by *Salmonella livingstone* (O:6,7). Biochem Cell Biol 67:278–280
21. Lindberg B, Leontein K, Lindquist U, Svenson SB, Wrangsell G, Dell A, Rogers M (1988) Structural studies of the O-antigen polysaccharide of *Salmonella thompson*, serogroup C_1 (6,7). Carbohydr Res 174:313–322
22. Di Fabio JL, Brisson J-R, Perry MB (1989) Structure of the lipopolysaccharide antigenic O-chain produced by *Salmonella ohio* (0:6,7). Carbohydr Res 189:161–168
23. Di Fabio JL, Perry MB, Brisson J-R (1988) Structure of the antigenic O-polysaccharide of the lipopolysaccharide produced by *Salmonella eimsbuttel*. Biochem Cell Biol 66:107–115
24. Hellerqvist CG, Hoffman J, Lindberg A, Lindberg B, Svensson S (1972) Sequence analysis of the polysaccharides from *Salmonella newport* and *Salmonella kentucky*. Acta Chem Scand 26:3282–3286
25. Torgov VI, Shibaev VN, Shashkov AS, Rozhnova SS (1990) Structural studies of the O-specific polysaccharide from *Salmonella kentucky* strain 98/39 (O:8, H:i, Z6). Carbohydr Res 208:293–300
26. Jann K, Westphal O (1975) Microbial polysaccharides. In: Sela M (ed) The antigens, vol III. Academic Press, New York, pp 1–125
27. Rahman MM, Guard-Petter J, Carlson RW (1997) A virulent isolate of *Salmonella enteritidis* produces a *Salmonella typhi*-like lipopolysaccharide. J Bacteriol 179:2126–2131
28. Brooks BW, Perry MB, Lutze-Wallace CL, MacLean LL (2008) Structural characterization and serological specificities of lipopolysaccharides from *Salmonella enterica* serovar Gallinarum biovar Pullorum standard, intermediate and variant antigenic type strains. Vet Microbiol 126:334–344
29. Hellerqvist CG, Lindberg B, Svensson S, Holme T, Lindberg AA (1969) Structural studies on the O-specific side chains of the cell wall lipopolysaccharides from *Salmonella typhi* and *S. enteridis*. Acta Chem Scand 23:1588–1596
30. Szafranek J, Gajdus J, Kaczynski Z, Dziadziuszko H, Kunikowska D, Glosnicka R, Yoshida T, Vihanto J, Pihlaja K (1998) Immunological and chemical studies of *Salmonella haarlem* somatic antigen epitopes. I. Structural studies of O-antigen. FEMS Immunol Med Microbiol 21:243–252
31. Nghiem HO, Himmelspach K, Mayer H (1992) Immunochemical and structural analysis of the O polysaccharides of *Salmonella zuerich* [1,9,27, (46)]. J Bacteriol 174:1904–1910
32. L'vov VL, Yakovlev AV, Shashkov AS (1989) Study of the structure of the O-specific polysaccharide from *Salmonella anatum* using ^{1}H and ^{13}C NMR spectroscopy. Bioorg Khim 9:1660–1663
33. Szafranek J, Kaczynska M, Kaczynski Z, Gajdus J, Czerwicka M, Dziadziuszko H, Glosnicka R (2003) Structure of the polysaccharide O-antigen of *Salmonella* Aberdeen (O:11). Pol J Chem 77:1135–1140

34. Perepelov AV, Liu B, Senchenkova SN, Shevelev SD, Feng L, Shashkov AS, Wang L, Knirel YA (2010) The O-antigen of *Salmonella enterica* O13 and its relation to the O-antigen of *Escherichia coli* O127. Carbohydr Res 345:1808–1811
35. Di Fabio JL, Brisson J-R, Perry MB (1988) Structure of the major lipopolysaccharide antigenic O-chain produced by *Salmonella carrau* (0:6, 14, 24). Carbohydr Res 179:233–244
36. Brisson J-R, Perry MB (1988) The structure of the two lipopolysaccharide O-chains produced by *Salmonella boecker*. Biochem Cell Biol 66:1066–1077
37. Di Fabio JL, Brisson J-R, Perry MB (1989) Structural analysis of the three lipopolysaccharides produced by *Salmonella madelia* (1,6,14,25). Biochem Cell Biol 67:78–85
38. Liu B, Perepelov AV, Guo D, Shevelev SD, Senchenkova SN, Shashkov AS, Feng L, Wang L, Knirel YA (2011) Structural and genetic relationships of two pairs of closely related O-antigens of *Escherichia coli* and *Salmonella enterica: E. coli* O11/*S. enterica* O16 and *E. coli* O21/*S. enterica* O38. FEMS Immunol Med Microbiol 61:258–268
39. Perepelov AV, Li D, Liu B, Senchenkova SN, Guo D, Shashkov AS, Feng L, Knirel YA, Wang L (2011) Structural and genetic characterization of the closely related O-antigens of *Escherichia coli* O85 and *Salmonella enterica* O17. Innate Immun 17:164–173
40. Vinogradov E, Nossova L, Radziejewska-Lebrecht J (2004) The structure of the O-specific polysaccharide from *Salmonella cerro* (serogroup K, O:6,14,18). Carbohydr Res 339: 2441–2443
41. Knirel YA, Perepelov AV, Senchenkova SN, Liu B, Feng L, Wang L (2010) New structures of *Salmonella enterica* O-antigens and their relationships with O-antigens of *Escherichia coli*. In: Abstracts of the 25th international carbohydrate symposium, Tokyo, Japan, 1–6 August 2010
42. Kumirska J, Dziadziuszko H, Czerwicka M, Lubecka EA, Kunikowska D, Siedlecka EM, Stepnowski P (2011) Heterogeneous structure of O-antigenic part of lipopolysaccharide of *Salmonella* Telaviv (serogroup O:28) containing 3-acetamido-3,6-dideoxy-D-glucopyranose. Biochemistry (Moscow) 76:780–790
43. Kumirska J, Szafranek J, Czerwicka M, Paszkiewicz M, Dziadziuszko H, Kunikowska D, Stepnowski P (2007) The structure of the O-polysaccharide isolated from the lipopolysaccharide of *Salmonella* Dakar (serogroup O:28). Carbohydr Res 342:2138–2143
44. Bundle DR, Gerken M, Perry MB (1986) Two-dimensional nuclear magnetic resonance at 500 MHz: the structural elucidation of a *Salmonella* serogroup N polysaccharide antigen. Can J Chem 64:255–264
45. Perry MB, Bundle DR, MacLean L, Perry JA, Griffith DW (1986) The structure of the antigenic lipopolysaccharide O-chains produced by *Salmonella urbana* and *Salmonella godesberg*. Carbohydr Res 156:107–122
46. Kenne L, Lindberg B, Söderholm E, Bundle DR, Griffith DW (1983) Structural studies of the O-antigens from *Salmonella greenside* and *Salmonella adelaide*. Carbohydr Res 111:289–296
47. Gajdus J, Kaczynski Z, Smietana J, Stepnowski P (2009) Structural determination of the O-antigenic polysaccharide from *Salmonella* Mara (O:39). Carbohydr Res 344:1054–1057
48. Perry MB, MacLean LL (1992) Structure of the polysaccharide O-antigen of *Salmonella riogrande* O40 (group R) related to blood group A activity. Carbohydr Res 232:143–150
49. Perepelov AV, Liu B, Senchenkova SN, Shashkov AS, Feng L, Knirel YA, Wang L (2010) Structure of the O-polysaccharide of *Salmonella enterica* O41. Carbohydr Res 345:971–973
50. Senchenkova SN, Perepelov AV, Shevelev SD, Shashkov AS, Knirel YA, Liu B, Feng L, Wang L (2010) The completed *Salmonella enterica* O-antigen structure elucidation. Paper presented at the 4th Baltic meeting on microbial carbohydrates, Hyytiälä, Finland, 19–22 September 2010
51. Perry MB, MacLean LL (1992) Structural characterization of the O-polysaccharide of the lipopolysaccharide produced by *Salmonella milwaukee* O:43 (group U) which possesses human blood group B activity. Biochem Cell Biol 70:49–55

52. Perepelov AV, Liu B, Senchenkova SN, Shashkov AS, Guo D, Feng L, Knirel YA, Wang L (2010) Structure and gene cluster of the O-polysaccharide of *Salmonella enterica* O44. Carbohydr Res 345:2099–2101
53. Shashkov AS, Vinogradov EV, Knirel YA, Nifant'ev NE, Kochetkov NK, Dabrowski J, Kholodkova EV, Stanislavsky ES (1993) Structure of the O-specific polysaccharide of *Salmonella arizonae* O45. Carbohydr Res 241:177–188
54. Perepelov AV, Wang Q, Senchenkova SN, Shashkov AS, Feng L, Wang L, Knirel YA (2009) Structure of O-antigen and characterization of O-antigen gene cluster of *Salmonella enterica* O47 containing ribitol phosphate and 2-acetimidoylamino-2,6-dideoxy-L-galactose. Biochemistry (Moscow) 74:416–420
55. Gamian A, Jones C, Lipinski T, Korzeniowska-Kowal A, Ravenscroft N (2000) Structure of the sialic acid-containing O-specific polysaccharide from *Salmonella enterica* serovar Toucra O48 lipopolysaccharide. Eur J Biochem 267:3160–3166
56. Feng L, Senchenkova SN, Tao J, Shashkov AS, Liu B, Shevelev SD, Reeves P, Xu J, Knirel YA, Wang L (2005) Structural and genetic characterization of enterohaemorrhagic *Escherichia coli* O145 O antigen and development of an O145 serogroup-specific PCR assay. J Bacteriol 187:758–764
57. Senchenkova SN, Shashkov AS, Knirel YA, Schwarzmüller E, Mayer H (1997) Structure of the O-specific polysaccharide of *Salmonella enterica* ssp. *arizonae* O50 (*Arizona* O9a,9b). Carbohydr Res 301:61–67
58. Perepelov AV, Liu B, Guo D, Senchenkova SN, Shashkov AS, Feng L, Wang L, Knirel YA (2011) Structure of the O-antigen of *Salmonella enterica* O51 and its structural and genetic relation to the O-antigen of *Escherichia coli* O23. Biochemistry (Moscow) 76:774–779
59. Perepelov AV, Liu B, Senchenkova SN, Shashkov AS, Feng L, Knirel YA, Wang L (2011) Structure of the O-polysaccharide and characterization of the O-antigen gene cluster of *Salmonella enterica* O53. Carbohydr Res 346:373–376
60. Keenleyside WJ, Perry M, MacLean L, Poppe C, Whitfield C (1994) A plasmid-encoded $rfb_{O:54}$ gene cluster is required for biosynthesis of the O:54 antigen in *Salmonella enterica* serovar Borreze. Mol Microbiol 11:437–448
61. Liu B, Perepelov AV, Svensson MV, Shevelev SD, Guo D, Senchenkova SN, Shashkov AS, Weintraub A, Feng L, Widmalm G, Knirel YA, Wang L (2010) Genetic and structural relationships of *Salmonella* O55 and *Escherichia coli* O103 O-antigens and identification of a 3-hydroxybutanoyltransferase gene involved in the synthesis of a Fuc3N derivative. Glycobiology 20:679–688
62. Perepelov AV, Liu B, Shevelev SD, Senchenkova SN, Hu B, Shashkov AS, Feng L, Knirel YA, Wang L (2010) Structural and genetic characterization of the O-antigen of *Salmonella enterica* O56 containing a novel derivative of 4-amino-4,6-dideoxy-D-glucose. Carbohydr Res 345:1891–1895
63. Perepelov AV, Liu B, Senchenkova SN, Guo D, Shevelev SD, Feng L, Shashkov AS, Wang L, Knirel YA (2011) O-antigen structure and gene clusters of *Escherichia* coli O51 and *Salmonella* enterica O57; another instance of identical O-antigens in the two species. Carbohydr Res 346:828–832
64. Perepelov AV, Liu B, Shevelev SD, Senchenkova SN, Shashkov AS, Feng L, Knirel YA, Wang L (2010) Relatedness of the O-polysaccharide structures of *Escherichia coli* O123 and *Salmonella enterica* O58, both containing 4,6-dideoxy-4-{*N*-(*S*)-3-hydroxybutanoyl]-D-alanyl}amino-D-glucose; revision of the *E. coli* O123 O-polysaccharide structure. Carbohydr Res 345:825–829
65. Perepelov AV, Liu B, Senchenkova SN, Shashkov AS, Guo D, Feng L, Knirel YA, Wang L (2011) Structures of the O-polysaccharides of *Salmonella enterica* O59 and *Escherichia coli* O15. Carbohydr Res 346:381–383
66. Perepelov AV, Liu B, Senchenkova SN, Shashkov AS, Feng L, Knirel YA, Wang L (2010) Structure and gene cluster of the O-antigen of *Salmonella enterica* O60 containing 3-formamido-3,6-dideoxy-D-galactose. Carbohydr Res 345:1632–1634

67. Knirel YA, Shashkov AS, Tsvetkov YE, Jansson P-E, Zähringer U (2003) 5,7-Diamino-3,5,7,9-tetradeoxynon-2-ulosonic acids in bacterial glycopolymers: chemistry and biochemistry. Adv Carbohydr Chem Biochem 58:371–417
68. Vinogradov EV, Knirel YA, Kochetkov NK, Schlecht S, Mayer H (1994) The structure of the O-specific polysaccharide of *Salmonella arizonae* O62. Carbohydr Res 253:101–110
69. Vinogradov EV, Knirel YA, Lipkind GM, Shashkov AS, Kochetkov NK, Stanislavsky ES, Kholodkova EV (1987) Antigenic polysaccharides of bacteria. 24. The structure of the O-specific polysaccharide chain of the *Salmonella arizonae* O63 (*Arizona* O8) lipopolysaccharide. Bioorg Khim 13:1399–1404
70. Liu B, Perepelov AV, Li D, Senchenkova SN, Han Y, Shashkov AS, Feng L, Knirel YA, Wang L (2010) Structure of the O-antigen of *Salmonella* O66 and the genetic basis for similarity and differences between the closely related O-antigens of *Escherichia coli* O166 and *Salmonella* O66. Microbiology 156:1642–1649
71. Kocharova NA, Vinogradov EV, Knirel YA, Shashkov AS, Kochetkov NK, Stanislavsky ES, Kholodkova EV (1988) The structure of the O-specific polysaccharide chains of the lipopolysaccharides of *Citrobacter* O32 and *Salmonella arizonae* O64. Bioorg Khim 14:697–700
72. Vinogradov EV, Knirel YA, Shashkov AS, Paramonov NA, Kochetkov NK, Stanislavsky ES, Kholodkova EV (1994) The structure of the O-specific polysaccharide of *Salmonella arizonae* O21 (*Arizona* 22) containing *N*-acetylneuraminic acid. Carbohydr Res 259:59–65
73. Gamian A, Lipinski T, Jones C, Hossam E, Korzeniowska-Kowal A, Rybka J Unpublished data
74. Author's unpublished data
75. Vinogradov EV, Knirel YA, Lipkind GM, Shashkov AS, Kochetkov NK, Stanislavsky ES, Kholodkova EV (1987) Antigenic polysaccharides of bacteria. 23. The structure of the O-specific polysaccharide chain of the lipopolysaccharide of *Salmonella arizonae* O59. Bioorg Khim 13:1275–1281
76. Kocharova NA, Knirel YA, Stanislavsky ES, Kholodkova EV, Lugowski C, Jachymek W, Romanowska E (1996) Structural and serological studies of lipopolysaccharides of *Citrobacter* O35 and O38 antigenically related to *Salmonella*. FEMS Immunol Med Microbiol 13:1–8
77. Hoffman J, Lindberg B, Glowacka M, Derylo M, Lorkiewicz Z (1980) Structural studies of the lipopolysaccharide from *Salmonella typhimurium* 902 (ColIb drd2). Eur J Biochem 105:103–107
78. Knirel YA, Kocharova NA, Bystrova OV, Katzenellenbogen E, Gamian A (2002) Structures and serology of the O-specific polysaccharides of bacteria of the genus *Citrobacter*. Arch Immunol Ther Exp 50:379–391
79. Kocharova NA, Mieszala M, Zatonsky GV, Staniszewska M, Shashkov AS, Gamian A, Knirel YA (2004) Structure of the O-polysaccharide of *Citrobacter youngae* O1 containing an α-D-ribofuranosyl group. Carbohydr Res 339:321–325
80. Mieszala M, Lipinski T, Kocharova NA, Zatonsky GV, Katzenellenbogen E, Shashkov AS, Gamian A, Knirel YA (2003) The identity of the O-specific polysaccharide structure of *Citrobacter* strains from serogroups O2, O20 and O25 and immunochemical characterisation of *C. youngae* PCM 1507 (O2a,1b) and related strains. FEMS Immunol Med Microbiol 36:71–76
81. Katzenellenbogen E, Zatonsky GV, Kocharova NA, Witkowska D, Bogulska M, Shashkov AS, Gamian A, Knirel YA (2003) Structural and serological studies on a new 4-deoxy-D-*arabino*-hexose-containing O-specific polysaccharide from the lipopolysaccharide of *Citrobacter braakii* PCM 1531 (serogroup O6). Eur J Biochem 270:2732–2738
82. Kocharova NA, Katzenellenbogen E, Zatonsky GV, Bzozovska E, Gamian A, Shashkov AS, Knirel YA (2010) Structure of the O-polysaccharide of *Citrobacter youngae* PCM 1503. Carbohydr Res 345:2571–2573

83. Ovchinnikova OG, Kocharova NA, Katzenellenbogen E, Zatonsky GV, Shashkov AS, Knirel YA, Lipinski T, Gamian A (2004) Structures of two O-polysaccharides of the lipopolysaccharide of *Citrobacter youngae* PCM 1538 (serogroup O9). Carbohydr Res 339:881–884
84. Katzenellenbogen E, Kocharova NA, Zatonsky GV, Bogulska M, Rybka J, Gamian A, Shashkov AS, Knirel YA (2003) Structure of the O-specific polysaccharide from the lipopolysaccharide of *Citrobacter gillenii* O11, strain PCM 1540. Carbohydr Res 338:1381–1387
85. Katzenellenbogen E, Kocharova NA, Korzeniowska-Kowal A, Bogulska M, Rybka J, Gamian A, Kachala VV, Shashkov AS, Knirel YA (2008) Structure of the glycerol phosphate-containing O-specific polysaccharide and serological studies on the lipopolysaccharides of *Citrobacter werkmanii* PCM 1548 and PCM 1549 (serogroup O14). FEMS Immunol Med Microbiol 54:255–262
86. Katzenellenbogen E, Kocharova NA, Toukach FV, Gorska S, Korzeniowska-Kowal A, Bogulska M, Gamian A, Knirel YA (2009) Structure of an abequose-containing O-polysaccharide from *Citrobacter freundii* O22 strain PCM 1555. Carbohydr Res 344:1724–1728
87. Katzenellenbogen E, Toukach FV, Kocharova NA, Korzeniowska-Kowal A, Gamian A, Shashkov AS, Knirel YA (2008) Structure of a phosphoethanolamine-containing O-polysaccharide of *Citrobacter freundii* strain PCM1443 from serogroup O39 and its relatedness to the *Klebsiella pneumoniae* O1 polysaccharide. FEMS Immunol Med Microbiol 53:60–64
88. Vinogradov E, Nossova L, Perry MB, Kay WW (2005) The structure of the antigenic O-polysaccharide of the lipopolysaccharide of *Edwardsiella ictaluri* strain MT104. Carbohydr Res 340:1509–1513
89. Vinogradov E, Nossova L, Perry MB, Kay WW (2005) Structural characterization of the O-polysaccharide antigen of *Edwardsiella tarda* MT 108. Carbohydr Res 340:85–90
90. Kocharova NA, Katzenellenbogen E, Toukach FV, Knirel YA, Shashkov AS (2009) Structures of the O-specific polysaccharides of the lipopolysaccharides of *Edwardsiella tarda*. In: Abstracts of the 15th European carbohydrate symposium, Vienna, 19–24 July 2009
91. Liu B, Knirel YA, Feng L, Perepelov AV, Senchenkova SN, Wang Q, Reeves P, Wang L (2008) Structure and genetics of *Shigella* O antigens. FEMS Microbiol Rev 32:627–653, Corrigendum in: FEMS Microbiol. Rev. **34**: 606 (2010)
92. Stenutz R, Weintraub A, Widmalm G (2006) The structures of *Escherichia coli* O-polysaccharide antigens. FEMS Microbiol Rev 30:382–403
93. Beynon LM, Bundle DR, Perry MB (1990) The structure of the antigenic lipopolysaccharide O-chain produced by *Escherichia hermannii* ATCC 33650 and 33652. Can J Chem 68:1456–1466
94. Perry MB, Richards JC (1990) Identification of the lipopolysaccharide O-chain of *Escherichia hermannii* (ATCC 33651) as a D-rhamnan. Carbohydr Res 205:371–376
95. Perry MB, Bundle DR (1990) Antigenic relationships of the lpopolysaccharides of *Escherichia hermannii* strains with those of *Escherichia coli* O157:H7, *Brucella melitensis*, and *Brucella abortus*. Infect Immun 58:1391–1395
96. Eserstam R, Rajaguru TP, Jansson P-E, Weintraub A, Albert MJ (2002) The structure of the O-chain of the lipopolysaccharide of a prototypal diarrheagenic strain of *Hafnia alvei* that has characteristics of a new species under the genus *Escherichia*. Eur J Biochem 269:3289–3295
97. West NP, Sansonetti P, Mounier J, Exley RM, Parsot C, Guadagnini S, Prevost MC, Prochnicka-Chalufour A, Delepierre M, Tanguy M, Tang CM (2005) Optimization of virulence functions through glucosylation of *Shigella* LPS. Science 307:1313–1317
98. Perepelov AV, Shevelev SD, Liu B, Senchenkova SN, Shashkov AS, Feng L, Knirel YA, Wang L (2010) Structures of the O-antigens of *Escherichia coli* O13, O129 and O135 related to the O-antigens of *Shigella flexneri*. Carbohydr Res 345:1594–1599
99. Perepelov AV, L'vov VL, Liu B, Senchenkova SN, Shekht ME, Shashkov AS, Feng L, Aparin PG, Wang L, Knirel YA (2009) A similarity in the O-acetylation pattern of the O-antigens of *Shigella flexneri* types 1a, 1b and 2a. Carbohydr Res 344:687–692

100. Kenne L, Lindberg B, Petersson K, Katzenellenbogen E, Romanowska E (1978) Structural studies of *Shigella flexneri* O-antigens. Eur J Biochem 91:279–284
101. Perepelov AV, L'vov VL, Liu B, Senchenkova SN, Shekht ME, Shashkov AS, Feng L, Aparin PG, Wang L, Knirel YA (2009) A new ethanolamine phosphate-containing variant of the O-antigen of *Shigella flexneri* type 4a. Carbohydr Res 344:1588–1591
102. Kenne L, Lindberg B, Petersson K, Katzenellenbogen E, Romanowska E (1977) Structural studies of the *Shigella flexneri* variant X, type 5a and 5b O-antigens. Eur J Biochem 76:327–330
103. Foster RA, Carlin NIA, Majcher M, Tabor H, Ng L-K, Widmalm G (2011) Structural elucidation of the O-antigen of the *Shigella flexneri* provisional serotype 88–893: structural and serological similarities with *Shigella flexneri* provisional serotype Y394 (1c). Carbohydr Res 346:872–876
104. Vinogradov E, Frirdich E, MacLean LL, Perry MB, Petersen BO, Duus JØ, Whitfield C (2002) Structures of lipopolysaccharides from *Klebsiella pneumoniae*. Eluicidation of the structure of the linkage region between core and polysaccharide O chain and identification of the residues at the non-reducing termini of the O chains. J Biol Chem 277:25070–25081
105. Clarke BR, Cuthbertson L, Whitfield C (2004) Nonreducing terminal modifications determine the chain length of polymannose O antigens of *Escherichia coli* and couple chain termination to polymer export via an ATP-binding cassette transporter. J Biol Chem 279:35709–35718
106. Whitfield C, Perry MB, MacLean LL, Yu SH (1992) Structural analysis of the O-antigen side chain polysaccharides in the lipopolysaccharides of *Klebsiella* serotypes O2(2a), O2(2a,2b), and O2(2a,2c). J Bacteriol 174:4913–4919
107. MacLean LL, Whitfield C, Perry MB (1993) Characterization of the polysaccharide antigen of *Klebsiella pneumoniae* O:9 lipopolysaccharide. Carbohydr Res 239:325–328
108. Kelly RF, Perry MB, MacLean LL, Whitfield C (1995) Structures of the O-antigens of *Klebsiella* serotypes O2(2a,2e), O2(2a,2e,2 h), and O2(2a,2f,2 g), members of a family of related D-galactan O-antigens in *Klebsiella* spp. J Endotoxin Res 2:131–140
109. Kelly RF, Severn WB, Richards JC, Perry MB, MacLean LL, Tomas JM, Merino S, Whitfield C (1993) Structural variations in the O-specific polysaccharides of *Klebsiella pneumoniae* serotype O1 and O8 lipopolysaccharide: evidence for clonal diversity in *rfb* genes. Mol Microbiol 10:615–625
110. Ansaruzzaman M, Albert MJ, Holme T, Jansson P-E, Rahman MM, Widmalm G (1996) A *Klebsiella pneumoniae* strain that shares a type-specific antigen with *Shigella flexneri* serotype 6. Characterization of the strain and structural studies of the O-antigenic polysaccharide. Eur J Biochem 237:786–791
111. Mertens K, Müller-Loennies S, Mamat U (2002) Analyses of the LPS O-antigens of non-typeable *Klebsiella* isolates: identification of two putative new O-serotypes. J Endotoxin Res 8:159–160
112. Mertens K, Müller-Loennies S, Stengel P, Podschun R, Hansen DS, Mamat U (2010) Antiserum against *Raoultella terrigena* ATCC 33257 identifies a large number of *Raoultella* and *Klebsiella* clinical isolates as serotype 012. Innate Immun 16:366–380
113. Leone S, Molinaro A, Dubery I, Lanzetta R, Parrilli M (2007) The O-specific polysaccharide structure from the lipopolysaccharide of the Gram-negative bacterium *Raoultella terrigena*. Carbohydr Res 342:1514–1518
114. Aucken HM, Wilkinson SG, Pitt TL (1998) Re-evaluation of the serotypes of *Serratia marcescens* and separation into two schemes based on lipopolysaccharide (O) and capsular polysaccharide (K) antigens. Microbiology 144:639–653
115. Vinogradov E, Petersen BO, Duus JØ, Radziejewska-Lebrecht J (2003) The structure of the polysaccharide part of the LPS from *Serratia marcescens* serotype O19, including linkage region to the core and the residue at the non-reducing end. Carbohydr Res 338:2757–2761
116. Aucken HM, Oxley D, Wilkinson SG (1993) Structural and serological characterisation of an O-specific polysaccharide from *Serratia plymuthica*. FEMS Microbiol Lett 111:295–300
117. Romanowska E (2000) Immunochemical aspects of *Hafnia alvei* O antigens. FEMS Immunol Med Microbiol 27:219–225

118. Katzenellenbogen E, Kocharova NA, Korzeniowska-Kowal A, Gamian A, Bogulska M, Szostko B, Shashkov AS, Knirel YA (2008) Immunochemical studies of the lipopolysaccharides of *Hafnia alvei* PCM 1219 and other strains with the O-antigens containing D-glucose 1-phosphate and 2-deoxy-2-[(*R*)-3-hydroxybutyramido]-D-glucose. Arch Immunol Ther Exp 56:347–352
119. Dag S, Niedziela T, Dzieciatkowska M, Lukasiewicz J, Jachymek W, Lugowski C, Kenne L (2004) The O-acetylation patterns in the O-antigens of *Hafnia alvei* strains PCM 1200 and 1203, serologically closely related to PCM 1205. Carbohydr Res 339:2521–2527
120. Jachymek W, Petersson C, Helander A, Kenne L, Niedziela T, Lugowski C (1996) Structural studies of the O-specific chain of *Hafnia alvei* strain 32 lipopolysaccharide. Carbohydr Res 292:117–128
121. Katzenellenbogen E, Kübler J, Gamian A, Romanowska E, Shashkov AS, Kocharova NA, Knirel YA, Kochetkov NK (1996) Structure and serological characterization of the O-specific polysaccharide of *Hafnia alvei* PCM 1185, another *Hafnia* O-antigen that contains 3-[(*R*)-3-hydroxybutyramido]-3,6-dideoxy-D-glucose. Carbohydr Res 293:61–70
122. Katzenellenbogen E, Kocharova NA, Zatonsky GV, Shashkov AS, Korzeniowska-Kowal A, Gamian A, Bogulska M, Knirel YA (2005) Structure of the O-polysaccharide of *Hafnia alvei* strain PCM 1189 that has hexa- to octa-saccharide repeating units owing to incomplete glucosylation. Carbohydr Res 340:263–270
123. Gamian A, Romanowska E, Dabrowski U, Dabrowski J (1993) Structure of the O-specific polysaccharide containing pentitol phosphate, isolated from *Hafnia alvei* strain PCM 1191 lipopolysaccharide. Eur J Biochem 213:1255–1260
124. Jachymek W, Petersson C, Helander A, Kenne L, Lugowski C, Niedziela T (1995) Structural studies of the O-specific chain and a core hexasaccharide of *Hafnia alvei* strain 1192 lipopolysaccharide. Carbohydr Res 269:125–138
125. Niedziela T, Kenne L, Lugowski C (2010) Novel O-antigen of *Hafnia alvei* PCM 1195 lipopolysaccharide with a teichoic acid-like structure. Carbohydr Res 345:270–274
126. Katzenellenbogen E, Zatonsky GV, Kocharova NA, Mieszala M, Gamian A, Shashkov AS, Romanowska E, Knirel YA (2001) Structure of the O-specific polysaccharide of *Hafnia alvei* PCM 1196. Carbohydr Res 330:523–528
127. Katzenellenbogen E, Romanowska E, Kocharova NA, Shashkov AS, Knirel YA, Kochetkov NK (1995) Structure of the O-specific polysaccharide of *Hafnia alvei* 1204 containing 3,6-dideoxy-3-formamido-D-glucose. Carbohydr Res 273:187–195
128. Jachymek W, Czaja J, Niedziela T, Lugowski C, Kenne L (1999) Structural studies of the O-specific polysaccharide of *Hafnia alvei* strain PCM 1207 lipopolysaccharide. Eur J Biochem 266:53–61
129. Katzenellenbogen E, Romanowska E, Dabrowski U, Dabrowski J (1991) O-Specific polysaccharide of *Hafnia alvei* lipopolysaccharide isolated from strain 1211. Structural study using chemical methods, gas-liquid chromatography/mass spectrometry and NMR spectroscopy. Eur J Biochem 200:401–407
130. Toukach FV, Shashkov AS, Katzenellenbogen E, Kocharova NA, Czarny A, Knirel YA, Romanowska E, Kochetkov NK (1996) Structure of the O-specific polysaccharide of *Hafnia alvei* strain 1222 containing 2-aminoethyl phosphate. Carbohydr Res 295:117–126
131. Katzenellenbogen E, Kocharova NA, Zatonsky GV, Kübler-Kielb J, Gamian A, Shashkov AS, Knirel YA, Romanowska E (2001) Structural and serological studies on *Hafnia alvei* O-specific polysaccharide of α-D-mannan type isolated from the lipopolysaccharide of strain PCM 1223. FEMS Immunol Med Microbiol 30:223–227
132. Katzenellenbogen E, Kocharova NA, Bogulska M, Shashkov AS, Knirel YA (2004) Structure of the O-polysaccharide from the lipopolysaccharide of *Hafnia alvei* strain PCM 1529. Carbohydr Res 339:723–727
133. Katzenellenbogen E, Kocharova NA, Zatonsky GV, Korzeniowska-Kowal A, Shashkov AS, Knirel YA (2003) Structure of the O-specific polysaccharide from the lipopolysaccharide of *Hafnia alvei* strain PCM 1546. Carbohydr Res 338:2153–2158

134. Karlsson C, Jansson PE, Wollin R (1997) Structure of the O-polysaccharide from the LPS of a *Hafnia alvei* strain isolated from a patient with suspect yersinosis. Carbohydr Res 300:191–197
135. Kubler-Kielb J, Vinogradov E, Garcia Fernandez JM, Szostko B, Zwiefka A, Gamian A (2006) Structure and serological analysis of the *Hafnia alvei* 481-L O-specific polysaccharide containing phosphate in the backbone chain. Carbohydr Res 341:2980–2985
136. MacLean LL, Vinogradov E, Pagotto F, Farber JM, Perry MB (2009) Characterization of the O-antigen in the lipopolysaccharide of *Cronobacter* (*Enterobacter*) *malonaticus* 3267. Biochem Cell Biol 87:927–932
137. MacLean LL, Pagotto F, Farber JM, Perry MB (2009) The structure of the O-antigen in the endotoxin of the emerging food pathogen *Cronobacter* (*Enterobacter*) *muytjensii* strain 3270. Carbohydr Res 344:667–671
138. Arbatsky NP, Wang M, Shashkov AS, Feng L, Knirel YA, Wang L (2010) Structure of the O-antigen of *Cronobacter sakazakii* serotype O1 containing 3-(*N*-acetyl-L-alanyl)amino-3,6-dideoxy-D-glucose. Carbohydr Res 345:2095–2098
139. MacLean LL, Pagotto F, Farber JM, Perry MB (2009) Structure of the antigenic repeating pentasaccharide unit of the LPS O-polysaccharide of *Cronobacter sakazakii* implicated in the Tennessee outbreak. Biochem Cell Biol 87:459–465
140. Arbatsky NP, Wang M, Shashkov AS, Chizhov AO, Feng L, Knirel YA, Wang L (2010) Structure of the O-antigen of *Cronobacter sakazakii* serotype O2 with a randomly O-acetylated L-rhamnose residue. Carbohydr Res 345:2090–2094
141. MacLean LL, Vinogradov E, Pagotto F, Farber JM, Perry MB (2010) The structure of the O-antigen of *Cronobacter sakazakii* HPB 2855 isolate involved in a neonatal infection. Carbohydr Res 345:1932–1937
142. Czerwicka M, Forsythe SJ, Bychowska A, Dziadziuszko H, Kunikowska D, Stepnowski P, Kaczycski Z (2010) Structure of the O-polysaccharide isolated from *Cronobacter sakazakii* 767. Carbohydr Res 345:908–913
143. Szafranek J, Czerwicka M, Kumirska J, Paszkiewicz M, Lojkowska E (2005) Repeating unit structure of *Enterobacter sakazakii* ZORB A 741 O-polysaccharide. Pol J Chem 79:287–295
144. Moule AL, Kuhl PMD, Galbraith L, Wilkinson SG (1989) Structure of the O-specific polysaccharide from *Enterobacter cloacae* strain N.C.T.C. 11579 (serogroup O10). Carbohydr Res 186:287–293
145. Cimmino A, Marchi G, Surico G, Hanuszkiewicz A, Evidente A, Holst O (2008) The structure of the O-specific polysaccharide of the lipopolysaccharide from *Pantoea agglomerans* strain FL1. Carbohydr Res 343:392–396
146. Staaf M, Urbina F, Weintraub A, Widmalm G (1999) Structure elucidation of the O-antigenic polysaccharide from the enteroaggregative *Escherichia coli* strain $62D_1$. Eur J Biochem 262:56–62
147. Karamanos Y, Kol O, Wieruszeski J-M, Strecker G, Fournet B, Zalisz R (1992) Structure of the O-specific polysaccharide chain of the lipopolysaccharide of *Enterobacter agglomerans*. Carbohydr Res 231:197–204
148. Knirel YA, Perepelov AV, Kondakova AN, Senchenkova SN, Sidorczyk Z, Rozalski A, Kaca W (2011) Structure and serology of O-antigens as the basis for classification of *Proteus* strains. Innate Immun 17:70–96
149. Kocharova NA, Torzewska A, Zatonsky GV, Blaszczyk A, Bystrova OV, Shashkov AS, Knirel YA, Rozalski A (2004) Structure of the O-polysaccharide of *Providencia stuartii* O4 containing 4-(*N*-acetyl-L-aspart-4-yl)amino-4,6-dideoxy-D-glucose. Carbohydr Res 339:195–200
150. Zatonsky GV, Bystrova OV, Kocharova NA, Shashkov AS, Knirel YA, Kholodkova EV, Stanislavsky ES (1999) Structure of a neutral O-specific polysaccharide of the bacterium *Providencia alcalifaciens* O5. Biochemistry (Moscow) 64:523–527
151. Ovchinnikova OG, Kocharova NA, Wykrota M, Shashkov AS, Knirel YA, Rozalski A (2007) Structure of a colitose-containing O-polysaccharide from the lipopolysaccharide of *Providencia alcalifaciens* O6. Carbohydr Res 342:2144–2148

152. Bystrova OV, Zatonsky GV, Borisova SA, Kocharova NA, Shashkov AS, Knirel YA, Kholodkova EV, Stanislavsky ES (2000) Structure of an acidic O-specific polysaccharide of the bacterium *Providencia alcalifaciens* O7. Biochemistry (Moscow) 65:677–684
153. Toukach FV, Kocharova NA, Maszewska A, Shashkov AS, Knirel YA, Rozalski A (2008) Structure of the O-polysaccharide of *Providencia alcalifaciens* O8 containing (2 *S*,4*R*)-2,4-dihydroxypentanoic acid, a new non-sugar component of bacterial glycans. Carbohydr Res 343:2706–2711
154. Kocharova NA, Ovchinnikova OG, Maszewska A, Shashkov AS, Arbatsky NP, Knirel YA, Rozalski A (2011) Elucidation of the full O-polysaccharide structure and identification of the oligosaccharide core type of the lipopolysaccharide of *Providencia alcalifaciens* O9. Carbohydr Res 346:644–650
155. Parkhomchuk AA, Kocharova NA, Bialczak-Kokot M, Shashkov AS, Chizhov AO, Knirel YA, Rozalski A (2010) Structure of the O-polysaccharide from the lipopolysaccharide of *Providencia alcalifaciens* O12. Carbohydr Res 345:1235–1239
156. Kocharova NA, Zatonsky GV, Torzewska A, Macieja Z, Bystrova OV, Shashkov AS, Knirel YA, Rozalski A (2003) Structure of the O-specific polysaccharide of *Providencia rustigianii* O14 containing N^{ε}-[(*S*)-1-carboxyethyl]-N^{α}-(D-galacturonoyl)-L-lysine. Carbohydr Res 338:1009–1016
157. Kondakova AN, Vinogradov EV, Lindner B, Kocharova NA, Rozalski A, Knirel YA (2007) Mass-spectrometric studies of *Providencia* SR-form lipopolysaccharides and elucidation of the biological repeating unit structure of *Providencia rustigianii* O14-polysaccharide. J Carbohydr Chem 26:497–512
158. Kocharova NA, Zatonsky GV, Bystrova OV, Ziolkowski A, Wykrota M, Shashkov AS, Knirel YA, Rozalski A (2002) Structure of the O-specific polysaccharide of *Providencia alcalifaciens* O16 containing *N*-acetylmuramic acid. Carbohydr Res 337:1667–1671
159. Kocharova NA, Blaszczyk A, Zatonsky GV, Torzewska A, Bystrova OV, Shashkov AS, Knirel YA, Rozalski A (2004) Structure and cross-reactivity of the O-antigen of *Providencia stuartii* O18 containing 3-acetamido-3,6-dideoxy-D-glucose. Carbohydr Res 339:409–413
160. Kocharova NA, Maszewska A, Zatonsky GV, Torzewska A, Bystrova OV, Shashkov AS, Knirel YA, Rozalski A (2004) Structure of the O-polysaccharide of *Providencia alcalifaciens* O19. Carbohydr Res 339:415–419
161. Kocharova NA, Vinogradov E, Kondakova AN, Shashkov AS, Rozalski A, Knirel YA (2008) The full structure of the carbohydrate chain of the lipopolysaccharide of *Providencia alcalifaciens* O19. J Carbohydr Chem 27:320–331
162. Shashkov AS, Kocharova NA, Zatonsky GV, Blaszczyk A, Knirel YA, Rozalski A (2007) Structure of the O-antigen of *Providencia stuartii* O20, a new polysaccharide containing 5,7-diacetamido-3,5,7,9-tetradeoxy-L-*glycero*-D-*galacto*-non-2-ulosonic acid. Carbohydr Res 342:653–658
163. Kocharova NA, Maszewska A, Zatonsky GV, Bystrova OV, Ziolkowski A, Torzewska A, Shashkov AS, Knirel YA, Rozalski A (2003) Structure of the O-polysaccharide of *Providencia alcalifaciens* O21 containing 3-formamido-3,6-dideoxy-D-galactose. Carbohydr Res 338:1425–1430
164. Ovchinnikova OG, Parkhomchuk NA, Kocharova NA, Kondakova AN, Shashkov AS, Knirel YA, Rozalski A (2009) Further progress in structural studies of lipopolysaccharides of bacteria of the genus *Providencia*. Paper presented at the 15th European carbohydrate symposium, Vienna, Austria, 19–24 July 2009
165. Kocharova NA, Vinogradov EV, Borisova SA, Shashkov AS, Knirel YA (1998) Identification of N^{ε}-[(*R*)-1-carboxyethyl]-L-lysine in, and the complete structure of, the repeating unit of the O-specific polysaccharide of *Providencia alcalifaciens* O23. Carbohydr Res 309:131–133
166. Kocharova NA, Ovchinnikova OG, Shashkov AS, Maszewska A, Knirel YA, Rozalski A (2011) Structure of the O-polysaccharide of *Providencia alcalifaciens* O25 containing an amide of D-galacturonic acid with N^{ε}-[(*S*)-1-carboxyethyl]-L-lysine. Biochemistry (Moscow) 76:707–712

167. Ovchinnikova OG, Bushmarinov IS, Kocharova NA, Toukach FV, Wykrota M, Shashkov AS, Knirel YA, Rozalski A (2007) New structure for the O-polysaccharide of *Providencia alcalifaciens* O27 and revised structure for the O-polysaccharide of *Providencia stuartii* O43. Carbohydr Res 342:1116–1121
168. Ovchinnikova OG, Kocharova NA, Kondakova AN, Shashkov AS, Knirel YA, Rozalski A (2003) New structures of L-fucose-containing O-polysaccharides of bacteria of the genus *Providencia* and the full lipopolysaccharide structure of *Providencia rustigianii* O34. In: Abstracts of the 24th international carbohydrate symposium, Oslo, Norway, 27 July-1 August 2008
169. Bushmarinov IS, Ovchinnikova OG, Kocharova NA, Toukach FV, Torzewska A, Shashkov AS, Knirel YA, Rozalski A (2006) Structure of the O-polysaccharide from the lipopolysaccharide of *Providencia alcalifaciens* O29. Carbohydr Res 341:1181–1185
170. Kocharova NA, Ovchinnikova OG, Torzewska A, Shashkov AS, Knirel YA, Rozalski A (2006) The structure of the O-polysaccharide from the lipopolysaccharide of *Providencia alcalifaciens* O30. Carbohydr Res 341:786–790
171. Ovchinnikova OG, Kocharova NA, Shashkov AS, Bialczak-Kokot M, Knirel YA, Rozalski A (2009) Structure of the O-polysaccharide from the lipopolysaccharide of *Providencia alcalifaciens* O31 containing an ether of D-mannose with (*2R,4R*)-2,4-dihydroxypentanoic acid. Carbohydr Res 344:683–686
172. Bushmarinov IS, Ovchinnikova OG, Kocharova NA, Toukach FV, Torzewska A, Shashkov AS, Knirel YA, Rozalski A (2007) Structure of the O-polysaccharide and serological cross-reactivity of the lipopolysaccharide of *Providencia alcalifaciens* O32 containing *N*-acetylisomuramic acid. Carbohydr Res 342:268–273
173. Torzewska A, Kocharova NA, Zatonsky GV, Blaszczyk A, Bystrova OV, Shashkov AS, Knirel YA, Rozalski A (2004) Structure of the O-polysaccharide and serological cross-reactivity of the *Providencia stuartii* O33 lipopolysaccharide containing 4-(*N*-acetyl-D-aspart-4-yl)amino-4,6-dideoxy-D-glucose. FEMS Immunol Med Microbiol 41:133–139
174. Kocharova NA, Kondakova AN, Vinogradov E, Ovchinnikova OG, Lindner B, Shashkov AS, Rozalski A, Knirel YA (2008) Full structure of the carbohydrate chain of the lipopolysaccharide of *Providencia rustigianii* O34. Chem Eur J 14:6184–6191
175. Kocharova NA, Ovchinnikova OG, Torzewska A, Shashkov AS, Knirel YA, Rozalski A (2007) The structure of the O-polysaccharide from the lipopolysaccharide of *Providencia alcalifaciens* O36 containing 3-deoxy-D-*manno*-oct-2-ulosonic acid. Carbohydr Res 342:665–670
176. Kocharova NA, Bushmarinov IS, Ovchinnikova OG, Toukach FV, Torzewska A, Shashkov AS, Knirel YA, Rozalski A (2005) The structure of the O-polysaccharide from the lipopolysaccharide of *Providencia stuartii* O44 containing L-quinovose, a 6-deoxy sugar rarely occurring in bacterial polysaccharides. Carbohydr Res 340:1419–1423
177. Ovchinnikova OG, Kocharova NA, Shashkov AS, Knirel YA, Rozalski A (2009) Antigenic polysaccharides of bacteria. 43. Structure of the O-specific polysaccharide of the bacterium *Providencia alcalifaciens* O46. Bioorg Khim 35:370–375
178. Ovchinnikova OG, Kocharova NA, Bakinovskiy LV, Torzewska A, Shashkov AS, Knirel YA, Rozalski A (2004) The structure of the O-polysaccharide from the lipopolysaccharide of *Providencia stuartii* O47. Carbohydr Res 339:2621–2626
179. Fedonenko YP, Egorenkova IV, Konnova SA, Ignatov VV (2001) Involvement of the lipopolysaccharides of *Azospirilla* in the interaction with wheat seedling roots. Microbiology 70:329–334
180. Bushmarinov IS, Ovchinnikova OG, Kocharova NA, Blaszczyk A, Toukach FV, Torzewska A, Shashkov AS, Knirel YA, Rozalski A (2004) Structure of the O-polysaccharide of *Providencia stuartii* O49. Carbohydr Res 339:1557–1560
181. Kocharova NA, Ovchinnikova OG, Bushmarinov IS, Toukach FV, Torzewska A, Shashkov AS, Knirel YA, Rozalski A (2005) The structure of the O-polysaccharide from the lipopolysaccharide of *Providencia stuartii* O57 containing an amide of D-galacturonic acid with L-alanine. Carbohydr Res 340:775–780

182. Ovchinnikova OG, Kocharova NA, Parkhomchuk AA, Bialczak-Kokot M, Shashkov AS, Knirel YA, Rozalski A (2011) Structure of the O-polysaccharide from the lipopolysaccharide of *Providencia alcalifaciens* O60. Carbohydr Res 346:377–380
183. Kilcoyne M, Shashkov AS, Senchenkova SN, Knirel YA, Vinogradov EV, Radziejewska-Lebrecht J, Galimska-Stypa R, Savage AV (2002) Structural investigation of the O-specific polysaccharides of *Morganella morganii* consisting of two higher sugars. Carbohydr Res 337:1697–1702
184. Shashkov AS, Torgov VI, Nazarenko EL, Zubkov VA, Gorshkova NM, Gorshkova RP, Widmalm G (2002) Structure of the phenol-soluble polysaccharide from *Shewanella putrefaciens* strain A6. Carbohydr Res 337:1119–1127
185. Bruneteau M, Minka S (2003) Lipopolysaccharides of bacterial pathogens from the genus *Yersinia*: a mini-review. Biochimie 85:145–152
186. Ovodov YS, Gorshkova RP (1988) Lipopolysaccharides of *Yersinia pseudotuberculosis*. Khim Prirod Soed 163–171
187. Ovodov YS, Gorshkova RP, Tomshich SV, Komandrova NA, Zubkov VA, Kalmykova EN, Isakov VV (1992) Chemical and immunochemical studies on lipopolysaccharides of some *Yersinia* species. A review of some recent investigations. J Carbohydr Chem 11:21–35
188. Holst O (2003) Lipopolysaccharides of *Yersinia*. An overview. Adv Exp Med Biol 529:219–228
189. Ho N, Kondakova AN, Knirel YA, Creuzenet C (2008) The biosynthesis and biological role of 6-deoxyheptoses in the lipopolysaccharide O-antigen of *Yersinia pseudotuberculosis*. Mol Microbiol 68:424–447
190. Kondakova AN, Shashkov AS, Komandrova NA, Anisimov AP, Skurnik M, Knirel YA Unpublished data
191. Kondakova AN, Shaikhutdinova RZ, Ivanov SA, Dentovskaya SV, Shashkov AS, Anisimov AP, Knirel YA (2009) Revision of the O-polysaccharide structure of *Yersinia pseudotuberculosis* O:1b. Carbohydr Res 344:2421–2423
192. De Castro C, Kenyon J, Cunneen MM, Reeves PR, Molinaro A, Holst O, Skurnik M (2011) Genetic characterization and structural analysis of the O-specific polysaccharide of *Yersinia pseudotuberculosis* serotype O:1c. Innate Immun 17:183–190
193. Kondakova AN, Ho N, Bystrova OV, Shashkov AS, Lindner B, Creuzenet C, Knirel YA (2008) Structural studies of the O-antigens of *Yersinia pseudotuberculosis* O:2a and mutants thereof with impaired 6-deoxy-D-*manno*-heptose biosynthesis pathway. Carbohydr Res 343:1383–1389
194. Kondakova AN, Bystrova OV, Shaikhutdinova RZ, Ivanov SA, Dentovskaya SV, Shashkov AS, Knirel YA, Anisimov AP (2009) Structure of the O-polysaccharide of *Yersinia pseudotuberculosis* O:2b. Carbohydr Res 344:405–407
195. Kondakova AN, Bystrova OV, Shaikhutdinova RZ, Ivanov SA, Dentovskaya SV, Shashkov AS, Knirel YA, Anisimov AP (2008) Reinvestigation of the O-antigens of *Yersinia pseudotuberculosis*: revision of the O2c and confirmation of the O3 antigen structures. Carbohydr Res 343:2486–2488
196. Kondakova AN, Bystrova OV, Shaikhutdinova RZ, Ivanov SA, Dentovskaya SV, Shashkov AS, Knirel YA, Anisimov AP (2009) Structure of the O-antigen of *Yersinia pseudotuberculosis* O:4a revised. Carbohydr Res 344:531–534
197. Kondakova AN, Bystrova OV, Shaikhutdinova RZ, Ivanov SA, Dentovskaya SV, Shashkov AS, Knirel YA, Anisimov AP (2009) Structure of the O-antigen of *Yersinia pseudotuberculosis* O:4b. Carbohydr Res 344:152–154
198. Zubkov VA, Gorshkova RP, Ovodov YS, Sviridov AF, Shashkov AS (1992) Synthesis of 3,6-dideoxy-4-*C*-(4^1-hydroxyethyl)hexopyranoses (yersinioses) from 1,6-anhydro-β-D-glycopyranose. Carbohydr Res 225:189–207
199. Beczala A, Ovchinnikova OG, Duda KA, Skurnik M, Radziejewska-Lebrecht J, Holst O (2009) Structure of *Yersinia pseudotuberculosis* O:9 O-specific polysaccharide repeating unit resolved. In: Abstracts of the 15th European carbohydrate symposium, Vienna, 19–24 July 2009

200. Kenyon JJ, De Castro C, Cunneen MM, Reeves PR, Molinaro A, Holst O, Skurnik M (2011) The genetics and structure of the O-specific polysaccharide of *Yersinia pseudotuberculosis* serotype O:10 and its relationship to *Escherichia coli* O111 and *Salmonella enterica* O35. Glycobiology, doi:10.1093/glycob/cwr010
201. Cunneen MM, De Castro C, Kenyon J, Parrilli M, Reeves PR, Molinaro A, Holst O, Skurnik M (2009) The O-specific polysaccharide structure and biosynthetic gene cluster of *Yersinia pseudotuberculosis* serotype O:11. Carbohydr Res 344:1533–1540
202. De Castro C, Skurnik M, Molinaro A, Holst O (2009) Characterization of the O-polysaccharide structure and biosynthetic gene cluster of *Yersinia pseudotuberculosis* serotype O:15. Innate Immun 15:351–359
203. Meikle PJ, Perry MB, Cherwonogrodzky JW, Bundle DR (1989) Fine structure of A and M antigens from *Brucella* biovars. Infect Immun 57:2820–2828
204. Gorshkova RP, Kalmykova EN, Isakov VV, Ovodov YS (1985) Structural studies on O-specific polysaccharides of lipopolysaccharides from *Yersinia enterocolitica* serovars O:1,2a,3, O:2a,2b,3 and O:3. Eur J Biochem 150:527–531
205. Gorshkova RP, Isakov VV, Kalmykova EN, Ovodov YS (1995) Structural stidies of O-specific polysaccharide chains of the lipopolysaccharide from *Yersinia enterocolitica* serovar O:10. Carbohydr Res 268:249–255
206. Marsden BJ, Bundle DR, Perry MB (1994) Serological and structural relationships between *Escherichia coli* O:98 and *Yersinia enterocolitica* O:11,23 and O:11,24 lipopolysaccharide O-antigens. Biochem Cell Biol 72:163–168
207. L'vov VL, Gur'yanova SV, Rodionov AV, Gorshkova RP (1992) Structure of the repeating unit of the O-specific polysaccharide of the lipopolysaccharide of *Yersinia kristensenii* strain 490 (O:12,25). Carbohydr Res 228:415–422
208. L'vov VL, Guryanova SV, Rodionov AV, Dmitriev BA, Shashkov AS, Ignatenko AV, Gorshkova RP, Ovodov YS (1990) The structure of the repeating unit of the glycerol phosphate-containing O-specific polysaccharide chain from the lipopolysaccharide of *Yersinia kristensenii* strain 103 (O:12,26). Bioorg Khim 16:379–389
209. Gorshkova RP, Isakov VV, Zubkov VA, Ovodov YS (1989) Structure of the O-specific polysaccharide of the lipopolysaccharide of *Yersinia frederiksenii* serovar O:16,29. Bioorg Khim 15:1627–1633
210. Gorshkova RP, Isakov VV, Nazarenko EL, Ovodov YS, Guryanova SV, Dmitriev BA (1993) Structure of the O-specific polysaccharide of the lipopolysaccharide from *Yersinia kristensenii* O:25,35. Carbohydr Res 241:201–208
211. Perry MB, MacLean LL (2000) Structural identification of the lipopolysaccharide O-antigen produced by *Yersinia enterocolitica* serotype O:28. Eur J Biochem 267:2567–2572
212. Zubkov VA, Gorshkova RP, Nazarenko EL, Shashkov AS, Ovodov YS (1991) Structure of the O-specific polysaccharide chain of lipopolysaccharide of *Yersinia aldovae*. Bioorg Khim 17:831–838
213. Gorshkova RP, Isakov VV, Zubkov VA, Ovodov YS (1994) Structure of O-specific polysaccharide of lipopolysaccharide from *Yersinia bercovieri* O:10. Bioorg Khim 20:1231–1235
214. Gorshkova RP, Isakov VV, Nazarenko EL, Shevchenko LS (1997) Structural study of the repeating unit of the O-specific polysaccharide from *Yersinia mollarettii* strain WS 42/90. Bioorg Khim 23:823–825
215. Zubkov VA, Nazarenko EL, Gorshkova RP, Ovodov YS (1993) Structure of O-specific polysaccharide of *Yersinia rohdei*. Bioorg Khim 19:729–732
216. Beynon LM, Richards JC, Perry MB (1994) The structure of the lipopolysaccharide O-antigen from *Yersinia ruckerii* serotype O1. Carbohydr Res 256:303–317
217. Bateman KP, Banoub JH, Thibault P (1996) Probing the microheterogeneity of O-specific chains from *Yersinia ruckeri* using capillary zone electrophoresis/electrospray mass spectrometry. Electrophoresis 17:1818–1828
218. Gorshkova RP, Kalmykova EN, Isakov VV, Ovodov YS (1986) Structural studies on O-specific polysaccharides of lipopolysaccharides from *Yersinia enterocolitica* servovars O:5 and O:5,27. Eur J Biochem 156:391–397

219. Pieretti G, Corsaro MM, Lanzetta R, Parrilli M, Canals R, Merino S, Tomas JM (2008) Structural studies of the O-chain polysaccharide from *Plesiomonas shigelloides* strain 302–73 (serotype O1). Eur J Org Chem 3149–3155
220. Pieretti G, Carillo S, Lindner B, Lanzetta R, Parrilli M, Jimenez N, Regue M, Tomas JM, Corsaro MM (2010) The complete structure of the core of the LPS from *Plesiomonas shigelloides* 302–73 and the identification of its O-antigen biological repeating unit. Carbohydr Res 345:2523–2528
221. Taylor DN, Trofa AC, Sadoff J, Chu C, Bryla D, Shiloach J, Cohen D, Ashkenazi S, Lerman Y, Egan W, Schneerson R, Robbins JB (1993) Synthesis, characterization, and clinical evaluation of conjugate vaccines composed of the O-specific polysaccharides of *Shigella dysenteriae* type 1, *Shigella flexneri* type 2a, and *Shigella sonnei* (*Pseudomonas shigelloides*) bound to bacterial toxoids. Infect Immun 61:3678–3687
222. Maciejewska A, Lukasiewicz J, Niedziela T, Szewczuk Z, Lugowski C (2009) Structural analysis of the O-specific polysaccharide isolated from *Plesiomonas shigelloides* O51 lipopolysaccharide. Carbohydr Res 344:894–900
223. Czaja J, Jachymek W, Niedziela T, Lugowski C, Aldova E, Kenne L (2000) Structural studies of the O-specific polysaccharides from *Plesiomonas shigelloides* strain CNCTC 113/92. Eur J Biochem 267:1672–1679
224. Niedziela T, Lukasiewicz J, Jachymek W, Dzieciatkowska M, Lugowski C, Kenne L (2002) Core oligosaccharides of *Plesiomonas shigelloides* O54:H2 (strain CNCTC 113/92). Structural and serological analysis of the lipopolysaccharide core region, the O-antigen biological repeating unit and the linkage between them. J Biol Chem 277:11653–11663
225. Niedziela T, Dag S, Lukasiewicz J, Dzieciatkowska M, Jachymek W, Lugowski C, Kenne L (2006) Complete lipopolysaccharide of *Plesiomonas shigelloides* O74:H5 (strain CNCTC 144/92). 1. Structural analysis of the highly hydrophobic lipopolysaccharide, including the O-antigen, its biological repeating unit, the core oligosaccharide, and the linkage between them. Biochemistry 45:10422–10433
226. Linnerborg M, Widmalm G, Weintraub A, Albert MJ (1995) Structural elucidation of the O-antigen lipopolysaccharide from two strains of *Plesiomonas shigelloides* that share a type-specific antigen with *Shigella flexneri* 6, and the common group 1 antigen with *Shigella flexneri* spp. and *Shigella dysenteriae*. Eur J Biochem 231:839–844
227. Jachymek W, Niedziela T, Petersson C, Lugowski C, Czaja J, Kenne L (1999) Structures of the O-specific polysaccharides from *Yokenella regensburgei* (*Koserella trabulsii*) strains PCM 2476, 2477, 2478, and 2494: high-resolution magic-angle spinning NMR investigation of the O-specific polysaccharides in native lipopolysaccharides and directly on the surface of living bacteria. Biochemistry 38:11788–11795
228. Zdorovenko EL, Varbanets LD, Brovarskaya OS, Valueva OA, Shashkov AS, Knirel YA (2011) Lipopolysaccharide of *Budvicia aquatica* 97U124: immunochemical properties and structure. Microbiology 80:372–377
229. Valueva OA, Zdorovenko EL, Kachala VV, Varbanets LD, Shubchinskiy VV, Arbatsky NP, Shashkov AS, Knirel YA (2011) Structure of the O-polysaccharide of *Pragia fontium* 27480 containing 2,3-diacetamido-2,3-dideoxy-D-mannuronic acid. Carbohydr Res 346:146–149
230. Zdorovenko EL, Valueva OA, Varbanets L, Shubchinskiy V, Shashkov AS, Knirel YA (2010) Structure of the O-polysaccharide of the lipopolysaccharide of *Pragia fontium* 97U116. Carbohydr Res 345:1812–1815
231. Zdorovenko EL, Varbanets LD, Zatonsky GV, Zdorovenko GM, Shashkov AS, Knirel YA (2009) Isolation and structure elucidation of two different polysaccharides from the lipopolysaccharide of *Rahnella aquatilis* 33071^{T}. Carbohydr Res 344:1259–1262
232. Zdorovenko EL, Varbanets LD, Zatonsky GV, Kachala VV, Zdorovenko GM, Shashkov AS, Knirel YA (2008) Structure of the O-specific polysaccharide of the lipopolysaccharide of *Rahnella aquatilis* 95 U003. Carbohydr Res 343:2494–2497
233. Zdorovenko EL, Varbanets LD, Zatonsky GV, Ostapchuk AN (2004) Structure of the O-polysaccharide of the lipopolysaccharide of *Rahnella aquatilis* 1-95. Carbohydr Res 339:1809–1812

234. Zdorovenko EL, Varbanets LD, Zatonsky GV, Ostapchuk AN (2006) Structures of two putative O-specific polysaccharides from the *Rahnella aquatilis* 3-95 lipopolysaccharide. Carbohydr Res 341:164–168
235. Ray TC, Smith ARW, Wait R, Hignett RC (1987) Structure of the sidechain of lipopolysaccharide from *Erwinia amylovara* T. Eur J Biochem 170:357–361
236. Senchenkova SN, Knirel YA, Shashkov AS, Ahmed M, Mavridis A, Rudolph K (2003) Structure of the O-polysaccharide of *Erwinia carotovora* ssp. *carotovora* GSPB 436. Carbohydr Res 338:2025–2027
237. Senchenkova SN, Shashkov AS, Knirel YA, Ahmed M, Mavridis A, Rudolph K (2005) Structure of the O-polysaccharide of *Erwinia carotovora* ssp. atroseptica GSPB 9205, containing a new higher branched monosaccharide. Rus Chem Bull Int Ed 54:1276–1281
238. Wang Z, Liu X, Garduno E, Garduno RA, Li J, Altman E (2009) Application of an immunoaffinity-based preconcentration method for mass spectrometric analysis of the O-chain polysaccharide of *Aeromonas salmonicida* from *in vitro*- and *in vivo*-grown cells. FEMS Microbiol Lett 295:148–155
239. Wang Z, Larocque S, Vinogradov E, Brisson JR, Dacanay A, Greenwell M, Brown LL, Li J, Altman E (2004) Structural studies of the capsular polysaccharide and lipopolysaccharide O-antigen of *Aeromonas salmonicida* strain 80204–1 produced under *in vitro* and *in vivo* growth conditions. Eur J Biochem 271:4507–4516
240. Corsaro MM, De Castro C, Molinaro A, Parrilli M (2001) Structure of lipopolysaccharides from phytopathogenic Gram-negative bacteria. Recent Res Dev Phytochem 5:119–138
241. Zdorovenko GM, Zdorovenko EL (2010) *Pseudomonas syringae* lipopolysaccharides: Immunochemical characteristics and structure as a basis for strain classification. Microbiology 79:47–57
242. Turska-Szewczuk A, Kozinska A, Russa R, Holst O (2010) The structure of the O-specific polysaccharide from the lipopolysaccharide of *Aeromonas bestiarum* strain 207. Carbohydr Res 345:680–684
243. Linnerborg M, Widmalm G, Rahman MM, Jansson P-E, Holme T, Qadri F, Albert MJ (1996) Structural studies of the O-antigenic polysaccharide from an *Aeromonas caviae* strain. Carbohydr Res 291:165–174
244. Wang Z, Liu X, Li J, Altman E (2008) Structural characterization of the O-chain polysaccharide of *Aeromonas caviae* ATCC 15468 lipopolysaccharide. Carbohydr Res 343:483–488
245. Shaw DH, Squires MJ (1984) O-Antigen structure in a virulent strain of *Aeromonas hydrophila*. FEMS Microbiol Lett 24:277–280
246. Knirel YA, Shashkov AS, Senchenkova SN, Merino S, Tomas JM (2002) Structure of the O-polysaccharide of *Aeromonas hydrophila* O:34; a case of random O-acetylation of 6-deoxy-L-talose. Carbohydr Res 337:1381–1386
247. Wang Z, Vinogradov E, Larocque S, Harrison BA, Li J, Altman E (2005) Structural and serological characterization of the O-chain polysaccharide of *Aeromonas salmonicida* strains A449, 80204 and 80204–1. Carbohydr Res 340:693–700
248. Wang Z, Liu X, Dacanay A, Harrison BA, Fast M, Colquhoun DJ, Lund V, Brown LL, Li J, Altman E (2007) Carbohydrate analysis and serological classification of typical and atypical isolates of *Aeromonas salmonicida*: a rationale for the lipopolysaccharide-based classification of *A. salmonicida*. Fish Shellfish Immunol 23:1095–1106
249. Shaw DH, Lee Y-Z, Squires MJ, Lüderitz O (1983) Structural studies on the O-antigen of *Aeromonas salmonicida*. Eur J Biochem 131:633–638
250. Knirel YA, Senchenkova SN, Jansson P-E, Weintraub A, Ansaruzzaman M, Albert MJ (1996) Structure of the O-specific polysaccharide of an *Aeromonas trota* strain cross-reactive with *Vibrio cholerae* O139 Bengal. Eur J Biochem 238:160–165
251. Nazarenko EL, Komandrova NA, Gorshkova RP, Tomshich SV, Zubkov VA, Kilcoyne M, Savage AV (2003) Structures of polysaccharides and oligosaccharides of some Gram-negative marine *Proteobacteria*. Carbohydr Res 338:2449–2457

252. Leone S, Silipo A, Nazarenko EL, Lanzetta R, Parrilli E, Molinaro A (2007) Molecular structure of endotoxins from Gram-negative marine bacteria: an update. Mar Drugs 5:85–112
253. Shashkov AS, Senchenkova SN, Chizhov AO, Knirel YA, Esteve C, Alcaide E, Merino S, Tomas JM (2009) Structure of a polysaccharide from the lipopolysaccharides of *Vibrio vulnificus* strains CECT 5198 and S3-I2-36, which is remarkably similar to the O-polysaccharide of *Pseudoalteromonas rubra* ATCC 29570. Carbohydr Res 344:2005–2009
254. Komandrova NA, Isakov VV, Tomshich SV, Romanenko LA, Perepelov AV, Shashkov AS (2010) Structure of an acidic O-specific polysaccharide of the marine bacterium *Pseudoalteromonas agarivorans* KMM 232 (R-form). Biochemistry (Moscow) 75:623–628
255. Perepelov AV, Shashkov AS, Torgov VI, Nazarenko EL, Gorshkova RP, Ivanova EP, Gorshkova NM, Widmalm G (2005) Structure of an acidic polysaccharide from the agar-decomposing marine bacterium *Pseudoalteromonas atlantica* strain IAM 14165 containing 5,7-diacetamido-3,5,7,9-tetradeoxy-L-*glycero*-L-*manno*-non-2-ulosonic acid. Carbohydr Res 340:69–74
256. Komandrova NA, Tomshich SV, Isakov VV, Romanenko LA (2001) O-specific polysaccharide of the marine bacterium "*Alteromonas marinoglutinosa*" NCIMB 1770. Biochemistry (Moscow) 66:894–897
257. Nazarenko EL, Perepelov AV, Shevchenko LS, Daeva ED, Ivanova EP, Shashkov AS, Widmalm G (2011) Structure of the O-specific polysaccharide from *Shewanella japonica* KMM 3601 containing 5,7-diacetamido-3,5,7,9-tetradeoxy-D-*glycero*-D-*talo*-non-2-ulosonic acid. Biochemistry (Moscow) 76:791–796
258. Kilcoyne M, Perepelov AV, Tomshich SV, Komandrova NA, Shashkov AS, Romanenko LA, Knirel YA, Savage AV (2004) Structure of the O-polysaccharide of *Idiomarina zobellii* KMM 231^{T} containing two unusual amino sugars with the free amino group, 4-amino-4,6-dideoxy-D-glucose and 2-amino-2-deoxy-L-guluronic acid. Carbohydr Res 339:477–482
259. Perry MB, Maclean LM, Brisson JR, Wilson ME (1996) Structures of the antigenic O-polysaccharides of lipopolysaccharides produced by *Actinobacillus actinomycetemcomitans* serotypes a, c, d and e. Eur J Biochem 242:682–688
260. Perry MB, MacLean LL, Gmür R, Wilson ME (1996) Characterization of the O-polysaccharide structure of lipopolysaccharide from *Actinobacillus actinomycetemcomitans* serotype b. Infect Immun 64:1215–1219
261. Kaplan JB, Perry MB, MacLean LL, Furgang D, Wilson ME, Fine DH (2001) Structural and genetic analyses of O polysaccharide from *Actinobacillus actinomycetemcomitans* serotype f. Infect Immun 69:5375–5384
262. Perry MB, Altman E, Brisson J-R (1990) Structural caharacteristics of the antigenic capsular polysaccharides and lipopolysaccharides involved in the serological classification of *Actinobacillus (Haemophilus) pleuropneumoniae* strains. Serodiagn Immunother Infect Dis 4:299–308
263. Beynon LM, Griffith DW, Richards JC, Perry MB (1992) Characterization of the lipopolysaccharide O antigens of *Actinobacillus pleuropneumoniae* serotype 9 and 11: antigenic relationships among serotypes 9, 11, and 1. J Bacteriol 174:5324–5331
264. Perry MB, MacLean LL, Vinogradov E (2005) Structural characterization of the antigenic capsular polysaccharide and lipopolysaccharide O-chain produced by *Actinobacillus pleuropneumoniae* serotype 15. Biochem Cell Biol 83:61–69
265. MacLean LL, Perry MB, Vinogradov E (2004) Characterization of the antigenic lipopolysaccharide O chain and the capsular polysaccharide produced by *Actinobacillus pleuropneumoniae* serotype 13. Infect Immun 72:5925–5930
266. Beynon LM, Perry MB, Richards JC (1991) Structure of the O-antigen of *Actinobacillus pleuropneumoniae* serotype 12 lipopolysaccharide. Can J Chem 69:218–224
267. Perry MB, MacLean LL (2004) Structural characterization of the antigenic O-polysaccharide in the lipopolysaccharide produced by *Actinobacillus pleuropneumoniae* serotype 14. Carbohydr Res 339:1399–1402

268. Monteiro MA, Slavic D, St Michael F, Brisson J-R, MacIinnes JI, Perry MB (2000) The first description of a (1 → 6)-β-D-glucan in prokaryotes: (1 → 6)-β-D-glucan is a common component of *Actinobacillus suis* and is the basis for a serotyping system. Carbohydr Res 329:121–130
269. Rullo A, Papp-Szabo E, Michael FS, Macinnes J, Monteiro MA (2006) The structural basis for the serospecificity of *Actinobacillus suis* serogroup O:2. Biochem Cell Biol 84:184–190
270. Severn WB, Richards JC (1993) Charactrerization of the O-polysaccharide of *Pasteurella haemolytica* serotype A1. Carbohydr Res 240:277–285
271. Leitch RA, Richards JC (1988) Structure of the O-chain of the lipopolysaccharide of *Pasteurella haemolytica* serotype T3. Biochem Cell Biol 66:1055–1065
272. Richards JC, Leitch RA (1989) Elucidation of the structure of the *Pasteurella haemolytica* serotype T10 lipopolysaccharide O-antigen by n.m.r. spectroscopy. Carbohydr Res 186:275–286
273. Hood DW, Randle G, Cox AD, Makepeace K, Li J, Schweda EK, Richards JC, Moxon ER (2004) Biosynthesis of cryptic lipopolysaccharide glycoforms in *Haemophilus influenzae* involves a mechanism similar to that required for O-antigen synthesis. J Bacteriol 186:7429–7439
274. Knirel YA, Vinogradov EV, Kocharova NA, Paramonov NA, Kochetkov NK, Dmitriev BA, Stanislavsky ES, Lányi B (1988) The structure of O-specific polysaccharides and serological classification of *Pseudomonas aeruginosa*. Acta Microbiol Hung 35:3–24
275. Knirel YA (1990) Polysaccharide antigens of *Pseudomonas aeruginosa*. CRC Crit Rev Microbiol 17:273–304
276. Knirel YA, Bystrova OV, Kocharova NA, Zähringer U, Pier GB (2006) Conserved and variable structural features in the lipopolysaccharide of *Pseudomonas aeruginosa*. J Endotoxin Res 12:324–336, Corrigendum in: Innate Immun. **16**: 274 (2010)
277. Knirel YA, Zdorovenko GM (1997) Structures of O-polysaccharide chains of lipopolysaccharides as the basis for classification of *Pseudomonas syringae* and related strains. In: Rudolph K, Burr TJ, Mansfield JW, Stead D, Vivian A, von Kietzell J (eds) *Pseudomonas syringae* pathovars and related pathogens. Kluwer Academic Publishers, Dordrecht, pp 475–480
278. Molinaro A, Newman M-A, Lanzetta R, Parrilli M (2009) The structures of lipopolysaccharides from plant-associated Gram-negative bacteria. Eur J Org Chem 5887–5896
279. Kooistra O, Lüneberg E, Lindner B, Knirel YA, Frosch M, Zähringer U (2001) Complex O-acetylation in *Legionella pneumophila* serogroup 1 lipopolysaccharide. Evidence for two genes involved in 8-O-acetylation of legionaminic acid. Biochemistry 40:7630–7640
280. Corsaro MM, Evidente A, Lanzetta R, Lavermicocca P, Parrilli M, Ummarino S (2002) 5,7-Diamino-5,7,9-trideoxynon-2-ulosonic acid: a novel sugar from a phytopathogenic *Pseudomonas* lipopolysaccharide. Carbohydr Res 337:955–959
281. Knirel YA, Zdorovenko GM, Paramonov NA, Veremeychenko SN, Toukach FV, Shashkov AS (1996) Somatic antigens of pseudomonads: structure of the O-specific polysaccharide of the reference strain for *Pseudomonas fluorescens* (IMV 4125, ATCC 13525, biovar A). Carbohydr Res 291:217–224
282. Knirel YA, Veremeychenko SN, Zdorovenko GM, Shashkov AS, Paramonov NA, Zakharova IY, Kochetkov NK (1994) Somatic antigens of pseudomonads: structure of the O-specific polysaccharide of *Pseudomonas fluorescens* biovar A strain IMV 472. Carbohydr Res 259:147–151
283. Knirel YA, Paramonov NA, Shashkov AS, Kochetkov NK, Zdorovenko GM, Veremeychenko SN, Zakharova IY (1993) Somatic antigens of pseudomonads: structure of the O-specific polysaccharide of *Pseudomonas fluorescens* biovar A strain IMV 1152. Carbohydr Res 243:205–210
284. Shashkov AS, Paramonov NA, Veremeychenko SN, Grosskurth H, Zdorovenko GM, Knirel YA, Kochetkov NK (1998) Somatic antigens of pseudomonads: structure of the O-specific

polysaccharide of *Pseudomonas fluorescens* biovar B, strain IMV 247. Carbohydr Res 306:297–303

285. Zatonsky GV, Kocharova NA, Veremeychenko SN, Zdorovenko EL, Shapovalova VY, Shashkov AS, Zdorovenko GM, Knirel YA (2002) Somatic antigens of pseudomonads: structure of the O-specific polysaccharide of *Pseudomonas fluorescens* IMV 2366 from (biovar C). Carbohydr Res 337:2365–2370
286. Khomenko VA, Naberezhnykh GA, Isakov VV, Solov'eva TF, Ovodov YS, Knirel YA, Vinogradov EV (1986) Structural study of O-specific polysaccharide chain of *Pseudomonas fluorescens* lipopolysaccharide. Bioorg Khim 12:1641–1648
287. Naberezhnykh GA, Khomenko VA, Isakov VV, El'kin YN, Solov'eva TF, Ovodov YS (1987) 3-(3-Hydroxy-2,3-dimethyl-5-oxoprolyl)amino-3,6-dideoxy-D-glucose: a novel amino sugar from the antigenic polysaccharide from *Pseudomonas fluorescens*. Bioorg Khim 13:1428–1429
288. Knirel YA, Zdorovenko GM, Veremeychenko SN, Shashkov AS, Zakharova IY, Kochetkov NK (1989) Antigenic polysaccharides of bacteria. 36. Structural study of O-specific polysaccharide chain of lipopolysaccharide of *Pseudomonas fluorescens* IMV 2763 (biovar G). Bioorg Khim 15:1538–1545
289. Knirel YA, Grosskurth H, Helbig JH, Zähringer U (1995) Structures of decasaccharide and tridecasaccharide tetraphosphates isolated by strong alkaline degradation of O-deacylated lipopolysaccharide of *Pseudomonas fluorescens* strain ATCC 49271. Carbohydr Res 279:215–226
290. Knirel YA, Zdorovenko GM, Veremeychenko SN, Lipkind GM, Shashkov AS, Zakharova IY, Kochetkov NK (1988) Antigenic polysaccharides of bacteria. 31. Structure of the O-specific polysaccharide chain of the *Pseudomonas aurantiaca* IMV 31 lipopolysaccharide. Bioorg Khim 14:352–358
291. Jiménez-Barbero J, De Castro C, Molinaro A, Nunziata R, Lanzetta R, Parrilli M, Holst O (2002) Structural determination of the O-specific chain of the lipopolysaccharide from *Pseudomonas cichorii*. Eur J Org Chem 1770–1775
292. Knirel YA, Shashkov AS, Senchenkova SN, Ajiki Y, Fukuoka S (2002) Structure of the O-polysaccharide of *Pseudomonas putida* FERM P-18867. Carbohydr Res 337:1589–1591
293. Molinaro A, Evidente A, Iacobellis NS, Lanzetta R, Cantore PL, Mancino A, Parrilli M (2002) O-specific chain structure from the lipopolysaccharide fraction of *Pseudomonas reactans*: a pathogen of the cultivated mushrooms. Carbohydr Res 337:467–471
294. Molinaro A, Bedini E, Ferrara R, Lanzetta R, Parrilli M, Evidente A, Lo CP, Iacobellis NS (2003) Structural determination of the O-specific chain of the lipopolysaccharide from the mushrooms pathogenic bacterium *Pseudomonas tolaasii*. Carbohydr Res 338:1251–1257
295. Leone S, Izzo V, Lanzetta R, Molinaro A, Parrilli M, Di Donato A (2005) The structure of the O-polysaccharide from *Pseudomonas stutzeri* OX1 containing two different 4-acylamido-4,6-dideoxy-residues, tomosamine and perosamine. Carbohydr Res 340:651–656
296. Leone S, Lanzetta R, Scognamiglio R, Alfieri F, Izzo V, Di Donato A, Parrilli M, Holst O, Molinaro A (2008) The structure of the O-specific polysaccharide from the lipopolysaccharide of *Pseudomonas* sp. OX1 cultivated in the presence of the azo dye Orange II. Carbohydr Res 343:674–684
297. Vinogradov EV, Pantophlet R, Haseley SR, Brade L, Holst O, Brade H (1997) Structural and serological characterisation of the O-specific polysaccharide from lipopolysaccharide of *Acinetobacter calcoaceticus* strain 7 (DNA group 1). Eur J Biochem 243:167–173
298. Galbraith L, Sharples JL, Wilkinson SG (1999) Structure of the O-specific polysaccharide for *Acinetobacter baumannii* serogroup O1. Carbohydr Res 319:204–208
299. Haseley SR, Wilkinson SG (1995) Structural studies of the putative O-specific polysaccharide of *Acinetobacter baumannii* O2 containing 3,6-dideoxy-3-*N*-(D-3-hydroxybutyryl) amino-D-galactose. Eur J Biochem 233:899–906
300. Haseley SR, Wilkinson SG (1996) Structure of the O-specific polysaccharide of *Acinetobacter baumannii* O5 containing 2-acetamido-2-deoxy-D-galacturonic acid. Eur J Biochem 237:229–233

301. Vinogradov EV, Pantophlet R, Dijkshoorn L, Brade L, Holst O, Brade H (1996) Structural and serological characterisation of two O-specific polysaccharides of *Acinetobacter*. Eur J Biochem 239:602–610
302. Haseley SR, Wilkinson SG (1998) Structure of the O-7 antigen from *Acinetobacter baumannii*. Carbohydr Res 306:257–263
303. Haseley SR, Wilkinson SG (1994) Structure of the putative O10 antigen from *Acinetobacter baumannii*. Carbohydr Res 264:73–81
304. Haseley SR, Wilkinson SG (1996) Structural studies of the putative O-specific polysaccharide of *Acinetobacter baumannii* O11. Eur J Biochem 237:266–271
305. Haseley SR, Diggle HJ, Wilkinson SG (1996) Structure of a surface polysaccharide from *Acinetobacter baumannii* O16. Carbohydr Res 293:259–265
306. Haseley SR, Traub WH, Wilkinson SG (1997) Structures of polymeric products isolated from the lipopolysaccharides of reference strains for *Acinetobacter baumannii* O23 and O12. Eur J Biochem 244:147–154
307. Haseley S, Wilkinson SG (1997) Structure of the O18 antigen from *Acinetobacter baumannii*. Carbohydr Res 301:187–192
308. Haseley SR, Galbraith L, Wilkinson SG (1994) Structure of a surface polysaccharide from *Acinetobacter baumannii* strain 214. Carbohydr Res 258:199–206
309. Haseley SR, Wilkinson SG (1997) Structural studies of the putative O-specific polysaccharide of *Acinetobacter baumannii* O24 containing 5,7-diamino-3,5,7,9-tetradeoxy-L-*glycero*-D-*galacto*-nonulosonic acid. Eur J Biochem 250:617–623
310. MacLean LL, Perry MB, Chen W, Vinogradov E (2009) The structure of the polysaccharide O-chain of the LPS from *Acinetobacter baumannii* strain ATCC 17961. Carbohydr Res 344:474–478
311. Haseley SR, Holst O, Brade H (1998) Structural studies of the O-antigen isolated from the phenol- soluble lipopolysaccharide of *Acinetobacter baumannii* (DNA group 2) strain 9. Eur J Biochem 251:189–194
312. Vinogradov EV, Brade L, Brade H, Holst O (2003) Structural and serological characterisation of the O-antigenic polysaccharide of the lipopolysaccharide from *Acinetobacter baumannii* strain 24. Carbohydr Res 338:2751–2756
313. Vinogradov EV, Brade L, Brade H, Holst O (2005) The structure of the O-specific polysaccharide of the lipopolysaccharide from *Acinetobacter* strain 44 (DNA group 3). Pol J Chem 79:267–273
314. Haseley SR, Holst O, Brade H (1997) Structural and serological characterisation of the O-antigenic polysaccharide of the lipopolysaccharide from *Acinetobacter haemolyticus* strain ATCC 17906. Eur J Biochem 244:761–766
315. Pantophlet R, Haseley SR, Vinogradov EV, Brade L, Holst O, Brade H (1999) Chemical and antigenic structure of the O-polysaccharide of the lipopolysaccharides from two *Acinetobacter haemolyticus* strains differing only in the anomeric configuration of one glycosyl residue in their O-antigens. Eur J Biochem 263:587–595
316. Haseley SR, Pantophlet R, Brade L, Holst O, Brade H (1997) Structural and serological characterisation of the O-antigenic polysaccharide of the lipopolysaccharide from *Acinetobacter junii* strain 65. Eur J Biochem 245:477–481
317. Arbatsky NP, Kondakova AN, Shashkov AS, Drutskaya MS, Belousov PV, Nedospasov SA, Petrova MA, Knirel YA (2010) Structure of the O-antigen of *Acinetobacter lwoffii* EK30A; identification of D-homoserine, a novel non-sugar component of bacterial polysaccharides. Org Biomol Chem 8:3571–3577
318. Arbatsky NP, Kondakova AN, Shashkov AS, Drutskaya MS, Belousov PV, Nedospasov SA, Petrova MA, Knirel YA (2010) Structure of the O-polysaccharide of *Acinetobacter* sp. VS-15 and *Acinetobacter lwoffii* EK67. Carbohydr Res 345:2287–2290
319. Haseley SR, Holst O, Brade H (1997) Structural and serological characterisation of the O-antigenic polysaccharide of the lipopolysaccharide from *Acinetobacter* strain 90 belonging to DNA group 10. Eur J Biochem 245:470–476

320. Haseley SR, Holst O, Brade H (1997) Structural studies of the O-antigenic polysaccharide of the lipopolysaccharide from *Acinetobacter* (DNA group 11) strain 94 containing 3-amino-3,6-dideoxy-D-galactose substituted by the previously unknown amide-linked L-2-acetoxypropionic acid or L-2-hydroxypropionic acid. Eur J Biochem 247:815–819
321. Vinogradov EV, Pantophlet R, Brade H, Holst O (2001) Structural and serological characterisation of the O-antigenic polysaccharide of the lipopolysaccharide from *Acinetobacter* strain 96 (DNA group 11). J Endotoxin Res 7:113–118
322. Chatterjee SN, Chaudhuri K (2003) Lipopolysaccharides of *Vibrio cholerae*. I. Physical and chemical characterization. Biochim Biophys Acta 1639:65–79
323. Kenne L, Lindberg B, Schweda E, Gustafsson B, Holme T (1988) Structural studies of the O-antigen from *Vibrio cholerae* O:2. Carbohydr Res 180:285–294
324. Chowdhury TA, Jansson P-E, Lindberg B, Gustavsson B, Holme T (1991) Structural studies of the *Vibrio cholerae* O:3 O-antigen polysaccharide. Carbohydr Res 215:303–314
325. Hermansson K, Jansson P-E, Holme T, Gustavsson B (1993) Structural studies of the *Vibrio cholerae* O:5 O-antigen polysaccharide. Carbohydr Res 248:199–211
326. Bergstrцm N, Nair GB, Weintraub A, Jansson P-E (2002) Structure of the O-polysaccharide from the lipopolysaccharide from *Vibrio cholerae* O6. Carbohydr Res 337:813–817
327. Kocharova NA, Perepelov AV, Zatonsky GV, Shashkov AS, Knirel YA, Jansson P-E, Weintraub A (2001) Structural studies of the O-specific polysaccharide of *Vibrio cholerae* O8 using solvolysis with triflic acid. Carbohydr Res 330:83–92
328. Kocharova NA, Knirel YA, Jansson P-E, Weintraub A (2001) Structure of the O-specific polysaccharide of *Vibrio cholerae* O9 containing 2-acetamido-2-deoxy-D-galacturonic acid. Carbohydr Res 332:279–284
329. Ansari AA, Kenne L, Lindberg B, Gustafsson B, Holme T (1986) Structural studies of the O-antigen from *Vibrio cholerae* O:21. Carbohydr Res 150:213–219
330. Perepelov AV, Kocharova NA, Knirel YA, Jansson P-E, Weintraub A (2011) Structure of the O-polysaccharide of *Vibrio cholerae* O43 containing a new monosaccharide derivative, 4-(*N*-acetyl-L-allothreonyl)amino-4,6-dideoxy-D-glucose. Carbohydr Res 346:70–96
331. Kondo S, Sano Y, Isshiki Y, Hisatsune K (1996) The O polysaccharide chain of the lipopolysaccharide from *Vibrio cholerae* O76 is a homopolymer of *N*-[(*S*)-(+)-2-hydroxypropionyl]-α-L-perosamine. Microbiology 142:2879–2885
332. Haishima Y, Kondo S, Hisatsune K (1990) The occurrence of α(1 → 2) linked *N*-acetylperosamine-homopolymer in lipopolysaccharides of non-O1 *Vibrio cholerae* possessing an antigenic factor in common with O1 *V. cholerae*. Microbiol Immunol 34: 1049–1054
333. Sano Y, Kondo S, Isshiki Y, Shimada T, Hisatsune K (1996) An *N*-[(*R*)-(−)-2-hydroxypropionyl]-α-L-perosamine homopolymer constitutes the O polysaccharide chain of the lipopolysaccharide from *Vibrio cholerae* O144 which has antigenic factor(s) in common with *V. cholerae* O76. Microbiol Immunol 40:735–741
334. Senchenkova SN, Zatonsky GV, Shashkov AS, Knirel YA, Jansson P-E, Weintraub A, Albert MJ (1998) Structure of the O-antigen of *Vibrio cholerae* O155 that shares a putative D-galactose-4,6-cyclophosphate-associated epitope with *V. cholerae* O139 Bengal. Eur J Biochem 254:58–62
335. Vinogradov EV, Holst O, Thomas-Oates JE, Broady KW, Brade H (1992) The structure of the O-antigenic polysaccharide from lipopolysaccharide of *Vibrio cholerae* strain H11 (non-O1). Eur J Biochem 210:491–498
336. Isshiki Y, Kondo S, Haishima Y, Iguchi T, Hisatsune K (1996) Identification of *N*-3-hydroxypropionyl-2-*O*-methyl-D-perosamine as a specific constituent of the lipopolysaccharide from *Vibrio* bio-serogroup 1875 which has Ogawa antigen factor B of *Vibrio cholerae* O1. J Endotoxin Res 3:143–149
337. Nazarenko EL, Shashkov AS, Knirel YA, Ivanova EP, Ovodov YS (1990) Uncommon acidic monosaccharides as components of O-specific polysaccharides of *Vibrio*. Bioorg Khim 16:1426–1429

338. Nazarenko EL, Zubkov VA, Ivanova EP, Gorshkova RP (1992) Structure of the O-specific polysaccharide of *Vibrio fluvialis* serovar 3. Bioorg Khim 18:418–421
339. Nazarenko EL, Gorshkova RP, Ovodov YS, Shashkov AS, Knirel YA (1989) Structure of the repeating unit of O-specific polysaccharide chain of *Vibrio fluvialis* lipopolysaccharide. Bioorg Khim 15:1100–1106
340. Nazarenko EL, Zubkov VA, Shashkov AS, Knirel YA, Komandrova NA, Gorshkova RP, Ovodov YS (1993) Structure of the repeating unit of the O-specific polysaccharide from *Vibrio fluvialis*. Bioorg Khim 19:989–1000
341. Kenne L, Lindberg B, Rahman MM, Mosihuzzaman M (1993) Structural studies of *Vibrio fluvialis* M-940 O-antigen polysaccharide. Carbohydr Res 242:181–189
342. Kondo S, Haishima Y, Ishida K, Isshiki Y, Hisatsune K (2000) The O-polysaccharide of lipopolysaccharide isolated from *Vibrio fluvialis* O19 is identical to that of *Vibrio* bioserogroup 1875 variant. Microbiol Immunol 44:941–944
343. Kondo S, Ishida K, Isshiki Y, Haishima Y, Iguchi T, Hisatsune K (1993) *N*-3-Hydroxypropionyl-α-D-perosamine homopolymer constituting the O-chain of lipopolysaccharides from *Vibrio* bioserogroup 1875 possessing antigenic factor(s) in common with O1 *Vibrio cholerae*. Biochem J 292:531–535
344. Kenne L, Lindberg B, Rahman MM, Mosihuzzaman M (1990) Structural studies of the O-antigen polysaccharide of *Vibrio fluvialis* AA-18239. Carbohydr Res 205:440–443
345. Landersjц C, Weintraub A, Ansaruzzaman M, Albert MJ, Widmalm G (1998) Structural analysis of the O-antigenic polysaccharide from *Vibrio mimicus* N-1990. Eur J Biochem 251:986–990
346. Kenne L, Lindberg B, Rahman MM, Mosihuzzman M (1993) Structural studies of the *Vibrio mimicus* W-26768 O-antigen polysaccharide. Carbohydr Res 243:131–138
347. Sadovskaya I, Brisson JR, Khieu NH, Mutharia LM, Altman E (1998) Structural characterization of the lipopolysaccharide O-antigen and capsular polysaccharide of *Vibrio ordalii* serotype O:2. Eur J Biochem 253:319–327
348. Kilcoyne M, Shashkov AS, Knirel YA, Gorshkova RP, Nazarenko EL, Ivanova EP, Gorshkova NM, Senchenkova SN, Savage AV (2005) The structure of the O-polysaccharide of the *Pseudoalteromonas rubra* ATCC 29570^{T} lipopolysaccharide containing a keto sugar. Carbohydr Res 340:2369–2375
349. Knirel YA, Senchenkova SN, Shashkov AS, Esteve C, Alcaide E, Merino S, Tomas JM (2009) Structure of a polysaccharide from the lipopolysaccharide of *Vibrio vulnificus* CECT4602 containing 2-acetamido-2,3,6-trideoxy-3-[(*S*)- and (*R*)-3-hydroxybutanoylamino]-L-mannose. Carbohydr Res 344:479–483
350. Senchenkova SN, Shashkov AS, Knirel YA, Esteve C, Alcaide E, Merino S, Tomas JM (2009) Structure of a polysaccharide from the lipopolysaccharide of *Vibrio vulnificus* clinical isolate YJ016 containing 2-acetimidoylamino-2-deoxy-L-galacturonic acid. Carbohydr Res 344:1009–1013
351. Sadovskaya I, Brisson J-R, Altman E, Mutharia LM (1996) Structural studies of the lipopolysaccharide O-antigen and capsular polysaccharide of *Vibrio anguillarum* serotype O:2. Carbohydr Res 283:111–127
352. Wang Z, Vinogradov E, Li J, Lund V, Altman E (2009) Structural characterization of the lipopolysaccharide O-antigen from atypical isolate of *Vibrio anguillarum* strain 1282. Carbohydr Res 344:1371–1375
353. Eguchi H, Kaya S, Araki Y, Kojima N, Yokota S (1992) Structure of the O-polysaccharide chain of the lipopolysaccharide of *Vibrio anguillarum* V-123. Carbohydr Res 231:159–169
354. Banoub JH, Michon F, Hodder HJ (1987) Structural elucidation of the O-specific polysaccharide of the phenol-phase soluble lipopolysaccharide of *Vibrio anguillarum*. Biochem Cell Biol 65:19–26
355. Molinaro A, Silipo A, Lanzetta R, Newman M-A, Dow JM, Parrilli M (2003) Structural elucidation of the O-chain of the lipopolysaccharide from *Xanthomonas campestris* strain 8004. Carbohydr Res 338:277–281
356. Senchenkova SN, Huang X, Laux P, Knirel YA, Shashkov AS, Rudolph K (2002) Structures of the O-polysaccharide chains of the lipopolysaccharides of *Xanthomonas campestris* pv.

phaseoli var. *fuscans* GSPB 271 and *Xanthomonas campestris* pv. *malvacearum* GSPB 1386 and GSPB 2388. Carbohydr Res 337:1723–1728
357. Molinaro A, Evidente A, Lo Cantore P, Iacobellis NS, Bedini E, Lanzetta R, Parrilli M (2003) Structural determination of a novel O-chain polysaccharide of the lipopolysaccharide from the bacterium *Xanthomonas campestris* pv. *pruni*. Eur J Org Chem 2254–2259
358. Wilkinson SG, Galbraith L, Anderton WJ (1983) Lipopolysaccharides from *Pseudomonas maltophilia*: Composition of the lipopolysaccharide and structure of the side-chain polysaccharide from strain N.C.I.B. 9204. Carbohydr Res 112:241–252
359. Winn AM, Wilkinson SG (1997) Structure of the O2 antigen of *Stenotrophomonas* (*Xanthomonas* or *Pseudomonas*) *maltophilia*. Carbohydr Res 298:213–217
360. Winn AM, Miles CT, Wilkinson SG (1996) Structure of the O3 antigen of *Stenotrophomonas* (*Xanthomonas* or *Pseudomonas*) *maltophilia*. Carbohydr Res 282:149–156
361. Winn AM, Wilkinson SG (2001) Structures of the O4 and O18 antigens of *Stenotrophomonas maltophilia*: a case of enantiomeric repeating units. Carbohydr Res 330:215–221
362. Winn AM, Wilkinson SG (1995) Structure of the O6 antigen of *Stenotrophomonas* (*Xanthomonas* or *Pseudomonas*) *maltophilia*. Carbohydr Res 272:225–230
363. Winn AM, Wilkinson SG (1998) The O7 antigen of *Stenotrophomonas maltophilia* is a linear D-rhamnan with a trisaccharide repeating unit that is also present in polymers from some *Pseudomonas* and *Burkholderia* species. FEMS Microbiol Lett 166:57–61
364. Neal DJ, Wilkinson SG (1982) Lipopolysaccharides from *Pseudomonas maltophilia*: Structural studies of the side-chain, core, and lipid A regions of the lipopolysaccharide from strain NCTC 10257. Eur J Biochem 128:143–149
365. Winn AM, Miller AM, Wilkinson SG (1995) Structure of the O10 antigen of *Stenotrophomonas* (*Xanthomonas*) *maltophilia*. Carbohydr Res 267:127–133
366. Di Fabio JL, Perry MB, Bundle DR (1987) Analysis of the lipopolysaccharide of *Pseudomonas maltophilia* 555. Biochem Cell Biol 65:968–977
367. Winn AM, Wilkinson SG (2001) Structure of the O16 antigen of *Stenotrophomonas maltophilia*. Carbohydr Res 330:279–283
368. Winn AM, Galbraith L, Temple GS, Wilkinson SG (1993) Structure of the O19 antigen of *Xanthomonas maltophilia*. Carbohydr Res 247:249–254
369. Winn AM, Wilkinson SG (1996) Structure of the O20 antigen of *Stenotrophomonas* (*Xanthomonas* or *Pseudomonas*) *maltophilia*. Carbohydr Res 294:109–115
370. Galbraith L, Wilkinson SG (2000) Structures of the O21 and O25 antigens of *Stenotrophomonas maltophilia*. Carbohydr Res 323:98–102
371. Gunn JS, Ernst RK (2007) The structure and function of *Francisella* lipopolysaccharide. Ann NY Acad Sci 1105:202–218
372. Kay W, Petersen BO, Duus J, Perry MB, Vinogradov E (2006) Characterization of the lipopolysaccharide and β-glucan of the fish pathogen *Francisella victoria*. FEBS J 273:3002–3013
373. Knirel YA, Senchenkova SN, Kocharova NA, Shashkov AS, Helbig JH, Zähringer U (2001) Identification of a homopolymer of 5-acetamidino-7-acetamido-3,5,7,9-tetradeoxy-D-*glycero*-D-*talo*-nonulosonic acid in the lipopolysaccharides of *Legionella pneumophila* non-1 serogroups. Biochemistry (Moscow) 66:1035–1041
374. Pieretti G, Corsaro MM, Lanzetta R, Parrilli M, Nicolaus B, Gambacorta A, Lindner B, Holst O (2008) Structural characterization of the core region of the lipopolysaccharide from the haloalkaliphilic *Halomonas pantelleriensis*: identification of the biological O-antigen repeating unit. Eur J Org Chem 721–728
375. Pieretti G, Nicolaus B, Poli A, Corsaro MM, Lanzetta R, Parrilli M (2009) Structural determination of the O-chain polysaccharide from the haloalkaliphilic *Halomonas alkaliantarctica* bacterium strain CRSS. Carbohydr Res 344:2051–2055
376. De Castro C, Molinaro A, Nunziata R, Grant W, Wallace A, Parrilli M (2003) The O-specific chain structure of the major component from the lipopolysaccharide fraction of *Halomonas magadii* strain 21 MI (NCIMB 13595). Carbohydr Res 338:567–570

377. De Castro C, Molinaro A, Wallace A, Grant WD, Parrilli M (2003) Structural determination of the O-specific chain of the lipopolysaccharide fraction from the alkaliphilic bacterium *Halomonas magadii* strain 21 MI. Eur J Org Chem 1029–1034
378. Pieretti G, Carillo S, Kim KK, Lee KC, Lee J-S, Lanzetta R, Parrilli M, Corsaro MM (2011) O-chain structure from the lipopolysaccharide of the human pathogen *Halomonas stevensii* strain S18214. Carbohydr Res 346:362–365
379. Zubkov VA, Nazarenko EL, Ivanova EP, Gorshkova NM, Gorshkova RP (1999) Structure of the repeating unit of the O-specific polysaccharide of *Marinomonas communis* strain ATCC 27118^{T}. Bioorg Khim 25:290–292
380. Vinogradov EV, Campos-Portuguez S, Yokota A, Mayer H (1994) The structure of the O-specific polysaccharide from *Thiobacillus ferrooxidans* IFO 14262. Carbohydr Res 261:103–109
381. Ormeno-Orrillo E (2005) Lipopolysaccharides of rhizobiaceae: structure and biosynthesis. Rev Latinoam Microbiol 47:165–175
382. De Castro C, Molinaro A, Lanzetta R, Silipo A, Parrilli M (2008) Lipopolysaccharide structures from *Agrobacterium* and *Rhizobiaceae* species. Carbohydr Res 343:1924–1933
383. D'Haeze W, Leoff C, Freshour G, Noel KD, Carlson RW (2007) *Rhizobium etli* CE3 bacteroid lipopolysaccharides are structurally similar but not identical to those produced by cultured CE3 bacteria. J Biol Chem 282:17101–17113
384. Forsberg LS, Bhat UR, Carlson RW (2000) Structural characterization of the O-antigenic polysaccharide of the lipopolysaccharide from *Rhizobium etli* strain CE3. A unique O-acetylated glycan of discrete size, containing 3-*O*-methyl-6-deoxy-L-talose and 2,3,4-tri-*O*-methyl-L-fucose. J Biol Chem 275:18851–18863
385. Turska-Szewczuk A, Pietras H, Borucki W, Russa R (2008) Alteration of O-specific polysaccharide structure of symbiotically defective *Mesorhizobium loti* mutant 2213.1 derived from strain NZP2213. Acta Biochim Pol 55:191–199
386. Turska-Szewczuk A, Palusinska-Szysz M, Russa R (2008) Structural studies of the O-polysaccharide chain from the lipopolysaccharide of symbiotically enhanced mutant Mlo-13 of *Mesorhizobium loti* NZP2213. Carbohydr Res 343:477–482
387. Forsberg LS, Carlson RW (2008) Structural characterization of the primary O-antigenic polysaccharide of the *Rhizobium leguminosarum* 3841 lipopolysaccharide and identification of a new 3-acetimidoylamino-3-deoxyhexuronic acid glycosyl component: a unique O-methylated glycan of uniform size, containing 6-deoxy-3-*O*-methyl-D-talose, *N*-acetylquinovosamine, and rhizoaminuronic acid (3-acetimidoylamino-3-deoxy-D-gluco-hexuronic acid). J Biol Chem 283:16037–16050
388. Muszynski A, Laus M, Kijne JW, Carlson RW (2011) The structures of the lipopolysaccharides from *Rhizobium leguminosarum* RBL5523 and its UDP-glucose dehydrogenase mutant (*exo5*). Glycobiology 21:55–68
389. Russa R, Urbanik-Sypniewska T, Shashkov AS, Banaszek A, Zamojski A, Mayer H (1996) Partial structure of lipopolysaccharides isolated from *Rhizobium leguminosarum* bv. *trifolii* 24 and its GalA-negative exo^{-} mutant AR20. Syst Appl Microbiol 19:1–8
390. Banaszek A (1998) Synthesis of the unique trisaccharide repeating unit, isolated from lipopolysaccharides *Rhizobium leguminosarum bv trifolii* 24, and its analogs. Carbohydr Res 306:379–385
391. Zdorovenko EL, Valueva OA, Kachala VV, Shashkov AS, Kocharova NA, Knirel YA, Kutkowska J, Turska-Szewczuk A, Urbanik-Sypniewska T, Choma A, Russa R (2009) Structure of the O-polysaccharides of the lipopolysaccharides of *Mesorhizobium loti* HAMBI 1148 and *Mesorhizobium amorphae* ATCC 19655 containing two methylated monosaccharides. Carbohydr Res 344:2519–2527
392. Russa R, Urbanik-Sypniewska T, Shashkov AS, Kochanowski H, Mayer H (1995) The structure of the homopolymeric O-specific chain from the phenol soluble LPS of the *Rhizobium loti* type strain NZP2213. Carbohydr Polym 27:299–303

393. Fernandez de Cordoba FJ, Rodriguez-Carvajal MA, Tejero-Mateo P, Corzo J, Gil-Serrano AM (2008) Structure of the O-antigen of the main lipopolysaccharide isolated from *Sinorhizobium fredii* SMH12. Biomacromolecules 9:678–685
394. Reuhs BL, Relic B, Forsberg LS, Marie C, Ojanen-Reuhs T, Stephens SB, Wong CH, Jabbouri S, Broughton WJ (2005) Structural characterization of a flavonoid-inducible *Pseudomonas aeruginosa* A-band-like O antigen of *Rhizobium* sp. strain NGR234, required for the formation of nitrogen-fixing nodules. J Bacteriol 187:6479–6487
395. Valueva OA, Zdorovenko EL, Kachala VV, Shashkov AS, Knirel YA, Komaniecka I, Choma A (2010) Structural investigation of the O-polysaccharide of *Azorhizobium caulinodans* HAMBI 216 consisting of rhamnose, 2-*O*-methylrhamnose and 3-*C*-methylrhamnose. In: Abstracts of the 4th Baltic meeting on microbial carbohydrates, Hyytiälä, Finland. 19–22 September 2010
396. Velasco J, Moll H, Vinogradov EV, Moriyyn I, Zähringer U (1996) Determination of the O-specific polysaccharide structure in the lipopolysaccharide of *Ochrobactrum anthropi* LMG 3331. Carbohydr Res 287:123–126
397. Shashkov AS, Campos-Portuguez S, Kochanowski H, Yokota A, Mayer H (1995) The structure of the O-specific polysaccharide from *Thiobacillus* sp. IFO 14570, with three different diaminopyranoses forming the repeating unit. Carbohydr Res 269:157–166
398. Previato JO, Jones C, Stephan MP, Almeida LPA, Mendonça-Previato L (1997) Structure of the repeating oligosaccharide from the lipopolysaccharide of the nitrogen-fixing bacterium *Acetobacter diazotrophicus* strain PAL 5. Carbohydr Res 298:311–318
399. Choma A, Komaniecka I, Sowinski P (2009) Revised structure of the repeating unit of the O-specific polysaccharide from *Azospirillum lipoferum* strain SpBr17. Carbohydr Res 344:936–939
400. Boiko AS, Smol'kina ON, Fedonenko YP, Zdorovenko EL, Kachala VV, Konnova SA, Ignatov VV (2010) O-Polysaccharide structure in serogroup I azospirilla. Microbiology 79:197–205
401. Fedonenko YP, Katsy EI, Petrova LP, Boyko AS, Zdorovenko EL, Kachala VV, Shashkov AS, Knirel YA (2010) The structure of the O-specific polysaccharide from a mutant of nitrogen-fixing rhizobacterium *Azospirillum brasilense* Sp245 with an altered plasmid content. Bioorg Khim 36:219–223
402. Wilkinson SG (1981) Structural studies of an acetylated mannan from *Pseudomonas diminuta* N.C.T.C. 8545. Carbohydr Res 93:269–278
403. Knirel YA, Paramonov NA, Shashkov AS, Kochetkov NK, Yarullin RG, Farber SM, Efremenko VI (1992) Structure of the polysaccharide chains of *Pseudomonas pseudomallei* lipopolysaccharides. Carbohydr Res 233:185–193
404. Perry MB, MacLean LL, Schollaardt T, Bryan LE, Ho M (1995) Structural characterization of the lipopolysaccharide O antigens of *Burkholderia pseudomallei*. Infect Immun 63:3348–3352
405. Brett PJ, Burtnick MN, Woods DE (2003) The wbiA locus is required for the 2-O-acetylation of lipopolysaccharides expressed by *Burkholderia pseudomallei* and *Burkholderia thailandensis*. FEMS Microbiol Lett 218:323–328
406. Burtnick MN, Brett PJ, Woods DE (2002) Molecular and physical characterization of *Burkholderia mallei* O antigens. J Bacteriol 184:849–852
407. Soldatkina MA, Knirel YA, Tanatar NV, Zakharova IY (1989) Immunological and structural studies of *Pseudomonas cepacia* lipopolysaccharide. Mikrobiol Zh 51:32–38
408. Vinion-Dubiel AD, Goldberg JB (2003) Lipopolysaccharide of *Burkholderia cepacia* complex. J Endotoxin Res 9:201–213
409. Ortega X, Hunt TA, Loutet S, Vinion-Dubiel AD, Datta A, Choudhury B, Goldberg JB, Carlson R, Valvano MA (2005) Reconstitution of O-specific lipopolysaccharide expression in *Burkholderia cenocepacia* strain J2315, which is associated with transmissible infections in patients with cystic fibrosis. J Bacteriol 187:1324–1333

410. Cérantola S, Montrozier H (1997) Structural elucidation of two polysaccharides present in the lipopolysaccharide of a clinical isolate of *Burkholderia cepacia*. Eur J Biochem 246:360–366
411. Ierano T, Silipo A, Cescutti P, Leone MR, Rizzo R, Lanzetta R, Parrilli M, Molinaro A (2009) Structural study and conformational behavior of the two different lipopolysaccharide O-antigens produced by the cystic fibrosis pathogen *Burkholderia multivorans*. Chem Eur J 15:7156–7166
412. Gaur D, Galbraith L, Wilkinson SG (1998) Structural characterisation of a rhamnan and a fucorhamnan, both present in the lipopolysaccharide of *Burkholderia vietnamiensis* strain LMG 10926. Eur J Biochem 258:696–701
413. Carillo S, Silipo A, Perino V, Lanzetta R, Parrilli M, Molinaro A (2009) The structure of the O-specific polysaccharide from the lipopolysaccharide of *Burkholderia anthina*. Carbohydr Res 344:1697–1700
414. Karapetyan G, Kaczynski Z, Iacobellis NS, Evidente A, Holst O (2006) The structure of the O-specific polysaccharide of the lipopolysaccharide from *Burkholderia gladioli* pv. *agaricicola*. Carbohydr Res 341:930–934
415. Mattos KA, Todeschini AR, Heise N, Jones C, Previato JO, Mendonca-Previato L (2005) Nitrogen-fixing bacterium *Burkholderia brasiliensis* produces a novel yersiniose A-containing O-polysaccharide. Glycobiology 15:313–321
416. De Castro C, Molinaro A, Lanzetta R, Holst O, Parrilli M (2005) The linkage between O-specific caryan and core region in the lipopolysaccharide of *Burkholderia caryophylli* is furnished by a primer monosaccharide. Carbohydr Res 340:1802–1807
417. Kocharova NA, Knirel YA, Shashkov AS, Nifant'ev NE, Kochetkov NK, Varbanets LD, Moskalenko NV, Brovarskaya OS, Muras VA, Young JM (1993) Studies of O-specific polysaccharide chains of *Pseudomonas solanacearum* lipopolysaccharides consisting of structurally different repeating units. Carbohydr Res 250:275–287
418. Galbraith L, George R, Wyklicky J, Wilkinson SG (1996) Structure of the O-specific polysaccharide from *Burkholderia pickettii* strain NCTC 11149. Carbohydr Res 282:263–269
419. Vinogradov E, Nossova L, Swierzko A, Cedzynski M (2004) The structure of the O-specific polysaccharide from *Ralstonia pickettii*. Carbohydr Res 339:2045–2047
420. Larocque S, Brisson JR, Therisod H, Perry MB, Caroff M (2003) Structural characterization of the O-chain polysaccharide isolated from *Bordetella avium* ATCC 5086: variation on a theme. FEBS Lett 535:11–16
421. Preston A, Petersen BO, Duus JO, Kubler-Kielb J, Ben Menachem G, Li J, Vinogradov E (2006) Complete structure of *Bordetella bronchiseptica* and *Bordetella parapertussis* lipopolysaccharides. J Biol Chem 281:18135–18144
422. Vinogradov E, King JD, Pathak AK, Harvill ET, Preston A (2010) Antigenic variation among *Bordetella*: *Bordetella bronchiseptica* strain MO149 expresses a novel O chain that is poorly immunogenic. J Biol Chem 285:26869–26877
423. Vinogradov E (2002) Structure of the O-specific polysaccharide chain of the lipopolysaccharide of *Bordetella hinzii*. Carbohydr Res 337:961–963
424. Vinogradov E, Caroff M (2005) Structure of the *Bordetella trematum* LPS O-chain subunit. FEBS Lett 579:18–24
425. Allen A, Maskell D (1996) The identification, cloning and mutagenesis of a genetic locus required for lipopolysaccharide biosynthesis in *Bordetella pertussis*. Mol Microbiol 19:37–52
426. Vinogradov E, MacLean LL, Brooks BW, Lutze-Wallace C, Perry MB (2008) The structure of the polysaccharide of the lipopolysaccharide produced by *Taylorella equigenitalis* type strain (ATCC 35865). Carbohydr Res 343:3079–3084
427. Vinogradov E, MacLean LL, Brooks BW, Lutze-Wallace C, Perry MB (2008) Structure of the O-polysaccharide of the lipopolysaccharide produced by *Taylorella asinigenitalis* type strain (ATCC 700933). Biochem Cell Biol 86:278–284
428. Knirel YA, Zdorovenko GM, Shashkov AS, Zakharova IY, Kochetkov NK (1986) Antigenic polysaccharides of bacteria. 19. Structure of O-specific polysaccharide chain of *Alcaligenes faecalis* lipopolysaccharide. Bioorg Khim 12:1530–1539

429. Silipo A, Molinaro A, Jiang CL, Jiang Y, Xu P, Xu LH, Lanzetta R, Parrilli M (2007) The O-chain structure from the LPS of the bacterium *Naxibacter alkalitolerans* YIM 31775 T. Carbohydr Res 342:757–761
430. Masoud H, Neszmŭlyi A, Mayer H (1991) Chemical characterization of the O-specific chain of *Sphaerotilus natans* ATCC 13338 lipopolysaccharide. Arch Microbiol 156:176–180
431. Vinogradov EV, Brade H, Holst O (1994) The structure of the O-specific polysaccharide of the lipopolysaccharide from *Chromobacterium violaceum* NCTC 9694. Carbohydr Res 264:313–317
432. Karlyshev AV, Champion OL, Churcher C, Brisson JR, Jarrell HC, Gilbert M, Brochu D, St Michael F, Li J, Wakarchuk WW, Goodhead I, Sanders M, Stevens K, White B, Parkhill J, Wren BW, Szymanski CM (2005) Analysis of *Campylobacter jejuni* capsular loci reveals multiple mechanisms for the generation of structural diversity and the ability to form complex heptoses. Mol Microbiol 55:90–103
433. Kilcoyne M, Moran AP, Shashkov AS, Senchenkova SN, Ferris JA, Corcoran AT, Savage AV (2006) Molecular origin of two polysaccharides of *Campylobacter jejuni* 81116. FEMS Microbiol Lett 263:214–222
434. Senchenkova SN, Shashkov AS, Knirel YA, McGovern JJ, Moran AP (1997) The O-specific polysaccharide chain of *Campylobacter fetus* serotype A lipopolysaccharide is a partially O-acetylated 1,3-linked α-D-mannan. Eur J Biochem 245:637–641
435. Senchenkova SN, Knirel YA, Shashkov AS, McGovern JJ, Moran AP (1996) The O-specific polysaccharide chain of *Campylobacter fetus* serotype B lipopolysaccharide is a linear D-rhamnan terminated with 3-*O*-methyl-D-rhamnose (D-acofriose). Eur J Biochem 239:434–438
436. Monteiro MA (2001) *Helicobacter pylori*: a wolf in sheep's clothing: the glycotype families of *Helicobacter pylori* lipopolysaccharides expressing histo-blood groups: structure, biosynthesis, and role in pathogenesis. Adv Carbohydr Chem Biochem 57:99–158
437. Moran AP (2008) Relevance of fucosylation and Lewis antigen expression in the bacterial gastroduodenal pathogen *Helicobacter pylori*. Carbohydr Res 343:1952–1965
438. Moran AP, Knirel YA, Senchenkova SN, Widmalm G, Hynes SO, Jansson P-E (2002) Phenotypic variation in molecular mimicry between *Helicobacter pylori* lipopolysaccharides and human gastric epithelial cell surface glycoforms. Acid-induced phase variation in Lewisx and Lewisy expression by *H. pylori* lipopolysaccharides. J Biol Chem 277: 5785–5795
439. Kocharova NA, Knirel YA, Widmalm G, Jansson P-E, Moran AP (2000) Structure of an atypical O-antigen polysaccharide of *Helicobacter pylori* containing a novel monosaccharide 3-*C*-methyl-D-mannose. Biochemistry 39:4755–4760
440. MacLean LL, Perry MB, Crump EM, Kay WW (2003) Structural characterization of the lipopolysaccharide *O*-polysaccharide antigen produced by *Flavobacterium columnare* ATCC 43622. Eur J Biochem 270:3440–3446
441. MacLean LL, Vinogradov E, Crump EM, Perry MB, Kay WW (2001) The structure of the lipopolysaccharide O-antigen produced by *Flavobacterium psychrophilum* (259–93). Eur J Biochem 268:2710–2716
442. Vinogradov E, MacLean LL, Crump EM, Perry MB, Kay WW (2003) Structure of the polysaccharide chain of the lipopolysaccharide from *Flexibacter maritimus*. Eur J Biochem 270:1810–1815
443. Perepelov AV, Shashkov AS, Tomshich SV, Komandrova NA, Nedashkovskaya OI (2007) A pseudoaminic acid-containing O-specific polysaccharide from a marine bacterium *Cellulophaga fucicola*. Carbohydr Res 342:1378–1381
444. Tomshich SV, Komandrova NA, Widmalm G, Nedashkovskaya OI, Shashkov AS, Perepelov AV (2007) Structure of acidic O-specific polysaccharide from the marine bacterium *Cellulophaga baltica*. Bioorg Khim 33:83–87
445. Hermansson K, Perry MB, Altman E, Brisson J-R, Garcia MM (1993) Structural studies of the O-antigenic polysaccharide of *Fusobacterium necrophorum*. Eur J Biochem 212:801–809

446. Senchenkova SN, Shashkov AS, Moran AP, Helander I, Knirel YA (1995) Structures of the O-specific polysaccharide chains of *Pectinatus cerevisiiphilus* and *Pectinatus frisingensis* lipopolysaccharides. Eur J Biochem 232:552–557
447. Paramonov N, Bailey D, Rangarajan M, Hashim A, Kelly G, Curtis MA, Hounsell EF (2001) Structural analysis of the polysaccharide from the lipopolysaccharide of *Porphyromonas gingivalis* strain W50. Eur J Biochem 268:4698–4707
448. Rangarajan M, Aduse-Opoku J, Paramonov N, Hashim A, Bostanci N, Fraser OP, Tarelli E, Curtis MA (2008) Identification of a second lipopolysaccharide in *Porphyromonas gingivalis* W50. J Bacteriol 190:2920–2932
449. Matsuo K, Isogai E, Araki O (2001) Structural characterization of the O-antigenic polysaccharide chain of *Porphyromonas circumdentaria* NCTC 12469. Microbiol Immunol 45:299–306
450. Hashimoto M, Kirikae F, Dohi T, Adachi S, Kusumoto S, Suda Y, Fujita T, Naoki H, Kirikae T (2002) Structural study on lipid A and the O-specific polysaccharide of the lipopolysaccharide from a clinical isolate of *Bacteroides vulgatus* from a patient with Crohn's disease. Eur J Biochem 269:3715–3721

Chemical Synthesis of Lipid A and Analogues

4

Shoichi Kusumoto

4.1 Introduction

Chemical syntheses of lipid A and its analogues have made substantial contributions to our current understanding of the endotoxic active structures by providing homogeneous products with definite structures. Precise information has been obtained from the use of synthetic lipid A derivatives, which exclude the possibility of influence by contaminants from bacterial or other sources. In this chapter, representative synthetic work on lipid A and analogues will be outlined with brief explanations of their significance for endotoxin research.

After "lipid A" was first described as the endotoxically active principle of bacterial lipopolysaccharide (LPS) by Westphal and Lüderitz [1], it soon became an important target of research in microbiology, immunology, and related fields. However, more time was required until this particular molecule became attractive to organic chemists. This was because little was known on the structural features of lipid A mainly owing to difficulties in the purification of this amphiphilic and intrinsically heterogeneous molecule for the chemical characterization. Advances in the early 1970s towards the understanding of the structural features of lipid A made it possible to use synthetic approaches to build the proposed structures aiming at chemical reproduction of endotoxic activities with synthetic homogeneous molecular species. This body of early research, in part summarized in several articles in a book [2], was based on the incomplete structural information available at that time. Although these efforts did not lead to the production of definite synthetic endotoxic compounds, they contributed to resolve many important issues for the chemical construction of complex glycoconjugate molecules. In fact, this

S. Kusumoto (✉)
Suntory Institute for Bioorganic Research, Wakayamadai 1-1-1, Shimamoto-cho, Mishima-gun, Osaka 618–8503, Japan
e-mail: skus@sunbor.or.jp

Y.A. Knirel and M.A. Valvano (eds.), *Bacterial Lipopolysaccharides*,
DOI 10.1007/978-3-7091-0733-1_4,

accumulated knowledge was soon utilized for the successful synthesis of lipid A derivatives once the lipid A structure was properly elucidated.

4.2 Early Chemical Syntheses of Endotoxic Lipid A

In the early attempts for chemical synthesis, the first molecule chosen as a target structure for synthesis was a tetraacylated glucosamine disaccharide bisphosphate, which corresponds to a disaccharide biosynthetic precursor to lipid A. This compound, designated precursor Ia or recently called more frequently lipid IV_A, was first isolated from a temperature sensitive mutant of *Escherichia coli* and shown to contain 4 mol of (*R*)-3-hydroxytetradecanoic acid (D-β-hydroxymyristic acid) linked to the hydrophilic backbone [3]. The originally proposed backbone structure, 1,4′-bisphosphate of β-(1 → 6) disaccharide of 2-amino-2-deoxy-D-glucose (D-glucosamine), was confirmed by structural studies of a purified major component of *E. coli* lipid A. However, the position of direct acylation on the backbone was revised to be the 2,2′-N- and 3,3′-O-positions of the disaccharide [4]. Based on this information the chemical structure of the precursor Ia is represented as **1**. Since precursor Ia exhibited full endotoxic activity in conventional in vivo systems with mice and rabbits, **1** was regarded as the best target of the next synthetic effort (Fig. 4.1).

Synthesis of **1** was successfully completed through a multi-step procedure as illustrated in Scheme 4.1 [5, 6]. Biological testing clearly proved that synthetic **1** exhibited the full range of toxic and nontoxic biological activities described for endotoxin [7, 8]. Thus, **1** was the first man-made endotoxic compound. At the same time, 100 years after its discovery by Pfeifer in 1983, it could be unequivocally established that lipid A is the active entity of endotoxin [1]. Because of the simple structural feature of **1**, which contains four identical fatty acids in a symmetrical distribution on the disaccharide, the synthetic strategy for **1** was rather simple. Two N-bound and O-bound acyl groups were introduced in one step, respectively, after formation of a protected β-(1 → 6) glucosamine disaccharide **3**. The chemically

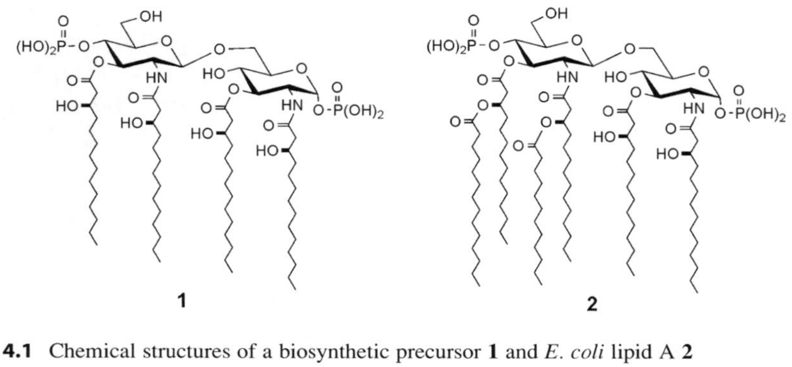

Fig. 4.1 Chemical structures of a biosynthetic precursor **1** and *E. coli* lipid A **2**

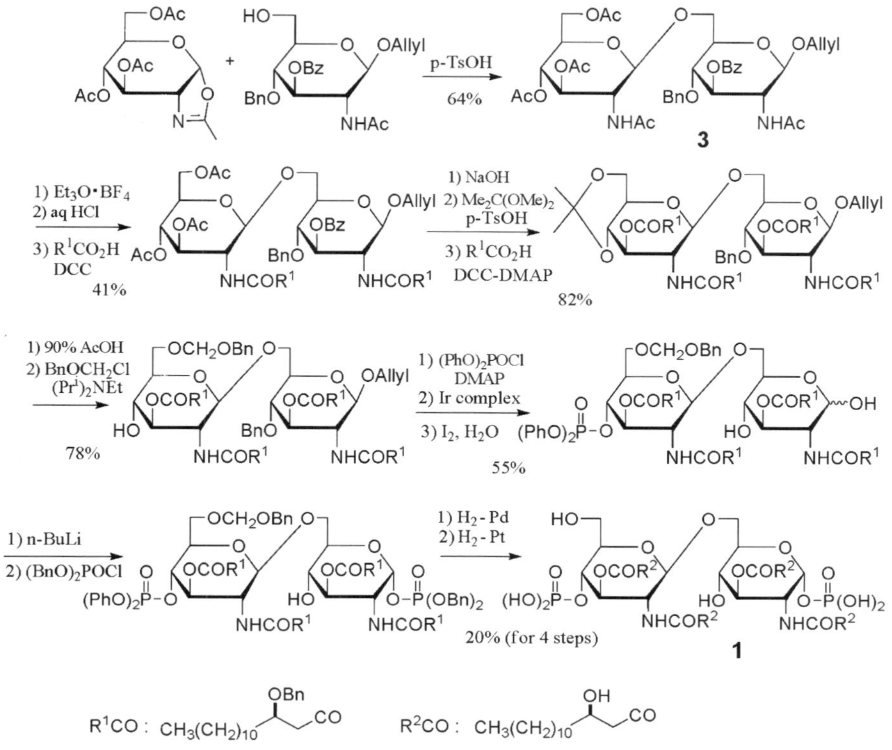

Bn : benzyl Bz : benzoyl Ir complex : $[Ir(cod)(PCH_3Ph_2)]PF_6$

Scheme 4.1 The first chemical synthesis of biosynthetic precursor to lipid A **1**

labile glycosyl phosphate moiety was introduced at the final synthetic stage just before the final deprotection. A novel method was conceived for the formation of α-glycosyl phosphates of 2-*N*-(3-benzyloxyacyl)-glucosamine with dibenzyl phosphorochloridate [9]. The persistent protecting groups were so designed that their final removal was achieved by hydrogenolysis after glycosyl phosphorylation to give the unprotected final product **1**. This strategy facilitated, as anticipated, the purification of the amphiphilic deprotected product. Optically pure (*R*)-3-hydroxytetradecanoic acid was obtained by enantioselective reduction of the corresponding keto ester [10] and its hydroxy group protected as benzyl ether during the synthesis.

In the meantime, the full structure of mature lipid A of *E. coli*, which contains additional dodecanoic and tetradecanoic acids, was deduced as **2** [11]. Because of the asymmetrical distribution of the acyl groups on the pair of glucosamine residues, synthesis of **2** required a more elaborated procedure than that for **1**. The following strategy was employed: (i) all the acyl groups and the protected 4′-phosphate were introduced before the disaccharide formation to reduce the number of protecting groups required, and (ii) only the *N*-acyl group of the distal glucosamine residue was introduced exceptionally after the formation of the disaccharide to avoid β-elimination of the *N*-3-acyloxyacyl group during the glycosylation reaction. 2,2,2-Trichloroethoxycarbonyl (Troc) group was used for the protection of the particular 2-amino group during the disaccharide formation. The Troc group assures selective β-glycosylation and can readily be removed later for N-acylation. For the remaining, the same strategy was employed as in the synthesis of the tetraacylated precursor **1**, e.g. introduction of the chemically labile glycosyl phosphate moiety at the final synthetic stage and subsequent deprotection by hydrogenolysis.

The outline of the first synthesis of hexaacylated *E. coli* lipid A **2** is illustrated in Scheme 4.2 [11, 12]. Coupling of a 2-*N*-Troc-glycosyl bromide with an acceptor proceeded with a selective reaction at the primary 6-hydroxy group of the latter to give the desired β-(1 → 6) disaccharide. Cleavage of the *N*- and *O*-Troc groups followed by N-acylation gave the fully acylated disaccharide **4**. The 6′-hydroxy group was then protected again and the 1-*O*-allyl group cleaved. Phosphorylation as descried above gave the desired 1-α-phosphate. Hydrogenolytic removal of all the benzyl-type protecting groups with a palladium catalyst followed by hydrogenolysis with a platinum catalyst to remove the phenyl esters of the 4′-phosphate afforded the desired free **2**.

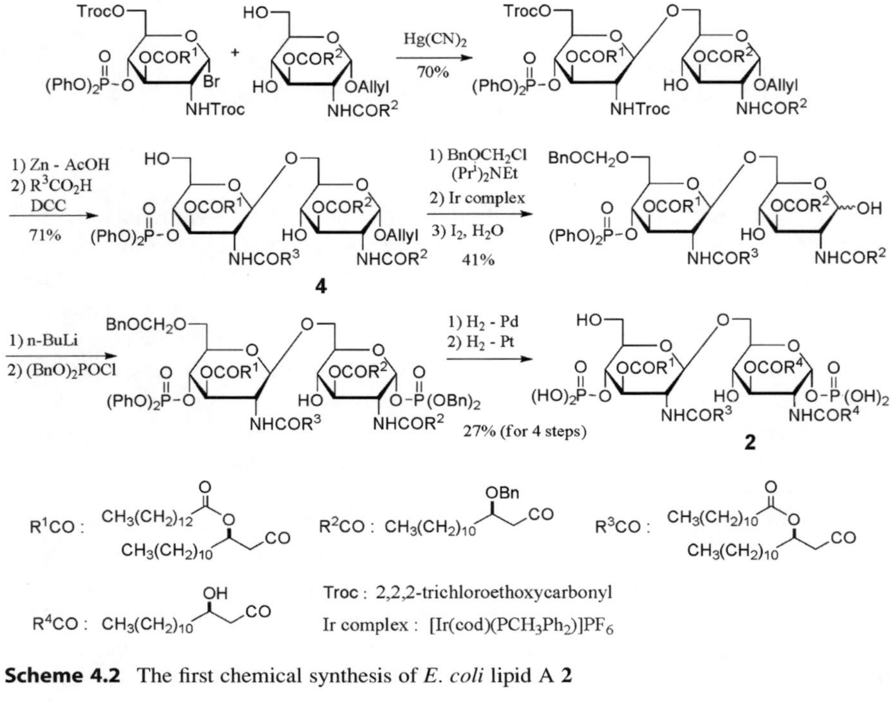

Scheme 4.2 The first chemical synthesis of *E. coli* lipid A **2**

Standard biological tests for endotoxic activity demonstrated that synthetic **2** exhibited the same activities as those of the natural counterpart isolated from *E. coli* [13, 14]. When cultured human cells became available for biological tests, the tetraacylated **1** proved to be inactive as endotoxin but active as an antagonist, while hexaacylated lipid A **2** remained as active as bacterial LPS [15]. This unexpected information resulted only possible by the use of pure synthetic compounds, and also suggested the presence of specific receptors for lipid A on animal cells. The mechanism of the antagonistic function of **1** was finally unveiled after Toll-like-receptor 4/MD-2 was identified as the receptor complex for lipid A [16].

The efficiency of the earlier syntheses of **1** and **2** was not high, many reaction steps were required, and some of the protecting groups had to be replaced during the syntheses. Nevertheless, these efforts provided unequivocal evidence supporting the notion that lipid A is the endotoxic principle of LPS, and that structural analogues of lipid A could be made synthetically available.

4.3 Improved Synthesis of Lipid A Analogues

After the synthetic identification of the endotoxic principle, the next point to be investigated was the relationship between chemical structures and biological activity of lipid A analogues. Concomitantly, isolation and structural study were facilitated by much improved methods of purification and advances in structural analyses by the use of nuclear magnetic resonance (NMR) spectroscopy and mass spectrometry (MS). Under such situations lipid A molecules from various bacterial species were isolated and their structures elucidated (see Chap. 1). Therefore, it became possible to contemplate the chemical synthesis of lipid A analogues with different biological activities using a structure-guided approach. Also, the efficiency of chemical synthesis was improved to prepare a sufficient number and amount of analogues. Some typical examples from such syntheses will be described below. Several articles have recently appeared which summarize syntheses of lipid A and its analogues [17–20].

Since the first synthesis of *E. coli* lipid A synthetic, chemical procedures were modified and refined to reduce the total reaction steps and improve conversion yields. The important basic strategies of the early synthesis were thereby retained in most of the new syntheses, as follows: (i) the use of benzyl-type groups for persistent protections which enable the final hydrogenolytic deprotection, (ii) the use of *N*-Troc glycosyl donors, and (iii) introduction of the glycosyl phosphate at the latest synthetic stage before the final deprotection.

As one of the typical examples, new synthesis of *E. coli* lipid A **2** is illustrated in Scheme 4.3 [21]. The major points improved were: (i) a 6-*O*-benzylated glucosamine derivative was prepared via regioselective reductive opening of a 4,6-*O*-benzylidene ring in one step, and then converted into *N*-Troc imidate **5** to be used as the glycosyl donor; and (ii) a cyclic benzyl-type xylidene diester [22] used for the protection of the 4′-phosphate. The xylidene ester of a phosphate is stable enough

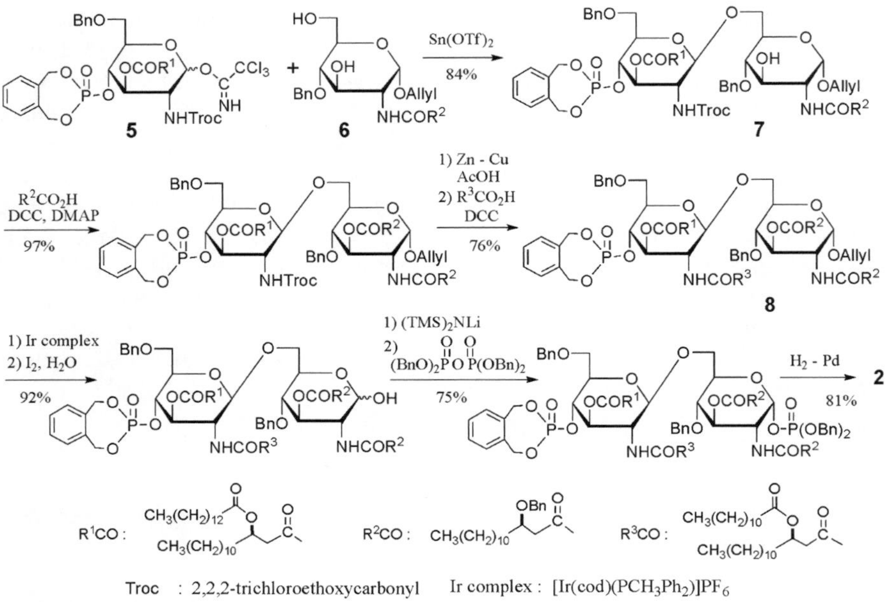

Scheme 4.3 An improved synthesis of *E. coli* lipid A **2**

to survive through the multistep conversion of total synthesis. The corresponding simple dibenzyl ester is not suitable for the same purpose because its partial cleavage cannot be avoided during the synthesis. The use of the xylidene protection for the 4′-phosphate also enabled the final hydrogenolytic deprotection in one step with a palladium catalyst. Glycosylation of a new acceptor **6** with the imidate **5** afforded the desired β-(1 → 6) disaccharide **7**. After stepwise acylation of 3-O- and 2′-N-positons of the disaccharide, the fully protected product **8** was then converted to 1-*O*-phosphate, which was subjected to hydrogenolytic deprotection in one step to give **2**.

New synthesis of tetraacyl precursor **1** was also achieved in a similar way [23]. Synthesis of an unnatural analogue of *E. coli*-type lipid A was reported, which contains enantiomeric (*S*)-3-hydroxytetradecanoic acids in place of the corresponding (*R*)-acids that are present in natural lipid A **2** [24]. The unnatural (*S*)-acid was prepared by novel procedures used in the preparation of optically pure β-hydroxy fatty acids [24–26]. Such unnatural structural analogues are only available by chemical synthesis.

In earlier work, purification of deprotected lipid A analogues after hydrogenolysis was not an easy step because of their amphiphilc property and strong tendency to aggregate in solution. Recovery yields from chromatographic purification were always low since free lipid A analogues are often strongly retained on both polar and lipophilic surfaces for column chromatographic separations. Application of the principle of liquid-liquid partition was then found to be efficient and practical for the purification of free lipid A analogues. For example, the biosynthetic precursor **1** prepared by a new route was purified successfully by partition chromatography with a two-phase chloroform : methanol : 2-propanol : water : triethylamine solvent system on a column of Sephadex LH-20 [23]. The method was also applicable to purification of other synthetic compounds [27].

Most lipid A analogues isolated from various bacteria share the same phosphorylated disaccharide as their hydrophilic backbone with varying numbers, chain lengths, and distributions of acyl groups depending on bacterial species and/or growth conditions. There are some analogues that lack one or both of the phosphates. The presence of hydrophilic substituents on the phosphates is also documented. Under such situation, various natural and unnatural lipid A analogues have been synthesized in order to confirm their structures and biological activities. Some natural lipid A species lack the 4′-phosphate. Lipid A form *Helicobacter pylori* belongs to this type but isolated preparations from usually grown cells of this bacterium consist of a mixture of 1-monophosphate with or without an ethanolamine substituent **9a** and **9b**. These structures were separately synthesized and their biological functions investigated [28].

The structures of lipid A from plant pathogenic and symbiotic bacteria were also studied. Among them, synthesis was reported of another unique lipid A analogue from *Rhizobium* sin-1. The backbone of this lipid A consists of a characteristic β-(1 → 6) disaccharide of glucosamine and 2-aminogluconolactone and lacks both phosphates. The location and composition of five fatty acyl groups containing one very long C_{28} acid are also unique. A divergent synthetic route was reported to this type of lipid A analogues and one of the products **10** was shown to have antagonistic activity, as reported for the natural counterparts, to suppress the effect of *E. coli* lipid A [29] (Fig. 4.2).

More complex structures containing additional sugar units linked to lipid A were also synthesized. The first compound synthesized corresponds to Re-type LPS

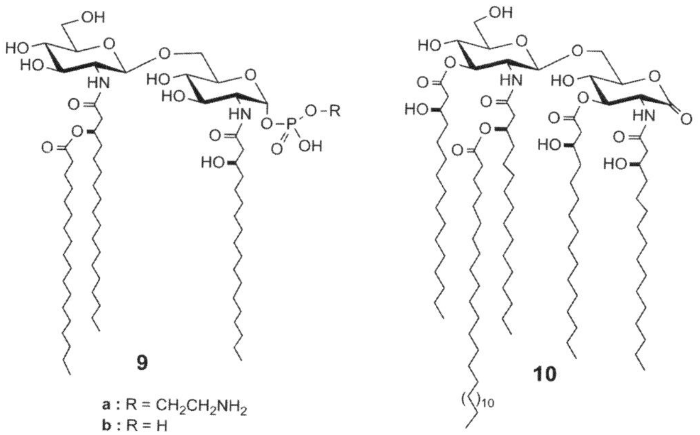

Fig. 4.2 Chemical structures of *Helicobacter pylori* lipid A **9** and a synthetic analogue **10** of *Rhizobium* sin-1 lipid A

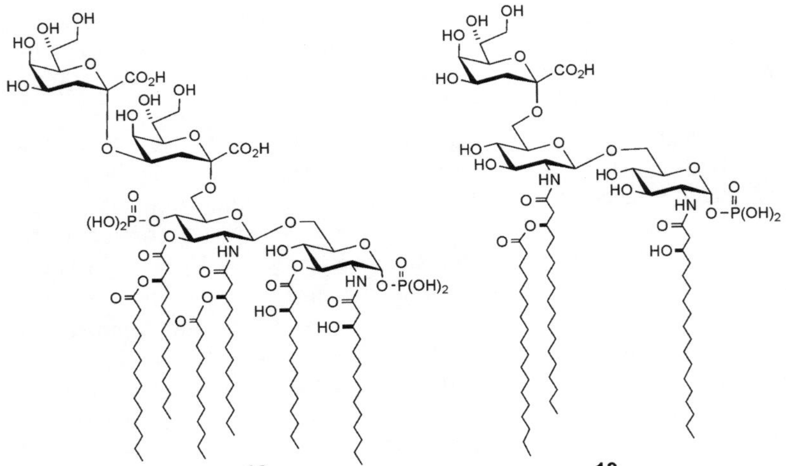

12 13

Fig. 4.3 Chemical structures of *E. coli* Re LPS **12** and a partial structure **13** of *Helicobacter pylori* LPS synthesized

12 isolated from an *E. coli* mutant. Re LPS contains two α-ketosidically linked 3-deoxy-*manno*-octulosonic acid (abbreviated Kdo according to its old name 3-keto-2-deoxy-octonic acid) residues linked to the 6′-position of *E. coli* lipid A **2**. A compound that contains a single Kdo linked to lipid A was also synthesized in a similar manner. Synthesis was achieved by stepwise condensation of Kdo donors to a lipid A part obtained by slight modifications of synthetic route to **2** [27]. Synthesis of another Kdo-containing compound **13** was also reported which corresponds to a partial structure of LPS of *Hericobactor pylori* [30]. Biological tests of these compounds showed the effect of additional Kdo residue(s) on the biological activity of lipid A (Fig. 4.3).

4.4 Synthesis of Unnatural Analogues of Lipid A

As mentioned above in relation to the *E. coli* lipid A analogue with (*S*)-fatty acids, chemical synthesis opened routes to analogues and derivatives that are not available from nature. Syntheses of several artificial structural analogues have already been reported for better understanding of the relationship between chemical structures and biological activities or for the purpose of obtaining more favorable compounds for certain practical purposes.

The labile glycosyl phosphate is one of the major reasons that make chemical synthesis of lipid A difficult. To avoid this problem, substitution with other acidic functionalities was attempted for the purpose to find new compounds, which could retain the beneficial potencies of lipid A such as antitumor activity.

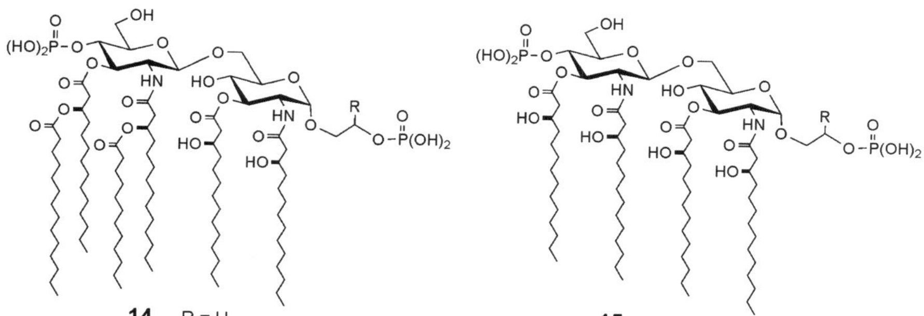

14 R = H
14a R = ^{3}H

15 R = H
15a R = ^{3}H

Fig. 4.4 Chemical structures of phosphonooxyethyl analogues of *E. coli*-type lipid A **14** and biosynthetic precursor **15**

Phosphonooxyethyl analogues **14** and **15** represent typical examples of this line of approach [31, 32]. The hexaacylated **14** exhibits potent endotoxic activity indistinguishable from that of the natural counterpart **2**, whereas **15** acts as an antagonist that inhibits the action of endotoxin as the natural precursor **1** does on human cells. Because of the chemical stability of the phosphonooxyethyl group as compared to the glycosyl phosphate, the synthesis and purification of **14** and **15** were much easier. With an improved synthetic route to a natural type of lipid A, as described in the previous section, these phosphonooxyethyl analogues can also be prepared quite efficiently via improved routes, though their first syntheses were achieved by a procedure similar to the early synthesis of natural type lipid A. The purification of the final deprotected products was also much easier than the natural type counterparts [21] (Fig. 4.4).

Discovery of phosphonooxyethyl derivatives led to the first successful synthesis of tritium-labeled endotoxic lipid A analogues **14a** and **15a** [33]. Radioactive lipid A has long been desired for the study of action mechanism of lipid A. Labeled lipid A of a very high specific radioactivity was required for that purpose because lipid A acts at very low concentrations on host cells. Preparation with such a high radioactivity can never be obtained by biosynthetic procedures. Labeling of the fatty acyl or phosphate moiety, which could be split off in living host cells, was not recommended to avoid false signals when used as a tracer.

Scheme 4.4 illustrates the synthesis of a phosphonooxyethyl analogue **14a** tritium-labeled at the ethylene glycol part. The same fully acylated disaccharide 4′-phosphate **8** used as a synthetic intermediate to *E. coli* lipid A (Scheme 4.3) served as the starting material of this synthesis. The allyl group of the disaccharide **8** was first oxidized to an aldehyde. Subsequent reduction of the aldehyde function with a tritium-labeled borohydride reagent smoothly gave the α-glycosidically linked radiolabeled hydroxyethyl group, which was then phosphorylated by the standard phosphoroamidite procedure followed by oxidation. The protected labeled product was intensively purified at this stage, and then subjected to hydrogenolysis to yield highly pure final product **14a** with high specific radioactivity.

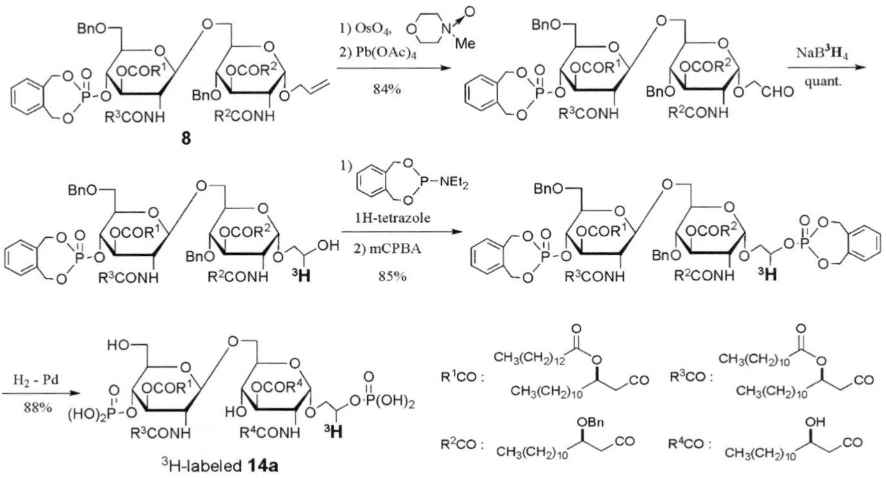

Scheme 4.4 Synthesis of a tritium-labeled phosphonooxyethyl analogue of *E. coli*-type lipid A **14a**

The corresponding tetraacylated biosynthetic precursor-type labeled derivative **15a** was also obtained similarly [33]. These radiolabeled compounds were utilized in precise analysis of the mode of interaction with their receptor proteins [16].

Substitution of the glycosyl phosphate with a carboxylic acid then proved to be another possibility for the synthesis of the carboxymethyl analogue of *E. coli* lipid A [21]. A series of carboxymethyl analogues with various distribution patterns of acyl groups were synthesized and their biological activities assessed in relation to their molecular conformations [34].

All the syntheses described above employed catalytic hydrogenolysis for the complete deprotection at the final stage to give the free lipid A analogues. This general strategy is satisfactory because most typical lipid A isolated from natural bacterial cells contain only saturated fatty acids as their components. Hydrogenolysis of benzyl, xylidene, and, in earlier syntheses, phenyl groups form only readily removable volatile by-products, so that the final free lipid A products were obtained in pure states without damaging other functionalities. This strategy is not applicable to lipid A analogues that contain unsaturated acyl groups. *Rhodobacter sphaeroides* lipid A **16** was reported to share the same hydrophilic backbone of the 1,4′-bisphosphorylated β-(1 → 6) glucosamine disaccharide as other so far known bacteria but contains unusual fatty acids: a 3-keto acid on the 2-amino group and an unsaturated acid in the 3-acyloxyacyl group linked to the 2′-amino group of the disaccharide [35, 36] (Fig. 4.5).

R. sphaeroides lipid A attracted attention from a pharmaceutical point of view owing to its potent antagonistic activity to suppress the endotoxic function of LPS. It may have a possible clinical application to therapy against sepsis and shock syndrome caused by LPS in the case of Gram-negative infections. A new synthetic route was elaborated to such unsaturated lipid A based on allyl-type protections

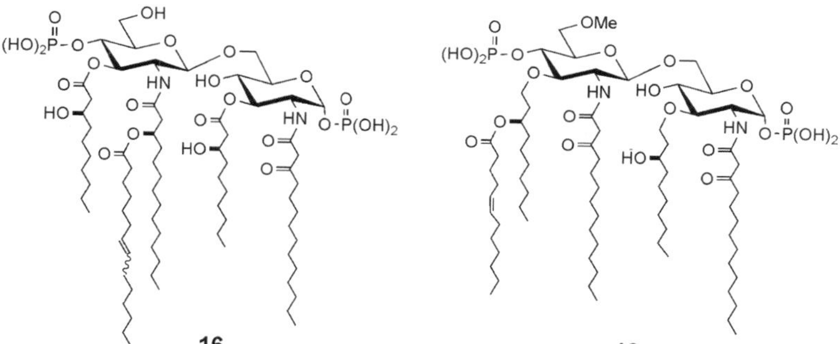

Fig. 4.5 Proposed chemical structure of *Rhodobacter sphaeroides* lipid A **16** and its artificial analogue **19** synthesized

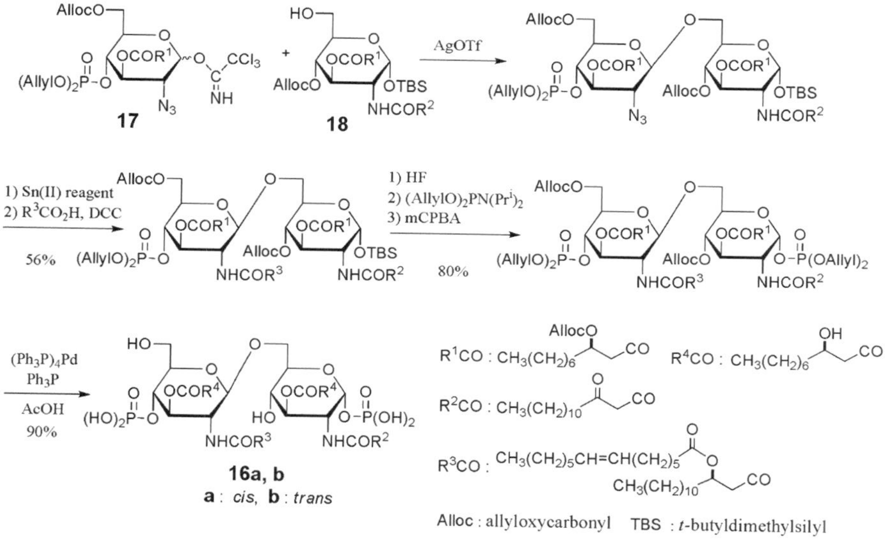

Scheme 4.5 Synthesis of proposed structure of *Rhodobacter sphaeroides* lipid A **16** containing an unsaturated fatty acid

[37]. Allyl-type protecting groups are known to be removable by transition metal-catalysed reactions leaving thereby the isolated double bond in the molecule intact.

In their synthesis, allyloxycarbonyl (Alloc) group and diallyl ester were employed for protection of hydroxy and the 4′-phosphate groups, respectively (Scheme 4.5). Coupling of a glycosyl trichloroacetimidate of a 2-azido sugar **17** with an acceptor **18** gave a β-(1 → 6) disaccharide. The azido group of the disaccharide was then reduced and the resulting amino group acylated. The *t*-butyldimethylsilyl (TBS) glycoside was then selectively cleaved and the glycosyl phosphate introduced as its diallyl ester by means of the phosphoroamidite method.

All the allyl-type protecting groups were cleanly removed in one step with a palladium(0) catalyst to give the first synthetic lipid A analogues **16a** and **16b** (*cis-* and *trans-*isomers of the double bond, respectively) containing one unsaturated acyl group. Though neither of them was identical with the natural lipid A obtained from *R. sphaeroides* cells, both showed potent antagonistic activities to suppress the toxic effect of LPS in a human monocyte system. The same research group then synthesized a novel artificial derivative **19** which was designed to be more resistant than **16** against biological degradation and expected to be of therapeutic value in near future [38].

4.5 Concluding Remarks

After the success in the first synthesis of *E. coli* lipid A and some of its structural analogues, efforts were directed towards understanding structure-function relationships. Indeed, synthetic targets automatically expanded to various analogues of both natural and unnatural structures. Some of such syntheses and review articles are cited above. As the result of critical methodological improvements structural analogues of lipid A are now synthetically feasible and can be obtained in highly pure forms. Syntheses were also extended to labeled compound such as regiospecifically ^{13}C-labeled lipid A derivatives for conformational study [39, 40], as well as radio- and fluorescence-labeled derivatives for functional studies [33, 38, 41]. Chemical synthesis can contribute significantly to a better understanding of the biological significance of endotoxin.

References

1. Rietschel ET, Westphal O (1999) Endotoxin: historical perspectives. In: Brade H, Opal SM, Vogel SN, Morrison DC (eds) Endotoxin in health and disease. Marcel Dekker, New York, pp 1–31
2. Anderson L, Unger FM (eds) (1983) Bacterial lipopolysaccharide, vol 231, ACS symposium series. American Chemical Society, Washington, DC
3. Lehman V (1977) Isolation, purification and properties of an intermediate in 3-deoxy-D-*manno*-octulosonic acid-lipid A biosynthesis. Eur J Biochem 75:257–266
4. Imoto M, Kusumoto S, Shiba T, Naoki H, Iwashita T, Rietschel ET, Wollenweber H-W, Galanos C, Lüderitz O (1983) Chemical structure of *E. coli* lipid A: linkage site of acyl groups in the disaccharide backbone. Tetrahedron Lett 24:4017–4020
5. Imoto M, Yoshimura H, Yamamoto M, Shimamoto T, Kusumoto S, Shiba T (1984) Chemical synthesis of phosphorylated tetraacyl disaccharide corresponding to a biosynthetic precursor of lipid A. Tetrahedron Lett 25:2667–2670
6. Imoto M, Yoshimura H, Yamamoto M, Shimamoto T, Kusumoto S, Shiba T (1987) Chemical synthesis of a biosynthetic precursor of lipid A with a phosphorylated tetraacyl disaccharide structure. Bull Chem Soc Jpn 60:2197–2204
7. Galanos C, Lehman V, Lüderitz O, Rietschel ET, Westphal O, Brade H, Brade L, Freudenberg MA, Hansen-Hagge T, Lüderitz T, McKenzie G, Schade U, Stritmatter W, Tanamoto K, Zähringer U, Imoto M, Yoshimura H, Yamamoto M, Shimamoto T, Kusumoto S, Shiba T

(1984) Endotoxic properties of chemically synthesized lipid A part structures – comparison of synthetic lipid A precursor and synthetic analogues with biosynthetic lipid A precursor and free lipid A. Eur J Biochem 140:221–227
8. Kotani S, Takada H, Tsujimoto M, Ogawa T, Harada K, Mori Y, Kawasaki A, Tanaka A, Nagao S, Tanaka S, Shiba T, Kusumoto S, Imoto M, Yoshimura H, Yamamoto M, Shimamoto T (1984) Immunobiologically active lipid A analogs synthesized according to a revised structural model of natural lipid A. Infect Immun 45:293–296
9. Inage M, Chaki H, Kusumoto S, Shiba T (1982) A convenient preparative method of carbohydrate phosphates with butyllithium and phosphorochloridate. Chem Lett 11: 1281–1284
10. Tai A, Nakahata M, Harada H, Izumi Y, Kusumoto S, Inage M, Shiba T (1980) A facile method for preparation of the optically pure 3-hydroxytetradecanoic acid by an application of asymmetrically modified nickel catalyst. Chem Lett 9:1125–1126
11. Imoto M, Yoshimura H, Kusumoto S, Shiba T (1984) Total synthesis of lipid A, active principle of bacterial endotoxin. Proc Jpn Acad Ser B Phys Biol Sci 60:285–288
12. Imoto M, Yoshimura H, Shimamoto T, Sakaguchi N, Kusumoto S, Shiba T (1987) Total synthesis of *Escherichia coli* lipid A, the endotoxically active principle of cell-surface lipopolysaccharide. Bull Chem Soc Jpn 60:2205–2214
13. Galanos C, Lüderitz O, Rietschel ET, Westphal O, Brade H, Brade L, Freudenberg M, Schade U, Imoto M, Yoshimura H, Kusumoto S, Shiba T (1985) Synthetic and natural *Escherichia coli* free lipid A express identical endotoxic activities. Eur J Biochem 148:1–5
14. Kotani S, Takada H, Tsujimoto M, Ogawa T, Takahashi I, Ikeda T, Harada K, Nagaki K, Kitamura H, Shiba T, Kusumoto S, Imoto M, Yoshimura H (1985) Synthetic lipid A with endotoxic and related biological activities comparable to those of a natural lipid A from an *Escherichia coli* re-mutant. Infect Immun 49:225–237
15. Loppnow H, Brade L, Brade H, Rietschel ET, Kusumoto S, Shiba T, Flad H-D (1986) Induction of human interleukin 1 by bacterial and synthetic lipid A. Eur J Immunol 16:1263–1267
16. Kobayashi M, Saitoh S, Tanimura N, Takahashi K, Kawasaki K, Nishijima M, Fujimoto Y, Fukase K, Akashi-Takamura S, Miyake K (2006) Lipid A antagonist, lipid lVa, is distinct from lipid A in interaction with Toll-like receptor 4 (TLR4)-MD-2 and ligand-induced TLR4 oligomerization. J Immunol 176:6211–6218
17. Kusumoto S, Fukase K, Oikawa M (1999) The chemical synthesis of lipid A. In: Brade H, Opal SM, Vogel SN, Morrison DC (eds) Endotoxin in health and disease. Marcel Dekker, New York, pp 243–256
18. Kusumoto S, Fukase K, Fujimoto Y (2007) Synthesis of lipopolysaccharide, peptidoglycan, and lipoteichoic acid fragments. In: Kamerling JP (ed) Comprehensive glycoscience, vol 1. Elsevier, Amsterdam, pp 685–711
19. Kusumoto S, Hashimoto M, Kawahara K (2009) Structure and synthesis of lipid A. In: Jeannin J-F (ed) Lipid A in cancer therapy. Landes Bioscience, Austin, TX, pp 5–23
20. Kusumoto S, Fukase K, Shiba T (2010) Key structures of bacterial peptidoglycan and lipopolysaccharide triggering the innate immune system of higher animals: chemical synthesis and functional studies. Proc J Acad B 86:322–337
21. Liu W-C, Oikawa M, Fukase K, Suda Y, Kusumoto S (1999) A divergent synthesis of lipid A and its chemically stable unnatural analogues. Bull Chem Soc Jpn 72:1377–1385
22. Watanabe Y, Kodama Y, Ebysuya K, Ozaki S (1990) An efficient phosphorylation method using a new phophitylating agent, 2-diethylamino-1,3,2-benzodioxaphosphepane. Tetrahedron Lett 31:255–256
23. Oikawa M, Wada A, Yoshizaki H, Fukase K, Kusumoto S (1997) New efficient synthesis of a biosynthetic precursor of lipid A. Bull Chem Soc Jpn 70:1435–1440
24. Liu W-C, Oikawa M, Fukase K, Suda Y, Winarno H, Mori S, Hashimoto M, Kusumoto S (1997) Enzymatic preparation of (*S*)-3-hydroxytetradecanoic acid and synthesis of unnatural analogues of lipid A containing the (*S*)-acid. Bull Chem Soc Jpn 70:1441–1450

25. Noyori R, Ohkuma T, Kitamura M, Takaya H, Sayo N, Kumobayashi H, Akutagawa S (1987) Asymmetric hydrogenation of β-keto carboxylic acid esters. A practical, purely chemical access to β-hydroxy esters in high enantiomeric purity. J Am Chem Soc 109:5856–5858
26. Oikawa M, Kusumoto S (1995) On a practical synthesis of β-hydroxy fatty acid derivatives. Tetrahedron: Asymmetry 6:961–966
27. Yoshizaki H, Fukuda N, Sato K, Oikawa M, Fukase K, Suda Y, Kusumoto S (2001) First total synthesis of the re-type lipopolysaccharide. Angew Chem Int Ed Engl 40:1475–1480
28. Sakai Y, Oikawa M, Yoshizaki H, Ogawa T, Suda Y, Fukase K, Kusumoto S (2000) Synthesis of *Helicobacter pylori* lipid A and its analogue using *p*-(trifluoromethyl)benzyl protecting group. Tetrahedron Lett 41:6843–6847
29. Demchenko AV, Wolfelt MA, Santhanam B, Moore JN, Boons G-J (2003) Synthesis and biological evaluation of *Rhizobium* sin-1 lipid A derivatives. J Am Chem Soc 125:6103–6112
30. Fujimoto Y, Iwata M, Imakita N, Shimoyama A, Suda Y, Kusumoto S, Fukase K (2007) Synthesis of immunoregulatory *Helicobacter pylori* lipopolysaccharide partial structures. Tetrahedron Lett 48:6577–6581
31. Kusama T, Soga T, Shioya E, Nakayama K, Nakajima H, Osada Y, Ono Y, Kusumoto S, Shiba T (1990) Synthesis and antitumor activity of lipid A analogs having a phosphonooxyethyl group with α- or β-configuration at position 1. Chem Pharm Bull 38:3366–3372
32. Kusama T, Soga T, Ono Y, Kumazawa E, Shioya E, Osada Y, Kusumoto S, Shiba T (1991) Synthesis and biological activity of a lipid A biosynthetic precursor: 1-*O*-phosphonooxyethyl-4′-*O*-phosphono-disaccharices with (*R*)-3-hydroxytetradecanoyl or tetradecanoyl groups at positions 2, 3, 2′, and 3′. Chem Pharm Bull 39:1994–1999
33. Fukase K, Kirikae T, Kirikae F, Liu W-C, Oikawa M, Suda Y, Kurosawa M, Fukase Y, Yoshizaki H, Kusumoto S (2001) Synthesis of [^{3}H]-labeled bioactive lipid A analogs and their use for detection of lipid A-binding proteins on murine macrophages. Bull Chem Soc Jpn 74:2189–2197
34. Fukase Y, Fujimoto Y, Adachi Y, Suda Y, Kusumoto S, Fukase K (2008) Synthesis of *Rubrivivax gelatinosus* lipid A and analogues for investigation of the structural basis for immunostimulating and inhibitory activities. Bull Chem Soc Jpn 81:796–819
35. Salimath PV, Weckesser J, Strittmatter W, Mayer H (1983) Structural studies on nontoxic lipid A from *Rhodobacter sphaeroides* ATCC 17023. Eur J Biochem 136:195–200
36. Takayama K, Qureshi N, Beutler B, Kirkland TH (1989) Diphosphoryl lipid A from *Rhodopseudomonas sphaeroides* ATCC 1702 blocks induction of cachectin in macrophage by lipopolysaccharide. Infect Immun 57:1336–1338
37. Christ WJ, McGuiness PD, Asano O, Wang Y, Mullarkey MA, Perez M, Hawkins LD, Blythe TA, Dubuc GR, Robidoux A (1994) Total synthesis of the proposed structure of *Rhodobacter sphaeroides* lipid A resulting in the synthesis of new potent lipopolysaccharide antagonists. J Am Chem Soc 116:3637–3638
38. Christ WJ, Asano O, Robidoux AL, Perez M, Wang Y, Dubuc GR, Gabin WE, Hawkins LD, McGuiness PD, Mullarkey MA, Lewis MD, Kishi Y, Kawata T, Bristol JR, Rose JR, Rossignor DP, Kobayashi S, Hashinuma I, Kimura A, Asakawa N, Katayama K, Yamatsu I (1995) E5531, a pure endotoxin antagonist of high potency. Science 268:80–83
39. Oikawa M, Shintaku T, Sekljic H, Fukase K, Kusumoto S (1999) Synthesis of ^{13}C-labeled biosynthetic precursor of lipid A and its analogue with shorter acyl chains. Bull Chem Soc Jpn 72:1857–1867
40. Fukase K, Oikawa M, Suda Y, Liu W-C, Fukase Y, Shintaku T, Sekljic H, Yoshizaki H, Kusumoto S (1999) New synthesis and conformational analysis of lipid A: biological activity and supramolecular assembly. J Endotoxin Res 5:46–51
41. Oikawa M, Furuta H, Suda Y, Kusumoto S (1999) Synthesis and bioactivity of a fluorescence-labeled lipid A analogue. Tetrahedron Lett 40:5199–5202

Chemical Synthesis of Lipopolysaccharide Core

5

Paul Kosma and Alla Zamyatina

5.1 Introduction

Genomic data and results compiled from analytical studies during the past decade have revealed a multitude of novel structural features within the core region of the lipopolysaccharide (LPS) of Gram-negative bacteria [1] (see also Chap. 2). Previous synthetic efforts have covered the basic structural units of the enterobacterial LPS core as well as biomedically relevant structures of O-antigens and capsular polysaccharides. These studies have already been summarized in close detail in the past decade. Hence this chapter will present an update of ongoing synthetic efforts using representative examples from the literature of the past decade [2–4]. The use of synthetic carbohydrate antigens as surrogates of the structures occurring in the core-region as well as repeating units of O-antigens and capsular polysaccharides serving as vaccine candidates has also been covered in excellent reviews [5, 6].

Synthetic approaches towards components of the inner core region have to deal with the elaboration of efficient protocols to prepare multigram amounts of the higher carbon aldoses L-*glycero*-D-*manno*-heptose and its 6-epimer as well as the octulosonic acids 3-deoxy-D-*manno*-oct-2-ulosonic acid (Kdo) and D-*glycero*-D-*talo*-oct-2-ulosonic acid (Ko) followed by transformation into suitable glycosyl donor and acceptor derivatives. Recent developments in the field with regard to these topics have therefore also been included in this review. Finally, the remarkable challenges associated with the need for orthogonal protecting groups and specific incorporation of additional groups such as phosphate, 2-aminoethyl phosphate or 4-deoxy-4-amino-L-arabinose substituents will also be discussed. Major research lines in the synthesis of core structures in the past few years covered truncated forms of LPS in order to elucidate their antigenic properties to be

P. Kosma (✉) • A. Zamyatina
Department of Chemistry, University of Natural Resources and Life Sciences, Muthgasse 18, A-1190 Vienna, Austria
e-mail: paul.kosma@boku.ac.at; alla.zamyatina@boku.ac.at

Y.A. Knirel and M.A. Valvano (eds.), *Bacterial Lipopolysaccharides*,
DOI 10.1007/978-3-7091-0733-1_5,

exploited as diagnostic markers and as lead structures for future vaccine development. Remarkable progress has been witnessed in defining the molecular basis for the interaction of Kdo-specific antibodies directed against synthetic antigens as well as oligosaccharides released and purified from bacterial LPS. Knowledge on the three-dimensional presentation of core oligosaccharides is still in its infancy and will undoubtedly benefit from the availability of defined synthetic products for structural studies by nuclear magnetic resonance (NMR) spectroscopy and X-ray crystallography to be complemented by molecular modeling approaches. In addition, these compounds may further be exploited as substrates for various enzymes acting in elongation of the core domain as well as for those adding highly specific and immunorelevant "decorating" groups onto the core units, respectively.

5.2 Recent Approaches for the Synthesis of 3-Deoxy-D-*manno*-oct-2-ulosonic Acid (Kdo)

5.2.1 Chemical Syntheses of Kdo

Considerable attention has been devoted to the development of efficient synthetic methodologies that enable the convenient preparation of 3-deoxy-2-ulosonic acids and their analogues in a minimum number of synthetic steps and the progress made in the synthesis of the major core constituent Kdo has been summarized in an in-depth review in 2003 [7]. A well-established method for Kdo synthesis developed by Cornforth involves a base-catalysed aldol condensation between D-arabinose and oxaloacetic acid followed by Ni^{2+} catalyzed decarboxylation under mildly acidic conditions. Whereas the diastereoselectivity of the Cornforth reaction is moderate, it nevertheless allows the synthesis of multigram amounts of the crystalline ammonium salt of Kdo, thus satisfying the demand for simplicity in combination with efficiency. Despite the utility of the overall transformation of the improved Cornforth reaction there are many cases in which it cannot be applied. Most of the modern methods for Kdo synthesis, with several exceptions, rely on either [6 + 2] atom incorporation starting from D-mannose and put special emphasis on the elaboration of the two-carbon unit, which is later converted into α-oxocarboxylic acid moiety, or on [5 + 3] elongation strategy applying D-arabinose as inexpensive starting material with successive addition of a three-carbon unit.

A concise synthesis of Kdo was achieved starting from D-mannose via elongation with a two-carbon unit and was based on a new β-elimination reaction of a cyclic sulfite as the key step [8]. Wittig olefination of diisopropylidene-α-D-mannofuranose **1** afforded C-6 unprotected **2** as the mixture of geometrical isomers (*E*/*Z* ratio 18:1), which were dihydroxylated to give a stereoisomeric mixture of **3** in 90% yield from **1** (Scheme 5.1). The C-6 hydroxyl group was well discriminated by formation of the five-membered cyclic sulfite by treatment with thionyl chloride and Et_3N to afford **4**. The β-elimination and the subsequent release of SO_2 from the unstable sulfite intermediate **4** under basic conditions (DBU) in the presence of

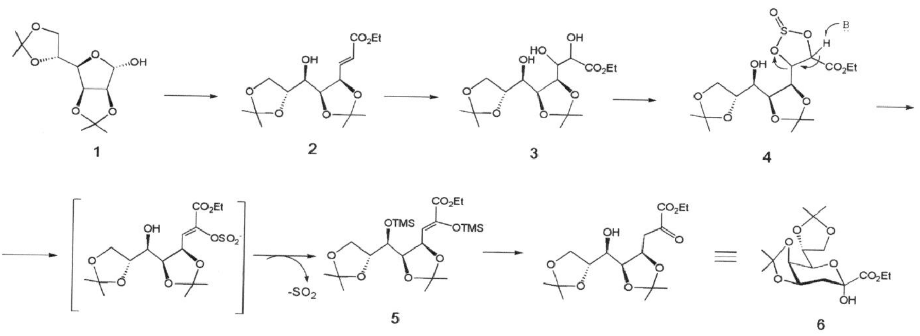

Scheme 5.1 Synthesis of Kdo via [6 + 2] atom incorporation strategy starting from D-mannose based on the β-elimination reaction of a cyclic sulfite as the key step

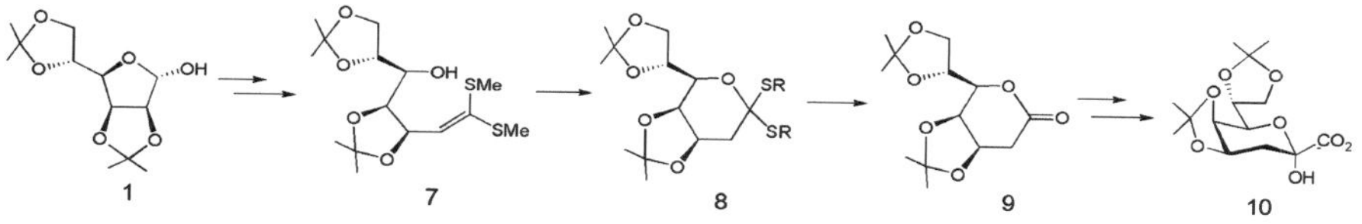

Scheme 5.2 Synthesis of Kdo via [6 + 2] atom incorporation starting from D-mannose based on cyclization of a ketene dithioacetal as the key step

TMSCl provided the enol **5** as bis-TMS ether. The acidic deprotection of the hydroxyl group on C-6 finally afforded Kdo ethyl ester **6** in 75% yield.

A novel short and efficient synthesis of the protected lactone precursor of Kdo, which involves cyclization of a ketene dithioacetal as the key step, has been recently disclosed [9]. Treatment of diisopropylidene-α-D-mannofuranose **1** with lithium derivative of bis(methylsulphanyl)trimethylsilylmethane provided ketene dithioacetal **7** in 90% yield (Scheme 5.2). After cyclization in acidic conditions (pyridinium *p*-toluenesulfonate) to afford **8**, the deprotection of the thioacetal group under classical conditions ($I_2/CaCO_3/H_2O$) provided the key lactone **9** in 80% overall yield.

An innovative, high-yielding 8-step synthetic pathway to Kdo and its 2-deoxy analogue, which implements ring-closing metathesis of highly functionalized α-alkoxyacrylates (such as **14**) and further functionalization of the enol ether double bond of the resulting oxygen heterocyclic intermediate (such as **16**), was recently described [10, 11] (Scheme 5.3). First, an essentially new two-step transformation to efficiently convert alcohol **11** into the corresponding α-alkoxy acrylate **14** has been developed. Thus, alcohol **11**, easily accessible from **1**, was treated with bromide **12** (prepared by reaction of commercially available 2,3-dibromopropionic acid methyl ester with pyrrolidine and Et_3N) in the presence of NaH to provide ester **13**. Reaction of **13** with MeI and Na_2CO_3 in refluxing methanol resulted in methylation of the pyrrolidine-nitrogen and subsequent base-induced elimination

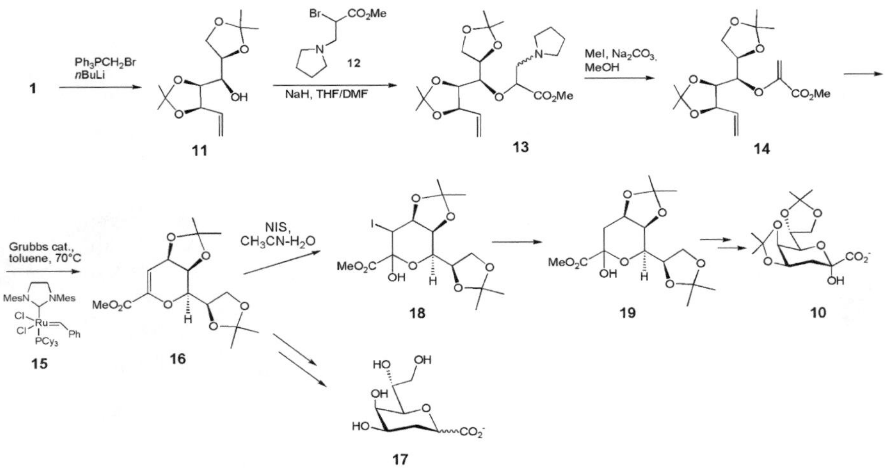

Scheme 5.3 Synthetic pathway to Kdo and 2-deoxy analogues via ring-closing metathesis of highly functionalized α-alkoxyacrylates

to give the desired ring-closing metathesis precursor **14**. Treatment of the α-enol ester **14** with second generation Grubbs' catalyst **15** under optimized reaction conditions led to a smooth conversion to the functionalized pyrene **16** without isomerization of the monosubstituted double bond. Thus, the glycal ester **16**, a known intermediate in the synthesis of Kdo and its derivatives, was prepared in only four steps from protected D-mannose in an overall yield of 64%. Quantitative hydrogenation of the double bond in **16** selectively yielded protected 2-deoxy-α-Kdo, which, after deprotection and anomerization, gave rise to 2-deoxy-β-Kdo **17**.

Reaction of glycal **16** with NIS in acetonitrile-water efficiently converted it into iodohydrin **18**, which was isolated as 1:1 mixture of diastereomers. Subsequent removal of iodide under hydrogenation conditions in the presence of Et_3N as HI-scavenger gave protected Kdo **19**, which, after common deprotection, provided Kdo **10** in 44% overall yield over eight steps. The reported NIS-based conversion represents a significantly more efficient alternative to the known methods, which relied on enolate formation of the hydrogenated form of **16**, followed by introduction of an oxygen or sulfur electrophile.

An attractive direct synthesis of furanosidic eight-membered ulosonic acid (such as **33**) via opening of the bicyclic precursors of octulosonic acids (such as **30**) was based on [5 + 3] carbon atom incorporation strategy and was achieved by treatment of differentially protected γ,δ-bis(silyloxy) *cis*-α,β-epoxy aldehyde **28** with ethyl 2-(trimethylsilyloxy)-2-propenoate **29** in the presence of boron trifluoride–diethyl ether [12] (Scheme 5.4). The synthesis of α,β-epoxy aldehyde was realized by glycol cleavage of diisopropylidene-D-mannitol **20** in the presence of sodium periodate and subsequent treatment of resulting D-glyceraldehyde **21** with methyl-(triphenylphosphoranylidene)acetate **22** which provided corresponding methyl ester **23** (*Z*/*E* ratio 6:1).

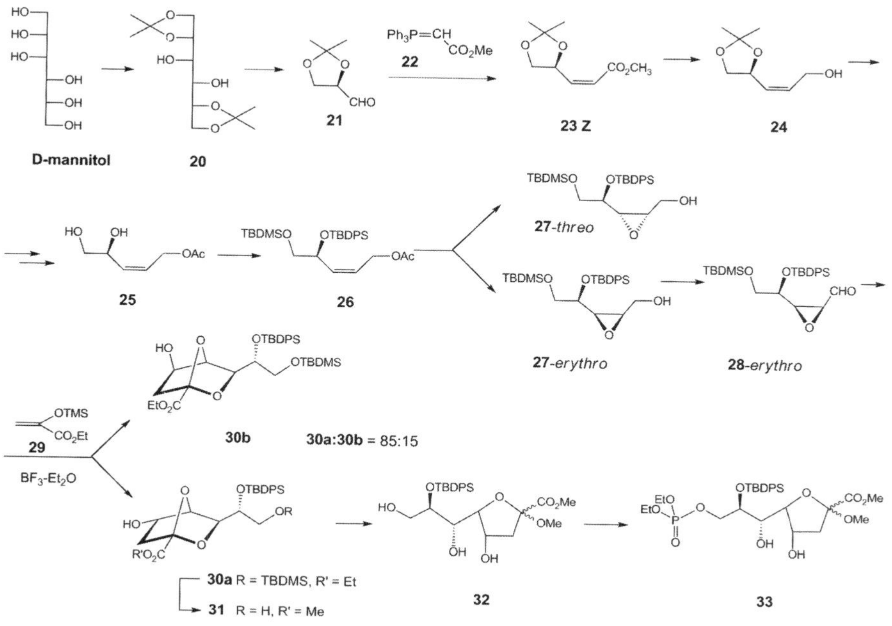

Scheme 5.4 Synthesis of Kdo via [5 + 3] carbon atom incorporation strategy by reaction of protected γ,δ-bis(silyloxy) *cis*-α,β-epoxy aldehydes with ethyl 2-(trimethylsilyloxy)-2-propenoate

Further reduction of the ester group in the presence of DIBAL-H gave the *Z*-allylic alcohol **24**. Subsequent acetate protection of the primary position and removal of the acetonide afforded diol **25**, which was selectively protected at the primary and secondary hydroxyl groups with TBDMS and TBDPS groups, respectively, affording compound **26**. After removal of the primary acetyl group, *m*CPBA in the presence of sodium bicarbonate was used to perform the epoxidation step. This resulted in the two diastereoisomers **27** in 87% yield. Finally, Swern oxidation of **27**-*erythro* gave the desired *cis*-α,β-epoxy aldehyde **28**. The Mukaiyama aldol condensation reaction between γ,δ-silyloxy *cis*-α,β-epoxy aldehyde **28**-*erythro* and ethyl 2-(trimethylsilyloxy)-2-propenoate **29** in the presence of $BF_3 \cdot Et_2O$ provided preferentially *syn*-aldol adducts that were cyclized *in situ* by an intramolecular process to provide bicycles **30a** and **30b**, which represent masked octulosonic acids, in excellent 78% yield. The bicyclic compound **30a** was opened by treatment with $SnCl_4$ in methanol, which resulted first in deprotection at C-8 and reesterification at C-1 positions to furnish intermediate **31**, which, after prolonged treatment with $SnCl_4$, gave the α- and β-methyl furanosides **32** (analogues of the C-4 epimer of the natural compound) in 62% yield. Selective phosphorylation of the primary hydroxyl group in **32** was achieved by reaction with triethyl phosphite and carbon tetrabromide in pyridine, resulting in the formation of the corresponding C-8 diethylphosphate ester **33**. Epimerization at C-4 position should afford the configuration of the natural Kdo.

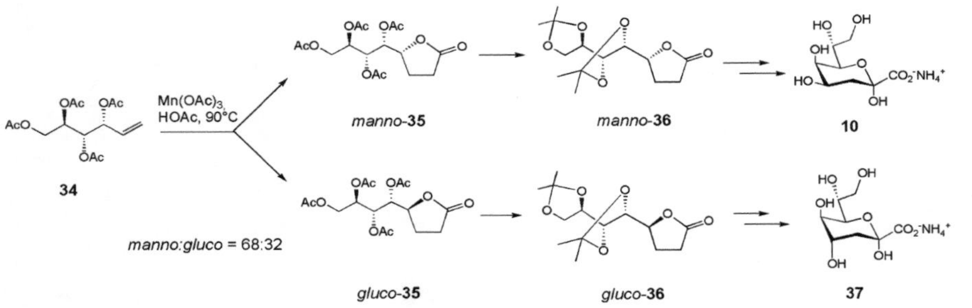

Scheme 5.5 Synthesis of Kdo and analogues via radical approach using carbohydrate-derived alkene as radical acceptor and acetic acid as C-2 building block

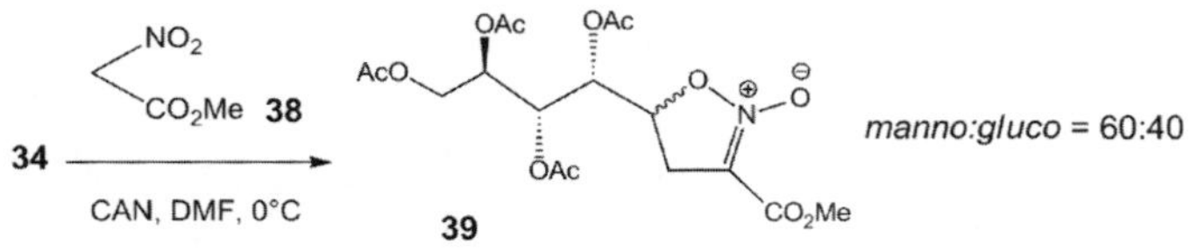

Scheme 5.6 Synthetic approach to Kdo via radical bond formation using carbohydrate-derived alkene as radical acceptor and methyl nitroacetate as C-2 building block

New radical approaches to Kdo and analogues were also recently revealed [13]. Using carbohydrate-derived alkene **34** as radical acceptor, and acetic acid as C-2 building block, under action of manganese (III) acetate, the oxidative radical bond formation was achieved in moderate stereoselectivity but in excellent 90% yield (Scheme 5.5).

The acetyl groups in **35** were exchanged for isopropylidene protecting groups giving rise to lactones **36**, which can be transformed into *manno*- and *gluco*-Kdo **10** and **37**, respectively, in only few steps. Thus, the addition of acetic acid to alkene **34** provided a convenient entry to 3-deoxy-2-ulosonic acids.

The second radical approach to Kdo, which comprised the addition of methyl nitroacetate **38** to alkene **34** in the presence of cerium (IV) ammonium nitrate, afforded isoxazoline *N*-oxides **39**, albeit in low yields and stereoselectivities (Scheme 5.6). Despite the low yields, the advantage of methyl nitroacetate as radical precursor for the total synthesis of Kdo pertains to the formation of C = N double bond in the addition products **39**, which allows the direct introduction of the required keto group by ozonolysis.

A completely different strategy, published by the same authors, envisaged an addition of a carbohydrate (diisopropylidene-D-arabinose **40**) as the radical precursor to an easily available alkene ethyl acrylate **41** in the presence of reducing electron-transfer reagent samarium (II) iodide, which generates radicals from the aldehyde functionality of a carbohydrate (Scheme 5.7). Thus, Sm-mediated radical reaction offers a simple one-step approach to the literature known lactones **36** [13].

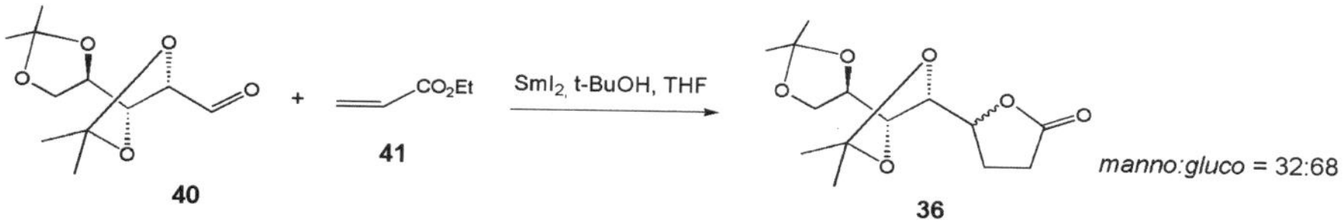

Scheme 5.7 Synthesis of Kdo by addition of diisopropylidene D-arabinose as radical precursor to alkene ethyl acrylate

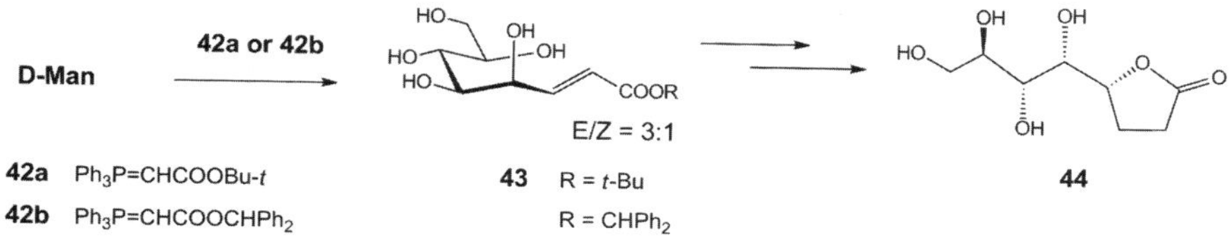

Scheme 5.8 Synthesis of Kdo from D-mannose by Wittig chain extension using highly substituted ylides

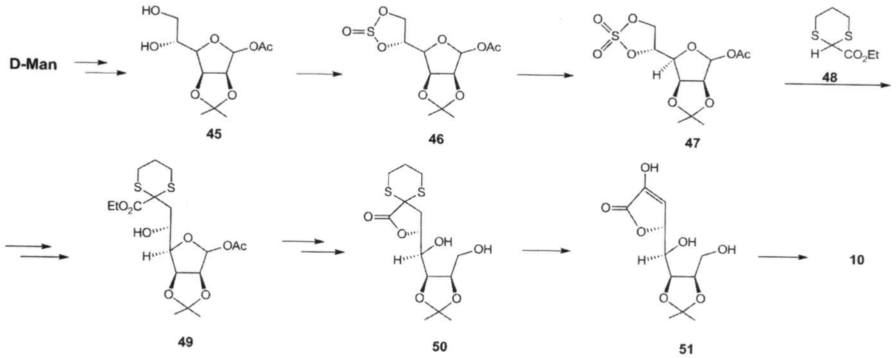

Scheme 5.9 Synthesis of Kdo by two-carbon chain extension at C-6 of D-mannose

The lactone precursor could also be prepared by homologation of unprotected mannose by Wittig chain extension using ylides **42a** or **42b** equipped with bulky *O*-alkyl groups [14]. Wittig reagents (**42a** or **42b**) reacted with unprotected mannose in hot dioxane to give α,β-unsaturated ester **43**, which, after hydrogenolysis and subsequent acid treatment, provided the known lactone **44** in 83% yield (Scheme 5.8).

Most of the reported methods for the chemical synthesis of Kdo starting from D-mannose derivatives are based on the two-carbon elongation at the anomeric position. A completely different approach to Kdo utilizing installation of glyoxylate dithioacetal unit onto C-6 of properly protected D-mannose was lately presented [15]. The key cyclic sulfate **47** was prepared from diol **45** by treatment with thionyl chloride in the presence of Et_3N to furnish cyclic sulfite **46**, which was subsequently oxidized ($RuCl_3$–$NaIO_4$) (Scheme 5.9). After alkylation of **47** with

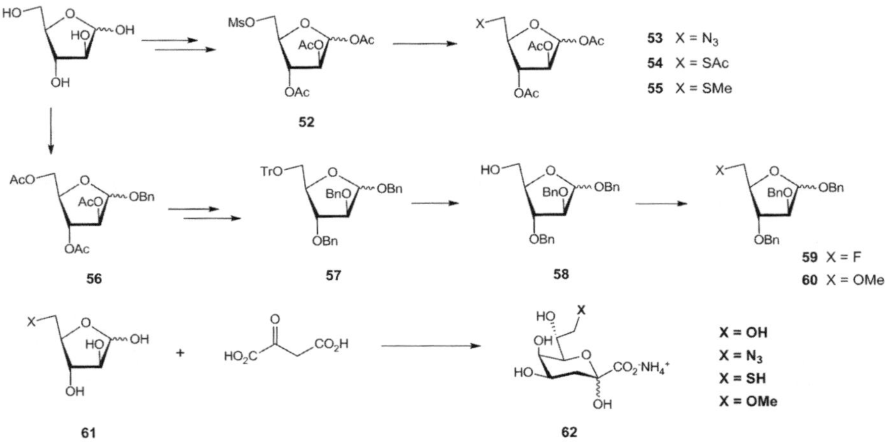

Scheme 5.10 Synthesis of C-8 Kdo analogues based on the modified Cornforth procedure starting from C-5 modified arabinose derivatives

ethyl 1,3-dithiane-2-carboxylate **48** followed by acidic hydrolysis, the hydroxyl ester **49** was obtained as the main product. Subsequent intramolecular lactonization, anomeric deacetylation and reductive ring opening of the hemiacetal with sodium borohydride provided diol **50**. The dithioacetal group in **50** was cleaved with NBS in aqueous acetone to give unsaturated lactone **51**, which, after deprotection, provided Kdo **10** (in seven steps from **45** with 26% overall yield).

A convenient and straightforward route to C-8 modified Kdo derivatives was suggested recently by Kiefel et al. [16]. The synthesis was based on the modified Cornforth procedure taking advantage of its inexpensiveness and simplicity. First, ready access to C-5 modified arabinose derivatives was attained by displacement of a leaving group in the key mesylate **52** (Scheme 5.10). Thus, 5-azido derivative **53**, 5-thioacetyl derivative **54** (by treatment of mesylate **52** with potassium thioacetate) and thiomethyl derivative **55** (by selective deprotection of **52** with hydrazine acetate and subsequent exposure to dimethyl sulfate) were synthesized. The C-5 fluorine-modified analogue **59** and methyl ether **60** were prepared by treatment of the primary hydroxyl group of the benzylated furanoside **58** with diethylamino sulfurtrifluoride (DAST) or with NaH/MeI, respectively. Next, the C-5 modified arabinose derivatives **61** were subjected to aldol condensation with a molar excess of oxaloacetic acid using appropriately adjusted reaction conditions which ensured a series of C-8 Kdo analogues **62** as epimeric mixtures at C-4 (with the ratio **62** to 4-*epi*-**62** approximately 5:1) in attractively high (65–85%) yields.

In spite of the development of all these promising approaches there is still no generally accepted high-yielding procedure for the stereoselective preparation of Kdo and analogues, therefore, systematic research and substantial know-how is required to meet the prerequisites for efficient chemical synthesis of 3-deoxy-2-ulosonic acids.

5.2.2 Chemoenzymatic Syntheses of Kdo

One of the most efficient chemo-enzymatic approaches to Kdo or Kdo-8-phosphate, which are based on the biosynthetic pathway of Kdo, involves specific aldol condensation of pyruvate or phosphoenolpyruvate (PEP) onto D-arabinose 5-phosphate catalysed by the appropriate aldolase or synthetase without recourse to protecting group chemistry [17–19]. In both cases, the C-3–C-4 bond is created with control of configuration at C-4 (Scheme 5.11). Due to the specificity of the enzymes for PEP or pyruvate and close analogues of D-arabinose, however, these methods do not allow the synthesis of diverse analogues of Kdo, especially the 4-deoxy-Kdo derivative.

To ensure an access to distinct Kdo-analogues, Bolte *et al.* developed a versatile approach based on the formation of the C-5–C-6 bond using fructose-1,6-bisphosphate aldolase from rabbit muscle (RAMA) [20, 21].This enzyme, which is not involved in the biosynthetic pathway of Kdo, catalyses condensation of dihydroxyacetone phosphate (DHAP) onto a variety of aldehydes, such as **63**, where the construction of C-5–C-6 bond and the configuration of these centers in compound **64** are controlled by the enzyme, whereas the configuration at C-4 and C-7 can be chosen to lead to Kdo or epimers (Scheme 5.12).

The substitution at C-4 in the aldehyde **63** can be omitted or, alternatively, the enantiomer of **63** at C-4 might be used in an aldolase-catalysed reaction to enable

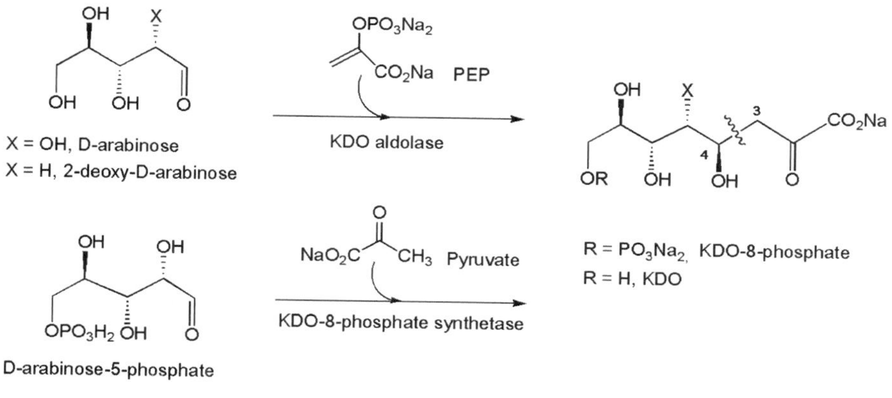

Scheme 5.11 Enzymatic synthesis of Kdo

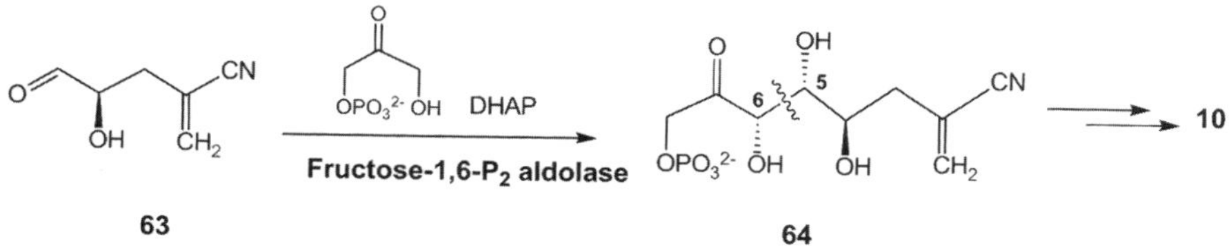

Scheme 5.12 Chemoenzymatic synthesis of Kdo using fructose-1,6-bis-phosphate aldolase

the synthesis of 4-deoxy-Kdo or C-4-epimer of Kdo, respectively. The configuration at C-7 of Kdo might be differentiated by chemical or enzymatic stereospecific reduction of the keto group to provide Kdo or respective analogues.

Another potent aldolase, recently described by Seeberger and Hilvert, is a macrophomate synthase (MPS) – a rare, unusually tolerant enzyme, accepting a range of protected (by ether, acetal, allyl, benzyl, silylether or ester groups) and unprotected aldehydes as substrates [22]. MPS acts through a mechanism that relies on a two-step Michael–aldol pathway, where an unprotected pyruvate, generated *in situ* by decarboxylation of oxaloacetate, serves as the nucleophile which then reacts with a suitable electrophile represented by the sugar aldehydes with three to six carbons, adding three carbons in each case. The MPS-catalysed Cornforth reaction conceptually offers multiple advantages; among others is the option to apply differentially protected sugars as starting aldehydes, so that the products of the enzyme-catalysed reaction can be carried successfully into standard chemical synthesis for subsequent transformations. Since the stereochemical preferences of MPS favour the formation of *4S*-configured alcohols, discovery and/or engineering of enantiocomplementary MPS mutants for the Cornforth-type synthesis of 3-deoxy sugars like Kdo that are *R*-configured at C-4 represents a challenge for the future.

5.2.3 Synthesis of Kdo Glycosyl Donors

Efficient Kdo glycosyl donors displaying good α-anomeric selectivity and sufficient reactivity have meanwhile been elaborated by introduction of electron donating groups such as silyl and benzyl ethers and employing fluoride as a leaving group [23]. The fluoride donor **66** was generated from the hemiketal **65** by reaction with DAST in 88% yield and was used for the synthesis of *Helicobacter pylori* LPS partial structures (Scheme 5.13). In addition, thioglycosides of Kdo have also

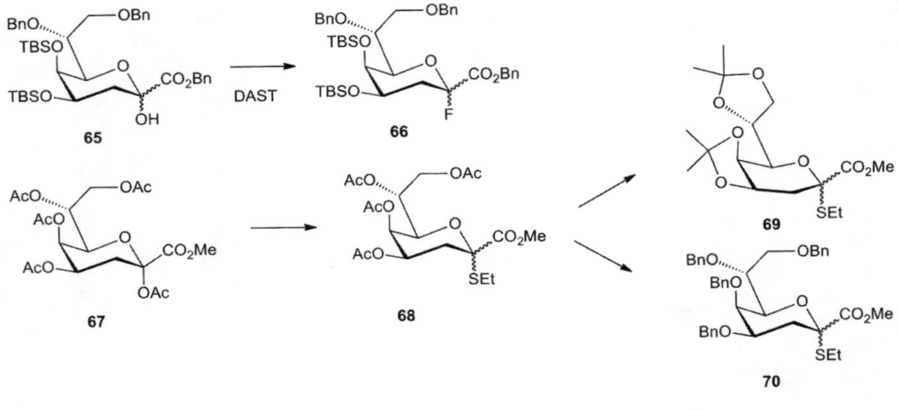

Scheme 5.13 Novel Kdo glycosyl donors

recently been utilized as Kdo donors accessible from the Kdo peracetate **67** [24]. The stability of the thioglycoside allows for the introduction of various protecting groups, giving rise to isopropylidene or benzyl-protected thioglycosides **69** and **70**, respectively, *via* Zemplén-deacetylation of the thioglycoside derivative **68**. These donors constitute improvements in comparison to previously employed Kdo bromide derivatives.

5.3 Synthesis and Antigenic Properties of Kdo Oligosaccharides Related to *Chlamydia* Core Structures

The bacterial family of *Chlamydiaceae* comprises a group of obligate intracellular pathogens which harbour a highly truncated LPS, being composed of Kdo units only. Remarkably, single CMP-Kdo transferases of several chlamydial species are highly promiscuous with respect to their substrate specificity enabling them to produce a variety of linear and branched Kdo oligosaccharides. In all chlamydial species, however, the linear trisaccharide α-Kdo-(2 → 8)-α-Kdo-(2 → 4)-α-Kdo has been found, thereby constituting a family specific antigen. As shown recently by the Brade group, this α-Kdo-(2 → 8)-α-Kdo-(2 → 4)-α-Kdo trisaccharide is still accessible for the heptosyl transferase WaaC. The lack of heptose elongation is explained by the fact that the underlying genes for heptosyl transferases have not been detected in the genomes of *C. trachomatis* and *C. pneumoniae* [25]. In addition to the family-specific trisaccharide antigen, also the (2 → 4)-interlinked trisaccharide α-Kdo-(2 → 4)-α-Kdo-(2 → 4)-α-Kdo and the branched tetrasaccharide α-Kdo-(2 → 4)-[α-Kdo-(2 → 8)]-α-Kdo-(2 → 4)-α-Kdo have been isolated and characterized from recombinant strains expressing the respective multifunctional CMP-Kdo transferase of *C. psittaci* [26]. The relevant Kdo oligosaccharides and part structures derived therefrom have been synthesized as allyl glycosides **71a–75a** as well as in the form of neoglycoconjugates **71b–75b** and have subsequently been used for the preparation and detailed analysis of the epitope specificities of monoclonal antibodies directed against chlamydial and enterobacterial LPS [27, 28] (Scheme 5.14).

Antibodies raised against these neoglycoconjugates displayed a wide range of specificities and affinities. In addition to immunochemical characterization of the antibodies using the BSA-conjugates, the allyl glycosides as well as oligosaccharides obtained in pure form *via* hydrolysis from LPS were used in crystallographic studies complexed to Fab fragments of various Kdo-specific antibodies [29]. Antibodies raised against the family-specific α-Kdo-(2 → 8)-α-Kdo-(2 → 4)-α-Kdo epitope reacted with the trisaccharide (mAb S25-23) as well as the inherent α-Kdo-(2 → 8)-α-Kdo and α-Kdo-(2 → 4)-α-Kdo disaccharide units (mAb S25-2). Thus, in the crystal structures of Kdo antigens complexed to the Fab fragments, specific binding was observed for the Kdo-(2 → 4)-Kdo epitope, whereas the Kdo-(2 → 8)-Kdo part was engaged in a more promiscuous binding mode. Within this series of antibodies, the terminal Kdo unit was invariably fixed in a highly

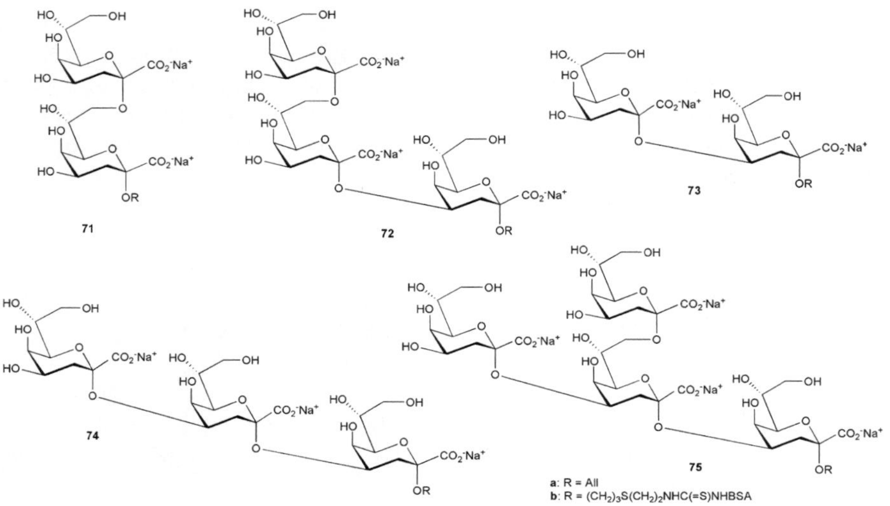

Scheme 5.14 Synthetic ligands and neoglycoconjugates corresponding to chlamydial LPS core units

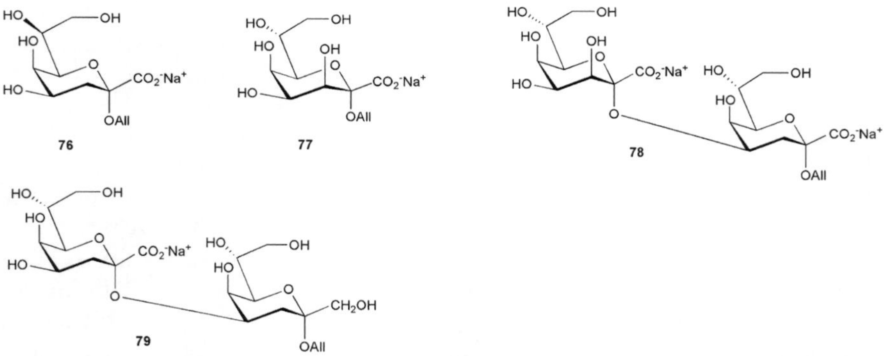

Scheme 5.15 Kdo analogues and Ko-containing ligands bound by monoclonal antibodies

conserved and germline-encoded binding pocket. By using a series of Kdo analogues, it could also be shown that modifications at C-7 (7-epi-Kdo, **76**), 3-hydroxy derivatives **77** and **78** of Kdo or carboxyl-reduction at the proximal Kdo unit as in compound **79** were tolerated by the mAb S25-2 [30] (Scheme 5.15).

Immunization of mice with the neoglycoconjugate derived from the *Burkholderia*-related core unit α-Ko-(2 → 4)-α-Kdo containing the 3-hydroxy analogue of Kdo, D-*glycero*-D-*talo*-oct-2-ulosonic acid (Ko), provided another near germ-line antibody (mAb S67-27) with relaxed binding preferences. Although an additional hydrogen bond was observed in the liganded complex of **78** with mAb S67-27, its binding affinity was equal in comparison to Kdo ligands [31].

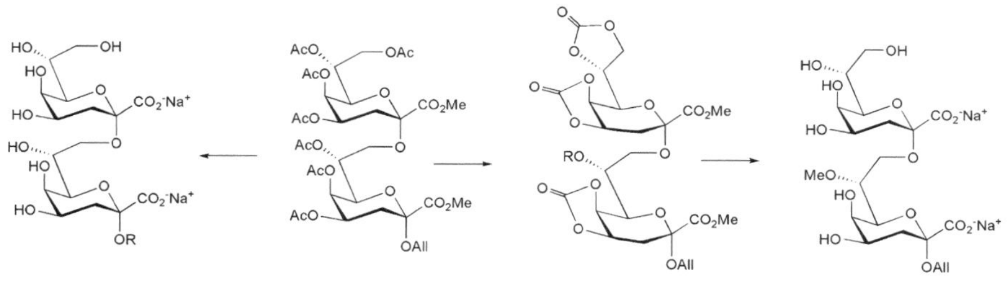

84 R = All
85 R = $(CH_2)_{1-4}CO_2^-Na^+$
86 R = $(CH_2)_2CH{=}CHCO_2^-Na^+$
87 R = $(CH_2)_3SCH_2CO_2^-Na^+$

80

81 R = H
82 R = Me

83

Scheme 5.16 Synthesis of a 7-*O*-methylated Kdo disaccharide and spacer-elongated haptens

The crystal structure also indicated a void in the binding pocket, allowing for binding of side-chain modified Kdo units. The synthesis of a methylated disaccharide analogue was straightforward using the peracetylated allyl glycoside precursor **80** [32]. After Zemplén deacetylation, the resulting intermediate was transformed into the tris-*O*-carbonyl derivative **81** by reaction with trichloromethyl chloroformate/*sym*-collidine in 84% yield. The remaining free hydroxyl group was amenable to further transformations and was methylated using TMS-diazomethane giving **82** in 76% yield. Deprotection under standard conditions afforded the target disaccharide **83** (Scheme 5.16). Notably, the 7-*O*-methyl disaccharide analogue **83** was bound by mAb S67-27 with a 30-fold higher affinity than to any other antigen tested, which was due to additional hydrophobic binding interactions of the methyl group with a proline and isoleucine residue in the binding site [33].

The impact of the spacer group on antibody reactivity was evaluated via the synthesis of modified spacer derivatives **85–87** generated from the α-Kdo-(2 → 8)-α-Kdo allyl disaccharide **84**, comprising shortened as well as chain-elongated linkers containing a terminal carboxyl acid group thereby mimicking the proximal Kdo moiety of the family specific trisaccharide [34].

The α-Kdo-(2 → 8)-α-Kdo allyl disaccharide **84** had previously been crystallized as a disodium salt and the crystal structure revealed an interresidue hydrogen bond extending from the terminal carboxylic group to OH-7 of the proximal Kdo unit indicating potential formation of interresidue lactones [35]. Whereas similar interresidue lactone formation has been extensively studied for neuraminic acid oligomers, the propensity of Kdo residues to form intramolecular lactones has not been addressed in close detail [36, 37]. Usage of neat acetic acid and extended reaction times, produced both 1′ → 7 as well as 1′ → 5 connected Kdo lactones from the respective α-Kdo-(2 → 8 or 4)-α-Kdo allyl disaccharides as evidenced by NMR measurements. Notably, the internal activation of the terminal carboxylic acid group in these disaccharides allows for a subsequent selective modification as demonstrated for the specific formation of the mono methyl

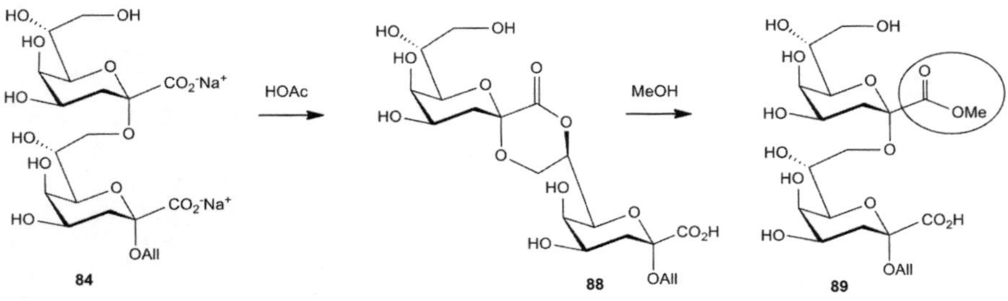

Scheme 5.17 Lactone formation of the (2 → 8)-linked Kdo disaccharide **84**

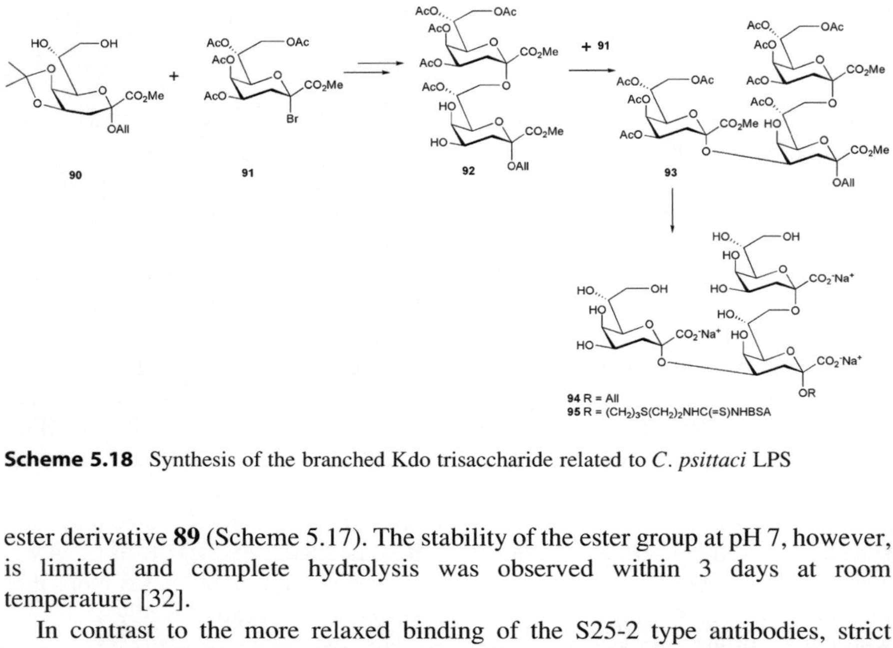

Scheme 5.18 Synthesis of the branched Kdo trisaccharide related to *C. psittaci* LPS

ester derivative **89** (Scheme 5.17). The stability of the ester group at pH 7, however, is limited and complete hydrolysis was observed within 3 days at room temperature [32].

In contrast to the more relaxed binding of the S25-2 type antibodies, strict preference for binding of α-Kdo-(2 → 4)-α-Kdo domains was introduced by the presence of a phenylalanine residue Phe_H99 as seen in the crystal structures of Fab fragments derived from mAb S45-18 [29] and S54-10, respectively [31]. Attempts to generate antibodies directed against the branched tetrasaccharide from *C. psittaci*, however, met with difficulties [38]. By using the tetrasaccharide neoglycoconjugate **75b** in immunization protocols, the α-Kdo-(2 → 4)-α-Kdo-(2 → 4)-α-Kdo part structure turned out as the immunodominant region. Thus, the synthesis of the shortened sequence α-Kdo-(2 → 4)-[α-Kdo-(2 → 8)]-α-Kdo has recently been accomplished [39]. Coupling of the Kdo bromide donor **91** to the diol derivative **90** afforded a mixture of (2 → 8)- and (2 → 7)-disaccharides (Scheme 5.18). Another approach for construction of the *Chlamydia*-specific 2 → 8 linkage was based on iodoalkoxylation of Kdo-glycal ester derivatives, which were efficiently coupled to ring-opened acceptor molecules substituted at C-2 with geminal sulfoxide/alkylsufanyl groups [40]. Following acetylation of the reaction products, the α-(2 → 8)-linked disaccharide could be isolated in 25% yield and subsequent removal of the 4,5-*O*-isopropylidene group gave the diol compound **92** in 82% yield to be subsequently coupled with donor **91** under Helferich conditions. This way, the trisaccharide **93** was isolated in 27% yield and deprotected under alkaline conditions producing the target allyl glycoside **94**. The material was further transformed into the corresponding BSA-conjugate **95** upon Michael addition of cysteamine to the allyl group and subsequent thiophosgene activation.

Immunization experiments using the neoglycoconjugate **95** containing the branched Kdo trisaccharide afforded murine mAb S73-2 and single chain fragments which were specific for *C. psittaci* LPS [41]. Recombinant antibody fragments obtained by bacteriophage display of scFv had a significantly enhanced binding affinity with a K_D in the range of 10 nM and readily discriminated between the branched Kdo trisaccharide and the α-Kdo-(2 → 4)-α-Kdo-(2 → 4)-α-Kdo trisaccharide epitope, respectively.

In conclusion, the results from the immunochemical and structural studies on Kdo specific antibodies clearly supported the immunorelevant role of Kdo determinants as well as the efficient usage of amino acid residues in the binding sites of several monoclonal antibodies recognizing Kdo residues in a broad spectrum of ligands with relaxed binding specificities, but also narrowing down the recognition of Kdo structures with high specificities upon minor adaptation in the heavy chain variable region. Specificity and binding affinities are driven mainly by hydrophobic interactions. It has to be pointed out, however, that the presentation of the Kdo core region when attached to a protein carrier is different to the native steric environment wherein Kdo is linked to the phosphorylated lipid A backbone.

5.4 Synthesis of Core Units Containing 4-Amino-4-deoxy-L-arabinose (Ara4N) Related to *Burkholderia* and *Proteus* Lipopolysaccharide

4-Amino-4-deoxy-L-arabinose (Ara4N) is a frequently detected sugar unit in the lipid A part of various bacterial LPs such as in *Salmonella*, *Proteus*, *Serratia*, *Burkholderia*, *Yersinia*, *Providencia*, *Chromobacterium* or *Pectinatus* spp. occurring in substoichiometric or stoichiometric amounts mainly at the nonreducing end, but also at the reducing end or at both positions of the lipid A backbone [1]. Substitution by Ara4N has been implicated as a major factor of bacterial resistance against polymyxin B and other cationic antimicrobial peptides (CAMP) by counteracting the anionic phosphate and carboxylate groups of the core-lipid A domain. In addition to the occurrence in lipid A, Ara4N residues have been found glycosidically linked to position 8 of Kdo as in the core region of *Proteus* strains but

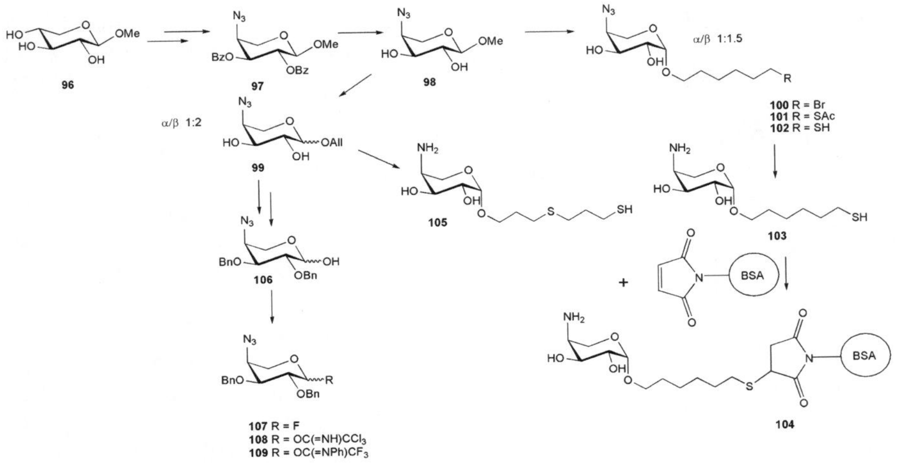

Scheme 5.19 Synthesis of a Ara4N neoglyconjugate and Ara4N glycosyl donors

also linked to carbon 8 of Ko as detected in the core regions of *Burkholderia* and *Serratia marcescens* strains [42–46] (Scheme 5.19).

For the synthesis of Ara4N glycosides, an efficient protocol has been elaborated in excellent overall yield utilizing methyl β-D-xylopyranoside **96** as starting material. Introduction of a good leaving group at O-4 (nosyl, tosyl, mesyl) *via* an intermediate stannylene acetal was followed by benzoylation to afford **97**. Azide displacement with inversion of configuration at C-4 gave the 4-azido-4-deoxy-α-L-arabinoside compound **98** in multigram amounts without recourse to chromatography [47]. To liberate the anomeric aglycon, further transformation involved transglycosylation with allyl alcohol as well as alcohols containing terminal halide substituents resulting in preferential formation of the axial products **99** and **100**, respectively. The β-bromohexyl glycoside **100** was eventually converted into the ω-thiol spacer glycoside **102** *via* the corresponding thioacetate **101**. Reduction of the azide group in the presence of 1,3-propanedithiol afforded **103**, which was subsequently reacted with maleimide-activated BSA to afford the neoglycoconjugate **104**. Notably, the allyl group could also be used as a hook to introduce a thiol-containing linker by reaction of the azide derivative with 1,3-propanedithiol giving the thio-ether bridged thiol spacer glycoside **105**. High-titre polyclonal rabbit and murine antisera were obtained using the neoglycoconjugate **104**. The sera were shown by inhibition experiments to be reactive with Ara4N-residues present in *Proteus* as well as *Burkholderia* LPS, but could not bind to Ara4N units linked to the lipid A domain [47].

The inner core unit of *Burkholderia* is composed of a trisaccharide of the sequence β-L-Ara4N-(1 → 8)-α-Ko-(2 → 4)-α-Kdo, whereas the branched trisaccharide β-L-Ara4N-(1 → 8)-[α-Kdo-(2 → 4)]-α-Kdo is present in the LPS of *Proteus* strains. Following di-*O*-benzylation of **99**, the allyl group was subsequently removed *via* Ir-catalyzed isomerization into the propenyl glycoside followed by hydrolysis to furnish the reducing sugar amenable to activation as glycosyl donor. Glycosyl donors of Ara4N were then elaborated from the hemiacetal **106** by treatment with DAST to give the fluoride donor **107** or by conversion into the corresponding trichloro- as well as the *N*-phenyltrifluoroacetimidate derivatives **108** and **109**, respectively [48]. As glycosyl acceptor, the α-Ko/Kdo allyl glycosides **100** were regioselectively silylated in ~85% yield using DABCO/TBSCl followed by acetylation and controlled treatment with 2% HF in MeCN.

Glycosylation of the alcohol **112** with the fluoride **107** or trichloroacetimidate derivative **108**, respectively, provided glycosides in only moderate yields, whereas the use of the *N*-phenyl trifluoroacetimidate donor **109** turned out as the method of choice producing an anomeric mixture of 8-linked disaccharides (α/β ratio ~ 1:2) and a small amount of 7-substituted material **115** in a combined yield of 80% (Scheme 5.20). Separation was achieved in two steps by chromatography of the crude glycosides to remove first the equatorial component followed by Zemplén deacetylation which gave pure disaccharide **114** and allowed the removal of the 7-linked byproduct **116**. Hydrogenolysis of the benzyl groups with concomitant reduction of the 4′-azido and the anomeric allyl group proved difficult and could be accomplished using $Pd(OH)_2$ in acetic acid affording the Ara4N-Kdo/Ko propyl glycosides **118**. Alternative removal of the benzyl groups using $TiCl_4$ followed by acetylation afforded the 4′-azido disaccharide derivative **119** amenable for subsequent introduction of spacer groups in order to generate neoglycoconjugates

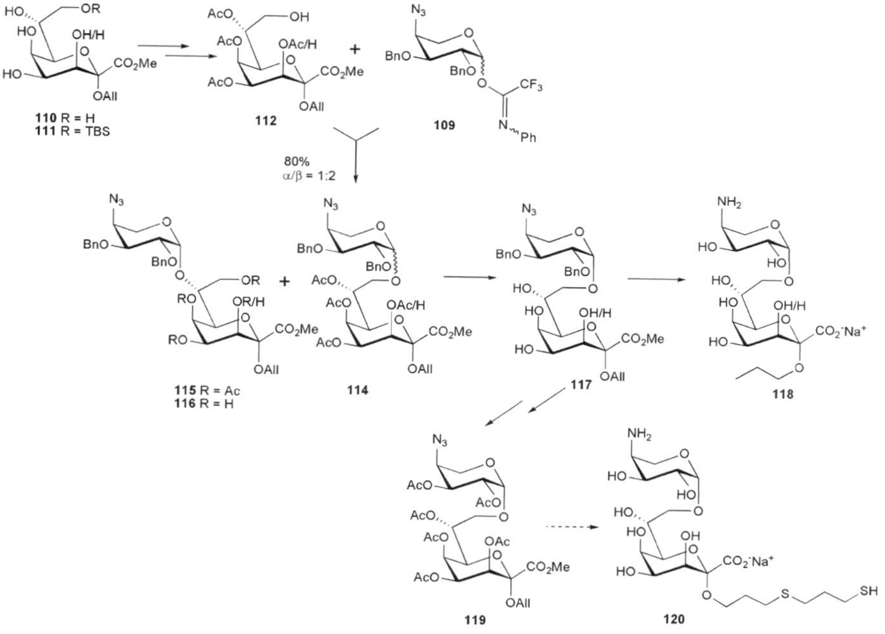

Scheme 5.20 Synthesis of *Burkholderia* and *Proteus* LPS core disaccharides

but also allowing access to disaccharide glycosyl donors *via* removal of the anomeric allyl group.

5.5 Synthesis of Core Oligosaccharides Containing L-*glycero*-D-*manno*-heptose Residues

Since the chemical synthesis of the major constituents of the heptose region comprising L-*glycero*-D-*manno*-heptose (L,D-Hep) and its 6-epimer has recently been summarized [2, 6, 49], it will not be addressed in this review. The main approaches relied on Sharpless asymmetric epoxidation or OsO_4-promoted dihydroxylation of an exocyclic double bond of mannopyranosides [50–52]. Further developments in heptose chemistry will be described herein within the context of oligosaccharide synthesis related to biomedically relevant core determinants.

5.5.1 Synthesis of Oligosaccharides Related to the Inner Core of *Haemophilus* and *Neisseria* LPS

Within the framework of a long-term research program, the group of Oscarson has undertaken the challenging synthesis of larger fragments derived from the heterogeneous core units from non-typeable *Haemophilus influenzae* strains and lipooligosaccharides (LOS) of *Neisseria meningitidis* aimed at the definition of the molecular basis of immune recognition of *Haemophilus* as well as *Neisseria* LPS [53–55]. The use of defined synthetic and pure oligosaccharides is of necessity in order to provide a rational basis for future vaccine development. Several syntheses of larger oligosaccharide fragments equipped with reactive spacer functions have been elaborated in the past and have already been summarized

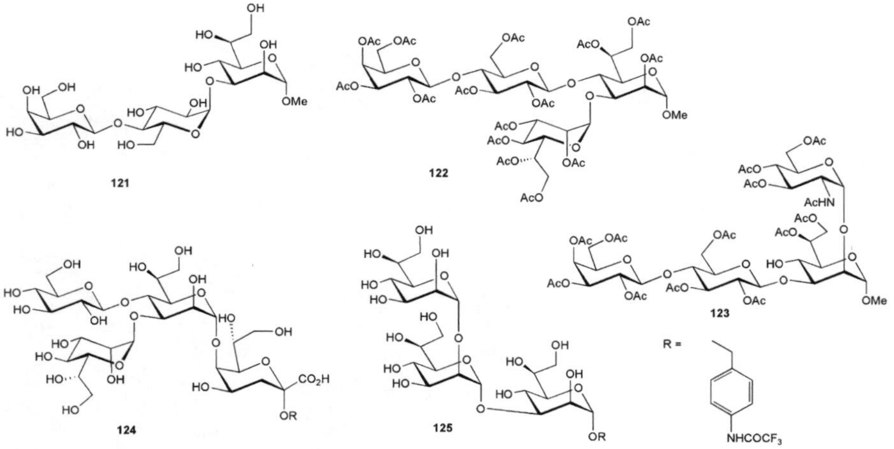

Scheme 5.21 Synthetic core fragments from *Haemophilus* and *Neisseria* LPS

in previous reviews (Scheme 5.21) [3]. Synthetic compounds prepared in the past also comprise the α-lactosyl-(1 → 3)-L-*glycero*-α-D-*manno*-heptopyranoside **121** as well as the 3,4- and 2,3-dibranched acetylated derivatives **122** and **123**, respectively, corresponding to part structures of *N. gonorrhoeae* strain 15,253 [56–58]. Recent studies employing LOS of this strain as affinity ligand indicated that human antibodies recognize several epitopes including the Kdo region but also the 3,4 branched as well as the 2,3;3,4 dibranched heptosyl epitopes [59].

Major accomplishments had previously been achieved in the synthesis of linear and branched spacer-equipped glycosides such as **124** and **125**, respectively, including also the α-(1 → 5) linkage of heptose to Kdo [54]. A complex synthetic issue in this context resides in the assembly of the 3,4-branched internal L,D-Hep moiety. This challenge could be successfully resolved by the use of a 1,6-anhydro heptosyl intermediate allowing for a release of steric congestion, which otherwise prevents further glycosylation at either 3- or 4-position of heptose once anyone of these sites has been glycosylated. The glycosylated 1,6-anhydro compounds could then be further transformed into suitable glycosyl donors following acetolysis of the 1,6-anhydro bridge and conversion of the resulting anomeric acetate into a thioglycoside moiety. This approach is illustrated in more detail by the recently described synthesis of a branched tetrasaccharide corresponding to the conserved inner core of *N. meningitidis* containing an α-linked GlcNAc residue at position 2 as well as a 2-aminoethyl phosphate substituent at position 6 of the α-(1 → 3)-linked heptose unit. In a first attempt, a 2 + 2 blockwise approach was tested using the previously described 1,6-anhydro acceptor derivative **126** and the preassembled α-(1 → 7) linked disaccharide thioglycoside donor **127** [60]. The donor had been equipped with a butane-2,3-dimethoxy acetal to secure α-selectivity in the ensuing glycosylation step (Scheme 5.22).

Despite several promoter and solvent conditions tested in glycosylations using both the ethyl-1-thio donor **127** as well as the corresponding, more reactive sulfoxide donor **128**, only small amounts of the tetrasaccharide product could be generated. Hence a stepwise approach was followed by first connecting the orthogonally protected heptosyl residue to O-3 of the 1,6-anhydro unit leading to **129**, followed by subsequent oxidative removal of the 2-*O*-methoxybenzyl protecting group to furnish trisaccharide acceptor **130**. Coupling of the trisaccharide acceptor **130** with the 2-azido-2-deoxy-β-D-glucopyranosyl ethyl-1-thio donor **131** was efficiently performed by activation with NIS/AgOTf resulting in an anomeric mixture of tetrasaccharides **132** in 95% yield. Other conditions tested such as TMSOTf, BF_3 · etherate, DMTST or dimethyldisulfide/triflic anhydride proved inefficient. The anomers were separated following removal of the 3,4-acetal and the material was further processed by $Sc(OTf)_3$ catalyzed acetolysis into the heptosyl acetate **133**. Subsequent conversion of the 2-azido function into the *N*-acetyl group and activation of the anomeric acetate in **134** as ethyl-1-thio glycoside giving **135** allowed the introduction of the Cbz-protected amino-spacer group. Eventually, the 6-*O*-chloroacetate was cleaved from tetrasaccharide **136** and the Boc-protected phosphodiester was installed using phosphoramidite chemistry. The differentially protected amino groups allow for selective conjugation of the spacer group in

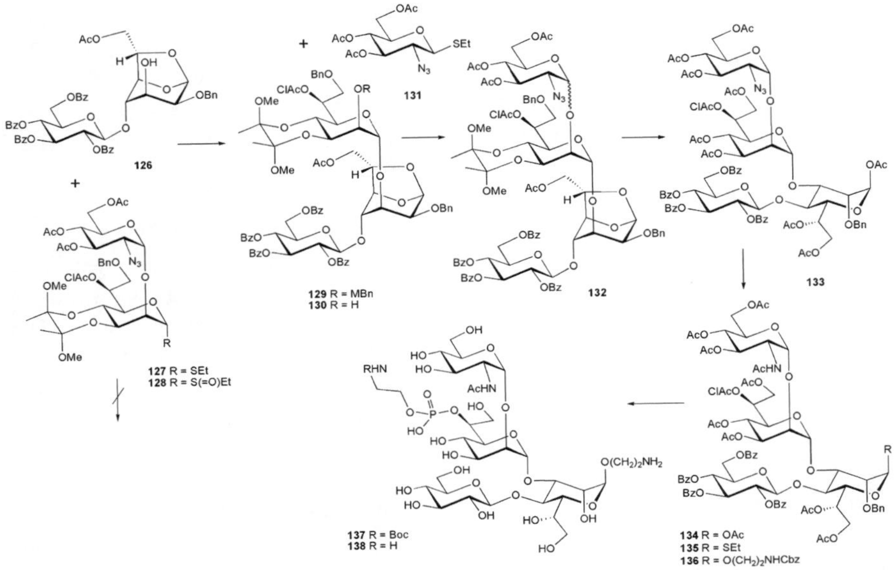

Scheme 5.22 Synthesis of a branched tetrasaccharide from *N. meningitidis* LPS

compound **137** to a protein carrier followed by final deprotection of the 2-aminoethyl phosphate moiety. Thus, deacylation and hydrogenolysis of the Cbz protecting group afforded the free spacer derivative **137**. Acid treatment was applied to release the Boc-protecting group to afford the target tetrasaccharide **138** [61].

Recently 2-aminoethyl phosphate-containing oligosaccharides have been elaborated within a novel strategy based on derivatization of de-*O*-acylated LPS from *N. meningitidis* [62]. The 2-aminoethyl phosphate moiety was preserved during chemical derivatization by introduction of a Boc protecting group. Notably, a deamidase from *Dictyostelium discoideum* was used to liberate the 2-amino group of the reducing glucosamine unit of lipid A, which was then covalently linked to a maleimide unit to generate **139**. Subsequent deblocking of the Boc group and coupling to SH-groups of a CRM-modified protein provided a vaccine conjugate with a high loading of antigen (Scheme 5.23).

A related synthetic approach had previously also been employed for the assembly of the phosporylated tetrasaccharide inner core of *Haemophilus influenzae* containing a terminal heptopyranosyl unit [63]. Again, orthogonal amino-protecting groups had been chosen to allow for selective modification of the spacer moiety while keeping the 2-aminoethyl phosphate unaffected. Chain extension of trisaccharide acceptor **130** with the heptosyl thioglycoside donor **140** afforded the α-(1 → 2)-connected tetrasaccharide **141**. Acetolysis and further transformations gave tetrasaccharide **142**. Manipulation of the anomeric center via the methyl-1-thio derivative **143** afforded the spacer compound **144**, which was phosphorylated and globally deprotected – while keeping the Boc group untouched – resulting in

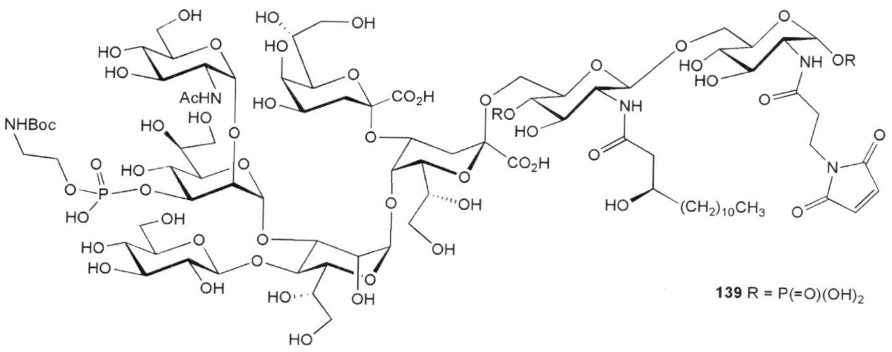

Scheme 5.23 Selective derivatization of partially de-O-acylated LOS from *N. meningitidis*

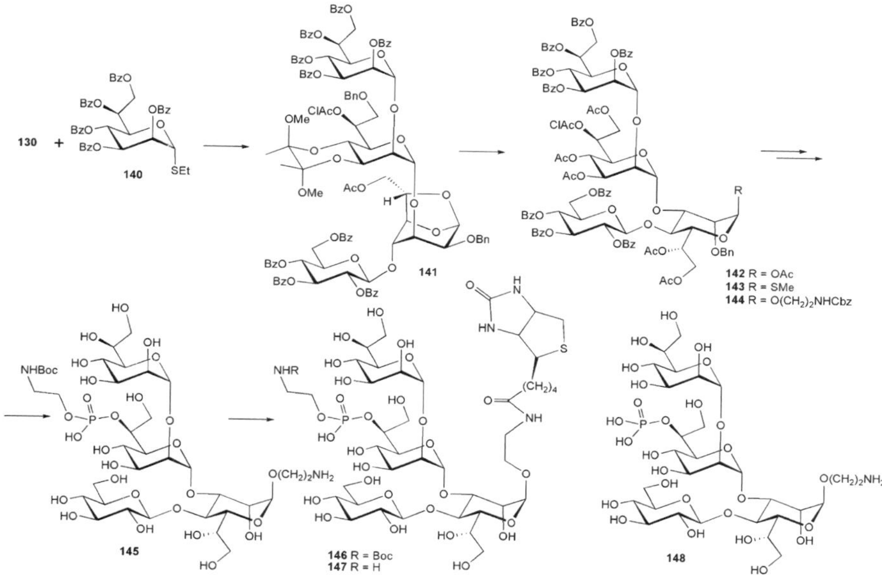

Scheme 5.24 Synthesis of the phosphorylated tetrasaccharide core unit from *H. influenzae*

formation of **145**. Finally, as test compound for the screening of monoclonal antibodies against nontypeable *H. influenzae*, a biotin label was installed at the spacer group providing compounds **146** and **147**, respectively (Scheme 5.24). For comparison, also the phosphomonoester **148** was synthesized.

Eventually this strategy has been further exploited using orthogonal protecting group patterns at the 6- and 3-position, respectively, allowing for the introduction of 2-aminoethyl phosphate residues at a later stage of the synthesis (Scheme 5.25). Along similar lines as described above, the 6-*O*-allyl-3-*O*-chloroacetyl protected trisaccharide **149** was converted into the spacer-equipped trisaccharide **152**,

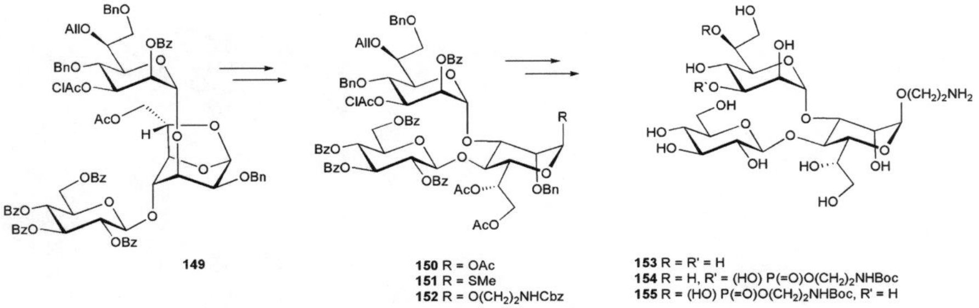

Scheme 5.25 Synthesis of 2-aminoethylphosphate containing trisaccharide units from *H. influenzae*

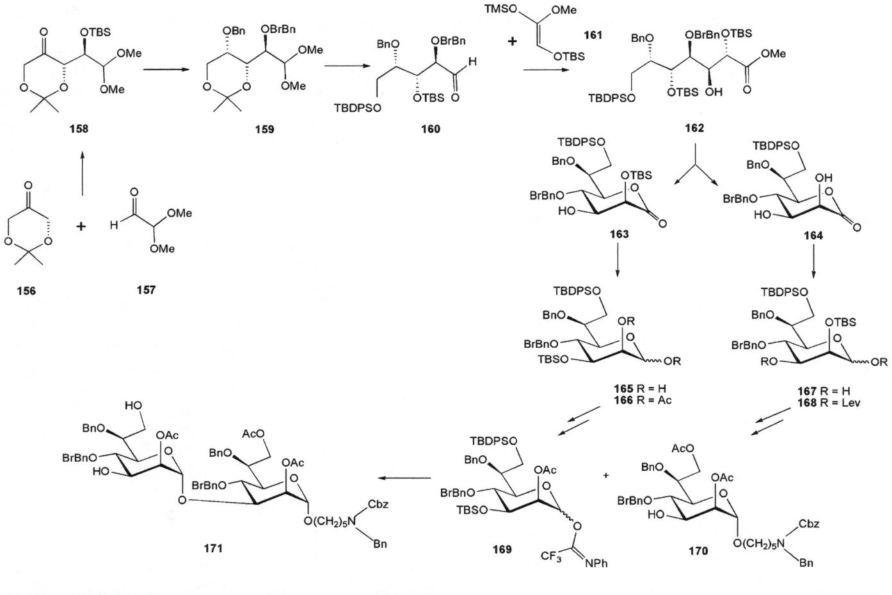

Scheme 5.26 Diastereoselective synthesis of L-*glycero*-D-*manno* heptose and conversion into a suitably protected disaccharide building block

selectively phosphorylated at either position and finally deprotected to give the Boc-protected 2-aminoethyl phosphodiester derivatives **154** and **155**, respectively [64]. In addition the nonphosphorylated compound **153** was also prepared.

Recently, a *de novo* synthesis of a central heptobiose building block for the future preparation of *Yersinia pestis* core structures has been communicated [65]. The enantioselective preparation of an orthogonally protected heptose precursor was elaborated *via* an *anti*-selective aldol reaction catalyzed by L-proline to give the ketone **158**. Selective reduction of the carbonyl group by selectride and further protecting group manipulation afforded the *p*-bromobenzyl protected intermediate **159**, which was further transformed into the silyl-protected dimethyl acetal **160**

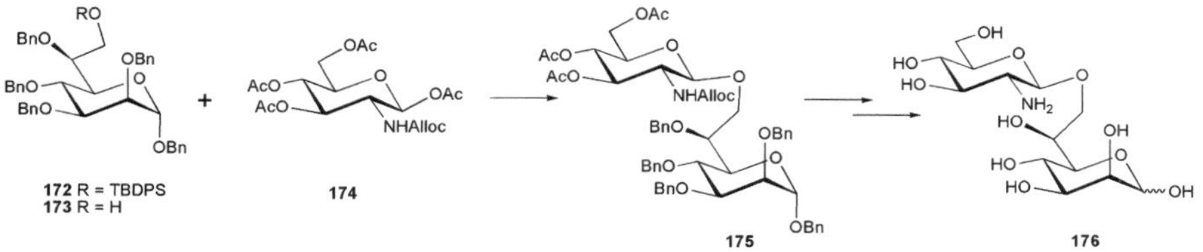

Scheme 5.27 Synthesis of a disaccharide corresponding to LPS core constituents from *Vibrio* and *Aeromonas* strains

(Scheme 5.26). Cleavage of the acetal was followed by a second highly selective aldol reaction with the silyl enol ether **161** to produce **162**, which was converted into a mixture of lactones **163** and **164**, respectively. Reduction of the heptonolactone **163** was followed by processing the intermediate acetate **166** into the *N*-phenyltrifluoroacetimidate donor **169**. Conversely, acceptor **170** was generated from the diol **167** *via* intermediate formation of levulinoyl ester **168**. Introduction of the Cbz-protected spacer group at the anomeric center and removal of the 3-*O* – levulinic ester group by hydrazinolysis furnished **170**. Eventually the coupling reaction – promoted by TMSOTf – followed by desilylation afforded the disaccharide building block **171**. The building block has been designed for further transformation into core structures related to *Yersinia pestis* LPS.

5.5.2 Synthesis of Core Units from *Vibrio* and *Aeromonas*

The disaccharide β-GlcN-(1 → 7)-L,D-Hep occurring in the LPS core of *Vibrio ordalii* and *Aeromonas salmonicida* had previously been synthesized by Paulsen using a heptofuranoside acceptor and a 2-azido-2-deoxy-glucopyranosyl bromide donor [66]. Recently, an alternative approach towards this disaccharide was reported [67]. The disaccharide was assembled using a TBDPS-protected benzyl heptopyranoside **172** as precursor. Removal of the 7-*O*-silyl ether was followed by coupling of the resulting alcohol **173** with an allyloxycarbonyl-protected glucosamine donor **174** in 89% yield. Deprotection of **175** *via* Zemplén deacetylation, Pd-catalyzed deallylation and hydrogenolysis provided the target compound **176** in high overall yield (Scheme 5.27).

5.6 Synthesis of Outer Core Oligosaccharides

5.6.1 Synthesis of a Distal Trisaccharide of the Core of *Escherichia coli* K-12

A similar approach for the preparation of the β-GlcN-(1 → 7)-L,D-Hep unit was followed in the synthesis of a trisaccharide corresponding to a distal unit in *Escherichia coli* K-12. First, heptosyl acceptor **173** was elongated at position 7

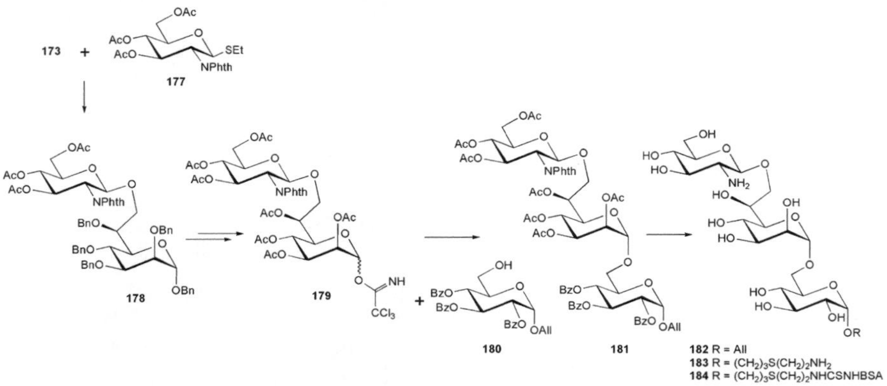

Scheme 5.28 Synthesis of an outer unit of *E.coli* K-12 LPS

by reaction with thioglycoside donor **177** to give the protected disaccharide **178** which was further developed into the disaccharide trichloroacetimidate donor **179** (Scheme 5.28). Condensation of donor **179** with the allyl glucopyranoside acceptor **180** provided the trisaccharide **181**. Alternatively, a stepwise approach was applied to prepare the trisaccharide followed by deprotection to afford **182**. The compound was further transformed *via* reaction of the anomeric allyl group with cysteamine to give **183**. Subsequent activation with thiosphosgene and coupling to BSA furnished the neoglycoconjugate **184** for immunochemical studies [68].

5.6.2 Synthesis of the α-Chain Pentasaccharide of the Lipooligosaccharide of *Neisseria gonorrhoeae* and *Neisseria meningitidis*

Pentasaccharide units corresponding to α-chain lipooligosaccharides of *Neisseria gonorrhoeae* and *N. meningitidis* have been synthesized comprising a common lacto-*N*-neotetraose core which had been extended with a D-GalNAc and Neu5Ac residue, respectively [69]. Di- and trisaccharide building blocks were preassembled using protected 2-amino-2-deoxy-*galacto*- and –*gluco* thioglycoside donors **187** and **190**, respectively, which were coupled to methoxyphenyl glycoside acceptors **186** and **191**, respectively, yielding the lactosamine derivative **188** and the trisaccharide **192** in good yields and anomeric selectivity (Scheme 5.29). The trisaccharide **192** served as central intermediate to act as glycosyl acceptor upon removal of the terminal 3- and 4-*O*-acetyl residues to afford **193**. Coupling of the acceptor derivative **193** with the disaccharide trichloroacetimidate donor **189** obtained from **188** *via* oxidative removal of the anomeric methoxyphenyl group afforded the pentasaccharide **194** in 84% yield. Compound **194** was deprotected and *N*-acetylated to eventually furnish the pentasaccharide **195** corresponding to the α-chain pentasaccharide of *N. gonorrhoeae* LOS. Stepwise elongation of the

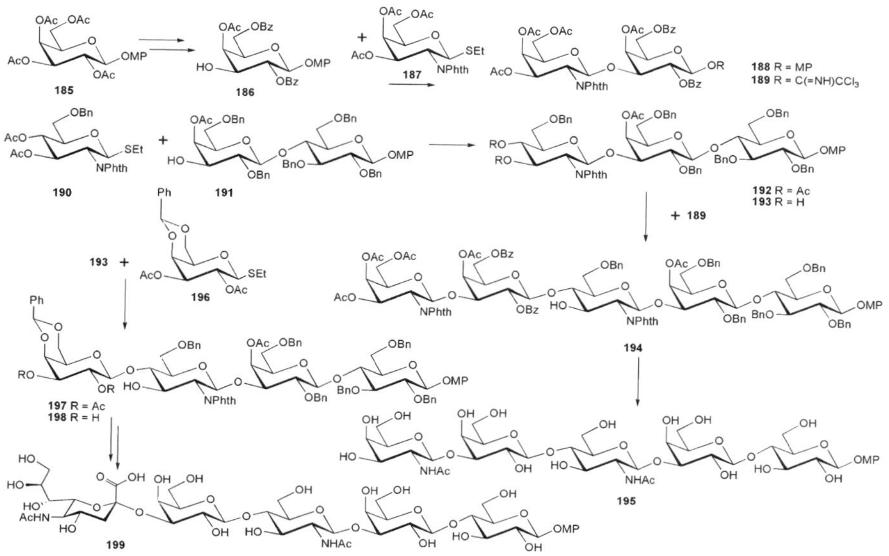

Scheme 5.29 Synthesis of lipooligosaccharides from *N. gonorrhoeae* and *N. meningitidis*

trisaccharide intermediate **193** with a 4,6-*O*-benzylidene protected thiogalactoside donor **196** produced the tetrasaccharide **197**, which after removal of the 2- and 3-*O*-acetyl groups afforded the diol **198**.

Subsequent condensation of **198** with a phenylthio glycosyl donor of *N*-acetylneuraminic acid gave the protected tetrasaccharide which was globally deprotected to afford the pentasaccharide glycoside **199**, corresponding to *N. meningitidis* LOS.

5.6.3 Synthesis of the Outer Core Region of the Lipopolysaccharide of *Pseudomonas aeruginosa*

The outer core from *P. aeruginosa* is composed of two glycoforms which modify the surface properties of the bacterial cell wall and which interact with the cystic fibrosis transmembrane conductance regulator, thereby enabling efficient clearing of the invading bacteria in healthy hosts [70]. For a detailed study on the binding interactions, several oligosaccharides corresponding to glycoform I have been synthesized including modifications of the GalN residue by *N*-acetylation as well as *N*-alanylation, respectively [71]. First, the protected disaccharide intermediate **200** was condensed at position 4 with the glucosyl *N*-phenyltrifluoroacetimidate donor **201** leading to the branched trisaccharide **202** in 62% yield (Scheme 5.30). Deprotection of **193** with reduction of the 2-azido group furnished the amine **203** which was coupled to the *N*-hydroxysuccinimide ester of Boc-protected L-alanine

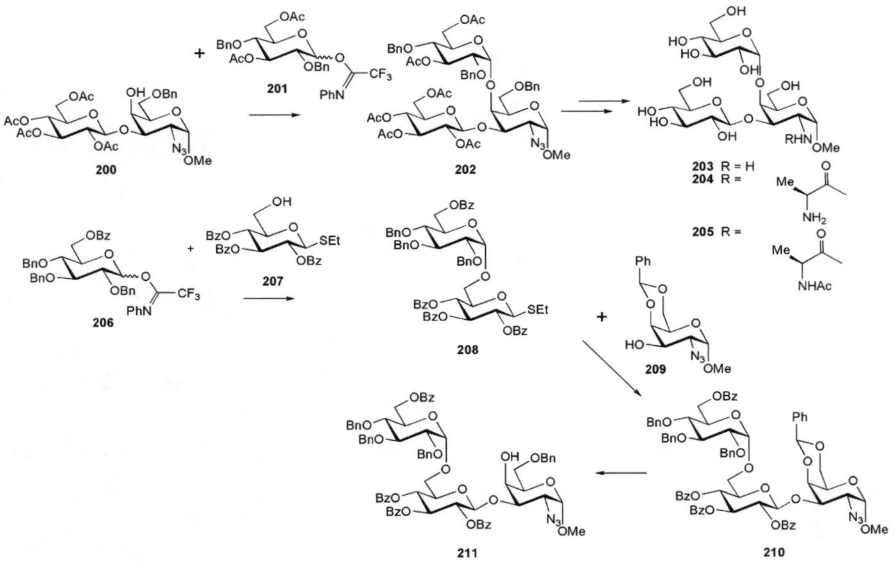

Scheme 5.30 Synthesis of outer core trisaccharide units from *P. aeruginosa*

followed by acid hydrolysis and *N*-acetylation to give the trisaccharide glycosides **204** and **205**, respectively.

Proceeding towards the pentasaccharide derivative, a series of α-selective glucosyl donors was tested and applied for the preparation of the α-(1 → 6)-linked disaccharide thioglycoside **208**. The *N*-phenyltrifluoroacetimidate donor **206** could thus be coupled to alcohol **207** in 63% yield [72]. Further chain elongation employing the 2-azido-2-deoxy-galactosyl acceptor **209** afforded the trisaccharide **210** in 70% yield. The further conversion into the 4-OH acceptor derivative **211**, however, met with difficulties in purification and hence, the synthetic scheme was redesigned. The assembly of the pentasaccharide was then based on a stepwise approach utilizing the 6-*O*-chloroacetyl-protected thioglycoside donor **212** which was connected to the 3-position of the methyl α-D-galactopyranoside **209**. Regioselective reductive opening of the 4,6-*O*-benzylidene acetal of disaccharide **213** was followed by condensation with a glucosyl donor to furnish the trisaccharide **214** in good yield (Scheme 5.31). Removal of the 6-*O*-chloroacetate protecting group in **214** was followed by glycosylation of acceptor **215** producing the tetrasaccharide **216** with good α-selectivity. Finally, selective de-*O*-acetylation of **216** gave alcohol **217** to be followed by silver triflate promoted coupling with the rhamnosyl donor **218** leading to the pentasaccharide **219**. Removal of the protecting groups was achieved by cleavage of the benzoyl ester groups to afford **220**. Dithiothreitol promoted reduction of the 2-azido group was followed either by conventional *N*-acetylation or introduction of a Boc protected L-alanine unit and catalytic hydrogenolysis of the benzyl groups leading eventually to

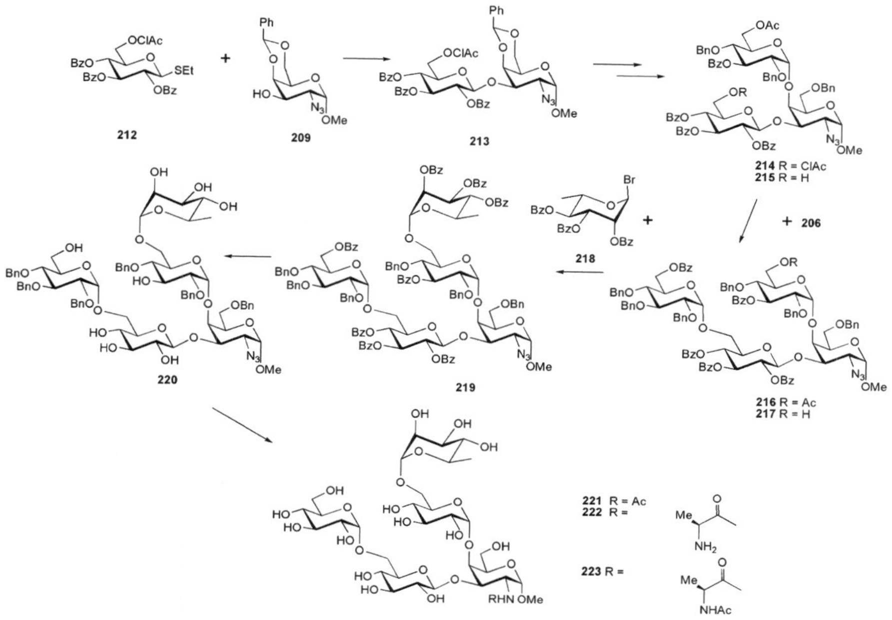

Scheme 5.31 Synthesis of outer core pentasaccharide fragments from *P. aeruginosa*

the pentasaccharide target compounds **221** and **222** (after removal of the Boc-protecting group). *N*-Acetylation of **222** finally gave pentasaccharide **223**.

The compounds will be tested for binding properties to the cystic fibrosis transmembrane conductance regulator [73].

References

1. Holst O (2007) The structures of core regions from enterobacterial lipopolysaccharides – an update. FEMS Microbiol Lett 271:3–11
2. Hansson J, Oscarson S (2000) Complex bacterial carbohydrate surface antigen structures: syntheses of Kdo- and heptose-containing lipopolysaccharide core structures and anomerically linked phosphodiester-linked oligosaccharide structures. Curr Org Chem 4:535–564
3. Kosma P (2009) Chemical synthesis of the core oligosaccharide of bacterial lipopolysaccharide. In: Moran A, Brennan P, Holst O, von Itzstein M (eds) Microbial glycobiology: structures, relevance and applications. Elsevier, San Diego, pp 429–454
4. Oscarson S (1997) Synthesis of oligosaccharides of bacterial origin containing heptoses, uronic acids and fructofuranoses as synthetic challenges. Top Curr Chem 186:171–202
5. Pozsgay V (2003) Chemical synthesis of bacterial carbohydrates. In: Wong SYC, Arsequell G (eds) Immunobiology of carbohydrates. Kluwer Academic/Plenum Publishers, New York, pp 192–273
6. Pozsgay V (2008) Recent developments in synthetic oligosaccharide-based bacterial vaccines. Curr Top Med Chem 8:126–140
7. Li LS, Wu YL (2003) Recent progress in the syntheses of higher 3-deoxy-octulosonic acids and their derivatives. Curr Org Chem 7:447–475

8. Kuboki A, Tajimi T, Tokuda Y, Kato Di, Sugai T, Ohira S (2004) Concise synthesis of 3-deoxy-D-*manno*-oct-2-ulosonic acid (KDO) as a protected form based on a new transformation of α,β-unsaturated ester to α-oxocarboxylic acid ester via diol cyclic sulfite. Tetrahedron Lett 45:4545–4548
9. Kikelj V, Plantier-Royon R, Portella C (2006) A new short and efficient route to 3-deoxy-D-*manno*-oct-2-ulosonic acid (KDO) and 3-deoxy-D-*arabino*-hept-2-ulosonic acid (DAH). Synthesis 1200–1204
10. Hekking KFW, van Delft FL, Rutjes FPJT (2003) Ring-closing metathesis of α-substituted enol ethers: application to the shortest synthesis of KDO. Tetrahedron 59:6751–6758
11. Hekking KFW, Moelands MAH, van Delft FL, Rutjes FPJT (2006) An in-depth study on ring-closing metathesis of carbohydrate-derived α-alkoxyacrylates: efficient syntheses of DAH, KDO, and 2-deoxy-β-KDO. J Org Chem 71:6444–6450
12. Sugisaki CH, Ruland Y, Baltas M (2003) Direct access to furanosidic eight-membered ulosonic esters from cis-α,β-epoxy aldehydes. Eur J Org Chem 672–688
13. Kim BG, Schilde U, Linker T (2005) New radical approaches to 3-deoxy-D-oct-2-ulosonic acids (KDO). Synthesis 1507–1513
14. Railton CJ, Clive DLJ (1996) Wittig chain extension of unprotected carbohydrates: formation of carbohydrate-derived α,β-unsaturated esters. Carbohydr Res 281:69–77
15. Ichiyanagi T, Sakamoto N, Ochi K, Yamasaki R (2009) A chemical synthesis of 3-deoxy-D-*manno*-2-octulosonic acid from D-mannose. J Carbohydr Chem 28:53–63
16. Winzar R, Philips J, Kiefel MJA (2010) Simple synthesis of C-8 modified 2-keto-3-deoxy-D-*manno*-octulosonic acid (KDO) derivatives. Synlett 583–586
17. Sugai T, Shen GJ, Ichikawa Y, Wong CH (1993) Synthesis of 3-deoxy-D-*manno*-2-octulosonic acid (KDO) and its analogs based on KDO aldolase-catalyzed reactions. J Am Chem Soc 115:413–421
18. Lamble HJ, Royer SF, Hough DW, Danson MJ, Taylor GL, Bull SDA (2007) Thermostable aldolase for the synthesis of 3-deoxy-2-ulosonic acids. Adv Synth Catal 349:817–821
19. Dean SM, Greenberg WA, Wong C-H (2010) Recent advances in aldolase-catalyzed asymmetric synthesis. Adv Synth Catal 349:1308–1320
20. Guerard C, Demuynck C, Bolte J (1999) Enzymatic synthesis of 3-deoxy-D-*manno*-2-octulosonic acid and analogues: a new approach by a non metabolic pathway. Tetrahedron Lett 40:4181–4182
21. Crestia D, Demuynck C, Bolte J (2004) Transketolase and fructose-1,6-bisphosphate aldolase, complementary tools for access to new ulosonic acid analogues. Tetrahedron 60: 2417–2425
22. Gillingham DG, Stallforth P, Adibekian A, Seeberger PH, Hilvert D (2010) Chemoenzymatic synthesis of differentially protected 3-deoxysugars. Nat Chem 2:102–105
23. Fujimoto Y, Iwata M, Imakita N, Shimoyama A, Suda Y, Kusumoto S, Fukase K (2007) Synthesis of immunoregulatory *Helicobacter pylori* lipopolysaccharide partial structures. Tetrahedron Lett 48:6577–6581
24. Mannerstedt K, Ekelöf K, Oscarson S (2007) Evaluation of Kdo as glycosyl donors. Carbohydr Res 342:631–637
25. Gronow S, Lindner B, Brade H, Müller-Loennies S (2008) Kdo-(2 → 8)-Kdo-(2 → 4)-Kdo but not Kdo-(2 → 4)-Kdo-(2 → 4)-Kdo is an acceptor for transfer of L-*glycero*-α-D-*manno*-heptose by *Escherichia coli* heptosyltransferase I (WaaC). Innate Immun 15:13–23
26. Rund S, Lindner B, Brade H, Holst O (2000) Structural analysis of the lipopolysaccharide from *Chlamydophila psittaci* strain 6BC. Eur J Biochem 267:5717–5726
27. Kosma P, Brade H, Evans S (2008) Lipopolysaccharide antigens of *Chlamydia*. In: Roy R (ed) Carbohydrate-based vaccines, vol 989, ACS symposium series. American Chemical Society, Washington DC, pp 239–257
28. Müller-Loennies S, MacKenzie R, Patenaude SI, Evans SV, Kosma P, Brade H, Brade L, Narang S (2000) Characterization of high affinity monoclonal antibodies specific for chlamydial lipopolysaccharide. Glycobiology 10:121–130

29. Nguyen HP, Seto NOL, MacKenzie CR, Brade L, Kosma P, Brade H, Evans SV (2003) Murine germline antibodies recognize multiple carbohydrate epitopes by flexible utilization of binding site residues. Nat Struct Biol 10:1019–1025
30. Brooks CL, Müller-Loennies S, Brade L, Kosma P, Hirama T, MacKenzie CR, Brade H, Evans SV (2008) Exploration of specificity in germline monoclonal antibody recognition of a range of natural and synthetic epitopes. J Mol Biol 377:450–468
31. Brooks CL, Müller-Loennies S, Borisova SN, Brade L, Kosma P, Hirama T, MacKenzie CR, Brade H, Evans SV (2010) Antibodies raised against chlamydial lipopolysaccharide antigens reveal convergence in germline gene usage and differential epitope recognition. Biochemistry 49:570–581
32. Sixta G, Wimmer K, Hofinger A, Brade H, Kosma P (2009) Synthesis and antigenic properties of C-7-modified Kdo mono- and disaccharide ligands and Kdo disaccharide interresidue lactones. Carbohydr Res 344:1660–1669
33. Brooks CL, Blackler RJ, Sixta G, Kosma P, Müller-Loennies S, Brade L, Hirama T, MacKenzie CR, Brade H, Evans SV (2010) The role of CDR H3 in antibody recognition of a synthetic analogue of a lipopolysaccharide antigen. Glycobiology 20:148–157
34. Sixta G, Hofinger A, Kosma P (2007) Synthesis of spacer-containing chlamydial disaccharides as analogues of the α-Kdo*p*-(2 → 8)-α-Kdo*p*-(2 → 4)-α-Kdo*p* trisaccharide epitope. Carbohydr Res 342:576–585
35. Mikol V, Kosma P, Brade H (1994) Crystal and molecular structure of allyl *O*-(sodium 3-deoxy-α-D-*manno*-2-octulopyranosylonate)-(2 → 8)-*O*-(sodium 3-deoxy-α-D-*manno*-2-octulopyranosidonate)-monohydrate. Carbohydr Res 263:35–42
36. Pudelko M, Lindgren A, Tengel T, Reis CA, Elofsson M, Kihlberg J (2006) Formation of lactones from sialylated MUC1 glycopeptides. Org Biomol Chem 4:713–720
37. Cheng M-C, Lin C-H, Khoo K-H, Wu S-H (1999) Regioselective lactonization of α-(2 → 8)-trisialic acid. Angew Chem Int Ed Engl 38:686–689
38. Müller-Loennies S, Gronow S, Brade L, MacKenzie R, Kosma P, Brade H (2006) A monoclonal antibody that recognizes an epitope present in the lipopolysaccharide of *Chlamydiales* differentiates *Chlamydophila psittaci* 6BC from *Chlamydophila pneumoniae* and *Chlamydia trachomatis*. Glycobiology 16:184–196
39. Kosma P, Hofinger A, Müller-Loennies S, Brade H (2010) Synthesis of a neoglycoconjugate containing a *Chlamydophila psittaci*-specific branched Kdo trisaccharide epitope. Carbohydr Res 345:704–708
40. Tanaka H, Takahashi D, Takahashi T (2006) Stereoselective synthesis of oligo-α(2,8)-3-deoxy-D-*manno*-2-octulosonic acid derivatives. Angew Chem Int Ed Engl 45:770–773
41. Gerstenbruch S, Brooks CL, Kosma P, Brade L, MacKenzie CR, Evans SV, Brade H, Müller-Loennies S (2010) Analysis of cross-reactive and specific anti-carbohydrate antibodies against lipopolysaccharide from *Chlamydophila psittaci*. Glycobiology 20:461–472
42. Sidorczyk Z, Kaca W, Brade H, Rietschel ET, Sinnwell V, Zähringer U (1987) Isolation and structural characterization of an 8-*O*-(4-amino-4-deoxy-β-L-arabinosyl)-3-deoxy-D-*manno*-octulosonic acid disaccharide in the lipopolysaccharide of a *Proteus mirabilis* deep rough mutant. Eur J Biochem 168:269–273
43. Vinion-Dubiel AD, Goldberg JB (2003) Lipopolysaccharide of *Burkholderia cepacia* complex. J Endotoxin Res 9:201–213
44. De Soyza A, Silipo A, Lanzetta R, Govan JR, Molinaro A (2008) Chemical and biological features of *Burkholderia cepacia* lipopolysaccharides. Innate Immun 14:127–144
45. Isshiki Y, Kawahara K, Zähringer U (1998) Isolation and characterization of disodium (4-amino-4-deoxy-β-L-arabinosyl)-(1 → 8)-(D-*glycero*-D-*talo*-2-octulosonate)-(2 → 4)-(methyl 3-deoxy-D-*manno*-2-octulopyranosid)onate from the lipopolysaccharide of *Burkholderia cepacia*. Carbohydr Res 313:21–27
46. Vinogradov E, Lindner B, Seltmann G, Radziejewska-Lebrecht J, Holst O (2006) Lipopolysaccharides from *Serratia marcescens* possess one or two 4-amino-4-deoxy-L-arabinopyranose

1-phosphate residues in the lipid A and D-*glycero*-D-*talo*-oct-2-ulosonic acid in the inner core region. Chem Eur J 12:6692–6700
47. Müller B, Blaukopf M, Hofinger A, Zamyatina A, Brade H, Kosma P (2010) Efficient synthesis of 4-amino-4-deoxy-L-arabinose and of spacer-equipped 4-amino-4-deoxy-L-arabinopyranosides by transglycosylation. Synthesis 3143–3151
48. Blaukopf M, Müller B, Brade H, Kosma P (2010) Chemical synthesis of the inner core of *Burkholderia* LPS. In: Abstracts of the 25th international carbohydrate symposium, Tokyo, Japan, 2–6 Aug 2010
49. Kosma P (2008) Occurrence, synthesis and biosynthesis of bacterial heptoses. Curr Org Chem 12:1021–1039
50. Gurjar MK, Talukdar A (2004) Synthesis of the terminal disaccharide unit of *Klebsiella pneumoniae* ssp. R20. Tetrahedron 60:3267–3271
51. Crich D, Banerjee A (2005) Synthesis and stereoselective glycosylation of D- and L-*glycero*-β-D-*manno*-heptopyranoses. Org Lett 7:1395–1398
52. Jaipuri FA, Collet YM, Pohl NL (2008) Synthesis and quantitative evaluation of *glycero*-D-*manno*-heptose binding to concanavalin A by fluorous-tag assistance. Angew Chem Int Ed Engl 47:1707–1710
53. Bernlind C, Oscarson S (1997) Synthesis of D-*glycero*-D-*manno*-heptopyranose-containing oligosaccharide structures found in lipopolysaccharides from *Haemophilus influenzae*. Carbohydr Res 297:251–260
54. Bernlind C, Oscarson S (1998) Synthesis of a branched heptose- and Kdo-containing common tetrasaccharide core structure of *Haemophilus influenzae* lipopolysaccharides via a 1,6-anhydro-L-*glycero*-β-D-*manno*-heptopyranose intermediate. J Org Chem 63: 7780–7788
55. Bernlind C, Bennett S, Oscarson S (2000) Synthesis of a d, d- and l, d-heptose-containing hexasaccharide corresponding to a structure from *Haemophilus ducrey* lipopolysaccharides. Tetrahedron Asymm 11:481–492
56. Ishii K, Kubo H, Yamasaki R (2002) Synthesis of α-lactosyl-(1 → 3)-L-*glycero*-α-D-*manno*-heptopyranoside, a partial oligosaccharide structure expressed within the lipooligosaccharide produced by *Neisseria gonorrhoeae* strain 15253. Carbohydr Res 337:11–20
57. Ishii K, Esumi Y, Iwasaki Y, Yamasaki R (2004) Synthesis of a 2,3-di-O-substituted heptose structure by regioselective 3-O-silylation of a 2-O-substituted heptose derivative. Eur J Org Chem 1214–1227
58. Kubo H, Ishii K, Koshino H, Toubetto K, Naruchi K, Yamasaki R (2004) Synthesis of a 3,4-di-O-substituted heptose structure: a partial oligosaccharide expressed in neisserial lipopolysaccharide. Eur J Org Chem 1202–1213
59. Yamasaki R, Yabe U, Kataoka C, Takeda U, Asuka S (2010) The oligosaccharides of gonococcal lipooligosaccharide contains several epitopes that are recognized by human antibodies. Infect Immun 78:3247–3257
60. Segerstedt E, Mannerstedt K, Johansson M, Oscarson S (2004) Synthesis of the branched trisaccharide L-*glycero*-α-D-*manno*-heptopyranosyl-(1 → 3)-[β-D-glucopyranosyl-(1 → 4)]-L-*glycero*-α-D-*manno*-heptopyranose, protected to allow flexible access to *Neisseria* and *Haemophilus* LPS inner core structures. J Carbohydr Chem 23:443–452
61. Olsson JDM, Oscarson S (2009) Synthesis of phosphorylated *Neisseria meningitidis* inner core lipopolysaccharide structures. Tetrahedron Asymm 20:879–886
62. Cox AD, St Michael F, Neelamegan D, Lacelle S, Cairns C, Richards J (2010) Investigating the candidacy of LPS-based glycoconjugates to prevent invasive meningococcal disease: chemical strategies to prepare glycoconjugates with good carbohydrate loading. Glycoconj J 27:401–417
63. Mannerstedt K, Segerstedt E, Olsson J, Oscarson S (2008) Synthesis of a common tetrasaccharide motif of *Haemophilus influenzae* LPS inner core structures. Org Biomol Chem 6:1087–1091

64. Olsson JDM, Oscarson S (2010) Synthesis of phosphorylated 3,4-branched trisaccharides corresponding to LPS inner core structures of *Neisseria meningitidis* and *Haemophilus influenzae*. Carbohydr Res 345:1331–1338
65. Ohara T, Adibekian A, Esposito D, Stallforth P, Seeberger PH (2010) Towards the synthesis of a *Yersinia pestis* cell wall polysaccharide: enantioselective synthesis of an L-*glycero*-D-*manno*-heptose building block. Chem Commun 46:4106–4108
66. Paulsen H, Wulff A, Heitmann AC (1988) Synthesis of disaccharides from L-*glycero*-D-*manno*-heptose and 2-amino-2-deoxy-D-glucose. Liebigs Ann Chem 1073–1078
67. Martin P, Lequart V, Cecchelli R, Boullanger P, Lafont D, Banoub J (2004) Novel synthesis of disaccharides containing the 2-amino-2-deoxy-β-D-glucopyranosyl unit and L-*glycero*-D-*manno*- and 7-deoxy-L-*glycero*-D-*galacto*-heptopyranoses. Chem Lett 33:696–697
68. Antonov KV, Backinowsky LV, Grzeszcyk B, Brade L, Holst O, Zamojski A (1998) Synthesis and serological characterization of L-*glycero*-α-D-*manno*-heptopyranose-containing di- and trisaccharides of the non-reducing terminus of the *Escherichia coli* K-12 LPS core oligosaccharide. Carbohydr Res 314:85–93
69. Mandal PK, Misra AK (2008) Concise synthesis of two pentasaccharides corresponding to the α-chain oligosaccharides of *Neisseria gonorrhoeae* and *Neisseria meningitidis*. Tetrahedron 64:8685–8691
70. Yokota S, Fuji N (2007) Contributions of the lipopolysaccharide outer core oligosaccharide region on the cell surface properties of *Pseudomonas aeruginosa*. Comp Immunol Microbiol Infect Dis 30:97–109
71. Komarova BS, Tsvetkov YE, Knirel YA, Zähringer U, Pier GB, Nifantiev NE (2006) Synthesis of common trisaccharide fragment of glycoforms of the outer core region of the *Pseudomonas aeruginosa* lipopolysaccharide. Tetrahedron Lett 47:3583–3587
72. Komarova BS, Tsvetkov YE, Pier GB, Nifantiev NE (2008) First synthesis of pentasaccharide glycoform I of the outer core region of the *Pseudomonas aeruginosa* lipopolysaccharide. J Org Chem 73:8411–8421
73. Campodónico VL, Gadjeva M, Paradis-Bleau C, Uluer A, Pier GB (2008) Airway epithelial control of *Pseudomonas aeruginosa* infection in cystic fibrosis. Trends Mol Med 14:120–133

6 Genetics and Biosynthesis of Lipid A

Christopher M. Stead, Aaron C. Pride, and M. Stephen Trent

6.1 Introduction

The defining feature of Gram-negative bacteria is the presence of an outer membrane, which comprises the outermost surface of the cell envelope and is, therefore, in constant contact with the surrounding environment. The Gram-negative outer membrane is unique as compared to most biological membranes in that it is an asymmetric bilayer composed of a phospholipid inner leaflet and a lipopolysaccharide (LPS) outer leaflet as opposed to a symmetrical phospholipid bilayer (Fig. 6.1). The presence of LPS in the outer leaflet confers unique properties to the membrane including an efficient permeability barrier that affords Gram-negative bacteria additional protection from their surrounding environment. LPS is generally organized into three structural domains – O-antigen, core, and lipid A (Fig. 6.1). Lipid A is a unique glycolipid that serves as the hydrophobic anchor of LPS. Extended from lipid A is the core oligosaccharide followed by the O-antigen polysaccharide. The core and O-antigen domains are typically not required for growth, but are critical for resistance to antibiotics, evasion of complement, and various other environmental stresses.

The lipid A of *Escherichia coli* K-12 is a β-1′,6-linked disaccharide of glucosamine that is both phosphorylated and fatty acylated (Fig. 6.2). Lipid A is

C.M. Stead
Georgia Health Sciences University, Department of Biochemistry and Molecular Biology, Augusta, GA 30912, USA
e-mail: cstead@georgiahealth.edu

A.C. Pride
Institute of Cellular and Molecular Biology, University of Texas at Austin, Austin, TX 78712, USA
e-mail: acpride@mail.utexas.edu

M.S. Trent (✉)
Section of Molecular Genetics and Microbiology and Institute of Cellular and Molecular Biology, University of Texas at Austin, Austin, TX 78712, USA
e-mail: strent@mail.utexas.edu

Y.A. Knirel and M.A. Valvano (eds.), *Bacterial Lipopolysaccharides*,
DOI 10.1007/978-3-7091-0733-1_6, © Springer-Verlag/Wien 2011

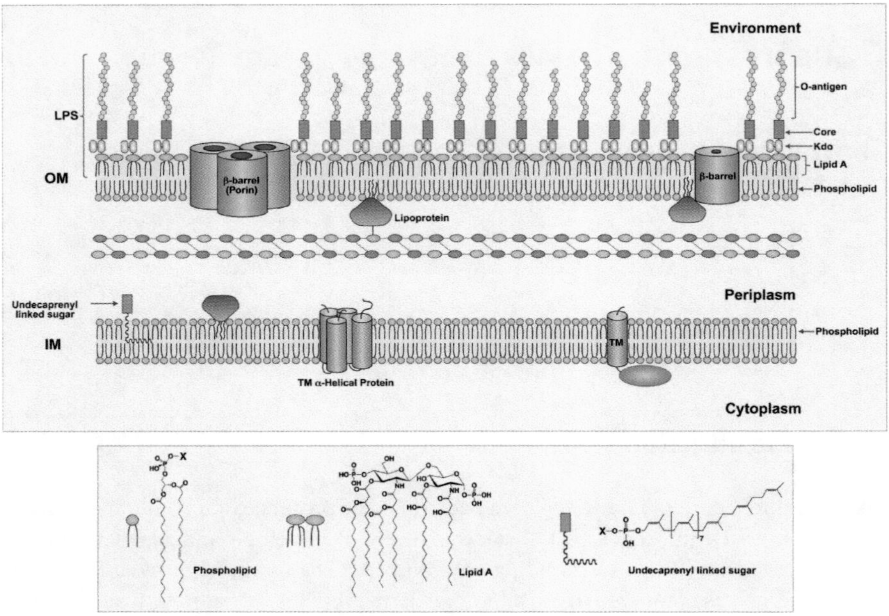

Fig. 6.1 Schematic of the Gram-negative cell envelope. The *upper panel* shows a representation of the Gram-negative cell envelope. The outer membrane (OM) is an asymmetric bilayer with the inner leaflet composed of phospholipids and the outer leaflet composed almost exclusively of lipopolysaccharide (LPS). Kdo, 3-deoxy-D-*manno*-octulosonic acid. The *lower panel* shows the chemical structures of lipid A, a phospholipid, and undecaprenyl-phosphate. The major phospholipids of *E. coli* are phosphatidylethanolamine and phosphatidylglycerol where X represents either a glycerol or ethanolamine group

glycosylated at the 6′-position with two Kdo (3-deoxy-D-*manno*-octulosonic acid) moieties with the inner Kdo serving as the point of attachment for the remaining core oligosaccharide. Both the lipid A domain and the core oligosaccharide, including the Kdo residues, are assembled on the cytoplasmic side of the inner membrane and subsequently translocated across the inner membrane by the ABC transporter MsbA. At the periplasmic side of the membrane, the O-antigen polysaccharide is ligated to the core-lipid A moiety completing LPS assembly [1]. Nascent LPS must then be shuttled across the periplasm and inserted into the outer leaflet of the outer membrane (see Chaps. 8–10). During Gram-negative infections, dissociated LPS is recognized by the innate immune system by Toll-like receptor 4 (TLR4) in complex with myeloid differentiation factor 2 (MD-2) that is present on many cell types including macrophages and dendritic cells (see Chap. 12). Notably, it is the lipid A domain that is recognized by the TLR4-MD-2 receptor. This chapter will focus on the biosynthesis and consequent modification of the Kdo-lipid A domain of LPS.

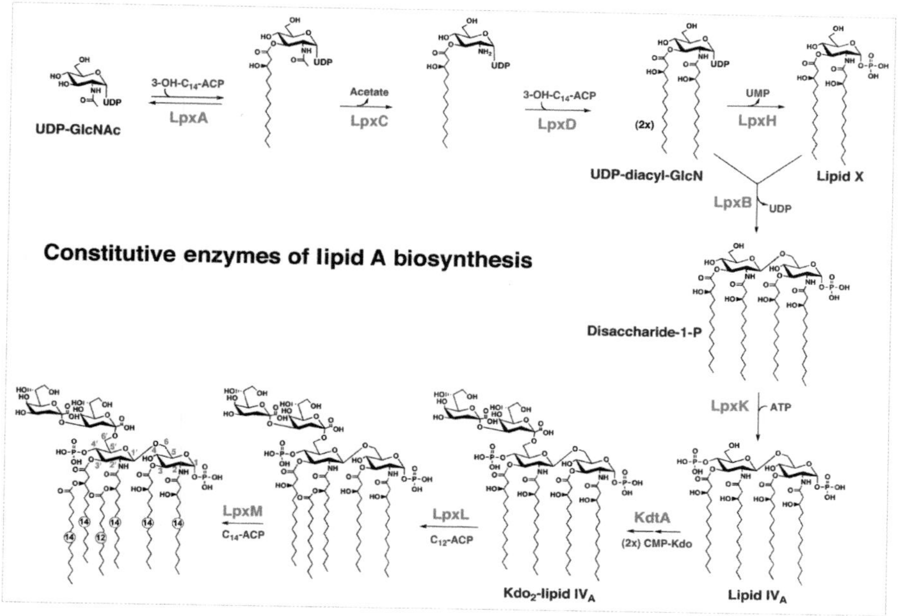

Fig. 6.2 Biosynthesis of the Kdo_2-lipid A. The structure of the intermediates in the biosynthesis pathway of Kdo_2-lipid A in *E. coli* K-12 and *S. typhimurium*, as well as their names and the nine enzymes catalyzing the reactions are shown. Acyl-ACP (acyl carrier protein) serves as the primary acyl donor for the various acyltransferases. The latter steps of pathway occur at the cytoplasmic face of the inner membrane beginning with LpxH

6.2 Kdo_2-Lipid A Biosynthesis

Kdo_2-lipid A biosynthesis has been well characterized in *E. coli* and shown to proceed via a nine-step enzymatic pathway known as the "Raetz Pathway" (Fig. 6.2) [1]. Bioinformatic analyses predict that homologues to each of these enzymes exist in almost all Gram-negative bacteria, suggesting a high degree of conservation with regards to lipid A biosynthesis. Lipid A structural studies have confirmed these bioinformatic observations by showing that many Gram-negative bacteria are in fact capable of producing at least a minor lipid A species which resembles that of *E. coli*. Each of these reactions takes place in the cytoplasm prior to transport of the molecule across the inner membrane.

The first step of lipid A biosynthesis is catalyzed by LpxA and involves the addition of an acyl chain to the 3-OH group of UDP-GlcNAc forming an ester linkage. LpxA has a strict dependence for an acyl-acyl carrier protein (acyl-ACP) donor [2]. The length of acyl chain attached to the acyl-ACP donor is also important with *E. coli* LpxA preferring a β-hydroxymyristate (3-OH-C14:0); therefore, LpxA is said to possess a "hydrocarbon ruler" [3]. The presence of the hydroxyl group on the myristate is also essential, although an LpxA homologue does exist in

Chlamydia trachomatis that utilizes a non-hydroxylated acyl chain [2, 4]. The large majority of LpxA homologues also display a hydrocarbon ruler, although the preferred acyl chain length can differ between bacterial species. For example, the LpxA homologue of *Pseudomonas aeruginosa* incorporates a β-hydroxydecanoate (3-OH-C10:0) acyl chain [3] while the LpxA of *Neisseria meningitidis* transfers a 3-OH-C12:0 acyl chain [5]. Some exceptions to this rule include *Porphrymonas gingivalis*, *Bordetella bronchiseptica* and *B. pertussis*, which are all capable of producing lipid A species with a large degree of acyl chain heterogeneity [6, 7].

Not all LpxA homologues transfer an acyl chain to a hydroxyl group due to the presence of *gnnA* and *gnnB* within their genomes. GnnA, an oxidoreductase, and GnnB, a transaminase, were first discovered and biochemically characterized in *Acidithiobacillus ferrooxidans* and shown to replace the 3-hydroxyl group of UDP-GlcNAc with an amino group to give UDP-GlcNAc3N, via a two-step enzymatic pathway (Fig. 6.3) [8]. GnnA and GnnB activities are also present in *Leptospira interrogans*, *Mesorhizobium loti* and *Campylobacter jejuni* [9, 10]. Interestingly the LpxA proteins of *L. interrogans* and *M. loti* are specific for UDP-GlcNAc3N (Fig. 6.3) leading to the production of a lipid A species with four N-linked primary acyl chains [9], yet the *C. jejuni* and *A. ferooxidans* LpxAs are more promiscuous, leading to a large degree of heterogeneity with regards to the primary acyl chain linkage [8, 10].

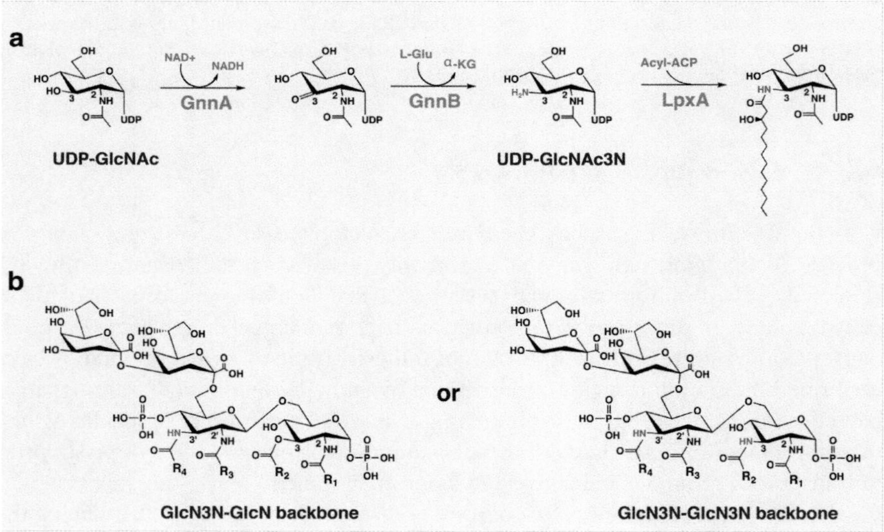

Fig. 6.3 Incorporation of diamino sugars into lipid A. (**a**) Prior to acylation by LpxA, UDP-GlcNAc is converted to its diamino-analog in which the 3-OH group is replaced with an NH_2 group by a two-step enzymatic process. GnnA, an oxidoreductase, catalyzes the formation of a ketone intermediate followed by a transamination event catalyzed by GnnB. (**b**) UDP-GlcNAc3N is acylated by LpxA and incorporated into the lipid A biosynthetic pathway resulting in amide-linked fatty acyl chains at the 3- and 3′-positions

The crystal structure of *E. coli* LpxA has been solved and shows a homotrimeric structure with a unique protein fold characterized by a left-handed helix of short parallel β-sheets [11]. Additional structures of *E. coli* LpxA in complex with its substrate, UDP-GlcNAc [12], and its reaction product UDP-3-*O*-acyl-GlcNAc [13] revealed three identical active sites at the subunit interfaces and corroborated the previously reported hydrocarbon ruler.

The next step involves deacetylation of UDP-3-*O*-acyl-GlcNAc by the Zn^{2+}-dependent enzyme LpxC to produce UDP-3-*O*-acylglucosamine. LpxC catalyzes the first committed step of lipid A biosynthesis due to the unfavourable equilibrium constant associated with UDP-3-*O*-acyl-GlcNAc production [14–16]. LpxC is currently an attractive and much studied target for a new class of antibiotics aimed at interrupting lipid A biosynthesis because it possesses no homology to other deacetylases or amidases and is essential for growth [17–19]. Indeed, compounds containing hydroxyamate or phosphonate zinc-binding motifs have been explored as potent LpxC inhibitors [18, 20–23]. One particular inhibitor, CHIR-090, is a potent inhibitor of LpxC displaying antibiotic activity comparable to that of commercial antibiotics [22]. Both nuclear magnetic resonance (NMR) and X-ray structures have been reported for unliganded LpxC [24] and LpxC-inhibitor complexes [25–27], which will greatly facilitate the design of broad-spectrum Gram-negative antibiotics with increased efficacy.

The action of LpxC then allows LpxD to transfer a second acyl chain to the newly generated amino group to form UDP-2,3-diacylglucosamine. *E. coli* LpxD shares the same properties as LpxA, including the presence of a hydrocarbon ruler specific for β-hydroxymyristate (3-OH-C14:0) and the utilization of ACP thioesters as the obligate acyl donor [28, 29]. The crystal structures of *E. coli* [29] and *C. trachomatis* LpxD [30] have been reported. Overall the structure of LpxD is similar to that of LpxA showing a homotrimer organization with three active sites located at the subunit interfaces. Comparison of the structures reveals differences in the hydrocarbon ruler that determines the selectively of the acyl-ACP substrate with the *C. trachomatis* protein capable of accommodating a larger fatty acyl chain. Chlamydial LpxD transfers a β-hydroxyarachidic acid (3-OH-C20:0) to UDP-3-*O*-acylglucosamine.

Following the formation of UDP-2,3-diacylglucosamine, lipid A biosynthesis proceeds with the pyrophosphatase LpxH, which produces UMP and 2,3-diacylglucosamine 1-phosphate, otherwise known as lipid X [31]. Accumulation of lipid X initiates the condensation of one molecule of UDP-2,3-diacylglucosamine with one molecule of lipid X catalyzed by LpxB, the disaccharide synthase [32, 33]. The strict order of these reactions was proven by the generation of an LpxH temperature sensitive mutant, which accumulated UDP-2,3-diacylglucosamine at the non-permissive temperature, demonstrating that LpxB could not condense two UDP-2,3-diacylglucosamine molecules [34]. As opposed to the earlier steps in the pathway, both LpxB and LpxH are peripheral membrane proteins.

Various Gram-negatives, including all α-proteobacteria, lack a homolog of LpxH yet they are capable of producing a mature lipid A species. These organisms produce a UDP-2,3-diacylglucosamine pyrophosphatase termed LpxI that shares no sequence similarity to LpxH. LpxI was discovered in *Caulobacter crescentus*, a bacterium with no LpxH homologue, because the location of its structural gene lies between the lipid A biosynthesis genes *lpxA* and *lpxB* [35]. Although LpxI produces the same products as LpxH, both enzymes have a different catalytic mechanism based upon which phosphate group is attacked by a water molecule. With LpxI a water molecule attacks the β-phosphate of the pyrophosphate, whereas with LpxH a water molecule attacks the α-phosphate [35].

Integral inner membrane proteins that require cytosolic factors for activity catalyze the latter steps of the pathway. The first of these, LpxK, catalyzes the addition of a phosphate group to the 4′-position of the tetraacylated monophosphorylated intermediate producing lipid IV_A [36–38] (Fig. 6.2). The reaction is ATP-dependent and precedes the addition of the Kdo sugars. Discovery of LpxK was a turning point in lipid A research because it enabled the production of high purity radiolabeled lipid A substrates, which could be used in in vitro assays to characterize lipid A biosynthesis and modification enzymes.

Although strictly part of the core oligosaccharide, Kdo residues are synonymous with lipid A biosynthesis because their presence is necessary for the addition of secondary acyl chains [39]. Two Kdo moieties are transferred to the distal glucosamine of lipid IV_A by the bifunctional enzyme WaaA (KdtA) [40–42]. WaaA homologues from different bacterial species can transfer between one and four Kdo residues, however, it is currently not possible to predict functionality by bioinformatic analyses alone. The ability to transfer more than two Kdo residues is only present in *Chlamydia* species with *C. trachomatis* and *C. pneumoniae* capable of transferring three Kdo residues and *C. psitticae* capable of transferring up to four Kdo residues [43–45]. A structure of the Kdo trisaccharide of *C. trachomatis* is shown in Fig. 6.4. *Vibrio cholerae* [46], *Haemophilus influenzae* [47], *B. pertussis* [48] and the hyperthermophile *Aquifex aeolicus* [49] all express a monofunctional WaaA. Recent studies utilizing *E. coli* and *H. influenzae* chimeric WaaA proteins were able to demonstrate that the N-terminal half of each protein was responsible for the observed differences in functionality [50]. Interestingly the non-thermophiles with a monofunctional WaaA all require the presence of a phosphate group attached to the Kdo to enable the consequent addition of secondary acyl chains [46]. The phosphate group is transferred to the Kdo by an enzyme known as KdkA [51].

In *E. coli* and *Salmonella* the final hexaacylated lipid A species is produced by the action of two acyltransferases, LpxL and LpxM, that catalyze the addition of secondary acyl chains to the distal glucosamine [39]. LpxL and LpxM display significant sequence similarity to each other and both require the presence of the Kdo residues for activity. LpxL and LpxM have a pre-determined order established by a strict substrate preference. LpxL first transfers a lauroyl (C12:0) group to the 2′-position of Kdo_2-lipid IV_A followed by the addition of a myristoyl (C14:0) group by LpxM to the 3′-position forming Kdo_2-lipid A. Although the secondary acyltransferases show no homology to LpxA and LpxD, they utilize acyl-ACPs as their preferred acyl donor [52, 53]. One exception to this rule is the *V. cholerae* LpxL homologue, which can use acyl-CoA with the same efficiency as acyl-ACP [46].

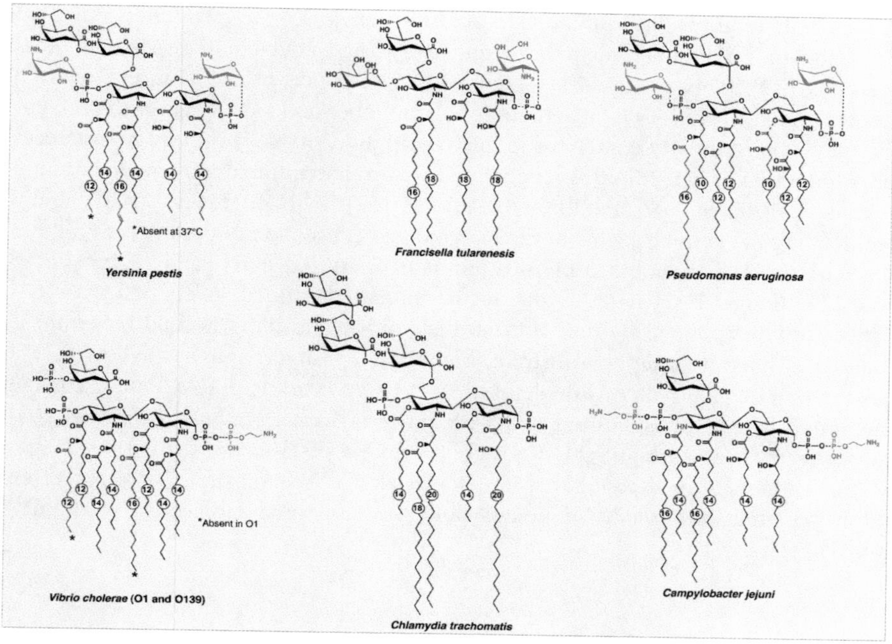

Fig. 6.4 Comparison of the Kdo-lipid A domains of Gram-negative bacteria. The chemical structure of the Kdo-lipid A domain found in the LPS of selected bacteria is shown. The *enclosed circles* show the lengths of the acyl chains and *dashed bonds* indicate partial substitution

The number of secondary acyltransferases and placement of secondary acyl chains differs throughout Gram-negative bacteria. For example, the genome of *C. trachomatis* contains a single homolog of the *E. coli* enzymes and synthesizes a lipid A with a single secondary fatty acyl chain (Fig. 6.4). On the other hand, *Helicobacter pylori* and *C. jejuni* are capable of synthesizing hexaacylated lipid A species, but contain only a single homolog of LpxL or LpxM. The LpxL homologue of *H. pylori* functions much like the *E. coli* enzyme, but transfers a stearoyl (C18:0) group to the 2′-position rather than a lauroyl group [54]. An activity for a second acyltransferase has been demonstrated in *H. pylori* membranes; however, this enzyme does not require the presence of the Kdo residues or the presence of the 2′-linked secondary acyl chain as seen with LpxM [54]. A relaxed requirement of the Kdo residues for secondary acylation is also seen with the *P. aeruginosa* [55, 56] and *N. meningitidis* [57] late acyltransferases.

The *E. coli* genome contains a third lipid A late acyltransferase homologous to LpxL, known as LpxP [58]. LpxP expression is turned on when *E. coli* are grown under cold shock conditions incorporating a palmitoleate (C16:1) in place of a laurate (C12:0) introducing an unsaturated acyl chain into the lipid A molecule [58]. The "kink" present in the unsaturated acyl chain may increase membrane fluidity, which would be advantageous in cold conditions. In the plant endosymbionts *Rhizobium leguminosarum* and *R. etli*, incorporation of a secondary acyl chain requires the unique late acyltransferase LpxXL, which is a distant ortholog to LpxL and transfers an unusually long acyl chain consisting of 28 carbons [59]. This process also requires the presence of the unique acyl-ACP termed ACP-XL [59–61]. Differences in the number and placement of the secondary acyl chains greatly contribute to the overall diversity seen in lipid A structures.

Production of Kdo_2-lipid A is regulated post-transcriptionally by the membrane-bound ATP-dependent metalloprotease FtsH [62]. FtsH degrades LpxC, thereby controlling the cellular levels of the enzyme and consequently lipid A biosynthesis because the deacetylation reaction is the first committed step of lipid A biosynthesis. Mutation of FtsH is lethal due to the increased cellular levels of LpxC [63]. This can be explained by the fact that the lipid A and phospholipid biosynthesis pathways have a common substrate, known as *R*-3-hydroxymyristoyl-ACP. The increase in LpxC effectively depletes the *R*-3-hydroxymyristoyl-ACP pool, causing an imbalance in phospholipid and LPS ratios, which is lethal to the cell [63]. More recent studies highlighted a second role for FtsH in the regulation of lipid A biosynthesis after showing that WaaA was also a FtsH substrate [64]. Therefore, FtsH post-transcriptionally regulates both the early and late stages of lipid A biosynthesis.

6.3 Lipid A and Innate Immunity

The human innate immune system has evolved various mechanisms for recognizing conserved microbial motifs including the action of TLR [65]. TLRs are constitutively present and, therefore, provide a rapid detection system for invading microbes. Once a microbe is detected a signalling cascade is induced, which eventually leads to the production of pro-inflammatory cytokines to help clear the infection [65]. TLR4 and its co-receptor MD-2 [66–68] are of particular interest because they evolved to recognize lipid A, which is an excellent candidate for a TLR because of the conserved lipid A biosynthesis pathway. To combat this line of defence Gram-negative bacteria modify their lipid A to prevent detection by TLR4-MD-2 [69]. The enzymes used for such modifications will be described in detail in the proceeding pages, whereas, the specific interactions between the TLR4-MD-2 receptor and lipid A ligand will be described in Chap. 13.

Another arm of the human innate immune system with a close association to lipid A is the production of cationic antimicrobial peptides (CAMPs). CAMPs are small positively charged peptides, which are responsible for killing a variety of invading microbes, including Gram-negative bacteria [70]. CAMPs are initially attracted to the negative charges present on the Gram-negative outer membrane before traversing the outer and inner membranes to exert their bactericidal effect, which occurs via a variety of mechanisms, including cell lysis and inhibition of protein synthesis [71]. Gram-negative bacteria commonly resist the bactericidal action of CAMPs by preventing the initial electrostatic interaction. A primary mechanism involved in this defence is the modification of lipid A phosphate groups to negate the strong negative charge associated with the phosphate groups. Gram-negative bacteria achieve this end by removing the phosphate groups or decorating them with positively charged moieties, such as phosphoethanolamine (PEtN) or 4-amino-4-deoxy-L-arabinose (Ara4N) [72–74], as described in the next section.

6.4 Lipid A Modifications

Despite the conservation of Kdo_2-lipid A biosynthesis a large degree of heterogeneity is seen between the Kdo_2-lipid A species generated by diverse Gram-negative bacteria (Fig. 6.4). This diversity is the product of lipid A modification enzymes. Nearly all of these modifications occur after conserved lipid A biosynthesis either in the periplasm or outer membrane and confer an advantage to the bacterium in evading the innate immune system as described above. In the majority of cases these modifications are only important for a portion of the bacterial life cycle and as a consequence are regulated. The most common form of regulation occurs via the PhoP/PhoQ and PmrA/PmrB bacterial two-component regulatory systems [75, 76]. Well studied examples of non-regulated lipid A modifications do exist in *H. pylori* [77] and is thought to occur because *H. pylori* exists in only one well defined niche, a concept which will be revisited in a future section.

6.4.1 Removal of Phosphate Groups

Various bacteria express enzymes catalyzing the removal of the 1- and 4′-phosphate groups of lipid A. The enzymes catalyzing these reactions are inner membrane proteins, highly specific for their respective positions, and function only after the transport of lipid A to the periplasmic face of the inner membrane (Fig. 6.5). The 1-phosphate group is cleaved by LpxE, which was first identified in *Rhizobium leguminosarum* [78]. LpxE exhibits sequence similarity to members of the phosphatidic acid-phosphatase superfamily (PAP2) characterized by a conserved phosphatase motif KX_6RP-(X_{12-54})-PSGH-(X_{31-54})-SRX_5HX_3D [79]. LpxE homologues have been characterized in *F. novicida* [80], *H. pylori* [81] and *P. gingivalis* [82]. Heterologous expression of LpxE in *E. coli* results in the production of 1-dephosphorylated lipid A. However, phosphatase activity is lost in conditional MsbA mutants unable to transport core-lipid A across the inner membrane. Studies on *H. pylori* LpxE demonstrated that LpxE was important for resistance to the cationic peptide polymyxin B [72], a common experimental substitute for human CAMPs, which has a similar mode of action.

The characterized lipid A structures of *Rhizobium* [83, 84], *Francisella* [85, 86], *P. gingivalis* [82, 87], *H. pylori* [77, 88] and *L. interrogans* [89] all predict the presence of a 4′-phosphatase. This prediction has been confirmed in each of the bacterial species with the exception of *L. interrogans*. The 4′-phosphatase of *F. novicida* was the first to be characterized and annotated as LpxF [90]. As with

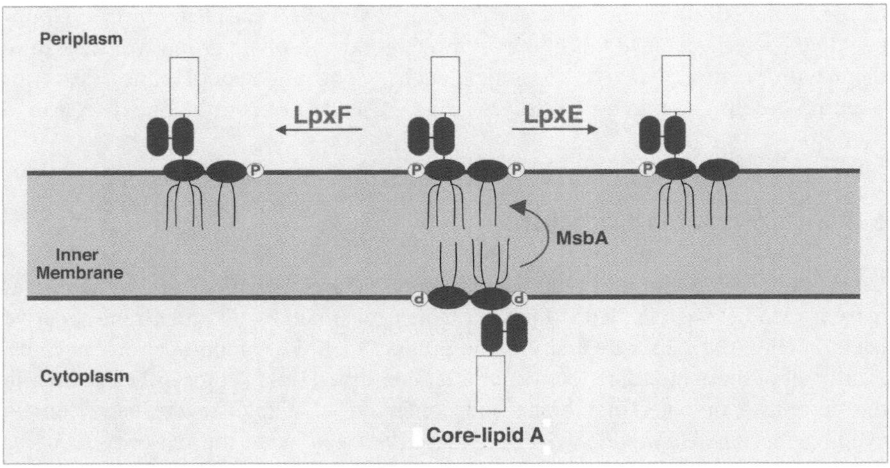

Fig. 6.5 Periplasmic dephosphorylation of lipid A. MsbA flips core-lipid A molecules assembled on the cytoplasmic side of the inner membrane to the periplasmic face of the membrane. In some organisms, dedicated phosphatases dephosphorylate the lipid A anchor of LPS often contributing to CAMP resistance. LpxE removes the 1-phosphate group and LpxF removes the 4′-phosphate group

LpxE, LpxF is a member of the PAP2 superfamily and depends on MsbA transport for activity. These lipid A modifications are important for virulence of *F. novicida*, since LpxF mutants are hypersensitive to polymyxin and also attenuated in a mouse model of infection [91]. Both *Rhizobium etli* [92] and *P. gingivalis* [82] LpxE and LpxF mutants have also been shown to be sensitive to the action of polymyxin, reinforcing negative charge reduction as an important defence against CAMPs. *P. gingivalis* phosphatase mutants changed the lipid A species produced by the bacterium from a TLR4 non-activator to a TLR4 activator [82]. This phenotype took an interesting turn when wild type *P. gingivalis* was grown in the presence of high haemin concentrations. The presence of haemin selectively turned off the 1-phosphatase, which led to the production of a lipid A species that now displayed TLR4 antagonism [82].

6.4.2 Decoration of Phosphate Groups

The modified forms of both *E. coli* and *Salmonella* lipid A includes decoration of the phosphate groups with Ara4N (Fig. 6.6) by the enzyme ArnT, an enzyme displaying distant similarity to yeast protein mannosyltransferases [73]. ArnT is under the control of the PmrA/PmrB two-component regulatory system, which is stimulated by mildly acidic pH, Fe^{3+} and PhoP/PhoQ [93, 94]. The transferase utilizes an undecaprenyl-linked Ara4N as the donor substrate (Fig. 6.7) [95] modifying lipid A on the periplasmic side of the inner membrane [96]. ArnT

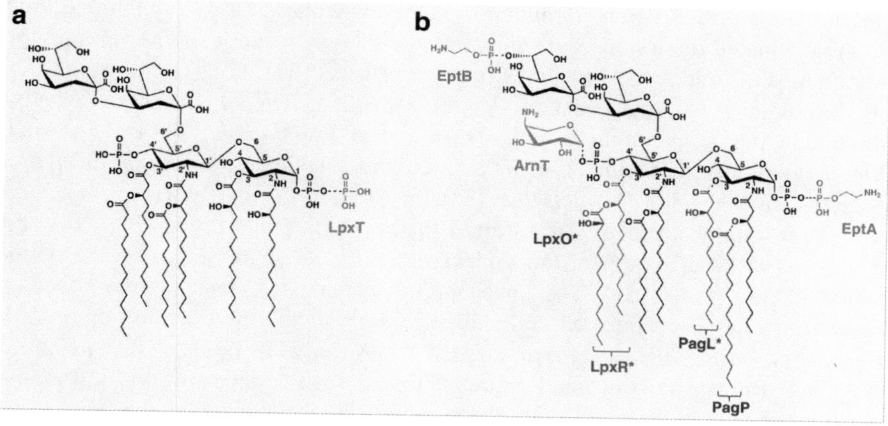

Fig. 6.6 Covalent modifications of Kdo$_2$-lipid A in *E. coli* and *Salmonella*. (**a**) The unmodified Kdo$_2$-lipid A of *E. coli* K-12 or *S. typhimurium* is hexaacylated and contains monophosphate groups at positions 1 and 4′. During growth in nutrient rich broth, approximately one-third of the lipid A molecules have an additional phosphate group at the 1-position forming a diphosphate species. LpxT catalyzes the formation of 1-diphosphate lipid A. (**b**) Possible modifications that are induced upon exposure to environmental stimuli that impact the bacterial membrane. LpxO, LpxR, and PagL are not present in *E. coli* K-12 but a homolog of LpxR can be found in some pathogenic strains of *E. coli*

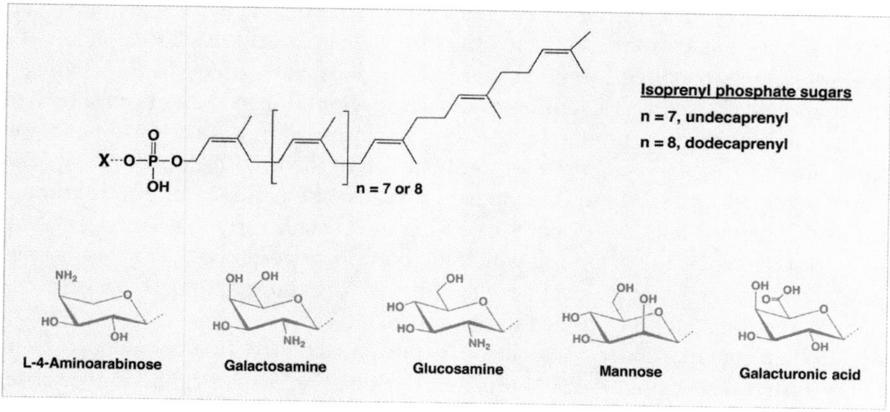

Fig. 6.7 Structures of isoprenyl-linked sugar donors utilized by lipid A modification enzymes. Undecaprenyl-phosphate sugar donors required for the modification of *E. coli* and *Francisella* lipid A with 4-amino-4-deoxy-L-arabinose and galactosamine, respectively. Dodecaprenyl-phosphate galacturonic acid is required for the modification of *Rhizobium* lipid A, and isoprenyl-linked glucosamine and mannose are required for the modification of *Bordetella* and *Francisella* lipid A, respectively

preferentially modifies the 4′-phosphate group and requires the presence of the secondary acyl chain at the 3′-position for optimal activity [97]. Ara4N is a positively charged sugar and its presence confers resistance to CAMPs including polymyxin B [98]. *Salmonella* mutants unable to synthesize Ara4N-modified lipid A are attenuated in virulence as compared to wild type bacteria, demonstrating the importance of lipid A modifications for pathogenesis [99].

Homologues to ArnT exist in several bacteria known to synthesize Ara4N-modified lipid A including *E. coli*, *Yersinia* and *Pseudomonas* (Fig. 6.4). ArnT homologues also exist which transfer a glucosamine residue (*Bordetella*) [100] or a galactosamine residue (*Francisella*) [85]. In *Francisella tularensis* ssp. *novicida* the galactosamine residue is transferred to the 1-phosphate group (Fig. 6.4) and analogous to Ara4N transfer, an undecaprenyl-linked sugar serves as the donor substrate (Fig. 6.7) [101]. The undecaprenyl phosphate-galactosamine donor of *Francisella* has been purified and shown to support the modification of lipid A precursors. *Francisella* can also modify the 4′-position of its lipid A with a mannose (Fig. 6.4) from a probable undecaprenyl linked sugar donor (Fig. 6.7) after the phosphate has been cleaved [102]. This modification is mediated by the same ArnT homologue, which decorates the 1-phosphate group [102]. *F. novicida* mutants unable to modify their lipid A with either galactosamine or mannose are attenuated in a mouse model [102]. The *Bordetella* ArnT homologs decorate both the 1- and 4′-phosphate groups with a glucosamine from a probable undecaprenyl-linked sugar donor (Fig. 6.7). In *B. pertussis*, presence of the glucosamines modulates human TLR4 activation by making the lipid A more stimulatory, leading to an increased production of proinflammatory cytokines in human THP-1-derived macrophages [103].

Another enzyme with similar properties to ArnT exists in *Rhizobium*; however, it transfers a galacturonic acid directly to the disaccharide backbone at the 4′-position after *Rhizobium* LpxF has cleaved the phosphate group. In *Rhizobium*, a dodecaprenyl phosphate galacturonic acid carrier lipid (Fig. 6.8) has been shown to be required for modification of the core oligosaccharide [104]. In all likelihood, the same donor substrate is required for modification of the lipid A domain [105]. The two elucidated *Aquifex* lipid A structures revealed the presence of a galacturonic acid substituent at both the 1- and 4′-positions attached directly to a hydroxyl group [49, 106]. Notably, in bacteria in which the phosphate groups are removed, a non-positively charged sugar decorates the lipid A backbone, perhaps because the phosphate groups no longer need to be masked.

Another enzyme under the control of the PmrA/PmrB two-component regulatory system is the lipid A PEtN transferase EptA [74, 107]. PEtN is a zwitterion including both positively (ethanolamine) and negatively (phosphate) charged groups. In contrast to Ara4N, coupling of PEtN to lipid A phosphate group does not neutralize its negative charge. However, even in this case incorporation of the positively charged group (EtN) may be advantageous at low pH when EtN is protonated.

EptA has been shown to contribute to polymyxin resistance in several organisms including *H. pylori* [72], *N. meningitidis* [108, 109], and *C. jejuni* [110]. Modification of lipid A with PEtN in *H. pylori* is unusual in that the EptA homologue transfers the PEtN directly to a hydroxyl group at the 1-position of the disaccharide backbone as opposed to a phosphate group [81]. The hydroxyl group is generated

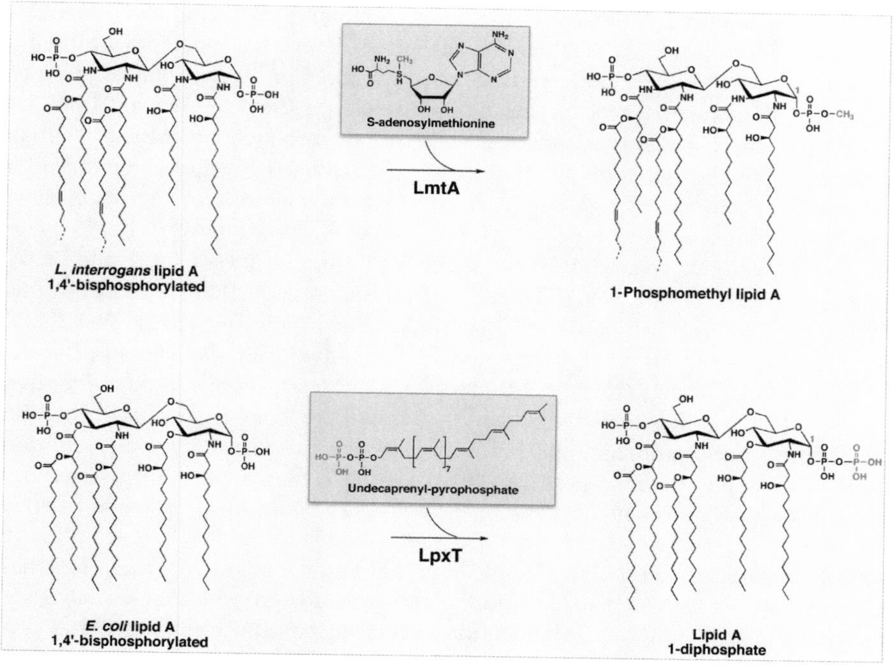

Fig. 6.8 Modification of the lipid A 1-phosphate group by LpxT and LmtA. The enzyme LmtA of *L. interrogans* transfers a methyl group from S-adenosylmethionine (SAM) to the 1-phosphate group of lipid A (*top panel*). Unlike *E. coli*, the lipid A of *L. interrogans* contains four N-linked acyl chains. LpxT acts as a kinase transferring the distal phosphate of undecaprenyl-pyrophosphate to the 1-phosphate group of lipid A (*bottom panel*)

by the prior action of the *H. pylori* lipid A 1-phosphatase. The *C. jejuni* EptA homologue is of particular interest because not only is it required for PEtN transfer to lipid A, but it was also shown to be necessary for motility [110]. This second seemingly unrelated phenotype was attributed to a PEtN modification of FlgG, which forms part of the flagellar basal body [110]. This is the first time that a lipid A modification enzyme has been shown to function on a second substrate, which is remarkable given how divergent each of the substrates are. The EptA homologue in *Neisseria* species has also been shown to be important for resistance to human complement-mediated killing [111] as well as adhesion to human endothelial and epithelial cells when the bacteria is unencapsulated [112], indicating a diverse role for the enzyme. To date no mechanism exists to explain these extraordinary phenotypes, however, it is tempting to speculate that a similar scenario exists as seen in *C. jejuni*, and the *Neisseria* EptA protein is modifying a secondary target.

A further two lipid A phosphate decorating enzymes have been documented; however, neither of the modifications provide resistance to CAMPs. LpxT phosphorylates lipid A at the 1-position forming a 1-diphosphate lipid A species (Fig. 6.8). In *E. coli* K-12, the 1-diphosphate form represents approximately one-third of the lipid A present in the outer membrane [113] (Fig. 6.6). Interestingly, LpxT does not use ATP as the phosphate donor, but rather undecaprenyl pyrophosphate (Fig. 6.8) [113] phosphorylating lipid A at the periplasmic face of the inner membrane. Undecaprenyl pyrophosphate is used as a carrier lipid for peptidoglycan, O-antigen and other bacterial surface carbohydrate polymers, which requires removal of the terminal phosphate to be recycled after releasing its cargo [114, 115]. Therefore, by removing a phosphate from undecaprenyl pyrophosphate and transferring it to lipid A (Fig. 6.8), LpxT links the biosynthesis of lipid A with the assembly of other essential bacterial envelope structures. Like LpxE and LpxF, LpxT is a member of the PAP2 family of phosphatases. Other members of this family are required in undecaprenyl pyrophosphate recycling [116]. The PAP2 phosphatase family is widespread in Gram-negative bacteria, opening up the possibility that formation of the 1-diphosphate species is more common than first thought. LpxT is negatively regulated by PmrA/PmrB; however, this regulation occurs post-transcriptionally [117]. This regulation is also linked to PEtN addition by EptA (Fig. 6.6) because EptA can only function in the absence of LpxT activity [117]. The exact mechanism used by *E. coli* to regulate LpxT remains elusive. The role of the 1-diphosphate species is also unknown; however, it is tempting to speculate that it could provide a source of energy at the outer membrane, which has no direct access to ATP. Finally, in *L. interrogans*, the enzyme LmtA methylates a lipid A phosphate group. Methylation occurs at the 1-position and LmtA was shown to use an S-adenosylmethionine donor during in vitro assays [118] (Fig. 6.8). The LmtA active site is cytoplasmic, unlike most lipid A modifying enzymes, which usually function after transport of Kdo_2-lipid A across the inner membrane by MsbA.

6.4.3 Acyl Chain Modifications

Following the conserved biosynthetic pathway, three enzymes have been identified which are involved in modulating the number of acyl chains present on a given lipid A species. PagL and LpxR are responsible for a reduction in acyl chain numbers [119, 120] whereas PagP is responsible for an increase in acyl chain numbers [121, 122]. Interestingly all three proteins are located in the outer membrane and are immersed in their own substrate. Each has a solved crystal structure (Fig. 6.9), giving a large amount of insight into the catalytic mechanisms of lipid A acyl chain rearrangements.

Transfer of a palmitoyl residue (C16:0) to lipid A was first demonstrated in an in vitro assay system using *E. coli* membranes as the enzyme source in 1987 [123]. It took another 11 years to discover the enzyme responsible for this activity in *S. enterica* serovar Typhimurium (herein *S. typhimurium*) [121]. The first clue as to the genes identity was garnered during studies of the PhoP/PhoQ two-component regulatory system. PhoP/PhoQ was shown to be activated in Mg^{2+} limiting conditions and as a consequence lipid A modifications were induced, including the addition of an acyl chain [124]. The PhoP/PhoQ activated gene responsible for

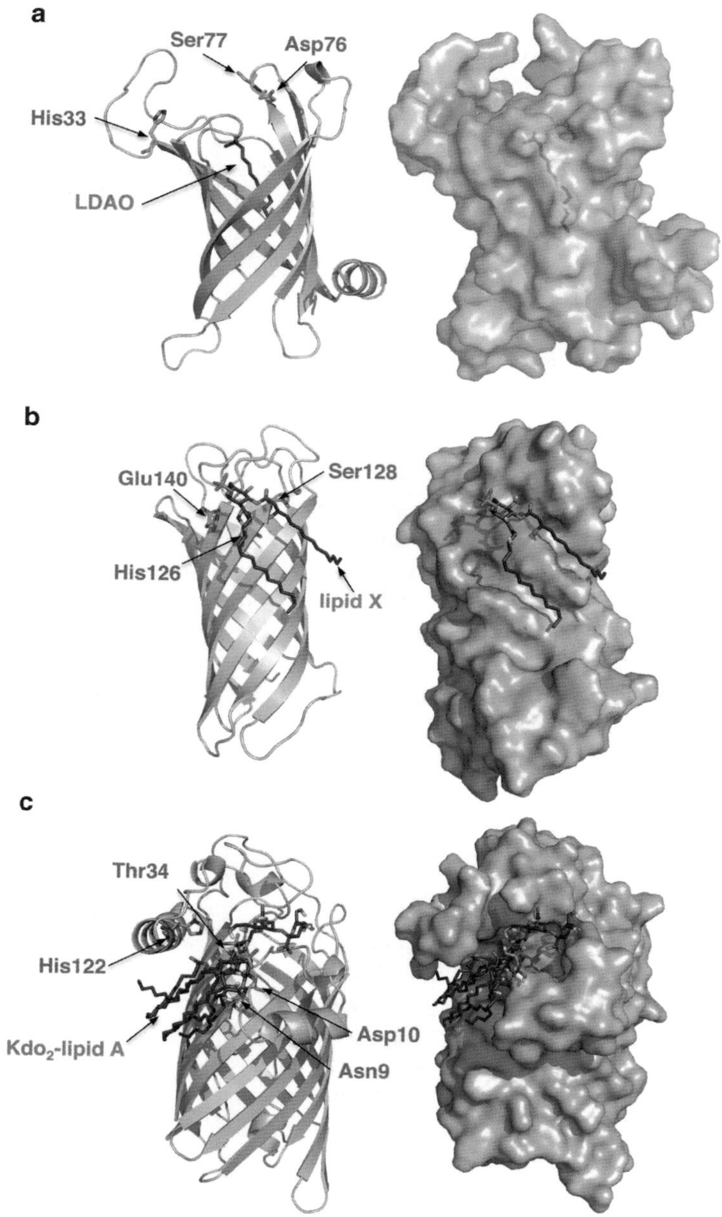

Fig. 6.9 Structures of outer membrane enzymes that modify lipid A. A ribbon (*left*) and surface (*right*) representation is shown for each protein in wheat color. Active-site residues are modeled in green and associated lipids are shown in dark gray. (**a**) The structure of the *E. coli* palmitate transferase, PagP (PDB code 1THQ). A single lauroyldimethylamine-N-oxide (LDAO) molecule present in the structure is shown in dark gray bound within the active site. The loop connecting the

the transfer of the acyl chain was consequently identified as PagP [121] and since then a plethora of research has been conducted on PagP concerning both its virulence properties and structure/function.

E. coli, *Salmonella*, *Shigella*, *Yersinia*, *Bordetella*, *Legionella* and *Pseudomonas* all have a PagP-like activity and several pathogenic traits have been attributed to PagP in many of these bacteria [125, 126]. One surprising phenotype was an increase in CAMP sensitivity seen with a *Salmonella* PagP mutant [121]. Normally resistance to CAMPs, generated by lipid A modifications, is provided by reducing the net negative charge present at the bacterial surface, thereby preventing the initial binding of CAMPs. It is believed that PagP provides resistance instead by inhibiting the translocation of CAMPs across the outer membrane, which is more tightly packed due to the presence of an additional acyl chain. Furthermore, PagP activity in *Salmonella* is related to a reduction in TLR4 activation [127]. Presumably the extra acyl chain interferes with lipid A binding to the TLR4-MD-2 complex, which is reviewed in Chap. 13. Therefore, the action of one enzyme provides resistance to two arms of the innate immune system.

The structure of *E. coli* PagP has been determined by NMR spectroscopy and X-ray crystallography. As with other integral membrane proteins found within the Gram-negative outer membrane, PagP exists as an antiparallel β-barrel. Specifically, PagP is an eight-stranded β-barrel with long extracellular loops and an amphipathic α-helix located at the N-terminus (Fig. 6.9) [128, 129]. Much like the acyltransferases required for synthesis of the conserved lipid A structure, the core of PagP contains a hydrocarbon ruler providing substrate selectivity [130]. In Fig. 6.9, the detergent lauroyldimethylamine-N-oxide (LDAO) is shown within the barrel highlighting the acyl-chain binding pocket. Events leading to disruption of the asymmetry of the outer membrane result in migration of phospholipids from the inner to the outer leaflet, which serve as acyl donors [131]. In *E. coli* and *Salmonella*, PagP specifically transfers a palmitate to the 2-position of the proximal glucosamine of lipid A resulting in a heptaacylated structure (Fig. 6.10) [122]. Biochemical and structural data support that disruptions of the hydrogen bonding between the strands of the barrel at two opposing sites provide a route for lateral access of lipid substrates [132]. Residues implicated in catalysis (highlighted in Fig. 6.9) are not organized into a prototypical catalytic triad. However, NMR data support that conformational changes within the large extracellular loop and exterior regions of the β-barrel occur to promote catalysis [129]. Based upon reported lipid A structures, PagP proteins from different bacterial species catalyze palmitate addition to different sites of lipid A. For example in *Bordetella* and *Pseudomonas*,

---

Fig. 6.9 (Continued) first and second β-strands is disordered and was introduced subsequently and energy minimized. Coordinates were provided by Chris Neale and Régis Pomès (University of Toronto). (**b**) Structure of the *P. aeruginosa* 3-O-deacylase PagL (PDB code 2ERV). (**c**) Structure of the *S. typhimurium* 3′-O-deacylase LpxR (PDB code 3FID). Lipid X and Kdo_2-lipid A were modelled into the active sites of PagL and LpxR, respectively. Coordinates for the PagL-lipid X complex were provided by Lucy Rutten and Jan Tommassen (Utrecht University)

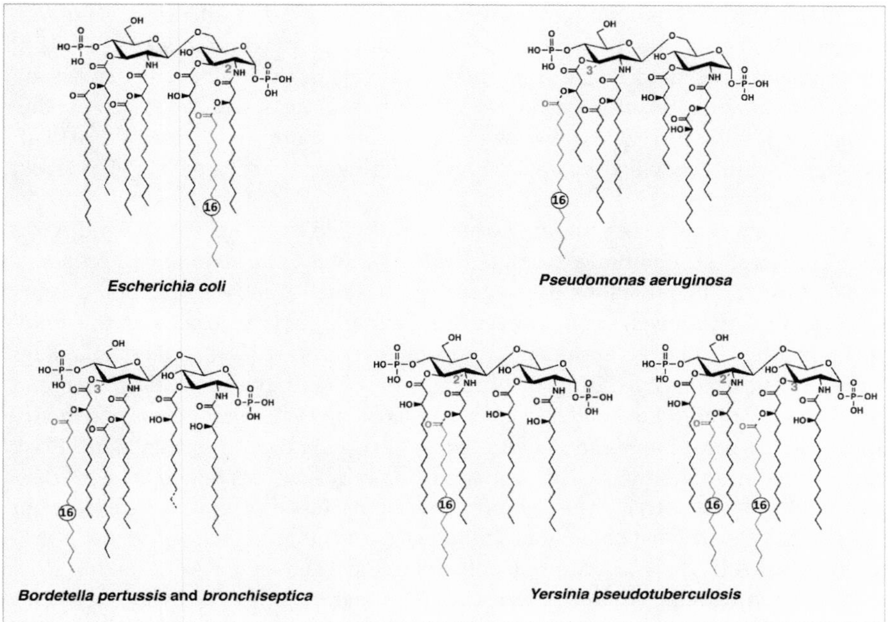

Fig. 6.10 Structural comparison of lipid A species containing palmitate. The lipid A structures of selected Gram-negative bacteria containing palmitate are shown. The location of the palmitate residue (C16:0) is highlighted

PagP catalyzes palmitate addition to the 3′-position, whereas in *Y. pseudotuberculosis* palmitate addition was proposed to occur at both 3- and 2′-positions (Fig. 6.10). How different PagP proteins select the site of palmitate addition is of interest, but remains unknown.

PagP activity in *Legionella pneumophila* and *B. bronchiseptica* also contributes to evasion from the innate immune system. An *L. pneumophila* PagP mutant was shown to have an increased sensitivity to a variety of CAMPs, which then correlated to a decrease in intracellular survival and a reduced efficiency for lung colonization in mice [133]. *B. bronchiseptica* PagP mutants display a slightly different phenotype in that PagP is not necessary for initial colonization but rather persistence of the organism once infection has been established [134]. One possible explanation for this observation is the increased sensitivity to complement mediated lysis seen with a *B. bronchiseptica* PagP mutant [134].

The pathogenic *Yersinia*, *Y. pestis*, *Y. enterocolitica* and *Y. pseudotuberculosis* all have a PagP homologue. Based upon structural analysis of lipid A, however, a palmitate (C16:0) residue is absent in *Y. pestis* [135]. This could be a consequence of a premature stop codon, which removes the final three amino acids [136]. The *Y. enterocolitica* PagP homologue is induced in Mg^{2+} limiting conditions [121] whereas the *Y. pseudotuberculosis* PagP homologue is induced at the body temperature of the mammalian host [135]. The addition of palmitate to *P. aeruginosa* lipid

A is thought to be important for survival in the cystic fibrosis lung since palmitate-modified lipid A is present in all *P. aeruginosa* cystic fibrosis isolates [137, 138]. This same modification is absent in all *P. aeruginosa* environmental isolates and non-cystic fibrosis infection isolates [138]. Currently the structural gene encoding *Pseudomonas* PagP remains unknown, but our laboratory has observed a PagP-like activity in the wild type membranes during in vitro assays (M.S. Trent, unpublished data).

An enzyme capable of removing of the 3-O-linked acyl chain from lipid A was first discovered in membrane extracts from *R. leguminosarum* and *P. aeruginosa* [139]. However, identification of the gene responsible for 3-O-deacylation was first achieved in *S. typhimurium* after discovering that the enzymatic activity was PhoP/PhoQ regulated and the resultant protein named PagL [120]. *Salmonella* PagL shows a robust activity during in vitro assays using *S. typhimurium* membranes grown under PhoP/PhoQ inducing conditions. However, PagL does not modify lipid A in the outer membrane in vivo when *S. typhimurium* is grown under PhoP/PhoQ inducing conditions. This apparent anomaly was explained when studies showed that PagL activity was inhibited by the presence of lipid A Ara4N modifications, a property that could not attributed to substrate preference because Ara4N addition is not 100% [140]. Subsequent studies have shown that PagL extracellular loops interact with the Ara4N moieties attached to the lipid A phosphates to silence the enzyme and that mutants which abolish this interaction can release PagL from latency in the presence of Ara4N modifications [141]. To date, a growth condition supporting modification of lipid A by PagL in wild type *Salmonella* has not been observed. As with PagP, modulation of acyl chain numbers by PagL in *Salmonella* decreases the endotoxic activity associated with PagL modified lipid A [142].

On the other hand, the *P. aeruginosa* PagL is active within the outer membrane and has been implicated in the adaptation of the bacterium within the cystic fibrosis lung. *P. aeruginosa* isolates from early or mild cystic fibrosis infections all have lipid A species lacking an acyl chain due to PagL activity; however, this modification is absent in *P. aeruginosa* isolates from non-cystic fibrosis infections [143]. Similar to PagP, the crystal structure of *Pseudomonas* PagL consists of an eight-stranded β-barrel [144]. The active site consists of a distinct Ser-His-Glu catalytic triad characteristic of serine esterases (Fig. 6.9) that is facing the outer surface of the outer membrane. Modelling the lipid A precursor lipid X onto the active site revealed hydrophobic groves on the exterior of the protein accommodating the 3-O-linked acyl chain (Fig. 6.9) [144]. However, unlike PagP it appears that PagL does not have a strict acyl chain preference [145]. A structure of *Salmonella* PagL has not been reported.

The last of the three enzymes known to modulate the number of lipid A acyl chains is LpxR. It was characterized in *S. typhimurium* and shown to remove both of the 3′-O-linked acyl chains in a single cleavage reaction, which was dependent upon Ca^{2+} [119]. Despite a robust activity seen during in vitro assays, *Salmonella* LpxR is not active in vivo, even in the presence of high Ca^{2+}, until the bacteria reaches stationary phase, indicating a growth phase dependent regulation [119, 146]. When stationary phase bacteria are used to infect macrophages a wild type strain shows increased intracellular growth as compared to an LpxR mutant, suggesting a role in pathogenesis for LpxR [146]. Whether or not other signals induce *Salmonella* LpxR within the outer membrane remains to be determined. LpxR is considerably larger than PagP or PagL existing as a 12-stranded β-barrel with a periplasmic plug that is formed by an unusually long periplasmic turn (Fig. 6.9). The active site of LpxR is located between the barrel wall and an α-helix of one of the extracellular loops. Thus, like PagP and PagL the active site is extracellular. Site-directed mutagenesis and modelling of Kdo_2-lipid A onto the predicted active site of the enzyme predicted a mechanism in which histidine-122 activates a water molecule leading to attack of the carbonyl oxygen of the scissile bond. A Ca^{2+} ion is required for the oxyanion hole explaining the requirement for Ca^{2+} during in vitro assay. Homologues to LpxR exist in *H. pylori*, *V. cholerae*, *E. coli* O157:H7 and *Y. enterocolitica*. Like *Salmonella* the LpxR homologue of each of these bacteria is inactive under normal growth conditions with the exception of *H. pylori*, which is constitutively active (Trent lab unpublished data). Understanding the difference in regulation is a very interesting subject and one that may be answered by solving the *H. pylori* LpxR crystal structure.

As well as the addition and removal of acyl chains, one other acyl chain modification has been documented in *S. typhimurium*, which involves the addition of a hydroxyl group to the 2-position of the 3′-linked secondary acyl chain. LpxO catalyzes this hydroxylation in an oxygen dependent manner at the cytoplasmic surface of the inner membrane [147]. Although initially thought to be regulated by PhoP/PhoQ [124] the basal levels of LpxO expression are sufficient to enable significant modification under PhoP/PhoQ non-inducing conditions [147]. As of yet no role in pathogenesis has been shown for LpxO. The elucidated lipid A structures of *Klebsiella* [148], *Pseudomonas* [149], *Bordetella* [150] and *Legionella* [151] all contain hydroxylated secondary acyl chains, which is supported by the presence of a LpxO homologue in each of the genomes [147].

6.4.4 Kdo Modifications

As mentioned previously, the number of Kdo residues transferred to lipid A by WaaA can vary between one and four sugars. Initially WaaA functionality was thought to be the only mechanism involved in determining how many Kdo residues were attached to a given lipid A species. However, an enzymatic activity catalyzing the removal of the outer Kdo residue was discovered in membranes isolated from *H. pylori* [152]. In the same investigation *H. pylori* WaaA was shown to be bifunctional, demonstrating that the Kdo hydrolase was in fact responsible for the number of Kdo residues present in the *H. pylori* mature LPS species. A Kdo hydrolase activity was also detected in *Francisella* membranes, which facilitated the discovery of a two-component protein complex responsible for removing the Kdo sugar. After a genomic comparison of *Francisella* and *Helicobacter*, a structural gene encoding a protein (KdoH1) with a bacterial sialidase domain was identified [88, 153]. A second integral membrane protein (KdoH2), encoded by a neighbouring gene, with no predicted function was also required for Kdo hydrolase activity. Removal of the Kdo residue occurs on the periplasmic side of the inner membrane [88].

The *H. pylori* Kdo hydrolase mutant had two unusual phenotypes related to O-antigen expression and CAMP resistance. Firstly, a Kdo hydrolase mutant was more sensitive to the action of polymyxin. This phenotype was likely due to that the 4′-phosphate group was no longer removed with 100% efficiency (see homogenous vs. heterogeneous lipid A profiles section below), increasing the net negative charge present at the outer surface. Secondly, transfer of O-antigen to core-lipid A was significantly diminished in a Kdo hydrolase mutant via a mechanism that still remains to be solved [88].

E. coli, *Salmonella* and *Rhizobium* all modify their outer Kdo sugar by transferring a PEtN (*E. coli* and *Salmonella*) or two galacturonic acid moieties (*Rhizobium*). In *E. coli* and *Salmonella* EptB, a PEtN transferase with high sequence homology to EptA, transfers PEtN to the outer Kdo [154]. EptB modifications are turned on in response to high Ca^{2+} concentrations in the growth media [154]. In *Rhizobium* two independent proteins, known as RgtA and RgtB, are responsible for decorating the outer Kdo sugar with two galacturonic acid residues [155]. RgtA and RgtB use dodecaprenyl linked sugar donors (Fig. 6.7) [104], similar to ArnT and its homologues.

6.4.5 Homogenous Versus Heterogeneous Lipid A Profiles

The large majority of the lipid A modifications described previously are regulated and only occur as a response to environmental changes. *S. typhimurium* serves as an excellent example of a bacterial species that regulates how its lipid A is presented. *S. typhimurium* has two two-component regulatory systems, which are intimately involved in modulating alterations to lipid A. The first, PhoP/PhoQ, responds to low Mg^{2+} concentrations or CAMP exposure turning on expression of *pagP* and *pagL*, modifying the lipid acylation pattern, providing resistance to the innate immune system. PhoP/PhoQ also upregulates *pmrD* expression, which has a stimulatory effect on the PmrA/PmrB two-component regulatory system. PmrA/PmrB can also be directly stimulated under mildly acidic conditions or in the presence of iron. Once activated, PmrA/PmrB upregulates the expression of *arnT* and *eptA*, which then decorate the lipid A phosphate groups also providing resistance to the innate immune system. Given that *S. typhimurium* has a variable lifestyle during which it colonizes diverse hosts and environmental niches, modification of the Kdo-lipid A domain is not always necessary.

Other important human pathogens with diverse lifestyles, such as *P. aeruginosa* and *Yersinia*, also modify their lipid A in response to environmental changes. *Yersinia* species modulate the number of acyl chains present on their lipid A in response to temperature changes. In *Y. pestis*, fewer acyl chains are present on its lipid A when grown at 37°C as compared to 21°C, a temperature shift that mimics the transition from flea to human. *P. aeruginosa* initiates lipid A modifications after infection of the cystic fibrosis lung; however, the same modifications are not seen in *P. aeruginosa* isolates from any other infection site indicating adaptation to a specific environment.

H. pylori has a very different lifestyle to *S. typhimurium*, *P. aeruginosa* and *Y. pestis* with only one known reservoir, the stomach of primates. Like *Salmonella* the lipid A of *H. pylori* is also highly modified; however, *H. pylori* produces a homogenous lipid A profile when grown in the laboratory under a diverse set of conditions. This suggests that *H. pylori* lipid A modifications are constitutively active when colonizing the stomach, although this has not been proven directly due to problems with isolating sufficient biomass to allow lipid A analysis. This lack of regulation is likely the result of a simpler lifestyle in which adaptation is not necessary. *H. pylori* lipid A is modified by the action of five enzymes (Fig. 6.11) to produce a lipid A species lacking the 3′-O-linked acyl chains, the 4′-phosphate group, a Kdo residue, and having PEtN directly attached to the disaccharide

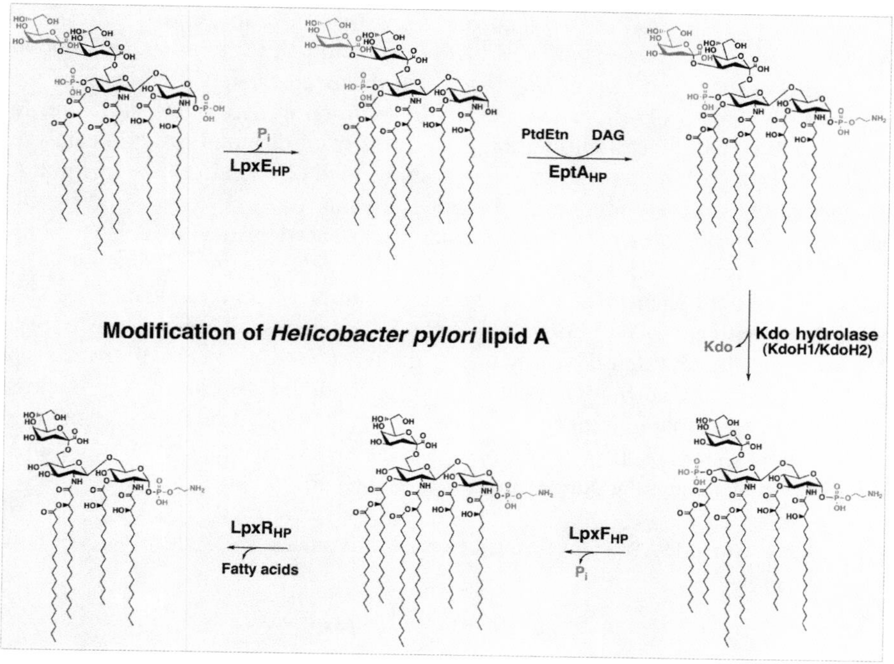

Fig. 6.11 Modification of *H. pylori* Kdo_2-lipid A. *H. pylori* produces a highly modified lipid A species via a five-step enzymatic pathway. The 1-phosphate is first cleaved by $LpxE_{HP}$, leaving a free hydroxyl group, followed by addition of a phosphoethanolamine by $EptA_{HP}$. Presumably, phosphatidylethanolmaine (PtdEtN) serves as the donor for the phosphoethanoalmine addition resulting in the formation of diacylglycerol (DAG). Next, a two-component protein complex (KdoH1 and KdoH2) work in concert to remove the terminal Kdo sugar. The 4′-phosphate group is removed by $LpxF_{HP}$ (M.S. Trent, unpublished data); however, the remaining hydroxyl group is not further modified as is the case at the 1 position. The final step involves the removal of the 3′-O-linked acyl chains by $LpxR_{HP}$, resulting in a tetraacylated lipid A

backbone at the 1-position. These modifications proceed via a strict order governed by substrate preference with the action of the proceeding enzyme enabling the following enzyme to function [54, 72, 81, 88, 152]. This ensures the production of a single lipid A species, highlighting the importance for a homogenous lipid A profile when only colonizing a very specific niche.

6.5 Concluding Remarks

On the surface, it may appear that our understanding of lipid A biosynthesis is nearing completion. However, the level of complexity associated with the synthesis of this remarkable glycolipid is impressive and requires further research. Furthermore, the majority of investigations have been carried out in *E. coli* and *S. typhimurium* and assumed to translate to other Gram-negative bacteria. Upon closer inspection this is not the case. For example, the late acylation stages of lipid A biosynthesis in several bacteria, such as *H. pylori*, *Y. pestis*, and *F. tularensis*, deviate from the "conserved pathway". Although common features can be found in the lipid A structure of Gram-negative bacteria, mechanistic differences of the enzymes catalyzing these reactions results in variation in the lipid A structure. In the instance of palmitate addition, PagP enzymes from different bacteria catalyze acyl transfer to different locations contributing to the diversity seen in lipid A structures. Although a common lipid A modification may exist between different bacteria, a clear homolog of the enzyme may be absent. A structural gene encoding a 3′-O-deacylase (LpxR) is not present in the *Francisella* genome although it lacks a 3′-O-linked acyl chain. Novel modifications and modification enzymes are still being discovered and as more lipid A structures are determined one can only expect that more will be revealed. Possibly one of the most exciting recent discoveries in lipid A research is the involvement of lipid A modification enzymes in distinct cellular processes, such as flagellar assembly in *C. jejuni*. LpxT involvement in undecaprenyl pyrophosphate recycling serves as another example of how lipid A modification systems may play a larger role in the cell.

Acknowledgement M.S. Trent is supported by National Institutes of Health Grants AI064184 and AI076322.

References

1. Raetz CR, Whitfield C (2002) Lipopolysaccharide endotoxins. Annu Rev Biochem 71:635–700
2. Anderson MS, Raetz CR (1987) Biosynthesis of lipid A precursors in *Escherichia coli*. A cytoplasmic acyltransferase that converts UDP-*N*-acetylglucosamine to UDP-3-*O*-(*R*-3-hydroxymyristoyl)-*N*-acetylglucosamine. J Biol Chem 262:5159–5169
3. Wyckoff TJ, Lin S, Cotter RJ, Dotson GD, Raetz CR (1998) Hydrocarbon rulers in UDP-*N*-acetylglucosamine acyltransferases. J Biol Chem 273:32369–32372

4. Sweet CR, Lin S, Cotter RJ, Raetz CR (2001) A *Chlamydia trachomatis* UDP-*N*-acetylglucosamine acyltransferase selective for myristoyl-acyl carrier protein. Expression in *Escherichia coli* and formation of hybrid lipid A species. J Biol Chem 276:19565–19574
5. Odegaard TJ, Kaltashov IA, Cotter RJ, Steeghs L, van der Ley P, Khan S, Maskell DJ, Raetz CR (1997) Shortened hydroxyacyl chains on lipid A of *Escherichia coli* cells expressing a foreign UDP-*N*-acetylglucosamine *O*-acyltransferase. J Biol Chem 272:19688–19696
6. Sweet CR, Preston A, Toland E, Ramirez SM, Cotter RJ, Maskell DJ, Raetz CR (2002) Relaxed acyl chain specificity of *Bordetella* UDP-*N*-acetylglucosamine acyltransferases. J Biol Chem 277:18281–18290
7. Bainbridge BW, Karimi-Naser L, Reife R, Blethen F, Ernst RK, Darveau RP (2008) Acyl chain specificity of the acyltransferases LpxA and LpxD and substrate availability contribute to lipid A fatty acid heterogeneity in *Porphyromonas gingivalis*. J Bacteriol 190:4549–4558
8. Sweet CR, Ribeiro AA, Raetz CR (2004) Oxidation and transamination of the 3″-position of UDP-*N*-acetylglucosamine by enzymes from *Acidithiobacillus ferrooxidans*. Role in the formation of lipid A molecules with four amide-linked acyl chains. J Biol Chem 279:25400–25410
9. Sweet CR, Williams AH, Karbarz MJ, Werts C, Kalb SR, Cotter RJ, Raetz CR (2004) Enzymatic synthesis of lipid A molecules with four amide-linked acyl chains. LpxA acyltransferases selective for an analog of UDP-*N*-acetylglucosamine in which an amine replaces the 3″-hydroxyl group. J Biol Chem 279:25411–25419
10. van Mourik A, Steeghs L, van Laar J, Meiring HD, Hamstra HJ, van Putten JP, Wosten MM (2010) Altered linkage of hydroxyacyl chains in lipid A of *Campylobacter jejuni* reduces TLR4 activation and antimicrobial resistance. J Biol Chem 285:15828–15836
11. Raetz CR, Roderick SL (1995) A left-handed parallel β helix in the structure of UDP-*N*-acetylglucosamine acyltransferase. Science 270:997–1000
12. Ulaganathan V, Buetow L, Hunter WN (2007) Nucleotide substrate recognition by UDP-*N*-acetylglucosamine acyltransferase (LpxA) in the first step of lipid A biosynthesis. J Mol Biol 369:305–312
13. Williams AH, Raetz CR (2007) Structural basis for the acyl chain selectivity and mechanism of UDP-*N*-acetylglucosamine acyltransferase. Proc Natl Acad Sci USA 104:13543–13550
14. Anderson MS, Bull HG, Galloway SM, Kelly TM, Mohan S, Radika K, Raetz CR (1993) UDP-*N*-acetylglucosamine acyltransferase of *Escherichia coli*. The first step of endotoxin biosynthesis is thermodynamically unfavorable. J Biol Chem 268:19858–19865
15. Anderson MS, Robertson AD, Macher I, Raetz CR (1988) Biosynthesis of lipid A in *Escherichia coli*: identification of UDP-3-O-[(R)-3-hydroxymyristoyl]-α-d-glucosamine as a precursor of UDP-N2, O3-bis[(R)-3-hydroxymyristoyl]-α-d-glucosamine. Biochemistry 27:1908–1917
16. Young K, Silver LL, Bramhill D, Cameron P, Eveland SS, Raetz CR, Hyland SA, Anderson MS (1995) The *envA* permeability/cell division gene of *Escherichia coli* encodes the second enzyme of lipid A biosynthesis. UDP-3-*O*-(*R*-3-hydroxymyristoyl)-*N*-acetylglucosamine deacetylase. J Biol Chem 270:30384–30391
17. Onishi HR, Pelak BA, Gerckens LS, Silver LL, Kahan FM, Chen MH, Patchett AA, Galloway SM, Hyland SA, Anderson MS, Raetz CR (1996) Antibacterial agents that inhibit lipid A biosynthesis. Science 274:980–982
18. Jackman JE, Fierke CA, Tumey LN, Pirrung M, Uchiyama T, Tahir SH, Hindsgaul O, Raetz CR (2000) Antibacterial agents that target lipid A biosynthesis in Gram-negative bacteria. Inhibition of diverse UDP-3-*O*-(*R*-3-hydroxymyristoyl)-*N*-acetylglucosamine deacetylases by substrate analogs containing zinc binding motifs. J Biol Chem 275:11002–11009
19. Clements JM, Coignard F, Johnson I, Chandler S, Palan S, Waller A, Wijkmans J, Hunter MG (2002) Antibacterial activities and characterization of novel inhibitors of LpxC. Antimicrob Agents Chemother 46:1793–1799
20. Kline T, Andersen NH, Harwood EA, Bowman J, Malanda A, Endsley S, Erwin AL, Doyle M, Fong S, Harris AL, Mendelsohn B, Mdluli K, Raetz CR, Stover CK, Witte PR,

Yabannavar A, Zhu S (2002) Potent, novel in vitro inhibitors of the *Pseudomonas aeruginosa* deacetylase LpxC. J Med Chem 45:3112–3129
21. Li X, Uchiyama T, Raetz CR, Hindsgaul O (2003) Synthesis of a carbohydrate-derived hydroxamic acid inhibitor of the bacterial enzyme (LpxC) involved in lipid A biosynthesis. Org Lett 5:539–541
22. McClerren AL, Endsley S, Bowman JL, Andersen NH, Guan Z, Rudolph J, Raetz CR (2005) A slow, tight-binding inhibitor of the zinc-dependent deacetylase LpxC of lipid A biosynthesis with antibiotic activity comparable to ciprofloxacin. Biochemistry 44: 16574–16583
23. Pirrung MC, Tumey LN, Raetz CR, Jackman JE, Snehalatha K, McClerren AL, Fierke CA, Gantt SL, Rusche KM (2002) Inhibition of the antibacterial target UDP-(3-*O*-acyl)-*N*-acetylglucosamine deacetylase (LpxC): isoxazoline zinc amidase inhibitors bearing diverse metal binding groups. J Med Chem 45:4359–4370
24. Whittington DA, Rusche KM, Shin H, Fierke CA, Christianson DW (2003) Crystal structure of LpxC, a zinc-dependent deacetylase essential for endotoxin biosynthesis. Proc Natl Acad Sci USA 100:8146–8150
25. Coggins BE, Li X, McClerren AL, Hindsgaul O, Raetz CR, Zhou P (2003) Structure of the LpxC deacetylase with a bound substrate-analog inhibitor. Nat Struct Biol 10:645–651
26. Mochalkin I, Knafels JD, Lightle S (2008) Crystal structure of LpxC from *Pseudomonas aeruginosa* complexed with the potent BB-78485 inhibitor. Protein Sci 17:450–457
27. Barb AW, Jiang L, Raetz CR, Zhou P (2007) Structure of the deacetylase LpxC bound to the antibiotic CHIR-090: time-dependent inhibition and specificity in ligand binding. Proc Natl Acad Sci USA 104:18433–18438
28. Kelly TM, Stachula SA, Raetz CR, Anderson MS (1993) The *firA* gene of *Escherichia coli* encodes UDP-3-*O*-(*R*-3-hydroxymyristoyl)-glucosamine *N*-acyltransferase. The third step of endotoxin biosynthesis. J Biol Chem 268:19866–19874
29. Bartling CM, Raetz CR (2009) Crystal structure and acyl chain selectivity of *Escherichia coli* LpxD, the *N*-acyltransferase of lipid A biosynthesis. Biochemistry 48:8672–8683
30. Buetow L, Smith TK, Dawson A, Fyffe S, Hunter WN (2007) Structure and reactivity of LpxD, the *N*-acyltransferase of lipid A biosynthesis. Proc Natl Acad Sci USA 104:4321–4326
31. Babinski KJ, Ribeiro AA, Raetz CR (2002) The *Escherichia coli* gene encoding the UDP-2,3-diacylglucosamine pyrophosphatase of lipid A biosynthesis. J Biol Chem 277: 25937–25946
32. Ray BL, Painter G, Raetz CR (1984) The biosynthesis of Gram-negative endotoxin. Formation of lipid A disaccharides from monosaccharide precursors in extracts of *Escherichia coli*. J Biol Chem 259:4852–4859
33. Radika K, Raetz CR (1988) Purification and properties of lipid A disaccharide synthase of *Escherichia coli*. J Biol Chem 263:14859–14867
34. Babinski KJ, Kanjilal SJ, Raetz CR (2002) Accumulation of the lipid A precursor UDP-2,3-diacylglucosamine in an *Escherichia coli* mutant lacking the *lpxH* gene. J Biol Chem 277:25947–25956
35. Metzger LE IV, Raetz CR (2010) An alternative route for UDP-diacylglucosamine hydrolysis in bacterial lipid A biosynthesis. Biochemistry 49:6715–6726
36. Garrett TA, Kadrmas JL, Raetz CR (1997) Identification of the gene encoding the *Escherichia coli* lipid A 4′-kinase. Facile phosphorylation of endotoxin analogs with recombinant LpxK. J Biol Chem 272:21855–21864
37. Garrett TA, Que NL, Raetz CR (1998) Accumulation of a lipid A precursor lacking the 4′-phosphate following inactivation of the *Escherichia coli lpxK* gene. J Biol Chem 273:12457–12465
38. Ray BL, Raetz CR (1987) The biosynthesis of Gram-negative endotoxin. A novel kinase in *Escherichia coli* membranes that incorporates the 4′-phosphate of lipid A. J Biol Chem 262:1122–1128

39. Brozek KA, Raetz CR (1990) Biosynthesis of lipid A in *Escherichia coli*. Acyl carrier protein-dependent incorporation of laurate and myristate. J Biol Chem 265:15410–15417
40. Brozek KA, Hosaka K, Robertson AD, Raetz CR (1989) Biosynthesis of lipopolysaccharide in *Escherichia coli*. Cytoplasmic enzymes that attach 3-deoxy-D-*manno*-octulosonic acid to lipid A. J Biol Chem 264:6956–6966
41. Belunis CJ, Raetz CR (1992) Biosynthesis of endotoxins. Purification and catalytic properties of 3-deoxy-D-*manno*-octulosonic acid transferase from *Escherichia coli*. J Biol Chem 267:9988–9997
42. Clementz T. Raetz CR (1991) A gene coding for 3-deoxy-D-*manno*-octulosonic-acid transferase in *Escherichia coli*. Identification, mapping, cloning, and sequencing. J Biol Chem 266:9687–9696
43. Belunis CJ, Clementz T, Carty SM, Raetz CR (1995) Inhibition of lipopolysaccharide biosynthesis and cell growth following inactivation of the *kdtA* gene in *Escherichia coli*. J Biol Chem 270:27646–27652
44. Brabetz W, Lindner B, Brade H (2000) Comparative analyses of secondary gene products of 3-deoxy-D-*manno*-oct-2-ulosonic acid transferases from *Chlamydiaceae* in *Escherichia coli* K-12. Eur J Biochem 267:5458–5465
45. Rund S, Lindner B, Brade H, Holst O (2000) Structural analysis of the lipopolysaccharide from *Chlamydophila psittaci* strain 6BC. Eur J Biochem 267:5717–5726
46. Hankins JV, Trent MS (2009) Secondary acylation of *Vibrio cholerae* lipopolysaccharide requires phosphorylation of Kdo. J Biol Chem 284:25804–25812
47. White KA, Kaltashov IA, Cotter RJ, Raetz CR (1997) A mono-functional 3-deoxy-D-*manno*-octulosonic acid (Kdo) transferase and a Kdo kinase in extracts of *Haemophilus influenzae*. J Biol Chem 272:16555–16563
48. Isobe T, White KA, Allen AG, Peacock M, Raetz CR, Maskell DJ (1999) *Bordetella pertussis waaA* encodes a monofunctional 2-keto-3-deoxy- D-*manno*-octulosonic acid transferase that can complement an *Escherichia coli waaA* mutation. J Bacteriol 181: 2648–2651
49. Mamat U, Schmidt H, Munoz E, Lindner B, Fukase K, Hanuszkiewicz A, Wu J, Meredith TC, Woodard RW, Hilgenfeld R, Mesters JR, Holst O (2009) WaaA of the hyperthermophilic bacterium Aquifex aeolicus is a monofunctional 3-deoxy-D-*manno*-oct-2-ulosonic acid transferase involved in lipopolysaccharide biosynthesis. J Biol Chem 284:22248–22262
50. Chung HS, Raetz CR (2010) Interchangeable domains in the Kdo transferases of *Escherichia coli* and *Haemophilus influenzae*. Biochemistry 49:4126–4137
51. White KA, Lin S, Cotter RJ, Raetz CR (1999) A *Haemophilus influenzae* gene that encodes a membrane bound 3-deoxy-D-*manno*-octulosonic acid (Kdo) kinase. Possible involvement of Kdo phosphorylation in bacterial virulence. J Biol Chem 274:31391–31400
52. Clementz T, Bednarski JJ, Raetz CR (1996) Function of the *htrB* high temperature requirement gene of *Escherchia coli* in the acylation of lipid A: HtrB catalyzed incorporation of laurate. J Biol Chem 271:12095–12102
53. Clementz T, Zhou Z, Raetz CR (1997) Function of the *Escherichia coli msbB* gene, a multicopy suppressor of *htrB* knockouts, in the acylation of lipid A. Acylation by MsbB follows laurate incorporation by HtrB. J Biol Chem 272:10353–10360
54. Stead CM, Beasley A, Cotter RJ, Trent MS (2008) Deciphering the unusual acylation pattern of *Helicobacter pylori* lipid A. J Bacteriol 190:7012–7021
55. Mohan S, Raetz CR (1994) Endotoxin biosynthesis in *Pseudomonas aeruginosa*: enzymatic incorporation of laurate before 3-deoxy-D-*manno*-octulosonate. J Bacteriol 176:6944–6951
56. Goldman RC, Doran CC, Kadam SK, Capobianco JO (1988) Lipid A precursor from *Pseudomonas aeruginosa* is completely acylated prior to addition of 3-deoxy-D-*manno*-octulosonate. J Biol Chem 263:5217–5223
57. Tzeng YL, Datta A, Kolli VK, Carlson RW, Stephens DS (2002) Endotoxin of *Neisseria meningitidis* composed only of intact lipid A: inactivation of the meningococcal 3-deoxy-D-*manno*-octulosonic acid transferase. J Bacteriol 184:2379–2388

58. Carty SM, Sreekumar KR, Raetz CR (1999) Effect of cold shock on lipid A biosynthesis in *Escherichia coli*. Induction at 12°C of an acyltransferase specific for palmitoleoyl-acyl carrier protein. J Biol Chem 274:9677–9685
59. Basu SS, Karbarz MJ, Raetz CR (2002) Expression cloning and characterization of the C28 acyltransferase of lipid A biosynthesis in *Rhizobium leguminosarum*. J Biol Chem 277:28959–28971
60. Brozek KA, Carlson RW, Raetz CR (1996) A special acyl carrier protein for transferring long hydroxylated fatty acids to lipid A in *Rhizobium*. J Biol Chem 271:32126–32136
61. Vedam V, Kannenberg EL, Haynes JG, Sherrier DJ, Datta A, Carlson RW (2003) A *Rhizobium leguminosarum* AcpXL mutant produces lipopolysaccharide lacking 27-hydroxyoctacosanoic acid. J Bacteriol 185:1841–1850
62. Narberhaus F, Obrist M, Fuhrer F, Langklotz S (2009) Degradation of cytoplasmic substrates by FtsH, a membrane-anchored protease with many talents. Res Microbiol 160:652–659
63. Ogura T, Inoue K, Tatsuta T, Suzaki T, Karata K, Young K, Su LH, Fierke CA, Jackman JE, Raetz CR, Coleman J, Tomoyasu T, Matsuzawa H (1999) Balanced biosynthesis of major membrane components through regulated degradation of the committed enzyme of lipid A biosynthesis by the AAA protease FtsH (HflB) in *Escherichia coli*. Mol Microbiol 31:833–844
64. Katz C, Ron EZ (2008) Dual role of FtsH in regulating lipopolysaccharide biosynthesis in *Escherichia coli*. J Bacteriol 190:7117–7122
65. Akira S, Uematsu S, Takeuchi O (2006) Pathogen recognition and innate immunity. Cell 124:783–801
66. Aderem A, Ulevitch RJ (2000) Toll-like receptors in the induction of the innate immune response. Nature 406:782–787
67. Hoshino K, Takeuchi O, Kawai T, Sanjo H, Ogawa T, Takeda Y, Takeda K, Akira S (1999) Cutting edge: Toll-like receptor 4 (TLR4)-deficient mice are hyporesponsive to lipopolysaccharide: evidence for TLR4 as the *Lps* gene product. J Immunol 162: 3749–3752
68. Shimazu R, Akashi S, Ogata H, Nagai Y, Fukudome K, Miyake K, Kimoto M (1999) MD-2, a molecule that confers lipopolysaccharide responsiveness on Toll-like receptor 4. J Exp Med 189:1777–1782
69. Trent MS, Stead CM, Tran AX, Hankins JV (2006) Diversity of endotoxin and its impact on pathogenesis. J Endotoxin Res 12:205–223
70. Jenssen H, Hamill P, Hancock RE (2006) Peptide antimicrobial agents. Clin Microbiol Rev 19:491–511
71. Brogden KA (2005) Antimicrobial peptides: pore formers or metabolic inhibitors in bacteria? Nat Rev Microbiol 3:238–250
72. Tran AX, Whittimore JD, Wyrick PB, McGrath SC, Cotter RJ, Trent MS (2006) The lipid A 1-phosphatase of *Helicobacter pylori* is required for resistance to the antimicrobial peptide polymyxin. J Bacteriol 188:4531–4541
73. Trent MS, Ribeiro AA, Lin S, Cotter RJ, Raetz CR (2001) An inner membrane enzyme in *Salmonella* and *Escherichia coli* that transfers 4-amino-4-deoxy-L-arabinose to lipid A: induction on polymyxin-resistant mutants and role of a novel lipid-linked donor. J Biol Chem 276:43122–43131
74. Lee H, Hsu FF, Turk J, Groisman EA (2004) The PmrA-regulated *pmrC* gene mediates phosphoethanolamine modification of lipid A and polymyxin resistance in *Salmonella enterica*. J Bacteriol 186:4124–4133
75. Gunn JS (2008) The *Salmonella* PmrAB regulon: lipopolysaccharide modifications, antimicrobial peptide resistance and more. Trends Microbiol 16:284–290
76. Kato A, Groisman EA (2008) The PhoQ/PhoP regulatory network of *Salmonella enterica*. Adv Exp Med Biol 631:7–21
77. Tran AX, Stead CM, Trent MS (2005) Remodeling of *Helicobacter pylori* lipopolysaccharide. J Endotoxin Res 11:161–166

78. Karbarz MJ, Kalb SR, Cotter RJ, Raetz CR (2003) Expression cloning and biochemical characterization of a *Rhizobium leguminosarum* lipid A 1-phosphatase. J Biol Chem 278:39269–39279
79. Stukey J, Carman GM (1997) Identification of a novel phosphatase sequence motif. Protein Sci 6:469–472
80. Wang X, Karbarz MJ, McGrath SC, Cotter RJ, Raetz CR (2004) MsbA transporter-dependent lipid A 1-dephosphorylation on the periplasmic surface of the inner membrane: topography of *Francisella novicida* LpxE expressed in *Escherichia coli*. J Biol Chem 279:49470–49478
81. Tran AX, Karbarz MJ, Wang X, Raetz CR, McGrath SC, Cotter RJ, Trent MS (2004) Periplasmic cleavage and modification of the 1-phosphate group of *Helicobacter pylori* lipid A. J Biol Chem 279:55780–55791
82. Coats SR, Jones JW, Do CT, Braham PH, Bainbridge BW, To TT, Goodlett DR, Ernst RK, Darveau RP (2009) Human Toll-like receptor 4 responses to *P. gingivalis* are regulated by lipid A 1- and 4′-phosphatase activities. Cell Microbiol 11:1587–1599
83. Que NL, Lin S, Cotter RJ, Raetz CR (2000) Purification and mass spectrometry of six lipid A species from the bacterial endosymbiont *Rhizobium etli*. Demonstration of a conserved distal unit and a variable proximal portion. J Biol Chem 275:28006–28016
84. Que NL, Ribeiro AA, Raetz CR (2000) Two-dimensional NMR spectroscopy and structures of six lipid A species from *Rhizobium etli* CE3. Detection of an acyloxyacyl residue in each component and origin of the aminogluconate moiety. J Biol Chem 275:28017–28027
85. Wang X, Ribeiro AA, Guan Z, McGrath SC, Cotter RJ, Raetz CR (2006) Structure and biosynthesis of free lipid A molecules that replace lipopolysaccharide in *Francisella tularensis* subsp. *novicida*. Biochemistry 45:14427–14440
86. Shaffer SA, Harvey MD, Goodlett DR, Ernst RK (2007) Structural heterogeneity and environmentally regulated remodeling of *Francisella tularensis* subspecies *novicida* lipid A characterized by tandem mass spectrometry. J Am Soc Mass Spectrom 18:1080–1092
87. Ogawa T (1993) Chemical structure of lipid A from *Porphyromonas* (*Bacteroides*) *gingivalis* lipopolysaccharide. FEBS Lett 332:197–201
88. Stead CM, Zhao J, Raetz CR, Trent MS (2010) Removal of the outer Kdo from *Helicobacter pylori* lipopolysaccharide and its impact on the bacterial surface. Mol Microbiol 78:837–852
89. Que-Gewirth NL, Ribeiro AA, Kalb SR, Cotter RJ, Bulach DM, Adler B, Girons IS, Werts C, Raetz CR (2004) A methylated phosphate group and four amide-linked acyl chains in leptospira interrogans lipid A. The membrane anchor of an unusual lipopolysaccharide that activates TLR2. J Biol Chem 279:25420–25429
90. Wang X, McGrath SC, Cotter RJ, Raetz CR (2006) Expression cloning and periplasmic orientation of the *Francisella novicida* lipid A 4′-phosphatase LpxF. J Biol Chem 281:9321–9330
91. Wang X, Ribeiro AA, Guan Z, Abraham SN, Raetz CR (2007) Attenuated virulence of a *Francisella* mutant lacking the lipid A 4′-phosphatase. Proc Natl Acad Sci USA 104:4136–4141
92. Ingram BO, Sohlenkamp C, Geiger O, Raetz CR (2010) Altered lipid A structures and polymyxin hypersensitivity of *Rhizobium etli* mutants lacking the LpxE and LpxF phosphatases. Biochim Biophys Acta 1801:593–604
93. Gunn JS, Miller SI (1996) PhoP-PhoQ activates transcription of *pmrAB*, encoding a two-component regulatory system involved in *Salmonella typhimurium* antimicrobial peptide resistance. J Bacteriol 178:6857–6864
94. Wosten MM, Kox LF, Chamnongpol S, Soncini FC, Groisman EA (2000) A signal transduction system that responds to extracellular iron. Cell 103:113–125
95. Trent MS, Ribeiro AA, Doerrler WT, Lin S, Cotter RJ, Raetz CR (2001) Accumulation of a polyisoprene-linked amino sugar in polymyxin-resistant *Salmonella typhimurium* and *Escherichia coli*: structural characterization and transfer to lipid A in the periplasm. J Biol Chem 276:43132–43144

96. Doerrler WT, Gibbons HS, Raetz CR (2004) MsbA-dependent translocation of lipids across the inner membrane of *Escherichia coli*. J Biol Chem 279:45102–45109
97. Tran AX, Lester ME, Stead CM, Raetz CR, Maskell DJ, McGrath SC, Cotter RJ, Trent MS (2005) Resistance to the antimicrobial peptide polymyxin requires myristoylation of *Escherichia coli* and *Salmonella typhimurium* lipid A. J Biol Chem 280:28186–28194
98. Gunn JS, Lim KB, Krueger J, Kim K, Guo L, Hackett M, Miller SI (1998) PmrA-PmrB-regulated genes necessary for 4-aminoarabinose lipid A modification and polymyxin resistance. Mol Microbiol 27:1171–1182
99. Gunn JS, Ryan SS, Van Velkinburgh JC, Ernst RK, Miller SI (2000) Genetic and functional analysis of a PmrA-PmrB-regulated locus necessary for lipopolysaccharide modification, antimicrobial peptide resistance, and oral virulence of *Salmonella enterica* serovar Typhimurium. Infect Immun 68:6139–6146
100. Marr N, Tirsoaga A, Blanot D, Fernandez R, Caroff M (2008) Glucosamine found as a substituent of both phosphate groups in *Bordetella* lipid A backbones: role of a BvgAS-activated ArnT ortholog. J Bacteriol 190:4281–4290
101. Wang X, Ribeiro AA, Guan Z, Raetz CR (2009) Identification of undecaprenyl phosphate-β-D-galactosamine in *Francisella novicida* and its function in lipid A modification. Biochemistry 48:1162–1172
102. Kanistanon D, Hajjar AM, Pelletier MR, Gallagher LA, Kalhorn T, Shaffer SA, Goodlett DR, Rohmer L, Brittnacher MJ, Skerrett SJ, Ernst RK (2008) A *Francisella* mutant in lipid A carbohydrate modification elicits protective immunity. PLoS Pathog 4:e24
103. Marr N, Hajjar AM, Shah NR, Novikov A, Yam CS, Caroff M, Fernandez RC (2010) Substitution of the *Bordetella pertussis* lipid A phosphate groups with glucosamine is required for robust NF-κB activation and release of proinflammatory cytokines in cells expressing human but not murine Toll-like receptor 4-MD-2-CD14. Infect Immun 78:2060–2069
104. Kanjilal-Kolar S, Raetz CR (2006) Dodecaprenyl phosphate-galacturonic acid as a donor substrate for lipopolysaccharide core glycosylation in *Rhizobium leguminosarum*. J Biol Chem 281:12879–12887
105. Raetz CR, Reynolds CM, Trent MS, Bishop RE (2007) Lipid A modification systems in Gram-negative bacteria. Annu Rev Biochem 76:295–329
106. Plotz BM, Lindner B, Stetter KO, Holst O (2000) Characterization of a novel lipid A containing D-galacturonic acid that replaces phosphate residues. The structure of the lipid A of the lipopolysaccharide from the hyperthermophilic bacterium Aquifex pyrophilus. J Biol Chem 275:11222–11228
107. Trent MS, Raetz CRH (2002) Cloning of EptA, the lipid A phosphoethanolamine transferase associated with polymyxin resistance. J Endotoxin Res 8:159
108. Cox AD, Wright JC, Li J, Hood DW, Moxon ER, Richards JC (2003) Phosphorylation of the lipid A region of meningococcal lipopolysaccharide: identification of a family of transferases that add phosphoethanolamine to lipopolysaccharide. J Bacteriol 185:3270–3277
109. Tzeng YL, Ambrose KD, Zughaier S, Zhou X, Miller YK, Shafer WM, Stephens DS (2005) Cationic antimicrobial peptide resistance in *Neisseria meningitidis*. J Bacteriol 187:5387–5396
110. Cullen TW, Trent MS (2010) A link between the assembly of flagella and lipooligosaccharide of the Gram-negative bacterium *Campylobacter jejuni*. Proc Natl Acad Sci USA 107:5160–5165
111. Lewis LA, Choudhury B, Balthazar JT, Martin LE, Ram S, Rice PA, Stephens DS, Carlson R, Shafer WM (2009) Phosphoethanolamine substitution of lipid A and resistance of *Neisseria gonorrhoeae* to cationic antimicrobial peptides and complement-mediated killing by normal human serum. Infect Immun 77:1112–1120
112. Takahashi H, Carlson RW, Muszynski A, Choudhury B, Kim KS, Stephens DS, Watanabe H (2008) Modification of lipooligosaccharide with phosphoethanolamine by LptA in *Neisseria meningitidis* enhances meningococcal adhesion to human endothelial and epithelial cells. Infect Immun 76:5777–5789

113. Touze T, Tran AX, Hankins JV, Mengin-Lecreulx D, Trent MS (2008) Periplasmic phosphorylation of lipid A is linked to the synthesis of undecaprenyl phosphate. Mol Microbiol 67:264–277
114. Valvano MA (2008) Undecaprenyl phosphate recycling comes out of age. Mol Microbiol 67:232–235
115. Bouhss A, Trunkfield AE, Bugg TD, Mengin-Lecreulx D (2008) The biosynthesis of peptidoglycan lipid-linked intermediates. FEMS Microbiol Rev 32:208–233
116. El Ghachi M, Derbise A, Bouhss A, Mengin-Lecreulx D (2005) Identification of multiple genes encoding membrane proteins with undecaprenyl pyrophosphate phosphatase (UppP) activity in *Escherichia coli*. J Biol Chem 280:18689–18695
117. Herrera CM, Hankins JV, Trent MS (2010) Activation of PmrA inhibits LpxT-dependent phosphorylation of lipid A promoting resistance to antimicrobial peptides. Mol Microbiol 76:1444–1460
118. Boon Hinckley M, Reynolds CM, Ribeiro AA, McGrath SC, Cotter RJ, Lauw FN, Golenbock DT, Raetz CR (2005) A *Leptospira interrogans* enzyme with similarity to yeast Ste14p that methylates the 1-phosphate group of lipid A. J Biol Chem 280:30214–30224
119. Reynolds CM, Ribeiro AA, McGrath SC, Cotter RJ, Raetz CR, Trent MS (2006) An outer membrane enzyme encoded by *Salmonella typhimurium lpxR* that removes the 3′-acyloxyacyl moiety of lipid A. J Biol Chem 281:21974–21987
120. Trent MS, Pabich W, Raetz CR, Miller SI (2001) A PhoP/PhoQ-induced lipase (PagL) that catalyzes 3-O-deacylation of lipid A precursors in membranes of *Salmonella typhimurium*. J Biol Chem 276:9083–9092
121. Guo L, Lim KB, Poduje CM, Daniel M, Gunn JS, Hackett M, Miller SI (1998) Lipid A acylation and bacterial resistance against vertebrate antimicrobial peptides. Cell 95:189–198
122. Bishop RE, Gibbons HS, Guina T, Trent MS, Miller SI, Raetz CR (2000) Transfer of palmitate from phospholipids to lipid A in outer membranes of Gram-negative bacteria. EMBO J 19:5071–5080
123. Brozek KA, Bulawa CE, Raetz CR (1987) Biosynthesis of lipid A precursors in *Escherichia coli*. A membrane-bound enzyme that transfers a palmitoyl residue from a glycerophospholipid to lipid X. J Biol Chem 262:5170–5179
124. Guo L, Lim KB, Gunn JS, Bainbridge B, Darveau RP, Hackett M, Miller SI (1997) Regulation of lipid A modifications by *Salmonella typhimurium* virulence genes *phoP-phoQ*. Science 276:250–253
125. Bishop RE (2005) The lipid A palmitoyltransferase PagP: molecular mechanisms and role in bacterial pathogenesis. Mol Microbiol 57:900–912
126. Bishop RE (2008) Structural biology of membrane-intrinsic β-barrel enzymes: sentinels of the bacterial outer membrane. Biochim Biophys Acta 1778:1881–1896
127. Kawasaki K, Ernst RK, Miller SI (2004) Deacylation and palmitoylation of lipid A by *Salmonellae* outer membrane enzymes modulate host signaling through Toll-like receptor 4. J Endotoxin Res 10:439–444
128. Hwang PM, Choy WY, Lo EI, Chen L, Forman-Kay JD, Raetz CR, Prive GG, Bishop RE, Kay LE (2002) Solution structure and dynamics of the outer membrane enzyme PagP by NMR. Proc Natl Acad Sci USA 99:13560–13565
129. Hwang PM, Bishop RE, Kay LE (2004) The integral membrane enzyme PagP alternates between two dynamically distinct states. Proc Natl Acad Sci USA 101:9618–9623
130. Ahn VE, Lo EI, Engel CK, Chen L, Hwang PM, Kay LE, Bishop RE, Prive GG (2004) A hydrocarbon ruler measures palmitate in the enzymatic acylation of endotoxin. EMBO J 23:2931–2941
131. Jia W, El Zoeiby A, Petruzziello TN, Jayabalasingham B, Seyedirashti S, Bishop RE (2004) Lipid trafficking controls endotoxin acylation in outer membranes of *Escherichia coli*. J Biol Chem 279:44966–44975

132. Khan MA, Bishop RE (2009) Molecular mechanism for lateral lipid diffusion between the outer membrane external leaflet and a b-barrel hydrocarbon ruler. Biochemistry 48:9745–9756
133. Robey M, O'Connell W, Cianciotto NP (2001) Identification of *Legionella pneumophila rcp*, a pagP-like gene that confers resistance to cationic antimicrobial peptides and promotes intracellular infection. Infect Immun 69:4276–4286
134. Pilione MR, Pishko EJ, Preston A, Maskell DJ, Harvill ET (2004) *pagP* is required for resistance to antibody-mediated complement lysis during *Bordetella bronchiseptica* respiratory infection. Infect Immun 72:2837–2842
135. Rebeil R, Ernst RK, Gowen BB, Miller SI, Hinnebusch BJ (2004) Variation in lipid A structure in the pathogenic yersiniae. Mol Microbiol 52:1363–1373
136. Bishop RE, Kim SH, El Zoeiby A (2005) Role of lipid A palmitoylation in bacterial pathogenesis. J Endotoxin Res 11:174–180
137. Ernst RK, Yi EC, Guo L, Lim KB, Burns JL, Hackett M, Miller SI (1999) Specific lipopolysaccharide found in cystic fibrosis airway *Pseudomonas aeruginosa*. Science 286:1561–1565
138. Ernst RK, Moskowitz SM, Emerson JC, Kraig GM, Adams KN, Harvey MD, Ramsey B, Speert DP, Burns JL, Miller SI (2007) Unique lipid A modifications in *Pseudomonas aeruginosa* isolated from the airways of patients with cystic fibrosis. J Infect Dis 196:1088–1092
139. Basu SS, White KA, Que NL, Raetz CR (1999) A deacylase in *Rhizobium leguminosarum* membranes that cleaves the 3-O-linked β-hydroxymyristoyl moiety of lipid A precursors. J Biol Chem 274:11150–11158
140. Kawasaki K, Ernst RK, Miller SI (2005) Inhibition of *Salmonella enterica* serovar Typhimurium lipopolysaccharide deacylation by aminoarabinose membrane modification. J Bacteriol 187:2448–2457
141. Manabe T, Kawasaki K (2008) Extracellular loops of lipid A 3-O-deacylase PagL are involved in recognition of aminoarabinose-based membrane modifications in *Salmonella enterica* serovar Typhimurium. J Bacteriol 190:5597–5606
142. Kawasaki K, Ernst RK, Miller SI (2004) 3-O-Deacylation of lipid A by PagL, a PhoP/PhoQ-regulated deacylase of *Salmonella typhimurium*, modulates signaling through Toll-like receptor 4. J Biol Chem 279:20044–20048
143. Ernst RK, Adams KN, Moskowitz SM, Kraig GM, Kawasaki K, Stead CM, Trent MS, Miller SI (2006) The *Pseudomonas aeruginosa* lipid A deacylase: selection for expression and loss within the cystic fibrosis airway. J Bacteriol 188:191–201
144. Rutten L, Geurtsen J, Lambert W, Smolenaers JJ, Bonvin AM, de Haan A, van der Ley P, Egmond MR, Gros P, Tommassen J (2006) Crystal structure and catalytic mechanism of the LPS 3-O-deacylase PagL from *Pseudomonas aeruginosa*. Proc Natl Acad Sci USA 103:7071–7076
145. Geurtsen J, Steeghs L, Hove JT, van der Ley P, Tommassen J (2005) Dissemination of lipid A deacylases (PagL) among Gram-negative bacteria: identification of active-site histidine and serine residues. J Biol Chem 280:8248–8259
146. Kawano M, Manabe T, Kawasaki K (2010) *Salmonella enterica* serovar Typhimurium lipopolysaccharide deacylation enhances its intracellular growth within macrophages. FEBS Lett 584:207–212
147. Gibbons HS, Lin S, Cotter RJ, Raetz CR (2000) Oxygen requirement for the biosynthesis of the *S*-2-hydroxymyristate moiety in *Salmonella typhimurium* lipid A. Function of LpxO, a new Fe^{2+}/α-ketoglutarate-dependent dioxygenase homologue. J Biol Chem 275:32940–32949
148. Sforza S, Silipo A, Molinaro A, Marchelli R, Parrilli M, Lanzetta R (2004) Determination of fatty acid positions in native lipid A by positive and negative electrospray ionization mass spectrometry. J Mass Spectrom 39:378–383

149. Kulshin VA, Zähringer U, Lindner B, Jäger KE, Dmitriev BA, Rietschel ET (1991) Structural characterization of the lipid A component of *Pseudomonas aeruginosa* wild-type and rough mutant lipopolysaccharides. Eur J Biochem 198:697–704
150. Kawai Y, Moribayashi A (1982) Characteristic lipids of *Bordetella pertussis*: simple fatty acid composition, hydroxy fatty acids, and an ornithine-containing lipid. J Bacteriol 151:996–1005
151. Zähringer U, Knirel YA, Lindner B, Helbig JH, Sonesson A, Marre R, Rietschel ET (1995) The lipopolysaccharide of *Legionella pneumophila* serogroup 1 (strain Philadelphia 1): chemical structure and biological significance. Prog Clin Biol Res 392:113–139
152. Stead C, Tran A, Ferguson D Jr, McGrath S, Cotter R, Trent S (2005) A novel 3-deoxy-D-*manno*-octulosonic acid (Kdo) hydrolase that removes the outer Kdo sugar of *Helicobacter pylori* lipopolysaccharide. J Bacteriol 187:3374–3383
153. Zhao J, Raetz CR (2010) A two-component Kdo hydrolase in the inner membrane of *Francisella novicida*. Mol Microbiol 78:820–836
154. Reynolds CM, Kalb SR, Cotter RJ, Raetz CR (2005) A phosphoethanolamine transferase specific for the outer 3-deoxy-D-*manno*-octulosonic acid residue of *Escherichia coli* lipopolysaccharide. Identification of the *eptB* gene and Ca^{2+} hypersensitivity of an *eptB* deletion mutant. J Biol Chem 280:21202–21211
155. Kanjilal-Kolar S, Basu SS, Kanipes MI, Guan Z, Garrett TA, Raetz CR (2006) Expression cloning of three *Rhizobium leguminosarum* lipopolysaccharide core galacturonosyltransferases. J Biol Chem 281:12865–12878

Pathways for the Biosynthesis of NDP Sugars

7

Youai Hao and Joseph S. Lam

7.1 Introduction

Bacterial lipopolysaccharide (LPS) is an important surface structure of Gram-negative bacteria for maintaining the integrity of the outer membrane. It is also a virulence factor in many bacteria, particularly those that are pathogens of plants and animals. Structurally, the LPS can be divided into three domains, lipid A, core oligosaccharide and O-polysaccharide (or O-antigen). Its polysaccharide constituents contain a great variety of sugars including neutral sugars, charged sugars that are acidic or amino substituted (see Chap. 3). Substitutions and enzymatic modifications of the basic sugar structure also lead to interesting deoxy or dideoxy sugars. To date, more than 100 new sugar moieties are found in bacterial polysaccharides. In contrast, eukaryotic glycoproteins and glycolipids are synthesized from only nine sugar donors [1, 2]. Since many of the LPS monosaccharide components are rare sugars and only present in certain pathogenic bacteria species, these unusual sugars and the enzymes involved in their synthesis can be targets for novel antimicrobial drug development. An in-depth understanding of the biosynthetic pathways of these sugars and the mechanisms of the encoded enzymes is an essential first step to undertake.

As demonstrated by Leloir in the 1950s, the sugar units must be converted into sugar nucleotides before they are recognized by specific glycosyltransferases and assembled into a sugar polysaccharide one by one [3]. Different sugars are activated by different nucleotide triphosphate (NTP) to form either nucleotide monophosphate (NMP) or nucleotide diphosphate (NDP) derivatives. Except for

Y. Hao • J.S. Lam (✉)
Department of Molecular and Cellular Biology, University of Guelph, 50 Stone Road E., Guelph, Canada, ON, N1G 2W1
e-mail: haoy@uoguelph.ca; jlam@uoguelph.ca

Y.A. Knirel and M.A. Valvano (eds.), *Bacterial Lipopolysaccharides*,
DOI 10.1007/978-3-7091-0733-1_7, © Springer-Verlag/Wien 2011

a few sugars such as glucose, galactose and *N*-acetylglucosamine which are common components of many other structural glycans and are utilized in other housekeeping metabolic functions, the genes responsible for the biosynthesis of sugars found in LPS are usually located and organized in gene clusters (e.g. core or O-antigen gene clusters, Chaps. 8 and 9).

In recent years, the advancements of molecular genetic knowledge and sequencing techniques, the availability of tools for manipulating and constructing recombinant DNA, and the rapid expansion of the databases for annotation of whole-genome sequences have made identification and sequencing of these gene clusters easier. The development of new methods for mass spectrometry (MS), nuclear magnetic resonance (NMR) spectroscopy, coupled with improvements of the sensitivity of these instruments allows the precise determination of the exact composition of a LPS structure. The knowledge of these sugar structures and the availability of sequenced polysaccharide biosynthesis gene clusters allow for the prediction of biosynthesis pathways and pave way for understanding the biochemistry of specific enzymatic-substrate reactions concerning microbial glycobiology. The possible biosynthesis genes involved in particular steps in the pathways could be identified from the corresponding gene cluster, and based on *in silico* comparisons of sequence similarity and identity, "putative" functions could be assigned to these genes. To determine the exact enzymatic functions, the proteins of interest could be over-expressed, purified, and used to develop enzymatic assays. Capillary electrophoresis (CE), high-performance liquid chromatography coupled with MS, and NMR spectroscopy have been used by our group and others to facilitate in vitro biochemical characterization of the enzymatic properties. Besides biochemical studies, a sufficiently high yield of the purified enzymes could also facilitate structural studies using X-ray crystallography or protein NMR methods. Such studies are important for determining the 3D structures of these proteins and the mechanisms of the enzymatic activities.

To date, the biosynthesis pathways of more than 30 of the NDP sugars precursors have been reported and discussed below. The majority of the sugars found in LPS are hexoses and their derivatives. There are also non-hexoses including 3-deoxy-D-*manno*-oct-2-ulosonic acid (Kdo) and L-*glycero*-D-*manno*-heptose (L,D-Hep), both of which are highly conserved in the core oligosaccharides of most LPS structures. Hexoses and hexose derivatives are generally derived from either glucose 6-phosphate (Glc-6-P) or fructose 6-phosphate (Fru-6-P), which can be obtained from the bacterial central metabolic pathway. In this review, we will summarize the current knowledge of the characterized NDP sugar biosynthesis pathways of these sugars. We have grouped the hexoses according to their original sugar sources (Glc-6-P or Fru-6-P) and the identity of the coupled NDP (dTDP, GDP or UDP), and we also highlight common rules among these pathways when appropriate. However, it should be noted that due to page limits, some rare hexoses and hexose derivatives that are also derived from Glc-6-P or Fru-6-P (such as UDP-2,4-diacetamido-2,4,6-trideoxy-D-glucose, dTDP-3-acetamido-3,6-dideoxy-D-glucose, CDP-3,6-dideoxyhexoses) are not covered in this chapter. We have also reviewed the biosynthesis pathways of the non-hexose sugars Kdo and L,D-Hep.

7.2 Sugars Derived From Glucose-6-P

7.2.1 Biosynthesis of UDP Sugars

7.2.1.1 UDP-D-Glucose (UDP-D-Glc)

D-Glc (1)[1] is a relatively common monosaccharide in the outer core of most LPS molecules, and is also a constituent of many O-antigens. The proposed active form of D-Glc recognized by glucosyltransferase is UDP-D-Glc (4). The initial substrate for biosynthesis of UDP-D-Glc is Glc-6-P (2) derived from the central metabolic pathway. Glc-6-P can either be directly transported into the bacteria cell or can be converted from D-Glc by the enzyme glucokinase (Glk) (EC 2.7.1.2). Three steps are required for the conversion of Glc-6-P to UDP-D-Glc (Scheme 7.1). In *Escherichia coli*, a phosphoglucomutase (EC 5.4.2.2) (encoded by *pgm*) catalyzes the reversible conversion of Glc-6-P to glucose 1-phosphate (Glc-1-P) (3) [4], a common intermediate for the synthesis of both UDP-D-Glc and dTDP-L-rhamnose (dTDP-L-Rha) (11) and related sugars. In *Pseudomonas aeruginosa*, a bifunctional enzyme AlgC is responsible for this reaction [5]. AlgC has both phosphoglucomutase (PGM) and phosphomannomutase (PMM) activity. While the PGM activity is required for the conversion of Glc-6-P to Glc-1-P, the PMM activity is involved in the biosynthesis of GDP-D-Rha (21). The *algC* mutant strains of *P. aeruginosa* produced LPS with truncated cores missing all Glc and Rha residues [6]. Mutational analysis also indicated that AlgC is the only PGM in *P. aeruginosa*, as the crude cell extract of *algC* mutant strain showed no detectable PGM activity [5].

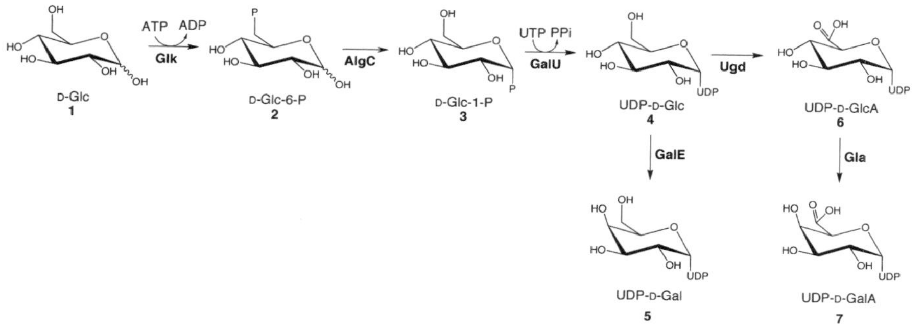

Scheme 7.1 Biosynthesis pathways of UDP-D-glucose, UDP-D-galactose, UDP-D-glucuronic acid and UDP-D-galacturonic acid

[1]Numbers in parentheses refer to the corresponding structures depicted in the figures.

The formation of UDP-D-Glc (4) from UTP and Glc-1-P is then catalyzed by Glc-1-P uridilyltransferase (also known as UDP-D-Glc pyrophosphorylase) (EC 2.7.7.9). In *E. coli*, this enzyme is encoded by the *galU* gene [7]. The 3D crystal structure of the *E. coli* GalU reveals that this protein is a member of the short chain dehydrogenase/reductase (SDR) superfamily, forms a tetramer, and shares remarkable structural similarity to Glc-1-P thymidylyltransferase involved in the biosynthesis of dTDP-L-Rha (see Sect. 7.2.2.1) [8]. Homologs of *galU* have also been isolated and genetically characterized in *P. aeruginosa* [9–11], *Streptococcus pneumoniae* [12] and many other bacterial species.

7.2.1.2 UDP-D-Glc Derived Sugars (UDP-D-Gal, UDP-D-GlcA and UDP-D-GalA)

UDP-D-Glc is a common precursor for the biosynthesis of several other UDP sugars commonly found in the bacterial surface glycans, including its C-4 epimer UDP-D-galactose (UDP-D-Gal) (5), UDP-D-glucuronic acid (UDP-D-GlcA) (6) and UDP-D-galacturonic acid (UDP-D-GalA) (7). The hexose D-Gal is a highly conserved constituent found in both LPS cores and O-antigens of bacteria. Its nucleotide activated precursor, UDP-D-Gal, is synthesized in bacteria by the catalytic activity of UDP-D-Glc 4-epimerase (also known as UDP-D-Gal 4-epimerase) GalE (EC 5.1.3.2) using UDP-D-Glc as the substrate (Scheme 7.1). This is a reversible reaction and it enables the catabolism of exogenous galactose via the glycolytic pathway [3]. GalE, another member of the SDR superfamily, has been well characterized from *E. coli* and other bacterial species [13, 14]. In *E. coli*, *galE* is not localized in LPS gene clusters; instead, it is found in the *gal* operon involved in galactose uptake and catabolism [15].

D-GlcA is found in both O-antigens and the exopolysaccharide (EPS) colanic acid of many serotypes of *E. coli* (such as O4, O5, O6 and O9) and *Salmonella enterica* [16, 17]. It is also present in the O-antigen of *Proteus vulgaris* O4 [18], the capsular polysaccharide (CPS) of *Vibrio cholerae* O139 [19] and *Streptococcus pneumoniae* types 1, 2, 3 and 8 [20–23]. The nucleotide-activated precursor UDP-D-GlcA (6) is synthesized from UDP-D-Glc by the dehydrogenase Ugd (EC 1.1.1.22) (Scheme 7.1). The *ugd* gene has been found in the colanic acid biosynthesis cluster of *E. coli* [16, 24], and in the CPS cluster of *V. cholerae* O139 and *S. pneumoniae* [25, 26]. In *P. aeruginosa*, *ugd* is found in the *PA4773-PA4775-pmrAB* and *pmrHFIJKLM-ugd* operons, and has been implicated in the resistance mechanism against cationic antimicrobial peptides such as polymyxin B. Interestingly, redundancy of *ugd* has been observed, and two copies of *ugd* have been found in *P. aeruginosa* [27] and *Burkholderia cenocepacia* [28].

UDP-D-GlcA is the precursor for the biosynthesis of UDP-D-GalA (7), the nucleotide-activated form of D-GalA. This sugar is also commonly present in a variety of surface glycans. To name a few, it has been found in the O-antigens of *E. coli* O113 [29] and *V. cholerae* O139 and O22 [30, 31], the core oligosaccharide of *Rhizobium leguminosarum* [32], *Proteus penneri* [33] and *Klebsiella pneumoniae* [34], and the CPS of *S. pneumoniae* [22]. UDP-D-GalA is converted from UDP-D-GlcA by a 4-epimerase. This enzyme was first characterized from the *S. pneumoniae* (Cap1J) [23], and recently in *K. pneumoniae* [35, 36]. The gene that encodes this enzyme was originally named *uge*, but was changed to *gla* as recommended by Reeves et al. [37]. The Gla enzymes from the two bacterial species apparently have different biochemical properties. For example, Gla from *S. pneumoniae* is highly specific for the interconversion of UDP-D-GlcA and UDP-D-GalA. However, the enzyme from *K. pneumoniae* could catalyze not only the interconversion of UDP-D-GlcA and UDP-D-GalA, but also the interconversion of UDP-D-Glc and UDP-D-Gal, as well as UDP-2-acetamido-2-deoxy-D-glucose and UDP-2-acetamido-2-deoxy-D-galactose [23, 36].

7.2.2 Biosynthesis of dTDP Sugars

7.2.2.1 dTDP-L-Rhamnose (dTDP-L-Rha)

L-Rha is widely distributed in the O-antigens of Gram-negative bacteria such as *S. enterica*, *V. cholerae*, *E. coli* [38, 39], and *P. aeruginosa* (serotypes O4, O6, O13, O14, O15 and O19) [40]. It is also a common constituent of the outer core of LPS in *P. aeruginosa* [41], and the CPS of Gram-positive bacteria including *S. pneumoniae* [42–44]. In *Mycobacterium* species, L-Rha links arabinogalactan to the peptidoglycan layer [45], which is vital to mycobacteria survival and growth [46]. As mammals do not produce or utilize L-Rha, the biosynthetic pathway of L-Rha and the enzymes involved represent potential targets against which new therapeutic drugs might be designed [47].

The biosynthesis of the nucleotide-activated precursor dTDP-L-Rha (11) [48] has been thoroughly characterized. Four enzymes RmlA, RmlB, RmlC and RmlD catalyze the conversion of D-Glc-1-P and dTTP to dTDP-L-Rha [49] (Scheme 7.2). RmlA (EC 2.7.7.24) is a glucose-1-phosphate thymidylyltransferase and catalyzes the transfer of dTMP from dTTP to Glc-1-P to form dTDP-D-Glc (8). The final product of the pathway dTDP-L-Rha shows feedback inhibition of the synthesis activity of RmlA [47, 50]. RmlB has been shown to have dTDP-D-glucose 4,6-dehydratase activity (EC 4.2.1.46). It is also a member of the SDR superfamily of proteins, requiring NAD^+ as a cofactor to catalyze the conversion of dTDP-D-Glc to dTDP-6-deoxy-4-keto-D-Glc (dTDP-6-deoxy-D-*xylo*-hexos-4-ulose) (9). The third enzyme RmlC (dTDP-6-deoxy-4-keto-D-Glc 3,5-epimerase, EC 5.1.3.13) then converts dTDP-6-deoxy-4-keto-D-Glc (9) to dTDP-4-keto-L-Rha (dTDP-6-deoxy-L-*lyxo*-hexos-4-ulose) (10) by catalyzing the epimerization at C-5 and C-3. Finally, another SDR superfamily protein, RmlD (dTDP-6-deoxy-L-*lyxo*-hexos-4-ulose 4-reductase, EC 1.1.1.133) carries out the reduction reaction at position 4 of dTDP-4-keto-L-Rha forming dTDP-L-Rha (11).

The 3D structures of RmlA [47, 51] and RmlC [52] from *P. aeruginosa*, as well as RmlB [53], RmlC [54] and RmlD [55] from *S. enterica* serovar Typhimurium have been solved recently. The knowledge gained from these structures has shed light on the mechanisms of their enzymatic reactions. Specific catalytic residues were identified based on their contact with the substrate and on sequence

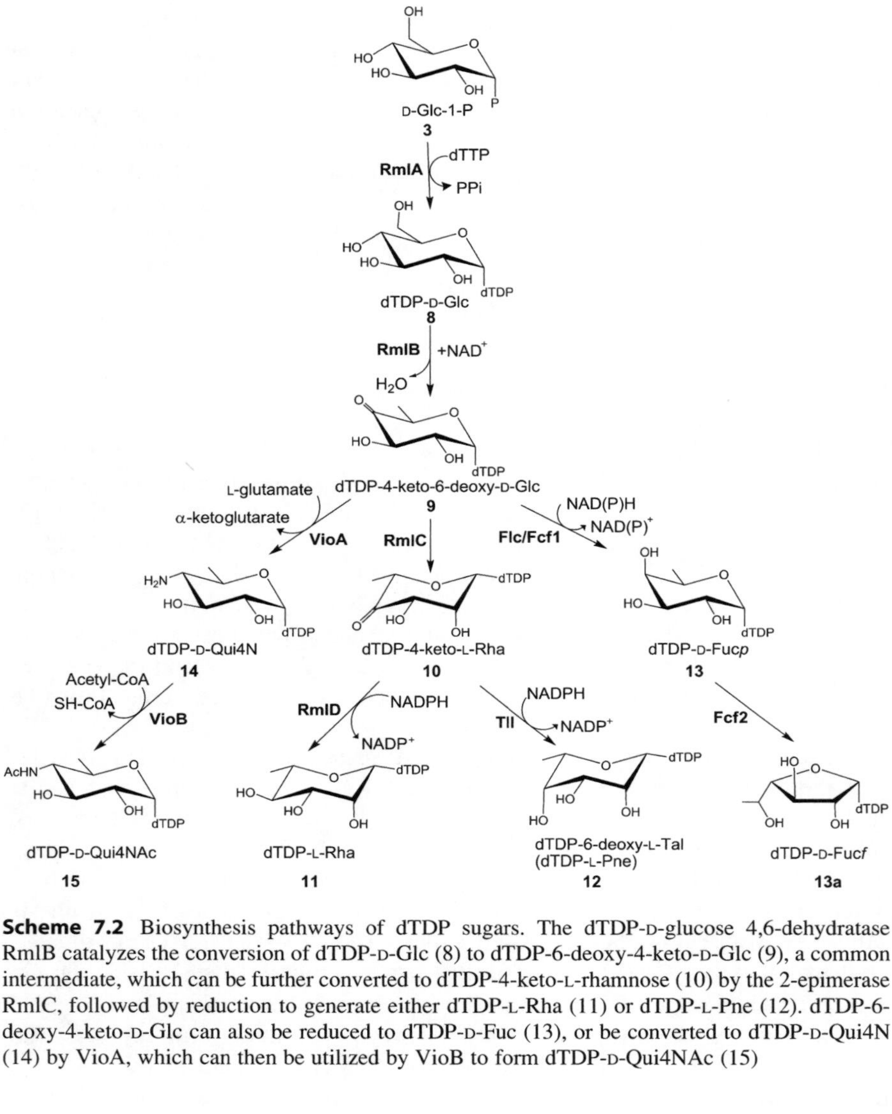

Scheme 7.2 Biosynthesis pathways of dTDP sugars. The dTDP-D-glucose 4,6-dehydratase RmlB catalyzes the conversion of dTDP-D-Glc (8) to dTDP-6-deoxy-4-keto-D-Glc (9), a common intermediate, which can be further converted to dTDP-4-keto-L-rhamnose (10) by the 2-epimerase RmlC, followed by reduction to generate either dTDP-L-Rha (11) or dTDP-L-Pne (12). dTDP-6-deoxy-4-keto-D-Glc can also be reduced to dTDP-D-Fuc (13), or be converted to dTDP-D-Qui4N (14) by VioA, which can then be utilized by VioB to form dTDP-D-Qui4NAc (15)

conservation. RmlA from *P. aeruginosa* is a tetramer (dimer of dimer) and its catalytic residues in the active sites have been identified to include R15, K25, D110, K162, and D225 [47]. RmlB from *S. enterica* is a homodimer. Its monomeric structure contains two domains, a N-terminal NAD^+ cofactor-binding domain, and a C-terminal sugar-nucleotide-binding domain. The N-terminal domain contains the highly conserved YXXXK catalytic couple and the GXXGXXG motif, which are characteristic of SDR extended family [53]. RmlD also contains these particular motifs, which characterized it as a SDR superfamily member.

Unlike other SDR enzymes, RmlD shows no strong preference for either NADH or NADPH as cofactor. It also requires the binding of another cofactor Mg^{2+} for dimerization [55]. RmlC represents a new class of epimerases that do not require any cofactor [54]. The four genes *rmlA*, *rmlB*, *rmlC* and *rmlD* are conserved and are clustered together (although the order of the genes in the cluster may differ) in all bacterial species studied. This observation has allowed the Reeves laboratory to use them as target genes for bioinformatics studies of lateral gene transfer of O-antigen gene clusters between species [56, 57].

The catalytic reaction product of RmlB, dTDP-6-deoxy-4-keto-D-Glc (9), is a common intermediate for the biosynthesis of several dTDP-activated sugar precursors found in the LPS biosynthesis in addition to dTDP-L-Rha. For example, dTDP-6-deoxy-L-talose (dTDP-L-pneumose) (12), dTDP-D-fucose (13), dTDP-4-amino-4,6-dideoxy-D-glucose (14) and dTDP-4-acetamido-4,6-dideoxy-D-glucose (15) (see below).

7.2.2.2 dTDP-L-Pneumose (dTDP-L-Pne)

The monosaccharide L-pneumose (L-Pne, 6-deoxy-L-talose) has been described by Gaugler and Gabriel as an unusual sugar [58]. It is a component of the O-antigens of *E. coli* O45 and O66 [59] and *Burkholderia plantarii* [60]. The O-specific chain of the LPS from *Rhizobium loti* NZP2213 is a homopolymer of L-Pne [61]. The serotype c-specific polysaccharide of the Gram-negative bacteria *Actinobacillus actinomycetemcomitans* is 6-deoxy-L-talan, which consists of 2-O-acetylated 1,3-linked L-Pne [62]. The activated nucleotide-sugar form of L-Pne is dTDP-L-Pne (12) and is very unstable [58, 63]. The biosynthetic pathway for dTDP-L-Pne has so far only been characterized in *A. actinomycetemcomitans* [63]. The chemical structures of L-Pne and L-Rha differ only in the stereochemistry of the C-4 carbon. Not surprisingly, the biosynthetic pathway of these two nucleotide sugar precursors only differs in the last step. A *tll* gene encoding dTDP-L-Pne synthase was identified and characterized from *A. actinomycetemcomitans* NCTC 9710 [63]. Like RmlD in the L-Rha biosynthesis pathway (see above), Tll is a dTDP-6-deoxy-L-*lyxo*-hexos-4-ulose reductase. Both RmlD and Tll catalyze the reduction at the 4-keto group of the same substrate dTDP-4-keto-L-Rha (10) (the RmlC catalyzed reaction product), but yield distinct products due to the stereospecificity of the reactions. While both are dTDP-6-deoxy-L-*lyxo*-4-hexulose reductase, RmlD and Tll from *A. actinomycetemcomitans* exhibit a low level of sequence similarity (20.6% in amino acid sequence), except for the consensus nucleotide-binding motif (GXXGXXG) [63].

7.2.2.3 dTDP-D-Fucose (dTDP-D-Fuc)

D-Fuc is a rare sugar that has been found in the LPS of several bacteria species. For example, the O-antigens of *Pectinatus cerevisiiphilus* [64] and *E. coli* O52 [65] contain D-Fuc in furanose form, while the O-antigen of *Stenotrophomonas maltophilia* O3 [66] and O19 [67] contains this sugar in pyranose form. D-Fuc is also present in the LPS of *Pseudomonas syringae*, *Pseudomonas fluorescens*, *Burkholderia cepacia*, *Burkholderia gladioli* and *Erwinia amylovora* [68]. The serotype b-specific polysaccharide of *A. actinomycetemcomitans* Y4 is composed of a disaccharide repeating unit → 3)-α-D-Fuc-(1 → 2)-α-L-Rha-(1 → [69]. The disaccharide repeating units of the S-layer glycoprotein of the recently affiliated *Geobacillus tepidamans* GS5-97^T is composed of α1,3-linked D-Fuc and L-Rha [70].

The biosynthesis of dTDP-D-fucopyranose (dTDP-D-Fuc*p*) (13) has been described in *A. actinomycetemcomitans* Y4, serotype b, [71] and most recently in *Geobacillus tepidamans* GS5-97^T [70]. In both cases, an *flc* gene, which encodes dTDP-6-deoxy-4-keto-D-Glc reductase was identified. The heterologously expressed and purified protein Flc was able to catalyze the conversion of dTDP-6-deoxy-4-keto-D-Glc (9) and NAD(P)H to dTDP-D-Fuc*p* (13) and NAD(P)$^+$ [70, 71] (Scheme 7.2). The dTDP-D-fucofuranose (dTDP-D-Fuc*f*) (13a) synthetic pathway has been characterized only in *E. coli* O52 and found to include one additional step, the conversion of dTDP-D-Fuc*p* (the dTDP-6-deoxy-4-keto-D-Glc reductase Fcf1 catalyzed product) to dTDP-D-Fuc*f* by dTDP-D-fucopyranose mutase Fcf2 [72] (Scheme 7.2).

7.2.2.4 dTDP-4-Amino-4,6-Dideoxy-D-Glucose (dTDP-D-Qui4N) and dTDP-4-Acetamido-4,6-Dideoxy-D-Glucose (dTDP-D-Qui4NAc)

The tetrasaccharide repeating unit of the O-antigen of *Shigella dysenteriae* type 7[73] and *E. coli* O121 [74] contains a residue of 4-(*N*-acetylglycyl)amino-4,6-dideoxy-D-glucose (D-Qui4NGlyAc). *E. coli* O7 antigen contains 4-acetamido-4,6-dideoxy-D-glucose (D-Qui4NAc) [75]. The biosyntheses of dTDP-D-Qui4N (14) (the nucleotide activated precursor of D-Qui4NGlyAc) and dTDP-D-Qui4NAc (15) have been characterized in *S. dysenteriae* type 7 and *E. coli* O7. VioA, an aminotransferase, transfers an amino group from L-glutamate to dTDP-6-deoxy-4-keto-D-Glc (9) (the RmlB catalyzed product) to form dTDP-D-Qui4N (14) and α-ketoglutarate; VioB, an acetyltransferase, catalyzes the conversion of dTDP-D-Qiu4N and acetyl-CoA to dTDP-D-Qui4NAc (15) [76] (Scheme 7.2).

7.3 Sugars Derived From Fructose-6-P

7.3.1 Biosynthesis of GDP Sugars

7.3.1.1 GDP-D-Mannose (GDP-D-Man)

D-Man is found in the LPS of many bacteria including *S. enterica* [77]. The O-antigens of *K. pneumoniae* serotype O3 and O5, as well as *E. coli* O8 and O9 are mannose homopolysaccharides [78–80]. Some host immune systems have evolved to become capable of interacting with the mannose-rich O-polysaccharide of pathogens, thereby triggering the host defence systems. For instance, the human mannose binding protein binds to virulent *Salmonella montevideo* that produces a mannose-rich O-polysaccharide, and results in attachment, uptake, and killing of the bacteria by phagocytes [81]. In another report, surfactant protein D, which plays

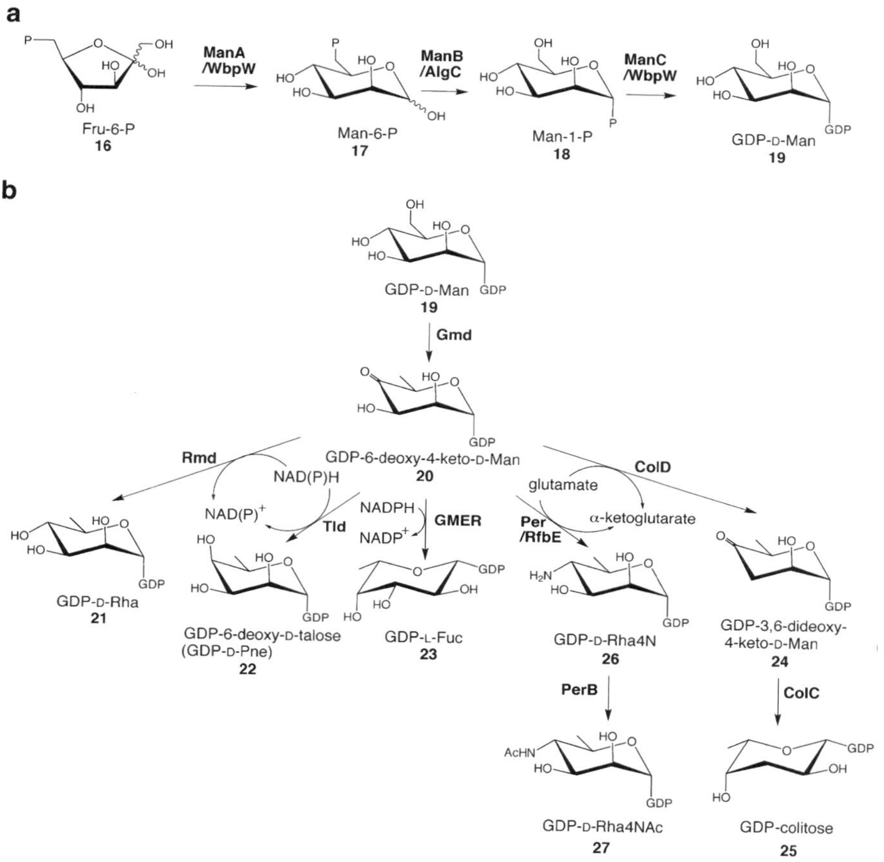

Scheme 7.3 Biosynthesis pathways of GDP sugars. (**a**) GDP-D-Man biosynthesis pathway. (**b**) Sugars derived from GDP-D-Man. The GDP-D-Man 4,6-dehydrotase Gmd catalyzes the formation of the common intermediate GDP-6-deoxy-4-keto-D-Man (20), which can then be converted to different GDP sugars

important roles in the regulation of innate immune responses in the lung, was found to selectively bind to LPS of clinical isolates of *Klebsiella* species with mannose-rich O-antigens [82].

The activated precursor of mannose GDP-D-Man (19) is synthesized from fructose 6-phosphate (Fru-6-P) (16) in three steps (Scheme 7.3a). In Enterobacteriaceae, such as *E. coli*, *S. enterica* and *K. pneumoniae*, ManA, ManB and ManC catalyze each of the three steps. ManA is a type I phosphomannose isomerase (PMI) (EC 5.3.1.8) that can catalyze the reversible interconversion of Fru-6-P (16) and mannose 6-phosphate (Man-6-P) (17), and is responsible for the first step, the conversion of Fru-6-P to Man-6-P. The reversible PMI reaction, the conversion of Man-6-P to Fru-6-P, enables the catabolism of exogenous mannose via the glycolytic pathway [83]. ManB is a PMM (EC 5.4.2.8) and catalyzes the second step of the GDP-D-Man biosynthesis pathway, the conversion of Man-6-P (17) to Man-1-P (18). The third enzyme ManC, mannose-1-phosphate guanylyltransferase (EC 2.7.7.22) (also called GDP-D-Man pyrophosphorylase [GMP]), catalyzes the synthesis of GDP-D-Man (19) from Man-1-P (18) and GTP. There are two types of PMIs, type I such as ManA are zinc-dependent monofunctional enzymes catalyzing only the isomerization reaction. Type II such as WbpW from *P. aeruginosa* is a bifunctional enzyme that has both PMI and GMP activities, catalyzing both the first and the last step of GDP-D-Man biosynthesis pathway. Our group has shown that *wbpW* could complement both *E. coli manA* and *manC* mutants to restore K30 capsule biosynthesis [84]. In *P. aeruginosa*, AlgC catalyzes the second step of GDP-D-Man biosynthesis. As presented earlier (Sect. 7.2.1.1), AlgC is a bifunctional enzyme which has both PMM and PGM activities. The PMM activity is involved in the GDP-D-Man biosynthesis pathway, while the PGM activity is required for the synthesis of the important intermediate Glc-1-P (3).

Due to its role in mannose catabolism, the gene *manA* is generally present in *E. coli* and *S. enterica* and is located outside of the O-antigen gene cluster [83]. The two genes *manB* and *manC* are usually transcribed from the same operon and located within relevant polysaccharide gene clusters [85]. In *P. aeruginosa* genome, in addition to *wbpW*, which is located in the common O-polysaccharide (formerly called A-band) gene cluster, two other homologs are present, *algA* which is located in the alginate biosynthesis gene cluster, and *pslB* (originally ORF488) from the locus responsible for synthesis of a cell surface polysaccharide Psl that are important for biofilm formation [86]. Both AlgA and PslB have been shown to exhibit PMI and GMP functions [87, 88].

7.3.1.2 GDP-D-Man-Derived Sugars

GDP-D-Man is a precursor for the biosynthesis of many other sugars found in LPS. The enzyme GDP-D-mannose 4,6-dehydratase (Gmd) (EC 4.2.1.47) catalyzes the conversion of GDP-D-Man (19) to GDP-6-deoxy-4-keto-D-Man (GDP-6-deoxy-D-*lyxo*-hexos-4-ulose) (20). Like the dTDP-6-deoxy-4-keto-D-Glc (9) in the biosynthesis pathways of dTDP sugars, GDP-6-deoxy-4-keto-D-Man (20) is an important common intermediate for the biosynthesis of many GDP sugars, including GDP-D-rhamnose (21), GDP-6-deoxy-D-talose (GDP-D-pneumose) (22), GDP-L-fucose (23), GDP-colitose (25) and GDP-4-acetamido-4,6-dideoxy-D-mannose (27), and acts as the branching point in the biosynthesis pathways (Scheme 7.3b). The gene *gmd* was first identified in *E. coli* in 1996 [16]; to date, it has been characterized at the biochemical and structural levels from several different sources [89–91]. The protein Gmd also belongs to the SDR superfamily, and the N-terminal domain binds to its cofactor NADP(H). The *E. coli* Gmd is a dimer [90], while the Gmd enzymes from *P. aeruginosa* and *Arabidopsis thaliana* are tetramers, or dimers of dimers [91].

GDP-D-Rhamnose (GDP-D-Rha) and GDP-D-Pneumose (GDP-D-Pne)

D-Rha is a rare sugar that has mainly been found in LPS or EPS of Gram-negative bacteria. The common O-polysaccharide (formerly called A-band polysaccharide) of *P. aeruginosa* is a homopolymer of D-Rha [92, 93]. D-Rha is also a constituent of the O-antigens of *Pseudomonas syringae* [94], *Xanthomonas campestris* [95], *Campylobacter fetus* [96] and *Helicobacter pylori* [97].

The biochemical pathway of D-Rha biosynthesis was first studied in *Aneurinibacillus thermoaerophilus* where D-Rha is present in the S-layer protein glycan [98]. A SDR superfamily protein GDP-6-deoxy-4-keto-D-mannose reductase Rmd (EC 1.1.1.281) catalyzes the stereospecific reduction of the 4-keto group of GDP-6-deoxy-4-keto-D-Man (20) (the Gmd catalyzed product) and results in the synthesis of GDP-D-Rha (21) [98]. This reaction is analogous to the RmlD catalyzed reduction of dTDP-4-keto-L-Rha to form dTDP-L-Rha (Scheme 7.2). The Gmd and Rmd from *P. aeruginosa* have also been well characterized [84, 99, 100]. The genes *gmd* and *rmd* are localized in the common O-polysaccharide gene cluster [84]. Heterologously expressed *P. aeruginosa* Gmd showed remarkable structural similarity to *A. thermoaerophilus* Rmd, and is also bifunctional, able to catalyze both GDP-D-mannose 4,6-dehydration and the subsequent reduction reaction to produce GDP-D-Rha [99]. *P. aeruginosa* Rmd, same as other reported Rmd, catalyzes the stereospecific reduction of GDP-6-deoxy-4-keto-D-Man and generates GDP-D-Rha [99, 100].

The C-4 epimer of D-Rha, 6-deoxy-D-talose (D-pneumose, D-Pne), is another rare sugar. It has been reported in the EPS of the *B. plantarii* [60] and in the serotype a-specific polysaccharide of *A. actinomycetemcomitans* [62, 101]. The biosynthesis pathway of GDP-D-Pne (22) has been determined in the latter organism. The gene *tld*, encoding another GDP-6-deoxy-4-keto-D-Man reductase that catalyzes the synthesis of GDP-D-Pne, has been identified and characterized [102, 103]. Both Rmd and Tld are SDR family proteins requiring the binding of the cofactor NAD(P)H. They use the same substrate GDP-6-deoxy-4-keto-D-Man but show different stereospecifity of the reaction.

GDP-L-Fucose (GDP-L-Fuc)

L-Fuc is a sugar commonly found in complex glycoconjugates of species ranging from bacteria to mammals. For example, L-Fuc is found in the human ABO blood group antigens and the Lewis (Le) antigens [104, 105]. The EPS colanic acid produced by most *E. coli* strains and other species within Enterobacteriaceae generally contains L-Fuc [106]. This sugar is also a component of various Nod factors (lipo-chitooligosaccharides) produced by the plant associated nitrogen fixing bacteria *Azorhizobium* and *Rhizobium* [107, 108]. L-Fuc is also present in the LPS of some human pathogens. For example, it is a structural constituent of the O-antigens of *E. coli* O157 [109], *Yersinia enterocolitica* O8 [110], *Yersinia pseudotuberculosis* O3 [111], *Campylobacter fetus* [112] and *Helicobacter pylori* [113, 114]. The O-antigens in most *H. pylori* strains contain fucosylated glycans, which are structurally similar to human LeX or LeY antigens and have been studied extensively.

In bacteria, the biosynthesis of GDP-L-Fuc (23) (the activated form of L-Fuc) from the intermediate GDP-6-deoxy-4-keto-D-Man (20) was first characterized in *E. coli*. A *fcl* gene, encoding a bifunctional GDP-6-deoxy-4-keto-D-mannose

3,5-epimerase/4-reductase (GMER, also called Fcl or WcaG, EC 1.1.1.271), was identified in the colanic acid gene cluster of *E. coli* [115]. The enzyme GMER first catalyzes the epimerization of the GDP-6-deoxy-4-keto-D-Man at C-3 and C-5, leading to the formation of GDP-6-deoxy-4-keto-L-Gal (GDP-6-deoxy-L-*xylo*-hexos-4-ulose), and then it catalyzes the NADPH-dependent reduction of the 4-keto group, finally resulting in the formation of GDP-L-Fuc (23). The GMER from *E. coli* and *H. pylori* are also members of the SDR superfamily of proteins and have been characterized at the biochemical and structural levels [116, 117].

GDP-Colitose (GDP-Col)

Colitose (3,6-dideoxy-L-*xylo*-hexose) is another rare sugar. It has been found in the O-antigens of some pathogens such as *S. enterica* O35 [118], *E. coli* O111 [119], *E. coli* O55 [120], *Vibrio cholerae* O139 [121] and *Yersinia pseudotuberculosis* O6 [122]. It is also a constituent of the O-antigens of some marine bacteria including *Pseudoalteromonas tetraodonis* [123] and *Pseudoalteromonas carrageenovora* [124]. Although colitose being synthesized as a GDP-derivative has been known as early as 1965 [125], its biosynthesis pathway has only been experimentally characterized recently [126, 127] (Scheme 7.3b).

A five-gene cluster (from *colA* to *colE*) was identified from *Y. pseudotuberculosis* VI, and *colE* and *colB* are *manC* and *gmd* homologs that catalyze the formation of GDP-D-Man (19) and the further 4,6-dehydration of GDP-D-Man to form the common intermediate GDP-6-deoxy-4-keto-D-Man (20), respectively. The gene *colD* was shown to encode a coenzyme B_6 (pyridoxal 5-phosphate, PLP)-dependent GDP-6-deoxy-4-keto-D-mannose 3-deoxygenase, which catalyzes the removal of the hydroxyl group at position 3 of GDP-6-deoxy-4-keto-D-Man (20) to form GDP-3,6-dideoxy-4-keto-D-Man (24) [105, 128]. The gene *colC* encodes a bifunctional enzyme that catalyzes both the C-5 epimerization and the 4-keto reduction of GDP-3,6-dideoxy-4-keto-D-Man to finally form GDP-colitose (25) [128]. Cook et al. [129] reported the structures of ColD and the enzyme/cofactor (ColD/PLP) complex from *E. coli* strain 5a (serotype O55:H7). It was found that two subunits of ColD form a tight dimer that shows a characteristic feature of the aspartate aminotransferase superfamily [129]. Unlike most PLP-dependent enzymes that contain a lysine in the active site, ColD utilizes a histidine residue in the active site as the catalytic base [130]. Results from site-directed mutagenesis showed that His188 is critical for the deoxygenase activity. Replacing His188 with a Lys or Asn abrogated the deoxygenase activity of ColD [126, 130].

GDP-4-Amino-4,6-Dideoxy-D-Mannose (GDP-D-Rha4N) and GDP-4-Acetamido-4,6-Dideoxy-D-Mannose (GDP-D-Rha4NAc)

4-Amino-4,6-dideoxy-D-mannose (D-Rha4N) is an unusual sugar found in the O-antigen of the human pathogen *V. cholerae* O1 [131]. The N-acetylated version of it (D-Rha4NAc) has been reported in the O-antigens of several Gram-negative bacteria, including *Caulobacter crescentus* CB15 [132], *E. coli* O157:H7 [133], *S. enterica* O30 [134] and *Citrobacter freundii* F90 [135].

The biosynthesis pathway of the nucleotide-activated sugar GDP-D-Rha4N (26) was first genetically and biochemically characterized in *V. cholerae* O1. A four-gene cluster (*rfbA*, *rfbB*, *rfbD*, *rfbE*) from *V. cholerae* O1 was proposed to encode enzymes for the conversion of Fru-6-P to GDP-D-Rha4N in five steps. RfbA, like WbpW from *P. aeruginosa*, is a bifunctional enzyme that has both ManA and ManC activity, and RfbB is a phosphomannomutase. The combined activities of RfbA and RfbB catalyze the conversion of Fru-6-P to GDP-D-Man. The other two genes *rfbD* and *rfbE* were predicated to encode proteins responsible for the synthesis of GDP-D-Rha4N from the GDP-D-Man [136]. Biochemical characterization of *rfbD* and *rfbE* of *V. cholerae* O1 was reported by Albermann and Piepersberg [137], who showed that *rfbD* encode Gmd, which catalyzes the formation of the intermediate GDP-6-deoxy-4-keto-D-Man, and *rfbE* encodes a GDP-D-Rha4N synthase which transferred an amino group from glutamate to the position 4 of GDP-6-deoxy-4-keto-D-Man (20) to form GDP-D-Rha4N (26) and α-ketoglutarate (Scheme 7.3b). Most recently, GDP-D-Rha4N synthase (Per) from *E. coli* O157:H7 was also characterized at the biochemical level, and was shown to have some different characteristics compared to RfbE from *V. cholerae* [138]. The differences include, first, Per is a decamer while RfbE is a tetramer, and second, Per uses only L-glutamate as an amino donor while RfbE uses both L-glutamate and L-glutamine. The structure of the GDP-D-Rha4N synthase from *Caulobacter crescentus* CB15 has been determined, and the overall structure places it into the aspartate aminotransferase superfamily [139]. It also shows remarkable structure similarity to another PLP-dependent enzyme, deoxygenase ColD from *E. coli* in the GDP-colitose pathway that was described earlier. The authors showed that by manipulating the protein with two site-directed mutations, ColD exhibited aminotransferase activity instead of its original deoxygenase activity [140]. A *perB* (also called *wbdR*) gene encoding an *N*-acetyltransferase that catalyzes the N-acetylation of GDP-D-Rha4N to form GDP-D-Rha4NAc (27) was characterized from *E. coli* O157:H7 [141] (Scheme 7.3b). However, no corresponding genes have been identified in the O-antigen gene cluster of *S. enterica* O30 or *C. freundii* F90, although both have the same O-antigen structure as *E. coli* O157:H7 [141].

7.3.2 Biosynthesis of UDP Sugars

7.3.2.1 UDP-2-Acetamido-2-Deoxy-D-Glucose (UDP-D-GlcNAc)

D-GlcNAc is an important component of LPS. It is the precursor for the disaccharide moiety of lipid A of most Gram-negative bacteria [142] (Chaps. 1 and 6), and also a constituent in the core and O-polysaccharides [39, 143] (Chaps. 2 and 3). The majority of the *E. coli* [17], *Shigella* [144] and *Proteus* [145] O-antigen structures contain D-GlcNAc at the reducing termini. D-GlcNAc is also required for the synthesis of peptidoglycan of the bacterial cell wall [146, 147].

The biosynthesis of UDP-D-GlcNAc (31) in bacteria has been well characterized (Scheme 7.4). It is synthesized from D-Fru-6-P (16) in four steps: the first being the formation of 2-amino-2-deoxy-D-glucose 6-phosphate (GlcN-6-P) (28) from

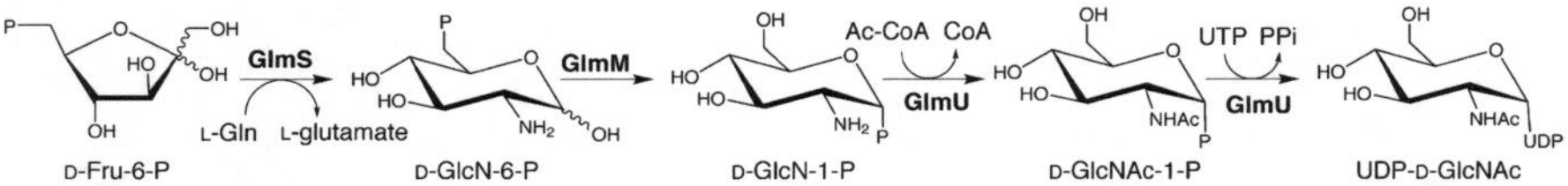

Scheme 7.4 UDP-D-GlcNAc biosynthesis pathway

D-Fru-6-P, catalyzed by the enzyme GlcN-6-P synthase (GlmS) (EC 2.6.1.26) [148]; in the second step, phosphoglucosamine mutase (GlmM) (EC 5.4.2.10) catalyzes the conversion of D-GlcN-6-P (28) to D-GlcN-1-P (29) [149]; a bifunctional enzyme GlmU (EC 2.3.1.57/2.7.7.23) that has both acetyltransferase and uridylyltransferase (pyrophosphorylase) activities catalyzes the third and the fourth steps [150, 151]. It first transfers an acetyl group from acetyl-CoA to D-GlcN-1-P (29) to form D-GlcNAc-1-P (30), followed by the transfer of the UMP group from UTP to D-GlcNAc-1-P (30) to form UDP-D-GlcNAc (31). The essential nature of this particular pathway in bacteria is apparent, especially since mutation of any of the genes that encode these enzymes or inhibition of the enzymes causes dramatic morphological changes in cell shape and finally results in cell lysis [149, 150, 152, 153].

The key enzymes in this pathway are considered attractive targets for new antimicrobial discovery [154]; hence, they have been studied extensively. The GlmM from *E. coli* has been characterized at the biochemical level and the mechanism of reaction has been elucidated [155, 156]. The enzyme is active only in a phosphorylated form, and acts in a classical ping-pong mechanism [155, 157]. Later, GlmM homologs from *H. pylori* [158], *P. aeruginosa* [159], *Streptococcus gordonii* [160] and *Staphylococcus aureus* [161] among others have been identified and characterized. The bifunctional enzyme GlmU has been characterized at the biochemical and structural levels from several different organisms and successful crystallization of this protein has provided structural and functional insights into GlmU activity and inhibition mechanisms [110, 162, 163]. Studies of the N-terminal and C-terminal truncation variants of GlmU showed that the C-terminal variant catalyzed acetyltransfer, and the N-terminus was capable of only uridylytransfer activity [164]. The crystal structure of GlmU and the complexes of GlmU binding with different substrates (acetyl-CoA, UTP, GlcNAc-1-P) were first studied in *E. coli* [162, 165, 166]. High-resolution 3D structures of GlmU based on X-ray crystallography studies have been obtained from *Streptococcus pneumoniae* [167], *Haemophilus influenzae* [168], and *Mycobacterium tuberculosis* [169].

7.3.2.2 UDP-D-GlcNAc-Derived Sugars

Besides being an important structural component, UDP-D-GlcNAc (31) is also a common precursor and an important convergent point for the metabolic pathways for the biosynthesis of many other sugars found in the LPS and other bacterial surface glycans [48]. Sugars that have an *N*-acetylamino group at position 2 are usually synthesized from UDP-D-GlcNAc by different combinations of epimerization, dehydration, oxidation, reduction, amino and acetyl group transfer. For example, both D and L enantiomers of FucNAc and QuiNAc were synthesized as UDP derivatives from UDP-D-GlcNAc. Other examples include UDP-D-GalNAc, UDP-D-GalNAcA, UDP-D-ManNAc, UDP-D-ManNAcA, UDP-D-Glc(2NAc3NAc)A, UDP-D-Man(2NAc3NAc)A and UDP-D-Man(2NAc3NAm)A.

UDP-2-Acetamido-2,6-Dideoxy-D-Glucose (UDP-D-QuiNAc), UDP-2-Acetamido-2,6-Dideoxy-D-Galactose (UDP-D-FucNAc) and UDP-2-Acetamido-2,6-Dideoxy-D-Xylo-Hexos-4-Ulose

2-Acetamido-2,6-dideoxy-D-glucose (D-QuiNAc) and its C-4 epimer 2-acetamido-2,6-dideoxy-D-galactose (D-FucNAc) are rare sugars in nature, but they are found in the O-antigens of many serotypes of *P. aeruginosa*. D-QuiNAc is found in serotypes O1, O4, O6, O9, O10, O12, O13, O14 and O19 of the *P. aeruginosa* International Antigenic Typing Scheme, and D-FucNAc is present in serotypes O1, O2, O5, O7, O8, O9, O16 and O18 [40, 170]. D-QuiNAc has also been reported in the outer core of *Rhizobium etli* [171, 172]. The biosynthesis of the nucleotide-activated sugars UDP-D-QuiNAc (33) and UDP-D-FucNAc (34) has been proposed to start from UDP-D-GlcNAc (31) involving two steps (Scheme 7.5). Similar to the previously presented dTDP and GDP pathways, the first step is the generation of the 6-deoxy-

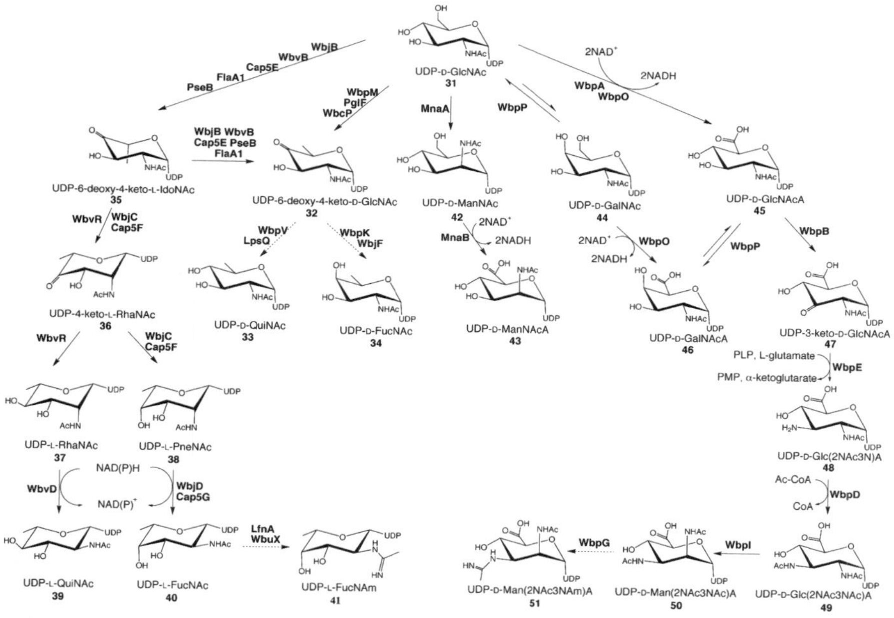

Scheme 7.5 Proposed divergent pathways for the biosynthesis of UDP-sugars derived from the common precursor substrate UDP-D-GlcNAc. Steps containing arrows with *dashed lines* indicate the lack of biochemical evidence to date. Note that in the reactions catalyzed by the 4-epimerase WbpP, the *larger arrows* indicate the kinetically favoured steps (thus more physiologically relevant) versus the *small arrows*, which show the kinetically less favoured steps

4-keto derivative UDP-2-acetamido-2,6-dideoxy-D-*xylo*-hexos-4-ulose (also called UDP-6-deoxy-4-keto-D-GlcNAc or UDP-4-keto-D-QuiNAc) (32) catalyzed by the UDP-D-GlcNAc 4,6-dehydratase. The 4-keto group can then be reduced to a 4-hydroxy group in different orientations by different stereospecific 4-reductases to form either UDP-D-QuiNAc or UDP-D-FucNAc.

The gene *wbpM* in *P. aeruginosa* encodes the enzyme UDP-D-GlcNAc 4,6-dehydratase (EC number is pending) involved in the first step and has been studied extensively [170, 173]. Mutation of *wbpM* from many serotypes of *P. aeruginosa* including O3, O5, O6, O7, O10 and O11 that contain D-FucNAc and/or D-QuiNAc in their LPS abrogated O-antigen biosynthesis [10, 174–177]. Interestingly, knockout of *wbpM* does not affect O-antigen biosynthesis of *P. aeruginosa* serotypes O15 or O17 since these two serotypes do not contain either D-FucNAc or D-QuiNAc in their O-polysaccharides [175]. The *wbpM* homologous genes from *Plesiomonas shigelloides* (*wbgZ*), *Bordetella pertussis* (*wlbL*) and *S. aureus* (*cap8D*) could complement the *wbpM* mutation in *P. aeruginosa* [175]. Recently, two other homologues of WbpM, PglF from *Campylobacter jejuni* in the UDP-2,4-diacetamido-2,4,6-trideoxy-D-glucose biosynthesis pathway [178], and WbcP from *Yersinia enterocolitica* serotype O:3 have been reported to catalyze the same reaction, e.g. the conversion of UDP-D-GlcNAc (31) to UDP-4-keto-D-QuiNAc (32) [179].

In *R. etli*, genetic evidence suggests that *lpsQ* encodes the UDP-6-deoxy-4-keto-D-GlcNAc 4-reductase, which catalyzes the second step of the D-QuiNAc biosynthesis pathway. Mutation of *lpsQ* results in the synthesis of LPS containing 4-keto-D-QuiNAc instead of D-QuiNAc [180]. Its homologue in *P. aeruginosa* O6, *wbpV*, is able to complement the above *lpsQ* mutant [181]. The gene *wbpV* is also essential for the biosynthesis of O-antigen in *P. aeruginosa* serotypes that contain D-QuiNAc in LPS, and mutation of *wbpV* abrogated O-antigen expression [177]. This information suggests that WbpV is the 4-reductase involved in UDP-D-QuiNAc biosynthesis in *P. aeruginosa*.

Bioinformatics and genetic evidence suggested that WbpK from *P. aeruginosa* O5 is the 4-reductase involved in the biosynthesis of D-FucNAc. It showed a high sequence similarity to the UDP-6-deoxy-4-keto-D-GlcNAc 4-reductase WbpV in *P. aeruginosa* O6 (36% identity in amino acid sequence). Putative homologs of WbpK with more than 50% identity in amino acid sequences were found in almost all *P. aeruginosa* serotypes containing D-FucNAc in their LPS [170]. Mutation of *wbpK* abrogates O-antigen biosynthesis in *P. aeruginosa* O5 [177]. Interestingly, despite such high level of sequence identity between *wbpV* (O6) and *wbpK* (O5), these two genes could not cross complement knockout mutants of each other [177], indicating the expected stereospecificity of the reduction activities by these two enzymes. However, there is no direct biochemical evidence so far about the proposed 4-reductase activities of these two enzymes.

The outer core of *Y. enterocolitica* serotype O:3 was recently reported to contain a residue of 6-deoxy-4-keto-D-GlcNAc (2-acetamido-2,6-dideoxy-D-*xylo*-hexos-4-ulose) [179] instead of FucNAc as reported earlier [182]. This keto sugar has also been reported in the LPS of *Vibrio ordalii* O:2 [183], *Flavobacterium columnare* [184] and *Pseudoalteromonas rubra* [185], and in the CPS of

S. pneumoniae type 5 [186]. The intermediate in the UDP-D-QuiNAc or UDP-D-FucNAc pathway, UDP-6-deoxy-4-keto-D-GlcNAc (32), is likely the activated sugar-nucleotide precursor. WbcP from *Y. enterocolitica* serotype O:3, as mentioned earlier, is the UDP-D-GlcNAc 4,6-dehydratase that catalyzes the conversion of UDP-D-GlcNAc to UDP-6-deoxy-4-keto-D-GlcNAc [179]. It has also been shown in that report that *Y. enterocolitica* serotype O:3 lacks UDP-6-deoxy-4-keto-D-GlcNAc 4-reductase. Introduction of an exogenous 4-reductase (either WbpK or WbpV from *P. aeruginosa*) caused the reduction of UDP-6-deoxy-4-keto-D-GlcNAc to UDP-D-FucNAc or UDP-D-QuiNAc, and changed the chemical composition of the outer core produced in the transconjugant strains [179].

UDP-2-Acetamido-2,6-Dideoxy-L-Glucose (UDP-L-QuiNAc), UDP-2-Acetamido-2,6-Dideoxy-L-Galactose (UDP-L-FucNAc) and UDP-2-Acetimidoylamino-2,6-Dideoxy-L-Galactose (UDP-L-FucNAm)

The L enantiomers of QuiNAc and FucNAc are also rare sugars found in the LPS of some Gram-negative bacteria or the surface polysaccharide of some Gram-positive bacteria. For example, L-QuiNAc has been found in the O-antigens of *E. coli* O98 [187], *Shigella boydii* type 13 [188], *Proteus* O1, O2, O31a, O31a,b and O55 [145], *Y. enterocolitica* O11,23 and O11,24 [187], and *S. enterica* O41 [189]. It is also a constituent of the CPS of *Bacteroides fragilis* NCTC 9343 [190]. L-FucNAc is a rare sugar that has only been reported in bacterial polysaccharide structures [191]. It is a constituent of the O-antigens of *P. aeruginosa* O4 and O11 [40], *E. coli* O4: K52, O4:K6, O25, O26 and O 172 [17], *Proteus* O6, O8, O12 O19a, O19a,b, O39, O67, O68, O70 and O76 [145] and *Salmonella arizonae* O59 [192], as well as the CPS of *S. aureus* type 5 [193] and type 8 [194], *S. pneumoniae* type 4 [195] and *B. fragilis* [190].

Recently, the biosynthesis of the nucleotide-sugar precursors UDP-L-QuiNAc (39) and UDP-L-FucNAc (40) has been investigated in *P. aeruginosa* O11, *S. aureus* capsular type 5 and *V. cholerae* O37. Three enzymes from *P. aeruginosa* O11 (WbjB, WbjC and WbjD) and *S. aureus* type 5 (Cap5E, Cap5F and Cap5G) are required for the conversion of UDP-D-GlcNAc (31) to UDP-L-FucNAc (40) [191, 196], while three enzymes WbvB, WbvR and WbvD from *V. cholerae* O37 were able to convert UDP-D-GlcNAc (31) to UDP-L-QuiNAc (39) [197]. The data obtained in these recent studies by our group suggest that the biosynthesis of UDP-L-FucNAc and UDP-L-QuiNAc from UDP-D-GlcNAc involved four shared and parallel steps (Scheme 7.5). An UDP-D-GlcNAc 5-inverting 4,6-dehydratase (EC 4.2.1.115) is a new type of dehydratase that catalyses the first step: the conversion of UDP-D-GlcNAc (31) to UDP-2-acetamido-2,6-dideoxy-L-*arabino*-hexos-4-ulose (UDP-6-deoxy-4-keto-L-IdoNAc) (35). The second step is the 3-epimerization of UDP-6-deoxy-4-keto-L-IdoNAc to form UDP-4-keto-L-RhaNAc (36), which is followed by the reduction of the 4-keto group to form either UDP-L-RhaNAc (37) or UDP-2-acetamido-2,6-dideoxy-L-talose (UDP-L-PneNAc) (38), depending on the orientation of the newly generated 4-hydroxy group. The last step is the 2-epimerization of UDP-L-RhaNAc or UDP-L-PneNAc to finally generate UDP-L-QuiNAc (39) or UDP-L-FucNAc (40).

The enzyme WbjB/Cap5E/WbvB is apparently the UDP-D-GlcNAc 5-inverting 4,6-dehydratase (EC 4.2.1.115) that catalyzes the first step reaction of the pathway. However, it should be noted that because of the labile nature of the product UDP-6-deoxy-4-keto-L-IdoNAc (35), it is not detected in earlier experiments [198], and the activity of this group of enzymes was previously misannotated as multifunctional containing 4,6-dehydratase, 5-epimerase and 3-epimerase activities [191, 197], and later as bifunctional with both 4,6-dehydratase and 5-epimerase activities [196]. With the development of a "real-time" NMR spectroscopy method, e.g. by placing an enzyme-substrate reaction mixture into an NMR tube and monitoring the yield of intermediates or products, the structure of the otherwise labile intermediate UDP-6-deoxy-4-keto-L-IdoNAc was unequivocally determined [198].

The 3D structures of the WjbB homologs from *Helicobacter pylori* (FlaA1, now renamed as PseB) [199] and from *Campylobacter jejuni* (PseB) [200] were determined and the structural information from these studies revealed that the reaction mechanism is different from other simple (or retaining) 4,6-dehydratases (such as dTDP-D-Glc 4,6-dehydratase and GDP-D-Man 4,6-dehydratase presented earlier). The activity of the inverting 4,6-dehydratase would remove the H-5 proton and then replace it on opposite face of the sugar ring, resulting in the inversion of the C-5 chiral center [200]. Sequence comparison of the UDP-D-GlcNAc inverting 4,6-dehydratase (WbjB/Cap5E/WbvB/FlaA1/PseB) with the UDP-D-GlcNAc retaining 4,6-dehydratase (WbpM/WbgZ/WlbL/WbcP) showed that they have high sequence similarity at the C-terminal region. However, the latter have longer sequences (around 600 amino acids) and are predicted to be membrane associated while the former are shorter (around 350 amino acids) and are predicted to be soluble proteins [170]. These 5-inverting 4,6-dehydratases can slowly catalyze the 5-epimerization of UDP-6-deoxy-4-keto-L-IdoNAc (35) to form UDP-6-deoxy-4-keto-D-GlcNAc (32) [198–200]. Compared to the dehydratase activity, the rate of the 5-epimerization is too slow and probably not physiologically relevant [200].

The second enzyme WbjC/Cap5F/WbvR is a bifunctional enzyme with both UDP-6-deoxy-4-keto-L-IdoNAc 3-epimerase and UDP-4-keto-L-RhaNAc 4-reductase activities and catalyzes both the second and the third steps. It first catalyzes the 3-epimerization of UDP-6-deoxy-4-keto-L-IdoNAc (35) to generate UDP-4-keto-L-RhaNAc (36), followed by stereospecific reduction of the 4-keto group to form either UDP-L-RhaNAc (37) (catalyzed by WbvR) or UDP-L-PneNAc (38) (catalyzed by WbjC/Cap5F) [196]. This group of enzymes also belongs to the SDR superfamily and requires the binding of the cofactor NADH or NADPH [191]. WbvD from *V. cholerae* O37 has been biochemically characterized as a UDP-L-RhaNAc 2-epimerase that converts UDP-L-RhaNAc (37) to UDP-L-QuiNAc (39) [197]. WbjD/Cap5G from *P. aeruginosa* O11/*S. aureus* O5 encodes UDP-L-PneNAc 2-epimerase that converts UDP-L-PneNAc (38) to UDP-L-FucNAc (40) [191, 196].

The rare sugar 2-acetimidoylamino-2,6-dideoxy-L-galactose (L-FucNAm) has been found in the LPS of a few pathogenic bacteria including *P. aeruginosa* serogroup O12 [201], *E. coli* O145 [202], *S. enterica* serovar Toucra O48 [202, 203] and *S. enterica* serovar Arizonae O21 [109]. Since homologs of *wbjB*, *wbjC* and *wbjD* were present in these organisms, it was proposed that UDP-L-FucNAc (40) was synthesized by the same scheme as described above, and UDP-L-FucNAm (41) was then derived from modification of the acetamido group of UDP-L-FucNAc (40) by an amidotransfer reaction [84, 176, 202]. In a recent report by our group, a putative amidotransferase encoding gene, *lfnA* from *P. aeruginosa* O12, was essential for the expression of L-FucNAm containing O-antigen. The *lfnA* mutant strain produces LPS containing L-FucNAc in the usual place of L-FucNAm, while its homolog in *E. coli* O145 (*wbuX*) was able to cross complement the mutant [204]. This provides genetic evidence that *lfnA*/*wbuX* encode putative amidotransferases that catalyze the conversion of UDP-L-FucNAc to UDP-L-FucNAm. At present, biochemical evidence of the putative enzymatic activities is lacking and investigation of the functions of these proteins is underway.

UDP-2-Acetamido-2-Deoxy-D-Mannose (UDP-D-ManNAc) and UDP-2-Acetamido-2-Deoxy-D-Mannuronic Acid (UDP-D-ManNAcA)

ManNAc is found in the O-antigen of *E. coli* serogroups O1A [205], O1B [206], O1C [206] and O64 [207], *Aeromonas salmonicida* strains 80204-1, 80204 and A449 [208] and *S. enterica* O:54 [209]. More importantly, ManNAcA is a constituent of the enterobacterial common antigen (ECA), a glycolipid found in all species of the family Enterobacteriaceae [210, 211]. It is also present in the CPS of *S. aureus* serotype 5 and 8 [212] and *S. pneumoniae* type 19F [213].

The biosynthesis pathway of UDP-D-ManNAc and UDP-D-ManNAcA were first characterized in *E. coli*, and later in the Gram-positive bacterium *S. aureus* [214–220]. UDP-D-ManNAc (42) is the 2-epimer of UDP-D-GlcNAc (31), not surprisingly it is synthesized from UDP-D-GlcNAc by UDP-D-GlcNAc 2-epimerase (EC 5.1.3.14). The uronic acid derivative UDP-D-ManNAcA (43) is formed by the oxidation of UDP-D-ManNAc (42) by the activity of UDP-D-ManNAc 6-dehydrogenase (EC 1.1.1.n3). The gene encoding the UDP-D-GlcNAc 2-epimerase has been identified and well characterized from several different organisms: *wecB* (originally called *rffE*) from *E. coli* [221], *rfbC* from *S. enterica* [209], *cps19fK* from *S. pneumoniae* [213] and *cap5P* from *S. aureus* [218, 220, 222].

The gene *wecC* (formerly *rffD*) from *E. coli* [221] and *cap5O* from *S. aureus* [220] encode UDP-D-ManNAc 6-dehydrogenase. Cap5O was biochemically characterized and its activity requires the binding of the cofactor NAD^+ [220]. Reeves proposed the renaming of the genes encoding UDP-D-GlcNAc 2-epimerase and UDP-D-ManNAc 6-dehydrogenase as *mnaA* and *mnaB*, respectively [37].

UDP-2-Acetamido-2-Deoxy-D-Galactose (UDP-D-GalNAc) and UDP-2-Acetamido-2-Deoxy-D-Galacturonic Acid (UDP-D-GalNAcA)

The sugar D-GalNAc is ubiquitous among O-antigens of Gram-negative bacteria. For example, 26 of the 87 established *E. coli* O-repeat unit structures contain at least one D-GalNAc residue [17]. This group of bacteria include pathogenic strains such as O55 [120], O86:B7 [223] and O157 [133]. D-GalNAc is also commonly found in the O-antigens of *Proteus* and *Shigella* species: 37 out of the 88 published *Proteus* O-repeat unit structures [145] and 15 out of the 41 *Shigella* serotypes with known O-antigen structures [144] contain at least one D-GalNAc. Like D-GlcNAc, the sugar D-GalNAc is usually located at the reducing termini of the O-repeat units. For example, of the 26 *E. coli* O-repeat unit structures that contain D-GalNAc, only 4 structures do not have the D-GalNAc residue at the reducing termini [17]. Compared to D-GalNAc, the uronic acid D-GalNAcA is less common in LPS. It has been reported in the O-antigens of a few Gram-negative bacteria including *E. coli* O98 [187], O121 [74] and O138 [224], *Acinetobacter haemolyticus* ATCC 17906 [225], *Proteus vulgaris* TG155 [226], *Aeromonas salmonicida* 80204-1 [227], *Pseudomonas fluorescens* IMV 247 [228] and *P. aeruginosa* O6, O13 and O14 [229].

D-GalNAc is the C-4 epimer of D-GlcNAc. Its precursor, UDP-D-GalNAc (44) was thought to arise from UDP-D-GlcNAc by an epimerization reaction, followed by dehydrogenation in the next step of the pathway to form UDP-D-GalNAcA (45). An UDP-D-GlcNAc 4-epimerase WbpP (EC 5.1.3.7) was isolated and characterized from *P. aeruginosa* O6. Data from experiments using CE and CE coupled with MS revealed unequivocally that WbpP catalyzes the reversible conversion between UDP-D-GlcNAc (31) and UDP-D-GalNAc (44), and between UDP-D-GlcNAcA (45) and UDP-D-GalNAcA (46), and at a much lower rate between the non-acetamido nucleotide sugars UDP-D-Glc (4) and UDP-D-Gal (5) [230]. In contrast to the well-known 4-epimerase GalE, WbpP is the first bacterial 4-epimerase that showed a stronger preference for the acetamido substrates (lower K_m) than the non-acetamido sugars.

Another gene *wbpO*, also from *P. aeruginosa* O6, encodes a 6-dehydrogenase and can convert UDP-D-GalNAc (44) to UDP-D-GalNAcA (46), as well as UDP-D-GlcNAc (31) to UDP-D-GlcNAcA (45) [231]. Due to the relaxed substrate specificity of WbpP and WbpO, there are two possible pathways for the biosynthesis of UDP-D-GalNAcA from UDP-D-GlcNAc. The first possible pathway is that WbpP would catalyze the conversion of UDP-D-GlcNAc (31) to UDP-D-GalNAc (44), followed by the WbpO oxidation of UDP-D-GalNAc (44) to form UDP-D-GalNAcA (46). An alternative pathway could be that WbpO would first convert UDP-D-GlcNAc (31) to UDP-D-GlcNAcA (45), which is then epimerized to UDP-D-GalNAcA (46) by WbpP. However, comparison of the kinetics of the enzyme-substrate reactions and the equilibrium parameters showed that the latter, i.e. first oxidation and then epimerization, is favoured and thus more physiologically relevant [232] (Scheme 7.5).

Another UDP-D-GlcNAc 4-epimerase, WbgU from *Plesiomonas shigelloides* O17, has also been characterized. Similar to WbpP, although it is capable of interconverting both acetamido (UDP-D-GlcNAc and UDP-D-GalNAc) and non-acetamido derivatives (UDP-D-Glc and UDP-D-Gal), the rate is much lower for the latter [233]. Two other UDP-D-GlcNAc 4-epimerases from *E. coli* O55:H7 (Gne) [234] and *E. coli* O86:B7 (Gne1) [235] have been reported. However, there is still controversy in the literature about the activity of *E. coli* Gne and Gne1. Recently,

Z3206 from *E. coli* O157, which exhibits 100% sequence identity to Gne and Gne1, was found to be incapable of converting UDP-D-GlcNAc to UDP-D-GalNAc [236]. *E. coli* O55:H7, O86:B7 and O157 all have D-GalNAc as the first residue (the reducing end) in the O-repeat units. Sugar 1-phosphate transferase WecA, the O-unit initiating enzyme that transfers the first residue to undecaprenyl phosphate (Und-P), was not able to recognize UDP-D-GalNAc and transfer D-GalNAc-P to Und-P to form GalNAc-PP-Und, but only able to transfer D-GlcNAc-P from UDP-D-GlcNAc to form GlcNAc-PP-Und [236]. Interestingly, Z3206 was capable of converting GlcNAc-PP-Und to GalNAc-PP-Und [236]. It is possible that the GalNAc residue at the reducing end of the O-repeat units is not derived from the epimerization of UDP-D-GlcNAc, but from the epimerization of GlcNAc-PP-Und. More in-depth investigation is warranted to clarify this controversy.

UDP-2,3-Diacetamido-2,3-Dideoxy-D-Mannuronic Acid (UDP-D-Man(2NAc3NAc)A), UDP-2,3-Diacetamido-2,3-Dideoxy-D-Glucuronic Acid (UDP-D-Glc(2NAc3NAc)A) and 2-Acetamido-3-Acetimidoylamino-2,3-Dideoxy-D-Mannuronic Acid (UDP-D-Man(2NAc3NAm)A)

The O-antigen of *P. aeruginosa* serotype O5 [40] and the band-A trisaccharide of *Bordetella pertussis* [237] contain a rare diacetamido uronic acid D-Man(2NAc3NAc)A. Bioinformatics and mutational studies first suggested that five genes from *P. aeruginosa* O5 (*wbpA*, *wbpB*, *wbpD*, *wbpE* and *wbpI*) [176, 238] and four genes from *B. pertussis* (*wlbA-D*) [239, 240] were involved in the biosynthesis of this sugar. Since WbpI and WbpA showed high sequence similarity to UDP-D-GlcNAc 2-epimerases (WecB and CapP) and UDP-D-ManNAc 6-dehydrogenase (WecC and Cap5O) in the ManNAcA biosynthesis pathway (see Sect. 7.3.2.2.3), respectively, the first two steps of the biosynthesis pathway of D-Man(2NAc3NAc)A and D-ManNAcA were originally proposed to be identical [176, 240]. However, attempts to cross complement between *wecB* and *wbpI*, and between *wecC* and *wbpA* were unsuccessful while complementation of knockout mutants with their respective homologous genes was positive. These observations indicate that these genes have different functions [175]. Thus, the biosynthetic pathways of D-Man(2NAc3NAc)A and D-ManNAcA are clearly different.

Recent results from our lab studying the biosynthesis of UDP-D-Man(2NAc3NAc)A from UDP-D-GlcNAc in *P. aeruginosa* PAO1 (O5) has provided unambiguous evidence that this pathway involves five steps (Scheme 7.5). WbpA from *P. aeruginosa* is the first biochemically characterized enzyme of this pathway. It encodes a UDP-D-GlcNAc 6-dehydrogenase (EC 1.1.1.136), together with its cofactor NAD^+, catalyzing the conversion of UDP-D-GlcNAc (31) to UDP-D-GlcNAcA (45), the first step of the pathway [241]. Intriguingly, as shown firstly by our group [242] and followed by Larkin and Imperiali [243], the second and third steps require coupling of two enzymes WbpB (UDP-D-GlcNAcA 3-dehydrogenase, EC 1.1.1.-) and WbpE (UDP-3-keto-D-GlcNAcA transaminase, EC 2.6.1.-). WbpB catalyzes the 3-dehydrogenation of UDP-D-GlcNAcA (45) to form UDP-

3-keto-D-GlcNAcA (47) (the second step), which is then used by WbpE as substrate for transamination to generate UDP-3-amino-3-deoxy-D-GlcNAcA (48) (the third step). In vitro experiments showed that these two enzymes were only active when both were present at the same time in a single reaction mixture with the initiating substrate, UDP-D-GlcNAcA, the product of the previous step catalyzed by WbpA, although the exact mechanism of the coupled reactions still needs clarification [242–244]. Most recently, the crystal structure of WbpE in complex with its cofactor PLP and product UDP-Glc(NAc3N)A has been solved in high resolution and revealed the key residues associated with the enzymatic activity of this enzyme [244].

WbpD is an *N*-acetyltransferase which transfers an acetyl group from acetyl-CoA to the 3-amino group of UDP-D-Glc(NAc3N)A (48) to create UDP-D-Glc(2NAc3NAc)A (49) (the fourth step) [242–244]. The chemical synthesis of the rare sugar UDP-D-Glc(2NAc3NAc)A by the Field laboratory [245] enables the characterization of another enzyme in the pathway, the UDP-D-Glc (2NAc3NAc)A 2-epimerase (EC 5.1.3.23) (WbpI from *P. aeruginosa* or WlbD from *B. pertussis*), which converts UDP-D-Glc(2NAc3NAc)A (49) to UDP-D-Man (2NAc3NAc)A (50), the last step of the pathway [246]. Homologs of the *P. aeruginosa* genes (*wbpA*, *wbpB*, *wbpE*, *wbpD* and *wbpI*) are also present in other bacterial species (such as *B. pertussis*, *B. parapertussis*, and *B. bronchiseptica*) that contain D-Man(2NAc3NAc)A and derivatives in their polysaccharides. The corresponding homolog genes from *B. pertussis* could fully complement *P. aeruginosa wbpA*, *wbpB*, *wbpE*, *wbpD* and *wbpI* mutants indicating that different organisms use the same scheme for biosynthesis of UDP-D-Man(2NAc3NAc)A [247].

The WbpD-catalyzed intermediate product UDP-D-Glc(2NAc3NAc)A is utilized as the NDP-activated sugar precursor for the assembly of the O-antigen of *P. aeruginosa* serotype O1, which contains D-Glc(2NAc3NAc)A in the O-repeat units [248]. In fact, *orf6*, *orf7*, *orf8* and *orf9* from the O-antigen gene cluster of *P. aeruginosa* serotype O1 showed high sequence similarity (>75%) to the *wbpA*, *wbpB*, *wbpD* and *wbpE* of serotype O5 and thus were proposed to encode the proteins responsible for the synthesis of UDP-D-Glc(2NAc3NAc)A using the same scheme [170].

2-Acetamido-3-acetimidoylamino-2,3-dideoxy-D-mannuronic acid (D-Man (2NAc3NAm)A), is another rare sugar constituent of the O-antigens of *P. aeruginosa* serogroups O2, O5, O16, O18 and O20 [40]. The sugar UDP-D-Man(2NAc3NAm) A (51) was proposed to arise from UDP-D-Man(2NAc3NAc)A (50) (the WbpI product) by an amidotransferase. The protein WbpG from *P. aeruginosa* PAO1 (O5) has conserved amidotransferase domain and showed high sequence similarity to LfnA of *P. aeruginosa* O12, an amidotransferase involved in the synthesis of L-FucNAm [204] (Sect. 7.3.2.2.2). A knockout mutant of *wbpG* was deficient in the O-antigen biosynthesis [249]. These information has led to the proposal that WbpG is the amidotransferase that converts UDP-D-Man(2NAc3NAc)A to UDP-D-Man (2NAc3NAm)A. To fully understand the biochemical function of WbpG, more work is warranted.

7.4 Biosynthesis of Non-hexose Sugar Precursors

7.4.1 Biosynthesis of CMP-3-Deoxy-D-Manno-Oct-2-Ulosonic Acid (CMP-Kdo)

Kdo (56) is an essential component of LPS of Gram-negative bacteria and has been found in all LPS inner core structures investigated so far [250] (see Chap. 2). In *E. coli*, the minimal LPS structure required for bacterial survival and growth is composed of two Kdo residues attached to lipid A [251, 252]. Although the majority of higher plants and some green algae also contain Kdo [253], it is not present in yeast and animals [252]. The fact that Kdo is essential for bacteria survival and absent in animals made the biosynthesis pathway of Kdo an attractive target for designing of novel antibacterial agents. As a result, the biosynthesis pathway of Kdo has been exceptionally well investigated, and the current progress in the antimicrobial drug design targeting the Kdo biosynthetic pathway has recently been reviewed [254].

CMP-Kdo is the activated sugar-nucleotide precursor of Kdo [255]. The biosynthesis of CMP-Kdo involves four steps (Scheme 7.6). The first step is the conversion of D-ribulose 5-phosphate (52) into D-arabinose 5-phosphate (53), catalyzed by the enzyme D-arabinose-5-phosphate isomerase (EC 5.3.1.6) [256]. In *E. coli* there are two such isomerases, KdsD and GutQ, which have almost the same biochemical properties [257, 258]. The KdsD homolog is present in all sequenced genomes of Gram-negative bacteria, while only a subset of Enterobacteriaceae encodes GutQ homologs [254]. The gene *gutQ* is a paralogue of *kdsD* deriving from a duplication event associated with other specific pathways but still capable of substituting for *kdsD* [254]. The second step of the CMP-Kdo pathway is the condensation of phosphoenolpyruvate (54) and D-arabinose 5-phosphate (53) into 3-deoxy-D-*manno*-oct-2-ulosonate 8-phosphate (Kdo-8-P) (55), catalyzed by Kdo-8-P synthase (EC 4.1.2.16) [259, 260]. This is the first committed step in the Kdo pathway [256]. In *E. coli*, Kdo-8-P synthase is encoded by *kdsA* and has been studied extensively, and crystal structures of KdsA homologs have been solved [261, 262]. Kdo-8-P phosphatase (EC 3.1.3.45) catalyzes the third step of CMP-Kdo biosynthesis pathway and hydrolyzes Kdo-8-P (55) to Kdo (56) and inorganic phosphate. In 1980, Kdo-8-P phosphatase was first purified and characterized from *E. coli* [260]. The encoding gene *kdsC* was identified and cloned later from *E. coli*

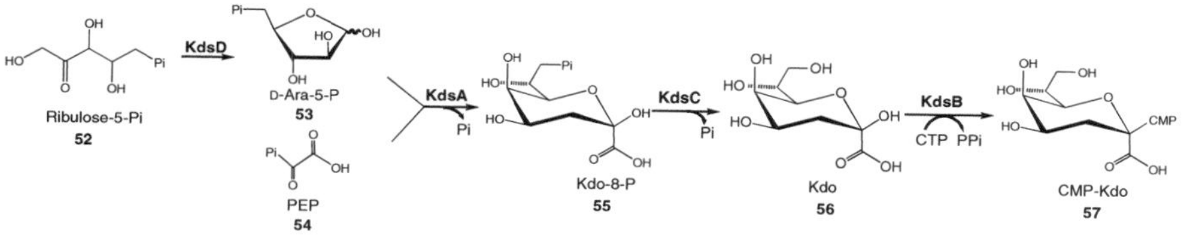

Scheme 7.6 CMP-Kdo biosynthesis pathway

[263]. Finally, the enzyme CMP-Kdo synthase (EC 2.7.7.38) encoded by *kdsB* in *E. coli* catalyzes the formation of the activated sugar CMP-Kdo (57) [264].

7.4.2 Biosynthesis of ADP-L-Glycero-D-Manno-Heptose (L,D-Hep)

L-*glycero*-D-*manno*-Heptose (L,D-Hep) is another highly conserved sugar moiety of the LPS inner core structure. Research by Eidels and Osborn indicated that D-sedoheptulose 7-phosphate (58), an intermediate in the pentose phosphate pathway, is the initial substrate for Hep biosynthesis [265, 266]. Later, ADP-L-*glycero*-D-*manno*-heptose (ADP-L,D-Hep) (63) was shown to be the activated sugar nucleotide form used by heptosyltransferase in *Shigella sonnei* and *S. enterica* [267]. A four-step biosynthesis pathway was previously proposed: (1) D-sedoheptulose 7-phosphate (58) is converted to D-*glycero*-D-*manno*-heptose 7-phosphate (D,D-Hep-7-P) (59) by a phosphoheptose isomerase; (2) conversion of D,D-Hep-7-P (59) to D-*glycero*-D-*manno*-heptose 1-phosphate (D,D-Hep-1-P) (61) by a mutase; (3) formation of ADP-D-*glycero*-D-*manno*-heptose (ADP-D,D-Hep) from D,D-Hep-1-P (62) and ATP catalyzed by an adenylyltransferase; and (4) conversion of ADP-D,D-Hep (62) to ADP-L,D-Hep (63) by an epimerase [265]. The enzymes catalyzing step (1) (GmhA) (EC 5.3.1.-) and (5) (GmhD) (EC 5.1.3.20) have been identified and biochemically characterized [268, 269]. A bifunctional two-domain enzyme HldE (formerly RfaE), involved in the intermediate steps of ADP-L,D-Hep biosynthesis, was isolated from *E. coli* [270]. One of the domains of HldE shows considerable structural similarity to members of the ribokinase family, while the other domain shows conserved features of nucleotidyltransferases. A phosphatase GmhB (EC 3.1.3.-) purified from *E. coli* uses D-*glycero*-D-*manno*-heptose 1,7-bisphosphate (D,D-Hep-1,7-PP) as substrate [271].

This information has led to the notion that a kinase/phosphatase cascade replaces the mutase step in the previously proposed ADP-L,D-Hep pathway. The newly

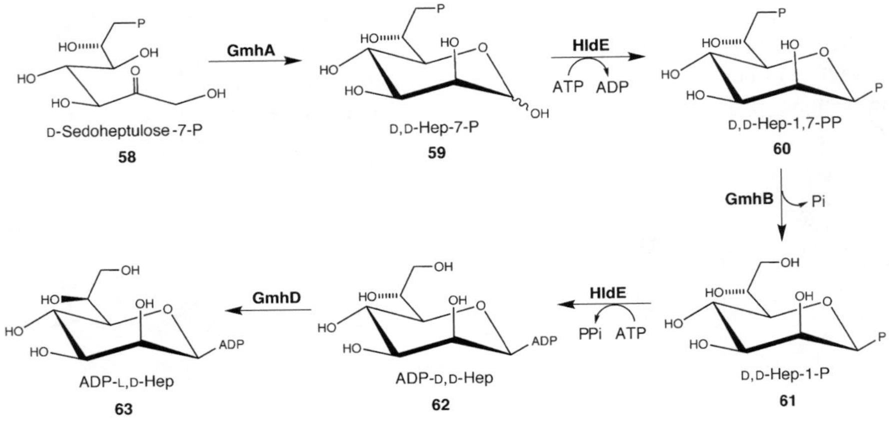

Scheme 7.7 ADP-L-D-Hep biosynthesis pathway

proposed five-step pathway is depicted as the following: (1) D-sedoheptulose 7-phosphate (58) is first converted to D,D-Hep-7-P (59) by GmhA, (2) D,D-Hep-1,7-PP (60) is then formed by the kinase activity of HldE, (3) the phosphatase GmhB converts the D,D-Hep-1,7-PP (60) to D,D-Hep-1-P (61), (4) ADP-D,D-Hep (62) is then synthesized from D,D-Hep-1-P and ATP by the bifunctional enzyme HldE, and finally (5) the epimerase GmhD catalyzes the conversion of ADP-D,D-Hep (62) to ADP-L,D-Hep (63) (Scheme 7.7) [272].

7.5 Conclusions

To date, the biosynthesis pathways of many of the sugar precursors utilized for bacterial LPS assembly have been elucidated. Here we have reviewed the current knowledge of the biosynthesis pathways of the UDP-D-Glc and related sugars, dTDP sugars, GDP sugars, UDP-D-GlcNAc and related sugars, as well as the CMP-Kdo and ADP-L,D-Hep. The initiating substrate for the biosynthesis of the majority of the NDP-hexoses is either Glc-6-P or Fru-6-P, which could be derived from the central metabolic pathway. Kinases, phosphatases, and phosphomutases usually act on the earlier steps to generate a sugar 1-phosphate intermediate (such as Glc-1-P, Fru-1-P and GlcNAc-1-P). Their activities are usually followed by another reaction step, the coupling of NMP to the sugar 1-phosphate by sugar nucleotidyltransferases (or NDP-sugar pyrophosphorylases) to generate NDP sugars that can serve as common precursors (such as UDP-D-Glc, dTDP-D-Glc, GDP-D-Man and UDP-D-GlcNAc).

The ensuing steps for modifying the common precursors would be through single or multiple enzymatic reactions such as epimerization, oxidation, dehydration, reduction, amino- and acetyl-transfer activities. These reactions generate a great variety of hexose derivatives. For example, Glc-6-P can be converted to two common NDP-sugar precursors, UDP-D-Glc and dTDP-D-Glc; while Fru-6-P can be converted to GDP-D-Man and UDP-D-GlcNAc. Subsequent oxidation, epimerization or a combination of both, would convert, for instance, UDP-D-Glc to UDP-D-GlcA, UDP-D-Gal and UDP-D-GalA, and UDP-D-GlcNAc to UDP-D-GalNAc, UDP-D-GalNAcA, UDP-D-ManNAc and UDP-D-ManNAcA. The biosynthesis of hexose derivatives deoxygenated at C6 first requires the generation of a 6-deoxy-4-keto derivative intermediate catalyzed by a 4,6-dehydratase (such as dTDP-6-deoxy-4-keto-D-Glc, GDP-6-deoxy-4-keto-D-Man, UDP-6-deoxy-4-keto-D-GlcNAc and UDP-6-deoxy-4-keto-L-IdoNAc), then by a combination of reduction, epimerization and amino and acetyl transfer, a variety of 6-deoxyhexose derivatives (dTDP-L-Rha, dTDP-D-Fuc, GDP-6-deoxy-L-Tal, GDP-D-Rha, GDP-L-Fuc, GDP-6-deoxy-D-Tal), 2-amino-2,6-dideoxyhexose derivatives (UDP-D-QuiNAc, UDP-L-QuiNAc, UDP-D-FucNAc UDP-L-FucNAc, UDP-L-FucNAm, UDP-L-RhaNAc), 3,6-dideoxyhexose derivatives (GDP-colitose) and 4-amino-4,6-dideoxyhexose derivatives (dTDP-D-Qui4N, dTDP-D-Qui4NAc, GDP-D-Rha4N and GDP-D-Rha4NAc) could be generated. Hexoses with the 2-acetamido group are generally obtained from UDP-D-GlcNAc, while hexoses with the 4-acetamido group (such as D-Qui4NAc and D-Rha4NAc) are obtained from other pathways.

Many of the enzymes involved in NDP-sugar biosynthesis are members of the short-chain dehydratase/reductase (SDR) superfamily with the highly-conserved signature motif GXXGXXG for binding of the cofactor $NAD(P)^+$/NAD(P)H. This family consists of enzymes with diverse functions including dehydratases (such as RmlB, Gmd, WbjB/Cap5E and WbpM), reductases (including RmlD, GMER, Tld and WbjC/Cap5F) and epimerases (such as GalE and WbpP). Structural and mechanistic studies of the key enzymes involved in NDP-sugar biosynthesis enabled the development of new inhibitors targeting these pathways. A thorough understanding of the biosynthesis pathways of natural NDP sugars as well as the catalytic mechanisms of the enzymes involved would make it possible to engineer bacteria and enzymes to perform in vivo or in vitro enzymatic glycodiversification for generating new glycoforms as reviewed recently by several groups [2, 273–275].

Many of the sugar biosynthesis pathways are conserved among different species. By sequence comparison with genes from well-characterized pathways, the functions of genes from newly sequenced LPS clusters could be predicated with sufficiently high level of confidence. However, it should be noted that proteins encoding the same type of enzyme from different organisms could show low sequence similarity, while proteins with high sequence similarity might exhibit totally different functions. Although bioinformatics is very useful in predicting the functions of unknown proteins, in many cases, biochemical characterization is still absolutely necessary to accurately decipher the function of the enzymes involved in each step of the nucleotide-sugar synthesis pathways.

Acknowledgements Research in the Lam laboratory is supported by operating grants from the Canadian Institute of Health Research (#MOP-14687), and the Canadian Cystic Fibrosis Foundation. J.S.L. holds a Canada Research Chair in Cystic Fibrosis and Microbial Glycobiology jointly funded by the Canadian Foundation of Innovation and the Ontario Research Fund.

References

1. Varki A, Cummings R, Esko J, Freeze H, Hart G, Marth J (1999) Essentials of glycobiology. Cold Spring Harbor, New York
2. Thibodeaux CJ, Melancon CE III, Liu HW (2008) Natural-product sugar biosynthesis and enzymatic glycodiversification. Angew Chem Int Ed Engl 47:9814–9859
3. Leloir LF (1951) The enzymatic transformation of uridine diphosphate glucose into a galactose derivative. Arch Biochem 33:186–190
4. Lu M, Kleckner N (1994) Molecular cloning and characterization of the *pgm* gene encoding phosphoglucomutase of *Escherichia coli*. J Bacteriol 176:5847–5851
5. Coyne MJ Jr, Russell KS, Coyle CL, Goldberg JB (1994) The *Pseudomonas aeruginosa algC* gene encodes phosphoglucomutase, required for the synthesis of a complete lipopolysaccharide core. J Bacteriol 176:3500–3507
6. Kooistra O, Bedoux G, Brecker L, Lindner B, Sanchez Carballo P, Haras D, Zähringer U (2003) Structure of a highly phosphorylated lipopolysaccharide core in the Delta *algC*

mutants derived from *Pseudomonas aeruginosa* wild-type strains PAO1 (serogroup O5) and PAC1R (serogroup O3). Carbohydr Res 338:2667–2677
7. Weissborn AC, Liu Q, Rumley MK, Kennedy EP (1994) UTP:α-D-glucose-1-phosphate uridylyltransferase of *Escherichia coli*: isolation and DNA sequence of the *galU* gene and purification of the enzyme. J Bacteriol 176:2611–2618
8. Thoden JB, Holden HM (2007) The molecular architecture of glucose-1-phosphate uridylyltransferase. Protein Sci 16:432–440
9. Priebe GP, Dean CR, Zaidi T, Meluleni GJ, Coleman FT, Coutinho YS, Noto MJ, Urban TA, Pier GB, Goldberg JB (2004) The *galU* gene of *Pseudomonas aeruginosa* is required for corneal infection and efficient systemic spread following pneumonia but not for infection confined to the lung. Infect Immun 72:4224–4232
10. Dean CR, Goldberg JB (2002) *Pseudomonas aeruginosa galU* is required for a complete lipopolysaccharide core and repairs a secondary mutation in a PA103 (serogroup O11) *wbpM* mutant. FEMS Microbiol Lett 210:277–283
11. Choudhury B, Carlson RW, Goldberg JB (2005) The structure of the lipopolysaccharide from a *galU* mutant of *Pseudomonas aeruginosa* serogroup-O11. Carbohydr Res 340:2761–2772
12. Mollerach M, Lopez R, Garcia E (1998) Characterization of the *galU* gene of *Streptococcus pneumoniae* encoding a uridine diphosphoglucose pyrophosphorylase: a gene essential for capsular polysaccharide biosynthesis. J Exp Med 188:2047–2056
13. Thoden JB, Holden HM (1998) Dramatic differences in the binding of UDP-galactose and UDP-glucose to UDP-galactose 4-epimerase from *Escherichia coli*. Biochemistry 37:11469–11477
14. Thoden JB, Frey PA, Holden HM (1996) Molecular structure of the NADH/UDP-glucose abortive complex of UDP-galactose 4-epimerase from *Escherichia coli*: implications for the catalytic mechanism. Biochemistry 35:5137–5144
15. Wilson DB, Hogness DS (1964) The enzymes of the galactose eperon in *Escherichia coli*. I. Purification and characterization of uridine diphosphogalactose 4-epimerase. J Biol Chem 239:2469–2481
16. Stevenson G, Andrianopoulos K, Hobbs M, Reeves PR (1996) Organization of the *Escherichia coli* K-12 gene cluster responsible for production of the extracellular polysaccharide colanic acid. J Bacteriol 178:4885–4893
17. Stenutz R, Weintraub A, Widmalm G (2006) The structures of *Escherichia coli* O-polysaccharide antigens. FEMS Microbiol Rev 30:382–403
18. Perepelov AV, Babicka D, Senchenkova SN, Shashkov AS, Moll H, Rozalski A, Zähringer U, Knirel YA (2001) Structure of the O-specific polysaccharide of *Proteus vulgaris* O4 containing a new component of bacterial polysaccharides, 4,6-dideoxy-4-{*N*-[(*R*)-3-hydroxybutyryl]-L-alanyl}amino-D-glucose. Carbohydr Res 331:195–202
19. Knirel YA, Paredes L, Jansson PE, Weintraub A, Widmalm G, Albert MJ (1995) Structure of the capsular polysaccharide of *Vibrio cholerae* O139 synonym Bengal containing D-galactose 4,6-cyclophosphate. Eur J Biochem 232:391–396
20. Munoz R, Mollerach M, Lopez R, Garcia E (1999) Characterization of the type 8 capsular gene cluster of *Streptococcus pneumoniae*. J Bacteriol 181:6214–6219
21. Lindberg B, Lindqvist B, Lönngren J, Powell DA (1980) Structural studies of the capsular polysaccharide from *Streptococcus pneumoniae* type 1. Carbohydr Res 78:111–117
22. Jansson PE, Lindberg B, Anderson M, Lindquist U, Henrichsen J (1988) Structural studies of the capsular polysaccharide from *Streptococcus pneumoniae* type 2, a reinvestigation. Carbohydr Res 182:111–117
23. Cartee RT, Forsee WT, Schutzbach JS, Yother J (2000) Mechanism of type 3 capsular polysaccharide synthesis in *Streptococcus pneumoniae*. J Biol Chem 275:3907–3914
24. Gottesman S, Stout V (1991) Regulation of capsular polysaccharide synthesis in *Escherichia coli* K12. Mol Microbiol 5:1599–1606

25. Arrecubieta C, Garcia E, Lopez R (1996) Demonstration of UDP-glucose dehydrogenase activity in cell extracts of *Escherichia coli* expressing the pneumococcal *cap3A* gene required for the synthesis of type 3 capsular polysaccharide. J Bacteriol 178:2971–2974
26. Jiang SM, Wang L, Reeves PR (2001) Molecular characterization of *Streptococcus pneumoniae* type 4, 6B, 8, and 18 C capsular polysaccharide gene clusters. Infect Immun 69:1244–1255
27. Hung RJ, Chien HS, Lin RZ, Lin CT, Vatsyayan J, Peng HL, Chang HY (2007) Comparative analysis of two UDP-glucose dehydrogenases in *Pseudomonas aeruginosa* PAO1. J Biol Chem 282:17738–17748
28. Loutet SA, Bartholdson SJ, Govan JR, Campopiano DJ, Valvano MA (2009) Contributions of two UDP-glucose dehydrogenases to viability and polymyxin B resistance of *Burkholderia cenocepacia*. Microbiology 155:2029–2039
29. Parolis H, Parolis LA (1995) The structure of the O-specific polysaccharide from *Escherichia coli* O113 lipopolysaccharide. Carbohydr Res 267:263–269
30. Hisatsune K, Kondo S, Isshiki Y, Iguchi T, Kawamata Y, Shimada T (1993) O-antigenic lipopolysaccharide of *Vibrio cholerae* O139 Bengal, a new epidemic strain for recent cholera in the Indian subcontinent. Biochem Biophys Res Commun 196:1309–1315
31. Isshiki Y, Kondo S, Iguchi T, Sano Y, Shimada T, Hisatsune K (1996) An immunochemical study of serological cross-reaction between lipopolysaccharides from *Vibrio cholerae* O22 and O139. Microbiology 142:1499–1504
32. Carlson RW, Garci F, Noel D, Hollingsworth R (1989) The structures of the lipopolysaccharide core components from *Rhizobium leguminosarum* biovar phaseoli CE3 and two of its symbiotic mutants, CE109 and CE309. Carbohydr Res 195:101–110
33. Vinogradov E, Sidorczyk Z, Knirel YA (2002) Structure of the core part of the lipopolysaccharides from *Proteus penneri* strains 7, 8, 14, 15, and 21. Carbohydr Res 337:643–649
34. Frirdich E, Bouwman C, Vinogradov E, Whitfield C (2005) The role of galacturonic acid in outer membrane stability in *Klebsiella pneumoniae*. J Biol Chem 280:27604–27612
35. Regue M, Hita B, Pique N, Izquierdo L, Merino S, Fresno S, Benedi VJ, Tomas JM (2004) A gene, *uge*, is essential for *Klebsiella pneumoniae* virulence. Infect Immun 72:54–61
36. Frirdich E, Whitfield C (2005) Characterization of Gla(KP), a UDP-galacturonic acid C4-epimerase from *Klebsiella pneumoniae* with extended substrate specificity. J Bacteriol 187:4104–4115
37. Reeves PR, Hobbs M, Valvano MA, Skurnik M, Whitfield C, Coplin D, Kido N, Klena J, Maskell D, Raetz CR, Rick PD (1996) Bacterial polysaccharide synthesis and gene nomenclature. Trends Microbiol 4:495–503
38. Erbing C, Svensson S, Hammarstrom S (1975) Structural studies on the *O*-specific side-chains of the cell-wall lipopolysaccharide from *Escherichia coli* O 75. Carbohydr Res 44:259–265
39. Stevenson G, Neal B, Liu D, Hobbs M, Packer NH, Batley M, Redmond JW, Lindquist L, Reeves P (1994) Structure of the O antigen of *Escherichia coli* K-12 and the sequence of its *rfb* gene cluster. J Bacteriol 176:4144–4156
40. Knirel YA, Bystrova OV, Kocharova NA, Zähringer U, Pier GB (2006) Conserved and variable structural features in the lipopolysaccharide of *Pseudomonas aeruginosa*. J Endotoxin Res 12:324–336
41. Bystrova OV, Knirel YA, Lindner B, Kocharova NA, Kondakova AN, Zähringer U, Pier GB (2006) Structures of the core oligosaccharide and O-units in the R- and SR-type lipopolysaccharides of reference strains of *Pseudomonas aeruginosa* O-serogroups. FEMS Immunol Med Microbiol 46:85–99
42. Moreau M, Richards JC, Perry MB, Kniskern PJ (1988) Structural analysis of the specific capsular polysaccharide of *Streptococcus pneumoniae* type 45 (American type 72). Biochemistry 27:6820–6829
43. Daoust V, Carlo DJ, Zeltner JY, Perry MB (1981) Specific capsular polysaccharide of type 45 *Streptococcus pneumoniae* (American type 72). Infect Immun 32:1028–1033

44. Lutticken R, Temme N, Hahn G, Bartelheimer EW (1986) Meningitis caused by *Streptococcus suis*: case report and review of the literature. Infection 14:181–185
45. McNeil M, Daffe M, Brennan PJ (1990) Evidence for the nature of the link between the arabinogalactan and peptidoglycan of mycobacterial cell walls. J Biol Chem 265: 18200–18206
46. Deng L, Mikusova K, Robuck KG, Scherman M, Brennan PJ, McNeil MR (1995) Recognition of multiple effects of ethambutol on metabolism of mycobacterial cell envelope. Antimicrob Agents Chemother 39:694–701
47. Blankenfeldt W, Asuncion M, Lam JS, Naismith JH (2000) The structural basis of the catalytic mechanism and regulation of glucose-1-phosphate thymidylyltransferase (RmlA). EMBO J 19:6652–6663
48. Shibaev VN (1986) Biosynthesis of bacterial polysaccharide chains composed of repeating units. Adv Carbohydr Chem Biochem 44:277–339
49. Koplin R, Wang G, Hotte B, Priefer UB, Puhler A (1993) A 3.9-kb DNA region of *Xanthomonas campestris* pv. campestris that is necessary for lipopolysaccharide production encodes a set of enzymes involved in the synthesis of dTDP-rhamnose. J Bacteriol 175: 7786–7792
50. Melo A, Glaser L (1965) The nucleotide specificity and feedback control of thymidine diphosphate D-glucose pyrophosphorylase. J Biol Chem 240:398–405
51. Blankenfeldt W, Giraud MF, Leonard G, Rahim R, Creuzenet C, Lam JS, Naismith JH (2000) The purification, crystallization and preliminary structural characterization of glucose-1-phosphate thymidylyltransferase (RmlA), the first enzyme of the dTDP-L-rhamnose synthesis pathway from *Pseudomonas aeruginosa*. Acta Crystallogr D Biol Crystallogr 56:1501–1504
52. Dong C, Major LL, Srikannathasan V, Errey JC, Giraud MF, Lam JS, Graninger M, Messner P, McNeil MR, Field RA, Whitfield C, Naismith JH (2007) RmlC, a C3′ and C5′ carbohydrate epimerase, appears to operate via an intermediate with an unusual twist boat conformation. J Mol Biol 365:146–159
53. Allard ST, Giraud MF, Whitfield C, Graninger M, Messner P, Naismith JH (2001) The crystal structure of dTDP-D-Glucose 4,6-dehydratase (RmlB) from *Salmonella enterica* serovar Typhimurium, the second enzyme in the dTDP-L-rhamnose pathway. J Mol Biol 307:283–295
54. Giraud MF, Leonard GA, Field RA, Berlind C, Naismith JH (2000) RmlC, the third enzyme of dTDP-L-rhamnose pathway, is a new class of epimerase. Nat Struct Biol 7: 398–402
55. Blankenfeldt W, Kerr ID, Giraud MF, McMiken HJ, Leonard G, Whitfield C, Messner P, Graninger M, Naismith JH (2002) Variation on a theme of SDR. dTDP-6-deoxy-L-*lyxo*-4-hexulose reductase (RmlD) shows a new Mg^{2+}-dependent dimerization mode. Structure 10:773–786
56. Li Q, Reeves PR (2000) Genetic variation of dTDP-L-rhamnose pathway genes in *Salmonella enterica*. Microbiology 146:2291–2307
57. Li Q, Hobbs M, Reeves PR (2003) The variation of dTDP-L-rhamnose pathway genes in *Vibrio cholerae*. Microbiology 149:2463–2474
58. Gaugler RW, Gabriel O (1973) Biological mechanisms involved in the formation of deoxy sugars. VII. Biosynthesis of 6-deoxy-L-talose. J Biol Chem 248:6041–6049
59. Jann B, Shashkov A, Torgov V, Kochanowski H, Seltmann G, Jann K (1995) NMR investigation of the 6-deoxy-L-talose-containing O45, O45-related (O45rel), and O66 polysaccharides of *Escherichia coli*. Carbohydr Res 278:155–165
60. Zähringer U, Rettenmaier H, Moll H, Senchenkova SN, Knirel YA (1997) Structure of a new 6-deoxy-α-D-talan from *Burkholderia* (*Pseudomonas*) *plantarii* strain DSM 6535, which is different from the O-chain of the lipopolysaccharide. Carbohydr Res 300:143–151
61. Russa R, Urbanik-Sypniewska T, Lindstrom K, Mayer H (1995) Chemical characterization of two lipopolysaccharide species isolated from *Rhizobium loti* NZP2213. Arch Microbiol 163:345–351

62. Shibuya N, Amano K, Azuma J, Nishihara T, Kitamura Y, Noguchi T, Koga T (1991) 6-Deoxy-D-talan and 6-deoxy-L-talan. Novel serotype-specific polysaccharide antigens from *Actinobacillus actinomycetemcomitans*. J Biol Chem 266:16318–16323
63. Nakano Y, Suzuki N, Yoshida Y, Nezu T, Yamashita Y, Koga T (2000) Thymidine diphosphate-6-deoxy-L-*lyxo*-4-hexulose reductase synthesizing dTDP-6-deoxy-L-talose from *Actinobacillus actinomycetemcomitans*. J Biol Chem 275:6806–6812
64. Senchenkova SN, Shashkov AS, Moran AP, Helander IM, Knirel YA (1995) Structures of the O-specific polysaccharide chains of *Pectinatus cerevisiiphilus* and *Pectinatus frisingensis* lipopolysaccharides. Eur J Biochem 232:552–557
65. Feng L, Senchenkova SN, Yang J, Shashkov AS, Tao J, Guo H, Cheng J, Ren Y, Knirel YA, Reeves PR, Wang L (2004) Synthesis of the heteropolysaccharide O antigen of *Escherichia coli* O52 requires an ABC transporter: structural and genetic evidence. J Bacteriol 186: 4510–4519
66. Winn AM, Miles CT, Wilkinson SG (1996) Structure of the O3 antigen of *Stenotrophomonas* (*Xanthomonas* or *Pseudomonas*) *maltophilia*. Carbohydr Res 282:149–156
67. Winn AM, Galbraith L, Temple GS, Wilkinson SG (1993) Structure of the O19 antigen of *Xanthomonas maltophilia*. Carbohydr Res 247:249–254
68. Knirel YA, Kochetkov NK (1994) The structure of lipopolysaccharides of gram-negative bacteria. III. The structure of O-antigens. Biochem Moscow 12:1325–1383
69. Amano K, Nishihara T, Shibuya N, Noguchi T, Koga T (1989) Immunochemical and structural characterization of a serotype-specific polysaccharide antigen from *Actinobacillus actinomycetemcomitans* Y4 (serotype b). Infect Immun 57:2942–2946
70. Kählig H, Kolarich D, Zayni S, Scheberl A, Kosma P, Schäffer C, Messner P (2005) *N*-Acetylmuramic acid as capping element of a-D-fucose-containing S-layer glycoprotein glycans from *Geobacillus tepidamans* GS5-97^{T}. J Biol Chem 280:20292–20299
71. Yoshida Y, Nakano Y, Nezu T, Yamashita Y, Koga T (1999) A novel NDP-6-deoxyhexosyl-4-ulose reductase in the pathway for the synthesis of thymidine diphosphate-D-fucose. J Biol Chem 274:16933–16939
72. Wang Q, Ding P, Perepelov AV, Xu Y, Wang Y, Knirel YA, Wang L, Feng L (2008) Characterization of the dTDP-D-fucofuranose biosynthetic pathway in *Escherichia coli* O52. Mol Microbiol 70:1358–1367
73. Knirel YA, Dashunin VV, Shashkov AS, Kochetkov NK, Dmitriev BA, Hofman IL (1988) Somatic antigens of *Shigella*: structure of the O-specific polysaccharide chain of the *Shigella dysenteriae* type 7 lipopolysaccharide. Carbohydr Res 179:51–60
74. Parolis H, Parolis LA, Olivieri G (1997) Structural studies on the *Shigella*-like *Escherichia coli* O121 O-specific polysaccharide. Carbohydr Res 303:319–325
75. L'vov VL, Shashkov AS, Dmitriev BA, Kochetkov NK, Jann B, Jann K (1984) Structural studies of the O-specific side chain of the lipopolysaccharide from *Escherichia coli* O:7. Carbohydr Res 126:249–259
76. Wang Y, Xu Y, Perepelov AV, Qi Y, Knirel YA, Wang L, Feng L (2007) Biochemical characterization of dTDP-D-Qui4N and dTDP-D-Qui4NAc biosynthetic pathways in *Shigella dysenteriae* type 7 and *Escherichia coli* O7. J Bacteriol 189:8626–8635
77. Lüderitz O, Staub AM, Westphal O (1966) Immunochemistry of O and R antigens of *Salmonella* and related Enterobacteriaceae. Bacteriol Rev 30:192–255
78. Jansson PE, Lönngren J, Widmalm G, Leontein K, Slettengren K, Svenson SB, Wrangsell G, Dell A, Tiller PR (1985) Structural studies of the O-antigen polysaccharides of *Klebsiella* O5 and *Escherichia coli* O8. Carbohydr Res 145:59–66
79. Ørskov I, Ørskov F (1984) Serotyping of *Klebsiella*. In: Bergan T (ed) Methods in microbiology, vol 14. Academic, London, pp 143–164
80. Prehm P, Jann B, Jann K (1976) The O9 antigen of *Escherichia coli*. Structure of the polysaccharide chain. Eur J Biochem 67:53–56
81. Kuhlman M, Joiner K, Ezekowitz RA (1989) The human mannose-binding protein functions as an opsonin. J Exp Med 169:1733–1745

82. Sahly H, Ofek I, Podschun R, Brade H, He Y, Ullmann U, Crouch E (2002) Surfactant protein D binds selectively to *Klebsiella pneumoniae* lipopolysaccharides containing mannose-rich O-antigens. J Immunol 169:3267–3274
83. Neidhardt FC, Curtiss R III, Ingraham JL, Lin ECC, Low BK, Magasanik B, Reznikoff WS, Riley M, Schaechter M, Umbarger HE (eds) (1996) *Escherichia coli* and *Salmonella*: cellular and molecular biology, 2nd edn. ASM Press, Washington, DC
84. Rocchetta HL, Pacan JC, Lam JS (1998) Synthesis of the A-band polysaccharide sugar D-rhamnose requires Rmd and WbpW: identification of multiple AlgA homologues, WbpW and ORF488, in *Pseudomonas aeruginosa*. Mol Microbiol 29:1419–1434
85. Jensen SO, Reeves PR (2001) Molecular evolution of the GDP-mannose pathway genes (*manB* and *manC*) in *Salmonella enterica*. Microbiology 147:599–610
86. Byrd MS, Sadovskaya I, Vinogradov E, Lu H, Sprinkle AB, Richardson SH, Ma L, Ralston B, Parsek MR, Anderson EM, Lam JS, Wozniak DJ (2009) Genetic and biochemical analyses of the *Pseudomonas aeruginosa* Psl exopolysaccharide reveal overlapping roles for polysaccharide synthesis enzymes in Psl and LPS production. Mol Microbiol 73:622–638
87. Shinabarger D, Berry A, May TB, Rothmel R, Fialho A, Chakrabarty AM (1991) Purification and characterization of phosphomannose isomerase-guanosine diphospho-D-mannose pyrophosphorylase. A bifunctional enzyme in the alginate biosynthetic pathway of *Pseudomonas aeruginosa*. J Biol Chem 266:2080–2088
88. Lee HJ, Chang HY, Venkatesan N, Peng HL (2008) Identification of amino acid residues important for the phosphomannose isomerase activity of PslB in *Pseudomonas aeruginosa* PAO1. FEBS Lett 582:3479–3483
89. Mulichak AM, Bonin CP, Reiter WD, Garavito RM (2002) Structure of the MUR1 GDP-mannose 4,6-dehydratase from *Arabidopsis thaliana*: implications for ligand binding and specificity. Biochemistry 41:15578–15589
90. Somoza JR, Menon S, Schmidt H, Joseph-McCarthy D, Dessen A, Stahl ML, Somers WS, Sullivan FX (2000) Structural and kinetic analysis of *Escherichia coli* GDP-mannose 4,6 dehydratase provides insights into the enzyme's catalytic mechanism and regulation by GDP-fucose. Structure 8:123–135
91. Webb NA, Mulichak AM, Lam JS, Rocchetta HL, Garavito RM (2004) Crystal structure of a tetrameric GDP-D-mannose 4,6-dehydratase from a bacterial GDP-D-rhamnose biosynthetic pathway. Protein Sci 13:529–539
92. Arsenault TL, Hughes DW, MacLean DB, Szarek WA, Kropinski AMB, Lam JS (1991) Structural studies on the polysaccharide portion of "A-band" lipopolysaccharide from a mutant (AK1401) of *Pseudomonas aeruginosa* strain PAO1. Can J Chem 69: 1273–1280
93. Yokota S, Kaya S, Sawada S, Kawamura T, Araki Y, Ito E (1987) Characterization of a polysaccharide component of lipopolysaccharide from *Pseudomonas aeruginosa* IID 1008 (ATCC 27584) as D-rhamnan. Eur J Biochem 167:203–209
94. Ovod V, Rudolph K, Knirel Y, Krohn K (1996) Immunochemical characterization of O polysaccharides composing the α-D-rhamnose backbone of lipopolysaccharide of *Pseudomonas syringae* and classification of bacteria into serogroups O1 and O2 with monoclonal antibodies. J Bacteriol 178:6459–6465
95. Molinaro A, Silipo A, Lanzetta R, Newman MA, Dow JM, Parrilli M (2003) Structural elucidation of the O-chain of the lipopolysaccharide from *Xanthomonas campestris* strain 8004. Carbohydr Res 338:277–281
96. Senchenkova SN, Shashkov AS, Knirel YA, McGovern JJ, Moran AP (1996) The O-specific polysaccharide chain of *Campylobacter fetus* serotype B lipopolysaccharide is a D-rhamnan terminated with 3-*O*-methyl-D-rhamnose (D-acofriose). Eur J Biochem 239:434–438
97. Kocharova NA, Knirel YA, Widmalm G, Jansson PE, Moran AP (2000) Structure of an atypical O-antigen polysaccharide of *Helicobacter pylori* containing a novel monosaccharide 3-*C*-methyl-D-mannose. Biochemistry 39:4755–4760

98. Kneidinger B, Graninger M, Adam G, Puchberger M, Kosma P, Zayni S, Messner P (2001) Identification of two GDP-6-deoxy-D-*lyxo*-4-hexulose reductases synthesizing GDP-D-rhamnose in *Aneurinibacillus thermoaerophilus* L420-91T. J Biol Chem 276:5577–5583
99. King JD, Poon KK, Webb NA, Anderson EM, McNally DJ, Brisson JR, Messner P, Garavito RM, Lam JS (2009) The structural basis for catalytic function of GMD and RMD, two closely related enzymes from the GDP-D-rhamnose biosynthesis pathway. FEBS J 276: 2686–2700
100. Maki M, Jarvinen N, Rabina J, Roos C, Maaheimo H, Renkonen R (2002) Functional expression of *Pseudomonas aeruginosa* GDP-6-deoxy-4-keto-D-mannose reductase which synthesizes GDP-rhamnose. Eur J Biochem 269:593–601
101. Perry MB, MacLean LM, Brisson JR, Wilson ME (1996) Structures of the antigenic O-polysaccharides of lipopolysaccharides produced by *Actinobacillus actinomycetem-comitans* serotypes a, c, d and e. Eur J Biochem 242:682–688
102. Maki M, Jarvinen N, Rabina J, Maaheimo H, Mattila P, Renkonen R (2003) Cloning and functional expression of a novel GDP-6-deoxy-D-talose synthetase from *Actinobacillus actinomycetemcomitans*. Glycobiology 13:295–303
103. Suzuki N, Nakano Y, Yoshida Y, Nezu T, Terada Y, Yamashita Y, Koga T (2002) Guanosine diphosphate-6-deoxy-4-keto-D-mannose reductase in the pathway for the synthesis of GDP-6-deoxy-D-talose in *Actinobacillus actinomycetemcomitans*. Eur J Biochem 269:5963–5971
104. Tonetti M, Sturla L, Bisso A, Zanardi D, Benatti U, De Flora A (1998) The metabolism of 6-deoxyhexoses in bacterial and animal cells. Biochimie 80:923–931
105. Becker DJ, Lowe JB (2003) Fucose: biosynthesis and biological function in mammals. Glycobiology 13:41R–53R
106. Whitfield C, Roberts IS (1999) Structure, assembly and regulation of expression of capsules in *Escherichia coli*. Mol Microbiol 31:1307–1319
107. Carlson RW, Price NP, Stacey G (1994) The biosynthesis of rhizobial lipo-oligosaccharide nodulation signal molecules. Mol Plant Microbe Interact 7:684–695
108. Stacey G, Luka S, Sanjuan J, Banfalvi Z, Nieuwkoop AJ, Chun JY, Forsberg LS, Carlson R (1994) *nodZ*, a unique host-specific nodulation gene, is involved in the fucosylation of the lipooligosaccharide nodulation signal of *Bradyrhizobium japonicum*. J Bacteriol 176: 620–633
109. Wang L, Reeves PR (1998) Organization of *Escherichia coli* O157 O antigen gene cluster and identification of its specific genes. Infect Immun 66:3545–3551
110. Zhang L, Radziejewska-Lebrecht J, Krajewska-Pietrasik D, Toivanen P, Skurnik M (1997) Molecular and chemical characterization of the lipopolysaccharide O-antigen and its role in the virulence of *Yersinia enterocolitica* serotype O:8. Mol Microbiol 23:63–76
111. Skurnik M, Zhang L (1996) Molecular genetics and biochemistry of *Yersinia* lipopolysaccharide. APMIS 104:849–872
112. Moran AP, O'Malley DT, Kosunen TU, Helander IM (1994) Biochemical characterization of *Campylobacter fetus* lipopolysaccharides. Infect Immun 62:3922–3929
113. Wang G, Ge Z, Rasko DA, Taylor DE (2000) Lewis antigens in *Helicobacter pylori*: biosynthesis and phase variation. Mol Microbiol 36:1187–1196
114. Appelmelk BJ, Vandenbroucke-Grauls CM (2000) *H. pylori* and Lewis antigens. Gut 47: 10–11
115. Andrianopoulos K, Wang L, Reeves PR (1998) Identification of the fucose synthetase gene in the colanic acid gene cluster of *Escherichia coli* K-12. J Bacteriol 180:998–1001
116. Rosano C, Bisso A, Izzo G, Tonetti M, Sturla L, De Flora A, Bolognesi M (2000) Probing the catalytic mechanism of GDP-6-deoxy-4-keto-D-mannose epimerase/reductase by kinetic and crystallographic characterization of site-specific mutants. J Mol Biol 303:77–91
117. Wu B, Zhang Y, Wang PG (2001) Identification and characterization of GDP-D-mannose 4,6-dehydratase and GDP-L-fucose snthetase in a GDP-L-fucose biosynthetic gene cluster from *Helicobacter pylori*. Biochem Biophys Res Commun 285:364–371

118. Xiang SH, Haase AM, Reeves PR (1993) Variation of the *rfb* gene clusters in *Salmonella enterica*. J Bacteriol 175:4877–4884
119. Edstrom RD, Heath EC (1965) Isolation of colitose-containing oligosaccharides from the cell wall lipopolysaccharide of *Escherichia coli*. Biochem Biophys Res Commun 21: 638–643
120. Lindberg B, Lindh F, Lönngren J (1981) Structural studies of the O-specific side-chain of the lipopolysaccharide from *Escherichia coli* O 55. Carbohydr Res 97:105–112
121. Cox AD, Brisson JR, Varma V, Perry M (1996) Structural analysis of the lipopolysaccharide from *Vibrio cholerae* O139. Carbohydr Res 290:43–58
122. Komandrova NA, Gorshkova RP, Zubkov VA, Ovodov IuS (1989) The structure of the O-specific polysaccharide chain of the lipopolysaccharide of *Yersinia pseudotuberculosis* serovar VII. Bioorg Khim 15:104–110
123. Muldoon J, Perepelov AV, Shashkov AS, Gorshkova RP, Nazarenko EL, Zubkov VA, Ivanova EP, Knirel YA, Savage AV (2001) Structure of a colitose-containing O-specific polysaccharide of the marine bacterium *Pseudoalteromonas tetraodonis* IAM 14160^{T}. Carbohydr Res 333:41–46
124. Silipo A, Molinaro A, Nazarenko EL, Gorshkova RP, Ivanova EP, Lanzetta R, Parrilli M (2005) The O-chain structure from the LPS of marine halophilic bacterium *Pseudoalteromonas carrageenovora*-type strain IAM 12662^{T}. Carbohydr Res 340:2693–2697
125. Elbein AD, Heath EC (1965) The biosynthesis of cell wall lipopolysaccharide in *Escherichia coli*. II. Guanosine diphosphate C-6-deoxy-4-keto-D-mannose, an intermediate in the biosynthesis of guanosine diphosphate colitose. J Biol Chem 240:1926–1931
126. Cook PD, Holden HM (2008) GDP-6-deoxy-4-keto-D-mannose 3-dehydratase, accommodating a sugar substrate in the active site. J Biol Chem 283:4295–4303
127. Beyer N, Alam J, Hallis TM, Guo Z, Liu HW (2003) The biosynthesis of GDP-L-colitose: C-3 deoxygenation is catalyzed by a unique coenzyme B_6-dependent enzyme. J Am Chem Soc 125:5584–5585
128. Alam J, Beyer N, Liu HW (2004) Biosynthesis of colitose: expression, purification, and mechanistic characterization of GDP-6-deoxy-4-keto-D-mannose-3-dehydrase (ColD) and GDP-L-colitose synthase (ColC). Biochemistry 43:16450–16460
129. Cook PD, Thoden JB, Holden HM (2006) The structure of GDP-6-deoxy-4-keto-D-mannose-3-dehydratase: a unique coenzyme B_6-dependent enzyme. Protein Sci 15:2093–2106
130. Cook PD, Holden HM (2007) A structural study of GDP-6-deoxy-4-keto-D-mannose-3-dehydratase: caught in the act of geminal diamine formation. Biochemistry 46: 14215–14224
131. Redmond JW (1975) 4-Amino-4,6-dideoxy-D-mannose (D-perosamine): a component of the lipopolysaccharide of *Vibrio cholerae* 569B (Inaba). FEBS Lett 50:147–149
132. Awram P, Smit J (2001) Identification of lipopolysaccharide O antigen synthesis genes required for attachment of the S-layer of *Caulobacter crescentus*. Microbiology 147:1451–1460
133. Perry MB, MacLean L, Griffith DW (1986) Structure of the O-chain polysaccharide of the phenol-phase soluble lipopolysaccharide of *Escherichia coli* O:157:H7. Biochem Cell Biol 64:21–28
134. Samuel G, Hogbin JP, Wang L, Reeves PR (2004) Relationships of the *Escherichia coli* O157, O111, and O55 O-antigen gene clusters with those of *Salmonella enterica* and *Citrobacter freundii*, which express identical O antigens. J Bacteriol 186:6536–6543
135. Bettelheim KA, Evangelidis H, Pearce JL, Sowers E, Strockbine NA (1993) Isolation of a *Citrobacter freundii* strain which carries the *Escherichia coli* O157 antigen. J Clin Microbiol 31:760–761
136. Stroeher UH, Karageorgos LE, Brown MH, Morona R, Manning PA (1995) A putative pathway for perosamine biosynthesis is the first function encoded within the *rfb* region of *Vibrio cholerae* O1. Gene 166:33–42

137. Albermann C, Piepersberg W (2001) Expression and identification of the RfbE protein from *Vibrio cholerae* O1 and its use for the enzymatic synthesis of GDP-D-perosamine. Glycobiology 11:655–661
138. Zhao G, Liu J, Liu X, Chen M, Zhang H, Wang PG (2007) Cloning and characterization of GDP-perosamine synthetase (Per) from *Escherichia coli* O157:H7 and synthesis of GDP-perosamine in vitro. Biochem Biophys Res Commun 363:525–530
139. Cook PD, Holden HM (2008) GDP-perosamine synthase: structural analysis and production of a novel trideoxysugar. Biochemistry 47:2833–2840
140. Cook PD, Kubiak RL, Toomey DP, Holden HM (2009) Two site-directed mutations are required for the conversion of a sugar dehydratase into an aminotransferase. Biochemistry 48:5246–5253
141. Albermann C, Beuttler H (2008) Identification of the GDP-*N*-acetyl-D-perosamine producing enzymes from *Escherichia coli* O157:H7. FEBS Lett 582:479–484
142. Raetz CR (1987) Structure and biosynthesis of lipid A. In: Neidhardt FC, Ingraham JL, Low KB, Magasanik B, Schaechter M, Umbarger HE (eds) *Escherichia coli* and *Salmonella typhimurium*: cellular and molecular biology, vol 1. ASM Press, Washington, DC, pp 498–503
143. Kuhn HM, Meier-Dieter U, Mayer H (1988) ECA, the enterobacterial common antigen. FEMS Microbiol Rev 4:195–222
144. Liu B, Knirel YA, Feng L, Perepelov AV, Senchenkova SN, Wang Q, Reeves PR, Wang L (2008) Structure and genetics of *Shigella* O antigens. FEMS Microbiol Rev 32:627–653
145. Knirel YA, Perepelov AV, Kondakova AN, Senchenkova SN, Sidorczyk Z, Rozalski A, Kaca W (2011) Structure and serology of O-antigens as the basis for classification of *Proteus* strains. Innate Immun 17:70–96
146. Park JT (1987) Murein synthesis. In: Neidhardt FC, Ingraham JL, Low KB, Magasanik B, Schaechter M, Umbarger HE (eds) *Escherichia coli* and *Salmonella typhimurium*: cellular and molecular biology, vol 1. ASM Press, Washington, DC, pp 663–671
147. Holtje JV, Schwarz U (1985) Biosynthesis and growth of the murein sacculus. In: Nanninga N (ed) Molecular cytology of *Escherichia coli*. Academic, London, pp 77–119
148. Dutka-Malen S, Mazodier P, Badet B (1988) Molecular cloning and overexpression of the glucosamine synthetase gene from *Escherichia coli*. Biochimie 70:287–290
149. Mengin-Lecreulx D, van Heijenoort J (1996) Characterization of the essential gene glmM encoding phosphoglucosamine mutase in *Escherichia coli*. J Biol Chem 271:32–39
150. Mengin-Lecreulx D, van Heijenoort J (1993) Identification of the *glmU* gene encoding *N*-acetylglucosamine-1-phosphate uridyltransferase in *Escherichia coli*. J Bacteriol 175: 6150–6157
151. Mengin-Lecreulx D, van Heijenoort J (1994) Copurification of glucosamine-1-phosphate acetyltransferase and *N*-acetylglucosamine-1-phosphate uridyltransferase activities of *Escherichia coli*: characterization of the *glmU* gene product as a bifunctional enzyme catalyzing two subsequent steps in the pathway for UDP-*N*-acetylglucosamine synthesis. J 7Bacteriol 176:5788–5795
152. Sarvas M (1971) Mutant of *Escherichia coli* K-12 defective in D-glucosamine biosynthesis. J Bacteriol 105:467–471
153. Wu HC, Wu TC (1971) Isolation and characterization of a glucosamine-requiring mutant of *Escherichia coli* K-12 defective in glucosamine-6-phosphate synthetase. J Bacteriol 105: 455–466
154. Green DW (2002) The bacterial cell wall as a source of antibacterial targets. Expert Opin Ther Targets 6:1–19
155. Jolly L, Ferrari P, Blanot D, Van Heijenoort J, Fassy F, Mengin-Lecreulx D (1999) Reaction mechanism of phosphoglucosamine mutase from *Escherichia coli*. Eur J Biochem 262: 202–210
156. Jolly L, Pompeo F, van Heijenoort J, Fassy F, Mengin-Lecreulx D (2000) Autophosphorylation of phosphoglucosamine mutase from *Escherichia coli*. J Bacteriol 182: 1280–1285

157. Segel IH (1975) Enzyme kinetics, behavior and analysis of rapid equilibrium and steady state enzyme systems. Wiley, New York
158. De Reuse H, Labigne A, Mengin-Lecreulx D (1997) The *Helicobacter pylori ureC* gene codes for a phosphoglucosamine mutase. J Bacteriol 179:3488–3493
159. Tavares IM, Jolly L, Pompeo F, Leitao JH, Fialho AM, Sa-Correia I, Mengin-Lecreulx D (2000) Identification of the *Pseudomonas aeruginosa glmM* gene, encoding phosphoglucosamine mutase. J Bacteriol 182:4453–4457
160. Shimazu K, Takahashi Y, Uchikawa Y, Shimazu Y, Yajima A, Takashima E, Aoba T, Konishi K (2008) Identification of the *Streptococcus gordonii glmM* gene encoding phosphoglucosamine mutase and its role in bacterial cell morphology, biofilm formation, and sensitivity to antibiotics. FEMS Immunol Med Microbiol 53:166–177
161. Jolly L, Wu S, van Heijenoort J, de Lencastre H, Mengin-Lecreulx D, Tomasz A (1997) The *femR315* gene from *Staphylococcus aureus*, the interruption of which results in reduced methicillin resistance, encodes a phosphoglucosamine mutase. J Bacteriol 179: 5321–5325
162. Verma SK, Jaiswal M, Kumar N, Parikh A, Nandicoori VK, Prakash B (2009) Structure of *N*-acetylglucosamine-1-phosphate uridyltransferase (GlmU) from *Mycobacterium tuberculosis* in a cubic space group. Acta Crystallogr F Struct Biol Cryst Commun 65:435–439
163. Zhang Z, Bulloch EM, Bunker RD, Baker EN, Squire CJ (2009) Structure and function of GlmU from *Mycobacterium tuberculosis*. Acta Crystallogr D Biol Crystallogr 65:275–283
164. Olsen LR, Roderick SL (2001) Structure of the *Escherichia coli* GlmU pyrophosphorylase and acetyltransferase active sites. Biochemistry 40:1913–1921
165. Gehring AM, Lees WJ, Mindiola DJ, Walsh CT, Brown ED (1996) Acetyltransfer precedes uridylyltransfer in the formation of UDP-*N*-acetylglucosamine in separable active sites of the bifunctional GlmU protein of *Escherichia coli*. Biochemistry 35:579–585
166. Olsen LR, Vetting MW, Roderick SL (2007) Structure of the *E. coli* bifunctional GlmU acetyltransferase active site with substrates and products. Protein Sci 16:1230–1235
167. Brown K, Pompeo F, Dixon S, Mengin-Lecreulx D, Cambillau C, Bourne Y (1999) Crystal structure of the bifunctional *N*-acetylglucosamine 1-phosphate uridyltransferase from *Escherichia coli*: a paradigm for the related pyrophosphorylase superfamily. EMBO J 18: 4096–4107
168. Kostrewa D, D'Arcy A, Takacs B, Kamber M (2001) Crystal structures of *Streptococcus pneumoniae* *N*-acetylglucosamine-1-phosphate uridyltransferase, GlmU, in apo form at 2.33 Å resolution and in complex with UDP-*N*-acetylglucosamine and Mg^{2+} at 1.96 Å resolution. J Mol Biol 305:279–289
169. Mochalkin I, Lightle S, Zhu Y, Ohren JF, Spessard C, Chirgadze NY, Banotai C, Melnick M, McDowell L (2007) Characterization of substrate binding and catalysis in the potential antibacterial target *N*-acetylglucosamine-1-phosphate uridyltransferase (GlmU). Protein Sci 16:2657–2666
170. King JD, Kocincova D, Westman EL, Lam JS (2009) Lipopolysaccharide biosynthesis in *Pseudomonas aeruginosa*. Innate Immun 15:261–312
171. Bhat UR, Krishnaiah BS, Carlson RW (1991) Re-examination of the structures of the lipopolysaccharide core oligosaccharides from *Rhizobium leguminosarum* biovar phaseoli. Carbohydr Res 220:219–227
172. Forsberg LS, Carlson RW (1998) The structures of the lipopolysaccharides from *Rhizobium etli* strains CE358 and CE359. The complete structure of the core region of *R. etli* lipopolysaccharides. J Biol Chem 273:2747–2757
173. Creuzenet C, Lam JS (2001) Topological and functional characterization of WbpM, an inner membrane UDP-GlcNAc C6 dehydratase essential for lipopolysaccharide biosynthesis in *Pseudomonas aeruginosa*. Mol Microbiol 41:1295–1310
174. DiGiandomenico A, Matewish MJ, Bisaillon A, Stehle JR, Lam JS, Castric P (2002) Glycosylation of *Pseudomonas aeruginosa* 1244 pilin: glycan substrate specificity. Mol Microbiol 46:519–530

175. Burrows LL, Urbanic RV, Lam JS (2000) Functional conservation of the polysaccharide biosynthetic protein WbpM and its homologues in *Pseudomonas aeruginosa* and other medically significant bacteria. Infect Immun 68:931–936
176. Burrows LL, Charter DF, Lam JS (1996) Molecular characterization of the *Pseudomonas aeruginosa* serotype O5 (PAO1) B-band lipopolysaccharide gene cluster. Mol Microbiol 22:481–495
177. Belanger M, Burrows LL, Lam JS (1999) Functional analysis of genes responsible for the synthesis of the B-band O antigen of *Pseudomonas aeruginosa* serotype O6 lipopolysaccharide. Microbiology 145:3505–3521
178. Schoenhofen IC, McNally DJ, Vinogradov E, Whitfield D, Young NM, Dick S, Wakarchuk WW, Brisson JR, Logan SM (2006) Functional characterization of dehydratase/aminotransferase pairs from *Helicobacter* and *Campylobacter*: enzymes distinguishing the pseudaminic acid and bacillosamine biosynthetic pathways. J Biol Chem 281:723–732
179. Pinta E, Duda KA, Hanuszkiewicz A, Kaczynski Z, Lindner B, Miller WL, Hyytiainen H, Vogel C, Borowski S, Kasperkiewicz K, Lam JS, Radziejewska-Lebrecht J, Skurnik M, Holst O (2009) Identification and role of a 6-deoxy-4-keto-hexosamine in the lipopolysaccharide outer core of *Yersinia enterocolitica* serotype O:3. Chem Eur J 15:9747–9754
180. Forsberg LS, Noel KD, Box J, Carlson RW (2003) Genetic locus and structural characterization of the biochemical defect in the O-antigenic polysaccharide of the symbiotically deficient *Rhizobium etli* mutant, CE166. Replacement of *N*-acetylquinovosamine with its hexosyl-4-ulose precursor. J Biol Chem 278:51347–51359
181. Miller WL, Lam JS (2007) Molecular biology of cell-surface polysaccharides in *Pseudomonas aeruginosa*: from gene to protein function. In: Cornelis P (ed) *Pseudomonas*: genomics and molecular biology. Horizon Scientific Press, Norfolk, pp 87–128
182. Radziejewska-Lebrecht J, Skurnik M, Shashkov AS, Brade L, Rozalski A, Bartodziejska B, Mayer H (1998) Immunochemical studies on R mutants of *Yersinia enterocolitica* O:3. Acta Biochim Pol 45:1011–1019
183. Sadovskaya I, Brisson JR, Khieu NH, Mutharia LM, Altman E (1998) Structural characterization of the lipopolysaccharide O-antigen and capsular polysaccharide of *Vibrio ordalii* serotype O:2. Eur J Biochem 253:319–327
184. MacLean LL, Perry MB, Crump EM, Kay WW (2003) Structural characterization of the lipopolysaccharide O-polysaccharide antigen produced by *Flavobacterium columnare* ATCC 43622. Eur J Biochem 270:3440–3446
185. Kilcoyne M, Shashkov AS, Knirel YA, Gorshkova RP, Nazarenko EL, Ivanova EP, Gorshkova NM, Senchenkova SN, Savage AV (2005) The structure of the O-polysaccharide of the *Pseudoalteromonas rubra* ATCC 29570^{T} lipopolysaccharide containing a keto sugar. Carbohydr Res 340:2369–2375
186. Jansson PE, Lindberg B, Lindquist U (1985) Structural studies of the capsular polysaccharide from *Streptococcus pneumoniae* type 5. Carbohydr Res 140:101–110
187. Marsden BJ, Bundle DR, Perry MB (1994) Serological and structural relationships between *Escherichia coli* O:98 and *Yersinia enterocolitica* O:11,23 and O:11,24 lipopolysaccharide O-antigens. Biochem Cell Biol 72:163–168
188. Feng L, Senchenkova SN, Yang J, Shashkov AS, Tao J, Guo H, Zhao G, Knirel YA, Reeves P, Wang L (2004) Structural and genetic characterization of the *Shigella boydii* type 13 O antigen. J Bacteriol 186:383–392
189. Perepelov AV, Liu B, Senchenkova SN, Shashkov AS, Feng L, Knirel YA, Wang L (2010) Structure of the O-polysaccharide of *Salmonella enterica* O41. Carbohydr Res 345:971–973
190. Kasper DL, Weintraub A, Lindberg AA, Lönngren J (1983) Capsular polysaccharides and lipopolysaccharides from two *Bacteroides fragilis* reference strains: chemical and immunochemical characterization. J Bacteriol 153:991–997
191. Kneidinger B, O'Riordan K, Li J, Brisson JR, Lee JC, Lam JS (2003) Three highly conserved proteins catalyze the conversion of UDP-*N*-acetyl-D-glucosamine to precursors for the biosynthesis of O antigen in *Pseudomonas aeruginosa* O11 and capsule in *Staphylococcus*

aureus type 5. Implications for the UDP-*N*-acetyl-L-fucosamine biosynthetic pathway. J Biol Chem 278:3615–3627

192. Vinogradov EV, Knirel' Iu A, Lipkind GM, Shashkov AS, Kochetkov NK (1987) Antigenic bacterial polysaccharides. 23. The structure of the O-specific polysaccharide chain of *Salmonella arizonae* O59 lipopolysaccharide. Bioorg Khim 13:1275–1281
193. Moreau M, Richards JC, Fournier JM, Byrd RA, Karakawa WW, Vann WF (1990) Structure of the type 5 capsular polysaccharide of *Staphylococcus aureus*. Carbohydr Res 201:285–297
194. Fournier JM, Vann WF, Karakawa WW (1984) Purification and characterization of *Staphylococcus aureus* type 8 capsular polysaccharide. Infect Immun 45:87–93
195. Jones C, Currie F, Forster MJ (1991) N.m.r. and conformational analysis of the capsular polysaccharide from *Streptococcus pneumoniae* type 4. Carbohydr Res 221:95–121
196. Mulrooney EF, Poon KK, McNally DJ, Brisson JR, Lam JS (2005) Biosynthesis of UDP-*N*-acetyl-L-fucosamine, a precursor to the biosynthesis of lipopolysaccharide in *Pseudomonas aeruginosa* serotype O11. J Biol Chem 280:19535–19542
197. Kneidinger B, Larocque S, Brisson JR, Cadotte N, Lam JS (2003) Biosynthesis of 2-acetamido-2,6-dideoxy-L-hexoses in bacteria follows a pattern distinct from those of the pathways of 6-deoxy-L-hexoses. Biochem J 371:989–995
198. McNally DJ, Schoenhofen IC, Mulrooney EF, Whitfield DM, Vinogradov E, Lam JS, Logan SM, Brisson JR (2006) Identification of labile UDP-ketosugars in *Helicobacter pylori*, *Campylobacter jejuni* and *Pseudomonas aeruginosa*: key metabolites used to make glycan virulence factors. Chembiochem 7:1865–1868
199. Ishiyama N, Creuzenet C, Miller WL, Demendi M, Anderson EM, Harauz G, Lam JS, Berghuis AM (2006) Structural studies of FlaA1 from *Helicobacter pylori* reveal the mechanism for inverting 4,6-dehydratase activity. J Biol Chem 281:24489–24495
200. Morrison JP, Schoenhofen IC, Tanner ME (2008) Mechanistic studies on PseB of pseudaminic acid biosynthesis: a UDP-*N*-acetylglucosamine 5-inverting 4,6-dehydratase. Bioorg Chem 36:312–320
201. Bystrova OV, Lindner B, Moll H, Kocharova NA, Knirel YA, Zähringer U, Pier GB (2003) Structure of the lipopolysaccharide of *Pseudomonas aeruginosa* O-12 with a randomly O-acetylated core region. Carbohydr Res 338:1895–1905
202. Feng L, Senchenkova SN, Tao J, Shashkov AS, Liu B, Shevelev SD, Reeves PR, Xu J, Knirel YA, Wang L (2005) Structural and genetic characterization of enterohemorrhagic *Escherichia coli* O145 O antigen and development of an O145 serogroup-specific PCR assay. J Bacteriol 187:758–764
203. Gamian A, Jones C, Lipinski T, Korzeniowska-Kowal A, Ravenscroft N (2000) Structure of the sialic acid-containing O-specific polysaccharide from *Salmonella enterica* serovar Toucra O48 lipopolysaccharide. Eur J Biochem 267:3160–3167
204. King JD, Mulrooney EF, Vinogradov E, Kneidinger B, Mead K, Lam JS (2008) *lfnA* from *Pseudomonas aeruginosa* O12 and *wbuX* from *Escherichia coli* O145 encode membrane-associated proteins and are required for expression of 2,6-dideoxy-2-acetamidino-L-galactose in lipopolysaccharide O antigen. J Bacteriol 190:1671–1679
205. Baumann H, Jansson PE, Kenne L, Widmalm G (1991) Structural studies of the *Escherichia coli* O1A O-polysaccharide, using the computer program CASPER. Carbohydr Res 211:183–190
206. Gupta DS, Shashkov AS, Jann B, Jann K (1992) Structures of the O1B and O1C lipopolysaccharide antigens of *Escherichia coli*. J Bacteriol 174:7963–7970
207. Perry MB, MacLean LL, Brisson JR (1993) The characterization of the O-antigen of *Escherichia coli* O64:K99 lipopolysaccharide. Carbohydr Res 248:277–284
208. Wang Z, Vinogradov E, Larocque S, Harrison BA, Li J, Altman E (2005) Structural and serological characterization of the O-chain polysaccharide of *Aeromonas salmonicida* strains A449, 80204 and 80204-1. Carbohydr Res 340:693–700
209. Keenleyside WJ, Perry M, Maclean L, Poppe C, Whitfield C (1994) A plasmid-encoded $rfb_{O:54}$ gene cluster is required for biosynthesis of the O:54 antigen in *Salmonella enterica* serovar Borreze. Mol Microbiol 11:437–448

210. Mäkela PH, Mayer H (1976) Enterobacterial common antigen. Bacteriol Rev 40:591–632
211. Rick PD, Silver RP (1996) Enterobacterial common antigen and capsular polysaccharides. In: Neidhardt FC, Curtiss R III, Ingraham JL, Lin ECC, Low BK, Magasanik B, Reznikoff WS, Riley M, Schaechter M, Umbarger HE (eds) *Escherichia coli* and *Salmonella*: cellular and molecular biology, 2nd edn. ASM Press, Washington, DC, pp 104–122
212. Karakawa WW, Fournier JM, Vann WF, Arbeit R, Schneerson R, Robbins JB (1985) Method for the serological typing of the capsular polysaccharides of *Staphylococcus aureus*. J Clin Microbiol 22:445–447
213. Morona JK, Morona R, Paton JC (1997) Characterization of the locus encoding the *Streptococcus pneumoniae* type 19 F capsular polysaccharide biosynthetic pathway. Mol Microbiol 23:751–763
214. Lew HC, Nikaido H, Makela PH (1978) Biosynthesis of uridine diphosphate *N*-acetylmannosaminuronic acid in *rff* mutants of *Salmonella tryphimurium*. J Bacteriol 136:227–233
215. Kawamura T, Ichihara N, Ishimoto N, Ito E (1975) Biosynthesis of uridine diphosphate N-acetyl-D-mannosaminuronic acid from uridine diphosphate *N*-acetyl-D-glucosamine in *Escherichia coli*: separation of enzymes responsible for epimerization and dehydrogenation. Biochem Biophys Res Commun 66:1506–1512
216. Kawamura T, Ishimoto N, Ito E (1979) Enzymatic synthesis of uridine diphosphate *N*-acetyl-D-mannosaminuronic acid. J Biol Chem 254:8457–8465
217. Kawamura T, Ishimoto N, Ito E (1982) UDP-*N*-acetyl-D-glucosamine 2′-epimerase from *Escherichia coli*. Meth Enzymol 83:515–519
218. Kawamura T, Kimura M, Yamamori S, Ito E (1978) Enzymatic formation of uridine diphosphate *N*-acetyl-D-mannosamine. J Biol Chem 253:3595–3601
219. Kiser KB, Lee JC (1998) *Staphylococcus aureus cap5O* and *cap5P* genes functionally complement mutations affecting enterobacterial common-antigen biosynthesis in *Escherichia coli*. J Bacteriol 180:403–406
220. Portoles M, Kiser KB, Bhasin N, Chan KH, Lee JC (2001) *Staphylococcus aureus* Cap5O has UDP-ManNAc dehydrogenase activity and is essential for capsule expression. Infect Immun 69:917–923
221. Meier-Dieter U, Starman R, Barr K, Mayer H, Rick PD (1990) Biosynthesis of enterobacterial common antigen in *Escherichia coli*. Biochemical characterization of Tn10 insertion mutants defective in enterobacterial common antigen synthesis. J Biol Chem 265: 13490–13497
222. Kiser KB, Bhasin N, Deng L, Lee JC (1999) *Staphylococcus aureus cap5P* encodes a UDP-*N*-acetylglucosamine 2-epimerase with functional redundancy. J Bacteriol 181:4818–4824
223. Andersson M, Carlin N, Leontein K, Lindquist U, Slettengren K (1989) Structural studies of the O-antigenic polysaccharide of *Escherichia coli* O86, which possesses blood-group B activity. Carbohydr Res 185:211–223
224. Linnerborg M, Weintraub A, Widmalm G (1997) Structural studies of the O-antigen polysaccharide from *Escherichia coli* O138. Eur J Biochem 247:567–571
225. Haseley SR, Holst O, Brade H (1997) Structural and serological characterisation of the O-antigenic polysaccharide of the lipopolysaccharide from *Acinetobacter haemolyticus* strain ATCC 17906. Eur J Biochem 244:761–766
226. Kondakova AN, Kolodziejska K, Zych K, Senchenkova SN, Shashkov AS, Knirel YA, Sidorczyk Z (2003) Structure of the *N*-acetyl-L-rhamnosamine-containing O-polysaccharide of *Proteus vulgaris* TG 155 from a new *Proteus* serogroup, O55. Carbohydr Res 338:1999–2004
227. Wang Z, Larocque S, Vinogradov E, Brisson JR, Dacanay A, Greenwell M, Brown LL, Li J, Altman E (2004) Structural studies of the capsular polysaccharide and lipopolysaccharide O-antigen of *Aeromonas salmonicida* strain 80204-1 produced under in vitro and in vivo growth conditions. Eur J Biochem 271:4507–4516
228. Veremeichenko SN, Zdorovenko GM (2000) The distinctive features of the structure of the *Pseudomonas fluorescens* IMV 247 (biovar II) lipopolysaccharide. Mikrobiologiia 69: 362–369

229. Knirel YA (1990) Polysaccharide antigens of *Pseudomonas aeruginosa*. Crit Rev Microbiol 17:273–304
230. Creuzenet C, Belanger M, Wakarchuk WW, Lam JS (2000) Expression, purification, and biochemical characterization of WbpP, a new UDP-GlcNAc C4 epimerase from *Pseudomonas aeruginosa* serotype O6. J Biol Chem 275:19060–19067
231. Zhao X, Creuzenet C, Belanger M, Egbosimba E, Li J, Lam JS (2000) WbpO, a UDP-*N*-acetyl-D-galactosamine dehydrogenase from *Pseudomonas aeruginosa* serotype O6. J Biol Chem 275:33252–33259
232. Miller WL, Matewish MJ, McNally DJ, Ishiyama N, Anderson EM, Brewer D, Brisson JR, Berghuis AM, Lam JS (2008) Flagellin glycosylation in *Pseudomonas aeruginosa* PAK requires the O-antigen biosynthesis enzyme WbpO. J Biol Chem 283:3507–3518
233. Kowal P, Wang PG (2002) New UDP-GlcNAc C4 epimerase involved in the biosynthesis of 2-acetamino-2-deoxy-L-altruronic acid in the O-antigen repeating units of *Plesiomonas shigelloides* O17. Biochemistry 41:15410–15414
234. Wang L, Huskic S, Cisterne A, Rothemund D, Reeves PR (2002) The O-antigen gene cluster of *Escherichia coli* O55:H7 and identification of a new UDP-GlcNAc C4 epimerase gene. J Bacteriol 184:2620–2625
235. Guo H, Li L, Wang PG (2006) Biochemical characterization of UDP-GlcNAc/Glc 4-epimerase from *Escherichia coli* O86:B7. Biochemistry 45:13760–13768
236. Rush JS, Alaimo C, Robbiani R, Wacker M, Waechter CJ (2010) A novel epimerase that converts GlcNAc-P-P-undecaprenol to GalNAc-P-P-undecaprenol in *Escherichia coli* O157. J Biol Chem 285:1671–1680
237. Caroff M, Brisson JR, Martin A, Karibian D (2000) Structure of the *Bordetella pertussis* 1414 endotoxin. FEBS Lett 477:8–14
238. Wenzel CQ, Daniels C, Keates RA, Brewer D, Lam JS (2005) Evidence that WbpD is an *N*-acetyltransferase belonging to the hexapeptide acyltransferase superfamily and an important protein for O-antigen biosynthesis in *Pseudomonas aeruginosa* PAO1. Mol Microbiol 57:1288–1303
239. Preston A, Thomas R, Maskell DJ (2002) Mutational analysis of the *Bordetella pertussis wlb* LPS biosynthesis locus. Microb Pathog 33:91–95
240. Allen A, Maskell D (1996) The identification, cloning and mutagenesis of a genetic locus required for lipopolysaccharide biosynthesis in *Bordetella pertussis*. Mol Microbiol 19:37–52
241. Miller WL, Wenzel CQ, Daniels C, Larocque S, Brisson JR, Lam JS (2004) Biochemical characterization of WbpA, a UDP-*N*-acetyl-D-glucosamine 6-dehydrogenase involved in O-antigen biosynthesis in *Pseudomonas aeruginosa* PAO1. J Biol Chem 279:37551–37558
242. Westman EL, McNally DJ, Charchoglyan A, Brewer D, Field RA, Lam JS (2009) Characterization of WbpB, WbpE, and WbpD and reconstitution of a pathway for the biosynthesis of UDP-2,3-diacetamido-2,3-dideoxy-D-mannuronic acid in *Pseudomonas aeruginosa*. J Biol Chem 284:11854–11862
243. Larkin A, Imperiali B (2009) Biosynthesis of UDP-GlcNAc(3NAc)A by WbpB, WbpE, and WbpD: enzymes in the Wbp pathway responsible for O-antigen assembly in *Pseudomonas aeruginosa* PAO1. Biochemistry 48:5446–5455
244. Larkin A, Olivier NB, Imperiali B (2010) Structural analysis of WbpE from *Pseudomonas aeruginosa* PAO1: a nucleotide sugar aminotransferase involved in O-antigen assembly. Biochemistry 49:7227–7237
245. Rejzek M, Sri Kannathasan V, Wing C, Preston A, Westman EL, Lam JS, Naismith JH, Maskell DJ, Field RA (2009) Chemical synthesis of UDP-Glc-2,3-diNAcA, a key intermediate in cell surface polysaccharide biosynthesis in the human respiratory pathogens *B. pertussis* and *P. aeruginosa*. Org Biomol Chem 7:1203–1210
246. Westman EL, McNally DJ, Rejzek M, Miller WL, Kannathasan VS, Preston A, Maskell DJ, Field RA, Brisson JR, Lam JS (2007) Identification and biochemical characterization of two novel UDP-2,3-diacetamido-2,3-dideoxy-α-D-glucuronic acid 2-epimerases from respiratory pathogens. Biochem J 405:123–130

247. Westman EL, Preston A, Field RA, Lam JS (2008) Biosynthesis of a rare di-*N*-acetylated sugar in the lipopolysaccharides of both *Pseudomonas aeruginosa* and *Bordetella pertussis* occurs via an identical scheme despite different gene clusters. J Bacteriol 190: 6060–6069
248. Bystrova OV, Lindner B, Moll H, Kocharova NA, Knirel YA, Zähringer U, Pier GB (2003) Structure of the biological repeating unit of the O-antigen of *Pseudomonas aeruginosa* immunotype 4 containing both 2-acetamido-2,6-dideoxy-D-glucose and 2-acetamido-2,6-dideoxy-D-galactose. Carbohydr Res 338:1801–1806
249. Rocchetta HL, Burrows LL, Lam JS (1999) Genetics of O-antigen biosynthesis in *Pseudomonas aeruginosa*. Microbiol Mol Biol Rev 63:523–553
250. Holst O (2002) Chemical structure of the core region of lipopolysaccharides – an update. Trends Glycosci Glyc 14:87–103
251. Raetz CR, Whitfield C (2002) Lipopolysaccharide endotoxins. Annu Rev Biochem 71: 635–700
252. Gronow S, Brade H (2001) Lipopolysaccharide biosynthesis: which steps do bacteria need to survive? J Endotoxin Res 7:3–23
253. Cosgrove DJ (1997) Assembly and enlargement of the primary cell wall in plants. Annu Rev Cell Dev Biol 13:171–201
254. Cipolla L, Polissi A, Airoldi C, Galliani P, Sperandeo P, Nicotra F (2009) The Kdo biosynthetic pathway toward OM biogenesis as target in antibacterial drug design and development. Curr Drug Discov Technol 6:19–33
255. Ghalambor MA, Heath EC (1966) The biosynthesis of cell wall lipopolysaccharide in *Escherichia coli*. IV. Purification and properties of cytidine monophosphate 3-deoxy-D-*manno*-octulosonate synthetase. J Biol Chem 241:3216–3221
256. Raetz CR (1990) Biochemistry of endotoxins. Annu Rev Biochem 59:129–170
257. Sperandeo P, Pozzi C, Deho G, Polissi A (2006) Non-essential KDO biosynthesis and new essential cell envelope biogenesis genes in the *Escherichia coli yrbG-yhbG* locus. Res Microbiol 157:547–558
258. Meredith TC, Woodard RW (2005) Identification of GutQ from *Escherichia coli* as a D-arabinose 5-phosphate isomerase. J Bacteriol 187:6936–6942
259. Hedstrom L, Abeles R (1988) 3-Deoxy-D-*manno*-octulosonate-8-phosphate synthase catalyzes the C-O bond cleavage of phosphoenolpyruvate. Biochem Biophys Res Commun 157:816–820
260. Ray PH, Benedict CD (1980) Purification and characterization of specific 3-deoxy-D-*manno*-octulosonate 8-phosphate phosphatase from *Escherichia coli* B. J Bacteriol 142:60–68
261. Radaev S, Dastidar P, Patel M, Woodard RW, Gatti DL (2000) Preliminary X-ray analysis of a new crystal form of the *Escherichia coli* KDO8P synthase. Acta Crystallogr D Biol Crystallogr 56:516–519
262. Duewel HS, Radaev S, Wang J, Woodard RW, Gatti DL (2001) Substrate and metal complexes of 3-deoxy-D-*manno*-octulosonate-8-phosphate synthase from *Aquifex aeolicus* at 1.9-Å resolution. Implications for the condensation mechanism. J Biol Chem 276: 8393–8402
263. Wu J, Woodard RW (2003) *Escherichia coli* YrbI is 3-deoxy-D-*manno*-octulosonate 8-phos phate phosphatase. J Biol Chem 278:18117–18123
264. Goldman RC, Kohlbrenner WE (1985) Molecular cloning of the structural gene coding for CTP:CMP-3-deoxy-*manno*-octulosonate cytidylyltransferase from *Escherichia coli* K-12. J Bacteriol 163:256–261
265. Eidels L, Osborn MJ (1971) Lipopolysaccharide and aldoheptose biosynthesis in transketolase mutants of *Salmonella typhimurium*. Proc Natl Acad Sci USA 68:1673–1677
266. Eidels L, Osborn MJ (1974) Phosphoheptose isomerase, first enzyme in the biosynthesis of aldoheptose in *Salmonella typhimurium*. J Biol Chem 249:5642–5648
267. Kocsis B, Kontrohr T (1984) Isolation of adenosine 5′-diphosphate-L-*glycero*-D-*manno*-heptose, the assumed substrate of heptose transferase(s), from *Salmonella minnesota* R595 and *Shigella sonnei* Re mutants. J Biol Chem 259:11858–11860

268. Coleman WG Jr (1983) The *rfaD* gene codes for ADP-L-*glycero*-D-*manno*-heptose-6-epimerase. An enzyme required for lipopolysaccharide core biosynthesis. J Biol Chem 258: 1985–1990
269. Brooke JS, Valvano MA (1996) Biosynthesis of inner core lipopolysaccharide in enteric bacteria identification and characterization of a conserved phosphoheptose isomerase. J Biol Chem 271:3608–3614
270. Valvano MA, Marolda CL, Bittner M, Glaskin-Clay M, Simon TL, Klena JD (2000) The *rfaE* gene from *Escherichia coli* encodes a bifunctional protein involved in biosynthesis of the lipopolysaccharide core precursor ADP-L-*glycero*-D-*manno*-heptose. J Bacteriol 182: 488–497
271. Kneidinger B, Marolda C, Graninger M, Zamyatina A, McArthur F, Kosma P, Valvano MA, Messner P (2002) Biosynthesis pathway of ADP-L-*glycero*-β-D-*manno*-heptose in *Escherichia coli*. J Bacteriol 184:363–369
272. Valvano MA, Messner P, Kosma P (2002) Novel pathways for biosynthesis of nucleotide-activated *glycero-manno*-heptose precursors of bacterial glycoproteins and cell surface polysaccharides. Microbiology 148:1979–1989
273. Mendez C, Luzhetskyy A, Bechthold A, Salas JA (2008) Deoxysugars in bioactive natural products: development of novel derivatives by altering the sugar pattern. Curr Top Med Chem 8:710–724
274. Salas JA, Mendez C (2007) Engineering the glycosylation of natural products in actinomycetes. Trends Microbiol 15:219–232
275. Thibodeaux CJ, Melancon CE, Liu HW (2007) Unusual sugar biosynthesis and natural product glycodiversification. Nature 446:1008–1016

Lipopolysaccharide Core Oligosaccharide Biosynthesis and Assembly

8

Uwe Mamat, Mikael Skurnik, and José Antonio Bengoechea

8.1 Introduction

Gram-negative bacteria express on their outer membrane lipopolysaccharide (LPS) that typically comprises of three structural components: lipid A, core oligosaccharide and the O-polysaccharide (OPS). The biosynthesis, on the other hand, takes place at the cytoplasmic face of the inner membrane via two separate pathways for lipid A-core oligosaccharide and OPS that converge physically in the periplasmic face of the inner membrane. There, the undecaprenyl-diphosphate-carried OPS is joined by a carbon-oxygen ligase onto lipid A core and the resulting completed LPS molecule is shuffled onto outer membrane by the recently delineated Lpt translocation pathway.

In this chapter we will review the biosynthesis and genetics of the core oligosaccharide, and discuss selected examples.

U. Mamat
Division of Structural Biochemistry, Research Center Borstel, Leibniz-Center for Medicine and Biosciences, Parkallee 4a/4c, D-23845 Borstel, Germany
e-mail: umamat@fz-borstel.de

M. Skurnik (✉)
Department of Bacteriology and Immunology, Haartman Institute, University of Helsinki, P.O. Box 21, Haartmaninkatu 3, FIN-00014 Helsinki, Finland
e-mail: mikael.skurnik@helsinki.fi

J.A. Bengoechea
Laboratory Microbial Pathogenesis, Consejo Superior Investigaciones Científicas, Fundación de Investigación Sanitaria Illes Balears, Recinto Hospital Joan March, Carretera Sóller Km12; 07110 Bunyola, Spain
e-mail: bengoechea@caubet-cimera.es

Y.A. Knirel and M.A. Valvano (eds.), *Bacterial Lipopolysaccharides*,
DOI 10.1007/978-3-7091-0733-1_8, © Springer-Verlag/Wien 2011

8.2 Overview on LPS Core Types in Different Bacteria

Historically, the core structures of *Salmonella* were studied first, followed by those of *Escherichia coli*. Certain "rules" were extrapolated from those early studies. The core oligosaccharide typically contains 8–15 sugar residues that are hexoses, heptoses: either L-*glycero*-D-*manno*-heptopyranose (L,D-Hep) or D-*glycero*-D-*manno*-heptopyranose (D,D-Hep), and octulosonic acids: either 3-deoxy-D-*manno*-oct-2-ulopyranosonic acid (Kdo) or D-*glycero*-D-*talo*-oct-2-ulopyranosonic acid (Ko) whereas others, like in *Rhizobium*, lack heptose entirely [1] (see also Chap. 2). In bacteria that produce smooth LPS (S-LPS), the core oligosaccharides are conceptually divided into two regions: inner core (lipid A proximal) and outer core.

In the enterobacterial LPS, the binding of the core region to lipid A always occurs via a Kdo residue, and, as in all other LPS structures, the core region is negatively charged (provided by phosphoryl substituents and/or sugar acids like Kdo and uronic acids), which is thought to contribute to the stability of the Gram-negative outer membrane through intermolecular cationic cross links [2].

One Kdo residue is linked by an acid-sensitive glycosidic bond to O-6′ of lipid A glucosamine (GlcN). To the first Kdo residue another Kdo or Ko residue may be linked as well as a chain of two to four heptoses to which two or more hexoses are attached.

More recent structural studies have revealed a relatively variable repertoire of core structures in different Gram-negative bacteria. In addition to the common hexoses and heptoses, more rare sugars and other compounds such as phosphate, ethanolamine, acetyl and amino acid residues have been detected in the core oligosaccharide. The core structure of a single species is not uniform either. For example, among the *E. coli* strains five (Fig. 8.1) and in *Salmonella* two different core types have been recognized [1]. Below we describe briefly this structural diversity using bacteria from different taxonomical groups as an example.

Helicobacter. One of the unique features of *H. pylori* LPS is the abundance of D,D-Hep residues in its core region. D,D-Hep residues form an integral component of the core oligosaccharide as well as the linking region that connects the outer-core oligosaccharide to the O-chain. In some *H. pylori* isolates, the linking region is composed of a long D,D-heptoglycan polymer, while in other strains a single D,D-Hep residue links the O-chain to the core. Another peculiar feature is that the inner-core domain of *H. pylori* LPS contains a single Kdo sugar [3, 4] due to the activity of a Kdo hydrolase [5].

Yokenella. *Y. regensburgei* expresses a unique undecasaccharide lacking any phosphate group, and the only negative charges are provided by carboxyl groups of Kdo and galacturonic acid (GalA) [6].

Neisseria. *N. meningitidis* LPS is based on a diheptose backbone, which is attached via one of two Kdo residues to the lipid A portion. Additions occur to the first heptose (Hep I), and extension past the proximal glucose (Glc) residue is classed as the outer-core structure. The second heptose (Hep II) is invariably substituted by an *N*-acetylglucosamine (GlcNAc) residue. Additions of Glc and phosphoethanolamine (PEtN) to Hep II within the inner-core region also vary among immunotypes [7]. The incorporation of glycine has also been reported.

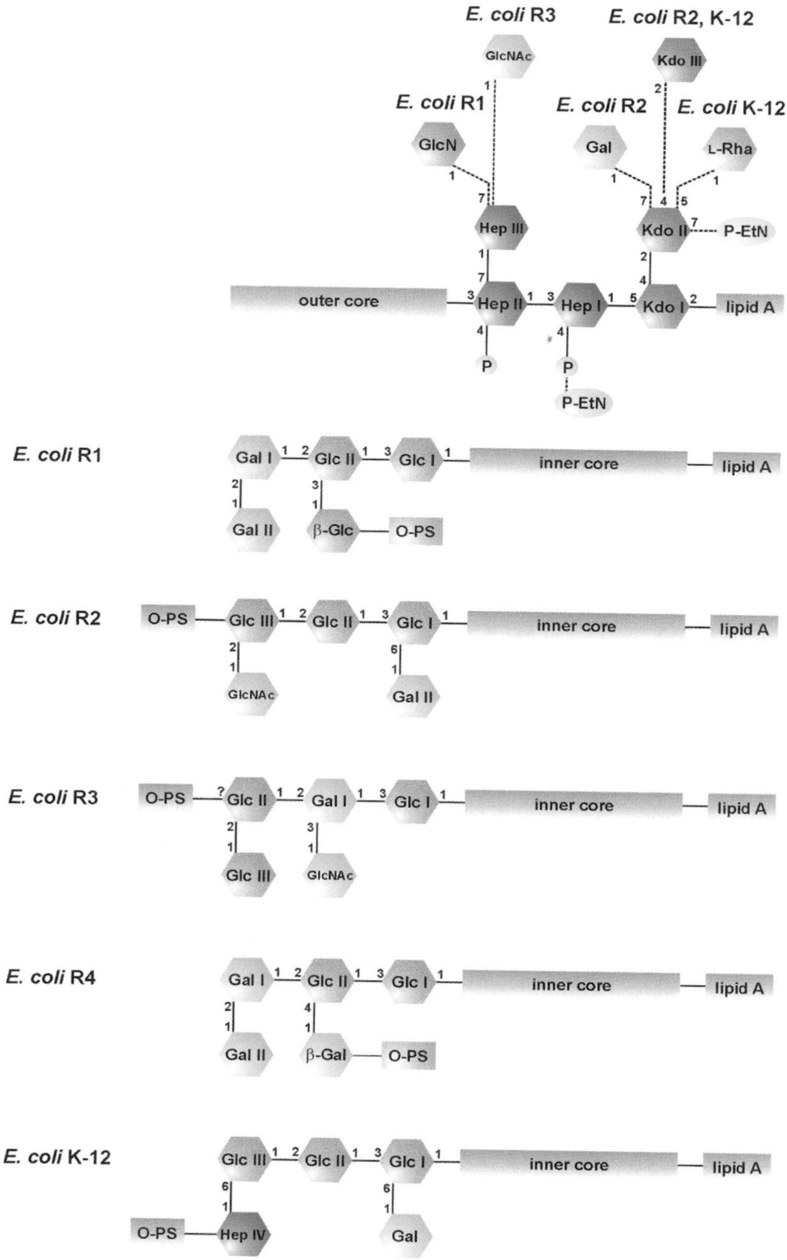

Fig. 8.1 Schematic architecture of the core types R1, R2, R3, R4 and K-12 of *E. coli*. The inner and outer core variations are illustrated at the top and bottom part, respectively

Haemophilus. Extensive structural studies of the LPS from *H. influenzae* have led to the identification of the conserved glucose-substituted triheptosyl inner-core moiety L-α-D-Hep*p*-(1 → 2)-[PEtN-6]-L-α-D-Hep*p*-(1 → 3)-[β-D-Glc*p*-(1 → 4)]-L-α-D-Hep*p* linked to lipid A via Kdo 4-phosphate. This inner-core unit provides the template for attachment of oligosaccharide- and non-carbohydrate substituents. Both inner- and outer-core glycosyl residues can be substituted by non-carbohydrate substituents. This adds considerably to the heterogeneity of LPS since certain substituents can be located at several positions in the same LPS molecule and in non-stoichiometric abundances. Nevertheless, certain sugars appear to carry phosphate-containing substituents stoichiometrically. Thus, in every strain investigated to date, Kdo is substituted at O-4 by 2-aminoethyl pyrophosphate (PPEtN), while Hep II carries PEtN at O-6. By contrast, Hep III is substituted by P or PEtN in only a limited number of strains [8, 9]. Phosphoryl choline substitution is a common feature of *H. influenzae* LPS. Ester-linked glycine is a prominent substituent in the inner-core region [10]. Recent studies would suggest that all strains are capable of expressing minor amounts of this amino acid on their LPS. Hep III is most frequently substituted by ester-linked glycine although it can also be found on Hep II or Kdo. Both outer- and inner-core residues can also be O-acetylated [11].

Pseudomonas. Among Gram-negative bacteria, *P. aeruginosa* has the most phosphorylated core. Phosphate substituents can be mono-, di- or even triphosphates, with most analyzed *P. aeruginosa* LPS having some triphosphate, which, to date, has only been detected in the LPS of this bacterial species [12]. Phosphorylation of LPS has been associated with intrinsic resistance to antibiotics. *P. aeruginosa* core has three phosphorylation sites: positions 2 and 4 of Hep I and position 6 of Hep II. In addition, the position 2 of Hep I is non-stoichiometrically occupied by diphosphoethanolamine. Another modification in the inner-core structure is the stoichiometric presence of an *O*-carbamoyl substituent at position 7 of Hep II. The outer core of the *P. aeruginosa* LPS is usually synthesized by an individual strain as two structurally similar isoforms (glycoforms 1 and 2), which are present in comparable amounts [12–15]. Both glycoforms contain three D-glucose (Glc) residues, one *N*-alanylated 2-amino-2-deoxy-D-galactose (GalN) residue, and one L-rhamnose (L-Rha) residue, the position of which differs in the two glycoforms. Glycoform $2_{(O+)}$, but not glycoform $1_{(O-)}$, can be further substituted by OPS; therefore, glycoform 1 has sometimes been called the "uncapped" core glycoform, whereas glycoform 2 is also known as the "capped" core glycoform [13]. While the basic core structure is conserved among various characterized *P. aeruginosa* strains, variation is seen among the peripheral structural features, including the presence of the terminal Glc IV and *O*-acetyl groups. Extensive *O*-acetylation of the outer-core sugars is relatively common and up to five *O*-acetyl groups have been found [13]. However, the acetates are not present in high amounts at any one position, making a clear structural determination of these substituents difficult. Also, *O*-acetyl groups are fairly labile under mild acid or base conditions and may be lost during LPS purification [12].

Rhizobium. A striking feature of the *Rhizobium* core oligosaccharide is the inner-core modification with three GalA moieties, two on the distal Kdo unit and one on the mannose residue. In addition, *Rhizobium* core regions differ from most LPS by the lack of heptoses and phosphate groups. This is also true for *Agrobacterium* LPS cores, thereby confirming the close taxonomical relationship between these two genera. Both *Agrobacterium* and *Rhizobium* LPS core regions share, in most cases, the same residue linked to the position O-5 of the first Kdo, namely α-D-Man or α-D-Glc [16].

Bordetella. The core oligosaccharides of *B. pertussis* and *B. bronchiseptica* possess an almost identical structure of a branched nonasaccharide with several free amino and carboxyl groups linked to a distal trisaccharide, called band A trisaccharide. *B. parapertussis* core comprises a heptasaccharide that lacks band A trisaccharide and two other monosaccharides [17].

Legionella. Although substitution of the lateral Kdo residues in LPS is known with various sugars and phosphate substituents, including substitution at position 8, occurrence of a Man*p*-(1 → 8)-Kdo disaccharide has been reported thus far only for *Legionella*. Together with an isomeric disaccharide Man*p*-(1 → 5)-Kdo it constitutes the inner, hydrophilic region of the LPS core of *Legionella*. In contrast, the outer region of the core is enriched with 6-deoxy sugars and *N*- and *O*-acetylated sugars [18].

8.3 Core Constituents and Their Biosynthesis

8.3.1 Kdo and Ko

All inner-core (lipid A proximal) structures investigated thus far contain at least one Kdo residue that links the carbohydrate domain via an α-(2 → 6) linkage to the distal GlcN of the lipid A backbone. In some cases, the lipid A-linked Kdo residue can be non-stoichiometrically replaced by Ko [19, 20]. A general feature of the core oligosaccharides is the addition of negatively charged substituents to position 4 of Kdo I (or Ko). While the inner core of the vast majority of Gram-negative bacteria contains an α-(2 → 4)-linked Kdo disaccharide, Ko has been found in place of Kdo II in *Burkholderia cepacia* [21, 22], *B. cenocepacia* [23], *Yersinia pestis* [24], *Serratia marcescens* [25], or members of the genus *Legionella* [26]. Other bacteria such as *H. influenzae* [27, 28], *B. pertussis* [29], *Pasteurella haemolytica* [30], and *Vibrio* spp. [31–33] show a substitution with a phosphate group at position 4 of Kdo I. Unique among core structures is the expression of an α-Kdo-(2 → 8)-α-Kdo-(2 → 4)-α-Kdo trisaccharide in *Chlamydophila pneumoniae* and *Chlamydia trachomatis* [34–36], as well as the synthesis of a branched tetrasaccharide of α-Kdo-(2 → 4)-[α-Kdo-(2 → 8)]-α-Kdo-(2 → 4)-α-Kdo in *C. psittaci* [37] as the only constituents of chlamydial rough-type LPS. The unusual α-(2 → 8)-linked Kdo disaccharide has also been identified outside the family *Chlamydiaceae* in the LPS core of an *Acinetobacter lwoffii* isolate [38].

The conserved pathway for biosynthesis of Kdo has now been well established (Fig. 8.2). Kdo is synthesized and activated for transfer to the lipid A moiety in a four-step enzymatic process that is initiated with the reversible 1,2-aldo/keto isomerization of the pentose pathway intermediate D-ribulose 5-phosphate to D-arabinose 5-phosphate (D-Ara5*P*) by the D-Ara5*P* isomerase (API) [39–42]. In

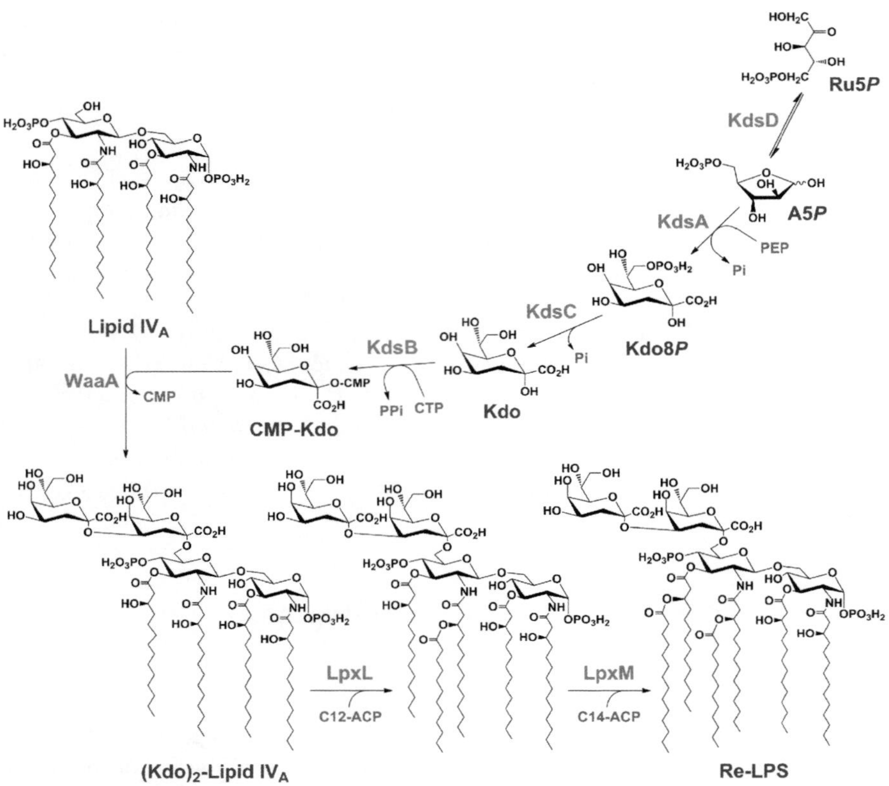

Fig. 8.2 Biosynthesis and incorporation of Kdo into the inner LPS core of *E. coli*. The Kdo pathway is initiated by the enzyme D-arabinose 5-phosphate isomerase (KdsD), which catalyzes the interconversion of D-ribulose 5-phosphate (Ru5*P*) and D-arabinose 5-phosphate (A5*P*). The Kdo 8-phosphate synthase KdsA subsequently condenses A5*P* with phosphoenolpyruvate (PEP) to form Kdo 8-phosphate (Kdo8*P*), followed by hydrolysis of Kdo8*P* to Kdo and inorganic phosphate (Pi) by the Kdo8*P* phosphatase KdsC, activation of Kdo to CMP-Kdo by the CMP-Kdo synthetase KdsB, before finally Kdo is transferred from CMP-Kdo to lipid IV$_A$ by the Kdo transferase WaaA. The Kdo-dependent late acyltransferases LpxL and LpxM subsequently transfer the fatty acids laurate and myristate, respectively, to Kdo$_2$-lipid IV$_A$ to generate the characteristic acyloxyacyl units of hexaacylated Re-LPS

E. coli K-12, API is encoded by two paralogous genes that are capable of complementing each other, e.g., *kdsD* for the first enzyme in the biosynthesis of Kdo and *gutQ* as part of the *gutAEBDMRQ* operon for a phosphoenolpyruvate:sugar phosphotransferase system that metabolizes D-glucitol [39, 43]. KpsF, a third paralogous copy of the API enzyme has been implicated in capsular polysaccharide expression in uropathogenic *E. coli* and *N. meningitidis* strains [42, 44, 45]. Consistent with multifunctional roles of API enzymes, YrbH of *Y. pestis* possesses a biofilm-related function apart from producing D-Ara5*P* for Kdo synthesis [41]. In fact, the API proteins share a similar domain architecture [40, 41, 43], with an N-terminal catalytic isomerase domain commonly found in phosphosugar isomerases [46], followed by a pair of cystathionine β-synthase domains of unknown function [47].

In the second step, D-Ara5*P* is condensed with phosphoenolpyruvate from the pentose pathway via an aldol-like condensation by the Kdo 8-phosphate synthase (KdsA) to form Kdo 8-phosphate (Kdo8*P*), the phosphorylated precursor of Kdo, and inorganic phosphate [48–51]. Kdo8*P* synthases have been separated into two classes of enzymes that differ primarily in their requirement for divalent metal ions [52, 53]. The Kdo8*P* synthases of *E. coli* [50, 54, 55] and *N. meningitidis* [56] belong to the metal ion-independent branch of this enzyme family, whereas the Kdo8*P* synthases from *Aquifex aeolicus* [53, 57–59], *A. pyrophilus* [60], *Helicobacter pylori* [61, 62] and *Acidithiobacillus ferrooxidans* [63] require a divalent metal ion for catalytic activity. Thus, Kdo8*P* synthases are unique enzymes with respect to their host-specific dependence on a metal cofactor. The tetrameric quaternary structures of both metal-dependent and metal-independent Kdo8*P* synthases are very similar, with most of the key residues of their active sites being conserved [50, 56, 57]. The only obvious difference between the structures of the two enzyme classes resides at the metal-binding site. Three of the four amino acid residues associated with metal binding (His, Glu and Asp) in metal-dependent Kdo8*P* synthases are retained in nearly identical positions in the metal-independent enzymes, whereas a Cys residue as the fourth metal ligand is replaced by an Asn residue in the metal-independent forms [52, 63]. In this context it is worth mentioning that single reciprocal substitutions of the Cys and Asn residues could at least partially interconvert metal-dependent and metal-independent catalytic activity [59, 64–66].

The Kdo8*P* phosphatase KdsC catalyzes the third step of the Kdo biosynthetic pathway, the dephosphorylation of Kdo8*P* to yield Kdo and inorganic phosphate [67, 68]. In *E. coli*, the *kdsC* and *kdsD* genes are organized in the *yrbG-yhbG* locus [69], the *yrbK* (now renamed *lptC*), *yhbN* (*lptA*) and *yhbG* (*lptB*) genes of which have now been identified to code for proteins required for the transport of LPS to the outer membrane [70–77] (see also Chap. 10). KdsC is an acid phosphatase that belongs to the large haloacid dehalogenase (HAD) superfamily of hydrolases [68, 78, 79]. The vast majority of catalytic activities of HAD superfamily members are directed at phosphoryl group transfer, with phosphatases and ATPases being the most prevalent [80]. Based on the presence and location of a so-called cap domain that moves as a rigid body by a hinge-like motion over the active site of the core domain, the HAD superfamily can be divided into three subfamilies [81]. Whereas members of the C1 and C2 subfamilies are usually monomeric and contain a cap domain, the Kdo8*P* phosphatase is a tetrameric enzyme of the subfamily C0 that does not possess the structural cap insertion [81, 82]. Structural and biochemical investigations indicated that both substrate specificity and catalytic efficiency of KdsC are the consequences of the tetramerization of the enzyme, the intersubunit contacts of which are mediated by a β-hairpin loop found in a location topologically equivalent to that of the cap domains [78, 79]. Furthermore, it has been suggested that the flexible C-terminal tail of KdsC binds to the active site and contributes to the catalytic efficiency of the enzyme by facilitating Kdo product release [78].

Prior to substitution of the lipid A moiety with Kdo, the sugar requires an activation into its transferable form, CMP-Kdo, which is a short-lived intermediate possessing a half-life of 34 min at 25°C [83]. Utilizing CTP, CMP-Kdo and pyrophosphate are generated by the activity of the CMP-Kdo synthetase (CKS) [84]. Kdo in solution exists predominantly in the α-pyranose form, with a minor fraction of about 2% being β-pyranosidic [85]. The latter Kdo anomer is the preferred CKS substrate for synthesis of CMP-Kdo containing Kdo in the β-configuration [86]. Two functionally distinct CKS paralogs of 44% amino acid sequence identity have been identified in *E. coli* strains with group II K antigens. One of the paralogs, KdsB (LPS-specific or L-CKS), is involved in LPS biosynthesis, whereas the other one, KpsU (capsule-specific or K-CKS), is necessary for expression of the capsular polysaccharide [87, 88]. Although both enzymes catalyze the same reaction, they display different kinetics under different conditions [88]. Due to temperature-regulated expression of capsule genes [89], elevated KpsU activity is only observed at permissive temperatures for capsule expression between 25°C and 37°C but not at capsule-restrictive temperatures below 20°C [87, 90]. Like *kdsA*, the *kdsB* gene of *E. coli* undergoes transcriptional regulation as a function of growth phase but not growth rate, with a rapid decline of mRNA but not protein levels when the bacterial cells enter the stationary growth phase [91]. The three-dimensional structures of KpsU from *E. coli* [92–94] and KdsB from *E. coli* [95] and *H. influenzae* [96] not only revealed a high degree of structural conservation among the dimeric CKS but also strong similarity of the Kdo-activating enzymes with DNA/RNA polymerases in terms of active site configuration and overall chemistry catalyzed, e.g., the formation of a sugar-phosphate linkage with release of pyrophosphate [95, 97, 98]. Based on structural and modelling data, and in analogy to the DNA/RNA polymerases, a two-metal-mechanism with recruitment of two magnesium ions to the active site has been proposed for the CKS-catalyzed activation of Kdo to CMP-Kdo [95].

8.3.2 Heptose

The majority of Gram-negative bacteria, including enterobacteria or strains of the genera *Vibrio*, *Pseudomonas*, *Helicobacter*, *Bordetella*, and *Haemophilus*, contain an L,D-Hep (Hep I) substitution at position 5 of Kdo I [99]. However, this position can be substituted with mannose in *Rhizobium* [100, 101] and *L. pneumophila* [18, 102], glucose in *Moraxella* [103] and *A. haemolyticus* [19], or even an additional Kdo residue in *A. baumannii* [104]. Most inner-core regions possess a second heptose residue (Hep II) linked to position 3 of Hep I, whereas a third heptose (Hep III) may be found in (1 → 7)-linkage to Hep II, forming, for example, in enterobacteria the common inner-core structural element L,D-Hep-(1 → 7)-L,D-Hep-(1 → 3)-L,D-Hep-(1 → 5)-[Kdo-(2 → 4)]-Kdo. The smaller L,D-Hep-(1 → 3)-L,D-Hep-(1 → 5)-Kdo unit has been identified in many Gram-negative bacteria. Frequently, Hep I is substituted at position 4 with either phosphate or PPEtN, while another phosphate residue can be attached to position 4 of Hep II [99]. Finally, position 3 of Hep II may serve as the attachment site for the outer core in different bacteria.

The biosynthesis of ADP-L-*glycero*-β-D-*manno*-heptose, the precursor to the final heptose residues in the inner core of the LPS molecule, involves five steps and four enzymes (Fig. 8.3). The sedoheptulose-7-phosphate isomerase GmhA catalyzes the conversion of D-sedoheptulose 7-phosphate into D-*glycero*-D-*manno*-heptose 7-phosphate, the first committed step in the pathway [105–108]. As gained from structural and functional investigations of the highly conserved tetrameric GmhA enzymes of *E. coli*, *P. aeruginosa*, *Vibrio cholerae* and *Campylobacter jejuni* [108, 109], the overall fold of each monomer is similar to the flavodoxin-type nucleotide-binding motif. It is worth to note that GmhA is likely to adopt an open and closed conformation for binding of the substrate and the product, respectively, consistent with the mechanism of the isomerase to catalyze both forward and reverse reactions [108].

The D-*glycero*-D-*manno*-heptose 7-phosphate exists as a mixture of α- and β-anomers. In the presence of ATP, the N-terminal kinase domain of the bifunctional D-β-D-heptose-phosphate kinase/D-β-D-heptose-1-phosphate adenyltransferase (HldE) selectively phosphorylates position 1 of the β-anomer to produce D-*glycero*-β-D-*manno*-heptose 1,7-bisphosphate in bacteria such as *E. coli* and *P. aeruginosa* [110–112]. HldE consists of two discrete functional domains. The N-terminal domain shows strong similarity to the ribokinase superfamily of kinases that phosphorylate a diverse spectrum of carbohydrate and non-carbohydrate substrates, while the C-terminal domain shares conserved features with the cytidylyltransferase superfamily [111, 112]. However, in some bacteria such as *B. cenocepacia*, the two different functions are accomplished by two separate enzymes, HldA and HldC [113, 114].

In the next step, the phosphate at O-7 is removed by the action of the D-α,β-D-heptose-1,7-bisphosphate phosphatase (GmhB) [114, 115], a member of the histidinol-phosphate phosphatase (HisB) subfamily of the HAD superfamily of phosphohydrolases with high catalytic efficiency and anomeric selectivity toward its physiological substrate [116–118]. The structures of the monomeric and capless GmhB enzymes from *E. coli* and *B. bronchiseptica* indicate that the cap is replaced by three peptide loops, designing the catalytic site in form of a concave, semicircular surface around the substrate leaving group, D-*glycero*-β-D-*manno*-heptose 1-phosphate [116]. The latter subsequently serves as the substrate for the C-terminal adenyltransferase function of HldE (or monofunctional HldC), which catalyzes the transfer of the AMP moiety from ATP to yield ADP-D-*glycero*-β-D-*manno*-heptose [112]. Finally, ADP-L-*glycero*-β-D-*manno*-heptose is generated by a reversible epimerization reaction catalyzed by the enzyme ADP-D-β-D-heptose 6-epimerase (HldD) [113, 119, 120]. Orthologs of ADP-L-*glycero*-β-D-*manno*-heptose pathway genes have been identified in various Gram-negative bacteria such as *H. influenzae* [105, 121, 122], *N. gonorrhoeae* [123, 124], *Actinobacillus pleuropneumoniae* [125], *S. enterica* sv. Typhimurium [126, 127], or *V. cholerae* [128], indicating a conservation of the pathway even in distantly related microorganisms.

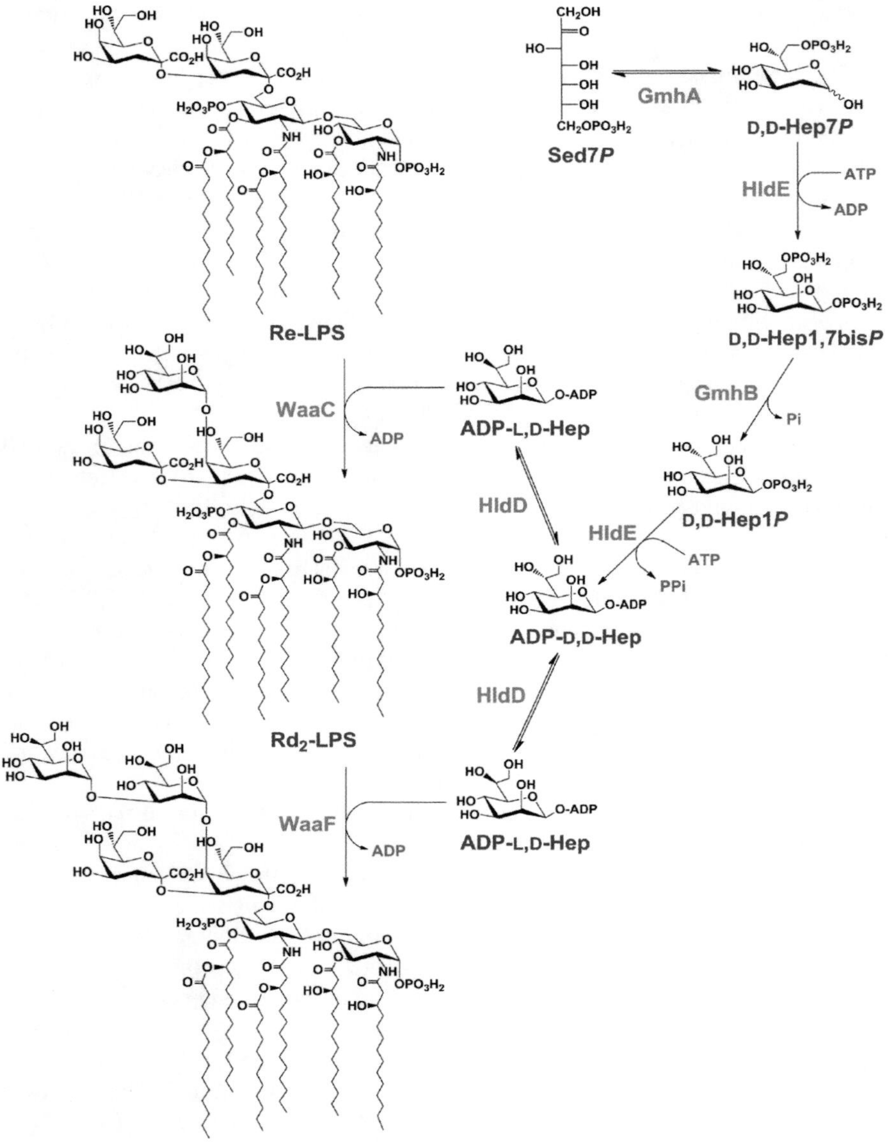

Fig. 8.3 Biosynthesis and incorporation of heptose into the inner LPS core of *E. coli*. The sedoheptulose 7-phosphate isomerase GmhA catalyzes the conversion of D-sedoheptulose 7-phosphate (Sed7*P*) into D-*glycero*-D-*manno*-heptose 7-phosphate (D,D-Hep7*P*). In the presence of ATP, the N-terminal kinase domain of the bifunctional D-β-D-heptosephosphate kinase/D-β-D-heptose-1-phosphate adenyl-transferase (HldE) phosphorylates position 1 of the β-anomer of D,D-Hep7*P* to yield D-*glycero*-β-D-*manno*-heptose 1,7-bisphosphate (D,D-Hep1,7*P*), followed by removal of the phosphate group at O-7 by the D-α-β-D-heptose-1,7-bisphosphate phosphatase

8.3.3 Phosphate Substitutions of Kdo and Heptose

The inner-core backbone structure usually carries several substituents in variable amounts. The substitutions may vary among strains even within a single species, probably depending on the specific genetic background of a given strain and its physiological demands under varying environmental conditions. Together with the carboxyl groups of Kdo, anionic substituents account for the negative charge of the inner core, which is believed to contribute to the integrity and biogenesis of the outer membrane by providing sites for electrostatic interactions with divalent cations, polyamines and positively charged groups on outer membrane proteins [129–133]. These ionic bridges minimize electrostatic repulsion while fostering lateral interactions between neighboring LPS molecules.

As demonstrated for bacteria such as *H. influenzae*, *V. cholerae* and *Pasteurella multocida*, the synthesis of LPS with a single phosphorylated Kdo residue in the inner-core region correlates with the presence of a Kdo kinase (KdkA) that specifically phosphorylates position 4 of Kdo [134–139]. It has been proposed that KdkA is distantly related to eukaryotic protein kinases [140]. In particular, analyses of KdkA residues essential for phosphorylation of *P. multocida* LPS and bioinformatic comparisons of Kdo kinases with the eukaryotic cyclic AMP-dependent protein kinase (cAPK) suggested similarities of the structures, ATP binding, and the catalytic residues of KdkA and eukaryotic protein kinases [137, 140]. Yet another LPS-phosphorylating enzyme, WaaP, is evolutionarily related to eukaryotic protein kinases [140]. In *E. coli* and *Salmonella*, the kinase WaaP transfers a phosphate group to position 4 of Hep I as a prerequisite for the sequential addition of Hep III to the inner-core oligosaccharide by WaaQ and another phosphate residue to position 4 of Hep II by WaaY [141–143].

Among Gram-negative bacteria, the LPS of *P. aeruginosa* is known to contain the most phosphorylated inner core, carrying three phosphate groups at positions 2 and 4 of Hep I and position 6 of Hep II [14, 144]. Like its enterobacterial orthologs, WaaP of *P. aeruginosa* catalyzes the transfer of a phosphate group to position 4 of Hep I [145]. Interestingly, in addition to being a sugar kinase associated with the biosynthesis of the LPS core, WaaP of *P. aeruginosa* is a self-phosphorylating phosphotyrosine kinase with significant similarities to eukaryotic protein-tyrosine kinase and Ser/Thr kinase families [146]. Furthermore, based on the presence of conserved kinase domains and the ability to complement an *S. enterica* sv. Typhimurium *waaP* mutant, the *P. aeruginosa* WapP and WapQ proteins may also be involved in the inner-core heptose phosphorylation [13, 145].

Fig. 8.3 (Continued) (GmhB). The resulting D-*glycero*-β-D-*manno*-heptose 1-phosphate (D,D-Hep1*P*) serves as the substrate for the C-terminal adenyltransferase activity of HldE, which catalyzes the transfer of AMP from ATP to yield ADP-D-*glycero*-β-D-*manno*-heptose (ADP-D,D-Hep). ADP-L-*glycero*-β-D-*manno*-heptose (ADP-L,D-Hep), the donor substrate of heptosyltransferase I (WaaC) and heptosyltransferase II (WaaF) is generated by a reversible epimerization reaction catalyzed by the enzyme ADP-D-β-D-heptose 6-epimerase (HldD)

As indicated before, PEtN residues attached to the inner-core structure and/or phosphate groups to form PPEtN are frequently found as substituents of LPS. The *lpt3* gene of *N. meningitidis* and *N. gonorrhoeae* was reported to code for a PEtN transferase that is required for modification of position 3 of Hep II with a PEtN residue [147–149]. Consistent with a PEtN substitution at position 6 of Hep II in strains of *N. meningitidis* and *H. influenzae*, another PEtN transferase gene, *lpt6*, is always present in those bacteria [150]. Moreover, the high degree of similarity of PEtN transferases suggests that they apparently form a separate family of transferases for the decoration of LPS molecules with PEtN moieties in a wide range of Gram-negative bacteria, including *E. coli*, *P. aeruginosa*, *Y. pestis*, *V. cholerae*, and others [147, 150]. The Lpt3 protein of *N. meningitidis* [148] has six significant orthologs in *E. coli*, the PEtN transferase EptB of which was shown to add specifically a PEtN group to Kdo II of a heptose-deficient mutant when grown in medium supplemented with $CaCl_2$ [151, 152]. In the presence of Ca^{2+}, the enzyme utilizes phosphatidylethanolamine to generate the product PEtN-Kdo_2-lipid A with the release of diacylglycerol as a by-product [152].

Some bacteria such as *Klebsiella pneumoniae*, *Rhizobium leguminosarum*, *R. etli*, and *Plesiomonas shigelloides* O:54 entirely lack phosphoryl groups in the inner-core region. Instead, several GalA units provide the negative charge required for outer membrane stability, thus playing an equivalent role to the phosphate substitutions [153–157]. Studies with *K. pneumoniae* mutants defective in the UDP-GalA C-4 epimerase of UDP-GalA precursor synthesis indicated that the GalA residues are indeed essential for maintenance of outer membrane integrity and permeability barrier functions [158, 159]. The GalA transferases WabG and WabO involved in transfer of an α-GalA residue to position 3 of Hep II and a terminal α-GalA moiety to position 7 of Hep III of the *K. pneumoniae* inner core, respectively, have been identified [160, 161]. Unlike WabG and WabO of *K. pneumoniae*, the membrane-associated dodecaprenyl-phosphate-β-D-GalA substrate but not a sugar nucleotide is utilized as the direct GalA donor for substitution of the inner core in *R. leguminosarum* [162]. The enzymes RgtA and RgtB catalyze the addition of two GalA units to Kdo II, whereas RgtC transfers another GalA residue to the inner-core mannose unit [163].

8.4 Glycosyltransferases Involved in the Core Assembly

8.4.1 Kdo Transferases

The WaaA-catalyzed transfer of Kdo from CMP-Kdo to an underacylated lipid A precursor is the first step in the biosynthesis of the inner core [164]. Following complete acylation of the substrate by usually Kdo-dependent late acyl transferases, the assembly of the remaining core oligosaccharide occurs on the Kdo-glycosylated lipid A acceptor at the inner face of the cytoplasmic membrane [1, 165]. The Kdo transfer reaction is characterized by an inversion of the stereochemistry at the anomeric reaction center of the donor substrate. While WaaA enzymes appear to develop high specificity toward the donor substrate, CMP-β-Kdo, they apparently can tolerate lipid A acceptor molecules with varying extent of acylation and different disaccharide backbones [164, 166–168]. However, the ability of WaaA to catalyze the transfer of Kdo strictly depends on the presence of a negatively charged phosphate group at position 4′ of the lipid A intermediate [164]. It has long been recognized that Kdo transferases, constituting the glycosyltransferase (GT) 30 family of the Carbohydrate-Active enZymes Database (CAZy; http://www.cazy.org/), are unusual GTs. Depending on the Gram-negative host and normally consistent with the number of Kdo residues present in the inner core of its LPS, Kdo transferases can be either mono-, bi-, tri-, or even tetra-functional. WaaA is usually bifunctional in bacteria with LPS that contains an α-(2 → 4)-linked Kdo disaccharide in the inner-core region such as *E. coli* [164], *K. pneumoniae* [169], *Legionella pneumophila* [167], *A. baumannii*, and *A. haemolyticus* [166]. Thus, the bifunctional enzymes are capable of catalyzing the formation of two different glycosidic bonds, resulting in an α-(2 → 6)-linkage between Kdo I and the carbohydrate backbone of the lipid A precursor, and an α-(2 → 4)-linkage between Kdo II and Kdo I. Consistent with the presence of a Kdo trisaccharide in *C. pneumoniae* and *C. trachomatis*, the Kdo transferases display at least trifunctional activity [170–174], whereas heterologous expression of the *waaA* gene from *C. psittaci* can direct the synthesis of the complete Kdo tetrasaccharide structure in *E. coli* [171]. Finally, the Kdo transferases of *H. influenzae*, *B. pertussis*, or *A. aeolicus* are monofunctional, which agrees with the presence of a single Kdo residue in the inner core of their LPS [134, 168, 175]. In rare cases, however, the number of Kdo sugars in the inner LPS core does not correlate with the functionality of WaaA. The LPS of *H. pylori* and *Francisella novicida* have been characterized as having single Kdo residues attached to lipid A while their Kdo transferases were demonstrated to act as bifunctional enzymes [4, 5, 176–178]. In both organisms, Kdo II is removed from the LPS core by an unusual heterodimeric Kdo hydrolase, the activity of which is absolutely dependent on the presence of two proteins that have to work in concert to perform the Kdo-trimming function [5, 178, 179]. In either case, the identification of the determinants conferring different functionality to the Kdo transferases must await further investigations. The construction of chimeras by swapping domains of the monofunctional Kdo transferase from *H. influenzae* and the bifunctional enzyme from *E. coli* provided recently the first experimental evidence that amino acid residues of the N-terminal part of WaaA play a critical role in determining whether one or two Kdo residues are transferred to the lipid A acceptor substrate [180].

8.4.2 Heptosyltransferases

In many Gram-negative bacteria, the assembly of the inner LPS core is completed by the addition of Hep I and Hep II. Accordingly, orthologs of the heptosyltransferase I, WaaC (GT9), and heptosyltransferase II, WaaF (GT9), have been identified in a variety of microorganisms, including *E. coli* and *S. enterica* [181–185], *K. pneumoniae* [169], *P. aeruginosa* [186], *N. gonorrhoeae* [187–189],

N. meningitidis [190, 191], *B. pertussis* [192], *Aeromonas hydrophila* [193], *A. salmonicida* [194], *B. cenocepacia* [23], *B. cepacia* [21], *Campylobacter* spp. [195, 196], *S. marcescens* [197], or *H. influenzae* [122, 198]. Both WaaC and WaaF preferentially utilize ADP-L-*glycero*-D-*manno*-heptose but not ADP-D-*glycero*-D-*manno*-heptose as their substrate, thus determining by their substrate specificity the conformation of L,D-Hep residues in the inner core [182, 199, 200]. Furthermore, as ADP-L-*glycero*-D-*manno*-heptose must be β-configured to serve as a substrate for both WaaC and WaaF, the Hep transfer reactions are characterized by inverting mechanisms to yield α-glycosidic bonds in the LPS [200].

The X-ray structures of WaaC [201] and WaaF (PDB code 1PSW) from *E. coli* have been determined. Despite a rather low identity at the primary amino acid sequence level, the structures of WaaC and WaaF are remarkably similar, displaying the classical fold of the glycosyltransferase superfamily GT-B, with an N-terminal acceptor-substrate binding domain and a C-terminal donor-substrate binding domain both adopting a Rossmann-like fold [201]. Of note, residue Lys192 of WaaC, which may be involved in discrimination between the ADP-L-*glycero*-D-*manno*-heptose and ADP-D-*glycero*-D-*manno*-heptose substrates, is strictly conserved among ADP-L-*glycero*-β-D-*manno*-heptosyltransferases but not, for example, in HP0479, an ADP-D-*glycero*-D-*manno*-heptosyltransferase of outer-core biosynthesis in *H. pylori* [202]. Together with the highly conserved motif D (S/T)(G/A)XXH, Lys192 could represent a characteristic signature of heptosyltransferases such as WaaC, WaaF, WaaQ, and OpsX [201]. OpsX from *H. influenzae* has been identified as a novel type of heptosyltransferase I with altered acceptor substrate specificity. The enzyme was shown to add Hep I to an acceptor molecule with a Kdo-phosphate but not to an acceptor substituted with a Kdo disaccharide [203], indicating that bacteria containing a single phosphorylated Kdo residue in the inner core may require a different heptosyltransferase I for LPS maturation. This assumption is further supported by the existence of OpsX orthologs in bacteria such as *Xanthomonas campestris* and *P. multocida*, which possess an inner-core glycoform with a single phosphorylated Kdo unit [136, 204].

The mannosyltransferase LpcC acts as a functional analog of WaaC in the assembly of the heptoseless inner-core region of *R. leguminosarum* [205–207]. The enzyme is highly selective for GDP-Man as the donor substrate in mannosylation of position 5 of Kdo I of the LPS precursor Kdo_2-lipid IV_A, with two Kdo moieties attached to the lipid acceptor substrate being crucial to LpcC activity [207]. In *Moraxella catarrhalis*, the transfer of a Glc residue to the Kdo_2-lipid A acceptor substrate is catalyzed by Lgt6, an α-1,5-glucosyltransferase replacing WaaC in the biosynthesis of the unusual heptoseless, glucose-containing inner-core structure [208].

8.4.3 Hexosyltransferases

GTs that transfer hexose units from their nucleotide-activated precursors to glycosylated lipid A acceptor molecules are central to the biosynthesis of many core oligosaccharide structures. However, detailed biochemical investigations aimed to elucidate the ill-defined mechanisms of glycosyl transfer have rarely been performed most likely due to problems associated with the expression and purification of the peripheral membrane proteins, as well as the lack of appropriate transferase assays. In the majority of cases, the functions of the GTs involved in the assembly of the outer core have been assigned by structural analyses of the LPS from defined mutants, heterologous gene complementation experiments and/or similarity searches of their deduced amino acid sequences [1]. Examples include the identification and characterization of hexose transferases in many distantly related organisms such as *B. pertussis* [209], *B. cenocepacia* [23], *C. jejuni* [210], *Haemophilus ducreyi* [211], *P. multocida* [212], or *Y. enterocolitica* [213].

The α-1,3-glucosyltransferase WaaG of *S. enterica* sv. Typhimurium and *E. coli* K-12 is currently one of the very few GTs of outer-core biosynthesis for which both biochemical and structural data are available [214–218]. WaaG belongs to the GT4 family and catalyzes the transfer of a Glc residue to position 3 of Hep II of the inner core. Since all five core types of *E. coli* (K-12 and R1 through R4) (Fig. 8.1) and the two different core oligosaccharides of *S. enterica*, represented by serovars Typhimurium (subspecies I) and Arizonae (subspecies IIIa), contain a Glc residue as the first sugar in the outer core, it is not surprising that the WaaG orthologs from each of the core types are highly conserved [184, 219–221]. The enzyme folds into two Rossmann-like (β/α/β) domains characteristic of GTs of the GT-B superfamily, and both the interactions with the nucleotide-sugar donor, UDP-Glc, and the proposed catalytic mechanism is similar to other retaining glucosyltransferases of the GT-B fold [216].

The α-1,3-galactosyltransferase WaaI of *S. enterica* and the α-1,3-glucosyltransferase WaaO of *E. coli* K-12 of the GT8 family are thought to add the second hexose to the outer-core backbone of the sequence α-Hex III-(1 → 2)-α-Hex II-(1 → 3)-α-Glc-(1→, which is Gal I in *Salmonella* and Glc II in *E. coli* strains of the K-12 core type [219, 220, 222–225]. WaaI and WaaO share significant similarity, and homologs of the proteins have been identified in *E. coli* R3, where Hex II is Gal I, as well as in R1, R2 and R4 strains, which contain a second Glc residue in Hex II position [226, 227].

The attachment of Hex III moieties to the backbone structure, Glc II in *S. enterica* and Glc III in *E. coli* K-12, requires the activity of another set of GTs, e.g., the α-1,2-glucosyltransferases WaaJ and WaaR, respectively [222, 224]. Again, homologs of WaaJ and WaaR are present in all *E. coli* R1-R4 strains, containing either Gal or Glc residues in Hex III position. To emphasize that the GTs utilize different UDP-sugar donor substrates and have different terminal sugars on their acceptor molecules, the enzymes were designated WaaR in R2, WaaJ in R3, and WaaT in R1 and R4 strains [220]. In fact, WaaR is a UDP-glucose:(glucosyl) LPS α-1,2-glucosyltransferase, whereas WaaJ and WaaT display UDP-glucose:(galactosyl) LPS α-1,2-glucosyltransferase and UDP-galactose:(glucosyl) LPS α-1,2-glucosyltransferase activities, respectively. When investigated *in vitro*, however, the enzymes displayed more flexibility for both donor and acceptor than was observed in complementation experiments *in vivo* [228].

According to the CAZy classification system, WaaR, WaaT, and WaaJ belong to the family GT8 of GTs, which are characterized by the typical fold of the glycosyltransferase GT-A superfamily, a retaining catalytic mechanism, and the presence of DXD sequence motifs [228, 229]. The latter are common to a wide range of GTs and have been studied in LgtC, a retaining α-1,4-galactosyltransferase of the GT8 family that catalyzes a key step in the biosynthesis of the LPS structure in *N. meningitidis* by transferring a Gal residue from UDP-Gal to a terminal lactose of the acceptor [230]. According to the proposed ordered bi-bi kinetic mechanism of LgtC, the donor sugar nucleotide binds to the protein first, followed by the acceptor substrate [230]. One DXD motif of the active site of the enzyme is predicted to bind the divalent metal ion, while the first aspartate of the second DXD motif may interact with the galactose ring of UDP-Gal. Except for some domains of high conservation such as the DXD sequence motifs, the overall similarity of the primary amino acid sequences of LgtC, WaaR, WaaT, and WaaJ is rather limited [228]. However, a common feature of LgtC, WaaJ and other GT8 members is the presence of a C-terminal domain with a high ratio of positively charged and hydrophobic amino acid residues, which are believed to contribute to interactions with the negatively charged, hydrophobic membrane lipids. As demonstrated for WaaJ, the C-terminal α-helix plays a critical role in catalytic activity and membrane association of the GT [229].

The elongation of the core backbone structures in *E. coli* K-12 and *S. enterica* was demonstrated to go along with the phenotypic appearance of multiple forms of the LPS core, consistent with the basic core structure being modified with additional residues as the synthesis of the main hexose chain progresses [184, 224]. WaaB of the GT4 family is the α-1,6-galactosyltransferase that adds a lateral Gal residue to Glc I of the core backbone. Of note, mutant strains defective in WaaB did not only lack the α-1,6-Gal substitution, but were also unable to transfer Hex II to the growing main chain, indicating that WaaO and WaaI presumably require the lateral Gal residue for an efficient recognition of the acceptor molecule so that the main hexose chain can be extended [224]. Among *E. coli* strains with R1-R4 core types, only R2 strains contain an α-1,6-Gal substitution at Hex I and hence a WaaB homolog [227].

While the outer cores of *E. coli* K-12 and *S. enterica* do not carry a side-branch residue at Hex II position, *E. coli* R1, R3, and R4 core types are substituted with β-1,3-Glc, α-1,3-GlcNAc, and β-1,4-Gal. The required activities for the glycosyl transfer may be provided by WaaV (GT2), WaaD (GT4), and WaaX (GT25), respectively [226]. Thus, the β-GTs WaaV and WaaX are the only inverting enzymes in the assembly of the *E. coli* and *S. enterica* core oligosaccharides, transferring different β-linked sugars to the same site of the otherwise identical R1 and R4 cores.

The substitution of the Hex III position with an α-1,2-linked GlcNAc residue in the *S. enterica* sv. Typhimurium Ra and *E. coli* R2 cores is catalyzed by WaaK (GT4) homologs [227, 231]. However, WaaK is replaced by the α-1,2-glucosyltransferase WaaH (GT52) in *S. enterica* sv. Arizonae, resulting in a core structure with a branched α-1,2-Glc but not an α-1,2-linked GlcNAc residue attached to Hex III [221]. In *E. coli* R1 and R4 strains, Gal II is transferred to the terminal Gal I residues by the α-1,2-galactosyltransferase WaaW (GT8), whereas it is not entirely clear yet which gene product catalyzes the attachment of the α-1,2-linked Glc III residue to the terminal Glc II in *E. coli* R3 [219, 226, 227].

Analyses of the LPS structures arising from defined mutants and/or individual enzymes have also led to the identification and characterization of hexose transferases involved in core biosynthesis in important human mucosal pathogens such as *N. gonorrhoeae* [232–234], *N. meningitidis* [235, 236], *H. influenzae* [11, 237–239], or *Moraxella catarrhalis* [240]. One of the strategies evolved to adjust LPS biosynthesis to microenvironmental changes in the host is that the outer core of LPS undergoes structural variation, being subject to both antigenic variation by changing the carbohydrate composition and phase variation by reversible on/off-switching of distinct outer-core constituents [241, 242]. This indicates that each bacterial strain can synthesize a set of LPS molecules simultaneously and regardless of the LPS forms being produced, the variable oligosaccharides exhibit the extraordinary feature to mimic functions of host molecules, which enables the bacteria to evade innate and adaptive immune responses of the host [243, 244].

In many human pathogens, the addition of Glc and/or Gal moieties to Hep residues of the inner core is a key factor in contributing to heterogeneity of the LPS. The addition of the sugars is a prerequisite for any further hexose extension from the inner core, thus having the potential to affect both the biology and the virulence of the bacteria [237]. A common feature of those LPS cores is the presence of a Glc-β-1,4-Hep linkage. LgtF (GT2), an inverting β-1,4-glucosyltransferase of the GT-A superfamily, catalyses the transfer of a Glc unit to the Hep I moiety in various organisms, including *N. meningitidis* [235], *H. influenzae* [11, 237], *H. ducreyi* [245], *B. pertussis* [209], and *C. jejuni* [210]. Interestingly, the *C. jejuni* enzyme has been suggested to act as a two-domain glucosyltransferase that is capable of transferring a β-1,4-Glc to Hep I and a β-1,2-Glc to Hep II of the inner core [210, 246]. Yet another glucosyltransferase, Lgt3 (GT2), with a bi-domain architecture and similarities to β-1,4-glucosyltransferases such as LgtF of *H. ducreyi* or WaaE (GT2) of *K. pneumoniae* was shown to extend the heptoseless inner-core Glc-Kdo_2-lipid A structure in *M. catarrhalis* [240, 245, 247, 248]. The LpsA (GT25) enzyme from *H. influenzae*, using an inverting reaction mechanism of glycosyl transfer, is rather unusual amongst bacterial GTs. It has been demonstrated that LpsA of the GT-A superfamily is capable of adding a hexose residue to Hep III of the inner LPS core, which can be either Glc or Gal linked by a β-1,2 or β-1,3 linkage [237, 249]. Intriguingly, a single key amino acid at position 151 of the protein apparently determines which hexose is added to Hep III, while the 3′ end of the gene appears to direct the anomeric linkage of the added hexose [249].

8.5 Genetic Basis of Core Biosynthesis

In *E. coli* and *Salmonella*, the core biosynthesis gene cluster consists of three operons located in the *waa* region (formerly *rfa*) of the chromosome, and mapping between *cysE* and *pyrE* at 81–82 min on the *E. coli* K-12 and *Salmonella* linkage maps. The *gmhD* (*hldD*), *waaQ* and *waaA* operons are defined by the first gene of the transcriptional unit. The *gmhD* (*hldD*)-*waaFC* genes are required for biosynthesis and transfer of L,D-heptose. The long central *waaQ* operon contains genes necessary for the biosynthesis of the outer core and for modification and/or decoration of core. In *E. coli* isolates with the R1 and R4 cores, this operon also contains the "ligase" structural gene (*waaL*), whose product is required to link the OPS to the completed core. The *waaQ* operon is preceded by a JUMPStart (*J*ust *U*pstream of *M*any *P*olysaccharide-associated gene *Start*s) sequence that includes the conserved 8-bp region known as *ops* (*o*peron *p*olarity *s*uppressor) that, together with RfaH (a NusG homolog), is required for operon polarity suppression (see below). The *waaA* transcript contains the structural gene (*waaA*, formerly *kdtA*) for the bifunctional Kdo transferase and a "non-LPS" gene encoding phosphopantetheine adenylyltransferase (*coaD*, formerly *kdtB*). Although some *waa*-encoded functions have been defined biochemically in *E. coli* K-12, assignment of others relies heavily on the construction of non-polar mutants in single genes followed by chemical analysis of LPS structure.

Whole-genome sequencing has led to the identification of homologs of core biosynthetic genes in a variety of enteric and non-enteric bacteria. Since many gene assignments currently rely heavily on *E. coli* and *Salmonella* prototypes, a consideration of the limitations of the functional assignments in these prototypes is critical. Examination of annotated genomes from more distantly related organisms indicates that some have clusters of subsets of core genes. Nevertheless it should be noted that in most cases gene functions are unknown. An approach to facilitate the characterization of these genes could be to test whether they complement known core deficiencies of *E. coli*, *Salmonella* or *Klebsiella*. In mucosal pathogens including *H. influenzae* and *N. meningitidis*, there is no significant clustering of inner-core biosynthesis genes, but many of the genes encoding GTs are usually found in the same locus.

8.6 Regulation of Core Expression and Role in Virulence

There is paucity of data on the regulation of core expression. This might be due to the fact that in most bacterial species loci involved in core synthesis do not form a single transcriptional unit being instead scattered through the genome. Nevertheless, early studies identified RfaH as an important component in the synthesis of LPS [250] and while originally thought to be an enzyme, RfaH is a positive regulator of core gene expression [251]. However, RfaH also regulates the expression of operons encoding structures targeted to the cell surface including OPS of LPS, F pilus, capsules and hemolysin [252]. Mutations in *rfaH* decrease promoter distal but not promoter proximal gene expression, a so-called elongation regulation. *In silico* analysis of the promoter regions of RfaH-regulated operons revealed a single common 8-bp motif termed the *ops* element and its deletion gives the same transcriptional phenotype as an *rfaH* mutant. The *ops* element functions even when placed within a non-native promoter, such as the p*tac* promoter [253]. The *ops* element is the second half of a direct repeat of 5′-GGTAGC-N_{15}-GGTAGC-3′

present in the 5′ region of operons that direct the synthesis among others of polysaccharides, including capsules, cores and OPS, and it was termed JUMPstart sequence [254].

Several studies support the notion that both RfaH and *ops* cooperate to control elongation of the initiated transcript, however the mechanism of this interaction remains to be fully characterized. It has been proposed that the *ops* element recruits RfaH and perhaps other factors to the transcription complex resulting in a modification of the RNA polymerase into a termination-resistant form [252]. Another possibility could be that the JUMPstart sequence is a place for the polymerase to pause due to stem-loop structures. However, in the presence of RfaH, *ops* element brings together RfaH, and perhaps also other regulatory proteins, with the RNA polymerase, and the binding of all these elements may prevent the formation of the stem-loop structures [255]. Shedding new light into the question, Artsimovitch et al. [256] have presented evidence indicating that RfaH recognized and binds both RNA polymerase and the *ops* element in the paused elongation complex hence acting as anti-terminator by reducing pausing and termination at factor-independent and Rho-dependent signals. Furthermore, there are no other cellular components involved thereby implying that RfaH indeed is the direct effector controlling elongation. At the structural level, it has been recently demonstrated that the N-terminal RfaH domain recognizes the *ops* element, binds to the RNA polymerase and reduces termination *in vitro* [257, 258].

There is virtually no data indicating that core expression could be affected by environmental signals. Available evidence supports the notion that temperature may affect core gene expression. Thus, heat shock response up-regulates the expression of some *E. coli* core genes [259] whereas the expression of *Y. enterocolitica* O:3 outer core is downregulated at 37°C (M. Skurnik, unpublished data).

Whereas the role of OPS in virulence was established quite soon in *Enterobacteriaceae*, it took some time to rigorously demonstrate the relative contribution of LPS core to virulence. This is due to the fact that in most of the Gram-negative species, the OPS is linked to the core. Therefore core mutants are also devoid of OPS, hence making impossible to dissect the relative contribution of each LPS section to virulence. However, the structure of *Y. enterocolitica* serotype O:3 LPS has some peculiarities rarely seen in other enterobacteria which provided the means to test the role of LPS core in virulence. The serotype O:3 OPS is a homopolymer of α-(1 → 2) linked 6-deoxy-L-altrose [260] that is attached to the inner-core region of the LPS. Of note, the outer-core hexasaccharide is also attached to the inner core thus forming a short branch in the LPS molecule. This peculiar structure has made it possible to construct mutants that are missing either the OPS, the outer core or both [261, 262], hence allowing experiments to determine the relative contribution of outer core to *Y. enterocolitica* serotype O:3 virulence. The LD_{50} of the outer-core mutant was approximately 1,000 times higher than that of the wild type in orogastrically infected mice [262]. However, in contrast to the mutant lacking OPS, co-infection experiments revealed that the outer-core mutant did colonize the Peyer's patches as efficiently as the wild type but it was much less efficient in colonizing deeper organs and at 5 days post-infection it was completely eliminated from Peyer's patches. Thus, it is clear that *Y. enterocolitica* outer core and OPS play different roles during infection. It seems that OPS is needed during the first hours of infection, whereas the outer core is required for prolonged survival of the bacteria in Peyer's patches and for invasion of deeper tissues like liver and spleen.

At present we can only speculate on the specific role of outer core in virulence. Among other possibilities, it could be that the outer core may prevent the access of harmful molecules into the outer membrane. In fact, several lines of evidence indicate that this could be the case because *Y. enterocolitica* outer-core mutants are more susceptible to antimicrobial peptides than the wild-type strain [262]. While antimicrobial peptides are ubiquitous in nature and in vertebrates, they are the front line of defence against infections in those areas exposed to pathogens. Therefore, resistance to antimicrobial peptides is a virulence trait for many bacterial pathogens [263]. The connection between LPS core and resistance to antimicrobial peptides may represent a general feature of this LPS section. Thus, core mutants of *Brucella abortus*, *Brucella ovis*, *B. cenocepacia* and *A. pleuropneumoniae* were more susceptible to antimicrobial peptides than the wild-type strains [23, 113, 264, 265]. Similar to the *Y. enterocolitica* O:3 outer-core mutants, the core mutants of the three bacterial species were also attenuated *in vivo* [23, 113, 264, 265].

However, it is also conceivable that some virulence factors located in the outer membrane require the presence of the core for its proper expression or functionality and hence the core role in virulence would be indirect. Studies with *K. pneumoniae* could be considered as an example of this possibility. It has been shown that *K. pneumoniae* capsule is associated with the bacterial surface by ionic interactions to the LPS negative charges present at the LPS core region [266]. As indicated before, several GalA units, added to the core heptoses by the transferases WabG and WabO, provide these negative charges [160, 161]. As expected, *wabG* or *wabO* mutants not only lack OPS and core residues but also showed reduced levels of cell-bound capsule compared to the wild-type [160, 161]. Although these mutants were strongly attenuated in the mouse model of pneumonia, it should be noted that the capsule is the single most important virulence factor necessary for *K. pneumoniae* survival in the lung [267]. Other core mutants were also attenuated in the same animal model but they also lacked capsule, hence pointing out that the role of *Klebsiella* core in virulence might be indirect.

Nevertheless, data support the notion that in *K. pneumoniae* there is a correlation between expression of a certain core and virulence, strongly suggesting that indeed *K. pneumoniae* core plays a significant role in virulence. Regue et al. [268] have demonstrated that *K. pneumoniae* strains express two types of core structures, termed type 1 and type 2. Both core types share the same inner core and the outer-core-proximal disaccharide, Glc*p*N-(1 → 4)-Gal*p*A, but they differ in the GlcN substituents. In core type 2, the Glc*p*N residue is substituted at O-4 by the disaccharide β-Glc*p*-(1 → 6)-α-Glc*p*, while in core type 1 the Glc*p*N residue is substituted at O-6 by either the disaccharide α-Hep-(1 → 4)-α-Kdo or a Kdo residue. This difference correlates with the presence of a two-gene region in the corresponding core biosynthetic clusters. Strains containing type 1 core contain *wabI* (encoding outer-core Kdo transferase) and *wabJ* (encoding outer-core heptosyltransferase), whereas type 2-expressing strains contain *wabK* and *wabM* (both encoding glucosyltransferases). Interestingly, epidemiological data indicates that type 2 core is prevalent in *K. pneumoniae* O1:K2 strains which are considered to be of high virulence potential for humans [269]. This may support the notion that type 1 and 2 cores are not equivalent in their virulence potential. The fact that the region specifying the expression of each core type is a unique and small DNA region allowed interchanging the two core types and experimentally tests the above hypothesis. The replacement of *Klebsiella* core type 1 in a highly virulent type 2 strain increased two-log the LD_{50} for mice inoculated intraperitoneally [268] pointing out to a direct role of LPS core structure in virulence. However, it cannot be rigorously ruled out that the difference in virulence could still be attributed to effects on proper location and/or expression of outer membrane components playing a direct role in virulence. It should be noted that the core type did not affect capsule expression.

8.7 Conclusions

Although analysis of the core structures of *Salmonella* and *E. coli* did allow to extrapolate certain rules regarding LPS core structures, recent structural studies have revealed that there is a variable repertoire of core structures in different Gram-negative bacteria which may even vary within a single species. In addition, more rare sugars and other compounds such as phosphate, ethanolamine, acetyl and amino acid residues have been detected from the core structures. Nevertheless, despite this structural diversity some biosynthetic pathways required for the synthesis of Kdo and heptose are conserved even in distantly related bacteria. We envision that next years will bring the identification of new enzymes and/or activites important for core biosynthesis in other bacteria than *Enterobactericeae*. Further, whole-genome sequencing of several non-enteric Gram-negative bacteria has paved the way to studies identifying core biosynthesis genes.

Still few studies have addressed in-depth the contribution of LPS core to virulence and we expect exciting studies on this topic. In any case, available data suggests that the core may play an important role in the interplay between the pathogen and the innate immune system. However, it should be noted that the proper expression or functionality of some virulence factors located in the outer membrane may require the presence of the core oligosaccharide and hence the role of core in virulence would be indirect.

References

1. Raetz CR, Whitfield C (2002) Lipopolysaccharide endotoxins. Annu Rev Biochem 71:635–700
2. Frirdich E, Whitfield C (2005) Lipopolysaccharide inner core oligosaccharide structure and outer membrane stability in human pathogens belonging to the *Enterobacteriaceae*. J Endotoxin Res 11:133–144

3. Aspinall GO, Monteiro MA, Shaver RT, Kurjanczyk LA, Penner JL (1997) Lipopolysaccharides of *Helicobacter pylori* serogroups O:3 and O:6. Structures of a class of lipopolysaccharides with reference to the location of oligomeric units of D-glycero-α-D-manno-heptose residues. Eur J Biochem 248:592–601
4. Monteiro MA, Appelmelk BJ, Rasko DA, Moran AP, Hynes SO, MacLean LL, Chan KH, Michael FS, Logan SM, O'Rourke J, Lee A, Taylor DE, Perry MB (2000) Lipopolysaccharide structures of *Helicobacter pylori* genomic strains 26695 and J99, mouse model *H. pylori* Sydney strain, *H. pylori* P466 carrying sialyl Lewis X, and *H. pylori* UA915 expressing Lewis B classification of *H. pylori* lipopolysaccharides into glycotype families. Eur J Biochem 267:305–320
5. Stead C, Tran A, Ferguson D Jr, McGrath S, Cotter R, Trent S (2005) A novel 3-deoxy-D-*manno*-octulosonic acid (Kdo) hydrolase that removes the outer Kdo sugar of *Helicobacter pylori* lipopolysaccharide. J Bacteriol 187:3374–3383
6. Niedziela T, Jachymek W, Lukasiewicz J, Maciejewska A, Andersson R, Kenne L, Lugowski C (2010) Structures of two novel, serologically nonrelated core oligosaccharides of *Yokenella regensburgei* lipopolysaccharides differing only by a single hexose substitution. Glycobiology 20:207–214
7. Tsai CM, Jankowska-Stephens E, Mizanur RM, Cipollo JF (2009) The fine structure of *Neisseria meningitidis* lipooligosaccharide from the M986 strain and three of its variants. J Biol Chem 284:4616–4625
8. Mansson M, Hood DW, Moxon ER, Schweda EK (2003) Structural diversity in lipopolysaccharide expression in nontypeable *Haemophilus influenzae*. Identification of L-glycerol-D-*manno*-heptose in the outer-core region in three clinical isolates. Eur J Biochem 270:610–624
9. Masoud H, Moxon ER, Martin A, Krajcarski D, Richards JC (1997) Structure of the variable and conserved lipopolysaccharide oligosaccharide epitopes expressed by *Haemophilus influenzae* serotype b strain Eagan. Biochemistry 36:2091–2103
10. Li J, Bauer SH, Mansson M, Moxon ER, Richards JC, Schweda EK (2001) Glycine is a common substituent of the inner core in *Haemophilus influenzae* lipopolysaccharide. Glycobiology 11:1009–1015
11. Schweda EK, Richards JC, Hood DW, Moxon ER (2007) Expression and structural diversity of the lipopolysaccharide of *Haemophilus influenzae*: implication in virulence. Int J Med Microbiol 297:297–306
12. Pier GB (2007) *Pseudomonas aeruginosa* lipopolysaccharide: a major virulence factor, initiator of inflammation and target for effective immunity. Int J Med Microbiol 297:277–295
13. King JD, Kocincova D, Westman EL, Lam JS (2009) Lipopolysaccharide biosynthesis in *Pseudomonas aeruginosa*. Innate Immun 15:261–312
14. Bystrova OV, Shashkov AS, Kocharova NA, Knirel YA, Lindner B, Zähringer U, Pier GB (2002) Structural studies on the core and the O-polysaccharide repeating unit of *Pseudomonas aeruginosa* immunotype 1 lipopolysaccharide. Eur J Biochem 269:2194–2203
15. Sadovskaya I, Brisson JR, Thibault P, Richards JC, Lam JS, Altman E (2000) Structural characterization of the outer core and the O-chain linkage region of lipopolysaccharide from *Pseudomonas aeruginosa* serotype O5. Eur J Biochem 267:1640–1650
16. De CC, Molinaro A, Lanzetta R, Silipo A, Parrilli M (2008) Lipopolysaccharide structures from *Agrobacterium* and *Rhizobiaceae* species. Carbohydr Res 343:1924–1933
17. Caroff M, Aussel L, Zarrouk H, Martin A, Richards JC, Therisod H, Perry MB, Karibian D (2001) Structural variability and originality of the *Bordetella* endotoxins. J Endotoxin Res 7:63–68
18. Moll H, Knirel YA, Helbig JH, Zähringer U (1997) Identification of an α-D-Man*p*-(1 → 8)-Kdo disaccharide in the inner core region and the structure of the complete core region of the *Legionella pneumophila* serogroup 1 lipopolysaccharide. Carbohydr Res 304:91–95

19. Vinogradov EV, Müller-Loennies S, Petersen BO, Meshkov S, Thomas-Oates JE, Holst O, Brade H (1997) Structural investigation of the lipopolysaccharide from *Acinetobacter haemolyticus* strain NCTC 10305 (ATCC 17906, DNA group 4). Eur J Biochem 247:82–90
20. Vinogradov EV, Bock K, Petersen BO, Holst O, Brade H (1997) The structure of the carbohydrate backbone of the lipopolysaccharide from *Acinetobacter* strain ATCC 17905. Eur J Biochem 243:122–127
21. Gronow S, Noah C, Blumenthal A, Lindner B, Brade H (2003) Construction of a deep-rough mutant of *Burkholderia cepacia* ATCC 25416 and characterization of its chemical and biological properties. J Biol Chem 278:1647–1655
22. Isshiki Y, Kawahara K, Zähringer U (1998) Isolation and characterisation of disodium (4-amino-4-deoxy-α-L-arabinopyranosyl)-(1 → 8)-(D-*glycero*-α-D-*talo*-oct-2-ulopyranosylonate)-(2 → 4)-(methyl 3-deoxy-D-*manno*-oct-2-ulopyranosid)onate from the lipopolysaccharide of *Burkholderia cepacia*. Carbohydr Res 313:21–27
23. Ortega X, Silipo A, Saldias MS, Bates CC, Molinaro A, Valvano MA (2009) Biosynthesis and structure of the *Burkholderia cenocepacia* K56-2 lipopolysaccharide core oligosaccharide: truncation of the core oligosaccharide leads to increased binding and sensitivity to polymyxin B. J Biol Chem 284:21738–21751
24. Vinogradov EV, Lindner B, Kocharova NA, Senchenkova SN, Shashkov AS, Knirel YA, Holst O, Gremyakova TA, Shaikhutdinova RZ, Anisimov AP (2002) The core structure of the lipopolysaccharide from the causative agent of plague, *Yersinia pestis*. Carbohydr Res 337:775–777
25. Vinogradov E, Lindner B, Seltmann G, Radziejewska-Lebrecht J, Holst O (2006) Lipopolysaccharides from *Serratia marcescens* possess one or two 4-amino-4-deoxy-L-arabinopyranose 1-phosphate residues in the lipid A and D-*glycero*-D-*talo*-oct-2-ulopyranosonic acid in the inner core region. Chem Eur J 12:6692–6700
26. Sonesson A, Jantzen E, Bryn K, Tangen T, Eng J, Zähringer U (1994) Composition of 2,3-dihydroxy fatty acid-containing lipopolysaccharides from *Legionella israelensis*, *Legionella maceachernii* and *Legionella micdadei*. Microbiology 140:1261–1271
27. Phillips NJ, Apicella MA, Griffiss JM, Gibson BW (1992) Structural characterization of the cell surface lipooligosaccharides from a nontypable strain of *Haemophilus influenzae*. Biochemistry 31:4515–4526
28. Phillips NJ, Apicella MA, Griffiss JM, Gibson BW (1993) Structural studies of the lipooligosaccharides from *Haemophilus influenzae* type b strain A2. Biochemistry 32:2003–2012
29. Lebbar S, Caroff M, Szabo L, Merienne C, Szilogyi L (1994) Structure of a hexasaccharide proximal to the hydrophobic region of lipopolysaccharides present in *Bordetella pertussis* endotoxin preparations. Carbohydr Res 259:257–275
30. Holst O (1999) Chemical structure of the core region of lipopolysaccharides. In: Brade H, Opal SM, Vogel SN, Morrison DC (eds) Endotoxin in health and disease. Marcel Dekker, New York, pp 115–154
31. Edebrink P, Jansson P-E, Bogwald J, Hoffman J (1996) Structural studies of the *Vibrio salmonicida* lipopolysaccharide. Carbohydr Res 287:225–245
32. Kondo S, Zähringer U, Seydel U, Sinnwell V, Hisatsune K, Rietschel ET (1991) Chemical structure of the carbohydrate backbone of *Vibrio parahaemolyticus* serotype O12 lipopolysaccharide. Eur J Biochem 200:689–698
33. Vinogradov EV, Bock K, Holst O, Brade H (1995) The structure of the lipid A-core region of the lipopolysaccharides from *Vibrio cholerae* O1 smooth strain 569B (Inaba) and rough mutant strain 95R (Ogawa). Eur J Biochem 233:152–158
34. Brade H, Brabetz W, Brade L, Holst O, Löbau S, Lucakova M, Mamat U, Rozalski A, Zych K, Kosma P (1997) Chlamydial lipopolysaccharide. J Endotoxin Res 4:67–84
35. Heine H, Müller-Loennies S, Brade L, Lindner B, Brade H (2003) Endotoxic activity and chemical structure of lipopolysaccharides from *Chlamydia trachomatis* serotypes E and L2 and *Chlamydophila psittaci* 6BC. Eur J Biochem 270:440–450

36. Rund S, Lindner B, Brade H, Holst O (1999) Structural analysis of the lipopolysaccharide from *Chlamydia trachomatis* serotype L2. J Biol Chem 274:16819–16824
37. Rund S, Lindner B, Brade H, Holst O (2000) Structural analysis of the lipopolysaccharide from *Chlamydophila psittaci* strain 6BC. Eur J Biochem 267:5717–5726
38. Hanuszkiewicz A, Hubner G, Vinogradov E, Lindner B, Brade L, Brade H, Debarry J, Heine H, Holst O (2008) Structural and immunochemical analysis of the lipopolysaccharide from *Acinetobacter lwoffii* F78 located outside *Chlamydiaceae* with a *Chlamydia*-specific lipopolysaccharide epitope. Chem Eur J 14:10251–10258
39. Meredith TC, Woodard RW (2003) *Escherichia coli* YrbH is a D-arabinose 5-phosphate isomerase. J Biol Chem 278:32771–32777
40. Sommaruga S, Gioia LD, Tortora P, Polissi A (2009) Structure prediction and functional analysis of KdsD, an enzyme involved in lipopolysaccharide biosynthesis. Biochem Biophys Res Commun 388:222–227
41. Tan L, Darby C (2006) *Yersinia pestis* YrbH is a multifunctional protein required for both 3-deoxy-D-*manno*-oct-2-ulosonic acid biosynthesis and biofilm formation. Mol Microbiol 61:861–870
42. Tzeng YL, Datta A, Strole C, Kolli VS, Birck MR, Taylor WP, Carlson RW, Woodard RW, Stephens DS (2002) KpsF is the arabinose-5-phosphate isomerase required for 3-deoxy-D-*manno*-octulosonic acid biosynthesis and for both lipooligosaccharide assembly and capsular polysaccharide expression in *Neisseria meningitidis*. J Biol Chem 277:24103–24113
43. Meredith TC, Woodard RW (2005) Identification of GutQ from *Escherichia coli* as a D-arabinose 5-phosphate isomerase. J Bacteriol 187:6936–6942
44. Cieslewicz M, Vimr E (1997) Reduced polysialic acid capsule expression in *Escherichia coli* K1 mutants with chromosomal defects in *kpsF*. Mol Microbiol 26:237–249
45. Meredith TC, Woodard RW (2006) Characterization of *Escherichia coli* D-arabinose 5-phosphate isomerase encoded by *kpsF*: implications for group 2 capsule biosynthesis. Biochem J 395:427–432
46. Bateman A (1999) The SIS domain: a phosphosugar-binding domain. Trends Biochem Sci 24:94–95
47. Bateman A (1997) The structure of a domain common to archaebacteria and the homocystinuria disease protein. Trends Biochem Sci 22:12–13
48. Dotson GD, Nanjappan P, Reily MD, Woodard RW (1993) Stereochemistry of 3-deoxyoctulosonate 8-phosphate synthase. Biochemistry 32:12392–12397
49. Dotson GD, Dua RK, Clemens JC, Wooten EW, Woodard RW (1995) Overproduction and one-step purification of *Escherichia coli* 3-deoxy-D-*manno*-octulosonic acid 8-phosphate synthase and oxygen transfer studies during catalysis using isotopic-shifted heteronuclear NMR. J Biol Chem 270:13698–13705
50. Radaev S, Dastidar P, Patel M, Woodard RW, Gatti DL (2000) Structure and mechanism of 3-deoxy-D-*manno*-octulosonate 8-phosphate synthase. J Biol Chem 275:9476–9484
51. Radaev S, Dastidar P, Patel M, Woodard RW, Gatti DL (2000) Preliminary X-ray analysis of a new crystal form of the *Escherichia coli* KDO8P synthase. Acta Crystallogr D Biol Crystallogr 56:516–519
52. Birck MR, Woodard RW (2001) *Aquifex aeolicus* 3-deoxy-D-*manno*-2-octulosonic acid 8-phosphate synthase: a new class of KDO 8-P synthase? J Mol Evol 52:205–214
53. Duewel HS, Woodard RW (2000) A metal bridge between two enzyme families. 3-Deoxy-D-*manno*-octulosonate-8-phosphate synthase from *Aquifex aeolicus* requires a divalent metal for activity. J Biol Chem 275:22824–22831
54. Vainer R, Belakhov V, Rabkin E, Baasov T, Adir N (2005) Crystal structures of *Escherichia coli* KDO8P synthase complexes reveal the source of catalytic irreversibility. J Mol Biol 351:641–652
55. Wagner T, Kretsinger RH, Bauerle R, Tolbert WD (2000) 3-Deoxy-D-*manno*-octulosonate-8-phosphate synthase from *Escherichia coli*. Model of binding of phosphoenolpyruvate and D-arabinose-5-phosphate. J Mol Biol 301:233–238

56. Cochrane FC, Cookson TV, Jameson GB, Parker EJ (2009) Reversing evolution: re-establishing obligate metal ion dependence in a metal-independent KDO8P synthase. J Mol Biol 390:646–661
57. Duewel HS, Radaev S, Wang J, Woodard RW, Gatti DL (2001) Substrate and metal complexes of 3-deoxy-D-*manno*-octulosonate-8-phosphate synthase from *Aquifex aeolicus* at 1.9-Å resolution. Implications for the condensation mechanism. J Biol Chem 276:8393–8402
58. Kona F, Tao P, Martin P, Xu X, Gatti DL (2009) Electronic structure of the metal center in the Cd^{2+}, Zn^{2+}, and Cu^{2+} substituted forms of KDO8P synthase: implications for catalysis. Biochemistry 48:3610–3630
59. Li J, Wu J, Fleischhacker AS, Woodard RW (2004) Conversion of *Aquifex aeolicus* 3-deoxy-D-*manno*-octulosonate 8-phosphate synthase, a metalloenzyme, into a nonmetalloenzyme. J Am Chem Soc 126:7448–7449
60. Shulami S, Yaniv O, Rabkin E, Shoham Y, Baasov T (2003) Cloning, expression, and biochemical characterization of 3-deoxy-D-*manno*-2-octulosonate-8-phosphate (KDO8P) synthase from the hyperthermophilic bacterium *Aquifex pyrophilus*. Extremophiles 7:471–481
61. Krosky DJ, Alm R, Berg M, Carmel G, Tummino PJ, Xu B, Yang W (2002) *Helicobacter pylori* 3-deoxy-D-*manno*-octulosonate-8-phosphate (KDO-8-P) synthase is a zinc-metalloenzyme. Biochim Biophys Acta 1594:297–306
62. Sau AK, Li Z, Anderson KS (2004) Probing the role of metal ions in the catalysis of *Helicobacter pylori* 3-deoxy-D-*manno*-octulosonate-8-phosphate synthase using a transient kinetic analysis. J Biol Chem 279:15787–15794
63. Allison TM, Yeoman JA, Hutton RD, Cochrane FC, Jameson GB, Parker EJ (2010) Specificity and mutational analysis of the metal-dependent 3-deoxy-D-*manno*-octulosonate 8-phosphate synthase from *Acidithiobacillus ferrooxidans*. Biochim Biophys Acta 1804:1526–1536
64. Kona F, Xu X, Martin P, Kuzmic P, Gatti DL (2007) Structural and mechanistic changes along an engineered path from metallo to nonmetallo 3-deoxy-D-*manno*-octulosonate 8-phosphate synthases. Biochemistry 46:4532–4544
65. Oliynyk Z, Briseno-Roa L, Janowitz T, Sondergeld P, Fersht AR (2004) Designing a metal-binding site in the scaffold of *Escherichia coli* KDO8PS. Protein Eng Des Sel 17:383–390
66. Shulami S, Furdui C, Adir N, Shoham Y, Anderson KS, Baasov T (2004) A reciprocal single mutation affects the metal requirement of 3-deoxy-D-*manno*-2-octulosonate-8-phosphate (KDO8P) synthases from *Aquifex pyrophilus* and *Escherichia coli*. J Biol Chem 279:45110–45120
67. Ray PH, Benedict CD (1982) 3-Deoxy-D-*manno*-octulosonate-8-phosphate (KDO-8-P) phosphatase. Method Enzymol 83:530–535
68. Wu J, Woodard RW (2003) *Escherichia coli* YrbI is 3-deoxy-D-*manno*-octulosonate 8-phosphate phosphatase. J Biol Chem 278:18117–18123
69. Sperandeo P, Pozzi C, Deho G, Polissi A (2006) Non-essential KDO biosynthesis and new essential cell envelope biogenesis genes in the *Escherichia coli yrbG-yhbG* locus. Res Microbiol 157:547–558
70. Chng SS, Gronenberg LS, Kahne D (2010) Proteins required for lipopolysaccharide assembly in *Escherichia coli* form a transenvelope complex. Biochemistry 49:4565–4567
71. Ma B, Reynolds CM, Raetz CR (2008) Periplasmic orientation of nascent lipid A in the inner membrane of an *Escherichia coli* LptA mutant. Proc Natl Acad Sci USA 105:13823–13828
72. Narita S, Tokuda H (2009) Biochemical characterization of an ABC transporter LptBFGC complex required for the outer membrane sorting of lipopolysaccharides. FEBS Lett 583:2160–2164
73. Ruiz N, Gronenberg LS, Kahne D, Silhavy TJ (2008) Identification of two inner-membrane proteins required for the transport of lipopolysaccharide to the outer membrane of *Escherichia coli*. Proc Natl Acad Sci USA 105:5537–5542

74. Sperandeo P, Cescutti R, Villa R, Di Benedetto C, Candia D, Deho G, Polissi A (2007) Characterization of *lptA* and *lptB*, two essential genes implicated in lipopolysaccharide transport to the outer membrane of *Escherichia coli*. J Bacteriol 189:244–253
75. Sperandeo P, Lau FK, Carpentieri A, De Castro C, Molinaro A, Deho G, Silhavy TJ, Polissi A (2008) Functional analysis of the protein machinery required for transport of lipopolysaccharide to the outer membrane of *Escherichia coli*. J Bacteriol 190:4460–4469
76. Suits MD, Sperandeo P, Deho G, Polissi A, Jia Z (2008) Novel structure of the conserved gram-negative lipopolysaccharide transport protein A and mutagenesis analysis. J Mol Biol 380:476–488
77. Tran AX, Trent MS, Whitfield C (2008) The LptA protein of *Escherichia coli* is a periplasmic lipid A-binding protein involved in the lipopolysaccharide export pathway. J Biol Chem 283:20342–20349
78. Biswas T, Yi L, Aggarwal P, Wu J, Rubin JR, Stuckey JA, Woodard RW, Tsodikov OV (2009) The tail of KdsC: conformational changes control the activity of a haloacid dehalogenase superfamily phosphatase. J Biol Chem 284:30594–30603
79. Parsons JF, Lim K, Tempczyk A, Krajewski W, Eisenstein E, Herzberg O (2002) From structure to function: YrbI from *Haemophilus influenzae* (HI1679) is a phosphatase. Proteins 46:393–404
80. Allen KN, Dunaway-Mariano D (2004) Phosphoryl group transfer: evolution of a catalytic scaffold. Trends Biochem Sci 29:495–503
81. Burroughs AM, Allen KN, Dunaway-Mariano D, Aravind L (2006) Evolutionary genomics of the HAD superfamily: understanding the structural adaptations and catalytic diversity in a superfamily of phosphoesterases and allied enzymes. J Mol Biol 361:1003–1034
82. Lu Z, Wang L, Dunaway-Mariano D, Allen KN (2009) Structure-function analysis of 2-keto-3-deoxy-D-*glycero*-D-*galacto*-nononate-9-phosphate phosphatase defines specificity elements in type C0 haloalkanoate dehalogenase family members. J Biol Chem 284:1224–1233
83. Lin CH, Murray BW, Ollmann IR, Wong CH (1997) Why is CMP-ketodeoxyoctonate highly unstable? Biochemistry 36:780–785
84. Ray PH, Benedict CD (1982) CTP:CMP-3-deoxy-D-*manno*-octulosonate cytidylyltransferase (CMP-KDO synthetase). Method Enzymol 83:535–540
85. Brade H, Zähringer U, Rietschel ET, Christian R, Schulz G, Unger FM (1984) Spectroscopic analysis of a 3-deoxy-D-*manno*-2-octulosonic acid (Kdo)-disaccharide from the lipopolysaccharide of a *Salmonella godesberg* Re mutant. Carbohydr Res 134:157–166
86. Kohlbrenner WE, Fesik SW (1985) Determination of the anomeric specificity of the *Escherichia coli* CTP:CMP-3-deoxy-D-*manno*-octulosonate cytidylyltransferase by ^{13}C NMR spectroscopy. J Biol Chem 260:14695–14700
87. Finke A, Roberts I, Boulnois G, Pzzani C, Jann K (1989) Activity of CMP-2-keto-3-deoxyoctulosonic acid synthetase in *Escherichia coli* strains expressing the capsular K5 polysaccharide implication for K5 polysaccharide biosynthesis. J Bacteriol 171:3074–3079
88. Rosenow C, Roberts IS, Jann K (1995) Isolation from recombinant *Escherichia coli* and characterization of CMP-Kdo synthetase, involved in the expression of the capsular K5 polysaccharide (K-CKS). FEMS Microbiol Lett 125:159–164
89. Jann B, Jann K (1990) Structure and biosynthesis of the capsular antigens of *Escherichia coli*. Curr Top Microbiol Immunol 150:19–42
90. Finke A, Jann B, Jann K (1990) CMP-KDO-synthetase activity in *Escherichia coli* expressing capsular polysaccharides. FEMS Microbiol Lett 57:129–133
91. Strohmaier H, Remler P, Renner W, Hogenauer G (1995) Expression of genes *kdsA* and *kdsB* involved in 3-deoxy-D-*manno*-octulosonic acid metabolism and biosynthesis of enterobacterial lipopolysaccharide is growth phase regulated primarily at the transcriptional level in *Escherichia coli* K-12. J Bacteriol 177:4488–4500
92. Jelakovic S, Jann K, Schulz GE (1996) The three-dimensional structure of capsule-specific CMP: 2-keto-3-deoxy-*manno*-octonic acid synthetase from *Escherichia coli*. FEBS Lett 391:157–161

93. Jelakovic S, Schulz GE (2001) The structure of CMP:2-keto-3-deoxy-*manno*-octonic acid synthetase and of its complexes with substrates and substrate analogs. J Mol Biol 312:143–155
94. Jelakovic S, Schulz GE (2002) Catalytic mechanism of CMP:2-keto-3-deoxy-*manno*-octonic acid synthetase as derived from complexes with reaction educt and product. Biochemistry 41:1174–1181
95. Heyes DJ, Levy C, Lafite P, Roberts IS, Goldrick M, Stachulski AV, Rossington SB, Stanford D, Rigby SE, Scrutton NS, Leys D (2009) Structure-based mechanism of CMP-2-keto-3-deoxy-*manno*-octulonic acid synthetase: convergent evolution of a sugar-activating enzyme with DNA/RNA polymerases. J Biol Chem 284:35514–35523
96. Yoon HJ, Ku MJ, Mikami B, Suh SW (2008) Structure of 3-deoxy-*manno*-octulosonate cytidylyltransferase from *Haemophilus influenzae* complexed with the substrate 3-deoxy-*manno*-octulosonate in the β-configuration. Acta Crystallogr D Biol Crystallogr 64:1292–1294
97. Nudler E (2009) RNA polymerase active center: the molecular engine of transcription. Annu Rev Biochem 78:335–361
98. Rothwell PJ, Waksman G (2005) Structure and mechanism of DNA polymerases. Adv Protein Chem 71:401–440
99. Holst O (2002) Chemical structure of the core region of lipopolysaccharides – an update. Trends Glycosci Glyc 14:87–103
100. Carlson RW, Reuhs B, Chen TB, Bhat UR, Noel KD (1995) Lipopolysaccharide core structures in *Rhizobium etli* and mutants deficient in O-antigen. J Biol Chem 270:11783–11788
101. Kannenberg EL, Reuhs B, Forsberg LS, Carlson RW (1998) Lipopolysaccharides and K-antigens: their structures, biosynthesis, and functions in *Rhizobium*-legume interactions. In: Spaink HH, Kondrosi A, Hooykaas PJJ (eds) The *Rhizobiaceae*. Kluwer, Amsterdam, pp 119–154
102. Knirel YA, Moll H, Zähringer U (1996) Structural study of a highly *O*-acetylated core of *Legionella pneumophila* serogroup 1 lipopolysaccharide. Carbohydr Res 293:223–234
103. Edebrink P, Jansson P-E, Widmalm G, Holme T, Rahman M (1996) The structures of oligosaccharides isolated from the lipopolysaccharide of *Moraxella catarrhalis* serotype B, strain CCUG 3292. Carbohydr Res 295:127–146
104. Vinogradov EV, Petersen BO, Thomas-Oates JE, Duus J, Brade H, Holst O (1998) Characterization of a novel branched tetrasaccharide of 3-deoxy-D-*manno*-oct-2-ulopyranosonic acid. The structure of the carbohydrate backbone of the lipopolysaccharide from *Acinetobacter baumannii* strain NCTC 10303 (ATCC 17904). J Biol Chem 273:28122–28131
105. Brooke JS, Valvano MA (1996) Molecular cloning of the *Haemophilus influenzae gmhA* (*lpcA*) gene encoding a phosphoheptose isomerase required for lipooligosaccharide biosynthesis. J Bacteriol 178:3339–3341
106. Brooke JS, Valvano MA (1996) Biosynthesis of inner core lipopolysaccharide in enteric bacteria identification and characterization of a conserved phosphoheptose isomerase. J Biol Chem 271:3608–3614
107. Eidels L, Osborn MJ (1974) Phosphoheptose isomerase, first enzyme in the biosynthesis of aldoheptose in *Salmonella typhimurium*. J Biol Chem 249:5642–5648
108. Taylor PL, Blakely KM, de Leon GP, Walker JR, McArthur F, Evdokimova E, Zhang K, Valvano MA, Wright GD, Junop MS (2008) Structure and function of sedoheptulose-7-phosphate isomerase, a critical enzyme for lipopolysaccharide biosynthesis and a target for antibiotic adjuvants. J Biol Chem 283:2835–2845
109. Seetharaman J, Rajashankar KR, Solorzano V, Kniewel R, Lima CD, Bonanno JB, Burley SK, Swaminathan S (2006) Crystal structures of two putative phosphoheptose isomerases. Proteins 63:1092–1096
110. Kneidinger B, Graninger M, Puchberger M, Kosma P, Messner P (2001) Biosynthesis of nucleotide-activated D-*glycero*-D-*manno*-heptose. J Biol Chem 276:20935–20944

111. McArthur F, Andersson CE, Loutet S, Mowbray SL, Valvano MA (2005) Functional analysis of the *glycero-manno*-heptose 7-phosphate kinase domain from the bifunctional HldE protein, which is involved in ADP-L-*glycero*-D-*manno*-heptose biosynthesis. J Bacteriol 187:5292–5300
112. Valvano MA, Marolda CL, Bittner M, Glaskin-Clay M, Simon TL, Klena JD (2000) The *rfaE* gene from *Escherichia coli* encodes a bifunctional protein involved in biosynthesis of the lipopolysaccharide core precursor ADP-L-*glycero*-D-*manno*-heptose. J Bacteriol 182:488–497
113. Loutet SA, Flannagan RS, Kooi C, Sokol PA, Valvano MA (2006) A complete lipopolysaccharide inner core oligosaccharide is required for resistance of *Burkholderia cenocepacia* to antimicrobial peptides and bacterial survival *in vivo*. J Bacteriol 188:2073–2080
114. Valvano MA, Messner P, Kosma P (2002) Novel pathways for biosynthesis of nucleotide-activated *glycero-manno*-heptose precursors of bacterial glycoproteins and cell surface polysaccharides. Microbiology 148:1979–1989
115. Kneidinger B, Marolda C, Graninger M, Zamyatina A, McArthur F, Kosma P, Valvano MA, Messner P (2002) Biosynthesis pathway of ADP-L-*glycero*-β-D-*manno*-heptose in *Escherichia coli*. J Bacteriol 184:363–369
116. Nguyen HH, Wang L, Huang H, Peisach E, Dunaway-Mariano D, Allen KN (2010) Structural determinants of substrate recognition in the HAD superfamily member D-*glycero*-D-*manno*-heptose-1,7-bisphosphate phosphatase (GmhB). Biochemistry 49:1082–1092
117. Taylor PL, Sugiman-Marangos S, Zhang K, Valvano MA, Wright GD, Junop MS (2010) Structural and kinetic characterization of the LPS biosynthetic enzyme d-α, β-d-heptose-1,7-bisphosphate phosphatase (GmhB) from Escherichia coli. Biochemistry 49:1033–1041
118. Wang L, Huang H, Nguyen HH, Allen KN, Mariano PS, Dunaway-Mariano D (2010) Divergence of biochemical function in the HAD superfamily: D-*glycero*-D-*manno*-heptose-1,7-bisphosphate phosphatase (GmhB). Biochemistry 49:1072–1081
119. Morrison JP, Read JA, Coleman WG Jr, Tanner ME (2005) Dismutase activity of ADP-L-*glycero*-D-*manno*-heptose 6-epimerase: evidence for a direct oxidation/reduction mechanism. Biochemistry 44:5907–5915
120. Morrison JP, Tanner ME (2007) A two-base mechanism for *Escherichia coli* ADP-L-*glycero*-D-*manno*-heptose 6-epimerase. Biochemistry 46:3916–3924
121. Lee NG, Sunshine MG, Apicella MA (1995) Molecular cloning and characterization of the nontypeable *Haemophilus influenzae* 2019 *rfaE* gene required for lipopolysaccharide biosynthesis. Infect Immun 63:818–824
122. Nichols WA, Gibson BW, Melaugh W, Lee NG, Sunshine M, Apicella MA (1997) Identification of the ADP-L-*glycero*-D-*manno*-heptose-6-epimerase (*rfaD*) and heptosyltransferase II (*rfaF*) biosynthesis genes from nontypeable *Haemophilus influenzae* 2019. Infect Immun 65:1377–1386
123. Drazek ES, Stein DC, Deal CD (1995) A mutation in the *Neisseria gonorrhoeae rfaD* homolog results in altered lipooligosaccharide expression. J Bacteriol 177:2321–2327
124. Levin JC, Stein DC (1996) Cloning, complementation, and characterization of an *rfaE* homolog from *Neisseria gonorrhoeae*. J Bacteriol 178:4571–4575
125. Provost M, Harel J, Labrie J, Sirois M, Jacques M (2003) Identification, cloning and characterization of *rfaE* of *Actinobacillus pleuropneumoniae* serotype 1, a gene involved in lipopolysaccharide inner-core biosynthesis. FEMS Microbiol Lett 223:7–14
126. Jin UH, Chung TW, Lee YC, Ha SD, Kim CH (2001) Molecular cloning and functional expression of the *rfaE* gene required for lipopolysaccharide biosynthesis in *Salmonella typhimurium*. Glycoconj J 18:779–787
127. Sirisena DM, MacLachlan PR, Liu SL, Hessel A, Sanderson KE (1994) Molecular analysis of the *rfaD* gene, for heptose synthesis, and the *rfaF* gene, for heptose transfer, in lipopolysaccharide synthesis in *Salmonella typhimurium*. J Bacteriol 176:2379–2385
128. Stroeher UH, Karageorgos LE, Morona R, Manning PA (1995) In *Vibrio cholerae* serogroup O1, *rfaD* is closely linked to the *rfb* operon. Gene 155:67–72

129. Andrä J, de Cock H, Garidel P, Howe J, Brandenburg K (2005) Investigation into the interaction of the phosphoporin PhoE with outer membrane lipids: physicochemical characterization and biological activity. Med Chem 1:537–546
130. de Cock H, Brandenburg K, Wiese A, Holst O, Seydel U (1999) Non-lamellar structure and negative charges of lipopolysaccharides required for efficient folding of outer membrane protein PhoE of *Escherichia coli*. J Biol Chem 274:5114–5119
131. Hagge SO, de Cock H, Gutsmann T, Beckers F, Seydel U, Wiese A (2002) Pore formation and function of phosphoporin PhoE of *Escherichia coli* are determined by the core sugar moiety of lipopolysaccharide. J Biol Chem 277:34247–34253
132. Nikaido H (2003) Molecular basis of bacterial outer membrane permeability revisited. Microbiol Mol Biol Rev 67:593–656
133. Qu J, Behrens-Kneip S, Holst O, Kleinschmidt JH (2009) Binding regions of outer membrane protein A in complexes with the periplasmic chaperone Skp. A site-directed fluorescence study. Biochemistry 48:4926–4936
134. Brabetz W, Müller-Loennies S, Brade H (2000) 3-Deoxy-D-*manno*-oct-2-ulosonic acid (Kdo) transferase (WaaA) and Kdo kinase (KdkA) of *Haemophilus influenzae* are both required to complement a *waaA* knockout mutation of *Escherichia coli*. J Biol Chem 275:34954–34962
135. Hankins JV, Trent MS (2009) Secondary acylation of *Vibrio cholerae* lipopolysaccharide requires phosphorylation of Kdo. J Biol Chem 284:25804–25812
136. Harper M, Boyce JD, Cox AD, St Michael F, Wilkie IW, Blackall PJ, Adler B (2007) *Pasteurella multocida* expresses two lipopolysaccharide glycoforms simultaneously, but only a single form is required for virulence: identification of two acceptor-specific heptosyl I transferases. Infect Immun 75:3885–3893
137. Harper M, Cox AD, St Michael F, Ford M, Wilkie IW, Adler B, Boyce JD (2010) Natural selection in the chicken host identifies Kdo kinase residues essential for phosphorylation of *Pasteurella multocida* LPS. Infect Immun 78:3669–3677
138. White KA, Kaltashov IA, Cotter RJ, Raetz CR (1997) A mono-functional 3-deoxy-D-*manno*-octulosonic acid (Kdo) transferase and a Kdo kinase in extracts of *Haemophilus influenzae*. J Biol Chem 272:16555–16563
139. White KA, Lin S, Cotter RJ, Raetz CR (1999) A *Haemophilus influenzae* gene that encodes a membrane bound 3-deoxy-D-*manno*-octulosonic acid (Kdo) kinase. Possible involvement of Kdo phosphorylation in bacterial virulence. J Biol Chem 274:31391–31400
140. Krupa A, Srinivasan N (2002) Lipopolysaccharide phosphorylating enzymes encoded in the genomes of gram-negative bacteria are related to the eukaryotic protein kinases. Protein Sci 11:1580–1584
141. Yethon JA, Heinrichs D, Monteiro MA, Perry MB, Whitfield C (1998) Involvement of *waaY*, *waaQ*, and *waaP* in the modification of *Escherichia coli* lipopolysaccharide and their role in the formation of a stable outer membrane. J Biol Chem 273:26310–26316
142. Yethon JA, Gunn JS, Ernst RK, Miller SI, Laroche L, Malo D, Whitfield C (2000) *Salmonella enterica* serovar *typhimurium waaP* mutants show increased susceptibility to polymyxin and loss of virulence *in vivo*. Infect Immun 68:4485–4491
143. Yethon JA, Whitfield C (2001) Purification and characterization of WaaP from *Escherichia coli*, a lipopolysaccharide kinase essential for outer membrane stability. J Biol Chem 276:5498–5504
144. Kooistra O, Bedoux G, Brecker L, Lindner B, Sanchez CP, Haras D, Zähringer U (2003) Structure of a highly phosphorylated lipopolysaccharide core in the Δ*algC* mutants derived from *Pseudomonas aeruginosa* wild-type strains PAO1 (serogroup O5) and PAC1R (serogroup O3). Carbohydr Res 338:2667–2677
145. Walsh AG, Matewish MJ, Burrows LL, Monteiro MA, Perry MB, Lam JS (2000) Lipopolysaccharide core phosphates are required for viability and intrinsic drug resistance in *Pseudomonas aeruginosa*. Mol Microbiol 35:718–727

146. Zhao X, Lam JS (2002) WaaP of *Pseudomonas aeruginosa* is a novel eukaryotic type protein-tyrosine kinase as well as a sugar kinase essential for the biosynthesis of core lipopolysaccharide. J Biol Chem 277:4722–4730
147. Cox AD, Wright JC, Li J, Hood DW, Moxon ER, Richards JC (2003) Phosphorylation of the lipid A region of meningococcal lipopolysaccharide: identification of a family of transferases that add phosphoethanolamine to lipopolysaccharide. J Bacteriol 185:3270–3277
148. Mackinnon FG, Cox AD, Plested JS, Tang CM, Makepeace K, Coull PA, Wright JC, Chalmers R, Hood DW, Richards JC, Moxon ER (2002) Identification of a gene (*lpt-3*) required for the addition of phosphoethanolamine to the lipopolysaccharide inner core of *Neisseria meningitidis* and its role in mediating susceptibility to bactericidal killing and opsonophagocytosis. Mol Microbiol 43:931–943
149. O'Connor ET, Piekarowicz A, Swanson KV, Griffiss JM, Stein DC (2006) Biochemical analysis of Lpt3, a protein responsible for phosphoethanolamine addition to lipooligosaccharide of pathogenic *Neisseria*. J Bacteriol 188:1039–1048
150. Wright JC, Hood DW, Randle GA, Makepeace K, Cox AD, Li J, Chalmers R, Richards JC, Moxon ER (2004) *lpt6*, a gene required for addition of phosphoethanolamine to inner-core lipopolysaccharide of *Neisseria meningitidis* and *Haemophilus influenzae*. J Bacteriol 186:6970–6982
151. Brabetz W, Müller-Loennies S, Holst O, Brade H (1997) Deletion of the heptosyltransferase genes *rfaC* and *rfaF* in *Escherichia coli* K-12 results in an Re-type lipopolysaccharide with a high degree of 2-aminoethanol phosphate substitution. Eur J Biochem 247:716–724
152. Reynolds CM, Kalb SR, Cotter RJ, Raetz CR (2005) A phosphoethanolamine transferase specific for the outer 3-deoxy-D-*manno*-octulosonic acid residue of *Escherichia coli* lipopolysaccharide. Identification of the *eptB* gene and Ca^{2+} hypersensitivity of an *eptB* deletion mutant. J Biol Chem 280:21202–21211
153. Forsberg LS, Carlson RW (1998) The structures of the lipopolysaccharides from *Rhizobium etli* strains CE358 and CE359. The complete structure of the core region of *R. etli* lipopolysaccharides. J Biol Chem 273:2747–2757
154. Niedziela T, Lukasiewicz J, Jachymek W, Dzieciatkowska M, Lugowski C, Kenne L (2002) Core oligosaccharides of *Plesiomonas shigelloides* O54:H2 (strain CNCTC 113/92): structural and serological analysis of the lipopolysaccharide core region, the O-antigen biological repeating unit, and the linkage between them. J Biol Chem 277:11653–11663
155. Severn WB, Kelly RF, Richards JC, Whitfield C (1996) Structure of the core oligosaccharide in the serotype O8 lipopolysaccharide from *Klebsiella pneumoniae*. J Bacteriol 178:1731–1741
156. Süsskind M, Brade L, Brade H, Holst O (1998) Identification of a novel heptoglycan of $\alpha 1 \rightarrow 2$-linked D-*glycero*-D-*manno*-heptopyranose. Chemical and antigenic structure of lipopolysaccharides from *Klebsiella pneumoniae ssp. pneumoniae* rough strain R20 ($O1^-$:$K20^-$). J Biol Chem 273:7006–7017
157. Vinogradov E, Perry MB (2001) Structural analysis of the core region of the lipopolysaccharides from eight serotypes of *Klebsiella pneumoniae*. Carbohydr Res 335:291–296
158. Frirdich E, Bouwman C, Vinogradov E, Whitfield C (2005) The role of galacturonic acid in outer membrane stability in *Klebsiella pneumoniae*. J Biol Chem 280:27604–27612
159. Regue M, Hita B, Pique N, Izquierdo L, Merino S, Fresno S, Benedi VJ, Tomas JM (2004) A gene, *uge*, is essential for *Klebsiella pneumoniae* virulence. Infect Immun 72:54–61
160. Fresno S, Jimenez N, Canals R, Merino S, Corsaro MM, Lanzetta R, Parrilli M, Pieretti G, Regue M, Tomas JM (2007) A second galacturonic acid transferase is required for core lipopolysaccharide biosynthesis and complete capsule association with the cell surface in *Klebsiella pneumoniae*. J Bacteriol 189:1128–1137
161. Izquierdo L, Coderch N, Pique N, Bedini E, Corsaro MM, Merino S, Fresno S, Tomas JM, Regue M (2003) The *Klebsiella pneumoniae wabG* gene: role in biosynthesis of the core lipopolysaccharide and virulence. J Bacteriol 185:7213–7221

162. Kanjilal-Kolar S, Raetz CR (2006) Dodecaprenyl phosphate-galacturonic acid as a donor substrate for lipopolysaccharide core glycosylation in *Rhizobium leguminosarum*. J Biol Chem 281:12879–12887
163. Kanjilal-Kolar S, Basu SS, Kanipes MI, Guan Z, Garrett TA, Raetz CR (2006) Expression cloning of three *Rhizobium leguminosarum* lipopolysaccharide core galacturonosyltransferases. J Biol Chem 281:12865–12878
164. Belunis CJ, Raetz CR (1992) Biosynthesis of endotoxins. Purification and catalytic properties of 3-deoxy-D-*manno*-octulosonic acid transferase from *Escherichia coli*. J Biol Chem 267:9988–9997
165. Raetz CR, Reynolds CM, Trent MS, Bishop RE (2007) Lipid A modification systems in gram-negative bacteria. Annu Rev Biochem 76:295–329
166. Bode CE, Brabetz W, Brade H (1998) Cloning and characterization of 3-deoxy-D-*manno*-oct-2-ulosonic acid (Kdo) transferase genes (*kdtA*) from *Acinetobacter baumannii* and *Acinetobacter haemolyticus*. Eur J Biochem 254:404–412
167. Brabetz W, Schirmer CE, Brade H (2000) 3-Deoxy-D-*manno*-oct-2-ulosonic acid (Kdo) transferase of *Legionella pneumophila* transfers two Kdo residues to a structurally different lipid A precursor of *Escherichia coli*. J Bacteriol 182:4654–4657
168. Mamat U, Schmidt H, Munoz E, Lindner B, Fukase K, Hanuszkiewicz A, Wu J, Meredith TC, Woodard RW, Hilgenfeld R, Mesters JR, Holst O (2009) WaaA of the hyperthermophilic bacterium *Aquifex aeolicus* is a monofunctional 3-deoxy-D-*manno*-oct-2-ulosonic acid transferase involved in lipopolysaccharide biosynthesis. J Biol Chem 284:22248–22262
169. Noah C, Brabetz W, Gronow S, Brade H (2001) Cloning, sequencing, and functional analysis of three glycosyltransferases involved in the biosynthesis of the inner core region of *Klebsiella pneumoniae* lipopolysaccharide. J Endotoxin Res 7:25–33
170. Belunis CJ, Mdluli KE, Raetz CR, Nano FE (1992) A novel 3-deoxy-D-*manno*-octulosonic acid transferase from *Chlamydia trachomatis* required for expression of the genus-specific epitope. J Biol Chem 267:18702–18707
171. Brabetz W, Lindner B, Brade H (2000) Comparative analyses of secondary gene products of 3-deoxy-D-*manno*-oct-2-ulosonic acid transferases from *Chlamydiaceae* in *Escherichia coli* K-12. Eur J Biochem 267:5458–5465
172. Löbau S, Mamat U, Brabetz W, Brade H (1995) Molecular cloning, sequence analysis, and functional characterization of the lipopolysaccharide biosynthetic gene *kdtA* encoding 3-deoxy-α-D-*manno*-octulosonic acid transferase of *Chlamydia pneumoniae* strain TW-183. Mol Microbiol 18:391–399
173. Mamat U, Löbau S, Persson K, Brade H (1994) Nucleotide sequence variations within the lipopolysaccharide biosynthesis gene *gseA* (Kdo transferase) among the *Chlamydia trachomatis* serovars. Microb Pathog 17:87–97
174. Nano FE, Caldwell HD (1985) Expression of the chlamydial genus-specific lipopolysaccharide epitope in *Escherichia coli*. Science 228:742–744
175. Isobe T, White KA, Allen AG, Peacock M, Raetz CR, Maskell DJ (1999) *Bordetella pertussis waaA* encodes a monofunctional 2-keto-3-deoxy-D-*manno*-octulosonic acid transferase that can complement an *Escherichia coli waaA* mutation. J Bacteriol 181:2648–2651
176. Moran AP, Knirel YA, Senchenkova SN, Widmalm G, Hynes SO, Jansson P-E (2002) Phenotypic variation in molecular mimicry between *Helicobacter pylori* lipopolysaccharides and human gastric epithelial cell surface glycoforms. Acid-induced phase variation in Lewisx and Lewisy expression by *H. pylori* lipopolysaccharides. J Biol Chem 277:5785–5795
177. Vinogradov E, Perry MB, Conlan JW (2002) Structural analysis of *Francisella tularensis* lipopolysaccharide. Eur J Biochem 269:6112–6118
178. Zhao J, Raetz CR (2010) A two-component Kdo hydrolase in the inner membrane of *Francisella novicida*. Mol Microbiol 78:820–836
179. Stead CM, Zhao J, Raetz CR, Trent MS (2010) Removal of the outer Kdo from *Helicobacter pylori* lipopolysaccharide and its impact on the bacterial surface. Mol Microbiol 78:837–852

180. Chung HS, Raetz CR (2010) Interchangeable domains in the Kdo transferases of *Escherichia coli* and *Haemophilus influenzae*. Biochemistry 49:4126–4137
181. Chen L, Coleman WG Jr (1993) Cloning and characterization of the *Escherichia coli* K-12 *rfa-2 (rfaC)* gene, a gene required for lipopolysaccharide inner core synthesis. J Bacteriol 175:2534–2540
182. Gronow S, Brabetz W, Brade H (2000) Comparative functional characterization *in vitro* of heptosyltransferase I (WaaC) and II (WaaF) from *Escherichia coli*. Eur J Biochem 267:6602–6611
183. Kadrmas JL, Raetz CR (1998) Enzymatic synthesis of lipopolysaccharide in *Escherichia coli*. Purification and properties of heptosyltransferase I. J Biol Chem 273:2799–2807
184. Schnaitman CA, Klena JD (1993) Genetics of lipopolysaccharide biosynthesis in enteric bacteria. Microbiol Rev 57:655–682
185. Sirisena DM, Brozek KA, MacLachlan PR, Sanderson KE, Raetz CR (1992) The *rfaC* gene of *Salmonella typhimurium*. Cloning, sequencing, and enzymatic function in heptose transfer to lipopolysaccharide. J Biol Chem 267:18874–18884
186. de Kievit TR, Lam JS (1997) Isolation and characterization of two genes, *waaC* (*rfaC*) and *waaF* (*rfaF*), involved in *Pseudomonas aeruginosa* serotype O5 inner-core biosynthesis. J Bacteriol 179:3451–3457
187. Petricoin EF III, Danaher RJ, Stein DC (1991) Analysis of the *lsi* region involved in lipooligosaccharide biosynthesis in *Neisseria gonorrhoeae*. J Bacteriol 173:7896–7902
188. Schwan ET, Robertson BD, Brade H, van Putten JP (1995) Gonococcal *rfaF* mutants express Rd2 chemotype LPS and do not enter epithelial host cells. Mol Microbiol 15:267–275
189. Zhou D, Lee NG, Apicella MA (1994) Lipooligosaccharide biosynthesis in *Neisseria gonorrhoeae*: cloning, identification and characterization of the α1,5 heptosyltransferase I gene (*rfaC*). Mol Microbiol 14:609–618
190. Jennings MP, Bisercic M, Dunn KL, Virji M, Martin A, Wilks KE, Richards JC, Moxon ER (1995) Cloning and molecular analysis of the *Isi1* (*rfaF*) gene of *Neisseria meningitidis* which encodes a heptosyl-2-transferase involved in LPS biosynthesis: evaluation of surface exposed carbohydrates in LPS mediated toxicity for human endothelial cells. Microb Pathog 19:391–407
191. Stojiljkovic I, Hwa V, Larson J, Lin L, So M, Nassif X (1997) Cloning and characterization of the *Neisseria meningitidis rfaC* gene encoding α-1,5 heptosyltransferase I. FEMS Microbiol Lett 151:41–49
192. Allen AG, Isobe T, Maskell DJ (1998) Identification and cloning of *waaF* (*rfaF*) from *Bordetella pertussis* and use to generate mutants of *Bordetella* spp. with deep rough lipopolysaccharide. J Bacteriol 180:35–40
193. Jimenez N, Canals R, Lacasta A, Kondakova AN, Lindner B, Knirel YA, Merino S, Regue M, Tomas JM (2008) Molecular analysis of three *Aeromonas hydrophila* AH-3 (serotype O34) lipopolysaccharide core biosynthesis gene clusters. J Bacteriol 190:3176–3184
194. Jimenez N, Lacasta A, Vilches S, Reyes M, Vazquez J, Aquillini E, Merino S, Regue M, Tomas JM (2009) Genetics and proteomics of *Aeromonas salmonicida* lipopolysaccharide core biosynthesis. J Bacteriol 191:2228–2236
195. Kanipes MI, Papp-Szabo E, Guerry P, Monteiro MA (2006) Mutation of *waaC*, encoding heptosyltransferase I in *Campylobacter jejuni* 81–176, affects the structure of both lipooligosaccharide and capsular carbohydrate. J Bacteriol 188:3273–3279
196. Klena JD, Gray SA, Konkel ME (1998) Cloning, sequencing, and characterization of the lipopolysaccharide biosynthetic enzyme heptosyltransferase I gene (*waaC*) from *Campylobacter jejuni* and *Campylobacter coli*. Gene 222:177–185
197. Coderch N, Pique N, Lindner B, Abitiu N, Merino S, Izquierdo L, Jimenez N, Tomas JM, Holst O, Regue M (2004) Genetic and structural characterization of the core region of the lipopolysaccharide from *Serratia marcescens* N28b (serovar O4). J Bacteriol 186:978–988
198. Hood DW, Deadman ME, Allen T, Masoud H, Martin A, Brisson JR, Fleischmann R, Venter JC, Richards JC, Moxon ER (1996) Use of the complete genome sequence information of

Haemophilus influenzae strain Rd to investigate lipopolysaccharide biosynthesis. Mol Microbiol 22:951–965

199. Gronow S, Oertelt C, Ervelä E, Zamyatina A, Kosma P, Skurnik M, Holst O (2001) Characterization of the physiological substrate for lipopolysaccharide heptosyltransferases I and II. J Endotoxin Res 7:263–270
200. Zamyatina A, Gronow S, Oertelt C, Puchberger M, Brade H, Kosma P (2000) Efficient chemical synthesis of the two anomers of ADP-L-*glycero*- and D-*glycero*-D-*manno*-heptopyranose allows the determination of the substrate specificities of bacterial heptosyltransferases. Angew Chem Int Ed Engl 39:4150–4153
201. Grizot S, Salem M, Vongsouthi V, Durand L, Moreau F, Dohi H, Vincent S, Escaich S, Ducruix A (2006) Structure of the *Escherichia coli* heptosyltransferase WaaC: binary complexes with ADP and ADP-2-deoxy-2-fluoro heptose. J Mol Biol 363:383–394
202. Hiratsuka K, Logan SM, Conlan JW, Chandan V, Aubry A, Smirnova N, Ulrichsen H, Chan KH, Griffith DW, Harrison BA, Li J, Altman E (2005) Identification of a D-*glycero*-D-*manno*-heptosyltransferase gene from *Helicobacter pylori*. J Bacteriol 187:5156–5165
203. Gronow S, Brabetz W, Lindner B, Brade H (2005) OpsX from *Haemophilus influenzae* represents a novel type of heptosyltransferase I in lipopolysaccharide biosynthesis. J Bacteriol 187:6242–6247
204. Kingsley MT, Gabriel DW, Marlow GC, Roberts PD (1993) The *opsX* locus of *Xanthomonas campestris* affects host range and biosynthesis of lipopolysaccharide and extracellular polysaccharide. J Bacteriol 175:5839–5850
205. Brozek KA, Kadrmas JL, Raetz CR (1996) Lipopolysaccharide biosynthesis in *Rhizobium leguminosarum*. Novel enzymes that process precursors containing 3-deoxy-D-*manno*-octulosonic acid. J Biol Chem 271:32112–32118
206. Kadrmas JL, Brozek KA, Raetz CR (1996) Lipopolysaccharide core glycosylation in *Rhizobium leguminosarum*. An unusual mannosyl transferase resembling the heptosyl transferase I of *Escherichia coli*. J Biol Chem 271:32119–32125
207. Kanipes MI, Ribeiro AA, Lin S, Cotter RJ, Raetz CR (2003) A mannosyl transferase required for lipopolysaccharide inner core assembly in *Rhizobium leguminosarum*. Purification, substrate specificity, and expression in *Salmonella waaC* mutants. J Biol Chem 278:16356–16364
208. Schwingel JM, St Michael F, Cox AD, Masoud H, Richards JC, Campagnari AA (2008) A unique glycosyltransferase involved in the initial assembly of *Moraxella catarrhalis* lipooligosaccharides. Glycobiology 18:447–455
209. Geurtsen J, Dzieciatkowska M, Steeghs L, Hamstra HJ, Boleij J, Broen K, Akkerman G, El Hassan H, Li J, Richards JC, Tommassen J, van der Ley P (2009) Identification of a novel lipopolysaccharide core biosynthesis gene cluster in *Bordetella pertussis*, and influence of core structure and lipid A glucosamine substitution on endotoxic activity. Infect Immun 77:2602–2611
210. Kanipes MI, Tan X, Akelaitis A, Li J, Rockabrand D, Guerry P, Monteiro MA (2008) Genetic analysis of lipooligosaccharide core biosynthesis in *Campylobacter jejuni* 81-176. J Bacteriol 190:1568–1574
211. Tullius MV, Phillips NJ, Scheffler NK, Samuels NM, Munson JR Jr, Hansen EJ, Stevens-Riley M, Campagnari AA, Gibson BW (2002) The *lbgAB* gene cluster of *Haemophilus ducreyi* encodes a β-1,4-galactosyltransferase and an α-1,6-D,D-heptosyltransferase involved in lipooligosaccharide biosynthesis. Infect Immun 70:2853–2861
212. St Michael F, Vinogradov E, Li J, Cox AD (2005) Structural analysis of the lipopolysaccharide from *Pasteurella multocida* genome strain Pm70 and identification of the putative lipopolysaccharide glycosyltransferases. Glycobiology 15:323–333
213. Pinta E, Duda KA, Hanuszkiewicz A, Salminen TA, Bengoechea JA, Hyytiäinen H, Lindner B, Radziejewska-Lebrecht J, Holst O, Skurnik M (2010) Characterization of the six glycosyltransferases involved in the biosynthesis of *Yersinia enterocolitica* serotype O:3 lipopolysaccharide outer core. J Biol Chem 285:28333–28342

214. Austin EA, Graves JF, Hite LA, Parker CT, Schnaitman CA (1990) Genetic analysis of lipopolysaccharide core biosynthesis by *Escherichia coli* K-12: insertion mutagenesis of the *rfa* locus. J Bacteriol 172:5312–5325
215. Kadam SK, Rehemtulla A, Sanderson KE (1985) Cloning of *rfaG*, *B*, *I*, and *J* genes for glycosyltransferase enzymes for synthesis of the lipopolysaccharide core of *Salmonella typhimurium*. J Bacteriol 161:277–284
216. Martinez-Fleites C, Proctor M, Roberts S, Bolam DN, Gilbert HJ, Davies GJ (2006) Insights into the synthesis of lipopolysaccharide and antibiotics through the structures of two retaining glycosyltransferases from family GT4. Chem Biol 13:1143–1152
217. Muller E, Hinckley A, Rothfield L (1972) Studies of phospholipid-requiring bacterial enzymes. Purification and properties of uridine diphosphate glucose:lipopolysaccharide glucosyltransferase I. J Biol Chem 247:2614–2622
218. Parker CT, Pradel E, Schnaitman CA (1992) Identification and sequences of the lipopolysaccharide core biosynthetic genes *rfaQ*, *rfaP*, and *rfaG* of *Escherichia coli* K-12. J Bacteriol 174:930–934
219. Heinrichs DE, Yethon JA, Whitfield C (1998) Molecular basis for structural diversity in the core regions of the lipopolysaccharides of *Escherichia coli* and *Salmonella enterica*. Mol Microbiol 30:221–232
220. Heinrichs DE, Whitfield C, Valvano MA (1999) Biosynthesis and genetics of lipopolysaccharide core. In: Brade H, Opal SM, Vogel SN, Morrison DC (eds) Endotoxin in health and disease. Marcel Dekker, New York, pp 305–330
221. Kaniuk NA, Monteiro MA, Parker CT, Whitfield C (2002) Molecular diversity of the genetic loci responsible for lipopolysaccharide core oligosaccharide assembly within the genus *Salmonella*. Mol Microbiol 46:1305–1318
222. Carstenius P, Flock JI, Lindberg A (1990) Nucleotide sequence of *rfaI* and *rfaJ* genes encoding lipopolysaccharide glycosyl transferases from *Salmonella typhimurium*. Nucleic Acids Res 18:6128
223. Endo A, Rothfield L (1969) Studies of a phospholipid-requiring bacterial enzyme. I. Purification and properties of uridine diphosphate galactose: lipopolysaccharide α-3-galactosyl transferase. Biochemistry 8:3500–3507
224. Pradel E, Parker CT, Schnaitman CA (1992) Structures of the *rfaB*, *rfaI*, *rfaJ*, and *rfaS* genes of *Escherichia coli* K-12 and their roles in assembly of the lipopolysaccharide core. J Bacteriol 174:4736–4745
225. Whitfield C, Kaniuk N, Frirdich E (2003) Molecular insights into the assembly and diversity of the outer core oligosaccharide in lipopolysaccharides from *Escherichia coli* and *Salmonella*. J Endotoxin Res 9:244–249
226. Heinrichs DE, Yethon JA, Amor PA, Whitfield C (1998) The assembly system for the outer core portion of R1- and R4-type lipopolysaccharides of *Escherichia coli*. The R1 core-specific β-glucosyltransferase provides a novel attachment site for O-polysaccharides. J Biol Chem 273:29497–29505
227. Heinrichs DE, Monteiro MA, Perry MB, Whitfield C (1998) The assembly system for the lipopolysaccharide R2 core-type of *Escherichia coli* is a hybrid of those found in *Escherichia coli* K-12 and *Salmonella enterica*. Structure and function of the R2 WaaK and WaaL homologs. J Biol Chem 273:8849–8859
228. Leipold MD, Vinogradov E, Whitfield C (2007) Glycosyltransferases involved in biosynthesis of the outer core region of *Escherichia coli* lipopolysaccharides exhibit broader substrate specificities than is predicted from lipopolysaccharide structures. J Biol Chem 282:26786–26792
229. Leipold MD, Kaniuk NA, Whitfield C (2007) The C-terminal domain of the *Escherichia coli* WaaJ glycosyltransferase is important for catalytic activity and membrane association. J Biol Chem 282:1257–1264
230. Persson K, Ly HD, Dieckelmann M, Wakarchuk WW, Withers SG, Strynadka NC (2001) Crystal structure of the retaining galactosyltransferase LgtC from *Neisseria meningitidis* in complex with donor and acceptor sugar analogs. Nat Struct Biol 8:166–175

231. MacLachlan PR, Kadam SK, Sanderson KE (1991) Cloning, characterization, and DNA sequence of the *rfaLK* region for lipopolysaccharide synthesis in *Salmonella typhimurium* LT2. J Bacteriol 173:7151–7163
232. Banerjee A, Wang R, Uljon SN, Rice PA, Gotschlich EC, Stein DC (1998) Identification of the gene (*lgtG*) encoding the lipooligosaccharide β chain synthesizing glucosyl transferase from *Neisseria gonorrhoeae*. Proc Natl Acad Sci USA 95:10872–10877
233. Erwin AL, Haynes PA, Rice PA, Gotschlich EC (1996) Conservation of the lipooligosaccharide synthesis locus *lgt* among strains of *Neisseria gonorrhoeae*: requirement for *lgtE* in synthesis of the 2C7 epitope and of the β chain of strain 15253. J Exp Med 184:1233–1241
234. Gotschlich EC (1994) Genetic locus for the biosynthesis of the variable portion of *Neisseria gonorrhoeae* lipooligosaccharide. J Exp Med 180:2181–2190
235. Kahler CM, Carlson RW, Rahman MM, Martin LE, Stephens DS (1996) Two glycosyltransferase genes, *lgtF* and *rfaK*, constitute the lipooligosaccharide ice (inner core extension) biosynthesis operon of *Neisseria meningitidis*. J Bacteriol 178:6677–6684
236. Wakarchuk W, Martin A, Jennings MP, Moxon ER, Richards JC (1996) Functional relationships of the genetic locus encoding the glycosyltransferase enzymes involved in expression of the lacto-*N*-neotetraose terminal lipopolysaccharide structure in *Neisseria meningitidis*. J Biol Chem 271:19166–19173
237. Hood DW, Deadman ME, Cox AD, Makepeace K, Martin A, Richards JC, Moxon ER (2004) Three genes, *lgtF*, *lic2C* and *lpsA*, have a primary role in determining the pattern of oligosaccharide extension from the inner core of *Haemophilus influenzae* LPS. Microbiology 150:2089–2097
238. Weiser JN, Lindberg AA, Manning EJ, Hansen EJ, Moxon ER (1989) Identification of a chromosomal locus for expression of lipopolysaccharide epitopes in *Haemophilus influenzae*. Infect Immun 57:3045–3052
239. Yildirim HH, Li J, Richards JC, Hood DW, Moxon ER, Schweda EK (2005) An alternate pattern for globoside oligosaccharide expression in *Haemophilus influenzae* lipopolysaccharide: structural diversity in nontypeable strain 1124. Biochemistry 44:5207–5224
240. Edwards KJ, Allen S, Gibson BW, Campagnari AA (2005) Characterization of a cluster of three glycosyltransferase enzymes essential for *Moraxella catarrhalis* lipooligosaccharide assembly. J Bacteriol 187:2939–2947
241. Preston A, Mandrell RE, Gibson BW, Apicella MA (1996) The lipooligosaccharides of pathogenic Gram-negative bacteria. Crit Rev Microbiol 22:139–180
242. van Putten JP, Robertson BD (1995) Molecular mechanisms and implications for infection of lipopolysaccharide variation in *Neisseria*. Mol Microbiol 16:847–853
243. Griffiss JM, Schneider H (1999) The chemistry and biology of lipooligosaccharides: the endotoxins of bacteria of the respiratory and genital mucosae. In: Brade H, Opal SM, Vogel SN, Morrison DC (eds) Endotoxin in health and disease. Marcel Dekker, New York, pp 179–193
244. Schneider H, Hale TL, Zollinger WD, Seid RC Jr, Hammack CA, Griffiss JM (1984) Heterogeneity of molecular size and antigenic expression within lipooligosaccharides of individual strains of *Neisseria gonorrhoeae* and *Neisseria meningitidis*. Infect Immun 45:544–549
245. Filiatrault MJ, Gibson BW, Schilling B, Sun S, Munson RS Jr, Campagnari AA (2000) Construction and characterization of *Haemophilus ducreyi* lipooligosaccharide (LOS) mutants defective in expression of heptosyltransferase III and β1,4-glucosyltransferase: identification of LOS glycoforms containing lactosamine repeats. Infect Immun 68:3352–3361
246. Gilbert M, Karwaski MF, Bernatchez S, Young NM, Taboada E, Michniewicz J, Cunningham AM, Wakarchuk WW (2002) The genetic bases for the variation in the lipo-oligosaccharide of the mucosal pathogen. *Campylobacter jejuni*. Biosynthesis of sialylated ganglioside mimics in the core oligosaccharide. J Biol Chem 277:327–337

247. Faglin I, Tiralongo J, Wilson JC, Collins PM, Peak IR (2010) Biochemical analysis of Lgt3, a glycosyltransferase of the bacterium *Moraxella catarrhalis*. Biochem Biophys Res Commun 393:609–613
248. Regue M, Climent N, Abitiu N, Coderch N, Merino S, Izquierdo L, Altarriba M, Tomas JM (2001) Genetic characterization of the *Klebsiella pneumoniae waa* gene cluster, involved in core lipopolysaccharide biosynthesis. J Bacteriol 183:3564–3573
249. Deadman ME, Lundstrom SL, Schweda EK, Moxon ER, Hood DW (2006) Specific amino acids of the glycosyltransferase LpsA direct the addition of glucose or galactose to the terminal inner core heptose of *Haemophilus influenzae* lipopolysaccharide via alternative linkages. J Biol Chem 281:29455–29467
250. Wilkinson RG, Stocker BA (1968) Genetics and cultural properties of mutants of *Salmonella typhimurium* lacking glucosyl or galactosyl lipopolysaccharide transferases. Nature 217:955–957
251. Lindberg AA, Hellerqvist C-G (1980) Rough mutants of *Salmonella typhimurium*: immunochemical and structural analysis of lipopolysaccarides from *rfaH* mutants. J Gen Microbiol 116:25–32
252. Bailey MJA, Hughes C, Koronakis V (1997) RfaH and the *ops* element, components of a novel system controlling bacterial transcription elongation. Mol Microbiol 26:845–851
253. Nieto JM, Bailey MJA, Hughes C, Koronakis V (1996) Suppression of transcription polarity in the *Escherichia coli* haemolysin operon by a short upstream element shared by polysaccharide and DNA transfer determinants. Mol Microbiol 19:705–713
254. Hobbs M, Reeves PR (1994) The JUMPstart sequence: a 39 bp element common to several polysaccharide gene clusters. Mol Microbiol 12:855–856
255. Marolda CL, Valvano MA (1998) Promoter region of the *Escherichia coli* O7-specific lipopolysaccharide gene cluster: structural and functional characterization of an upstream untranslated mRNA sequence. J Bacteriol 180:3070–3079
256. Artsimovitch I, Landick R (2002) The transcriptional regulator RfaH stimulates RNA chain synthesis after recruitment to elongation complexes by the exposed nontemplate DNA strand. Cell 109:193–203
257. Belogurov GA, Mooney RA, Svetlov V, Landick R, Artsimovitch I (2009) Functional specialization of transcription elongation factors. EMBO J 28:112–122
258. Belogurov GA, Sevostyanova A, Svetlov V, Artsimovitch I (2010) Functional regions of the N-terminal domain of the antiterminator RfaH. Mol Microbiol 76:286–301
259. Karow M, Raina S, Georgopoulos C, Fayet O (1991) Complex phenotypes of null mutations in the *htr* genes, whose products are essential for *Escherichia coli* growth at elevated temperatures. Res Microbiol 142:289–294
260. Hoffman J, Lindberg B, Brubaker RR (1980) Structural studies of the O-specific side-chains of the lipopolysaccharide from *Yersinia enterocolitica* Ye 128. Carbohydr Res 78:212–214
261. Skurnik M, Venho R, Toivanen P, Al-Hendy A (1995) A novel locus of *Yersinia enterocolitica* serotype O:3 involved in lipopolysaccharide outer core biosynthesis. Mol Microbiol 17:575–594
262. Skurnik M, Venho R, Bengoechea J-A, Moriyón I (1999) The lipopolysaccharide outer core of *Yersinia enterocolitica* serotype O:3 is required for virulence and plays a role in outer membrane integrity. Mol Microbiol 31:1443–1462
263. Nizet V (2006) Antimicrobial peptide resistance mechanisms of human bacterial pathogens. Curr Issues Mol Biol 8:11–26
264. Monreal D, Grillo MJ, Gonzalez D, Marin CM, De Miguel MJ, Lopez-Goni I, Blasco JM, Cloeckaert A, Moriyon I (2003) Characterization of *Brucella abortus* O-polysaccharide and core lipopolysaccharide mutants and demonstration that a complete core is required for rough vaccines to be efficient against *Brucella abortus* and *Brucella ovis* in the mouse model. Infect Immun 71:3261–3271
265. Ramjeet M, Deslandes V, St MF, Cox AD, Kobisch M, Gottschalk M, Jacques M (2005) Truncation of the lipopolysaccharide outer core affects susceptibility to antimicrobial

peptides and virulence of *Actinobacillus pleuropneumoniae* serotype 1. J Biol Chem 280:39104–39114

266. Fresno S, Jimenez N, Izquierdo L, Merino S, Corsaro MM, De CC, Parrilli M, Naldi T, Regue M, Tomas JM (2006) The ionic interaction of *Klebsiella pneumoniae* K2 capsule and core lipopolysaccharide. Microbiology 152:1807–1818
267. Cortes G, Borrell N, de Astorza B, Gomez C, Sauleda J, Alberti S (2002) Molecular analysis of the contribution of the capsular polysaccharide and the lipopolysaccharide O side chain to the virulence of *Klebsiella pneumoniae* in a murine model of pneumonia. Infect Immun 70:2583–2590
268. Regue M, Izquierdo L, Fresno S, Pique N, Corsaro MM, Naldi T, De CC, Waidelich D, Merino S, Tomas JM (2005) A second outer-core region in *Klebsiella pneumoniae* lipopolysaccharide. J Bacteriol 187:4198–4206
269. Sahly H, Keisari Y, Crouch E, Sharon N, Ofek I (2008) Recognition of bacterial surface polysaccharides by lectins of the innate immune system and its contribution to defense against infection: the case of pulmonary pathogens. Infect Immun 76:1322–1332

Genetics, Biosynthesis and Assembly of O-Antigen

9

Miguel A. Valvano, Sarah E. Furlong, and Kinnari B. Patel

9.1 Introduction

Lipopolysaccharide (LPS), a major component of the outer leaflet of the Gram-negative bacterial outer membrane [1], consists of lipid A, core oligosaccharide (OS), and O-specific polysaccharide or O-antigen [1, 2]. LPS is a surface molecule unique to Gram-negative bacteria that plays a key role as an elicitor of innate immune responses, ranging from localized inflammation to disseminated sepsis [3]. The O-antigen, which is the most surface-exposed LPS moiety, also contributes to pathogenicity by protecting invading bacteria from bactericidal host responses [2]. A detailed understanding of the biosynthesis of the LPS O-antigen may contribute to identifying new means to curtail infections by interfering with its assembly.

LPS is synthesized at the cytoplasmic membrane followed by the transit of the molecule to the outer leaflet of the outer membrane, where it becomes surface exposed. The O-antigen is synthesized as a lipid-linked saccharide intermediate. The lipid component is undecaprenyl phosphate (Und-P), a C_{55} polyisoprenol (Fig. 9.1). The biogenesis of O-antigen in bacteria and lipid-linked saccharide moieties in eukaryotic protein N-glycosylation are remarkably similar (Fig. 9.2) [4–8], underscoring the general biological relevance of O-antigen biosynthesis pathways.

M.A. Valvano (✉) • S.E. Furlong • K.B. Patel
Centre for Human Immunology and Department of Microbiology and Immunology, University of Western Ontario, London, ON N6A 5C1, Canada
e-mail: mvalvano@uwo.ca; sfurlon@uwo.ca; kpatel59@uwo.ca

Y.A. Knirel and M.A. Valvano (eds.), *Bacterial Lipopolysaccharides*,
DOI 10.1007/978-3-7091-0733-1_9, © Springer-Verlag/Wien 2011

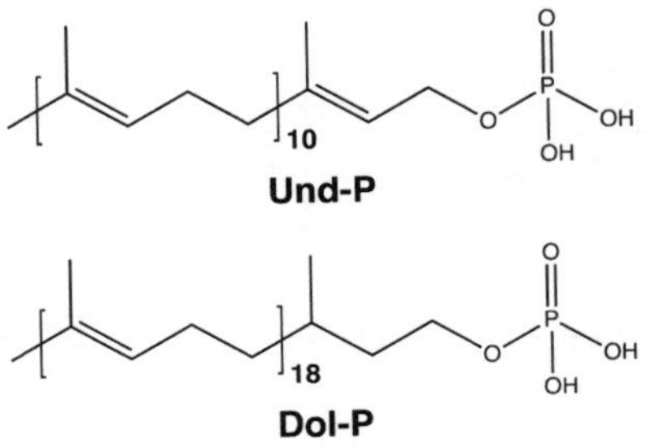

Fig. 9.1 Undecaprenyl phosphate (Und-P) and dolichyl phosphate (Dol-P) structures

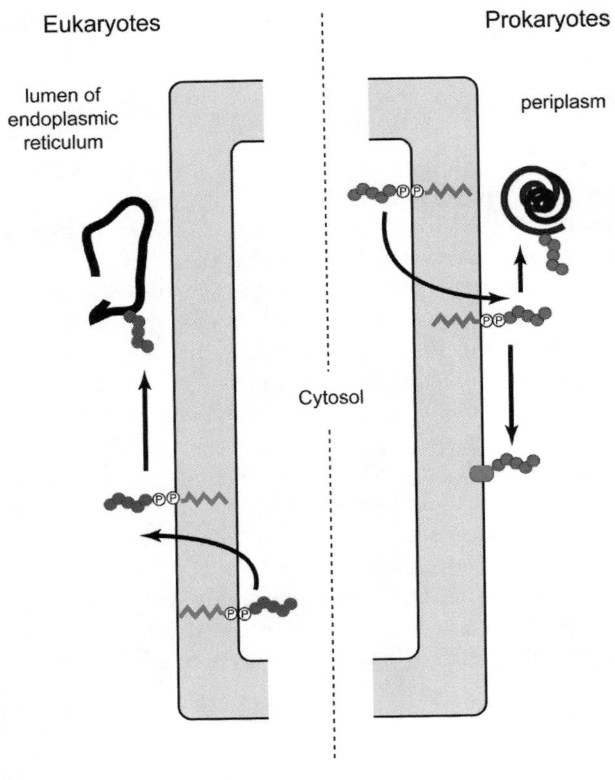

Fig. 9.2 Comparison of the lipid-linked glycan biosynthesis in eukaryotes and prokaryotes. In both systems, lipid-linked glycans are formed in association with isoprenoid lipids, and they must cross the corresponding membranes (endoplasmic reticulum or cellular membrane) before the glycan can be further transferred to donor molecules such as proteins (for N- and O-glycosylation) or lipids (lipid A-core OS)

LPS biosynthesis requires a large number of enzymes encoded by more than 40 genes [8–10]. O-antigens are polymers of OS repeats or O-units. The chemical composition, structure, and antigenicity of O-antigens vary widely among Gram-negative bacteria, giving rise to a large number of O-serotypes [11]. Many O-antigens are also virulence factors as they enable pathogenic bacteria to escape killing by phagocytosis and serum complement [12–14]. Lipid A, the membrane embedded portion of LPS, forms the majority of the outer lipid leaflet of the outer membrane. The core OS, which is made of hexoses, *glycero-manno*-heptose, and 2-keto-3-deoxyoctulosonic acid (Kdo) [15], is assembled on pre-formed lipid A by sequential glycosyltransfer reactions. In a separate pathway, the O-antigen is assembled on Und-P (Fig. 9.1) forming an Und-PP-linked saccharide. Both pathways converge by the ligation of O-antigen onto the lipid A-core OS, with the release of Und-PP [8, 9, 15, 16]. Und-PP is recycled into Und-P by a poorly characterized but conserved pathway [5, 8, 17] that involves the hydrolysis of the terminal phosphate [18–21]. The C_{55} Und-P is also the lipid carrier for the biosynthesis of enterobacterial common antigen (ECA) [22], peptidoglycan [23], and teichoic acid [24]. ECA is a surface glycolipid similar to the O-antigen but not commonly attached to lipid A-core OS [22], which can be assembled as a cyclic periplasmic soluble form [25, 26] or anchored to a diacylglycerophosphate [27]. In eukaryotes, the polyisoprenoid lipid carrier is C_{95} dolichyl phosphate (Dol-P) (Fig. 9.1) [5].

Polyisoprenols are lipid carrier intermediates used for the biogenesis of complex carbohydrate structures in all cells. In particular, cells employ phosphoisoprenol-linked saccharides for the early stages of protein glycosylation in eukaryotes and prokaryotes, as well as in the synthesis of bacterial cell walls and surface polysaccharides [4]. Nucleotide sugars, which donate carbohydrates for the synthesis of the saccharide moiety, are available as soluble molecules in cytosolic compartments (see Chap. 7). In contrast, phosphoisoprenol lipids are embedded within lipid membrane bilayers. Once assembled, phosphoisoprenol-linked saccharide molecules must cross the lipid bilayer for further processing. Thus, transmembrane movement of phosphoisoprenol-linked saccharides is an obligatory, conserved step of significant biologic importance in all types of cells (Fig. 9.2) [4, 28].

In the subsequent sections we will discuss the current mechanistic understanding of the synthesis and assembly of the O-antigen. Other chapters in this book deal with the synthesis and assembly of other regions of the LPS molecule (Chaps. 6, 8, and 10) and the evolution of LPS biosynthesis genes (Chap. 11).

9.2 Initiation of O-Antigen Biosynthesis

O-antigen biosynthesis initiation occurs at the cytoplasmic face of the bacterial cytosolic membrane. The cytosolic membrane contains Und-P that serves as a substrate for the initiation enzymes and as a lipid carrier for the synthesis of O-antigen, ECA, and cell wall synthesis [8, 29, 30]. The initiation enzymes are membrane proteins that catalyze the formation of a phosphoanhydride bond between Und-P and the first sugar 1-phosphate of the O-antigen unit transferred from uridine diphosphate (UDP), resulting in the release of uridine monophosphate (UMP). The enzymes initiating O-antigen synthesis form the Und-PP-saccharide precursor, which serves as a carrier glycolipid for other sugar transferases that sequentially add additional sugars to make the O-antigen polysaccharide. Unlike the initiating transferases, glycosyltransferases involved in the formation and elongation of the O-unit catalyze glycoside bond formation.

Two superfamilies of enzymes catalyze the initiation reaction: polyisoprenyl-phosphate *N*-acetylaminosugar-1-phosphate transferases (PNPT) and polyisoprenyl-phosphate hexose-1-phosphate transferases (PHPT) [8].

9.2.1 Polyisoprenyl-Phosphate *N*-Acetylaminosugar-1-Phosphate Transferases

The PNPT superfamily consists of multiple transmembrane domain (TMD) proteins that catalyze the transfer of *N*-acetylaminosugars. The family was previously thought to handle only *N*-acetylhexosamines [8], but it is now apparent that some members of the PNPT can transfer other *N*-acetylaminosugars (see below). PNPTs can be found in prokaryotes and eukaryotes. The eukaryotic members are UDP-GlcNAc:Dol-P GlcNAc-1-phosphate transferases (GPT) and reside in the rough endoplasmic reticulum (ER) membrane. The transfer reaction initiates the formation of Dol-PP-linked OSs, an essential step for protein N-glycosylation in eukaryotes [31]. Dol-P serves as a lipid carrier for C- and O-mannosylation, N-glycosylation of proteins, and the biosynthesis of glycosylphosphatidylinositol anchors [32].

Like GPT, prokaryotic members catalyze the transfer of the sugar 1-phosphate moiety of an UDP-sugar to a lipid carrier. However the specificities for the sugar and lipid carrier are different. Bacterial PNPTs are specific for Und-P, which contains 11 isoprenoid units with an unsaturated α-isoprenoid unit [30], while the eukaryotic GPTs are specific for Dol-P [33]. Und-P contains 11 isoprene units all of which are fully unsaturated, while Dol-P can be made of 15–19 isoprene units that have a saturated α-isoprene (Fig. 9.1) [34]. The α-isoprene is the phosphorylated end of the molecule, which participates in the phosphoanhydride bond formation with the *N*-acetylaminosugar-1-P. The ability of eukaryotic and bacterial enzymes to exquisitely discriminate their lipid substrate may reflect their evolutionary divergence.

While eukaryotic GPTs react with UDP-GlcNAc, the nucleotide sugar specificities of the prokaryotic PNPTs allow their classification into at least four subgroups [35]. The prototypic members of these subgroups are MraY, WecA, WbcO/WbpL and RgpG. The differences in their sugar specificities are proposed to depend on specific amino acid residues in their predicted fifth cytoplasmic domain (between IX and X TMDs) [35, 36]. MraY is an essential bacterial protein and shares the conserved residues that play functional roles in PNPTs. MraY is a key enzyme involved in the initiation of cell wall peptidoglycan synthesis by catalysing the transfer of *N*-acetylmuramyl-pentapeptide to Und-P. Despite similarities, MraY and the other PNPTs differ in their susceptibilities to various inhibitors such as tunicamycin (general PNPT inhibitor), and mureidomycin and liposidomycin (specific MraY inhibitors) [37]. Also, the MraY sugar substrate is a 9-carbon sugar, suggesting the key feature shared by the PNPT superfamily members is the ability to transfer 2-acetamido-2-deoxy-D-sugar 1-phosphates.

WecA-like transferases are specific for GlcNAc and initiate the synthesis of ECA [38] and O-antigens [29, 39–43]. Some O-antigens that require WecA for synthesis have *N*-acetylgalactosamine (GalNAc) at their reducing end. This led to the notion that WecA can indistinctly recognize UDP-GlcNAc or UDP-GalNAc as substrates. However, recent evidence shows that Und-PP-GalNAc is synthesized reversibly by a novel Und-PP-GlcNAc epimerase after the formation of Und-PP-GlcNAc by WecA [44].

WbcO/WbpL is proposed to utilize UDP-*N*-acetyl-D-fucosamine (UDP-FucNAc) and/or UDP-*N*-acetyl-D-quinovosamine (UDP-QuiNAc) [36, 45, 46]. WbpL in *Pseudomonas aeruginosa* initiates A- and B-band LPS biosynthesis through the addition of FucNAc (for B-band) or either GlcNAc or GalNAc (for A-band), suggesting that WbpL is bifunctional [46]. RgpG from *Streptococcus mutans* is an apparent WecA-WbcO hybrid [35] that can utilize either UDP-FucNAc or UDP-GlcNAc, which is required for rhamnose-glucose polysaccharide production [47]. However, the nucleotide sugar specificity of these proteins has not been directly confirmed biochemically.

9.2.1.1 Topology and Functional Motifs

The topology of the *Escherichia coli* and *Staphylococcus aureus* MraY transferases has been predicted and experimentally refined using β-lactamase fusions. Both proteins contain ten transmembrane segments [48]. The topology of the *E. coli* WecA, established by a combination of bioinformatics and the substituted cysteine accessibility method, reveals that the protein consists of 11 TMDs with the N-terminus residing in the periplasm and the C-terminus residing in the cytoplasm (Fig. 9.3) [49]. There are significant amino acid sequence similarities among PNPT family members, especially in regions predicted to face the cytoplasmic space, which suggests that these residues are involved in substrate or cofactor binding [36]. The PNPT family members have five cytosolic loops; 2, 3, and 4 are conserved among family members, while loops 1 and 5 show more diversity. Loop 5 is the least conserved among the subgroups [35].

The aspartic-acid rich DDxxD motif, which resides in the cytosolic loop II region of PNPT members, was postulated to play a role in binding the Mg^{2+} cofactor (Fig. 9.3). This was suggested since a similar motif is conserved in prenyltransferases that bind diphosphate-containing substrates via Mg^{2+} bridges [50, 51]. Amino acid replacements of the DD of the DDxxD motif in *E. coli* and *Bacillus subtilis* MraY, and *E. coli* WecA, result in a protein with reduced enzymatic activity [49, 52–54]. However, only the D90 residue of the *E. coli* WecA motif ($D_{90}D_{91}XXD_{94}$) was critical for binding Mg^{2+} [49].

PNPT members have proposed carbohydrate recognition (CR) domains in cytosolic loop 5 [35], which are variable among the various subfamilies (Fig. 9.3). The proposed CR domains are accessible to soluble nucleotide sugars and are highly basic (pI = 11.0), suggesting an interaction with an acidic ligand, such as a sugar

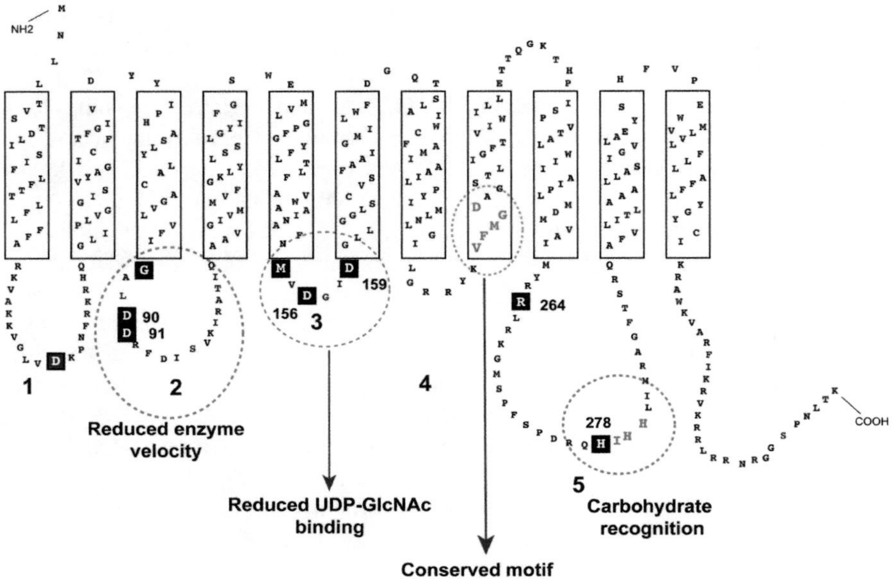

Fig. 9.3 Topology and functional region of the *E. coli* WecA GlcNAc-1-P transferase. The model was originally derived as described [49]. *Boldface* numbers indicate cytosolic loops 1–5. The residues spanning predicted transmembrane segments are enclosed in boxes. The residues with a *black square* background indicate residues that when mutated affect the function of WecA, as determined by genetic complementation of O7 LPS synthesis and by in vitro transferase assays [49]. *Dotted circles* indicate functional regions of WecA. The VFMGD motif (which is universally conserved in all members of the family; M.A. Furlong and M.A. Valvano, unpublished data) and the HIHH motif (which is important in UDP-sugar recognition; Ref. [55]) are indicated in *red*

substrate. Each subfamily of PNPT transferases shares highly conserved internal sequences that are unique to that subgroup, which have been postulated as having a role in UDP-sugar substrate binding [35]. Experimental evidence that the cytosolic loop 5 carries the CR domain was provided by Amer and Valvano [55] by showing that amino acid replacement of the conserved and highly basic $H_{278}IHH_{281}$ motif resulted in a loss of function of the enzyme and inability to bind tunicamycin, a nucleoside antibiotic which resembles the UDP-GlcNAc-polyisoprenoid lipid reaction intermediate [56].

9.2.1.2 Proposed Catalytic Mechanisms

PNPT family members share conserved amino acid sequences and function. Thus, it is conceivable they have a common enzymatic mechanism. Two alternative models exist to explain the reaction mechanism of PNPT members based on studies of the *B. subtilis* and *E. coli* MraY proteins. Al-Dabbagh et al. proposed a one-step reaction model whereby the enzyme uses an aspartic acid residue (residing in the DDxxD motif of cytosolic loop 2; Fig. 9.3) to remove a hydrogen atom from the phosphate of Und-P, which directly performs a nucleophilic attack on the

UDP-sugar [53]. This notion is based on mutational analyses of various conserved amino acids in the *B. subtilis* MraY, which demonstrated that no single residue substitution could lead to a completely nonfunctional enzyme when tested in an in vitro assay using purified protein. The first D of the DDxxD motif was the only residue mutant that could restore enzymatic activity with increasing pH, suggesting this aspartic acid can deprotonate the phosphate group of Und-P. This model is inconsistent with data from Lloyd et al. [52] on *E. coli* MraY, as well as with data suggesting a role for this aspartic acid residue in coordinating Mg^{2+} cofactor in the *E. coli* WecA [49]. The latter assays were conducted using membrane protein fractions; therefore the differences observed compared to the results by Al-Dabbagh et al. [53] could be due to the nature of the enzyme source.

The two-step reaction or double-displacement model proposes that an aspartic acid residue, residing in the highly conserved V/IFMGD motif (in the cytosolic loop IV of MraY) performs the nucleophilic attack on the UDP-sugar substrate forming a covalent-acyl enzyme intermediate and the release of UMP [52]. The second step involves the nucleophilic attack on the acyl-enzyme intermediate by Und-P, thereby completing the reaction [52, 57]. The highly conserved NxxNxxDGIDGL motif [36], present in all eukaryotic and prokaryotic PNPT members has been postulated to be involved in catalysis. Mutations at positions D156 and D159 of this motif in the *E. coli* WecA resulted in a severe reduction (less than 10% of the wild-type) in enzymatic activity and loss of tunicamycin and UDP-GlcNAc binding, suggesting that these residues may play a role in catalysis [49, 54]. Studying the catalytic mechanisms of the PNPT family members in detail has been hampered by the lack of purified enzyme in a purified in vitro system. The purification of the *Thermatoga maritima* WecA and *B. subtilis* MraY proteins [53, 58] will contribute to a better understanding of the kinetic parameters of these enzymes as well as potentially leading to structural analysis.

9.2.2 Polyisoprenyl-Phosphate Hexose-1-Phosphate Transferases

Members of the PHPT superfamily are integral membrane proteins that catalyze the synthesis of O-antigen, exopolysaccharide (EPS), capsular polysaccharide (CPS), and/or glycans for surface layer (S-layer) protein glycosylation and also general protein glycosylation [8, 59, 60]. To our knowledge, no eukaryotic proteins belong to the PHPT family. Therefore, these enzymes could be useful targets for the design of novel antimicrobial compounds. WbaP, a prototypic member of the PHPT family, is a cytosolic membrane protein that initiates O-antigen synthesis in *S. enterica* and capsule synthesis in *E. coli* K-30 by transferring galactose 1-phosphate from UDP-galactose to a phosphorylated lipid carrier (Und-P). Other galactose-1-phosphate transferases include WsaP, involved in S-layer glycosylation in *Geobacillus stearothermophilus* [60], and *Erwinia amylovora* AmsG [61], which initiates EPS synthesis. WbaP homologs such as *S. pneumoniae* Cps2E and *E. coli* K-12 WcaJ utilize UDP-glucose and initiate capsular and colanic acid synthesis, respectively [62, 63]. Similar to PNPT proteins, PHPT proteins studied in vitro require divalent cations Mg^{2+} or Mn^{2+} [60, 64, 65]. While all the previously mentioned members transfer galactose-1-P or glucose-1-P some outliers exist in the PHPT family that transfer sugars that are not simple hexoses. Examples of these proteins are PglC in *Campylobacter jejuni*, which transfers 2,4-diacetamido-2,4,6-trideoxyglucose-1-P [66], and PglB in *Neisseria* sp., which presumably transfers a 2(4)-acetamido-4(2)-glyceramido-2,4,6-trideoxyhexose-1-P [67, 68]. PglC and PglB are part of bacterial protein glycosylation systems.

9.2.2.1 Topology and Functional Domains

PHPT members can be large, hydrophobic, basic proteins with a predicted mass of around 56 kDa and five predicted transmembrane helices that are separated into three distinct domains: an N-terminal domain encompassing TM-I to TM-IV, a large periplasmic domain between TM-IV and TM-V and a large C-terminal cytoplasmic domain (Fig. 9.4) [64]. The N-terminal and periplasmic domains are conserved among a small group of O-antigen and LPS biosynthesis proteins in *Salmonella*, *E. coli*, *Haemophilus influenzae* and *Actinobacillus* whereas the C-terminal domain is highly conserved in a large number of PHPTs that are involved in the synthesis of polysaccharides from Gram-negative, Gram-positive and archaeal species [64]. The only region of WbaP required for galactosyltransferase activity is the C-terminal domain. Therefore, the presence of the C-terminal domain provides a characteristic signature to the members of this family. PHPT members such as PssY and PssZ of *Caulobacter crescentus*, required for holdfast synthesis [69], and the PglC and PglB proteins, discussed above, are well known examples of shorter PHPTs.

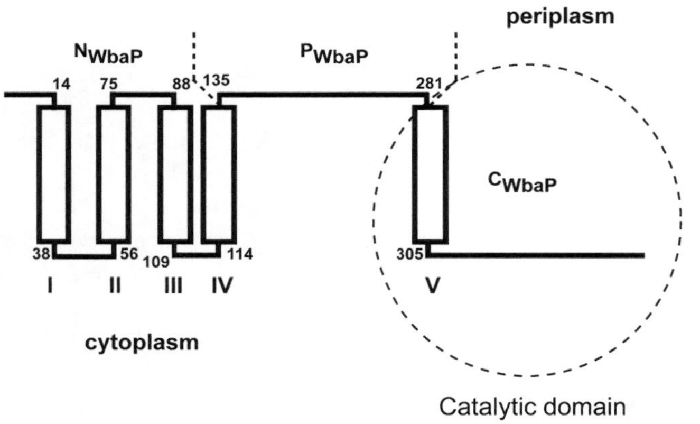

Fig. 9.4 Topological model of the WbaP galactose-1-P transferase. The model was generated as described [64, 70]. Open rectangles represent predicted TMs (indicated by *roman numerals*). The *numbers* indicate the amino acid positions of the boundary of each TM. *Dashed lines* indicate the boundaries of each of the three predicted domains of WbaP: the N-terminal domain (N_{WbaP}), the periplasmic domain (P_{WbaP}), and the cytosolic domain (C_{WbaP}). *Dotted circle* indicates that the C_{WbaP} carries the catalytic domain responsible for glycosyltransferase activity, which is the most conserved region in all members of this protein family

The function of the N-terminus and periplasmic domains of WbaP is not well understood. PHPT proteins with all five TMDs may be more stable [64]. The large periplasmic domain may also contribute to the stability of the protein. This periplasmic domain has regions of predicted secondary structure and many of the conserved residues in this domain are found in putative α-helices and β-strands [64]. Recent work, utilizing trypsin protease accessibility experiments, showed no cleavage in this region, despite the presence of multiple trypsin cleavage sites [70]. This work suggests that the predicted periplasmic domain may be located in the cytoplasm or it may have a packed secondary structure preventing the access of trypsin. Trypsin experiments also revealed that the first and second predicted loops are cytoplasmic and periplasmic respectively and that approximately the last 20 kDa of the protein are cytoplasmic, which was confirmed by fusing a green fluorescent protein reporter to the C-terminus of WbaP [70].

Deletion or overexpression of the periplasmic loop of WbaP in *S. enterica* affects the distribution of O-antigen chain length and the mutant proteins show reduced in vitro transferase activity. This was initially attributed to potential associations of this region to either the polymerase Wzy and/or the copolymerase Wzz [64]. In the Gram-positive bacterium *S. pneumoniae*, the extracellular loop of the glucosyltransferase Cps2E (equivalent to the periplasmic domain in WbaP) is important to modulate for the release of capsule from the cell and plays a role in modulating polymer assembly [65]. Therefore, it is possible that the central region in WbaP can exert an allosteric effect on the sugar transfer reaction and therefore affects the rate of polymer synthesis by regulating the rate of the initiation reaction.

The C-terminal domain, including TM-V, is required for function in all PHPT members studied to date. Residues that are involved in catalysis must reside in the cytoplasm as the initiation reaction occurs on the cytoplasmic face of the cytosolic membrane. Highly conserved residues within the 20-kDa region in WbaP, which was experimentally confirmed to be cytoplasmic, were investigated by introducing alanine replacements [70]. The function of the mutant proteins was assessed by in vivo complementation of O-antigen biosynthesis and an in vitro galactosyltransferase assay. Seven residues were identified as essential for activity [70], but the specific function of these amino acids, whether they interact with UDP-galactose, metal co-factor, or Und-P remains to be determined.

9.3 Assembly of O-Antigen

The Und-PP-linked sugar arising from the initiation reaction becomes an acceptor for subsequent additions of sugars to complete the formation of the O-unit. These reactions involve glycosidic bond formations and are catalyzed by specific glycosyltransferases. These proteins are classified in different families (see Refs. [71–73] for reviews) and they are typically associated with the cytoplasmic side of the membrane by ionic interactions. The cytoplasmic location of these enzymes is consistent with the notion that the assembly of Und-PP-linked O-units occurs at the cytosolic face of the plasma membrane. However, the ligation of Und-PP-linked

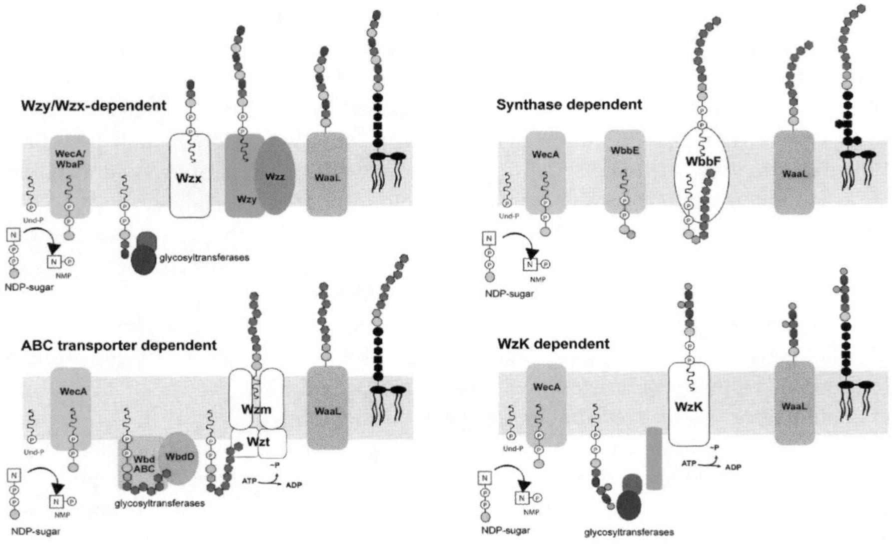

Fig. 9.5 Diagrams illustrating the known pathways for O-polysaccharide assembly

O-polysaccharides to lipid A-core OS takes place at the periplasmic face of the plasma membrane [74, 75], imposing the need for an obligatory mechanism whereby Und-PP-linked saccharides are translocated across the plasma membrane.

Four different mechanisms for the assembly of O-specific polysaccharides have been described: Wzy/Wzx-, ABC-transporter-, synthase-, and Wzk-dependent pathways (Fig. 9.5). The Wzy/Wzx pathway involves the synthesis of O-units by the sequential addition of monosaccharides at the non-reducing end of the molecule, a process that takes place on the cytosolic side of the plasma membrane [76]. These O-units are flipped across the plasma membrane, and are subsequently polymerized by Wzy using a mechanism involving the successive addition of the reducing end of the growing polymer to the non-reducing end of Und-PP-linked units (Fig. 9.5). The Und-PP-linked polymer is then ligated *en bloc* to the lipid A-core OS by reactions occurring on the periplasmic face of the membrane [74, 75, 77]. This pathway occurs in the synthesis of the majority of O-antigens, especially in those with repeating units made of different sugars (heteropolymeric O-antigens) [78]. The pathway also involves another protein, Wzx, which is a putative flippase, and it is always present in the gene clusters containing the *wzy* gene. Therefore, it would be more appropriate to refer to this pathway as the Wzy/Wzx-dependent pathway.

A second mechanism for O-antigen biosynthesis involves the formation of a polymeric O-antigen by reactions taking place on the cytosolic face of the plasma membrane (Fig. 9.5), which are mediated by the sequential action of glycosyltransferases elongating the polysaccharide at the non-reducing end [79]. The nascent polysaccharide is transported across the plasma membrane by a two-component

ATP-binding cassette transporter [78, 80], and subsequently ligated to lipid A-core OS. This pathway has been observed especially in the synthesis of O-antigens with repeating units made of the same sugar (homopolymeric O-antigens) such as those from *E. coli* O8 and O9 [76] and *Klebsiella pneumoniae* O1 [81], as well as in the synthesis of group 2 and 3 EPS capsules from *E. coli* [79].

9.3.1 Wzy/Wzx-Dependent Pathway

In contrast to the ABC transporter-dependent pathway for O-antigen biogenesis (see below), no obvious ABC transporters have been identified in Wzy/Wzx-dependent systems. At least three proteins (Wzx, Wzy, and Wzz) are involved in this export pathway but currently, there is no information concerning the manner in which these proteins interact with one another to facilitate the formation of predicted functional complexes. Once the individual Und-PP-linked O-units are formed, they must be exported to the site of polymerization at the periplasmic face of the plasma membrane (Fig. 9.5).

9.3.1.1 Und-PP-O-Antigen Translocation

All Wzx/Wzy-dependent O-antigen clusters studied to date contain a gene that encodes a plasma membrane protein designated Wzx, postulated as a candidate for the O-unit flippase or translocase [82]. Based on comparisons of hydrophobicity and secondary structure predictions, Wzx proteins are classified within a family of integral membrane proteins with 12 predicted transmembrane helices [83]. The membrane topology was experimentally confirmed for Wzx proteins of *S. enterica* serovar Typhimurium [84], *R. leguminosarum* (PssL) [85], *P. aeruginosa* [86], and *E. coli* O157 [87]. However, Wzx proteins share little primary amino acid sequence similarity, and their genes also have poor nucleotide sequence homology, to the extent that they can be used as genetic markers for molecular typing of *E. coli* strains expressing specific O-antigens [43, 88]. The involvement of Wzx proteins in the translocation of Und-PP-linked O-units was suggested based in experiments using a heterologous O-antigen expressing system in *S. enterica*. In vivo radio-labeling of O-antigen precursors in the presence of a *wzx* mutation indicated the accumulation of Und-PP-linked O-units with an apparent location on the cytoplasmic face of the plasma membrane [82]. However, the definitive localization was difficult since the *wzx* mutation only led to approximately 50% of the radiolabeled material being accumulated on the cytoplasmic face of the plasma membrane.

Wzx may facilitate the transit of Und-PP-linked O-units to the periplasmic face of the plasma membrane in a similar manner as a permease, and with the help of proton motive force as a source of energy. This could explain the absence of typical amino acid motifs indicative of ATP or GTP binding sites in the primary sequence of Wzx. Extensive mutagenesis of residues in the Wzx protein from *E. coli* O157 identified only four charged residues that yielded non-functional proteins when replaced by alanine [87]. Two of these residues were located in transmembrane helices, and each of the other two in internal and external soluble loops.

Conservative replacements at these positions demonstrated that the charge, but not the nature of the targeted amino acid is critical to retain the ability of Wzx to support O-antigen production. The results suggest that residues at these positions could be involved in making contacts with substrates directly or via water molecules. The presence of transmembrane helices with charged amino acids in Wzx proteins [84–87], together with the demonstration that at least some of these residues are functionally important [87], suggest the possibility of a tertiary structure whereby some of the helices may interact with each other and locate further away from the lipid bilayer. This is similar to the arrangement found for the LacY permease [89]. In agreement with this notion, the functional requirement of charged residues at both sides of the membrane and in two TM helices [87] could be important to create an electrostatic cavity [89, 90] and perhaps even electrostatic interactions with the phosphate groups of Und-PP-linked sugars, which may allow localized perturbation of the lipid bilayer to facilitate the movement of the Und-PP-linked saccharide substrate across the membrane. Remarkably, in vitro studies of interactions between hydrophobic peptides with lipid vesicles containing isoprenoid phosphates combined with molecular modelling studies of isoprenoid phosphates in artificial membranes [91, 92] support these conclusions.

The direct measurement of O-antigen translocase activity is critical to provide biochemical evidence on the mechanism of translocation, but addressing the topological orientation of Und-PP-linked sugars in the membrane is difficult. Some of these approaches also require purified Wzx protein and this is hard to achieve without using detergents that perturb native membrane structures. Therefore, although a translocase assay may reveal protein-dependent lipid flipping, the major challenge relies on maintaining the specificity of the assay for a particular protein. For example, an assay using a soluble analogue of Und-PP-GlcNAc was linked to a genetically identified translocase, WzxE, of *E. coli* K-12 [93]. However, the same strain contains another translocase, Wzx_{O16}, and despite that both WzxE and Wzx_{O16} are interchangeable [94], that assay did not detect an activity for Wzx_{O16} [93]. The biochemical assays to date require detergents to prepare membrane extracts, which generates denatured membrane proteins. This can provide continuity between the leaflets of the membranes and thereby serve as artefactual conduits for polar moieties [95–97]. The lack of a definitive biochemical assay to unequivocally determine "flipping" of the Und-PP-linked O-units complicates the functional analysis of Wzx proteins. Supporting a role for O-antigen translocation, recent work has demonstrated that PglK and Wzx are interchangeable [98]. However, PglK is an ABC transporter implicated in the flipping of Und-PP-linked saccharides from *C. jejuni* [98], which are the glycan component of a bacterial N-glycosylation machinery [99, 100].

Since the membrane translocation of Und-PP-linked O-units must be a conserved process, the absence of any obvious conserved motifs in the primary amino acid sequences of Wzx proteins is intriguing. One possible explanation for the abundant differences among Wzx proteins could be the requirement for the recognition of specific O-units, which are highly variable in terms of structure and sugar composition. However, a complete Und-PP-linked O-unit is not required for "flipping" since a single sugar, GlcNAc, can be incorporated into the lipid A-core OS of *E. coli* K-12 by a process requiring *wecA* and *wzx* [101]. This demonstrates that Und-PP-linked with the first sugar of the O-unit is the minimal substrate for Wzx-dependent translocation. Using a genetic system based on reconstructing O16 antigen synthesis in *E. coli* K-12, we showed that Wzx homologues from different bacteria complement an *E. coli* K-12 Δ*wzx* mutant [102]. More specifically, Wzx proteins from O-antigen systems that use Und-PP-GlcNAc or Und-PP-GalNAc for the initiation of the biosynthesis of the O-unit repeat can fully complement the formation of O16 LPS. Partial complementation was seen with Wzx from *P. aeruginosa*, a system that uses Und-PP-FucNAc in the initiation reaction, while the complementation with the Wzx protein from *S. enterica* (that uses Und-PP-Gal) was only possible under high levels of protein expression. Therefore, it would appear that Wzx proteins, like the initiating Und-PP-sugar transferases, occur in at least two functional classes that distinguish among Und-PP-bound *N*-acetylhexosamines or hexoses. However, it is not clear if this is due to specific recognition of the initiating sugar in the context of the phosphoisoprenoid lipid, an interaction between Wzx and the corresponding initiating transferase, or a combination of both.

9.3.1.2 Polymerization and Chain Length Regulation

The Wzy protein is required for the polymerization of Und-PP-linked O-units at the periplasmic face of the cytoplasmic membrane. The biochemical reaction involves transfer of nascent polymer from its Und-PP carrier to the reducing end of the new Und-PP-linked O-unit [103, 104], with the concomitant formation of a novel glycosidic bond joining the O-antigen units. The polymerase-mediated linkage brings another level of immunochemical variability to the O-antigens, as new serotypes can arise with the same O-units but different positions or/and anomeric forms of the glycosidic bond mediated by Wzy. For example, in the *S. enterica* serovar Anatum, one Wzy protein is responsible for joining the O-antigen repeats forming α-glycosidic linkages, while the other protein mediates β-glycosidic linkages, resulting in a different serotype [105]. Serotype conversion dependent on variations in the polymerase-mediated linkage of the O-units determined by lysogenic bacteriophages has been documented in *Salmonella* [106] and *P. aeruginosa* [107].

Mutants with defects in the *wzy* gene produce LPS with only a lipid A-core OS joined to a single O-unit [108]. Wzy proteins are integral membrane proteins with multiple predicted TMDs, and they exhibit little primary sequence similarity [109, 110]. Wzy proteins were examined with robust computer programs to predict topology and several were found to possess a relatively large predicted periplasmic loop, which may be important in the recognition of the O-unit or for catalytic activity. This topology has been experimentally confirmed in at least two Wzy proteins [86, 110]. In contrast to Wzx, Wzy proteins from different O-types are not interchangeable and display specificity for the cognate O-unit or for structures containing the same linkage between O-antigen units [2, 76]. This is consistent with the notion that Wzy proteins, like general glycosyltransferases, are highly specific for the O-units and it is likely that O-unit sugars provide a specific context for recognition by a cognate Wzy protein. The enzymatic mechanism of Wzy has not been resolved, in part because of the absence of distinguishing features in Wzy proteins, but also due to difficulties in expressing sufficient amounts for in vitro studies, which complicates the identification of catalytic and binding residues in Wzy. Recently, Wzy and Wzz proteins have been purified and the polymerization reaction was reconstituted in vitro, which represents a major achievement that will facilitate the better understanding of the polymerization mechanism as well as structural analysis of both proteins in association [111].

A curious but functionally important characteristic of polymerized O-antigens is the distribution of molecules with various chain lengths. This is not a random event but rather a strain-specific chain length distribution reflected in characteristic clusters of bands following gel electrophoresis of LPS samples [112]. Chain length distribution of O-antigen polysaccharides depends on Wzz. Wzz proteins appear not to be specific for a given O-repeat unit structure and they are not required for bacterial viability in the laboratory setting. However, there are numerous examples in the literature demonstrating that the distribution of the O-polysaccharide chain length is critical for virulence [113–119].

Wzz proteins reside in the plasma membrane, and all have two transmembrane helices flanking a periplasmic loop with a predicted coiled-coil structure [120]. Wzz belongs to a family of proteins called "polysaccharide co-polymerases" [120]. Some members can participate in the synthesis of capsules and have larger cytosolic C-terminal regions containing ATP-binding sites and several tyrosine residues that can become phosphorylated [121–125]. Phosphorylation and dephosphorylation of these proteins are important for the export of CPS (reviewed in Ref. [126]), but the specific mechanism of export remains unclear. More than one *wzz* gene has been observed in some microorganisms like *P. aeruginosa* [127] and *S. flexneri* [128]. It is not clear if the presence of additional Wzz activities would have an additive effect in the overall length of the O-polysaccharides or alternatively, they would be differentially required under varying physiological conditions.

Several models have been proposed to explain the modality in the polymerization process. Wzz was hypothesized to act as a timing clock, interacting with the Wzy polymerase and modulating its activity between two states that favour either chain elongation or chain termination caused by transfer of the O-polymer to the ligase [129]. An alternative model implicates Wzz as a molecular chaperone to assemble a complex consisting of Wzy, the WaaL ligase, and Und-PP-linked O-specific polysaccharide [130]. The specific modality would then be determined by different kinetics resulting from a Wzz-dependent ratio of Wzy relative to WaaL. However, it has been shown that ligation is not required for modality [100, 131]. The coiled-coil domains were proposed to be important for interactions of Wzz with Wzy, WaaL, or both. However, only Wzz oligomers have been identified by chemical cross-linking experiments [132], while a definitive evidence of cross-linking of Wzz to either Wzy or WaaL is lacking. Also, the regions of Wzz required for modality are not well defined [132, 133].

Recent work has shown that the periplasmic loop of the *E. coli* K-12 Wzz has extended conformation, and mutagenesis experiments suggest that the regions predicted as coiled coils are important for Wzz function by maintaining the native conformation of the protein, but these experiments do not support the existence of coiled-coils per se [134]. The elucidation of the crystal structures of the periplasmic domains of three Wzz homologues that impart substantially different chain length distributions to surface polysaccharides shows that they share a common protomer structure with a long extended central α-helix [135]. The protomers self-assemble into bell-shaped oligomers of variable sizes with a large internal cavity. Functional studies suggest that the top of the PCP oligomers is an important region for determining polysaccharide modal length. These observations have led to a new model for Wzz function in which the oligomers would organize the polymerization such that the size of the oligomer would determine the number of associated Wzy molecules and ultimately the length of the polymer [135]. A difficulty with this model is that Wzy is poorly expressed [110, 136], which would affect the stoichiometry of putative macromolecular complex. Also, there is no agreement in the number of protomers found in the complexes, which may vary depending on the methodology used to their isolation and characterization [135, 137, 138]. Regardless of the actual mechanism, a theme is emerging whereby a periplasmic protein (like Wzz or Wzc in capsule export) assembles into oligomeric structures that extend into the periplasmic space [139]. In the case of capsule export it is proposed that these complexes interact with an outer membrane protein channel required for the surface assembly of the polymer. For O-antigen synthesis, the putative complex may deliver the Und-PP-linked polysaccharide to the ligation step within the periplasmic space, but it still remains unclear how this process occurs, as efforts by several laboratories to demonstrate a complex have been unsuccessful to date (see Sect. 3.1.3). Recent work by Papadopoulos and Morona [140] shows a positive correlation between the stability of the Wzz oligomers and O-antigen chain length, suggesting that the ability of Wzz to form higher-order oligomers provides a scaffold that modulates the activity of Wzy, which is presumably associated to the complex. Although strong genetic evidence supports these associations [94], novel methods are needed to reveal the direct interactions between Wzy and Wzz by biochemical or microscopic means.

9.3.1.3 Membrane Complexes Formed by Wzy-Dependent Pathway Components

Several authors have suggested that the proteins of the Wzy-dependent pathway function as multi-protein complexes [130–132, 141, 142]. Also, it is possible that protein components for the assembly of the ECA, which is similar to the Wzy-dependent pathway, exist together in the plasma membrane as a complex [26]. Direct evidence exists for oligomerization in vivo of at least one of these proteins, Wzz, in *S. flexneri* [132], *E. coli* K-12/O16 [143], and *P. aeruginosa* [131]. However, efforts to provide biochemical evidence for the existence of a complex involving other proteins have been unsuccessful. Compelling genetic data support the notion that Wzy, Wzz, and Wzx work in concert as a functional complex [94].

O-antigen synthesis reconstitution experiments in *E. coli* K-12 reveal that the *wzxE* gene (encoding the translocase for ECA) can fully complement a wzx_{O16} deletion mutant only if the majority of the ECA gene cluster is deleted. Conversely, plasmids expressing either the WzyE polymerase or the WzzE chain-length regulator protein from the ECA cluster drastically reduce the O16 LPS complementing activity of WzxE. Similar results were observed with the O7 system and Wzx_{O7} can cross complement translocase defects in the O16 and O7 antigen clusters only in the absence of their cognate Wzz and Wzy proteins [94]. These genetic data strongly suggest that translocation of O-antigen across the plasma membrane and the subsequent assembly of periplasmic Und-PP-linked O-polysaccharide depends on interactions among Wzx, Wzz, and Wzy, which presumably form a multi-protein complex. Therefore, it is possible that multi-protein complexes at the plasma membrane exist for the translocation and assembly of the O-antigen. Additional evidence for the possible existence of complexes is that WecA, the initiating Und-PP-GlcNAc transferase, is located in discrete regions of the plasma membrane [49]. Early work in *Salmonella* has shown that new O-antigen LPS molecules appear on the cell surface at a limited number of sites ("adhesion zones") [144, 145] and more recently, experimental evidence supports the existence of multiprotein complexes for the assembly of CPSs, which may provide a molecular "scaffold" across the periplasm [146, 147]. Fractionation experiments have also shown that WecA is not only in the low-density plasma membrane fraction but also in a fraction of intermediate density which contains markers of both outer and plasma membrane proteins (Tatar and Valvano, unpublished data). These membrane fractions contain newly synthesized material that is exported to the outer membrane and are considered to be the biochemical equivalent of the "adhesion zones" [148–151].

9.3.1.4 Parallels of the Wzy-Dependent Pathway and Protein N-Glycosylation

The protein N-glycosylation pathway in eukaryotes has remarkable parallels with the biogenesis of Wzx/Wzy-dependent O-specific polysaccharides (Fig. 9.2). As in bacteria, the process can be divided into similar steps involving the assembly of a lipid-linked OS (analogous to the O-unit) on the cytosolic side of the ER, the translocation of this molecule across the ER membrane (analogous the flipping reaction mediated by Wzx), and the transfer of the OS from its lipid anchor to selected asparagines of nascent glycoproteins (analogous to the ligation reaction of the O-specific polymer with the lipid A-core OS molecules [5]). The initiation reaction that results in the biosynthesis of Dol-PP-GlcNAc is followed by subsequent reactions on the cytosolic side of the ER membrane that involve the addition of another GlcNAc and three mannose residues. These reactions are mediated by specific glycosyltransferases and result in the formation of a heptasaccharide-lipid linked intermediate, Dol-PP-$GlcNAc_2Man_5$. These reactions are analogous to those taking place on the cytosolic side of the bacterial plasma membrane, which result in the formation of the O-antigen units as well as other precursor molecules for peptidoglycan and cell surface polysaccharides in general.

The Rft1 protein [7] carries out the translocation of the lipid-linked heptasaccharide intermediate across the ER membrane in *Saccharomyces cerevisiae*. The *rft1* gene is highly conserved in eukaryotic genomes and a mutation in this gene is lethal in yeast as it results in protein glycosylation defects [152]. Rft1 is an integral membrane protein with 12 predicted TMDs, which in contrast to other lipid flippases [153], lacks any motifs indicative of ABC-type transporters. Interestingly, *Plasmodium falciparum*, an organism that lacks protein N-glycosylation also lacks a *rft1* gene homologue [154]. The function of Rft1 is analogous to that of bacterial Wzx. Both proteins share similarities in size and hydrophobicity plots, and both also lack identifiable motifs. However, the two families do not share any obvious similarities in their primary amino acid sequences. Whether these two protein families share a similar translocation mechanism or their structural differences reflect the nature of the different substrates it is presently unknown. The role of Rft1 in the flipping of Dol-PP-$GlcNAc_2Man_5$ has recently been challenged based strictly on biochemical reconstitution experiments [155], which unfortunately suffer from the same shortcomings described above (see Sect. 3.1.1) concerning the loss of specificity in these types of assays.

9.3.2 ABC Transporter-Dependent Pathway

The most significant features of this pathway are that the completion of the O-specific polysaccharide occurs at the cytosolic side of the cytoplasmic membrane and the export of the polymer to the outer face prior to ligation requires an ABC (ATP-binding cassette) transporter (Fig. 9.5). ABC transporters are not only involved in export of O-specific polysaccharides [156] but they also function in the export of lipid-linked glycans for the assembly of CPSs [157–159], teichoic acids [160], and glycoproteins [98]. A very comprehensive and recent review of ABC transporters for export of bacterial cell surface glycoconjugates, designated as glyco-ABC transporters, provides a classification of these transporters based on phylogenetic and functional features [161]. Seven different classes can be identified, six of which consist of independent pairs of TMD polypeptides and polypeptides containing the nucleotide-binding domains (NBDs). NBD and TMD containing proteins associate on the cytosolic face of the membrane. The seventh group (group G) consists of proteins that function as homodimers with each monomer containing one TMD and one NBD [161]. This group, which contains homologs of PglK [98] and MsbA [162], is required for the export of N-linked protein glycan in *C. jejuni* and lipid A, respectively. Furthermore, groups A and B contain NBD proteins with an extended C-terminal domain, which is absent in the other groups, and they can be found in all classes of prokaryotes, as well as in both O-antigen and glycoprotein biosynthesis systems. Group E is relatively homogeneous and contains NBDs involved in the export of the polyolphosphate teichoic acids in Gram-positive bacteria [161].

The biosynthesis of *E. coli* O8 and O9 has been used as a model system for the ABC-dependent transporter pathway [9]. Both O8 and O9 antigens are homopolymers of mannose and their structures and antigenicity depend on the different linkages formed between the mannose residues. One salient feature of the O-polysaccharides formed by this pathway is the participation of a primer Und-PP-GlcNAc intermediate followed by the addition of a sugar adaptor molecule. Although these O-polysaccharides are all initiated by the activity of WecA, they differ from Wzy/Wzx O-polysaccharides in that the GlcNAc residue transferred to lipid A-core OS during ligation occurs only once per chain, and thus it is not found within the O-repeat unit structure itself [163, 164]. Next, an O-polysaccharide-specific glycosyltransferase adds an adaptor sugar residue between the Und-PP-GlcNAc primer and the O-repeat unit domain, and this reaction also occurs only once per chain. Different enzymes are involved in adding the adaptor. In *E. coli* O9, adaptor formation involves the addition of a single mannose residue by WbdC [165]. In the other cases, such as some serotypes of *K. pneumoniae* LPS [164], the adaptor is added by the bifunctional galactosyltransferase, WbbO, which also participates in subsequent chain extension reactions [166, 167]. The O-polysaccharide is assembled in the chain extension phase by the processive transfer of residues to the nonreducing terminus of the Und-PP-linked acceptor [165, 166, 168], which may be mediated by either monofunctional or multifunctional transferases [165, 169, 170].

Due to the processive nature of the polymerization, an intriguing aspect about the polymers assembled by the ABC transport dependent pathway is their mode of termination. In the case of the Wzy/Wzx-exported O-polysaccharides, this process results from the interactions involving Wzy and Wzz. Despite that the ABC-transport dependent pathway does not involve a Wzz protein, the O-specific polysaccharides formed by this pathway display strain-specific chain length (modal) distributions [171]. The chain-terminating WbdD protein controls the length of these polymers by modifying their nonreducing end [172], causing termination of polymerization. The terminating residues are a methyl group in *E. coli* O8 and a phosphate plus a methyl group in *E. coli* O9 [172, 173]. The C-terminal region of WbdD physically interacts with WbdA6, which facilitates the association of a soluble glycosyltransferase to the membrane [174], and WbdD is also essential to couple biosynthesis and export [172]. Other types of termination includes the addition of a Kdo residue at the nonreducing end, as is the case in *K. pneumoniae* O4 and O12 [175]. However, the lack of detailed structural information in other polysaccharides assembled by this pathway precludes a comprehensive database of potential terminating residues.

The TMD component of the ABC-2 transporters is an integral membrane protein, Wzm, with an average of six TMDs, while the hydrophilic NBD protein is designated as Wzt. Genes encoding these two components are present within the O-polysaccharide biosynthesis clusters. As with other ABC-transporters involved in transmembrane export, Wzm homologues for O-polysaccharide biosynthesis display little primary sequence identity, but Wzt homologues are much more highly conserved, especially in the NBD domain. However, Wzm proteins are functionally interchangeable between different O-antigens, while Wzt proteins are not [176]. The extended C-terminal region of Wzt proteins of groups A and B determines the specificity [176] and recent structural data revealed that this domain forms a β-sandwich with an immunoglobulin-like fold that contains the O-polysaccharide-binding pocket [177]. Presumably, binding of the polysaccharide to this region would promote a conformational change in the NBD driving ATP hydrolysis, in turn promoting the interactions with Wzm which would result in the export of the Und-PP-linked O-polysaccharide.

The absence of an obvious C-terminal CR domain in groups C and D of ABC transporters suggests that a different mechanism exists for coupling polymerization with export. Very little mechanistic information is available for these two classes of ABC transporters. In the case of the D-galactan I found in O-serotypes of *K. pneumoniae*, capping occurs by an additional polysaccharide domain with a different structure, which can also define a new serotype [178, 179]. However, this may not be the case when additional "capping" polysaccharides are not obvious from available structural data.

9.3.3 Minor Export Pathways: Synthase- and Wzk-Mediated Pathways

The plasmid-encoded O:54 antigen of *S. enterica* serovar Borreze is the only known example of a synthase-dependent O-polysaccharide (Fig. 9.5) [180–182]. The O:54-specific polysaccharide is a homopolymer made of *N*-acetylmannosamine (ManNAc). WecA initiates the synthesis of the O:54 unit and the first ManNAc residue is transferred to the Und-PP-GlcNAc primer by the nonprocessive ManNAc transferase WbbE [183]. The second transferase, WbbF, belongs to the HasA (hyaluronan synthase) family of glycosaminoglycan glycosyltransferases [184], and it is proposed that this enzyme performs the chain-extension steps. Synthases are integral membrane proteins [184, 185], which appear to catalyze a vectorial polymerization reaction by a processive mechanism resulting in the extension of the polysaccharide chain with the simultaneous extrusion of the nascent polymer across the plasma membrane [185]. Although the exported polymer is presumably Und-PP-linked, there is little information on the exact mechanism of export mediated by WbbF as well as on the process leading to chain termination. The synthase family also has other members including the enzymes involved in biosynthesis of cellulose, chitin, and hyaluronan [184, 185], the type 2 and 3 capsules of *S. pneumoniae* [62, 186], as well as alginate [187] and poly-β-D-GlcNAc [188]. Two conserved domains, one likely involved in the glycosyltransfer reaction and the other implicated in the translocation of the nascent polymer characterize these enzymes. In the past 10 years, however, there has been little new information on the synthase mechanism.

Helicobacter pylori is the only known organism that synthesizes O-antigen by the recently discovered Wzk pathway [189]. The synthesis of this O-antigen begins with the formation of Und-PP-GlcNAc, mediated by WecA, and continues by the action of processive glycosyltransferases that add Gal and GlcNAc residues in an alternating fashion (Fig. 9.5). The Lewis antigen is generated by fucosyltransferases that attach fucose residues at select positions on the O-antigen backbone. The unique component of this pathway is the flippase Wzk, a homolog of *C. jejuni* PglK [98]. *C. jejuni* PglK mediates the membrane translocation of the Und-PP-linked heptasaccharide glycan used for protein glycosylation in the periplasmic space [98, 190]. To determine the role of WzK in *H. pylori*, complementation experiments in *E. coli* were performed where the N-glycosylation machinery was expressed in a glycan flippase *E. coli* mutant. Expression of a *C. jejuni* acceptor protein, AcrA, in *E. coli* was used to test for glycosylation. Wzx was also shown to restore flippase activity in a Wzx mutant of *E. coli*. Together these results indicate that WzK in *H. pylori* translocates the complete Und-PP- linked O-antigen polymer to the periplasmic space. Further studies are needed to determine whether Wzk, like PglK, has ATPase activity [98].

9.4 O-Antigen Ligation

Irrespective of the export and polymerization modes of the saccharide molecules, assembled Und PP-linked O-antigens are ligated in the periplasmic space to terminal sugar residues of the lipid A-core OS [74]. This is a specific glycosyl-transfer reaction mediated by the *waaL* gene product, which encodes an integral membrane protein. The *waaL* gene maps within the *waa* gene cluster that also encodes other enzymes for the biosynthesis and assembly of core OS [9, 15]. Mutant strains devoid of a functional *waaL* gene cannot ligate O-antigen molecules to lipid A-core OS resulting in the production of LPS lacking O-antigen polysaccharide and accumulation of membrane-bound Und-PP-linked O-antigen molecules [74, 75]. Remarkably, ligase mutants are viable, in contrast to mutants in the *wzx* flippase [93, 94]. It is not clear why the accumulation of Und-PP-linked O-antigen precursors would be lethal if it takes place in the cytosolic side of the membrane but not in the periplasmic side. Although it has not been demonstrated, it is possible that the periplasmic accumulation of unprocessed Und-PP-saccharides somehow provides a signal that results in downregulation of O-antigen biosynthesis.

The mechanism of ligation is still unresolved. Although the ligase catalyzes the formation of a glycosidic bond, WaaL proteins share no similarities with any of the glycosyltransferases that use sugar nucleotides. A requirement for a specific lipid A-core OS acceptor structure has been established in several model systems [191–194], which has led to the generalized notion that the specific WaaL protein can recognize a specific lipid A-core OS terminal structure. For example, in *E. coli* there are five chemically distinct core OS types, K-12, R1, R2, R3, and R4 [15], while only two types are found in *S. enterica* [15, 195, 196]. Both *Salmonella* and *E. coli* K-12, and presumably all Gram-negative bacteria, can ligate any number of Und-PP-linked recombinant O-antigens [9, 15, 196]. Donor Und-PP-linked glycans for the ligation reaction can originate from various biosynthesis pathways. For example, a small portion of *E. coli* K-12 colanic acid, a cell surface capsular material that is usually loosely associated with the bacterial cell, can be covalently linked to lipid A-core

OS by WaaL at the same attachment site position for O-antigen [197]. Recently, Tang and Mintz [198] have suggested that glycosylation of the collagen adhesin EmaA, a non-fimbrial surface protein of *Aggregatibacter actinomycetemcomitans*, depends on WaaL. Glycosylation of other prokaryote components is carried out by functional homologs of WaaL such as *N. meningitidis* PglL [199, 200] and *P. aeruginosa* PilO [201], which are responsible for O-glycosylation of pili in these organisms. A homologue of WaaL is also encoded by a cluster of genes for S-layer protein glycosylation in *G. stearothermophilus* [202].

It is unclear how WaaL recognizes the Und-PP-linked O-antigen and in particular, which part of the donor molecule participates in the enzymatic reaction. Most studies on O-antigen ligases have been limited to establishing the topology of the protein [86, 191, 193, 194, 203]. Unfortunately, WaaL proteins show significant divergence in their primary amino acid sequence, even for members from the same species [204]. The extremely low sequence conservation among O-antigen ligases makes comparative analyses difficult. Therefore, a detailed knowledge of the residues involved in ligase activity or the chemical characteristics of the ligation reaction are unknown. It is also difficult to establish relationships in WaaL proteins based on potential core OS acceptor structures. For example, the *E. coli* R2 and *S. enterica* WaaL proteins share ~80% amino acid sequence similarity and are functionally interchangeable, as both link the O-antigen polysaccharide to a terminal glucose in the core OS that has an α-1,2-linked *N*-acetylglucosamine [193]. In contrast, *E. coli* R3 WaaL is ~66% similar to the *Salmonella* protein but links the O-antigen polysaccharide to a different attachment site in the core OS that resembles a similar site in the K-12 core OS. However, the R3 WaaL shares little identity with the *E. coli* K-12 ligase [16]. More recent evidence suggests that the specificity of the ligation reaction for a particular lipid A-core OS structure does not solely depend on the WaaL protein, but on an additional factor or factors that have not yet been identified [196].

Recently, Abeyrathne and Lam [205] reported that highly purified WaaL from *P. aeruginosa* has ATPase activity and ATP hydrolysis is required for the in vitro ligation reaction. This is an intriguing finding since ATP is not present in the periplasmic space [206]. The requirement for ATP in the ligation reaction could not be confirmed in other WaaL proteins. In the *E. coli* K-12 WaaL, an extensive mutagenesis analysis of amino acid motifs that are putatively involved in ATP binding or hydrolysis did not afford ligation-defective proteins [207] and ATP is not required for the in vitro ligation assay (X. Ruan and M. A. Valvano, unpublished data). Also, the WaaL protein from *H. pylori* does not require ATP for activity [189].

Conceivably, WaaL activity requires amino acids exposed to the periplasmic space where they could interact with the donor and acceptor molecules. A critical histidine, which is somewhat conserved in many WaaL proteins, was identified in a periplasmic loop of the *Vibrio cholerae* WaaL [194], and a potentially common motif is emerging not only in WaaL proteins but also in proteins that ligate Und-PP-linked O-antigen precursors to pili [208]. A tri-dimensional structural model of the WaaL large periplasmic loop was proposed, which consists of two pairs of almost perpendicular α-helices. In this model, all the conserved residues in other WaaL proteins cluster within a putative catalytic region, as demonstrated by site-directed amino acid replacements [207]. The model predicts that Arginine-288 and Histidine-337, two residues that are critical for WaaL function, face each other and are exposed to the solvent in a spatial arrangement that suggests interactions with substrate molecules. In addition, a conserved arginine, also critical for WaaL function, is invariably present in the short periplasmic loop preceding the large loop [207]. These results support the notion that a positively charged region exposed to the periplasmic face of WaaL plays a critical role in either catalysis or binding of the Und-PP-linked O-antigen substrate.

9.5 Synthesis and Recycling of Und-PP

Und-P is a universal lipid carrier for the synthesis of bacterial cell surface glycans. Und-P arises from the dephosphorylation of Und-PP, which in turns results from the condensation of five-carbon building blocks, isopentenyl diphosphate and its isomer dimethylallyl diphosphate [209]. Two distinct biosynthetic pathways synthesize these precursors: (1) the methyl-erythritol phosphate pathway in eubacteria, green algae, and plant chloroplasts, and (2) the mevalonate pathway in eukaryotes and Archaea [210–215]. The UppS synthase catalyzes the de novo biosynthesis of Und-PP in a reaction that adds isoprene units onto farnesyl pyrophosphate [216, 217], which occurs at the cytosolic side of the plasma membrane. Und-PP is also regenerated at the periplasmic side of the membrane (or the external side of the membrane in Gram-positive bacteria). Free Und-PP is released from Und-PP-linked glycans upon glycan transfer reactions to complete the synthesis of LPS O-antigen, peptidoglycan, and bacterial glycoproteins (Fig. 9.6). Cellular availability of Und-P is a limiting factor in the biosynthesis of glycans, since this lipid carrier is made in very small amounts and is also required for the biosynthesis of multiple carbohydrate polymers. Therefore, the recycling pathway for Und-P synthesis from preformed Und-PP released at the external side of the plasma membrane contributes to the total Und-P pool available for the initiation of lipid-linked glycan biosynthesis (Fig. 9.6).

The de novo synthesis and regeneration of Und-PP at both sides of the membrane suggests a requirement for membrane pyrophosphatases whose active sites are in cytosolic and periplasmic environments, respectively. The specific enzymes involved in Und-PP dephosphorylation have recently been identified in *E. coli* K-12 and they are UppP (formerly BacA), which accounts for ~75% of the cellular Und-PP pyrophosphatase activity [18], YbjG, YeiU, and PgpB [19]. UppP lacks any known features of a typical acid phosphatase and catalyzes the conversion of de novo synthesized Und-PP into Und-P [18].

In eukaryotic cells, released Dol-PP after the transfer of the Dol-PP-linked glycan to glycoproteins is also recycled to Dol-P by dephosphorylation [218]. A dolichyl pyrophosphate phosphatase, Cwh8, was discovered in *S. cerevisiae* [17] and in mammalian cells [219]. Membrane proteins of the Cwh8 family are in the ER and their pyrophosphate phosphatase activity is mediated by a luminally-oriented

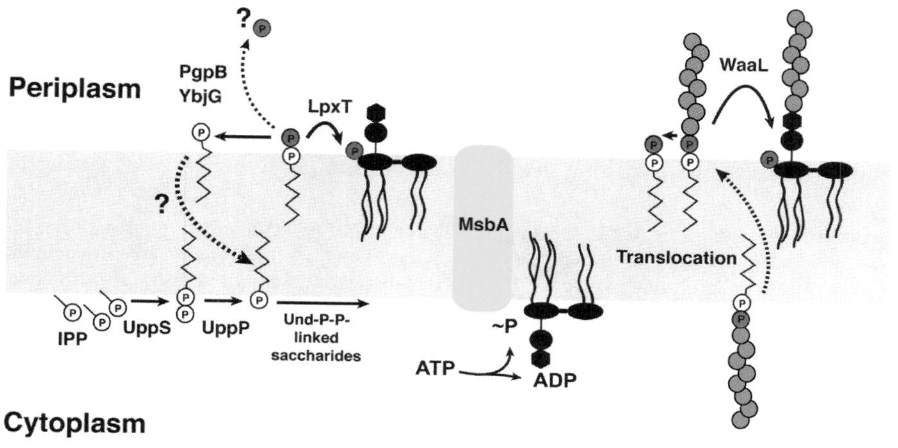

Fig. 9.6 Und-P/Und-PP biogenesis in *E. coli* (modified from Ref. [20]). The topography of Und-P synthesis by de novo and recycling pathways is indicated using the synthesis of O-antigen polysaccharide (*blue circles*) as an example of a surface polymer that requires an Und-PP-linked intermediary prior to its transfer to the lipid A-core OS at the periplasmic side of the plasma membrane, which is mediated by the WaaL O-antigen ligase. The terminal phosphate molecule transferred by LpxT from periplasmic Und-PP to the 1-position of lipid A [21] is indicated in *red*. *Question marks* indicate that it is presently unknown whether or not the other Und-PP phosphatases, PgpB and YbjG, transfer the terminal phosphate to unknown acceptor molecules, and also how the released periplasmic Und-P becomes available at the cytosolic side of the membrane to reinitiate the lipid-linked glycan synthesis. The de novo synthesis involves the condensation of isopentenyl diphosphate and/or dimethylallyl diphosphate (IPP) by UppS resulting in Und-PP followed by the dephosphorylation by UppP yielding Und-P

active site [17, 219]. Cwh8 has a segment of approximately 122 amino acids that corresponds to an acid phosphatase domain. This domain is shared among lipid phosphatases and non-specific acid phosphatases [220–222] and it faces the luminal side of the ER membrane, the same side of the membrane where the N-linked oligosaccharyl transferase reaction takes place [17]. The *E. coli* K-12 proteins YbjG, YeiU, and PgpB have a similar motif to that found in Cwh8 [19]. The presence of Cwh8 homologues in prokaryotes suggests that they might have an analogous function in the recycling of Und-PP. Tatar et al. [223] have shown that the acid phosphatase motifs of YbjG and YeiU face the periplasmic space. These authors also demonstrated that YbjG and, to a lesser extent, YeiU are implicated in the recycling of periplasmic Und-PP molecules. Furthermore, it has been shown that YeiU (subsequently renamed as LpxT) catalyzes the transfer of a phosphate group from the Und-PP donor to the 1-position of lipid A to form lipid A 1-diphosphate, and this reaction occurs at the periplasmic side of the cytosolic membrane [21]. The majority of lipid A molecules in *E. coli* K-12 are phosphorylated at 1 and 4′ positions, but approximately one-third of the molecules contain a diphosphate at the 1-position (lipid A 1-diphosphate) [204]. Further research is required to elucidate the biochemical functions of YbjG and PgpB. It is conceivable that these enzymes, like LpxT, may transfer the distal phosphate of Und-PP to other acceptor molecules in the periplasm (Fig. 9.6). These phosphotransfer reactions could be important in the fine-tuning of periplasmic and outer membrane molecules and could also play a role in bacterial adaptation to stress conditions. An important question that also deserves attention is how the released Und-P in the external side of the membrane regains access to the cytosolic side for its reutilization in Und-PP-glycan synthesis.

Acknowledgments Research in the authors' laboratory has been supported by grants from the Canadian Institutes of Health Research, Natural Sciences and Engineering Research Council, the Mizutani Foundation for Glycoscience, Cystic Fibrosis Canada, Ontario Ministry of Innovation, and the Canadian Foundation for Innovation. S.E.F. and K.B.P. hold Ontario Graduate Science and Technology scholarships. M.A.V. holds a Canada Research Chair in Infectious Diseases and Microbial Pathogenesis.

References

1. Nikaido H (1996) Outer membrane. In: Neidhardt FC, Curtiss R III, Ingraham JL, Lin ECC, Low KB, Magasanik B, Reznikoff WS, Riley M, Schaechter M, Umbarger HE (eds) *Escherichia coli* and *Salmonella*: cellular and molecular biology. ASM Press, Washington, pp 29–47
2. Whitfield C, Valvano MA (1993) Biosynthesis and expression of cell-surface polysaccharides in gram-negative bacteria. Adv Microb Physiol 35:135–246
3. Opal SM (2007) The host response to endotoxin, antilipopolysaccharide strategies, and the management of severe sepsis. Int J Med Microbiol 297:365–377
4. Bugg TD, Brandish PE (1994) From peptidoglycan to glycoproteins: common features of lipid-linked oligosaccharide biosynthesis. FEMS Microbiol Lett 119:255–262
5. Burda P, Aebi M (1999) The dolichol pathway of N-linked glycosylation. Biochim Biophys Acta 1426:239–257
6. Helenius J, Aebi M (2002) Transmembrane movement of dolichol linked carbohydrates during N-glycoprotein biosynthesis in the endoplasmic reticulum. Semin Cell Dev Biol 13: 171–178
7. Helenius J, Ng DT, Marolda CL, Walter P, Valvano MA, Aebi M (2002) Translocation of lipid-linked oligosaccharides across the ER membrane requires Rft1 protein. Nature 415: 447–450
8. Valvano MA (2003) Export of O-specific lipopolysaccharide. Front Biosci 8:s452–s471
9. Raetz CRH, Whitfield C (2002) Lipopolysaccharide endotoxins. Annu Rev Biochem 71: 635–700
10. Samuel G, Reeves P (2003) Biosynthesis of O-antigens: genes and pathways involved in nucleotide sugar precursor synthesis and O-antigen assembly. Carbohydr Res 338: 2503–2519
11. Jansson P-E (1999) The chemistry of O-polysaccharide chains in bacterial lipopolysaccharides. In: Brade H, Opal SM, Vogel SN, Morrison DC (eds) Endotoxin in health and disease. Marcel Dekker, New York, pp 155–178
12. Pluschke G, Mercer A, Kusecek B, Pohl A, Achtman M (1983) Induction of bacteremia in newborn rats by *Escherichia coli* K1 is correlated with only certain O (lipopolysaccharide) antigen types. Infect Immun 39:599–608
13. Pluschke G, Achtman M (1984) Degree of antibody-independent activation of the classical complement pathway by K1 *Escherichia coli* differs with O-antigen type and correlates with virulence of meningitis in newborns. Infect Immun 43:684–692

14. Joiner KA (1988) Complement evasion by bacteria and parasites. Annu Rev Microbiol 42:201–230
15. Heinrichs DE, Valvano MA, Whitfield C (1999) Biosynthesis and genetics of lipopolysaccharide core. In: Brade H, Opal SM, Vogel SN, Morrison DC (eds) Endotoxin in health and disease. Marcel Dekker, New York, pp 305–330
16. Heinrichs DE, Yethon JA, Whitfield C (1998) Molecular basis for structural diversity in the core regions of the lipopolysaccharides of *Escherichia coli* and *Salmonella enterica*. Mol Microbiol 30:221–232
17. Fernandez F, Rush JS, Toke DA, Han G, Quinn JE, Carman GM, Choi J-Y, Voelker DR, Aebi M, Waechter CJ (2001) The *CWH8* gene encodes a dolichyl pyrophosphate phosphatase with a luminally oriented active site in the endoplasmic reticulum of *Saccharomyces cerevisiae*. J Biol Chem 276:41455–41464
18. El Ghachi M, Bouhss A, Blanot D, Mengin-Lecreulx D (2004) The *bacA* gene of *Escherichia coli* encodes an undecaprenyl pyrophosphate phosphatase activity. J Biol Chem 279: 30106–30113
19. El Ghachi M, Derbise A, Bouhss A, Mengin-Lecreulx D (2005) Identification of multiple genes encoding membrane proteins with undecaprenyl pyrophosphate phosphatase (UppP) activity in *Escherichia coli*. J Biol Chem 280:18689–18695
20. Valvano MA (2008) Undecaprenyl phosphate recycling comes out of age. Mol Microbiol 67:232–235
21. Touze T, Tran AX, Hankins JV, Mengin-Lecreulx D, Trent MS (2008) Periplasmic phosphorylation of lipid A is linked to the synthesis of undecaprenyl phosphate. Mol Microbiol 67:264–277
22. Rick PD, Silver RP (1996) Enterobacterial common antigen and capsular polysaccharides. In: Neidhardt FC, Curtiss R III, Ingraham JL, Lin ECC, Low KB, Magasanik B, Reznikoff WS, Riley M, Schaechter M, Umbarger HE (eds) *Escherichia coli* and *Salmonella*: cellular and molecular biology. ASM Press, Washington, pp 104–122
23. Bouhss A, Trunkfield AE, Bugg TDH, Mengin-Lecreulx D (2008) The biosynthesis of peptidoglycan lipid-linked intermediates. FEMS Microbiol Rev 32:208–233
24. Swoboda JG, Campbell J, Meredith TC, Walker S (2010) Wall teichoic acid function, biosynthesis, and inhibition. Chembiochem 11:35–45
25. Erbel PJ, Barr K, Gao N, Gerwig GJ, Rick PD, Gardner KH (2003) Identification and biosynthesis of cyclic enterobacterial common antigen in *Escherichia coli*. J Bacteriol 185: 1995–2004
26. Kajimura J, Rahman A, Rick PD (2005) Assembly of cyclic enterobacterial common antigen in *Escherichia coli* K-12. J Bacteriol 187:6917–6927
27. Rick PD, Hubbard GL, Kitaoka M, Nagaki H, Kinoshita T, Dowd S, Simplaceanu V, Ho C (1998) Characterization of the lipid-carrier involved in the synthesis of enterobacterial common antigen (ECA) and identification of a novel phosphoglyceride in a mutant of *Salmonella typhimurium* defective in ECA synthesis. Glycobiology 8:557–567
28. Higgins CF (1994) Flip-flop: the transmembrane translocation of lipids. Cell 79:393–395
29. Alexander DC, Valvano MA (1994) Role of the *rfe* gene in the biosynthesis of the *Escherichia coli* O7-specific lipopolysaccharide and other O-specific polysaccharides containing *N*-acetylglucosamine. J Bacteriol 176:7079–7084
30. Rush JS, Rick PD, Waechter CJ (1997) Polyisoprenyl phosphate specificity of UDP-GlcNAc:undecaprenyl phosphate *N*-acetylglucosaminyl 1-P transferase from *E. coli*. Glycobiology 7:315–322
31. Lehrman MA (1994) A family of UDP-GlcNAc/MurNAc: polyisoprenol-P GlcNAc/MurNAc-1-P transferases. Glycobiology 4:768–771
32. Schenk B, Fernandez F, Waechter CJ (2001) The ins(ide) and outs(ide) of dolichyl phosphate biosynthesis and recycling in the endoplasmic reticulum. Glycobiology 11:61R–70R

33. Mankowski T, Sasak W, Chojnacki T (1975) Hydrogenated polyprenol phosphates – exogenous lipid acceptors of glucose from UDP glucose in rat liver microsomes. Biochem Biophys Res Commun 65:1292–1297
34. Pennock JF, Hemming FW, Morton RA (1960) Dolichol: a naturally occurring isoprenoid alcohol. Nature 186:470–472
35. Anderson MS, Eveland SS, Price NP (2000) Conserved cytoplasmic motifs that distinguish sub-groups of the polyprenol phosphate:*N*-acetylhexosamine-1-phosphate transferase family. FEMS Microbiol Lett 191:169–175
36. Price NP, Momany FA (2005) Modeling bacterial UDP-HexNAc: polyprenol-P HexNAc-1-P transferases. Glycobiology 15:29R–42R
37. Brandish P, Kimura K, Inukai M, Southgate R, Lonsdale J, Bugg T (1996) Modes of action of tunicamycin, liposidomycin B, and mureidomycin A: inhibition of phospho-*N*-acetylmuramyl-pentapeptide translocase from *Escherichia coli*. Antimicrob Agents Chemother 40:1640–1644
38. Meier-Dieter U, Starman R, Barr K, Mayer H, Rick PD (1990) Biosynthesis of enterobacterial common antigen in *Escherichia coli*. J Biol Chem 265:13490–13497
39. Klena JD, Schnaitman CA (1993) Function of the *rfb* gene cluster and the *rfe* gene in the synthesis of O-antigen by *Shigella dysenteriae* 1. Mol Microbiol 9:393–402
40. Yao Z, Valvano MA (1994) Genetic analysis of the O-specific lipopolysaccharide biosynthesis region (*rfb*) of *Escherichia coli* K-12W3110: identification of genes that confer group 6 specificity to *Shigella flexneri* serotypes Y and 4a. J Bacteriol 176:4133–4143
41. Wang L, Huskic S, Cisterne A, Rothemund D, Reeves PR (2002) The O-antigen gene cluster of *Escherichia coli* O55:H7 and identification of a new UDP-GlcNAc C4 epimerase gene. J Bacteriol 184:2620–2625
42. Zhang L, Radziejewska-Lebrecht J, Krajewska-Pietrasik D, Toivanen P, Skurnik M (1997) Molecular and chemical characterization of the lipopolysaccharide O-antigen and its role in the virulence of *Yersinia enterocolitica* serotype O:8. Mol Microbiol 23:63–76
43. Wang L, Reeves PR (1998) Organization of the *Escherichia coli* O157 O-antigen cluster and identification of its specific genes. Infect Immun 66:3545–3551
44. Rush JS, Alaimo C, Robbiani R, Wacker M, Waechter CJ (2010) A novel epimerase that converts GlcNAc-P-P-undecaprenol to GalNAc-P-P-undecaprenol in *Escherichia coli* O157. J Biol Chem 285:1671–1680
45. Ortega X, Hunt TA, Loutet S, Vinion-Dubiel AD, Datta A, Choudhury B, Goldberg JB, Carlson R, Valvano MA (2005) Reconstitution of O-specific lipopolysaccharide expression in the *Burkholderia cenocepacia* strain J2315 that is associated with transmissible infections in patients with cystic fibrosis. J Bacteriol 187:1324–1333
46. Rocchetta HL, Burrows LL, Pacan JC, Lam JS (1998) Three rhamnosyltransferases responsible for assembly of the A-band D-rhamnan polysaccharide in *Pseudomonas aeruginosa*: a fourth transferase, WbpL, is required for the initiation of both A-band and B-band lipopolysaccharide synthesis. Mol Microbiol 28:1103–1119
47. Yamashita Y, Shibata Y, Nakano Y, Tsuda H, Kido N, Ohta M, Koga T (1999) A Novel gene required for rhamnose-glucose polysaccharide synthesis in *Streptococcus mutans*. J Bacteriol 181:6556–6559
48. Bouhss A, Mengin-Lecreulx D, Le Beller D, Van Heijenoort J (1999) Topological analysis of the MraY protein catalysing the first membrane step of peptidoglycan synthesis. Mol Microbiol 34:576–585
49. Lehrer J, Vigeant KA, Tatar LD, Valvano MA (2007) Functional characterization and membrane topology of *Escherichia coli* WecA, a sugar-phosphate transferase initiating the biosynthesis of enterobacterial common antigen and O-antigen lipopolysaccharide. J Bacteriol 189:2618–2628
50. Ashby MN, Edwards PA (1990) Elucidation of the deficiency in teo yeast coenzyme Q mutants. Characterization of the structural gene encoding hexaprenyl pyrophospohate synthetase. J Biol Chem 265:13157–13164

51. Tarshis LC, Yan M, Poulter CD, Sacchettini JC (1994) Crystal structure of recombinant farnesyl diphosphate synthase at 2.6-Å resolution. Biochemistry 33:10871–10877
52. Lloyd AJ, Brandish PE, Gilbey AM, Bugg TD (2004) Phospho-*N*-acetyl-muramyl-pentapeptide translocase from *Escherichia coli*: catalytic role of conserved aspartic acid residues. J Bacteriol 186:1747–1757
53. Al-Dabbagh B, Henry X, El Ghachi M, Auger G, Blanot D, Parquet C, Mengin-Lecreulx D, Bouhss A (2008) Active site mapping of MraY, a member of the polyprenyl phosphate *N*-acetylhexosamine 1-phosphate transferase superfamily, catalyzing the first membrane step of peptidoglycan biosynthesis. Biochemistry 47:8919–8928
54. Amer AO, Valvano MA (2002) Conserved aspartic acids are essential for the enzymic activity of the WecA protein initiating the biosynthesis of O-specific lipopolysaccharide and enterobacterial common antigen in *Escherichia coli*. Microbiology 148:571–582
55. Amer AO, Valvano MA (2001) Conserved amino acid residues found in a predicted cytosolic domain of WecA (UDP-*N*-acetyl glucosamine:undecaprenol-phosphate *N*-acetylglucosamine-1-phosphate transferase) are implicated in the recognition of UDP-*N*-acetylglucosamine. Microbiology 147:3015–3025
56. Heifetz A, Keenan RW, Elbein AD (1979) Mechanism of action of tunicamycin on the UDP-GlcNAc:dolichyl phosphate GlcNAc-1-phosphate transferase. Biochemistry 18:2186–2192
57. Heydanek MG, Struve WG, Neuhaus FC (1969) On the initial stages of peptidoglycan synthesis. III. Kinetics and uncoupling of phospho-*N*-acetylomuramyl-pentapeptide translocase (uridine 5′-phosphate). Biochemistry 8:1214–1221
58. Al-Dabbagh B, Mengin-Lecreulx D, Bouhss A (2008) Purification and characterization of the bacterial UDP-GlcNAc:undecaprenyl-phosphate GlcNAc-1-phosphate transferase WecA. J Bacteriol 190:7141–7146
59. Wang L, Liu D, Reeves PR (1996) C-Terminal half of *Salmonella enterica* WbaP (RfbP) is the galactosyl-1-phosphate transferase domain catalyzing the first step of O-antigen synthesis. J Bacteriol 178:2598–2604
60. Steiner K, Novotny R, Patel K, Vinogradov E, Whitfield C, Valvano MA, Messner P, Schaffer C (2007) Functional characterization of the initiation enzyme of S-layer glycoprotein glycan biosynthesis in *Geobacillus stearothermophilus* NRS 2004/3a. J Bacteriol 189:2590–2598
61. Bugert P, Geider K (1995) Molecular analysis of the *ams* operon required for exopolysaccharide synthesis in *Erwinia amylovora*. Mol Microbiol 15:917–933
62. Cartee RT, Forsee WT, Bender MH, Ambrose KD, Yother J (2005) CpsE from type 2 *Streptococcus pneumoniae* catalyzes the reversible addition of glucose-1-phosphate to a polyprenyl phosphate acceptor, initiating type 2 capsule repeat unit formation. J Bacteriol 187:7425–7433
63. Stevenson G, Andrianopoulos K, Hobbs M, Reeves PR (1996) Organization of the *Escherichia coli* K-12 gene cluster responsible for production of the extracellular polysaccharide colanic acid. J Bacteriol 178:4885–4893
64. Saldías MS, Patel K, Marolda CL, Bittner M, Contreras I, Valvano MA (2008) Distinct functional domains of the *Salmonella enterica* WbaP transferase that is involved in the initiation reaction for synthesis of the O-antigen subunit. Microbiology 154:440–453
65. Xayarath B, Yother J (2007) Mutations blocking side chain assembly, polymerization, or transport of a Wzy-dependent *Streptococcus pneumoniae* capsule are lethal in the absence of suppressor mutations and can affect polymer transfer to the cell wall. J Bacteriol 189: 3369–3381
66. Glover KJ, Weerapana E, Chen MM, Imperiali B (2006) Direct biochemical evidence for the utilization of UDP-bacillosamine by PglC, an essential glycosyl-1-phosphate transferase in the *Campylobacter jejuni* N-linked glycosylation pathway. Biochemistry 45:5343–5350
67. Power PM, Roddam LF, Dieckelmann M, Srikhanta YN, Tan YC, Berrington AW, Jennings MP (2000) Genetic characterization of pilin glycosylation in *Neisseria meningitidis*. Microbiology 146:967–979

68. Chamot-Rooke J, Rousseau B, Lanternier F, Mikaty G, Mairey E, Malosse C, Bouchoux G, Pelicic V, Camoin L, Nassif X, Duménil G (2007) Alternative *Neisseria* spp. type IV pilin glycosylation with a glyceramido acetamido trideoxyhexose residue. Proc Natl Acad Sci USA 104:14783–14838
69. Toh E, Kurtz HD Jr, Brun YV (2008) Characterization of the *Caulobacter crescentus* holdfast polysaccharide biosynthesis pathway reveals significant redundancy in the initiating glycosyltransferase and polymerase steps. J Bacteriol 190:7219–7231
70. Patel KB, Furlong SE, Valvano MA (2010) Functional analysis of the C-terminal domain of the WbaP protein that mediates initiation of O-antigen synthesis in *Salmonella enterica.* Glycobiology 20:1389–1401
71. Campbell JA, Davies GJ, Bulone V, Henrissat B (1997) A classification of nucleotide-diphospho-sugar glycosyltransferases based on amino acid sequence similarities. Biochem J 326:929–939
72. Coutinho PM, Deleury E, Davies GJ, Henrissat B (2003) An evolving hierarchical family classification for glycosyltransferases. J Mol Biol 328:307–317
73. Lairson LL, Henrissat B, Davies GJ, Withers SG (2008) Glycosyltransferases: structures, functions, and mechanisms. Annu Rev Biochem 77:521–555
74. McGrath BC, Osborn MJ (1991) Localization of the terminal steps of O-antigen synthesis in *Salmonella typhimurium.* J Bacteriol 173:649–654
75. Mulford CA, Osborn MJ (1983) An intermediate step in translocation of lipopolysaccharide to the outer membrane of *Salmonella typhimurium.* Proc Natl Acad Sci USA 80:1159–1163
76. Whitfield C (1995) Biosynthesis of lipopolysaccharide O-antigens. Trends Microbiol 3:178–185
77. Marino PA, McGrath BC, Osborn MJ (1991) Energy dependence of O-antigen synthesis in *Salmonella typhimurium.* J Bacteriol 173:3128–3133
78. Keenleyside WJ, Whitfield C (1999) Genetics and biosynthesis of lipopolysaccharide O-antigens. In: Brade H, Opal SM, Vogel SN, Morrison DC (eds) Endotoxin in health and disease. Marcel Dekker, New York, pp 331–358
79. Whitfield C, Roberts IS (1999) Structure, assembly and regulation of expression of capsules in *Escherichia coli.* Mol Microbiol 31:1307–1319
80. Bronner D, Clarke BR, Whitfield C (1994) Identification of an ATP-binding cassette transport system required for translocation of lipopolysaccharide O-antigen side-chains across the cytoplasmic membrane of *Klebsiella pneumoniae* serotype O1. Mol Microbiol 14:505–519
81. Clarke BR, Whitfield C (1992) Molecular cloning of the *rfb* region of *Klebsiella pneumoniae* serotype O1:K20: the *rfb* gene cluster is responsible for synthesis of the D-galactan I O-polysaccharide. J Bacteriol 174:4614–4621
82. Liu D, Cole RA, Reeves PR (1996) An O-antigen processing function for Wzx (RfbX): a promising candidate for O-unit flippase. J Bacteriol 178:2102–2107
83. Paulsen IT, Beness AM, Saier MH (1997) Computer-based analyses of the protein constituents of transport systems catalysing export of complex carbohydrates in bacteria. Microbiology 143:2685–2699
84. Cunneen MM, Reeves PR (2008) Membrane topology of the *Salmonella enterica* serovar Typhimurium group B O-antigen translocase Wzx. FEMS Microbiol Lett 287:76–84
85. Mazur A, Marczak M, Król JE, Skorupska A (2005) Topological and transcriptional analysis of *pssL* gene product: a putative Wzx-like exopolysaccharide translocase in *Rhizobium leguminosarum* bv. *trifolii* TA1. Arch Microbiol 184:1–10
86. Islam ST, Taylor VL, Qi M, Lam JS (2010) Membrane topology mapping of the O-antigen flippase (Wzx), polymerase (Wzy), and ligase (WaaL) from *Pseudomonas aeruginosa* PAO1 reveals novel domain architectures. mBio 1:e00189–10
87. Marolda CL, Li B, Lung M, Yang M, Hanuszkiewicz A, Rosales AR, Valvano MA (2010) Membrane topology and identification of critical amino acid residues in the Wzx O-antigen translocase from *Escherichia coli* O157:H4. J Bacteriol 192:6160–6171

88. Marolda CL, Feldman MF, Valvano MA (1999) Genetic organization of the O7-specific lipopolysaccharide biosynthesis cluster of *Escherichia coli* VW187 (O7:K1). Microbiology 145:2485–2496
89. Sorgen PL, Hu Y, Guan L, Kaback HR, Girvin ME (2002) An approach to membrane protein structure without crystals. Proc Natl Acad Sci USA 99:14037–14040
90. Abramson J, Smirnova I, Kasho V, Verner G, Kaback HR, Iwata S (2003) Structure and mechanism of the lactose permease of *Escherichia coli*. Science 301:610–615
91. Zhou GP, Troy FA (2005) NMR study of the preferred membrane orientation of polyisoprenols (dolichol) and the impact of their complex with polyisoprenyl recognition sequence peptides on membrane structure. Glycobiology 15:347–359
92. Zhou GP, Troy FA 2nd (2005) NMR studies on how the binding complex of polyisoprenol recognition sequence peptides and polyisoprenols can modulate membrane structure. Curr Protein Pept Sci 6:399–411
93. Rick PD, Barr K, Sankaran K, Kajimura J, Rush JS, Waechter CJ (2003) Evidence that the *wzxE* gene of *Escherichia coli* K-12 encodes a protein involved in the transbilayer movement of a trisaccharide-lipid intermediate in the assembly of enterobacterial common antigen. J Biol Chem 278:16534–16542
94. Marolda CL, Tatar LD, Alaimo C, Aebi M, Valvano MA (2006) Interplay of the *wzx* translocase and the corresponding polymerase and chain length regulator proteins in the translocation and periplasmic assembly of lipopolysaccharide O-antigen. J Bacteriol 188: 5124–5135
95. Kol MA, de Kroon AI, Rijkers DT, Killian JA, de Kruijff B (2001) Membrane-spanning peptides induce phospholipid flop: a model for phospholipid translocation across the inner membrane of *E. coli*. Biochemistry 40:10500–10506
96. Kol MA, van Dalen A, de Kroon AI, de Kruijff B (2003) Translocation of phospholipids is facilitated by a subset of membrane-spanning proteins of the bacterial cytoplasmic membrane. J Biol Chem 278:24586–24593
97. Kol MA, van Laak AN, Rijkers DT, Killian JA, de Kroon AI, de Kruijff B (2003) Phospholipid flop induced by transmembrane peptides in model membranes is modulated by lipid composition. Biochemistry 42:231–237
98. Alaimo C, Catrein I, Morf L, Marolda CL, Callewaert N, Valvano MA, Feldman MF, Aebi M (2006) Two distinct but interchangeable mechanisms for flipping of lipid-linked oligosaccharides. EMBO J 25:967–976
99. Wacker M, Linton D, Hitchen PG, Nita-Lazar M, Haslam SM, North SJ, Panico M, Morris HR, Dell A, Wren B, Aebi M (2002) N-Linked glycosylation in *Campylobacter jejuni* and its functional transfer into *E. coli*. Science 298:1790–1793
100. Feldman MF, Wacker M, Hernandez M, Hitchen PG, Marolda CL, Kowarik M, Morris HR, Dell A, Valvano MA, Aebi M (2005) Engineering N-linked protein glycosylation with diverse O-antigen lipopolysaccharide structures in *Escherichia coli*. Proc Natl Acad Sci USA 102:3016–3021
101. Feldman MF, Marolda CL, Monteiro MA, Perry MB, Parodi AJ, Valvano MA (1999) The activity of a putative polyisoprenol-linked sugar translocase(Wzx) involved in *Escherichia coli* O-antigen assembly is independent of the chemical structure of the O-repeat. J Biol Chem 274:35129–35138
102. Marolda CL, Vicarioli J, Valvano MA (2004) Wzx proteins involved in O-antigen biosynthesis function in association with the first sugar of the O-specific lipopolysaccharide subunit. Microbiology 150:4095–4105
103. Bray D, Robbins PW (1967) The direction of chain growth in *Salmonella anatum* O-antigen biosynthesis. Biochem Biophys Res Commun 28:334–339
104. Robbins PW, Bray D, Dankert BM, Wright A (1967) Direction of chain growth in polysaccharide synthesis. Science 158:1536–1542

105. McConnell MR, Oakes KR, Patrick AN, Mills DM (2001) Two functional O-polysaccharide polymerase *wzy* (*rfc*) genes are present in the *rfb* gene cluster of group E1 *Salmonella enterica* serovar Anatum. FEMS Microbiol Lett 199:235–240
106. Losick R (1969) Isolation of a trypsin-sensitive inhibitor of O-antigen synthesis involved in lysogenic conversion by bacteriophage ε^{15}. J Mol Biol 42:237–246
107. Newton GJ, Daniels C, Burrows LL, Kropinski AM, Clarke AJ, Lam JS (2001) Three-component-mediated serotype conversion in *Pseudomonas aeruginosa* by bacteriophage D3. Mol Microbiol 39:1237–1247
108. Collins LV, Attridge S, Hackett J (1991) Mutations at rfc or pmi attenuate *Salmonella typhimurium* virulence for mice. Infect Immun 59:1079–1085
109. Morona R, Mavris M, Fallarino A, Manning PA (1994) Characterization of the *rfc* region of *Shigella flexneri*. J Bacteriol 176:733–747
110. Daniels C, Vindurampulle C, Morona R (1998) Overexpression and topology of the *Shigella flexneri* O-antigen polymerase (Rfc/Wzy). Mol Microbiol 28:1211–1222
111. Woodward R, Yi W, Li L, Zhao G, Eguchi H, Sridhar PR, Guo H, Song JK, Motari E, Cai L, Kelleher P, Liu X, Han W, Zhang W, Ding Y, Li M, Wang PG (2010) In vitro bacterial polysaccharide biosynthesis: defining the functions of Wzy and Wzz. Nat Chem Biol 6: 418–423
112. Batchelor RA, Haraguchi GE, Hull RA, Hull SI (1991) Regulation by a novel protein of the bimodal distribution of lipopolysaccharide in the outer membrane of *Escherichia coli*. J Bacteriol 173:5699–5704
113. Murray GL, Attridge SR, Morona R (2003) Regulation of *Salmonella typhimurium* lipopolysaccharide O-antigen chain length is required for virulence; identification of FepE as a second Wzz. Mol Microbiol 47:1395–1406
114. Murray GL, Attridge SR, Morona R (2005) Inducible serum resistance in *Salmonella typhimurium* is dependent on *wzz* (*fepE*)-regulated very long O-antigen chains. Microb Infect 7:1296–1304
115. Murray GL, Attridge SR, Morona R (2006) Altering the length of the lipopolysaccharide O-antigen has an impact on the interaction of *Salmonella enterica* serovar Typhimurium with macrophages and complement. J Bacteriol 188:2735–2739
116. Bengoechea JA, Zhang L, Toivanen P, Skurnik M (2002) Regulatory network of lipopolysaccharide O-antigen biosynthesis in *Yersinia enterocolitica* includes cell envelope-dependent signals. Mol Microbiol 44:1045–1062
117. Hoare A, Bittner M, Carter J, Alvarez S, Zaldívar M, Bravo D, Valvano MA, Contreras I (2006) The outer core lipopolysaccharide of *Salmonella enterica* serovar Typhi is required for bacterial entry into epithelial cells. Infect Immun 74:1555–1564
118. Jimenez N, Canals R, Salo MT, Vilches S, Merino S, Tomas JM (2008) The *Aeromonas hydrophila wb** O34 gene cluster: genetics and temperature regulation. J Bacteriol 190: 4198–4209
119. Kintz E, Scarff JM, DiGiandomenico A, Goldberg JB (2008) Lipopolysaccharide O-antigen chain length regulation in *Pseudomonas aeruginosa* serogroup O11 strain PA103. J Bacteriol 190:2709–2716
120. Morona R, Van Den Bosch L, Daniels C (2000) Evaluation of Wzz/MPA1/MPA2 proteins based on the presence of coiled-coil regions. Microbiology 146:1–4
121. Vincent C, Doublet P, Grangeasse C, Vaganay E, Cozzone AJ, Duclos B (1999) Cells of *Escherichia coli* contain a protein-tyrosine kinase, Wzc, and a phosphotyrosine-protein phosphatase, Wzb. J Bacteriol 181:3472–3477
122. Ilan O, Bloch Y, Frankel G, Ullrich H, Geider K, Rosenshine I (1999) Protein tyrosine kinases in bacterial pathogens are associated with virulence and production of exopolysaccharide. EMBO J 18:3241–3248
123. Cozzone AJ, Grangeasse C, Doublet P, Duclos B (2004) Protein phosphorylation on tyrosine in bacteria. Arch Microbiol 181:171–181

124. Doublet P, Grangeasse C, Obadia B, Vaganay E, Cozzone AJ (2002) Structural organization of the protein-tyrosine autokinase Wzc within *Escherichia coli* cells. J Biol Chem 277: 37339–37348
125. Doublet P, Vincent C, Grangeasse C, Cozzone AJ, Duclos B (1999) On the binding of ATP to the autophosphorylating protein, Ptk, of the bacterium *Acinetobacter johnsonii*. FEBS Lett 445:137–143
126. Whitfield C (2006) Biosynthesis and assembly of capsular polysaccharides in *Escherichia coli*. Annu Rev Biochem 75:39–68
127. Rocchetta HL, Burrows LL, Lam JS (1999) Genetics of O-antigen biosynthesis in *Pseudomonas aeruginosa*. Microbiol Mol Biol Rev 63:523–553
128. Stevenson G, Kessler A, Reeves PR (1995) A plasmid-borne O-antigen chain length determinant and its relationship to other chain length determinants. FEMS Microbiol Lett 125: 23–30
129. Bastin DA, Stevenson G, Brown PK, Haase A, Reeves PR (1993) Repeat unit polysaccharides of bacteria: a model for polymerization resembling that of ribosomes and fatty acid synthetase, with a novel mechanism for determining chain length. Mol Microbiol 7:725–734
130. Morona R, van den Bosch L, Manning PA (1995) Molecular, genetic, and topological characterization of O-antigen chain length regulation in *Shigella flexneri*. J Bacteriol 177: 1059–1068
131. Daniels C, Griffiths C, Cowles B, Lam JS (2002) *Pseudomonas aeruginosa* O-antigen chain length is determined before ligation to lipid A core. Environ Microbiol 4:883–897
132. Daniels C, Morona R (1999) Analysis of *Shigella flexneri* Wzz (Rol) function by mutagenesis and cross-linking: Wzz is able to oligomerize. Mol Microbiol 34:181–194
133. Franco AV, Liu D, Reeves PR (1998) The Wzz (Cld) protein in *Escherichia coli*: amino acid sequence variation determines O-antigen chain length specificity. J Bacteriol 180:2670–2675
134. Marolda CL, Haggerty ER, Lung M, Valvano MA (2008) Functional analysis of predicted coiled coil regions in the *Escherichia coli* K-12 O-antigen polysaccharide chain length determinant Wzz. J Bacteriol 190:2128–2137
135. Tocilj A, Munger C, Proteau A, Morona R, Purins L, Ajamian E, Wagner J, Papadopoulos M, Van Den Bosch L, Rubinstein JL, Fethiere J, Matte A, Cygler M (2008) Bacterial polysaccharide co-polymerases share a common framework for control of polymer length. Nat Struct Mol Biol 15:130–138
136. Lukomski S, Hull RA, Hull SI (1996) Identification of the O-antigen polymerase (*rfc*) gene in *Escherichia coli* O4 by insertional mutagenesis using a nonpolar chloramphenicol resistance cassette. J Bacteriol 178:240–247
137. Larue K, Kimber MS, Ford RC, Whitfield C (2009) Biochemical and structural analysis of bacterial O-antigen chain length regulator proteins reveals a conserved quaternary structure. J Biol Chem 284:7395–7403
138. Bechet E, Gruszczyk J, Terreux R, Gueguen-Chaignon V, Vigouroux A, Obadia B, Cozzone AJ, Nessler S, Grangeasse C (2010) Identification of structural and molecular determinants of the tyrosine-kinase Wzc and implications in capsular polysaccharide export. Mol Microbiol 77:1315–1325
139. Whitfield C, Naismith J (2008) Periplasmic export machines for outer membrane assembly. Curr Opin Struct Biol 18:1–9
140. Papadopoulos M, Morona R (2010) Mutagenesis and chemical cross-linking suggest that Wzz dimer stability and oligomerisation affect lipopolysaccharide O-antigen modal chain length control. J Bacteriol 192:3385–3393
141. Bengoechea JA, Pinta E, Salminen T, Oertelt C, Holst O, Radziejewska-Lebrecht J, Piotrowska-Seget Z, Venho R, Skurnik M (2002) Functional characterization of Gne (UDP-*N*-acetylglucosamine-4-epimerase), Wzz (chain length determinant), and Wzy (O-antigen polymerase) of *Yersinia enterocolitica* serotype O:8. J Bacteriol 184:4277–4287

142. Gaspar JA, Thomas JA, Marolda CL, Valvano MA (2000) Surface expression of O-specific lipopolysaccharide in *Escherichia coli* requires the function of the TolA protein. Mol Microbiol 38:262–275
143. Stenberg F, Chovanec P, Maslen SL, Robinson CV, Ilag L, von Heijne G, Daley DO (2005) Protein complexes of the *Escherichia coli* cell envelope. J Biol Chem 280:34409–34419
144. Kulpa CF Jr, Leive L (1976) Mode of insertion of lipopolysaccharide into the outer membrane of *Escherichia coli*. J Bacteriol 126:467–477
145. Muhlradt PF, Menzel J, Golecki JR, Speth V (1973) Outer membrane of *Salmonella*. Sites of export of newly synthesised lipopolysaccharide on the bacterial surface. Eur J Biochem 35:471–481
146. Collins RF, Beis K, Clarke BR, Ford RC, Hulley M, Naismith JH, Whitfield C (2006) Periplasmic protein-protein contacts in the inner membrane protein Wzc form a tetrameric complex required for the assembly of *Escherichia coli* group 1 capsules. J Biol Chem 281: 2144–2150
147. McNulty C, Thompson J, Barrett B, Lord L, Andersen C, Roberts IS (2006) The cell surface expression of group 2 capsular polysaccharides in *Escherichia coli*: the role of KpsD, RhsA and a multi-protein complex at the pole of the cell. Mol Microbiol 59:907–922
148. Ishidate K, Creeger ES, Zrike J, Deb S, Glauner B, MacAlister TJ, Rothfield LI (1986) Isolation of differentiated membrane domains from *Escherichia coli* and *Salmonella typhimurium*, including a fraction containing attachment sites between the inner and outer membranes and the murein skeleton of the cell envelope. J Biol Chem 261:428–443
149. Bouveret E, Derouiche R, Rigal A, Lloubes R, Lazdunski C, Benedetti H (1995) Peptidoglycan-associated lipoprotein-TolB interaction. A possible key to explaining the formation of contact sites between the inner and outer membranes of *Escherichia coli*. J Biol Chem 270:11071–11077
150. Cascales E, Gavioli M, Sturgis JN, Lloubés R (2000) Proton motive force drives the interaction of the inner membrane TolA and outer membrane Pal proteins in *Escherichia coli*. Mol Microbiol 38:904–915
151. Guihard G, Boulanger P, Benedetti H, Lloubes R, Besnard M, Letellier L (1994) Colicin A and the Tol proteins involved in its translocation are preferentially located in the contact sites between the inner and outer membranes of *Escherichia coli* cells. J Biol Chem 269: 5874–5880
152. Ng DTW, Spear ED, Walter P (2000) The unfolded protein response regulates multiple aspects of secretory and membrane protein biogenesis and endoplasmic reticulum quality control. J Cell Biol 150:77–88
153. Sprong H, van der Sluijs P, van Meer G (2001) How proteins move lipids and lipid move proteins. Nat Rev Mol Cell Biol 2:504–513
154. Davidson EA, Gowda DC (2001) Glycobiology of *Plasmodium falciparum*. Biochimie 83:601–604
155. Frank CG, Sanyal S, Rush JS, Waechter CJ, Menon AK (2008) Does Rft1 flip an N-glycan lipid precursor? Nature 454:E3–E4
156. Zhang L, Al-Hendy A, Toivanen P, Skurnik M (1993) Genetic organization and sequence of the *rfb* gene cluster of *Yersinia enterocolitica* serotype O:3: similarities to the dTDP-L-rhamnose biosynthesis pathway of *Salmonella* and to the bacterial polysaccharide transport systems. Mol Microbiol 9:309–321
157. Smith AN, Boulnois GJ, Roberts IS (1990) Molecular analysis of the *Escherichia coli* K5 *kps* locus: identification and characterization of an inner-membrane capsular polysaccharide transport system. Mol Microbiol 4:1863–1869
158. Pavelka MS, Wright LF, Silver RP (1991) Identification of two genes, *kpsM* and *kpsT*, in region 3 of the polysialic acid gene cluster of *Escherichia coli* K1. J Bacteriol 173: 4603–4610

159. Kroll JS, Loynds B, Brophy LN, Moxon ER (1990) The *bex* locus in encapsulated *Haemophilus influenzae:* a chromosomal region involved in capsule polysaccharide export. Mol Microbiol 4:1853–1862
160. Lazarevic V, Karamata D (1995) The *tagGH* operon of *Bacillus subtilis* 168 encodes a two-component ABC transporter involved in the metabolism of two wall teichoic acids. Mol Microbiol 16:345–355
161. Cuthbertson L, Kos V, Whitfield C (2010) ABC transporters involved in export of cell surface glycoconjugates. Microbiol Mol Biol Rev 74:341–362
162. Doerrler WT, Raetz CRH (2002) ATPase activity of the MsbA lipid flippase of *Escherichia coli*. J Biol Chem 277:36697–36705
163. Rick PD, Hubbard GL, Barr K (1994) Role of the *rfe* gene in the synthesis of the O8 antigen in *Escherichia coli* K-12. J Bacteriol 176:2877–2884
164. Süsskind M, Brade L, Brade H, Holst O (1998) Identification of a novel heptoglycan of α1→2-linked D-*glycero*-D-*manno*-heptopyranose. Chemical and antigenic structure of lipopolysaccharides from *Klebsiella pneumoniae* ssp. *pneumoniae* rough strain R20 (O1-: K20-). J Biol Chem 273:7006–7017
165. Kido N, Torgov VI, Sugiyama T, Uchiya K, Sugihara H, Komatsu T, Kato N, Jann K (1995) Expression of the O9 polysaccharide of *Escherichia coli*: sequencing of the *E. coli* O9 *rfb* gene cluster, characterization of mannosyl transferases, and evidence for an ATP-binding cassette transport system. J Bacteriol 177:2178–2187
166. Guan S, Clarke AJ, Whitfield C (2001) Functional analysis of the galactosyltransferases required for biosynthesis of D-galactan I, a component of the lipopolysaccharide O1 antigen of *Klebsiella pneumoniae*. J Bacteriol 183:3318–3327
167. Clarke BR, Bronner D, Keenleyside WJ, Severn WB, Richards JC, Whitfield C (1995) Role of Rfe and RfbF in the initiation of biosynthesis of D-galactan I, the lipopolysaccharide O-antigen from *Klebsiella pneumoniae* serotype O1. J Bacteriol 177:5411–5418
168. Weisgerber C, Jann K (1982) Glucosyldiphosphoundecaprenol, the mannose acceptor in the synthesis of the O9 antigen of *Escherichia coli*. Biosynthesis and characterization. Eur J Biochem 127:165–168
169. Kido N, Kobayashi H (2000) A single amino acid substitution in a mannosyltransferase, WbdA, converts the *Escherichia coli* O9 polysaccharide into O9a: generation of a new O-serotype group. J Bacteriol 182:2567–2573
170. Kido N, Sugiyama T, Yokochi T, Kobayashi H, Okawa Y (1998) Synthesis of *Escherichia coli* O9a polysaccharide requires the participation of two domains of WbdA, a mannosyltransferase encoded within the *wb** gene cluster. Mol Microbiol 27:1213–1221
171. Whitfield C, Amor PA, Koplin R (1997) Modulation of the surface architecture of gram-negative bacteria by the action of surface polymer:lipid A-core ligase and by determinants of polymer chain length. Mol Microbiol 23:629–638
172. Clarke B, Cuthbertson L, Whitfield C (2004) Nonreducing terminal modifications determine the chain length of polymannose O-antigens of *Escherichia coli* and couple chain termination to polymer export via an ATP-binding cassette transporter. J Biol Chem 279: 35709–35718
173. Lindberg B, Lönngren J, Nimmich W (1972) Structural studies on *Klebsiella* O group 5 lipopolysaccharides. Acta Chem Scand 26:2231–2236
174. Clarke BR, Greenfield LK, Bouwman C, Whitfield C (2009) Coordination of polymerization, chain termination, and export in assembly of the *Escherichia coli* lipopolysaccharide O9a antigen in an ATP-binding cassette transporter-dependent pathway. J Biol Chem 284: 30662–30672
175. Vinogradov E, Frirdich E, MacLean LL, Perry MB, Petersen BO, Duus JØ, Whitfield C (2002) Structures of lipopolysaccharides from *Klebsiella pneumoniae*. Elucidation of the structure of the linkage region between core and polysaccharide O chain and identification of the residues at the non-reducing termini of the O chains. J Biol Chem 277:25070–25081

176. Cuthbertson L, Powers J, Whitfield C (2005) The C-terminal domain of the nucleotide-binding domain protein Wzt determines substrate specificity in the ATP-binding cassette transporter for the lipopolysaccharide O-antigens in *Escherichia coli* serotypes O8 and O9a. J Biol Chem 280:30310–30319
177. Cuthbertson L, Kimber MS, Whitfield C (2007) Substrate binding by a bacterial ABC transporter involved in polysaccharide export. Proc Natl Acad Sci USA 104:19529–19534
178. Whitfield C, Richards JC, Perry MB, Clarke BR, MacLean LL (1991) Expression of two structurally distinct D-galactan O-antigens in the lipopolysaccharide of *Klebsiella pneumoniae* serotype O1. J Bacteriol 173:1420–1431
179. Whitfield C, Perry MB, MacLean LL, Yu SH (1992) Structural analysis of the O-antigen side chain polysaccharides in the lipopolysaccharides of *Klebsiella* serotypes O2(2a), O2(2a,2b), and O2(2a,2c). J Bacteriol 174:4913–4919
180. Keenleyside WJ, Whitfield C (1996) A novel pathway for O-polysaccharide biosynthesis in *Salmonella enterica* serovar Borreze. J Biol Chem 271:28581–28592
181. Keenleyside WJ, Whitfield C (1995) Lateral transfer of *rfb* genes: a mobilizable ColE1-type plasmid carries the *rfb* O:54 (O:54 antigen biosynthesis) gene cluster from *Salmonella enterica* serovar Borreze. J Bacteriol 177:5247–5253
182. Keenleyside WJ, Perry M, Maclean L, Poppe C, Whitfield C (1994) A plasmid-encoded rfbO:54 gene cluster is required for biosynthesis of the O:54 antigen in *Salmonella enterica* serovar Borreze. Mol Microbiol 11:437–448
183. Keenleyside WJ, Clarke AJ, Whitfield C (2001) Identification of residues involved in catalytic activity of the inverting glycosyl transferase WbbE from *Salmonella enterica* serovar borreze. J Bacteriol 183:77–85
184. DeAngelis PL (2002) Microbial glycosaminoglycan glycosyltransferases. Glycobiology 12:9R–16R
185. DeAngelis PL (1999) Hyaluronan synthases: fascinating glycosyltransferases from vertebrates, bacterial pathogens, and algal viruses. Cell Mol Life Sci 56:670–682
186. Forsee WT, Cartee RT, Yother J (2000) Biosynthesis of type 3 capsular polysaccharide in *Streptococcus pneumoniae*. Enzymatic chain release by an abortive translocation process. J Biol Chem 275:25972–25978
187. Remminghorst U, Rehm BH (2006) Bacterial alginates: from biosynthesis to applications. Biotechnol Lett 28:1701–1712
188. Itoh Y, Rice JD, Goller C, Pannuri A, Taylor J, Meisner J, Beveridge TJ, Preston JF, Romeo T (2008) Roles of *pgaABCD* genes in synthesis, modification, and export of the *Escherichia coli* biofilm adhesin poly-β-1,6-*N*-acetyl-D-glucosamine. J Bacteriol 190:3670–3680
189. Hug I, Couturier MR, Rooker MM, Taylor DE, Stein M, Feldman MF (2010) *Helicobacter pylori* lipopolysaccharide is synthesized via a novel pathway with an evolutionary connection to protein N-glycosylation. PLoS Pathog 6:e1000819
190. Kelly J, Jarrell H, Millar L, Tessier L, Fiori LM, Lau PC, Allan B, Szymanski CM (2006) Biosynthesis of the N-linked glycan in *Campylobacter jejuni* and addition onto protein through block transfer. J Bacteriol 188:2427–2434
191. Abeyrathne P, Daniels C, Poon KK, Matewish MJ, Lam J (2005) Functional characterization of WaaL, a ligase associated with linking O-antigen polysaccharide to the core of *Pseudomonas aeruginosa* lipopolysaccharide. J Bacteriol 187:3002–3012
192. Heinrichs DE, Yethon JA, Amor PA, Whitfield C (1998) The assembly system for the outer core portion of R1- and R4-type lipopolysaccharides of *Escherichia coli*. The R1 core-specific β-glucosyltransferase provides a novel attachment site for O-polysaccharides. J Biol Chem 273:29497–29505
193. Heinrichs DE, Monteiro MA, Perry MB, Whitfield C (1998) The assembly system for the lipopolysaccharide R2 core-type of *Escherichia coli* is a hybrid of those found in *Escherichia coli* K-12 and *Salmonella enterica*. Structure and function of the R2 WaaK and WaaL homologs. J Biol Chem 273:8849–8859

194. Schild S, Lamprecht AK, Reidl J (2005) Molecular and functional characterization of O-antigen transfer in *Vibrio cholerae*. J Biol Chem 280:25936–25947
195. Olsthoorn MM, Petersen BO, Schlecht S, Haverkamp J, Bock K, Thomas-Oates JE, Holst O (1998) Identification of a novel core type in *Salmonella* lipopolysaccharide. Complete structural analysis of the core region of the lipopolysaccharide from *Salmonella enterica* sv. Arizonae O62. J Biol Chem 273:3817–3829
196. Kaniuk NA, Vinogradov E, Whitfield C (2004) Investigation of the structural requirements in the lipopolysaccharide core acceptor for ligation of O-antigens in the genus *Salmonella*: WaaL "ligase" is not the sole determinant of acceptor specificity. J Biol Chem 279: 36470–36480
197. Meredith TC, Mamat U, Kaczynski Z, Lindner B, Holst O, Woodard RW (2007) Modification of lipopolysaccharide with colanic acid (M-antigen) repeats in *Escherichia coli*. J Biol Chem 282:7790–7798
198. Tang G, Mintz KP (2010) Glycosylation of the collagen adhesin EmaA of *Aggregatibacter actinomycetemcomitans* is dependent upon the lipopolysaccharide biosynthetic pathway. J Bacteriol 192:1395–1404
199. Power PM, Seib KL, Jennings MP (2006) Pilin glycosylation in *Neisseria meningitidis* occurs by a similar pathway to *wzy*-dependent O-antigen biosynthesis in *Escherichia coli*. Biochem Biophys Res Commun 347:904–908
200. Faridmoayer A, Fentabil MA, Haurat MF, Yi W, Woodward R, Wang PG, Feldman MF (2008) Extreme substrate promiscuity of the *Neisseria* oligosaccharyl transferase involved in protein O-glycosylation. J Biol Chem 283:34596–34604
201. Castric P (1995) *pilO*, a gene required for glycosylation of *Pseudomonas aeruginosa* 1244 pilin. Microbiology 141:1247–1254
202. Novotny R, Schäffer C, Strauss J, Messner P (2004) S-layer glycan-specific loci on the chromosome of *Geobacillus stearothermophilus* NRS 2004/3a and dTDP-L-rhamnose biosynthesis potential of G. *stearothermophilus* strains. Microbiology 150:953–965
203. Nesper J, Kraiss A, Schild S, Blass J, Klose KE, Bockemuhl J, Reidl J (2002) Comparative and genetic analyses of the putative *Vibrio cholerae* lipopolysaccharide core oligosaccharide biosynthesis (*wav*) gene cluster. Infect Immun 70:2419–2433
204. Raetz CR, Reynolds CM, Trent MS, Bishop RE (2007) Lipid A modification systems in gram-negative bacteria. Annu Rev Biochem 76:295–329
205. Abeyrathne P, Lam J (2007) WaaL of *Pseudomonas aeruginosa* utilizes ATP in in vitro ligation of O-antigen onto lipid A-core. Mol Microbiol 65:1345–1359
206. Pugsley AP (1993) The complete general secretory pathway in gram-negative bacteria. Microbiol Rev 57:50–108
207. Pérez JM, McGarry MA, Marolda CL, Valvano MA (2008) Functional analysis of the large periplasmic loop of the *Escherichia coli* K-12 WaaL O-antigen ligase. Mol Microbiol 70: 1424–1440
208. Qutyan M, Paliotti M, Castric P (2007) PilO of *Pseudomonas aeruginosa* 1244: subcellular location and domain assignment. Mol Microbiol 66:1444–1458
209. White RH (1996) Biosynthesis of isoprenoids in bacteria. In: Neidhardt FC, Curtiss R III, Ingraham JL, Lin ECC, Low KB, Magasanik B, Reznikoff WS, Riley M, Schaechter M, Umbarger HE (eds) *Escherichia coli* and *Salmonella*: cellular and molecular biology. ASM Press, Washington, pp 637–641
210. Rohmer M (1999) The discovery of a mevalonate-independent pathway for isoprenoid biosynthesis in bacteria, algae and higher plants. Nat Prod Rep 16:565–574
211. Rohmer M, Knani M, Simonin P, Sutter B, Sahm H (1993) Isoprenoid biosynthesis in bacteria: a novel pathway for the early steps leading to isopentenyl diphosphate. Biochem J 295:517–524
212. Kuzuyama T (2002) Mevalonate and nonmevalonate pathways for the biosynthesis of isoprene units. Biosci Biotechnol Biochem 66:1619–1627

213. Sprenger GA, Schorken U, Wiegert T, Grolle S, de Graaf AA, Taylor SV, Begley TP, Bringer-Meyer S, Sahm H (1997) Identification of a thiamin-dependent synthase in *Escherichia coli* required for the formation of the 1-deoxy-D-xylulose 5-phosphate precursor to isoprenoids, thiamin, and pyridoxol. Proc Natl Acad Sci USA 94:12857–12862
214. Lois LM, Campos N, Putra SR, Danielsen K, Rohmer M, Boronat A (1998) Cloning and characterization of a gene from *Escherichia coli* encoding a transketolase-like enzyme that catalyzes the synthesis of D-1-deoxyxylulose 5-phosphate, a common precursor for isoprenoid, thiamin, and pyridoxol biosynthesis. Proc Natl Acad Sci USA 95:2105–2110
215. Takahashi S, Kuzuyama T, Watanabe H, Seto H (1998) A 1-deoxy-D-xylulose 5-phosphate reductoisomerase catalyzing the formation of 2-*C*-methyl-D-erythritol 4-phosphate in an alternative nonmevalonate pathway for terpenoid biosynthesis. Proc Natl Acad Sci USA 95:9879–9884
216. Apfel CM, Takacs B, Fountoulakis M, Stieger M, Keck W (1999) Use of genomics to identify bacterial undecaprenyl pyrophosphate synthetase: cloning, expression, and characterization of the essential *uppS* gene. J Bacteriol 181:483–492
217. Shimizu N, Koyama T, Ogura K (1998) Molecular cloning, expression, and purification of undecaprenyl diphosphate synthase. No sequence similarity between E- and Z-prenyl diphosphate synthases. J Biol Chem 273:19476–19481
218. Abeijon C, Hirschberg CB (1992) Topography of glycosylation reactions in the endoplasmic reticulum. Trends Biochem Sci 17:32–36
219. Rush JS, Cho SK, Jiang S, Hofmann SL, Waechter CJ (2002) Identification and characterization of a cDNA encoding a dolichyl pyrophosphate phosphatase located in the endoplasmic reticulum of mammalian cells. J Biol Chem 277:45226–45234
220. Stukey J, Carman GM (1997) Identification of a novel phosphatase sequence motif. Protein Sci 6:469–472
221. Neuwald AF (1997) An unexpected structural relationship between integral membrane phosphatases and soluble haloperoxidases. Protein Sci 6:1764–1767
222. Ishikawa K, Mihara Y, Gondoh K, Suzuki E, Asano Y (2000) X-ray structures of a novel acid phosphatase from *Escherichia blattae* and its complex with the transition-state analog molybdate. EMBO J 19:2412–2423
223. Tatar LD, Marolda CL, Polischuk AN, van Leeuwen D, Valvano MA (2007) An *Escherichia coli* undecaprenyl-pyrophosphate phosphatase implicated in undecaprenyl-phosphate recycling. Microbiology 153:2518–2529

Lipopolysaccharide Export to the Outer Membrane

10

Paola Sperandeo, Gianni Dehò, and Alessandra Polissi

10.1 Introduction

In this chapter we will discuss how lipopolysaccharide (LPS) is transported and assembled from its site of synthesis (the cytoplasm and the inner membrane) to the cell surface. This is a remarkably complex process, as LPS must traverse three different cellular compartments to reach its final destination.

Details on the last steps of LPS biogenesis have only recently emerged and several factors implicated in LPS transport have been identified. Nevertheless, the molecular details of this process need to be clarified, as many questions still remain unanswered. Most of our understanding of this problem derives, not surprisingly, from studies performed in the Gram-negative model organism *Escherichia coli*. Interestingly, the pathogenic *Neisseria meningitidis* can survive without LPS under laboratory conditions [1]. This has made easier the disruption of LPS biogenesis genes and thus the genetic and biochemical analysis of this process.

10.2 The Peculiar Envelope of Gram-Negative Bacteria

The staining procedure developed by Hans Christian Gram in 1884 [2] created a divide in the bacterial world. Although the taxonomic subdivision in Gram-positives and the Gram-negatives transcends the actual result of the procedure (e.g., the Gram-positive *Mycoplasma* species do not stain), the Gram staining is

P. Sperandeo • A. Polissi (✉)
Dipartimento di Biotecnologie e Bioscienze, Università di Milano-Bicocca, Piazza della Scienza 2, 20126 Milan, Italy
e-mail: paola.sperandeo@unimib.it; alessandra.polissi@unimib.it

G. Dehò
Dipartimento di Scienze biomolecolari e Biotecnologie, Università di Milano, Via Celoria 26, 20133 Milan, Italy
e-mail: gianni.deho@unimi.it

Y.A. Knirel and M.A. Valvano (eds.), *Bacterial Lipopolysaccharides*,
DOI 10.1007/978-3-7091-0733-1_10, © Springer-Verlag/Wien 2011

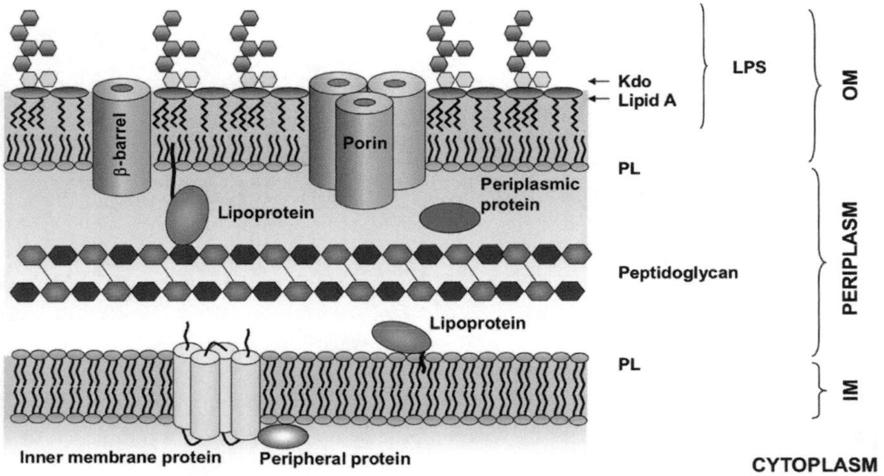

Fig. 10.1 The Gram-negative cell envelope. The envelope of Gram-negative bacteria is composed of an inner membrane (*IM*), the periplasm and an outer membrane (*OM*). The IM is a symmetric lipid bilayer composed of phospholipids (*PL*), integral proteins that span the membrane by α-helical transmembrane domains (*cyan cylinders*) and peripheral proteins associated to the inner leaflet of the IM (*light red ovals*). The periplasm is an aqueous compartment located between IM and OM containing a layer of peptidoglycan. The OM is an asymmetric bilayer composed of phospholipid in the inner leaflet and lipopolysaccharide (*LPS*) towards the outside. The lipid A, two molecules of Kdo (*yellow hexagons*) decorated by a core of non-repeated sugar units (*brown hexagons*) are shown; the variable repeated sugar unit (O-antigen) is omitted in the figure. The OM also contains integral proteins folded in β-barrel conformations (*orange barrels*). Both IM and OM contain lipoproteins (*green ovals*) anchored to their periplasmic faces. See text for details

still widely used by microbiologists as an important primary diagnostic tool. The Gram stain reflects differences in structure and composition of the bacterial cell envelope, although the basis for the differential Gram staining did not become immediately apparent when the nature of the bacterial envelope was elucidated [3].

Early biochemical approaches combined with imaging techniques such as electron microscopy led to our present understanding of the bacterial envelope as consisting of several differentially extractable components organized in physically distinct layers. These studies clearly illustrated the structural difference between Gram-positive and Gram-negative bacteria [4, 5]. By integrating various complementary approaches including genetics, molecular biology, structural biology and biophysics, spectacular progress has been made allowing a more refined and detailed picture of the structure and composition of the bacterial cell envelope [6, 7].

The envelope structure of a typical Gram-negative bacterium is illustrated in Fig. 10.1. In contrast to most Gram-positives, which are typically surrounded by the cytoplasmic membrane and by additional cell wall layers of variable complexity, Gram-negative bacteria are uniquely surrounded by two lipid membranes: the cytoplasmic (or inner) membrane (IM) and the outer membrane (OM) separated by the periplasmic space that also contains a thin peptidoglycan layer.

The IM is a classical symmetric phospholipid bilayer in which integral membrane proteins are typically embedded with hydrophobic α-helices spanning the membrane. Lipoproteins may also be anchored at the outer leaflet of the IM via an *N*-acyl-diacylglyceryl modification of a N-terminal cysteine [8, 9]. The IM represents the cell boundary and is primarily responsible for regulating all chemical exchanges and physical interactions between the cell and the environment; it is therefore involved in almost every aspect of bacterial homeostasis, growth, and metabolism [10].

The periplasmic space is an aqueous compartment between the IM and OM. Various processes that are vital to growth and viability of the cell occur in this compartment. Proteins residing in the periplasmic space fulfil important functions in the detection, processing and transport of nutrients into the cell. Periplasmic chaperones (including proteins involved in disulfide bonds formation) promote the biogenesis of periplasmic and OM proteins and external appendages such as pili and fimbriae. Detoxifying enzymes (such as β-lactamases) preserve the cell from toxic chemicals. It is important to note that the cell cannot export outside the IM high-energy phosphate bond molecules (ATP, GTP, PEP, etc.) and thus all endergonic processes occurring in the periplasm (and in the OM) must be energized from the cytoplasm or through the IM [11].

In the periplasmic space of Gram-negative resides the peptidoglycan (murein) layer, a chemically unique rigid structural component of the cell wall, which confers to the cell its shape and preserves its integrity in low osmolarity environments [12].

10.2.1 Outer Membrane Lipid Bilayer as a Diffusion Barrier

The OM is an asymmetric bilayer consisting of phospholipids and LPS in the inner and outer leaflet, respectively [7]. Proteins are associated to the lipid bilayer as either integral OM proteins or lipoproteins. The OM proteins are implicated in several functions: nutrients uptake, transport and secretion of various molecules (proteins, polysaccharides, drugs), assembly of proteins or proteinaceous structures at the OM. Integral OM proteins most often consist of antiparallel amphipatic β-strands that fold into cylindrical β-barrels with an hydrophilic interior and hydrophobic residues pointing outward to face the membrane lipids [13, 14]. However, the recent crystal structure of an integral OM protein (Wza) implicated in the secretion of capsular polysaccharide revealed a new structural paradigm for OM proteins. Wza shows a novel transmembrane α-helical barrel and a large central cavity, which is predicted to accommodate the secreted assembled capsular macromolecules; this is likely to be a common feature in other OM proteins involved in secretion processes [15, 16]. Many lipoproteins are associated at the OM: in *E. coli* the protein moiety usually faces the periplasm and the N-terminal *N*-acyl-diacylglycerylcysteine anchors the lipoprotein at the inner leaflet of the OM [9]. In other Gram-negative bacteria however, the OM lipoproteins may also extend in the extracellular medium. Examples are the *N. meningitidis* LbpB and TbpB components of the lactoferrin and transferrin receptor, respectively [17].

The asymmetry of the OM relies on the presence of LPS in the outer leaflet whereas phospholipids are confined to the inner leaflet of the bilayer. The absence of phospholipids in the outer leaflet of the OM was initially demonstrated by Nikaido and Kamio [18], who showed that chemical labelling of amino groups in intact cells of *Salmonella enterica* serovar Typhimurium by an OM-impermeable macromolecular reagent failed to label any phosphatidylethanolamine molecules.

This second lipid bilayer with an additional external hydrophilic region of long polysaccharide chains, endows Gram-negative bacteria with an additional diffusion barrier, which accounts for the generally higher resistance of Gram-negative bacteria, as compared to most Gram-positives, to many toxic chemicals such as antibiotics and detergents (e.g. bile salts) and to survive hostile environments such as the gastrointestinal tracts of mammals, encountered during host colonization or infection [7, 19].

The hydrophobicity of lipid A together with the strong lateral interactions between the LPS molecules contributes to the effectiveness OM permeability barrier [7]. The high number of fully saturated fatty acyl substituents per molecule of lipid A is thought to create a gel-like lipid interior of very low fluidity that contributes to the low permeability of hydrophobic solutes across the OM. The very strong lateral interaction between LPS molecules is mediated by the bridging action of divalent cations and also by the strong association of LPS to OM proteins such as FhuA, a ferric hydroxamate uptake receptor. Indeed, FhuA binding to lipid A offers an additional mode of interaction between neighbouring LPS molecules [20].

LPS organization is disrupted by defects in assembly of OM components [21], in mutants producing LPS severely truncated in sugar chains ("deep rough" mutants) [22] or by exposure to antimicrobial peptides and chelating agents such as EDTA which displace divalent cations needed to reduce the repulsive charges between LPS molecules [7]. In all these cases the consequence is that much of the LPS layer is shed and phospholipids from the inner leaflet migrate into the breached areas of the outer leaflet. These locally symmetric bilayer rafts are more permeable to hydrophobic molecules, which can now gain access to the periplasm while the OM continues to retain the more polar periplasmic contents [23]. Therefore, appreciable levels of phospholipids in the outer leaflet of the OM are detrimental to the cell.

The presence of phospholipids on the outer cell surface in stressed cells is detected by the integral OM β-barrel enzymes PagP and PldA, which act as sentinel of OM integrity and modify phospholipids to restore OM functionality through different mechanisms [24, 25]. PldA is an OM phospholipase that hydrolyzes a wide range of phospholipid substrates. The enzyme normally resides as an inactive monomer at the OM but in the presence of phospholipids a catalytically active PldA dimer is formed that destroys the invading lipid substrates thus restoring the asymmetry of the stressed OM [25]. PagP is also dormant in non-stressed cells and is activated through the PhoP/Q two component system, which responds to the limitation of divalent cations [26]. The enzyme transfers a palmitate moiety from phospholipids to lipid A forming a phospholipid by-product and a heptaacylated

LPS molecule with increased hydrophobicity [27] thus contributing to restore the permeability barrier function of the OM.

A third mechanism, the Mla pathway, which influences cell-surface phospholipid composition, has recently been discovered. This pathway comprises six proteins (MlaA-MlaF) forming an ABC (ATP-binding cassette) transport system that prevents accumulation of phospholipids in the outer leaflet of the OM [28]. Core components of the Mla pathway are conserved in Gram-negative bacteria and in the chloroplasts of plants [29]. Mutations in the Mla pathway are not lethal but lead to phospholipid accumulation in the outer leaflet of the OM, suggesting that this pathway plays a key role in preserving OM lipid asymmetry under non-stress conditions. Therefore it is proposed that the Mla proteins constitute a bacterial intermembrane phospholipids trafficking system [28].

10.2.2 Maintenance of the Outer Membrane Integrity

As the OM is a crucial structure for viability, it is not surprising that Gram-negative bacteria have evolved signalling pathways to monitor its integrity and to respond to envelope damage. In *E. coli* the σ^E regulon helps maintaining envelope integrity (reviewed by Alba and Gross [30] and Ades [31]). σ^E is an alternative sigma factor normally present in an inactive form by the membrane spanning anti-σ factor RseA. The σ^E pathway is activated by the accumulation of misfolded β-barrel OM proteins that trigger a proteolytic cascade leading to RseA degradation [32–34]. This frees σ^E to transcribe genes whose products (periplasmic folding catalysts, proteases, lipoproteins, enzymes involved in biosynthesis of lipid A, small regulatory RNAs) [35, 36] help restoring envelope homeostasis. Three genes *lptA*, *lptB* and *lptD* whose products are part of the machinery that transport LPS to the OM, are regulated by σ^E. The *lptAB* genes are organized in a di-cistronic operon controlled by σ^E and accordingly, their expression increases approximately ten-fold in an *rseA* gene deletion mutant [37]; *lptD* is co-transcribed with the periplasmic chaperone *surA* from a σ^E dependent promoter [35, 38]. The composition of the σ^E regulon suggests that the major, if not essential, physiological role of σ^E is to maintain the integrity and the functions of the envelope. Interestingly, σ^E activity is also sensitive to the status of LPS although the mechanism has not been elucidated [39]. Therefore, it may be relevant that the σ^E-dependent promoter upstream of *lptA* does not respond to any type of stress that activates the σ^E-dependent extracytoplasmic stress pathway except a subset of stressful conditions affecting LPS (A. Martorana et al., unpublished results).

10.3 Lipid Trafficking to the Outer Membrane

Transport of amphipatic molecules (such as phospholipids, lipoproteins and LPS) through membranes and in aqueous phases poses several problems. Movement through aqueous phases such as the periplasm would expose the hydrophobic moieties of amphipatic molecules to water molecules, whereas moving across membranes (intramembrane transport) would expose their hydrophilic moieties to the hydrophobic interior of the membranes.

Intramembrane transport of phospholipids is likely to be mediated by dedicated proteins [40]. The peptidoglycan layer, sandwiched between the IM and OM of Gram-negative bacteria, may effectively prevent membrane exchanges by vesicular trafficking [12]. Thus, lipid trafficking to the OM likely depends on soluble periplasmic lipid-transfer proteins and/or localized bridges connecting IM and OM [41].

10.3.1 Transport of Phospholipids and Lipoproteins

The transport of glycerophospholipids to the OM is poorly understood. The inner leaflet of the IM is the site of the late steps in the synthesis of phospholipids [42]. Reversible lipid transfer across model lipid bilayers (flip-flop) is extremely slow with half lives in the order of hours to days, but is very fast in biological membranes where it occurs in seconds or tenths of seconds [43]. This led to the conclusion that lipid flip-flop in biological membranes is catalyzed by proteins. Although various integral IM proteins may promote the rapid flipping of glycerophospholipids between the cytoplasmic and periplasmic leaflet of the IM [44], the precise mechanism for their transport across the IM bilayer remains largely not understood. The ATP-binding cassette transporter MsbA, which is responsible for the unidirectional flipping of the lipid A-core oligosaccharide (OS) across IM (see below), is also required for intramembrane exchange of glycerophospholipids in *E. coli* [45, 46], thus suggesting that MsbA also mediates phospholipid flipping across the IM. However, in *N. meningitidis* MsbA seems not required for glycerophospholipids biogenesis [47]. The role of MsbA in phospholipid movement across IM is controversial: it is likely that MsbA is primarily a flippase for lipid A-core OS translocation and may indirectly affect glycerophospholipids transport in *E. coli* (see below).

Contrary to glycerophospholipids, OM lipoprotein transport is well characterized [9]. Lipoproteins, as the bulk of secreted bacterial proteins, cross the IM via the Sec system [48]. In the periplasm their N-terminal cysteine is modified and anchored to the periplasmic leaflet of the IM; lipoproteins with an aspartate in position 2 will remain anchored to the IM whereas all the other will be sorted to the OM. This process requires an ABC transporter, LolCDE that mediates the detachment from the IM of OM-specific lipoproteins and delivers them to the periplasmic carrier protein LolA. Structure of LolA reveals a hydrophobic cavity, which is thought to accept the lipid moiety of lipoproteins. The LolA-lipoprotein complex moves across the periplasm, probably by diffusion, and interacts with a lipoprotein-specific receptor, LolB, which is itself a lipoprotein anchored to the OM. The lipoprotein is then transferred from LolA to LolB and incorporated into the inner leaflet of the OM (reviewed in Refs. [8, 9]). According to this model, the formation of the soluble complex LolA-lipoprotein with the LolA chaperone masking the hydrophobic moiety of the lipoprotein, makes the transport of an amphipatic molecule across the periplasm thermodynamically favourable. Notably, ATP hydrolysis mediated by the IM ABC transporter LolCDE provides the energy for the transport to the inner leaflet of the OM.

10.3.2 LPS Transport to the Cell Surface

10.3.2.1 Identification of LPS Biogenesis Factors

LPS is an essential structural component of the OM in most Gram-negative bacteria [6]. Many genes implicated in functions that are essential for bacterial viability have been identified through the study of conditional lethal (mostly temperature sensitive) and/or antibiotic resistant mutants. However, somewhat surprisingly, none of the temperature sensitive mutants isolated in over 50 years of bacterial genetics has been implicated in LPS transport, although many are implicated in LPS biosynthesis or in other aspects of OM biogenesis such as β-barrel OMP or lipoprotein transport [49]. On the other hand, chemical genetics (an approach which uses specific mutants in combination with antibiotics to search for the corresponding cellular targets) has proved successful for the identification of factors involved in OMPs biogenesis but not for the identification of LPS biogenesis functions (see for example Ref. [50]). Straight biochemical approaches do not seem suitable for the isolation of LPS transport components, as biochemical assays are not obvious.

A powerful genetic approach to identify genes involved in specific cellular functions is to devise screens for specific phenotypes. It is well known that viable mutants with an altered OM are more permeable to hydrophobic compounds, detergents, bile salts, dyes and large hydrophilic antibiotics [7]. Thus, screening for increased sensitivity to such compounds may lead to the identification of mutants in OM biogenesis. However, such properties are shared by viable mutants in late steps of LPS biosynthesis or in OMPs including drug efflux systems [6, 7], making the screening less specific. For instance, *E. coli* and *Salmonella* "deep rough" mutants that produce a truncated LPS molecule lacking the heptose region of the inner core OS exhibit increased sensitivity to hydrophobic dyes, antibiotics, detergents, phenols, fatty acids and polycyclic hydrocarbons [22]. Nevertheless, the first of the eight genes implicated in LPS transport known so far, *lptD* (originally known as *imp* or *ostA*; see below) has been identified for increased OM permeability in two independent screens [51, 52]. Other genes have been identified by random approaches such as multicopy suppression of a thermosensitive mutant, screening of conditional lethal expression mutants, and by direct conditional expression mutagenesis of genes adjacent to other genes implicated in LPS transport (*lpt* genes). Genetic and biochemical information obtained by studying these *lpt* genes allowed more direct screenings using bioinformatic analysis and mutagenesis of candidate genes, as well as copurification of proteins associated to a known Lpt machine component. These approaches, summarized in Table 10.1 and discussed in more detail below, led to the identification of eight genes involved in LPS biogenesis.

Table 10.1 Components of the LPS transport machine

Gene name	Synonyms	Chromosomal location	Method of identification and/or implication in LPS transport	Protein MW (kDa)	Protein localization	Protein properties/ function	References
lptA	*yhbN*	*yrbG-lptB* locus	Screening of conditional expression mutants upon Tn-SS2 mutagenesis; gene-specific conditional expression mutagenesis	18.6	Periplasm/associates to IM and OM Lpt components	Binds LPS	[37, 56, 76, 78, 79, 82]
lptB	*yhbG*	*yrbG-lptB* locus	As *lptA*	26.7	Cytoplasm/IM-associated	ABC protein; component of IM ABC transporter	[37, 56, 77, 82]
lptC	*yrbK*	*yrbG-lptB* locus	Gene-specific conditional expression mutagenesis of *lptA*	21.6	IM/periplasm exposed	Component of IM ABC transporter	[56, 82]
lptD	*imp*; *ostA*	*lptD-apaH* operon	Two independent screens for increased permeability: (i) suppression of *lamB* for large maltodextrins utilization, and (ii) increased sensitivity to organic solvents	87	OM	β-barrel component of OM complex for LPS assembly	[51, 52]
lptE	*rlpB*	*leuS-cobC* operon	Study of peptidoglycan biosynthesis genes; co-purification with His-tagged LptD	21.2	OM	LptD-associated lipoprotein; binds LPS	[55, 58, 116]
lptF	*yjgP*	*lptF-lptG* operon	Bioinformatics and gene specific mutagenesis	40.2	IM	Component of IM ABC transporter	[57]
lptG	*yjgQ*	*lptF-lptG* operon	Bioinformatics and gene specific mutagenesis	39.5	IM	Component of IM ABC transporter	[57]
msbA		*ycaI-ycaQ* operon	Suppressor of *htrB* mutants	64.3	IM	ABC transporter; LPS flipping over IM	[61, 65, 70]

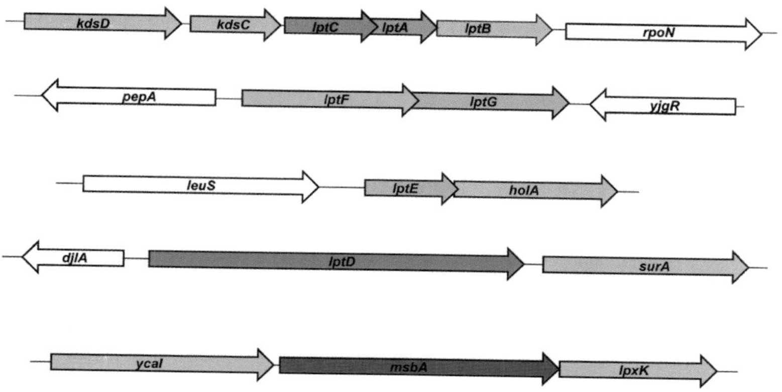

Fig. 10.2 Organization of genetic loci implicated in LPS transport in *E. coli*. Annotation is based on that of the *E. coli* strain MG1655 (*E. coli*) (http://ecocyc.org/). Lpt ORF lengths are to scale. In grey are represented neighbouring unrelated genes belonging to the same transcriptional units of Lpt genes

The genetic organization of the *lpt* genes in *E. coli* is shown in Fig. 10.2. It is now believed that these eight proteins constitute the machinery that transports LPS from its site of synthesis, the cytoplasmic side of the IM, to the cell surface. MsbA carries out the first step of LPS transport, namely the transbilayer movement of lipid A-core OS across the IM [53]. In strains producing O-antigen, ligation of the O-antigen polysaccharide to the lipid A-core OS acceptor molecule is catalyzed by the WaaL ligase (see Chap. 9) and occurs at the periplasmic face of the IM. Transport downstream of MsbA of lipid A-core OS and presumably full length LPS with O-antigen is carried out by the "Lpt machinery." The proteins implicated in this process are located at the IM (LptB, LptC, LptF and LptG), in the periplasm (LptA) and at the OM (LptD and LptE) and constitute a transenvelope complex that transports LPS across the periplasm and inserts it in the OM (Fig. 10.3) [37, 54–58].

10.3.2.2 Transport of LPS Across IM

The MsbA protein shows sequence similarity to a class of ABC transporters involved in multidrug resistance in bacteria and eukaryotic cells by extrusion of several different drugs [59, 60] (Fig. 10.3). MsbA is a "half-transporter," comprising a transmembrane domain with six membrane-spanning helices, which are believed to contain the substrate-binding site, and a nucleotide-binding domain, for a total molecular mass of 64.5 kDa [59]. The functional MsbA transporter is presumed to be a homodimer. Substrate transport is driven by the energy provided by ATP hydrolysis.

MsbA was originally identified in *E. coli* as a multicopy suppressor of the thermosensitive phenotype of an *htrB* deletion mutant [61]. *htrB* (later renamed *lpxL*) encodes a Kdo-dependent acyltransferase responsible for the addition of a lauroyl group to the tetraacylated Kdo_2-lipid IV_A thus forming the pentaacylated

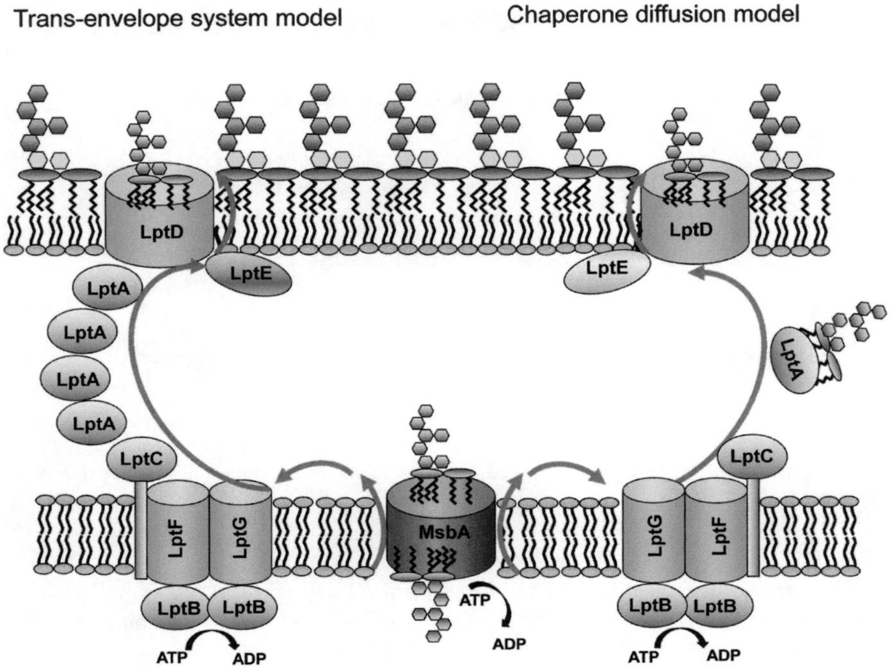

Fig. 10.3 Models for LPS transport through the cell envelope. After its synthesis at the inner leaflet of the IM, lipid A-core OS molecules are flipped across the IM by MsbA. If made, the O-antigen is ligated to the rest of the molecule by WaaL (not shown). Then LPS is extracted from the IM by the ABC transporter LptBCFG. In the "trans-envelope complex" model (shown on the *left*) LptA, LptE and LptD constitute a multiprotein complex with LptBCFG, which spans the cell envelope by bridging IM and OM components. In the "chaperone diffusion" model (shown on the *right*) LptA is the soluble chaperone protein, which accepts LPS from the IM ABC transporter and delivers it to the OMP complex LptDE

Kdo_2-lipid A [62] (see Chap. 6). Mutants in *htrB* are not viable at temperatures above 33°C and produce underacylated LPS that is not efficiently transported to the OM [45]. Under non-permissive conditions, the *htrB*/*lpxL* null mutant shows alterations in cell morphology (such as formation of bulges and filaments) and accumulates phospholipids [63] and the tetraacylated LPS precursor in the IM [45, 64]. Expression of *msbA* from a plasmid vector, which suppresses the thermo-sensitive growth defect of *htrB*/*lpxL* null mutant cells, restores phospholipids and tetraacylated LPS precursor translocation to the OM. Thus the higher expression of MsbA at higher temperature does not restore lipid IV_A acylation to give lipid A but seems to facilitate the transport of the immature LPS form to the OM [45]. By contrast, MsbA depleted cells accumulate hexaacylated lipid A at the IM [45], thus further implicating MsbA in LPS transport. Using an *E. coli* thermosensitive *msbA* mutant carrying a single amino acid substitution (A270T) in a transmembrane region of the protein, Doerrler et al. [65] observed a key role for MsbA in lipid trafficking. Transport of all major lipids (both LPS and phospholipids) to the OM is inhibited in these *msbA* mutant cells shifted to non-permissive temperature suggesting that *E. coli* MsbA is needed for export of all major membrane lipids [65].

In *N. meningitidis* the *msbA* gene is not essential for cell viability as this bacterium can survive without LPS [1]. *N. meningitidis msbA* mutants produce reduced amounts of LPS, a feature typical of mutants in LPS transport in this organism, but possess an OM mostly composed of phospholipids, indicating that phospholipid transport to the OM is not impaired and suggesting a difference in general lipid transport with respect to *E. coli* [47].

A topological analysis of lipids in vivo can be demonstrated using as markers covalent modifications catalyzed by compartment-specific enzymes. For the analysis of the topology of newly synthesised lipid A the temperature sensitive $msbA_{A270T}$ allele was analyzed in a polymyxin-resistant background [46]. In *E. coli* and *Salmonella* polymyxin resistance depends on enzymes acting at the periplasmic side of the IM that covalently modify lipid A with cationic substituents [66]. Upon MsbA inactivation at high temperature, newly synthesized lipid A was not modified, suggesting that the molecule accumulates in the IM facing the cytoplasm [46]. This is consistent with a model of MsbA-mediated translocation between membrane leaflets, rather than ejection from the bilayer.

Several in vitro studies have been performed to evaluate MsbA substrate specificity. The basal ATPase activity of purified MsbA reconstituted into liposomes is stimulated by hexaacylated lipid A, Kdo_2-lipidA, or LPS but not by underacylated lipid A precursors, suggesting that hexaacylated LPS is the substrate required for the transport [67] in line with previous genetic and biochemical evidence [45]. This work was further expanded by functional reconstitution of the protein into proteoliposomes of *E. coli* lipids to estimate MsbA binding affinities for nucleotides and putative transport substrates [68]. Using purified labeled MsbA simultaneous high affinity binding of lipid A and daunorubicin was demonstrated [69]. These results indicate that MsbA contains two substrate-binding sites that communicate with both the nucleotide-binding domain and with each other. One is a high affinity-binding site for the physiological substrate, lipid A, and the other site interacts with drugs with comparable affinity. Thus, MsbA may function as both a lipid flippase and a multidrug transporter [69]. Early attempts to demonstrate MsbA-mediated lipid flipping in vitro failed [44]. However, a direct measurement of the lipid flippase activity of purified MsbA in a reconstituted system has been recently reported [70].

The X-ray crystal structures of MsbA in different conformations from the three closely related orthologs from *E. coli*, *Vibrio cholerae*, and *S. enterica* (serovar Typhimurium) were recently reported [71], after the original MsbA structures were withdrawn due to the discovery of a flaw in the software used to solve them [72]. The overall shape and domain organization of MsbA resemble that of the 3.0-Å structure of the putative bacterial multidrug transporter Sav1866 [73] and the 8-Å cryo-EM structure of Pgp [74]. The analyses of crystal structures of MsbA trapped in different conformations indicate that this molecule may undergo large ranges of motion that may be required for substrate transport [71].

Collectively, these results show that MsbA has the potential, at least in vitro, to handle a variety of substrates as expected for a protein belonging to the sub-family of drug-efflux transporters. However, in vivo MsbA displays a remarkable selectivity towards the LPS substrates being capable to translocate only hexaacylated but not penta or tetraacylated LPS. This observation together with data that will be discussed in the following paragraphs suggests that MsbA may play the role of "quality control system" for LPS export to the OM.

10.3.2.3 Transport of LPS Across the Periplasm to the Cell Surface

After MsbA-mediated translocation across the IM, the lipid A-core OS or the full length LPS containing O-antigen is transported through the aqueous periplasm compartment and assembled at the outer leaflet of the OM. This process takes places outside the cell membrane in an environment that lacks high-energy phosphate bond molecules such as ATP as energy sources. However, energy is required to extract an amphipatic molecule such as LPS from a lipid membrane and to transport it trough the periplasm. The *E. coli* Lpt (*LP*S *t*ransport) machinery consists of seven essential proteins (LptABCDEFG) that accomplish LPS transport across the periplasm to its final assembly at the cell surface [37, 56–58, 75, 76] (Fig. 10.3). The components involved in this process have been recently discovered employing a combination of genetic, biochemical, and bioinformatic approaches but, to date, there is no mechanistic information on how this system operates to transport and assemble LPS at the OM.

The seven proteins of the Lpt machinery are located in each and every compartment of the cell. The LptBFGC proteins constitute an atypical IM ABC transporter. LptB is a 27-kDa cytoplasmic protein possessing the typical nucleotide-binding fold of ABC transporters. Earlier, LptB was found associated to an uncharacterized IM high molecular weight protein complex, of approximately 140 kDa, but the other components of the complex were not identified [77]. LptF (40.4 kDa) and LptG (39.6 kDa) are the transmembrane subunits of the IM ABC transporter and are encoded by two essential genes (previously known as *yjgP* and *yjgQ*, respectively) that form an operon distantly located from *lptB* in the chromosome [57]. LptC is a bitopic 21 kDa IM protein with a single transmembrane domain and a larger soluble periplasmic portion ([56]; A. Polissi and P. Sperandeo, unpublished results). The ABC transporters of Gram-negative bacteria are often associated to a periplasmic binding protein. In the Lpt machinery LptA is the putative periplasmic binding component. LptA is a small protein of 185 amino acids possessing a 23 amino acid signal sequence that is processed in the mature form (mature molecular weight of 18.6 kDa; [78]). The *E. coli* LptA expressed from a plasmid by an IPTG inducible promoter has been localized in the periplasm as a soluble protein [37, 78], whereas in *N. meningitidis* the majority of the corresponding protein are associated to the membrane fraction [41].

The components of the IM-associated ABC transporter were identified by different screenings. Polissi et al. [79] identified LptA, LptB and LptC by using a genetic screen designed to discover new essential functions. In this early work, random transposition with a Tn5 minitransposon derivative carrying the arabinose-inducible *araBp* promoter oriented outward at one end was used to generate mutants that were assayed for conditional lethal phenotypes. *lptA*, *lptB* and *lptC* (formerly *yhbN*, *yhbG* and *yrbK*, respectively) are genetically linked with two genes involved in LPS biosynthesis (*kdsC* and *kdsD* coding for enzymes involved in Kdo biosynthesis [80, 81]), which are located immediately upstream *lptC* in the same cluster [82]. Their sequence and genomic organization are highly conserved among Gram negative-bacteria: this strongly suggested that they might have a role in LPS biogenesis. Interestingly *lptA* and *lptB* form an operon expressed from a σ^E dependent promoter [37], which is induced upon extracytoplasmic stress [35], thus further implicating these genes in envelope biogenesis.

The direct evidence of the involvement of LptA, LptB and LptC in LPS biogenesis came from membrane fractionation experiments using sucrose density-gradient centrifugation of depleted mutant cells. These experiments showed that depletion of LptA, LptB and LptC results in a similar phenotype, namely: cessation of growth after about seven generations; accumulation of abnormal membrane structures in the periplasm; production of an anomalous LPS form characterized by a ladder-like banding of higher molecular weight species; and, more importantly, failure to transport to the OM de novo synthesized LPS, which accumulated in a novel membrane fraction with intermediate density between IM and OM [37, 56]. The anomalous LPS that accumulated upon LptA and LptB depletion turned out to be LPS decorated by repeated units of colanic acid transferred to the lipid A-core OS by the WaaL enzyme [56]. WaaL is the enzyme responsible of O-antigen ligation to lipid A-core OS at the periplasmic side of the IM [6] (see Chap. 9). Therefore, this evidence suggests that upon LptA, LptB or LptC depletion LPS largely accumulate at the periplasmic side of the IM where it can be modified by the WaaL ligase. This observation was confirmed by Raetz et al. exploiting the ectopic expression of the lipid A 1-phosphatase LpxE from *Francisella* and the lipid A 3-O-deacylase PagL from *Salmonella* as periplasmic and OM markers, respectively, of LPS topology in a novel temperature sensitive LptA mutant. These authors confirmed that upon LptA inactivation at the non-permissive temperature, the lipid A-core OS moiety of LPS is blocked at the periplasmic side of the IM where it becomes substrate for LpxE, and it fails to be modified by PagL, whose active site is localized at the OM [83].

This body of work strongly suggested that LptA, LptB and LptC are part of an ABC transporter involved in LPS biogenesis. LptC is a bitopic protein possessing only one transmembrane domain and cannot fulfil the role of the integral membrane components of typical ABC transporters, which consist of either one IM protein with 12 transmembrane domains or two proteins with 6 transmembrane domains each [84]. Ruiz et al. identified the missing transmembrane components of the transporter by bioinformatics [57]. Their approach exploited the high conservation of OM biogenesis genes among Gram-negative bacteria, including endosymbionts. The authors selected *Blochmannia floridanus* as the model organism because its genome is the smallest one among endosymbionts [85] and like *E. coli* belongs to the Enterobacteriaceae family. Despite that the *B. floridanus* proteome is only the 14% the size of that of *E. coli*, BLAST searches revealed that it contains most of the

OM biogenesis factors known in *E. coli*. This comparative analysis suggested that this organism could be a suitable candidate to search for missing OM biogenesis factors. This approach led to the identification of two paralogous six-transmembrane domain IM proteins, of unknown function (YjgP and YjgQ), as the missing transmembrane component of the Lpt ABC transporter. Both proteins, renamed LptF and LptG, respectively, turned out to be essential. Analyses of conditional expression mutants depleted of one or both LptF and LptG revealed that the proteins are involved in LPS biogenesis, as upon depletion LPS could not be modified by PagP and instead accumulated at the periplasmic side of the IM as a modified form [57]. This was reminiscent of the phenotype observed in LptA, LptB or LptC depleted mutants and strongly suggested that LptF and LptG are implicated in the same LPS transport pathway [57].

At the OM, the LptD/E complex is required to assemble LPS in the outer leaflet of the OM. LptD is an essential β-barrel protein of 87 kDa whereas LptE (formerly RlpB) is an essential lipoprotein of 21.1 kDa. LptD was the first Lpt factor identified in a genetic selection for OM mutants [51] but its function was revealed more than a decade later. The original genetic screening aimed at selecting suppressors of a maltodextrin-specific channel gene *lamB* deletion mutant that could grow using maltose as a sole carbon source. This led to the isolation of a mutant with increased OM permeability not only to maltodextrins but also to many hydrophobic and hydrophilic antibiotics. The gene affected by the mutation was originally named *imp* (for *i*ncreased *m*embrane *p*ermeability). Only several years later, *imp* was characterized and turned out to be essential, highly conserved in Gram-negative bacteria and involved in OM biogenesis. Initial analysis using a conditional *lptD* (*imp*) mutant showed that LptD depleted cells accumulated folded proteins and lipids in a novel membrane fraction with higher density than the OM [38]. This phenotype was ascribed to OMP mislocalization and a role for LptD in OMP biogenesis was postulated. The observation that *lptD* is located immediately upstream of *surA*, encoding for a periplasmic chaperone involved in OMP assembly [41] and that the *surA* and *lptD* are co-transcribed from a σ^E dependent promoter was considered an additional evidence in support of this role for LptD [35]. LptD, however, was implicated in LPS transport by exploiting the ability of *N. meningitidis* to survive without LPS [1]. The authors demonstrated that in mutants lacking the neisserial *lptD* ortholog, which are viable, LPS is not accessible to extracellularly added neuraminidase, an enzyme that modifies LPS by adding sialic acid residues, and its lipid A moiety is not deacylated by the ectopically expressed OM deacylase PagL, thus suggesting that in *lptD* mutants LPS is not transported to the cell surface [54].

The OM β-barrel protein LptD possesses a periplasmic N-terminal domain belonging to the same OstA superfamily as LptA and LptC [41]. The N-terminal domain of LptD is essential for its function, as a plasmid copy of *lptD* missing the periplasmic domain is not sufficient to complement the conditional *lptD* mutant under depleting conditions [58]. Initial studies on LptD revealed that this protein exists in a higher molecular weight complex in the OM [38]. The interacting protein was purified by affinity chromatography to a tagged LptD and identified by tandem mass spectrometry (MS) as RlpB (*r*are *l*ipo*p*rotein B, now referred as LptE), a previously identified OM lipoprotein of very low abundance [55]. The essential lipoprotein LptE is functional even without its N-terminal lipid anchor [58]. Conditional expression mutants depleted of LptE and LptD share the same phenotypes: newly synthesized LPS are not accessible to the OM LPS modification enzyme PagP [55]. Interestingly, in depleted cells the bulk of previously synthesized LPS becomes heptaacylated consistent with the notion that when LPS transport is blocked phospholipids reach the outer leaflet of OM thus activating the PagP enzyme [55].

Collectively, the data presented implicate the seven Lpt proteins in LPS transport to the cell surface. Are these proteins working in a concerted way? Sperandeo et al. showed that upon depletion of LptA, LptB, LptC LptD, and LptE the LPS assembly pathway is blocked in nearly the same fashion, which results in very similar phenotypes [56] and provided a first strong evidence of functional and/or physical interaction between the Lpt proteins.

10.4 The Lpt Machinery and Models of LPS Transport

The LPS transport from the IM to the cell surface is a thermodynamically unfavourable process that cannot occur by simple diffusion but needs to be energized from the cytoplasm. The identification of the Lpt proteins pointed out several similarities with the Lol system and suggested that periplasmic LPS transport may be analogous to the machinery that transports lipoproteins to the OM where the periplasmic carrier LolA escorts lipoproteins across the periplasm and delivers its cargo to its specific OM receptor [9] (Fig. 10.3). According to this model LptA may function as a soluble periplasmic chaperone that binds LPS, diffuses across the periplasm and delivers it to the LptD/E complex at the OM. The finding that LptA binds LPS in vitro is in line with this model [78]. However, some differences exist between the Lpt and the Lol systems: primarily, the Lol ABC transporter lacks an LptC analogous protein. Moreover, the "chaperone diffusion" model was challenged by Tommassen et al. who demonstrated that LPS transport to the OM still occurs in *E. coli* spheroplasts, namely cells effectively drained of periplasmic content [86]. It is worth mention that the same approach was successful in identifying LolA as the soluble periplasmic component implicated in lipoprotein transport to the OM [87].

A second model of LPS transport across the periplasm implicates a proteinaceous or a membrane bridge that would physically connect the IM and OM allowing for direct efflux of LPS to the cell surface [41] (Fig. 10.3). Bayer proposed this model more than 40 years ago [88, 89]. Indeed newly synthesised LPS appear in patches in the OM close to these "Bayer junctions" [90]. Finally LPS appears to transiently accumulate in a novel OM fraction of lower density (OM_L, "light" OM) that can be isolated by isopicnic density gradient centrifugation [91]. This OM_L fraction contains IM and OM components and is reminiscent of Bayer junctions.

In line with this second model, evidence exists of direct physical interaction between the seven Lpt proteins [76]. Indeed, all Lpt proteins co-fractionate with the

OM_L fraction and can be co-purified together, suggesting that these proteins can form a single complex connecting the IM and the OM. In these co-purification experiments LptA associates with both IM and OM Lpt proteins, possibly acting as the protein bridging the two membranes [76]. That all Lpt proteins can be found in the OM_L fraction (that contains IM and OM components) and that they directly interact strongly support the model that the Lpt proteins not only functionally interact in the same LPS transport pathway, but also physically interact to form a transenvelope complex.

In Gram-negative bacteria several transport processes are based on trans-envelope complexes connecting directly the cytoplasm with the exterior of the cell. For instance, Type III secretion systems [92], which inject toxins directly in the host cells during the infections process, or efflux pumps, which extrude noxious chemicals from the bacterial cell into the surrounding medium and are responsible for antibiotic resistance [93]. The Lpt system could operate in a similar fashion as one of these transenvelope complexes.

The crystal structure of LptA was solved in the presence and absence of LPS (Fig. 10.4). Crystals obtained in the presence of LPS revealed that multiple LptA

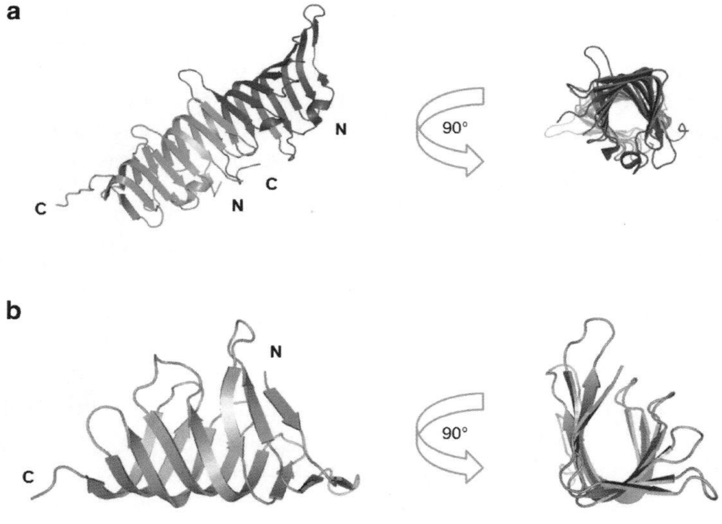

Fig. 10.4 Crystal structure of LptA and LptC. (**a**) Ribbon diagram of LptA with two molecules in the asymmetric unit at 2.15-Å resolution. The LptA molecules are organized in a head-to-tail fashion. LptA was rotated by 90° to highlight the channel formed along the length of LptA as a result of the β-jellyroll structure. (**b**) Ribbon diagram of a single His_6-LptC (24–191) molecule at 2.2-Å resolution. The structure of the periplasmic domain of LptC is composed of a series of 15 antiparallel β-strands that wind back along the path of the preceding peptide stretch throughout the length of the protein, resembling the structure of LptA. LptC was rotated by 90° to highlight the channel formed along the length of the molecule. The figures were generated using Pymol version 1.3 (www.pymol.org) and the Research Collaboratory for Structural Bioinformatics Protein Data Bank Accession codes 2R19 and 3MY2 [94, 95]

molecules can stack in a head-to-tail fashion to form a fibril containing a hydrophobic groove running through its entire length [94]. This result is in line with the hypothesis that LptA functions to bridge the IM and the OM spatially facilitating LPS export.

Both LptC and LptA belong to the same OstA superfamily as the N-terminal periplasmic domain of LptD [41]. LptC may connect with LptD through one or more copies of LptA [94]. Very recent data indicate that LptA and LptC physically interact thus supporting the above hypothesis (Sperandeo et al., submitted). This would be also consistent with the finding that LptA can interact with both the IM and the OM [76]. Overall these data support a model in which the OstA-like domains of the three proteins would create a continuous hydrophobic groove that may shield the lipid A portion of LPS molecules from the aqueous environment as they traverse the periplasm. It is not clear whether the Lpt proteins transiently associate at each transport cycle or form a stable complex that permanently spans the two membranes of the cell.

That the Lpt complex spans two separate membranes poses a major challenge for studying the mechanism of LPS transport. Presently an in vitro assay for LPS transport is missing and thus our current knowledge of the process is based on partial reconstitutions of the Lpt complex. The LptBCFG proteins physically interact and, as predicted, display ATPase activity [75]. The IM LptBCFG transporter therefore constitutes the engine for LPS export outside the cytoplasm. The subunit ratio of the complex is LptB:LptF:LptG:LptC = 2:1:1:1 [75]. The LptBFG and LptBCFG proteins were over-expressed, purified as membrane complexes and the respective molecular masses determined by size exclusion chromatography. The observed molecular masses of the LptBFG and LptBCFG were 280 and 330 kDa, respectively, in contrast with the predicted molecular masses of 135.5 and 157.2 kDa, respectively. These observations suggest either that the complexes exist as dimers under the selected experimental condition or that the discrepancy observed between predicted and experimentally determined molecular masses might be an artefact due to the detergent used to reconstitute the complexes, as already observed for other ABC transporters [75]. In vitro and in vivo data indicate that LptC may form dimers (P. Sperandeo et al., unpublished data) thus supporting the observations by Tokuda et al.

The ATPase activity of the LptBFG and LptBCFG complexes exhibits the same K_m and V_{max} values, revealing that LptC does not affect the kinetic parameters of the ATPase activity, despite its essential role in LPS transport [37]. However, Tokuda et al. reported that neither LPS nor lipid A, the putative substrates of the LptBCEF transporter, are able to stimulate the ATPase activity, suggesting that some component was missing in their in vitro assay [75].

LptA appears a key protein in the Lpt machinery as several direct and indirect evidences suggest that the protein bridges IM and OM and binds LPS in vitro. Despite the crystal structure of the protein is known [94] the requirements for LPS binding remain unknown. LptA presents a novel fold consisting of 16 antiparallel β-strands folded to resemble a semiclosed β-jellyroll; the structure in not completely symmetrical and it opens slightly at the N- and C-termini [94]. It is not yet clear how this structure can accommodate LPS. Based on the observation that LptA forms oligomers in the presence of LPS [94] it may be postulated that in vivo LptA oligomerization may be induced by the MsbA-translocated lipid A-core OS. The recently solved crystal structure of LptC reveals a striking structural similarity to LptA despite the two protein do not share sequence similarity. Like LptA, LptC binds LPS in vitro and LptA can displace it from LptC consistent with their proposed placement in the unidirectional LPS export pathway [95].

Several lines of evidence indicate that the LptD/E complex is responsible of LPS assembly at the cell surface [55, 58]. A proteolysis protection assay revealed that LptE strongly interacts with the C-terminal β-barrel domain of LptD and this interaction stabilizes LptD [58]. However, the role of LptE in LPS transport/assembly is not simply structural, as this protein is able to bind specifically LPS [58]. Based on the results of proteolysis protection experiments on the LptD/E complex and on the reported crystal structure of LptE orthologs [96, 97], LptE may form a plug that sits at the base of, or is somewhat buried within, the lumen of the β-barrel formed by the C-terminal domain of LptD, by analogy with the plug domains of other transporters (as P pili assembly usher or OM iron-chelate transporter FhuA), although in these cases the plug domains are part of the same polypeptide of the β-barrel domain [58]. According to these observations, recognition of LPS by LptE could trigger a conformational change of the protein that could be transmitted to LptD to allow LPS translocation.

A critical issue for transport machinery is the substrate specificity. The finding that LptA binds hexa and tetraacylated lipid A [78] in vitro supports the hypothesis, as mentioned earlier in this chapter, that transport occurring downstream of MsbA has relaxed substrate specificity and suggests that the quality control step in LPS transport is performed by MsbA. This notion is further supported by several genetic conditions in which mutants that assemble an OM devoid of Kdo_2-lipid A (the minimal LPS precursor normally required for the viability of most Gram-negative bacteria) and containing the Kdo_2-lipid IV_A or the lipid IV_A precursors are viable. For example, as discussed above, overexpression of MsbA is sufficient to transport the tetraacylated Kdo_2-lipid IV_A precursor that accumulates in *htrB/lpxL* mutants [61]. In addition, Woodard et al. showed that over-expression of the MsbA flippase from a multicopy plasmid enables growth of mutants unable to synthesize Kdo, thus suggesting that in vivo MsbA can flip the tetraacylated lipid IV_A precursor molecules, although with low efficiency [98], and that these precursors can be targeted to the OM by the Lpt machinery. Moreover these data suggest that the previously observed and well-documented dependence of cell viability on the synthesis of Kdo stems from lethal pleiotropic effects following the depletion of this molecule, rather than an intrinsic need for Kdo itself as an indispensable structural component of the OM LPS layer. More recently the same laboratory isolated several suppressor mutations able to bypass the normally essential Kdo_2-lipid A requirement. The first class of suppressors harboured single amino acid substitutions in *msbA* able to rescue viability of Kdo-deficient mutants that assembled lipid IV_A in the OM, thus suggesting a more relaxed substrate specificity of the mutant protein for tetraacylated lipid A [99]. The second class of suppressor mutations mapped in *yhjD*, which encodes a putative conserved transmembrane IM protein with unknown function [99]. The YhjD R134C amino acid substitution (*yhjD400* allele) suppresses the lethal phenotype of a *waaA* deletion mutant (see Chap. 6), which cannot ligate Kdo to lipid A and thus accumulates lipid IV_A. Interestingly, deletion of *msbA* is viable in a *yhjD400* genetic background. Thus suppression by *yhjD400* seems not associated to an abnormal MsbA activity, but with the activation of an independent transport pathway. This notion was further supported by the isolation of a suppressor-free *waaC lpxL lpxM lpxP* mutant [100]. This mutant is defective in heptosyltransferase I and late acyl-transferases activities (see Chaps. 6 and 8), produces a Kdo_2-lipid IV_A LPS derivative and, although viable under slow growth conditions at low temperatures, shows constitutively activated envelope stress response. Normal growth of *waaC lpxL lpxM lpxP* could be restored by extragenic chromosomal MsbA-D498V suppressor mutation or by over-expression of the *msbA* wild-type gene product [100].

Together, these observations suggest that the Lpt transport machine has relaxed substrate specificity and that MsbA provides the quality control step in LPS export to the OM.

10.5 LPS Export to Outer Membrane as Target for Novel Antibacterial Molecules

The discovery, development and clinical exploitation of antibiotics are among the most significant medical advances of the last century. However, the increasingly and alarming onset and spread of antibiotic resistant strains among pathogenic bacteria together with the unfavourable economics of antibiotic development poses as an urgent need the identification of new antibacterial agents that have a novel mode of action [101]. Multidrug-resistant (MDR) and pandrug-resistant (PDR) Gram-negative bacteria represent a serious threat, as these antibiotic resistant pathogens cause infections that are becoming truly untreatable. Strains resistant to some (MDR) or all (PDR) the antibiotics commonly used clinically have been isolated in *Acinetobacter baumannii*, *E. coli*, *Klebsiella pneumoniae*, and *Pseudomonas aeruginosa* and other pathogenic species [102]. Moreover, due to their OM that is impermeable to many drugs [7] and to their efflux pumps that actively expel many of the remainder [93], the prospects to find new antibiotics acting against these pathogens are especially poor.

Almost all drugs currently used to treat bacterial infections target one of the following four processes: protein, cell wall, nucleic acid or folate synthesis [103]. Since 1962 only three new classes of antibacterial agents have been approved for clinical use: oxazolidinones [104] and retapamulin [105], both targeting protein synthesis, and daptomycin, a narrow spectrum cyclic lipopeptide disrupting Gram-positive cytoplasmic membranes in a manner and with effects similar to those of the cationic antimicrobial peptides [106]. The availability of complete genome sequences has provided both the academia and pharmaceutical companies with a wealth of information that led to development of global technologies aiming at identifying virulence genes [107] or essential genes [108, 109] as potential new targets for the development of new antibacterial molecules. Despite the enormous amount of information provided by these global analyses, the feedback on drug development has been marginal [110]. Based on these considerations, novel molecules targeting different essential cellular pathways are urgently needed.

Target-driven discovery of novel antibacterials offers the advantage of prior knowledge of the protein/pathway target function thus potentially expediting the drug discovery process. Moreover the recent development of novel computational strategies to exploit target structural information and combining it with the existing chemical matter for another target lead to the identification of compounds that have the potential to become novel antibacterial leads [111, 112]. On the other end the whole-cell screening strategy, although empirical and less sensitive than molecular screens, may offer the advantage to identify a potential novel lead that is able to permeate the cell as it might be easier to find the cellular target of an antibacterial compound than it is to engineer a compound to increase its permeability without modifying its inhibitory activity [110].

A crucial point for success target-driven antibacterial drug discovery approaches is the identification/selection of the appropriate molecular target. The LPS export to the OM represents an attractive underexploited bacterial pathway. Several features of the Lpt transport proteins suggest that they could be good candidates as antibacterial targets. In fact, the components of the Lpt machinery are essential (LPS is an essential structure in most Gram-negative bacteria) and the proteins are conserved in many relevant bacterial pathogens. Importantly, as LPS is a molecule exclusively present in bacteria, the Lpt proteins do not have human counterparts.

Despite the Lpt proteins have been only very recently discovered and characterized, two inhibitors of the machinery have already been identified using two different approaches. The first compound targets LptB, the ATPase component of the IM ABC transporter [113]. This molecule has been identified by in vitro screening with 224 compounds from two commercially available kinase inhibitors libraries composed mostly of ATP-competitive inhibitors and, as expected, it is competitive with respect to ATP. The inhibitor binding constant was found to be in the micromolar range ($Ki = 5\ \mu M$); however, it does not display antibacterial activity against a wild type strain of *E. coli* whereas it shows a minimum inhibitory concentration (MIC), consistent with its Ki value, against a strain of *E. coli* with a leaky OM.

A peptidomimetic antibiotic specifically targeting LptD of *P. aeruginosa* is the second example of LPS transport inhibitors [114]. The starting point of this work is the synthesis of libraries of β-hairpin-shaped peptidomimetics based on the membranolytic host defence peptide protegrin I [115]. Following whole-cell screening, a lead showing low but significant broad-spectrum antibacterial activity was found. This lead was optimized for improved antimicrobial activity through iterative cycles of synthesis and screenings. This effort produced two peptidomimetics POL7001 and POL7080 with potent and selective action only against *P. aeruginosa* (MIC 0.13 and 0.25 $\mu\ mL^{-1}$, respectively). Interestingly, POL7001 and POL7080 also show activity against more than 90% of *P. aeruginosa* clinical isolated tested. A genetic approach was used to define the mechanism of action: spontaneous resistant *P. aeruginosa* mutants were selected and the mutations mapped. In all isolates the mutations mapped in the *P. aeruginosa* LptD homologue. Photoaffinity labelling experiments demonstrated that POL7001 and POL7080 bind LptD in intact cells and LPS is modified by PagP in cells grown under sub-lethal concentrations of the two peptidomimetics providing further evidence that LptD is the target of POL7001 and POL7080 and that LPS transport is inhibited. However, it is not clear whether the peptides need to permeate the bacteria to inhibit LptD function or whether they act from outside the cells. Nevertheless this work represents the "proof of concept" that LPS transport is indeed a good antibacterial target.

Overall, we believe that the Lpt machinery represents a composite cellular target that offers the opportunity not only to inhibit the function of any single protein but also to exploit different aspects of LPS biogenesis, namely the assembly of the complex and its ability to bind LPS.

10.6 Conclusions and Perspectives

Recent advances have led to the identification of the LPS transport protein machine that extracts LPS from the intracellular site of synthesis to the environment-exposed final destination. Strong genetic and biochemical evidences indicate that the Lpt machinery functions as a single device and that the seven proteins composing the system physically interact to form a transenvelope complex. However, neither the detailed mechanisms of LPS translocation across the periplasm and its insertion at the OM nor the molecular requirements for LPS binding to LptA and LptE are known. That the Lpt transport system consists of a single apparatus spanning IM and OM and the structural complexity of LPS, the transported molecule, pose a major challenge for studying the mechanism of LPS transport. The development of new tools may be needed to dissect the molecular mechanism of transport and to define what individual role each of these seven proteins play in the process. The identification of inhibitors that specifically target LPS transport in vitro and more importantly in vivo may represent important tools to dissect the transport pathway.

Acknowledgements This work was in part supported by Regione Lombardia "Cooperazione scientifica e tecnologica internazionale" grant 16876 SAL-18 (to A.P) and "Fondazione per la Ricerca sulla Fibrosi Cistica" grant FFC#13/2010 (to A.P.).

References

1. Steeghs L, den Hartog R, den Boer A, Zomer B, Roholl P, van der Ley P (1998) Meningitis bacterium is viable without endotoxin. Nature 392:449–450
2. Gram HCJ (1884) Über die isolierte Färbung der Schizomyceten in Schnitt- und Trockenpräparaten. Fortschr Med 2:185–189

3. Beveridge TJ, Davies JA (1983) Cellular responses of *Bacillus subtilis* and *Escherichia coli* to the Gram stain. J Bacteriol 156:846–858
4. Glauert AM, Thornley MJ (1969) The topography of the bacterial cell wall. Annu Rev Microbiol 23:159–198
5. Kellenberger E, Ryter A (1958) Cell wall and cytoplasmic membrane of *Escherichia coli*. J Biophys Biochem Cytol 4:323–326
6. Raetz CR, Whitfield C (2002) Lipopolysaccharide endotoxins. Annu Rev Biochem 71:635–700
7. Nikaido H (2003) Molecular basis of bacterial outer membrane permeability revisited. Microbiol Mol Biol Rev 67:593–656
8. Narita S, Matsuyama S, Tokuda H (2004) Lipoprotein trafficking in *Escherichia coli*. Arch Microbiol 182:1–6
9. Tokuda H (2009) Biogenesis of outer membranes in Gram-negative bacteria. Biosci Biotechnol Biochem 73:465–473
10. Kadner RJ (1996) Cytoplasmic membrane. In: Neidhardt FC, Curtiss R III, Ingraham JL, Lin ECC, Low KB, Magasanik B, Reznikoff WS, Riley M, Schaechter M, Umbarger HE (eds) *Escherichia coli* and *Salmonella*: cellular and molecular biology. ASM Press, Washington, DC, pp 58–87
11. Oliver DB (1996) Periplasm. In: Neidhardt FC, Curtiss R III, Ingraham JL, Lin ECC, Low KB, Magasanik B, Reznikoff WS, Riley M, Schaechter M, Umbarger HE (eds) *Escherichia coli* and *Salmonella*: cellular and molecular biology. ASM Press, Washington, DC, pp 88–103
12. Holtje JV (1998) Growth of the stress-bearing and shape-maintaining murein sacculus of *Escherichia coli*. Microbiol Mol Biol Rev 62:181–203
13. Koebnik R, Locher KP, Van Gelder P (2000) Structure and function of bacterial outer membrane proteins: barrels in a nutshell. Mol Microbiol 37:239–253
14. Schulz GE (2002) The structure of bacterial outer membrane proteins. Biochim Biophys Acta 1565:308–317
15. Dong C, Beis K, Nesper J, Brunkan-Lamontagne AL, Clarke BR, Whitfield C, Naismith JH (2006) Wza the translocon for *E. coli* capsular polysaccharides defines a new class of membrane protein. Nature 444:226–229
16. Collins RF, Beis K, Dong C, Botting CH, McDonnell C, Ford RC, Clarke BR, Whitfield C, Naismith JH (2007) The 3D structure of a periplasm-spanning platform required for assembly of group 1 capsular polysaccharides in *Escherichia coli*. Proc Natl Acad Sci USA 104:2390–2395
17. Pettersson A, Poolman JT, van der Ley P, Tommassen J (1997) Response of *Neisseria meningitidis* to iron limitation. Antonie Leeuwenhoek 71:129–136
18. Kamio Y, Nikaido H (1976) Outer membrane of *Salmonella typhimurium*: accessibility of phospholipid head groups to phospholipase c and cyanogen bromide activated dextran in the external medium. Biochemistry 15:2561–2570
19. Gunn JS (2000) Mechanisms of bacterial resistance and response to bile. Microbes Infect 2:907–913
20. Ferguson AD, Welte W, Hofmann E, Lindner B, Holst O, Coulton JW, Diederichs K (2000) A conserved structural motif for lipopolysaccharide recognition by procaryotic and eucaryotic proteins. Structure 8:585–592
21. Ruiz N, Kahne D, Silhavy TJ (2006) Advances in understanding bacterial outer-membrane biogenesis. Nat Rev Microbiol 4:57–66
22. Young K, Silver LL (1991) Leakage of periplasmic enzymes from envA1 strains of *Escherichia coli*. J Bacteriol 173:3609–3614
23. Nikaido H (2005) Restoring permeability barrier function to outer membrane. Chem Biol 12:507–509
24. Jia W, El Zoeiby A, Petruzziello TN, Jayabalasingham B, Seyedirashti S, Bishop RE (2004) Lipid trafficking controls endotoxin acylation in outer membranes of *Escherichia coli*. J Biol Chem 279:44966–44975

25. Dekker N (2000) Outer-membrane phospholipase A: known structure, unknown biological function. Mol Microbiol 35:711–717
26. Groisman EA (2001) The pleiotropic two-component regulatory system PhoP-PhoQ. J Bacteriol 183:1835–1842
27. Bishop RE (2008) Structural biology of membrane-intrinsic β-barrel enzymes: sentinels of the bacterial outer membrane. Biochim Biophys Acta 1778:1881–1896
28. Malinverni JC, Silhavy TJ (2009) An ABC transport system that maintains lipid asymmetry in the Gram-negative outer membrane. Proc Natl Acad Sci USA 106:8009–8014
29. Casali N, Riley LW (2007) A phylogenomic analysis of the *Actinomycetales mce* operons. BMC Genomics 8:60–83
30. Alba BM, Gross CA (2004) Regulation of the *Escherichia coli* sigma-dependent envelope stress response. Mol Microbiol 52:613–619
31. Ades SE (2008) Regulation by destruction: design of the σ^E envelope stress response. Curr Opin Microbiol 11:535–540
32. Ades SE, Connolly LE, Alba BM, Gross CA (1999) The *Escherichia coli* σ^E-dependent extracytoplasmic stress response is controlled by the regulated proteolysis of an anti-σ factor. Genes Dev 13:2449–2461
33. Alba BM, Leeds JA, Onufryk C, Lu CZ, Gross CA (2002) DegS and YaeL participate sequentially in the cleavage of RseA to activate the σ^E-dependent extracytoplasmic stress response. Genes Dev 16:2156–2168
34. Inaba K, Suzuki M, Maegawa K, Akiyama S, Ito K, Akiyama Y (2008) A pair of circularly permutated PDZ domains control RseP, the S2P family intramembrane protease of *Escherichia coli*. J Biol Chem 283:35042–35052
35. Dartigalongue C, Missiakas D, Raina S (2001) Characterization of the *Escherichia coli* σ^E regulon. J Biol Chem 276:20866–20875
36. Johansen J, Eriksen M, Kallipolitis B, Valentin-Hansen P (2008) Down-regulation of outer membrane proteins by noncoding RNAs: unraveling the cAMP-CRP- and σ^E-dependent CyaR-*ompX* regulatory case. J Mol Biol 383:1–9
37. Sperandeo P, Cescutti R, Villa R, Di Benedetto C, Candia D, Dehò G, Polissi A (2007) Characterization of *lptA* and *lptB*, two essential genes implicated in lipopolysaccharide transport to the outer membrane of *Escherichia coli*. J Bacteriol 189:244–253
38. Braun M, Silhavy TJ (2002) Imp/OstA is required for cell envelope biogenesis in *Escherichia coli*. Mol Microbiol 45:1289–1302
39. Tam C, Missiakas D (2005) Changes in lipopolysaccharide structure induce the σ^E-dependent response of *Escherichia coli*. Mol Microbiol 55:1403–1412
40. Kol MA, de Kroon AI, Killian JA, de Kruijff B (2004) Transbilayer movement of phospholipids in biogenic membranes. Biochemistry 43:2673–2681
41. Bos MP, Robert V, Tommassen J (2007) Biogenesis of the Gram-negative bacterial outer membrane. Annu Rev Microbiol 61:191–214
42. Cronan JE (2003) Bacterial membrane lipids: where do we stand? Annu Rev Microbiol 57:203–224
43. Rothman JE, Kennedy EP (1977) Rapid transmembrane movement of newly synthesized phospholipids during membrane assembly. Proc Natl Acad Sci USA 74:1821–1825
44. Kol MA, van Dalen A, de Kroon AI, de Kruijff B (2003) Translocation of phospholipids is facilitated by a subset of membrane-spanning proteins of the bacterial cytoplasmic membrane. J Biol Chem 278:24586–24593
45. Zhou Z, White KA, Polissi A, Georgopoulos C, Raetz CR (1998) Function of *Escherichia coli MsbA*, an essential ABC family transporter, in lipid A and phospholipid biosynthesis. J Biol Chem 273:12466–12475
46. Doerrler WT, Gibbons HS, Raetz CR (2004) MsbA-dependent translocation of lipids across the inner membrane of *Escherichia coli*. J Biol Chem 279:45102–45109
47. Tefsen B, Bos MP, Beckers F, Tommassen J, de Cock H (2005) MsbA is not required for phospholipid transport in *Neisseria meningitidis*. J Biol Chem 280:35961–35966

48. Driessen AJ, Nouwen N (2008) Protein translocation across the bacterial cytoplasmic membrane. Annu Rev Biochem 77:643–667
49. Raetz CR (1990) Biochemistry of endotoxins. Annu Rev Biochem 59:129–170
50. Wu T, Malinverni J, Ruiz N, Kim S, Silhavy TJ, Kahne D (2005) Identification of a multicomponent complex required for outer membrane biogenesis in *Escherichia coli*. Cell 121:235–245
51. Sampson BA, Misra R, Benson SA (1989) Identification and characterization of a new gene of *Escherichia coli* K-12 involved in outer membrane permeability. Genetics 122:491–501
52. Aono R, Negishi T, Aibe K, Inoue A, Horikoshi K (1994) Mapping of organic solvent tolerance gene *ostA* in *Escherichia coli* K-12. Biosci Biotechnol Biochem 58:1231–1235
53. Doerrler WT (2006) Lipid trafficking to the outer membrane of Gram-negative bacteria. Mol Microbiol 60:542–552
54. Bos MP, Tefsen B, Geurtsen J, Tommassen J (2004) Identification of an outer membrane protein required for the transport of lipopolysaccharide to the bacterial cell surface. Proc Natl Acad Sci USA 101:9417–9422
55. Wu T, McCandlish AC, Gronenberg LS, Chng SS, Silhavy TJ, Kahne D (2006) Identification of a protein complex that assembles lipopolysaccharide in the outer membrane of *Escherichia coli*. Proc Natl Acad Sci USA 103:11754–11759
56. Sperandeo P, Lau FK, Carpentieri A, De Castro C, Molinaro A, Dehò G, Silhavy TJ, Polissi A (2008) Functional analysis of the protein machinery required for transport of lipopolysaccharide to the outer membrane of *Escherichia coli*. J Bacteriol 190:4460–4469
57. Ruiz N, Gronenberg LS, Kahne D, Silhavy TJ (2008) Identification of two inner-membrane proteins required for the transport of lipopolysaccharide to the outer membrane of *Escherichia coli*. Proc Natl Acad Sci USA 105:5537–5542
58. Chng SS, Ruiz N, Chimalakonda G, Silhavy TJ, Kahne D (2010) Characterization of the two-protein complex in *Escherichia coli* responsible for lipopolysaccharide assembly at the outer membrane. Proc Natl Acad Sci USA 107:5363–5368
59. Davidson AL, Dassa E, Orelle C, Chen J (2008) Structure, function, and evolution of bacterial ATP-binding cassette systems. Microbiol Mol Biol Rev 72:317–364
60. Rees DC, Johnson E, Lewinson O (2009) ABC transporters: the power to change. Nat Rev Mol Cell Biol 10:218–227
61. Karow M, Georgopoulos C (1993) The essential *Escherichia coli msbA* gene, a multicopy suppressor of null mutations in the *htrB* gene, is related to the universally conserved family of ATP-dependent translocators. Mol Microbiol 7:69–79
62. Clementz T, Bednarski JJ, Raetz CR (1996) Function of the *htrB* high temperature requirement gene of *Escherchia coli* in the acylation of lipid A: HtrB catalyzed incorporation of laurate. J Biol Chem 271:12095–12102
63. Karow M, Fayet O, Georgopoulos C (1992) The lethal phenotype caused by null mutations in the *Escherichia coli htrB* gene is suppressed by mutations in the *accBC* operon, encoding two subunits of acetyl coenzyme A carboxylase. J Bacteriol 174:7407–7418
64. Polissi A, Georgopoulos C (1996) Mutational analysis and properties of the *msbA* gene of *Escherichia coli*, coding for an essential ABC family transporter. Mol Microbiol 20:1221–1233
65. Doerrler WT, Reedy MC, Raetz CR (2001) An *Escherichia coli* mutant defective in lipid export. J Biol Chem 276:11461–11464
66. Raetz CR, Reynolds CM, Trent MS, Bishop RE (2007) Lipid A modification systems in Gram-negative bacteria. Annu Rev Biochem 76:295–329
67. Doerrler WT, Raetz CR (2002) ATPase activity of the MsbA lipid flippase of *Escherichia coli*. J Biol Chem 277:36697–36705
68. Eckford PD, Sharom FJ (2008) Functional characterization of *Escherichia coli* MsbA: interaction with nucleotides and substrates. J Biol Chem 283:12840–12850
69. Siarheyeva A, Sharom FJ (2009) The ABC transporter MsbA interacts with lipid A and amphipathic drugs at different sites. Biochem J 419:317–328

70. Eckford PD, Sharom FJ (2010) The reconstituted *Escherichia coli* MsbA protein displays lipid flippase activity. Biochem J 429:195–203
71. Ward A, Reyes CL, Yu J, Roth CB, Chang G (2007) Flexibility in the ABC transporter MsbA: alternating access with a twist. Proc Natl Acad Sci USA 104:19005–19010
72. Chang G, Roth CB, Reyes CL, Pornillos O, Chen YJ, Chen AP (2006) Retraction. Science 314:1875
73. Dawson RJ, Locher KP (2006) Structure of a bacterial multidrug ABC transporter. Nature 443:180–185
74. Rosenberg MF, Callaghan R, Modok S, Higgins CF, Ford RC (2005) Three-dimensional structure of P-glycoprotein: the transmembrane regions adopt an asymmetric configuration in the nucleotide-bound state. J Biol Chem 280:2857–2862
75. Narita S, Tokuda H (2009) Biochemical characterization of an ABC transporter LptBFGC complex required for the outer membrane sorting of lipopolysaccharides. FEBS Lett 583:2160–2164
76. Chng SS, Gronenberg LS, Kahne D (2010) Proteins required for lipopolysaccharide assembly in *Escherichia coli* form a transenvelope complex. Biochemistry 49:4565–4567
77. Stenberg F, Chovanec P, Maslen SL, Robinson CV, Ilag LL, von Heijne G, Daley DO (2005) Protein complexes of the *Escherichia coli* cell envelope. J Biol Chem 280:34409–34419
78. Tran AX, Trent MS, Whitfield C (2008) The LptA protein of *Escherichia coli* is a periplasmic lipid A-binding protein involved in the lipopolysaccharide export pathway. J Biol Chem 283:20342–20349
79. Serina S, Nozza F, Nicastro G, Faggioni F, Mottl H, Dehò G, Polissi A (2004) Scanning the *Escherichia coli* chromosome by random transposon mutagenesis and multiple phenotypic screening. Res Microbiol 155:692–701
80. Meredith TC, Woodard RW (2003) *Escherichia coli* YrbH is a D-arabinose 5-phosphate isomerase. J Biol Chem 278:32771–32777
81. Wu J, Woodard RW (2003) *Escherichia coli* YrbI is 3-deoxy-D-*manno*-octulosonate 8-phos phate phosphatase. J Biol Chem 278:18117–18123
82. Sperandeo P, Pozzi C, Dehò G, Polissi A (2006) Non-essential KDO biosynthesis and new essential cell envelope biogenesis genes in the *Escherichia coli yrbG-yhbG* locus. Res Microbiol 157:547–558
83. Ma B, Reynolds CM, Raetz CR (2008) Periplasmic orientation of nascent lipid A in the inner membrane of an *Escherichia coli* LptA mutant. Proc Natl Acad Sci USA 105:13823–13828
84. Linton KJ, Higgins CF (2007) Structure and function of ABC transporters: the ATP switch provides flexible control. Pflugers Arch 453:555–567
85. Gil R, Silva FJ, Zientz E, Delmotte F, Gonzalez-Candelas F, Latorre A, Rausell C, Kamerbeek J, Gadau J, Holldobler B, van Ham RC, Gross R, Moya A (2003) The genome sequence of *Blochmannia floridanus*: comparative analysis of reduced genomes. Proc Natl Acad Sci USA 100:9388–9393
86. Tefsen B, Geurtsen J, Beckers F, Tommassen J, de Cock H (2005) Lipopolysaccharide transport to the bacterial outer membrane in spheroplasts. J Biol Chem 280:4504–4509
87. Matsuyama S, Tajima T, Tokuda H (1995) A novel periplasmic carrier protein involved in the sorting and transport of *Escherichia coli* lipoproteins destined for the outer membrane. EMBO J 14:3365–3372
88. Bayer ME (1968) Areas of adhesion between wall and membrane of *Escherichia coli*. J Gen Microbiol 53:395–404
89. Bayer ME (1991) Zones of membrane adhesion in the cryofixed envelope of *Escherichia coli*. J Struct Biol 107:268–280
90. Muhlradt PF, Menzel J, Golecki JR, Speth V (1973) Outer membrane of *Salmonella*. Sites of export of newly synthesised lipopolysaccharide on the bacterial surface. Eur J Biochem 35:471–481
91. Ishidate K, Creeger ES, Zrike J, Deb S, Glauner B, MacAlister TJ, Rothfield LI (1986) Isolation of differentiated membrane domains from *Escherichia coli* and *Salmonella*

typhimurium, including a fraction containing attachment sites between the inner and outer membranes and the murein skeleton of the cell envelope. J Biol Chem 261:428–443
92. Hueck CJ (1998) Type III protein secretion systems in bacterial pathogens of animals and plants. Microbiol Mol Biol Rev 62:379–433
93. Nikaido H, Zgurskaya HI (2001) AcrAB and related multidrug efflux pumps of *Escherichia coli*. J Mol Microbiol Biotechnol 3:215–218
94. Suits MD, Sperandeo P, Dehò G, Polissi A, Jia Z (2008) Novel structure of the conserved Gram-negative lipopolysaccharide transport protein A and mutagenesis analysis. J Mol Biol 380:476–488
95. Tran AX, Dong C, Whitfield C (2010) Structure and functional analysis of LptC, a conserved membrane protein involved in the lipopolysaccharide export pathway in *Escherichia coli*. J Biol Chem 285:33529–33539
96. Rossi P, Xiao R, Acton TB, Montelione GT (2007) Solution NMR structure of uncharacterized lipoprotein B from *Nitrosomonas europaea*. PDB ID: 2JXP doi:10.2210/pdb2jxp/pdb
97. Vorobiev SM, Abashidze M, Seetharaman J, Cunningham K, Maglaqui M, Owens L, Fang Y, Xiao R, Acton TB, Montelione GT, Tong L, Hunt JF (2007) Crystal structure of the A1KSW9_NEIMF protein from *Neisseria meningitidis*. PDB ID: 3BF2 doi:10.2210/pdb3bf2/pdb
98. Meredith TC, Aggarwal P, Mamat U, Lindner B, Woodard RW (2006) Redefining the requisite lipopolysaccharide structure in *Escherichia coli*. ACS Chem Biol 1:33–42
99. Mamat U, Meredith TC, Aggarwal P, Kuhl A, Kirchhoff P, Lindner B, Hanuszkiewicz A, Sun J, Holst O, Woodard RW (2008) Single amino acid substitutions in either YhjD or MsbA confer viability to 3-deoxy-D-*manno*-oct-2-ulosonic acid-depleted *Escherichia coli*. Mol Microbiol 67:633–648
100. Klein G, Lindner B, Brabetz W, Brade H, Raina S (2009) *Escherichia coli* K-12 suppressor-free mutants lacking early glycosyltransferases and late acyltransferases: minimal lipopolysaccharide structure and induction of envelope stress response. J Biol Chem 284:15369–15389
101. Fischbach MA, Walsh CT (2009) Antibiotics for emerging pathogens. Science 325:1089–1093
102. Falagas ME, Bliziotis IA, Kasiakou SK, Samonis G, Athanassopoulou P, Michalopoulos A (2005) Outcome of infections due to pandrug-resistant (PDR) Gram-negative bacteria. BMC Infect Dis 5:24
103. Walsh C (2003) Where will new antibiotics come from? Nat Rev Microbiol 1:65–70
104. Patel U, Yan YP, Hobbs FW Jr, Kaczmarczyk J, Slee AM, Pompliano DL, Kurilla MG, Bobkova EV (2001) Oxazolidinones mechanism of action: inhibition of the first peptide bond formation. J Biol Chem 276:37199–37205
105. Yan K, Madden L, Choudhry AE, Voigt CS, Copeland RA, Gontarek RR (2006) Biochemical characterization of the interactions of the novel pleuromutilin derivative retapamulin with bacterial ribosomes. Antimicrob Agents Chemother 50:3875–3881
106. Jung D, Rozek A, Okon M, Hancock RE (2004) Structural transitions as determinants of the action of the calcium-dependent antibiotic daptomycin. Chem Biol 11:949–957
107. Saenz HL, Dehio C (2005) Signature-tagged mutagenesis: technical advances in a negative selection method for virulence gene identification. Curr Opin Microbiol 8:612–619
108. Arigoni F, Talabot F, Peitsch M, Edgerton MD, Meldrum E, Allet E, Fish R, Jamotte T, Curchod ML, Loferer H (1998) A genome-based approach for the identification of essential bacterial genes. Nat Biotechnol 16:851–856
109. Akerley BJ, Rubin EJ, Camilli A, Lampe DJ, Robertson HM, Mekalanos JJ (1998) Systematic identification of essential genes by in vitro mariner mutagenesis. Proc Natl Acad Sci USA 95:8927–8932
110. Payne DJ, Gwynn MN, Holmes DJ, Pompliano DL (2007) Drugs for bad bugs: confronting the challenges of antibacterial discovery. Nat Rev Drug Discov 6:29–40

111. Weber A, Casini A, Heine A, Kuhn D, Supuran CT, Scozzafava A, Klebe G (2004) Unexpected nanomolar inhibition of carbonic anhydrase by COX-2-selective celecoxib: new pharmacological opportunities due to related binding site recognition. J Med Chem 47:550–557
112. Kinnings SL, Liu N, Buchmeier N, Tonge PJ, Xie L, Bourne PE (2009) Drug discovery using chemical systems biology: repositioning the safe medicine Comtan to treat multi-drug and extensively drug resistant tuberculosis. PLoS Comput Biol 5:e1000423
113. Gronenberg LS, Kahne D (2010) Development of an activity assay for discovery of inhibitors of lipopolysaccharide transport. J Am Chem Soc 132:2518–2519
114. Srinivas N, Jetter P, Ueberbacher BJ, Werneburg M, Zerbe K, Steinmann J, Van der MB, Bernardini F, Lederer A, Dias RL, Misson PE, Henze H, Zumbrunn J, Gombert FO, Obrecht D, Hunziker P, Schauer S, Ziegler U, Kach A, Eberl L, Riedel K, DeMarco SJ, Robinson JA (2010) Peptidomimetic antibiotics target outer-membrane biogenesis in *Pseudomonas aeruginosa*. Science 327:1010–1013
115. Kokryakov VN, Harwig SS, Panyutich EA, Shevchenko AA, Aleshina GM, Shamova OV, Korneva HA, Lehrer RI (1993) Protegrins: leukocyte antimicrobial peptides that combine features of corticostatic defensins and tachyplesins. FEBS Lett 327:231–236
116. Takase I, Ishino F, Wachi M, Kamata H, Doi N, Asoh S, Matsuzawa H, Ohta T, Matsuhashi M (1987) Genes encoding two lipoproteins in the *leuS-dacA* region of the *Escherichia coli* chromosome. J Bacteriol 169:5692–5699

Evolution of Lipopolysaccharide Biosynthesis Genes

11

Monica M. Cunneen and Peter R. Reeves

11.1 Introduction

Lipopolysaccharide (LPS) is a highly polymorphic structure that differs within and between genera, and contains three main components: lipid A, core oligosaccharide (OS), and O-specific antigen in the order in which they occur in LPS, which correlates with increasing structural diversity for each component. In *Escherichia coli*, for example, there are five core OS types known and over 180 O-antigen forms (including *Shigella*), and in *Salmonella enterica*, 2 and 46 respectively. The diversity of O-antigen forms has been widely studied for some species although the forms known may be underestimates as most of the isolates typed are from humans or domestic animals and their associated environments. Further examples for well-documented species are 20 O-antigen forms recognised in *Pseudomonas aeruginosa* [1], 21 in *Yersinia pseudotuberculosis* [2] and about 200 in *Vibrio cholerae* [3]. Such structural diversity is linked with genetic diversity, and in this chapter, the evolution and diversity of the genes required for the synthesis of these LPS structural components will be explored.

We start with an overview of the structure and genetics of each LPS component and discussion of the characteristics of the genes involved in biosynthesis, followed by case studies to highlight particular evolutionary aspects, such as the variation between genetic loci among species, how the clusters involved can be grouped by species or the pathways involved, and also the evidence for gene transfer events on a whole-cluster, gene-block and individual gene scale. We will then conclude with a discussion on the evolutionary forces driving the immense diversity of LPS.

M.M. Cunneen • P.R. Reeves (✉)
Division of Microbiology, School of Molecular and Microbial Biosciences, University of Sydney, Sydney, NSW 2006, Australia
e-mail: monica.cunneen@sydney.edu.au; peter.reeves@sydney.edu.au

Y.A. Knirel and M.A. Valvano (eds.), *Bacterial Lipopolysaccharides*,
DOI 10.1007/978-3-7091-0733-1_11,

11.2 Overview of LPS Structure and Gene Clusters

Biosynthesis of the basic 2-keto-3-deoxy-octulosonic acid (Kdo)-lipid A structure is generally well conserved among Gram-negative bacteria, with nine genes (*lpxA-D*, *lpxH*, *waaA* and *lpxK-M*), required for its synthesis (see also Chaps. 6 and 8). These genes are scattered throughout the genome: some individually, and others in clusters, in contrast to the genes for core OS and O-antigen, which are generally in specific gene clusters for each component. The genes for lipid A are in effect part of the core OS gene cluster, and a study in *Campylobacter* indicated sequence variation in the *lpxA* gene correlated with species divisions [4], as is expected for housekeeping genes.

The variation in lipid A involves for example, the type (length and number) of fatty acids present, the degree of phosphorylation, and the presence of substitutes like phosphoethanolamine and 4-amino-4-deoxy-L-arabinose. Genes that are not part of the core OS gene cluster may encode such modifications, and the presence or absence of these varies between strains, or differ in the substrate specificity of shared enzymes [5,6]. However, the evolutionary context for Kdo-lipid A genes is within the range for housekeeping functions and requires no special attention in this chapter, but this is not the case for the more variable components of LPS.

Synthesis of the core OS and O-antigen structures generally involves three major groups of proteins, those for biosynthesis of precursors, those for transfer of sugars or other components, and those for processing or export. Unlike those for Kdo-lipid-A synthesis, many of the genes encoding these proteins group together in what are known as the core OS and O-antigen gene clusters respectively (see also Chaps. 8 and 9). The higher structural diversity in O-antigens frequently involves sugars and other components not otherwise present in the cell, and the gene clusters often include a proportion of genes for biosynthesis of the precursors for these components. Over 80 monosaccharides and over 50 other components are given for O-antigens in Chap. 3 of this book. The genes for the three groups of proteins have different patterns of diversity, which has influenced nomenclature [7]. Genes for biosynthesis of precursors (see Chap. 7) are generally conserved across gene clusters and once the genes for a pathway have been identified for one gene cluster, they can usually be recognised in other gene clusters by sequence alone, particularly where the structure is also known, and can be named in a consistent manner based on their specific function. The glycosyltransferases (GTs), however, are very diverse and only a small proportion of those identified have been studied at all. It is commonly the case that such genes can be identified in a generic sense, but it is often not possible to allocate each GT to a specific linkage using only the DNA sequence. They are usually given names that do not imply the specific functions. For the processing genes, such as the *wzx* and *wzy* genes discussed below, it is usually possible to identify them from inferred secondary structure of the proteins, and often by the best hits in a BLAST search. But there can be enormous sequence diversity among genes with the same name, and diversity in specificity of the processing or export undertaken.

These diverse core OS and O-antigen gene clusters share several features. They often have a 39-bp sequence known as the "just upstream of many polysaccharide starts" (JUMPSTART) sequence [8] associated with the promoter. Core gene clusters tend to have three main operons with genes often in different orientations, whereas with few exceptions, O-antigen gene clusters are transcribed in one direction only. We will not discuss regulation further but readers are referred to a detailed study of the promoter region of *E. coli* O7 [9]. For both core OS and O-antigen, the genes commonly have a lower than genomic average GC content (see Fig. 11.1 for an example of an O-antigen gene cluster and structure). These "rules" will be used as a basis for discussion, using particular case studies for detail as required, but should be considered generalisations, and exceptions to these will also be discussed. Much of the discussion will be focussed on *E. coli* (including *Shigella*), *S. enterica*, *P. aeruginosa*, and *Y. pseudotuberculosis* as for each of them extensive structure and sequence data is available.

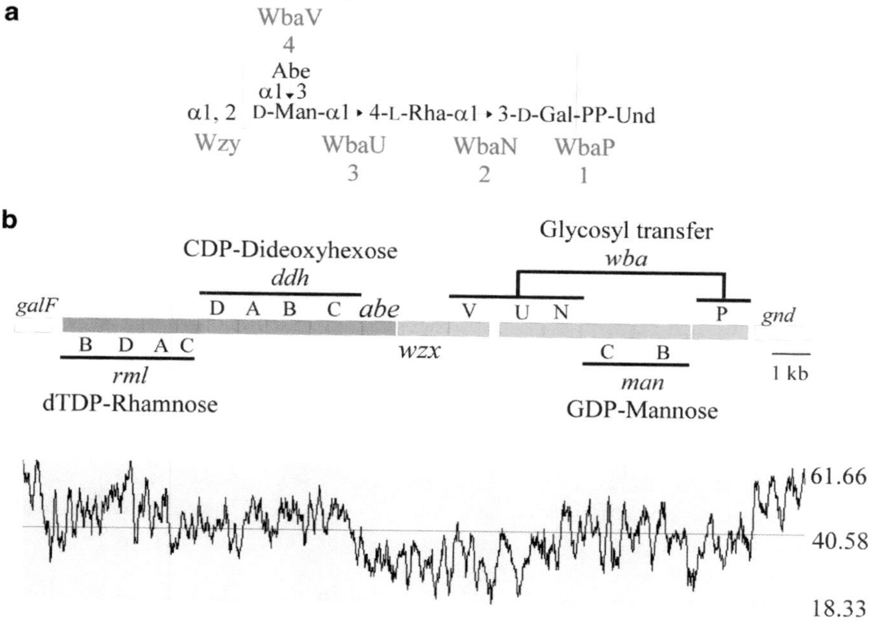

Fig. 11.1 The O-antigen structure and gene cluster of *S. enterica* group B1 O-antigen. (**a**) The structure of the group B repeat unit with the proteins responsible for each linkage shown in *red*. (**b**) The gene cluster with genes colour coded for the sugar involved; *purple* (rhamnose), *pink* (dideoxyhexose), *green* (mannose), *orange* (galactose). Below is a plot of the %GC content over the cluster using a 120 bp window

11.3 Evolution of O-Antigen Gene Clusters

O-antigen gene clusters exhibit enormous diversity both within and between species, and the same is true to a lesser extent for the LPS core OS gene cluster. Extensive lateral gene transfer (LGT) is the best conceivable explanation for this pattern of diversity. Only 9 of the 180 and 46 known O-antigens in *E. coli* and *S. enterica* respectively are shared by the two species. Shared O-antigens may still be discovered but that is not likely to remove the need to explain the origins of many new ones in both species. While there are pairs or groups of gene clusters that are related, with differences that could have evolved during species divergence, this does not apply to most of the genes, and the only realistic explanation is that many of the gene clusters have entered *E. coli* or *S. enterica* since divergence. This of course deprives us of access to the ancestors, as the evolution took place in unidentified species.

11.3.1 Initiating Transferases

The initiating transferase (IT) genes define groups of O-antigens. Despite the great diversity of GT genes in O-antigen gene clusters there are considerably fewer different ITs. The ITs WecA and WbaP are prototypes of the two known IT protein families and have been well studied in *E. coli* and *S. enterica* [10]. In the Enterobacteriaceae the *wecA* gene is in the gene cluster responsible for enterobacterial common antigen (ECA) synthesis, but WecA is also the major O-antigen IT for many species in the Enterobacteriaceae, including *E. coli* and *S. enterica* and several related genera such as *Citrobacter*, and O-antigens encoded at the *hemH/gsk* locus in *Yersinia*. WecA is again the probable IT for O-antigens encoded at the *cpxA/secB* locus in several *Proteus* species, as there is no IT gene in the sequenced gene clusters [11], the structures all contain GlcNAc, and there is a functional ECA locus [12]. WecA is also the usual IT for Wzm/Wzt O-antigens in the Enterobacteriaceae, where the complete O-antigen chain is built on one GlcNAc residue. The Enterobacteriaceae are unusual in this regard as gene clusters for repeat unit polysaccharides commonly include an IT gene: for example the *wbpL* gene in *Pseudomonas aeruginosa* O-antigen gene clusters discussed below. There is however within the Enterobacteriaceae a group of *S. enterica* O-antigens that have in their gene clusters the IT gene *wbaP*, which codes for a Gal-P transferase that initiates synthesis for this group and will be discussed below. There are also sporadic occurrences of other ITs in the Enterobacteriaceae such as WbyG for FucNAc4N-P in *Shigella sonnei* [13].

It seems that *wecA* predates the other ITs in the Enterobacteriaceae, as it is involved in this role in *E. coli* and *Yersinia*, and is also implicated in this role in *Proteus* [11], that diverges from *Escherichia* near the base of the family tree [14]. Also the majority of *S. enterica* serotypes do not have the *wbaP* gene, and in these cases WecA is the presumed IT responsible for initiating O-antigen synthesis with a GlcNAc or GalNAc residue as it was shown recently that in GalNAc initiated

O-antigens, UndPP-GlcNAc can be converted to UndPP-GalNAc after formation by WecA [15]. It seems likely therefore, that the ancestral IT in Enterobacteriaceae is WecA, and that WbaP and others were gained at a later stage, and this is also supported by the other IT genes being cluster associated, and probably gained at some point with the gene cluster. There is limited overlap in the structure of the *E. coli* and *S. enterica* O-antigens, but there are now nine cases where an *E. coli* and a *S. enterica* serotype have the same basic O-antigen structure, all being WecA initiated [16–22]. Sequence comparisons were first reported for the *E. coli* O55, O111 and O157 gene clusters in relation to the *S. enterica* O50, O35 and O30 gene clusters respectively [22]. The three pairs showed similar patterns of divergence, which was higher than usual for the core OS gene cluster, but given the consistency, and lack of extreme level, the divergence was taken to support their ancestry in the common ancestor with the sequences diverging as the two species diverged. For the additional pairs now known, the data fits the same pattern, and supports that conclusion. The higher rate of divergence in the O-antigen genes relative to core OS genes could be due to the much smaller population size if only one serotype is considered, which will affect rates of fixation, and/or be because the genes are still adapting to function in the *E. coli*/*S. enterica* background after transfer from an unrelated source. The data support the role of WecA in O-antigen initiation in the common ancestor of *E. coli* and *S. enterica*.

11.3.2 O-Antigen Gene Clusters

There are few cases where all or most of the gene clusters for identified serotypes of a species have been sequenced, for example *E. coli*/*Shigella*, *Y. pseudotuberculosis* and *P. aeruginosa*. Such sets provide information on how gene clusters within a species are related, with potential for the source of gene or gene blocks to be traceable through the set and will be discussed in the following sections.

11.3.2.1 The *S. enterica* Galactose-Initiated O-Antigens

There are 46 *S. enterica* O-antigens, of which 37 contain GlcNAc and/or GalNAc and are likely initiated by WecA, while eight are initiated by WbaP addition of galactose. The synthesis of *S. enterica* B1 is a well-studied example of the latter group, and the B1 O-unit structure and associated gene cluster are shown in Fig. 11.1. These Gal-initiated O-antigens (A, B1, B2, C2–C3, D1, D2, D3 and E) have closely related structures (Fig. 11.2) and gene clusters (Fig. 11.3), and also defines O-serogroups that are very commonly isolated for *S. enterica*. As discussed later, this group of O-antigens are probably of relatively recent origin in *Salmonella*, as the O-antigens in related species have almost entirely O-antigens initiated by WecA. As the structures, gene cluster sequences, and the functions of all gene products are known for all eight groups, the structural differences can be traced to variations in the respective gene clusters [23].

Most of these Gal-initiated O-antigens have the same Man-Rha-Gal backbone and one of three dideoxyhexoses (DDHs), paratose, tyvelose or abequose as a side

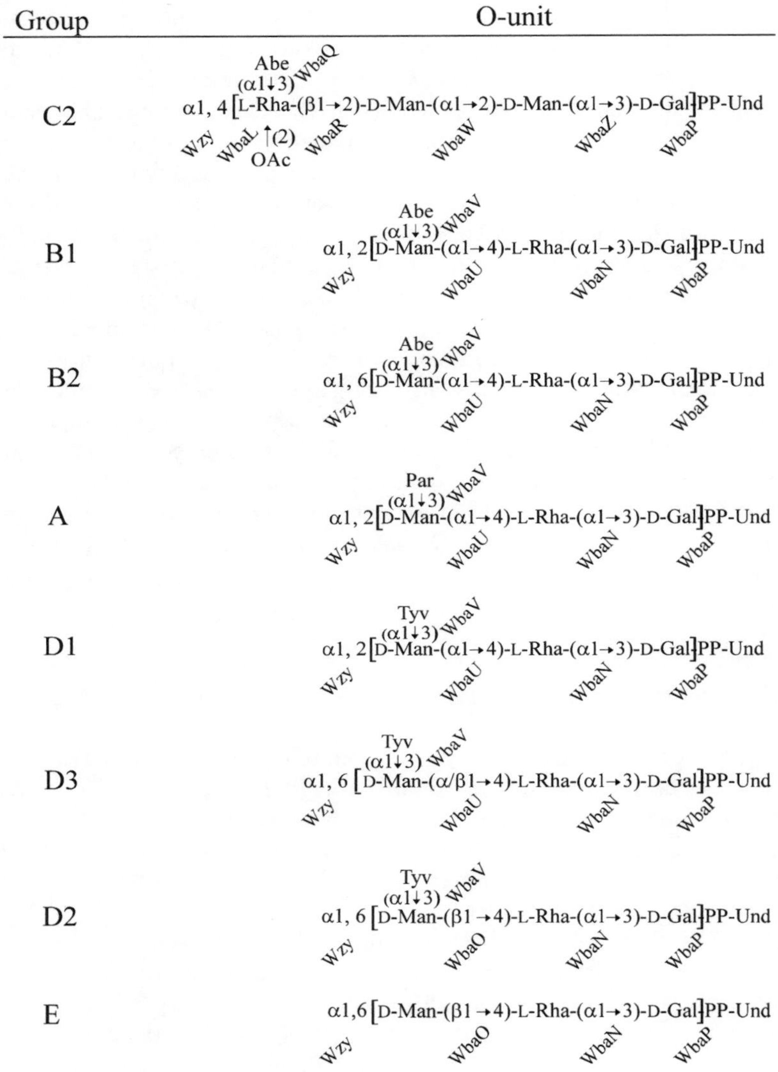

Fig. 11.2 The structures of the eight Galactose-initiated O-antigens of *S. enterica*. One O unit for each serotype is shown on the lipid carrier undecaprenol pyrophosphate (PP-Und) with the glycosyltransferase responsible for each linkage indicated. *Gal* galactose, *Rha* rhamnose, *Man* mannose, *Abe* abequose, *Par* paratose, *Tyv* tyvelose, *OAc O*-acetyl

branch sugar on the Man residue. Group E lacks the DDH side branch, while group C2–C3 backbone has the same three sugars in a different order, and its abequose side branch is on a Rha residue. As a result the C2–C3 gene cluster does not share a single GT gene with any of the others. The sequence comparisons (Fig. 11.3) show

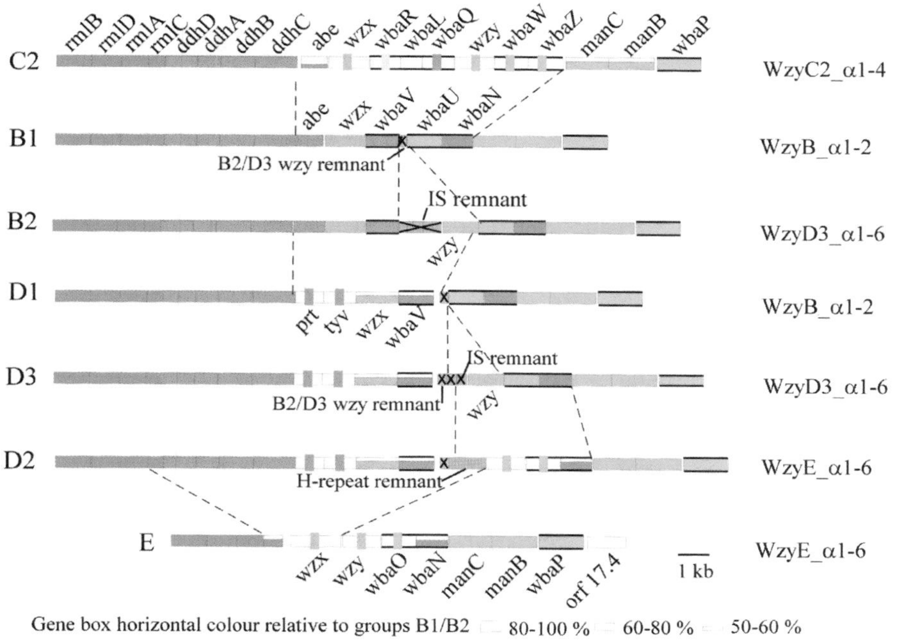

Fig. 11.3 The gene clusters for the eight Galactose-initiated O-antigens of *S. enterica*. Gene names are indicated. Genes colour coded for the sugar involved; *purple* (rhamnose), *pink* (dideoxyhexose), *green* (mannose), *orange* (galactose). The height of the coloured part of the *box* indicates level of amino acid identity of the gene product with that of group B1 (as shown): those with negligible homology are given a vertical band of colour. *Dashed lines* between pairs indicate boundary points between common and divergent regions in the clusters. Wzy variant and linkage in final structure indicated on the right. The group A gene cluster is not shown, but is related to that of group D1, except that the *tyv* gene is non-functional

that the shared genes are often nearly identical, while the other genes are only very distantly related to homologues in other strains of the group.

Groups B1, D1 and A differ structurally only in the DDH side branch, and each has the four gene block, *ddhDABC* required for synthesis of the immediate precursor of the basic DDH structure on CDP, while *prt* and *abe* are the genes for the final reduction step to give CDP-paratose or CDP-abequose respectively. CDP-tyvelose is made from CDP-paratose by epimerisation at C2. The *prt* and *abe* genes are only distantly related and their ancestry is not known. The three DDH sugars are related but give rise to major antigenic differences, as the three DDH sugars are immunodominant in these structures [24].

Groups B1 and D1 are sister groups with near identical genes apart from the presence of either the *abe* gene or the *prt* and *tyv* genes respectively. The sequences are otherwise near identical apart from the *wzx* and *wbaV* genes that follow the *abe* and *prt*/*tyv* genes, for which about 35% of amino acids differ. The junctions between *ddhC* and either *abe* or *prt* are very sharp, and likewise for the junctions in the intergenic region between *wbaV* and *wbaU*. These sharp boundaries mask the evolutionary origins of this pair of related gene clusters, but are probably currently maintained by concerted evolution of the shared genes at the two ends of the gene clusters. Any mutation that arises is either lost (usually) or (rarely) becomes established in groups B1 and D1 by random genetic drift in the genome, which can extend into the gene cluster up to the divergent segment in the center. However, mutations in the divergent regions will be fixed (or not) in either group B1 or group D1, but not in both as the sequence divergence is preventing recombination extending into this divergent segment. In this way the boundaries get sharper over time as only one side is diverging. It is interesting that Wzx and WbaV are the two proteins that could be affected by the difference between abequose and tyvelose, as WbaV adds the DDH to the O-unit and Wzx translocates the completed O-unit to the periplasmic surface. It is not known if the different sequence forms of *wzx* and *wbaV* provide selection for their maintenance, or if the selection is only for the *abe* gene versus the *prt/tyv* pair of genes in the two sequence blocks.

The distribution of near identical and divergent segments seen in the comparison of the B1 and D1 sequences is found in any pairwise comparison of the gene clusters shown in Fig. 11.3. In each case the divergent genes are in the center flanked by the near identical segments of the gene cluster, which are themselves flanked by shared genes. The genes that distinguish any two of the gene clusters can be transferred by a recombination event with the ends of the recombinant segment in shared DNA. Because all of the divergent sequence is in a single segment, there will be no reassortment of the genes in a normal homologous recombination event. This arrangement facilitates substitution of one O-antigen by another with little risk of generating other gene combinations. This pattern will be strongly supported by enabling "successful" gene clusters to move through the population, as must have happened for these O-antigens. This also has the effect that random genetic drift in the genome as a whole can extend into the gene cluster up to any of the junctions with the divergent segment as for B1 and D1. The divergent region can include genes that probably do not differ in function, for example *manC* and *manB* in the C2–C3 gene cluster. What we do not see is pairs that have three divergent segments separated by near identical sequence, which would allow two segments to be transferred independently.

Group A is interesting as the four group A serovar gene clusters have the *tyv* gene in a mutant form and it seems that they are recent derivatives of group D1 strains [25]. We can infer from this that the group A structure is only rarely advantageous, as it does not seem to have been maintained long term, but instead to have arisen occasionally from group D1 isolates. In each case there are one or two group D1 serovars with the same H1 and H2 flagellar antigens and the group D1 source serovar is identifiable. The *tyv* gene in serovar Paratyphi A is inactivated by a single base deletion [25], and would be expected to have accumulated further mutations if it had been non-functional for a significant time. This early frameshift generates a stop mutation in the group A *tyv* gene blocking CDP-paratose conversion to CDP-tyvelose, and paratose is the DDH-sugar that is present in the O-unit.

Groups B1, D1 and the derived group A all lack a *wzy* gene in the gene cluster, but have a *wzy* gene at a separate (*rfc*) locus. However between *wbaV* and *wbaU* there is a *wzy* gene remnant. It seems clear that groups B1 and D1 had ancestral forms with a different polymerisation linkage, and that their *wzy* genes lost function when the current *wzy* gene was acquired at the *rfc* locus. Group D3 resembles group D1, but has a α1–6 polymerisation linkage in place of the α1–2 linkage of groups B1 and D1. The remnant *wzy* genes in groups D1 and B1 are very similar to the corresponding segments of the D3 gene, so presumably D3 has that ancestral *wzy* gene, making D3 the ancestral form for group D1. Note that group D3 was reported to have two structures present simultaneously, with different Man-Rha linkages, but the gene cluster had only the gene for the linkage found in D1, B and A, and it is not known where the gene for the other linkage is located [26].

The group E structure lacks the DDH side branch, has a different β1–4 linkage between the Man and Rha residues, in place of the α1–4 linkage in the other groups, and has an a α1–6 polymerisation linkage. These differences from the group B and D1 structures are reflected in the absence of genes for any part of the DDH sugar pathway, the presence of a *wbaO* GT gene for the β1–4 linkage in place of *wbaU*, and a new *wzy* gene between *wbaV* and *wbaO*. The β1–4 linkage made by WbaO is the same linkage as reported for the second D3 linkage, however this *wbaO* gene was not found by PCR screening of a D3 strain [26].

The group D2 structure has the main chain of group E with a tyvelose side branch. The 5′ end of the D2 gene cluster is near identical to that of group D1, and the 3′ end near identical to that of the E1 gene cluster, which includes the *wbaO* and *wzy* genes (Fig. 11.3). There is a remnant H repeat at the junction of these segments, which is proposed to have been involved in a recombination event that gave rise to D2 [27]. In this case we see a rare recombination event which breaks the pattern described above for homologous recombination, and generates a new gene cluster which meets the pattern of having a central divergent segment in any pairwise comparison (see Fig. 11.4 for a model for the recombination event). Once such a gene cluster arises it is subject to natural selection driven by any benefit its O-antigen confers. There must be many such products that fail to survive.

The gene clusters of these Gal-initiated O-antigens (Fig. 11.3) have given us several cases where evolutionary origins can be inferred, but even here where the relationships are clear, we need to see intermediates to infer the processes involved in much of the diversification.

11.3.2.2 *Y. pseudotuberculosis* O-Antigen Gene Clusters

Y. pseudotuberculosis has a set of O-antigens that resembles in some ways the Gal-initiated set in *S. enterica*. There are 21 reported serotypes and 11 of the 13, for which we have both structure and gene cluster sequence, fit into this related set [28–30]. They each have the four DDH genes at the 5′ end of the gene cluster and main chain-related genes at the 3′ end of the gene cluster, but are significantly more diverse than those in *S. enterica*. There are several variants for the main chain and there are two additional DDH sugars (ascarylose and colitose, the latter with a quite different biosynthetic pathway), and also two serotypes have 6-deoxyaltrose, which

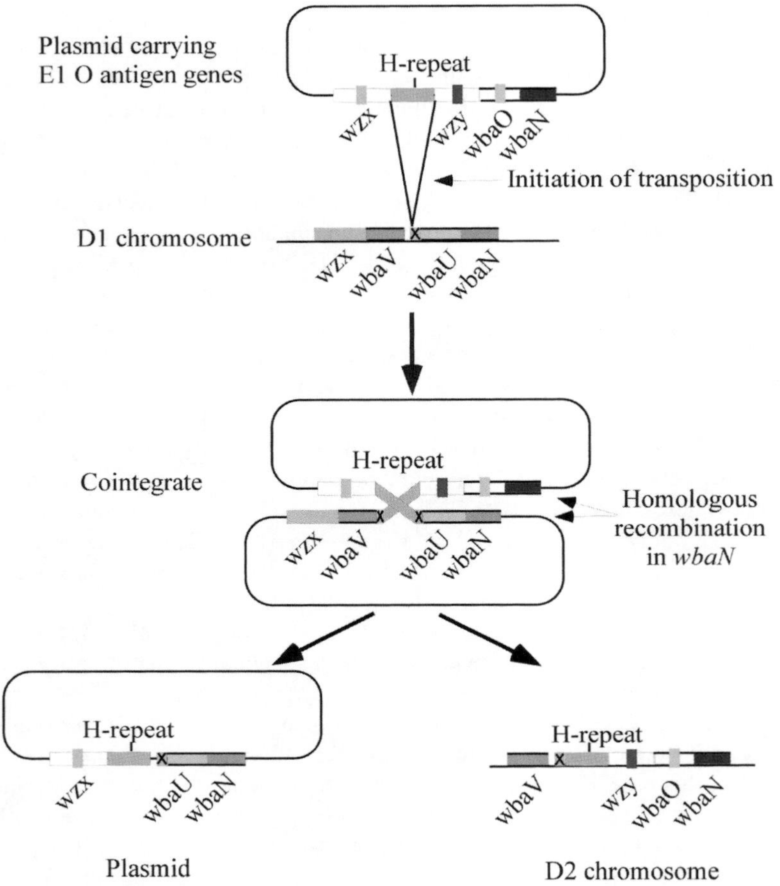

Fig. 11.4 A model for the recombination event that generated the *S. enterica* group D2 gene cluster. *Top*: hypothesised intermediate with incoming segment of the *S. enterica* group E gene cluster with H-repeat insertion prior to transposition into the D1 chromosome. *Center*: the hypothesised cointegrate of the transposition intermediate before resolution. *Bottom*: The product after resolution by homologous recombination (Modified from Ref. [27])

is not a DDH sugar, but its biosynthesis pathway branches off the major DDH pathway. There is the case of the *Y. pseudotuberculosis* O:1a gene cluster that combines sequences from two others in a similar way that *S. enterica* D2 has segments from serogroups D1 and E, to give an O-unit with the main chain of O:4b and the side branch DDH sugar of O:1b [31]. It is presumed to have arisen by recombination as for the *S. enterica* group D2 gene cluster. In general, the same pattern is found as for the Gal-initiated O-antigens of *S. enterica*, with the divergent genes in each pairwise comparison being in a central block of genes.

11.3.2.3 *E. coli* O-Antigen Gene Clusters

There are sequences and/or structures for many of the 180 O-antigens reported for *E. coli*, including both for all 34 *Shigella* O-antigens [32], and those traditional *E. coli* O-antigens closely related to any of the *Shigella* O-antigens. It is clear that *E. coli* has no major groups with related structures and sequences, as found in *S. enterica* and *Y. pseudotuberculosis*, but in several cases there are patterns suggesting that gene clusters are related, as discussed below.

One striking example is the gene clusters of *E. coli* O56, O24, O157, and O152 [33]. The O24 and O56 structures differ only in two linkages, reflected in the gene clusters by presence of shared genes for shared properties, and two serotype specific GT genes in each for the linkages that differ (Fig. 11.5). The *wfaP* and *wfaQ* GT genes specific to O56 are in the middle of its gene cluster, which has all of the genes required for O-unit synthesis. The only unusual feature is the presence of an H-repeat element between *galF* and the first gene of the cluster. The O24 gene cluster shares seven genes with O56, which are in the same order with about 80% sequence identity. Within this group of seven shared genes are the two O56-specific genes that are replaced in O24 by remnants of three IS elements. It appears that the O56-specific genes were lost by insertion of IS elements, but nothing recognisable is left of them and the IS are now reduced in size. The O24 replacement genes, *wbdN* and *wfaO*, are in two blocks (Blocks 1 and 3) at the ends of the shared segment (Block 2 in Fig. 11.5), and separated from the shared segment by IS and H-repeat elements. Putative sources for Blocks 1 and 3 are found in the *E. coli* O157 and O152 gene clusters respectively, again with about 80% identity. In the case of Block 1, there is a remnant of the O157 *wzy* gene next to *wbdN*, which presumably travelled together. It seems clear that the O24 gene cluster is derived from that of O56 by loss of two genes and gain of two genes from different sources, with these gains apparently mediated by IS or H-repeat elements.

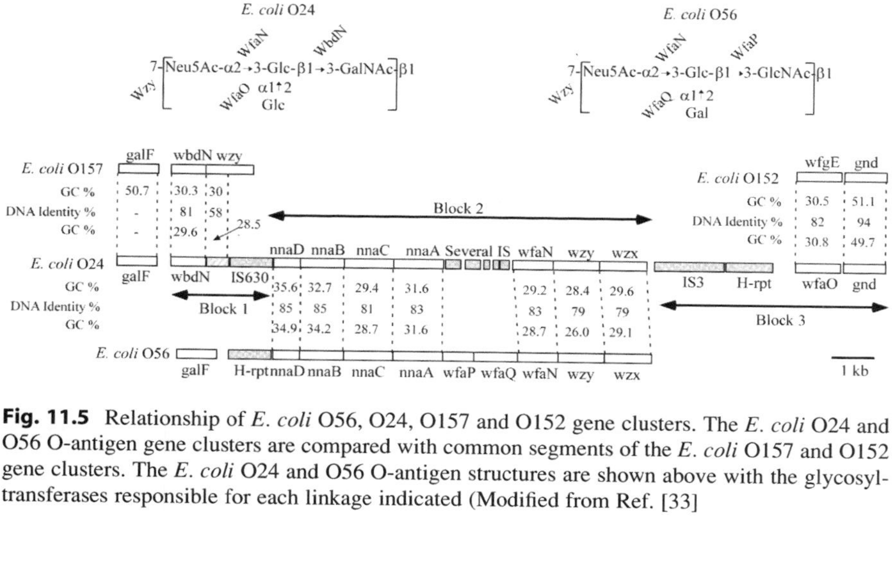

Fig. 11.5 Relationship of *E. coli* O56, O24, O157 and O152 gene clusters. The *E. coli* O24 and O56 O-antigen gene clusters are compared with common segments of the *E. coli* O157 and O152 gene clusters. The *E. coli* O24 and O56 O-antigen structures are shown above with the glycosyltransferases responsible for each linkage indicated (Modified from Ref. [33]

Another interesting pair are *E. coli* O148 and *Shigella dysenteriae* O1 [34]. In this case, *E. coli* O148 was the precursor, with Glc as the second residue and an appropriate GT gene, *wbbG*, for its addition. The *S. dysenteriae* O1 structure has a Gal residue in place of the Glc residue of O148, but is otherwise identical. The gene clusters are near identical but the *wbbG* Glc GT gene of O148 has a deletion in *S. dysenteriae* O1, and there is a plasmid-born Gal GT gene, *wbbP*, that functionally replaces the *wbbG* gene. It seems clear that *S. dysenteriae* O1 gained its *wbbP* gene on a plasmid and then lost *wbbG* function by mutation. Other cases involving *Shigella* and *E. coli* strains are given in a recent review [32]. The review also includes structures and gene cluster sequences for the 34 distinct *Shigella* O-antigens and their related *E. coli* O-antigens that highlight the diversity of structures and gene clusters in this selection from what is effectively the single species *E. coli*.

11.3.2.4 *P. aeruginosa* O-Antigen Gene Clusters

P. aeruginosa has two forms of O-antigen, known as A-band and B-band. B-band resembles the O-antigens of other species in being variable, with multiple structures and corresponding multiple forms of the gene cluster downstream of *ihfB*. However, there is only one structural form of A-band, which is present in almost all isolates. There are 20 B-band O-antigens recognised, and all structures are known [35] and also all gene cluster sequences except for that for O15, which must be at a different locus as the site has only a remnant [1]. Eleven of the 20 structures in this set fall into four groups of related structures, and each such group has only a single gene cluster sequence. Indeed there are only ten distinctive gene sequences at the locus, as there are two other pairs with identical gene clusters (some with genes inactivated by IS insertion). There has as yet been no detailed analysis of the relationships of the gene clusters, but each has the *wbpL* IT gene referred to above.

11.4 LPS Core OS

The LPS core OS is commonly subdivided into the inner and outer core, with the inner core more strongly conserved. In *E. coli* and *S. enterica* the genes for basic core synthesis are in a single gene cluster at the *waa* locus. This locus differs from those for O-antigens, in that most of the genes are for GTs. The most significant variation is in the outer core OS with five known types in *E. coli* (K-12, R1, R2, R3 and R4) and two in *S. enterica* (Typhimurium and Arizonae IIIa) [36,37].

There are three operons in the *waa* loci for the core OS structures of *E. coli* and *S. enterica*, coding for incorporation of both conserved and variable residues [37]. The first operon is the *hldD* (formerly *gmhD*) operon, starting with the conserved *hldD* gene for heptose synthesis, followed by two conserved inner core OS GT genes found in all seven variants. The last gene is *waaL*, for the ligase that attaches O-antigen to core OS, which is very variable in sequence as the ligase is specific for core OS structure. Next is the central *waaQ* operon, transcribed in the opposite direction towards the *hldD* operon. The third "operon" is the *waaA* gene for attachment of the two Kdo residues that starts inner core OS synthesis. All of the variation in gene content is in the central *waaQ* operon, relating to variation in the outer core OS. There are however four conserved genes in the *waaQ* operon, three of them at the start of the operon and thus part of the conserved region at the right-hand end of the *waa* locus. The three are *waaG* for the conserved GT attaching the first outer core OS residue to the inner core OS, *waaQ*, a conserved GT gene adding the last Hep residue of the inner core OS and *waaP* coding for a conserved kinase adding P to HepI. The fourth conserved gene, *waaY*, is an inner core OS gene in the middle of the operon, and to either side of it are genes with one to three occurrences in the seven outer core OS types. The effect then is similar to that for O-antigen gene clusters, with the variation between core OS types due to genes in the center of the gene cluster. However, in pairwise comparisons the genes that differ fall into two blocks separated by *waaY*. As for O-antigen gene clusters, this concentration of the non-conserved genes in the middle of the gene cluster will facilitate movement of core OS types by recombination in the conserved flanking DNA. The conserved *waaY* gene would be expected to allow new combinations of the gene blocks to the left and right of *waaY*, although this does not appear to occur. Perhaps most or all such combinations would not be advantageous, but this is an exception to the "rule" that the non-conserved genes form a single central block. Core types have not been studied as much as O-antigens, but the distribution of core OS types among the 72 *E. coli* strains of the ECOR set suggests that the core OS gene clusters are also mobile but not to the same extent as O-antigen gene clusters. All but the K-12 core OS gene cluster was found in more than one of the A, B1, B2, D and E groups of *E. coli* that were first distinguished by MLST, and the K-12 core OS type is found in two divergent subgroups within group A. The other core OS types are found in two or more of the *E. coli* groups, with R1, the most common core OS type, found in all five groups.

11.5 Loci for LPS Genes

The locations (map positions) of loci for the gene clusters for LPS components are themselves more variable than those for core OS genes of the genome, and in this section we will again focus on the four genera discussed above. All are members of γ-proteobacteria, with three from the Enterobacteriaceae and one from the Pseudomonaceae. The distribution of the loci involved is shown in Fig. 11.6.

For Kdo-lipid A synthesis, the *lpx*DAB genes cluster together, with a conserved genomic context, in *E. coli*, *S. enterica*, *Y. pseudotuberculosis* and *P. aeruginosa*. However, the loci for the *lpxCHKLM* genes are scattered in these species, although they share genomic context in all but *P. aeruginosa*, with *lpxK* and *lpxL* at unique loci, and *lpxM* missing though a second *lpxL*-like homologue is present [35]. Core genes, in contrast, are clustered together at the *waa* locus, and in *E. coli*, *S. enterica* and *Y. pseudotuberculosis* this is at the same site, between *htrL* and *mutM*, although many of the *waa* genes responsible for outer core OS synthesis are missing in *Y. pseudotuberculosis*, which presumably has species-specific core OS genes

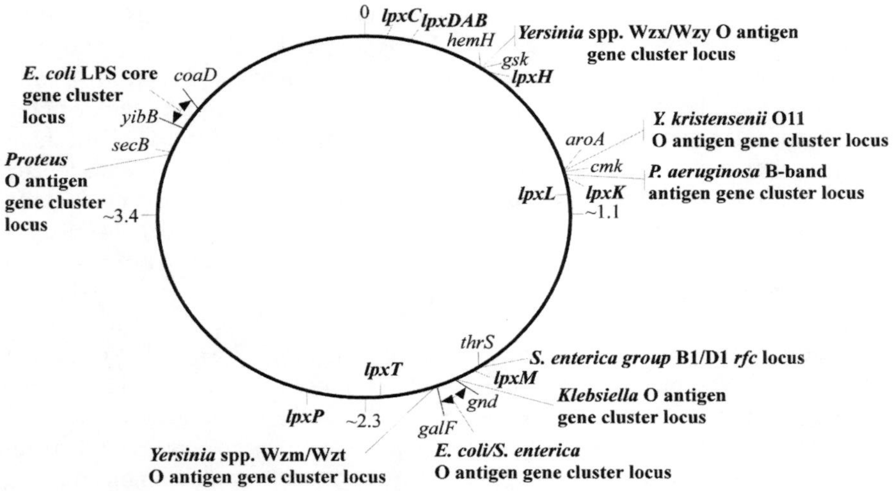

Fig. 11.6 The location of LPS gene loci for *Escherichia*, *Salmonella*, *Yersinia* and *Proteus*. *E. coli* K-12 chromosome (accession U00096) with positions given in Mb. The genes for lipid A, LPS core and O-antigen biosynthesis are indicated (*bold*) and neighbouring genes are also marked (*not bold*). The O-antigen loci for *S. enterica*, *Yersinia* spp, *Proteus*, and *P. aeruginosa* are shown at the site where the shared genes associated with the gene cluster map in *E. coli*. Note that the position of the *S. enterica rfc* locus is closer to the *galF*/*gnd* locus than in the *S. enterica* LT2 genome due to a major inversion difference

located elsewhere. The *P. aeruginosa waa* genes are not at this locus, although the cluster shares conserved inner core OS genes with the other species.

The major O-antigen loci are generally conserved within species and even genera, but tend to vary at higher taxonomic levels. It appears that these loci have changed over time to generate the loci now observed. Most O-antigens are synthesised by the Wzx/Wzy or Wzm/Wzt (ABC transporter) pathways, with only one O-antigen (*S. enterica* O54) having the synthase system, and its gene cluster is plasmid-borne [38,39]. The Wzx/Wzy clusters of *E. coli* and *S. enterica* and several related genera are between *galF* and *gnd*, those of *Y. pseudotuberculosis* between *hemH* and *gsk*, those of *P. aeruginosa* between *ihfR* and *tRNA-asn* and those of *P. mirabilis* between *cpxA* and *secB* [11]. For Wzm/Wzt systems, those in *E. coli* are between *gnd* and *his*, adjacent to the Wzx/Wzy clusters, and where present those of *Yersinia* spp. are upstream of *galU*. The locus for the *P. aeruginosa* A-band is upstream of *pseCoA* and *urvD*.

Generally, there is only one O-antigen expressed in any given strain, although for the *E. coli* O8 and O9/9a serotypes both types of gene clusters may be present, with the Wzm/Wzt form expressed as an O-antigen, and the Wzx/Wzy form expressed as a capsule. However, in *P. aeruginosa* both forms are co-expressed as O-antigen, with the A-band Wzm/Wzt gene cluster having only one form, while B-band Wzx/Wzy gene cluster is variable as discussed above. It is clear that as species diverge there is not only major change in the LPS genes and structures, but also in the genetic loci involved.

11.5.1 *Escherichia* Genome Sequences

Three of the *Escherichia* species have genome sequences for comparison: *E. coli* (many complete genomes, including K-12 MG1655; accession U00096.2), *E. fergusonii* (complete; accession CU928158.2), *E. albertii* (in progress; accession NZ_ABKX01000013.1). In *E. coli* the loci described above are also associated with the polysaccharide gene cluster for colanic acid, which is upstream of *galF*, and so it is evidently a "hot-spot" in the *E. coli* genome for these clusters. *E. fergusonii* resembles *E. coli* in having a colanic acid gene cluster upstream of *galF*, but this is missing in *E. albertii*, and both have the Wzx/Wzy O-antigen gene cluster between *galF* and *gnd*. Many of the O-antigen gene clusters in *E. coli* have now been sequenced, and the number of Wzx/Wzy O-antigens at the *galF*/*gnd* site far outweighs the two Wzm/Wzt type O-antigens that map between *gnd* and *his*. Two exceptions to the loci "rule" for *E. coli* are found in *E. coli* O52 and O99, which have Wzm/Wzt gene clusters between *galF* and *gnd* [40,41].

11.5.2 *Klebsiella* Genome Sequences

There are three *Klebsiella* genomes available (all *K. pneumonia*), and each has a Wzm/Wzt gene cluster between *gnd* and the *his* operon which is consistent with the location of the preciously studied Wzm/Wzt O-antigen gene clusters [42]. These genomes also have a capsule gene cluster between *galF* and *gnd* as shown previously for 12 capsule types [43]. It appears that *Klebsiella* has a capsule gene cluster at the locus usually occupied by the O-antigen gene cluster in *E. coli*, but that in some *E. coli* strains (O8 and O9), this site is occupied instead by a group 1 capsule gene cluster (for example, K-40; see Ref. [44] for review). *E. coli* O9 and O8 O-antigen gene clusters are between *gnd* and the *his* operon, and are related to those found in *Klebsiella*, indicating likely transfer between strains [45,46].

11.5.3 *Yersinia* Genome Sequences

There are currently ten species of *Yersinia*, and at least one genome sequence (partial or complete) for each (Table 11.1). The reported gene clusters for Wzx/Wzy systems are all between *hemH* and *gsk*, and those for Wzm/Wzt systems are associated with or upstream of *galU*/*galF* and *gnd*, with one isolated instance of a Wzx/Wzy O-antigen gene cluster between *aroA* and *cmk*, as is discussed below. Where genome sequence data is available for these regions, the Wzx/Wzy locus is occupied, with the locus for Wzm/Wzt systems occupied in only three of these species.

Table 11.1 O-antigen loci in *Yersinia*

Species	Accession	*hemH/gsk*	*galU*	*aroA/cmk*
Y. pseudotuberculosis O:1b[a]	CP000720.1	Yes	No	No
Y. pestis C092[a,b]	AL590842.1	Yes	No	No
Y. enterocolitica O:8[a]	AM286415.1	Yes	No	No
Y. enterocolitica O:9[c]	AJ605741.1	n/a	Yes	n/a
Y. enterocolitica O:3[c]	Z47767.1 Z18920.1	Yes, outer core OS	Yes, O-antigen	n/a
Y. aldovae[d]	NZ_ACCB01000002.1 NZ_ACCB01000005.1	n/a	No	No
Y. aldovae A125[c]	AJ871364.1	Yes	n/a	n/a
Y. bercovieri ATCC 43970[d]	NZ_AALC02000011.1 NZ_AALC02000079.1 NZ_AALC02000024.1	Yes	n/a	No
Y. frederiksenii ATCC 33641[d]	NZ_AALE02000006.1 NZ_AALE02000021.1 NZ_AALE02000015.1	Yes	Yes	No
Y. intermedia ATCC 29909[d]	NZ_AALF02000002.1 NZ_AALF02000015.1 NZ_AALF02000006.1	Yes	No	No
Y. kristensenii ATCC 33638[d]	NZ_ACCA01000010.1 NZ_ACCA01000001.1 NZ_ACCA01000014.1	Yes	No	Yes
Y. mollaretii ATCC 43969[d]	NZ_AALD02000001.1 NZ_AALD02000002.1 NZ_AALD02000003.1	Yes, with genes for both O forms	No	No
Y. rohdei ATCC 43380[d]	NZ_ACCD01000010.1 NZ_ACCD01000002.1 NZ_ACCD01000001.1	Yes	Yes	No
Y. ruckeri ATCC 29473[d]	NZ_ACCC01000007.1 NZ_ACCC01000001.1 NZ_ACCC01000001.1	Only *wzz* present	No	No

[a]Multiple genomes known in this species
[b]*Y. pestis* has only one reported O-antigen gene cluster and is an inactivated form derived from the *Y. pseudotuberculosis* O:1b O-antigen gene cluster [89]
[c]Only gene cluster sequence region known
[d]Genome in progress/incomplete; n/a indicates sequence is not available for analysis

For *Y. enterocolitica* the only genome sequence available is of the serotype O:8 that had been known before to have Wzx/Wzy gene cluster at the *hemH/gsk* locus and is now shown to have only that locus occupied. The other two serotypes that have been studied (O:3 and O:9) have Wzm/Wzt structures and gene clusters. The *Y. enterocolitica* O:9 gene cluster is at the *galU* locus but is atypical as it is in two segments, one upstream of *galU* and the other between *galF* and *gnd*.

The *hemH*/*gsk* locus has not been examined in the O:9 strain and we do not have genome sequences for these serotypes to compare to O:8. *Y. enterocolitica* O:3 has a Wzm/Wzt gene cluster at the *galU* locus for its polymeric O-antigen, and also a gene cluster at the *hemH*/*gsk* locus that lacks a *wzy* gene, and only a single "repeat unit" is made that is added to the inner core OS to form what has been called the outer core OS. It is likely that the O:9 serotype also produces an outer core OS structure based on serotyping studies, and is presumed to have the same outer core OS cluster at the *hemH*/*gsk* locus as O:3 [47]. The *Y. ruckeri* genome had a remnant gene cluster between *hemH* and *gsk*, with only the *wzz* gene, suggesting that no O-antigen is produced in that isolate. However, this scenario is not representative of the species, as O-antigen structures have been reported for other isolates [48,49].

It seems clear that in *Yersinia* the common pattern is to have only a Wzx/Wzy structure with the gene cluster between *hemH* and *gsk*, but some strains have an additional Wzm/Wzt gene cluster upstream of *galU*. As yet we do not know if both can be expressed as O-antigens, or if the pattern seen in the *Y. enterocolitica* O:3, and probably O:9, with a gene cluster at the *hemH*/*gsk* locus expressed as an outer core OS comprising a single "repeat unit", is at all common in the presence of a Wzm/Wzt O-antigen.

It is interesting that the major locus for O-antigen gene clusters in *Yersinia* (*hemH*/*gsk*) is at a different site than that for *E. coli* and *S. enterica* (*galF*/*gnd*), in both cases for Wzx/Wzy-processed O-antigens. However, both species groups have a major alternative site for Wzm/Wzt processed O-antigens in the *galF* region, although not at exactly the same locus, but there is too little data for any evolutionary implications from sharing this site.

11.5.4 *Pseudomonas* Genome Sequences

There are genome sequences for seven species of *Pseudomonas* and all have the Wzx/Wzy B-band synthesis locus downstream of *ihfB* (Table 11.2). The A-band gene cluster is present in all *P. aeruginosa* genomes but is otherwise found only in *P. fluorescens* O1, though at a different locus and is not present in *P. fluorescens*

Table 11.2 O-antigen loci in *Pseudomonas* spp

Species	Accession	A-band	B-band
P. aeruginosa PAO1[a]	AE004091.2	Yes	Yes
P. fluorescens Pf0-1[a]	CP000094.2	Yes, but different loci	Yes
P. entomophila L48	CT573326.1	No	Yes
P. mendocina ymp	CP000680.1	No	Yes
P. putida[a] F1	CP000712.1	No	Yes
P. stutzeri A1501	CP000304.1	No	No
P. syringae pv. Phaseolicola[a] 1448A	CP000058.1	No	Yes, but only few genes present

[a]Multiple genomes available in this species

O5. Both A-band and B-band are incorporated into LPS by the same ligase, WaaL [50], with its gene located close to LPS core OS genes. The site for addition of B-band is known but although the ligation site on the core OS for the A-band is not known, both are presumably added at the same site as alternative additions, as the same ligase is involved. The initiating GT gene *wbpL* is in the B-band gene cluster and also required for A-band initiation.

It is intriguing that the GC content of A-band genes is similar to that for the *P. aeruginosa* genome whereas those for B-band are lower [51]. The requirement of a B-band gene for A-band initiation and finding the A-band in only two species, suggests that A-band is evolutionally a recent addition, but the GC content suggests otherwise. The presence of the A-band gene cluster at different loci in the two species also suggests that the A-band cluster is not long established in the genus. The genome of *P. aeruginosa* is about 30% larger than that of *E. coli*, with many insertions, deletions and rearrangements relative to *E. coli*, that hamper a direct comparison [52]. Indeed, many of the genes that flank the LPS gene clusters in *E. coli* and *Yersinia* spp. are not adjacent, or are absent in *Pseudomonas*, making it impossible to find the corresponding regions in the two families for loci comparison. However, it is intriguing to observe that the *E. coli* O-antigen gene cluster is adjacent to the *his* operon, and that homologues of some of these *his* genes (*hisH* and *hisF*) are often present within the B-band O-antigen gene clusters of *P. aeruginosa* although the implications of this for their evolution in *E. coli* and *P. aeruginosa* is not clear.

11.5.5 LPS Gene Clusters in Other Genera

There is sporadic information for O-antigen loci for other species. *Enterobacter sakazakii* has an O-antigen gene cluster between *galF* and *gnd* [53]. A probable capsule gene cluster is also present between *galF* and *gnd* in the genome of *Serratia proteamaculans* 568 (CP000826.1), but there is no apparent O-antigen gene cluster in this species downstream of *gnd*, so the locus is currently unknown. However, another Wzx/Wzy O-antigen locus has been identified in several *Proteus* spp. and is between the genes *cpxA* and *secB* [11], which is markedly different than the loci previously discussed in this chapter. The *cpxA* and *secB* genes are not adjacent in *E. coli* K-12 or *S. enterica*, but are closely linked in *Y. pseudotuberculosis*. *P. aeruginosa* does not have a *cpxA* gene for locus comparisons. The number of isolates and species studied suggests this is a well-established locus in *Proteus*.

11.6 Genes not Linked to the Major Gene Clusters

There are many cases of LPS genes that are not in the relevant gene cluster. An example is the *wzy* gene of *S. enterica* groups B1 and D1 discussed above. In this case we are dealing with a function that is normally encoded in the gene cluster and it is thought that this is a relatively recently gained gene, as also discussed above.

There is no obvious advantage of having the *wzy* gene at a separate locus, as it is essential for O-antigen polymerisation, which itself is essential for survival for most Gram-negative bacteria. Having two loci for full O-antigen expression restricts movement of the B1 and D1 gene clusters to organisms that have this *wzy* gene, or relies on the chance transfer of both loci, which must be rare as the loci are about 750 kb apart (in strain LT2). In other cases, such as where the *E. coli* and *S. enterica* *wzz* gene is only a few kilobases from the O-antigen gene cluster, the arrangement may be beneficial. The *wzz* gene is close enough that it can be transferred with the gene cluster, but the few kilobases of DNA between the two loci, that is generally present in *E. coli*, still allows O-antigen structure and its variation in modal chain length, to be inherited independently.

In other cases the separate genes code for variations in the O-antigen, and genes for such "modifications" are often bacteriophage encoded. Several examples are known in *Shigella flexneri*, where different side-branch glucosylation or O-acetylation modifications of a common basic structure, result in different serotypes (Fig. 11.7) [54,55]. The same basic O-antigen structure is found in three traditional *E. coli* serotypes (O13, O129 and O135) that differ from each other by modifications similar to those found in *S. flexneri* [56]. The O129 and O135 modifications were the same as found in types 5a and 4b respectively, and O13 had a new glycosylation site. In *S. flexneri* these modifications are immunodominant, and the same must apply to the three traditional *E. coli*, as they were put in separate serogroups.

The *S. flexneri* side-branch glucosylation involves three genes as the glucosylation takes place after translocation of the O-unit by Wzx. Two genes are common to all of these *gtr* loci, with the third gene encoding the transferase itself and that confers site specificity. The single O-acetylation gene is also in a

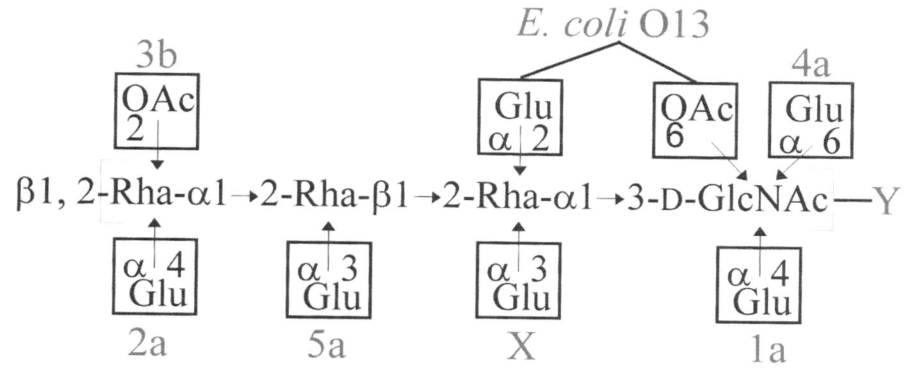

Fig. 11.7 O-antigen modification in *Shigella flexneri* and *E. coli*. The common backbone in *Shigella flexneri* is shown, and represents that of serotype Y. The sites for glucosylation and O-acylation in other serotypes are shown, using the name of the serotype that has only that modification (in *red*). Single additions are indicated with the linkages to the main backbone shown and the resulting serotype in *red*. Further serotypes (not shown) exist with combinations of the modifications shown in the figure. The *E. coli* O13 glucosylation and O-acetylation sites are also shown

bacteriophage genome. The immunity to infection by *S. flexneri* is specific to the serotype as shown by lack of significant cross-reaction on vaccination. In these cases involving immunodominant antigens, the selection is at least in part on the bacteriophage genome. Indeed, the bacteriophage gets a selective advantage from carrying the *gtr* locus by conferring selective benefit on the bacterium in certain circumstances as discussed below. The situation resembles that in the *S. enterica* groups discussed above, but in this case the side branch residues involved are determined by genes on bacteriophages, which appears to have allowed more rapid development as the strains involved are very closely related [57].

There is a parallel situation for *E. coli* O77, O17, O44, and O73 and *S. enterica* group O6, 14, which have the same backbone structure and gene clusters, but differ in glucosylation patterns, potentially due to presence of bacteriophage genomes carrying different *gtr* gene sets, and the role of a bacteriophage was shown for the O44 strain [21].

There are other examples of O-antigen modification genes being on bacteriophages, which received considerable attention in the early days of bacterial genetics. In several cases, it was shown that the bacteriophage uses the unmodified O-antigen as its receptor, and the modification reduced the ability of the bacteriophage to adsorb (see Ref. [58] for review), giving selective advantage to the bacteriophage after infection by blocking access to other bacteriophages. There are no similar studies on the *Shigella* converting bacteriophage, but it is possible that the same situation applies.

Glucosylation and O-acetylation are commonly encoded by genes outside of the gene cluster, as in many cases no appropriate gene is found in the gene cluster sequence. This is not invariable as, for example, the *S. enterica* C2–C3 acetylase gene (*wbaL*, previously known as *rfbL*) is in the gene cluster, with the acetylation step required for synthesis of the complete repeat unit [59].

11.7 Evolutionary Processes

In this section, we will discuss first the mobility of gene clusters, as there is a body of data showing that strains can gain new antigenicity by recombination in which an incoming gene cluster replaces an existing gene cluster at one of the polymorphic loci discussed above. We will then look at evolution of the gene clusters themselves, for which we have more circumstantial evidence.

11.7.1 Mobility of LPS Gene Clusters

The variety of gene clusters at the polymorphic loci discussed above may be maintained by selection. The gene clusters are able to transfer within a species by recombination involving homologous recombination in conserved flanking loci, and this mobility can be observed as discussed below. There are two hypotheses for the benefits of gene cluster transfer. One hypothesis is that the selection is simply for antigenic diversity, based on the benefits of novelty at the time of transfer, for example avoidance of immune response, predation by amoebae or resistance to bacteriophage adsorption etc., with no advantage for the specific structure. The alternative is that there are intrinsic benefits of some structures in specific circumstances (for example, a specific host species or the immune history of the host). In effect, the difference is selection against the current structure, or for the incoming structure, that drives the changes. It is likely that both apply from time to time, and the balance is not known. There would need to be selection at the time of transfer, and one can imagine that this would be very strong during a period of an immune response eliminating a colonisation, or when a bacteriophage is propagating through a clone, but the new form would then have to be competitive in future colonisation events. It certainly appears that such events are frequent enough to give higher rates of recombination at these loci, and turnover of much of the structural repertoire over the speciation timeframe.

Much of the evidence is based on the clonal nature of bacterial populations, ironically first deduced from repeated isolation of specific serotypes of *E. coli*. However, when population structures were studied using multilocus enzyme electrophoresis (MLEE) [60–63], and later sequence based typing, both based on housekeeping genes, it was found that surface antigens could vary in a clone that was recognised by stability in what is now seen as the core OS gene cluster [64]. Thus while the clonal nature of bacterial populations was confirmed, it was shown that antigen genes were particularly subject to change within a clone. This has now been observed in many studies, and is attributed to replacement of the antigen genes by homologous recombination in adjacent shared genes. An example is the diversification within *Shigella* groups 1 and 2 by change in the O-antigens, while the housekeeping genes vary very little [57].

11.7.2 Effect of Gene Cluster Transfer on Flanking DNA Sequences

By analysis of genes that flank the O-antigen loci in *E. coli*, it has been shown [65] that there is a recombination hot spot in these, with much more recombination than elsewhere, and is presumably driven by repeated selection for a new O-antigen. This was confirmed by comparison of a range of whole genome sequences [66], which showed that this was one of two regions that stood out in this regard. The recombination discussed above was by homologous recombination in shared genes that flank the area concerned, and further evidence for this can be found within these genes. For example, the *gnd* gene, a housekeeping gene located downstream of the O-antigen gene cluster in *E. coli* and *S. enterica*, has a relatively high level of variation, which is attributed to selection for O-antigen diversity [67,68]. However, the ends of the recombinant segment can be much further away. Milkman [69] observed increased diversity in groups of related strains extending up to about 160 kb in both directions from the O-antigen locus, and Touchon et al. [70], using

20 *E. coli* genome sequences, put the overall length for the region with increased recombination due the O-antigen gene cluster at 150 kb.

Recombination between subspecies is uncommon in *S. enterica* [71] and this is helpful for analysis of selection driven movement of O-antigen gene clusters across subspecies boundaries. A study of variation in the *gnd* gene gives an example of this and provides a view on a more local scale [72]. In that study, the *gnd* regions of 34 strains from the then seven subspecies of *S. enterica* were sequenced. Most gene sequences fell into groups corresponding to the subspecies, but four sequences had segments from two subspecies, which was attributed to recombination in which the segment adjacent to the O-antigen gene cluster was taken to represent the donor subspecies, and the other segment the subspecies of the recipient. In another strain of subspecies VI the *gnd* gene had the subspecies I sequence, and presumably the recombination occurred outside of the *gnd* gene. In interpreting these results we have to remember that recombination is much more common within than between subspecies. Therefore it is likely that following transfer of the gene cluster between subspecies there will have been further transfer by recombination within the recipient subspecies before we sequence a specific isolate. If there had been several such recombination events, then the boundary between the segments of donor and recipient subspecies sequence would represent the recombination site that was closest to the O-antigen gene cluster.

Another study using 13 isolates, representing *S. enterica* serotypes with rhamnose in the O-antigen, showed a similar pattern for the *rml* genes [73]. Rhamnose is present in about 25% of serotypes, but if present in both donor and recipient, the recombination junction can occur within this set of four *rml* genes. The six subspecies I strains studied had very similar sequences, while three of the four subspecies II strains shared a different sequence pattern. It appears that the *rml* genes can also be subspecies specific. The other subspecies II strains had a different sequence, probably derived from another subspecies. In each case the sequence at the 3′end, where it abuts the *ddh* genes, had a more complex pattern that was not subspecies related. This is consistent with the *gnd* data and shows that recombination events involved in movement of the O-antigen gene cluster can occur within the O-antigen gene cluster where there are shared genes. A similar study on *rml* genes in *V. cholerae* [74], in which the *rml* genes are at the other end of the gene cluster and in a different order, gave a comparable result.

11.7.3 Two Examples of Gene Cluster Mobility

Specific examples of O-antigen substitution are the origin of the O157:H7 clone from an O55:H7 clone [75], and the origin of the O139 variant of the generally O1 seventh pandemic clone of *V. cholerae* [76]. The event involving the *E. coli* O157 gene cluster was shown by genome sequencing to involve a 131 kb DNA segment [77]. These changes may well be driven by the selective advantage of a new antigen, as the antibody to O-antigen is often protective and can be a major contributor to immunity after infection on immunisation. It is interesting that the

V. cholerae O139 variant was able to infect older people that were immune to the usual O1 form [78], which would have provided a very strong selective advantage for the O139 form. It was successful for a few years and displaced the O1 form in some endemic areas, but later lost ground [78], presumably because those in endemic cholera areas became immune to both forms and some intrinsic advantage of the O1 form allowed it to come back.

11.7.3.1 Inter-Species Transfer

We have thus far considered only mobility within species, but as discussed earlier, the observation that most O-antigens found in *E. coli* have not been found in *S. enterica* and vice versa, indicates that there is turnover, and the majority have appeared in the two species since divergence. There is comparable data for other groups that have been studied, and this seems to be a general phenomenon. In some cases the new gene clusters seem to come from related species by homologous recombination. For example, the *E. coli* O8 and O9 structures have identical counterparts in *K. pneumoniae* O5 and O3 respectively, and based on sequence similarity the gene clusters in *E. coli* may arise by recombination from *K. pneumoniae* [45] where the genes are at the same locus. These are Wzm/Wzt gene clusters, which are normal in *K. pneumoniae* but rare in *E. coli*, and these are two of the four reported examples. It should be noted that this does not apply to the gene clusters shared by *E. coli* and *S. enterica*, as the sequence divergence is within the range expected if they were inherited from the common ancestor (see above).

A second means of gaining an O-antigen is by direct transfer from another species without homologous recombination. This is likely the only possible mechanism as sequences diverge. An example is *Y. kristensenii* O11, which has an O-antigen identical to that of *E. coli* O98, and gives an indication of how such a transition could occur. The *Y. kristensenii* gene cluster is at a novel locus between *aroA* and *cmk*, that has remnants of *E. coli*-like donor fragments of *galF* and *gnd* at the ends of the gene cluster [29]. It appears that the gene cluster was transferred from an *E. coli*-like organism (not *E. coli* as the sequences are too divergent for the presumed donor to have been *E. coli* itself). There were fragments of IS-elements associated with the junctions supporting the suggestion that the gene cluster was transferred from outside. Consistent with this hypothesis, the *aroA*/*cmk* genes are adjacent in the other *Yersinia* genomes. However, most of the O-antigen gene clusters are at one or two loci in any species, and the situation in *Y. kristensenii* O11 must be followed by movement of the new gene cluster to such a locus if it is to survive for long.

11.7.3.2 Variation in GC Content

Most of the genes in a bacterial genome have the same, or similar, G + C content throughout and directional mutation pressure is believed to account for this [79,80]. A major exception is that O-antigen gene clusters generally have a lower GC content, and the same applies to some other polysaccharide gene clusters, including for example the *E. coli waa* gene cluster. The simplest explanation is that these gene clusters were derived from a species with a low GC content, as is often done [81], and this may be so, but as the gene cluster sequences build up it is becoming very surprising that the divergence from normal is almost always to have a lower GC content. The GC content is also usually variable within the gene cluster and this is illustrated by *S. enterica* in Fig. 11.1. On the hypothesis that GC content reflects the source of the genes, this would indicate assembly from several sources, and this is reasonable. However, the GC content of *E. coli* O24 and O56 gene clusters is fairly consistent at about 30%, and there are some polysaccharide gene clusters that do not follow this pattern. For example, the ECA and *P. aeruginosa* A-band gene clusters have GC contents which is normal for the species involved, and the colanic acid gene cluster has a higher than normal GC content. We still have no better explanation for the low GC content, but as yet it does not appear to have been useful in identifying the source species. It certainly tells us that there is something unusual about these gene clusters. Within the gene clusters the *wzx* and *wzy* genes tend to have the lowest GC content, although the *E. coli* ECA gene cluster *wzx* and *wzy* genes have normal GC contents. In some cases the low GC content can be correlated with use of rare codons that reduce expression level of a *wzy* gene [82], but this does not seem sufficient to account for the low GC content overall.

11.7.4 Origins of New O-Antigens

We discussed above groups of O-antigen gene clusters that are related, with evidence for new clusters arising by reassortment of genes by recombination. There were examples in the *S. enterica* Gal-initiated set, and the origin of the *E. coli* O24 gene cluster from the O56 gene cluster. In those cases there was considerable diversity of strains with both ancestral and derived gene clusters, and we are not aware of any cases where one can identify pairs of closely related strains with ancestral and derived forms that could represent the before and after situations in their original context. It appears that in each case that we know of, the derived form has been distributed by gene cluster transfer as discussed in the previous section.

The examples we have considered are those that have survived selection pressures, but there are still clear indications of their history in remnant genes. What we do not see is the potential for such recombinant constructs arising. In this regard there are examples of interaction that arose during cloning experiments that are informative and we give an example below.

When the *E. coli* O4 gene cluster was cloned into *E. coli* K-12 it expressed well, but after making a deletion in the clone, the O-antigen structure changed [83]. This could not be interpreted until the *E. coli* K-12 O-antigen structure was determined and the K-12 and O4 O-antigen gene clusters were sequenced [84,85]. The novel structure is a combination of the O4 and O16 structures (Fig. 11.8), and the gene clusters show what has happened. The novel structure starts biosynthesis as an O4 structure, but this cannot be completed because the final GT gene (*wbuG*), for addition of the terminal rhamnose residue, was lost in the deletion, and the *wzy* gene for polymerisation was also affected. However, the K-12 WbbI GT adds the K-12 terminal Gal residue and polymerisation has to be by the K-12 Wzy polymerase.

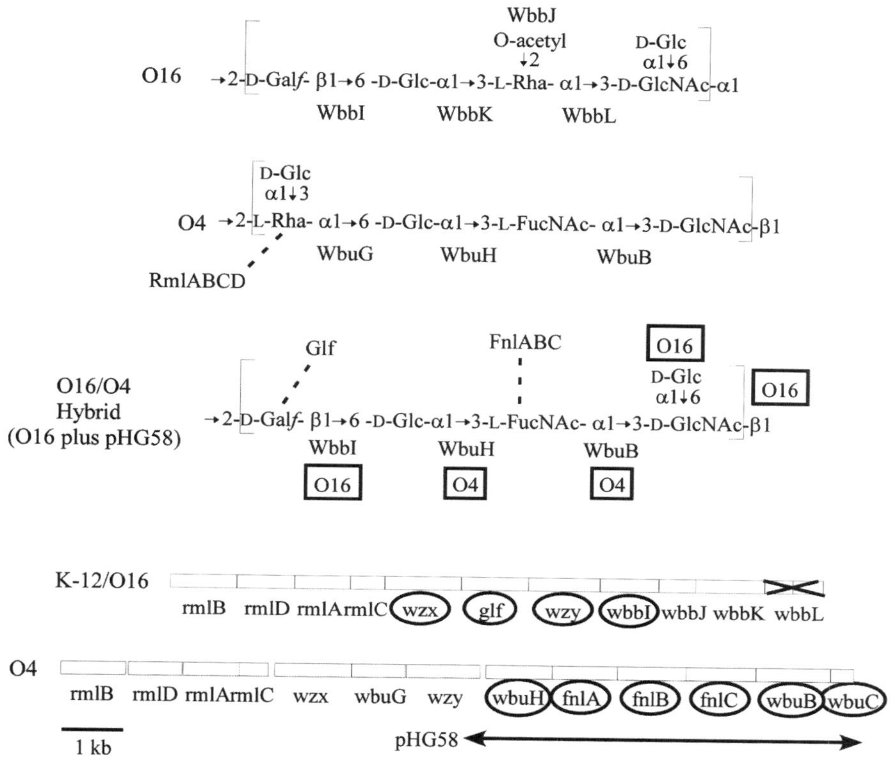

Fig. 11.8 O-antigens structures of *E. coli* O16 and O4 and a hybrid form, with corresponding gene clusters. The O16 and O4 structures are from [84] and [88] and the hybrid structure from [83]. Transferases responsible for each linkage are indicated. Some of the biosynthesis pathways for sugars in the structures are indicated with *dashed lines*. The original O-antigen source of each sugar in the hybrid structure is indicated. The K-12 and O4 gene clusters are shown below and are from [84] and [85] respectively, and the genes involved in synthesis of the hybrid structure are *circled*. The O4 genes present on plasmid pHG58 present in the K-12 strain producing the hybrid structure are shown [83]. The cross in K-12 indicates the inactivation of the rhamnosyltransferase gene *wbbL*

K-12 is rough strain with an insertion in the GT for the second sugar, so that O-unit biosynthesis usually cannot begin. The novel structure only appears when the O4 gene cluster is damaged, and there is no reason to believe that it has any selective advantages. But its appearance shows how easily such novel structures can arise. This novel structure involves nine genes for expression leaving six K-12 genes and ten O4 genes redundant. Of course the inactivation of the *wbbL* gene in the chromosome and the absence of part of the O4 gene cluster in the case of the novel O-antigen would prevent movement by homologous recombination in flanking DNA, but if the new antigen arose in nature and was beneficial, then there would be ongoing selection for the incoming segment to be incorporated close to or within the resident gene cluster, so that the whole could move as one gene cluster.

Although we do not have the intermediates, we can speculate that the *E. coli* O24 structure arose by two such events with O157 and O152 strains as donors, each starting with all or part of the two donor gene clusters starting on plasmids or bacteriophage genomes, and moving in stages to the current situation. There were probably many steps for incorporation of each segment. This must be a drawn out process and for O24 still has some way to go before the redundant DNA, including IS remnants, is removed. It is likely that the two GT substitutions arose sequentially but no intermediate have been reported. The O24 gene cluster is also unusual in that recombination with the O56 gene cluster can reassort the genes to give one or other of the new GT genes to the O56 cluster or remove it from O24 cluster. The alignment does not meet the "rule" that the genes that distinguish two gene clusters are in the center of the gene clusters. This may be a minor disadvantage as recombination is rare, but it is unusual, and perhaps would be resolved over time, though would involve substantial rearrangement. It is perhaps a reflection of the time required for perfection of a new gene cluster.

Some other examples of work in progress are seen in the gene clusters of *S. enterica* groups A, B1 and D1, all lacking a *wzy* gene, and dependant on a *wzy* gene elsewhere in the chromosome. There is another example in *S. enterica* O66, which also lacks a *wzy* gene and must have one elsewhere [17]. There is a similar situation in the *E. coli* O55 which has the *colA* and *colB* genes (previously known as *wbdJ* and *wbdK*) for the last two steps in GDP-colitose synthesis located just outside the gene cluster, and so are able to move with it [86] but are apparently not part of the original gene cluster. The list is by no means exhaustive. Finally, there are cases in *Shigella sonnei* and *S. enterica* Borreze O54 where the O-antigen is on a plasmid, which may represent the first step in acquisition of a new O-antigen.

11.8 Concluding Comments

We have seen that LPS gene clusters do not follow traditional vertical inheritance, which provides homologues in related species and allows traditional approaches to the study of evolution. The LPS genes, and genes for other polymorphic surface polysaccharides such as capsules, are not the only genes subjected to extensive LGT. Indeed such genes are now quite commonplace with those involved in pathogenicity perhaps best known. What is peculiar to surface polysaccharides is the pattern of multiple forms encoded by specific gene clusters at a polymorphic common locus. This allows periodic substitution of one structure by another and such substitutions repeated many times has lead to the patterns that we observe. We are in the early stages of understanding the processes involved, and have to infer the nature of the events from a limited number of examples. However, it seems that unlike the patterns of inheritance observed for the usual mobile genetic elements, the evolution of surface polysaccharides may have been greatly influenced by selection against an existing structure, with the loss achieved by recombination at a polymorphic locus. This would allow the loss to be offset by gain of a different gene cluster and retention (in our case) of complete LPS, but with a change in specificity.

LPS is remarkably diverse, with multiple forms of O-antigen and significant variation in core OS and lipid A. Some of the gene clusters involved must be among the most polymorphic of genetic loci, and as each Gram-negative species has its own repertoire, with almost complete turnover of O-antigens at least at genus level, the total repertoire must be truly enormous.

The finding that the pathways for the sugar precursors and other LPS components are generally conserved implies that the genes for each pathway have a single origin, with the corollary that pathway genes in a new gene cluster must come from other clusters that have that pathway. This probably accounts for the pathway genes being clustered within the gene cluster (see Fig. 11.1), as that makes it more likely that they will be transferred as a set of genes, thereby providing selection for this arrangement. The extreme diversity of sequences, for Wzy and Wzx in particular, implies very long periods since gene divergence, and again has implications for the evolution of new gene clusters. For the classes of genes in a cluster it is becoming realistic to relate most to a specific function when both sequence and structure are known, but this is often difficult for the GT genes. There are some cases of related genes having similar functions and this can help in assignment, and where there are several closely related structures with associated sequences, then function can be allocated by correlation of a specific gene with a specific linkage, as was done for the GT genes in *Y. pseudotuberculosis* (see above), but this is most unusual. As a result it is often the case that specific functions can be allocated to all except the GT genes. It can be a major undertaking to assay GTs experimentally and this is impeding our understanding of the evolution of GTs, including the relationships between sequence and substrate or linkage specificity. A possible approach is to mutate each GT gene separately and determine the size of the truncated OS [87], which if the structure is known should often associate each GT gene to a linkage without direct assay, and this may provide some relief.

References

1. Raymond CK, Sims EH, Kas A, Spencer DH, Kutyavin TV, Ivey RG, Zhou Y, Kaul R, Clendenning JB, Olson MV (2002) Genetic variation at the O-antigen biosynthetic locus in *Pseudomonas aeruginosa*. J Bacteriol 184:3614–3622
2. Bogdanovich T, Carniel E, Fukushima H, Skurnik M (2003) Use of O-antigen gene cluster-specific PCRs for the identification and O-genotyping of *Yersinia pseudotuberculosis* and *Yersinia pestis*. J Clin Microbiol 41:5103–5112
3. Shimada T, Arakawa E, Itoh K, Okitsu T, Matsushima A, Asai Y, Yamai S, Nakazato T, Nair G, Albert MJ, Takeda Y (1994) Extended serotyping scheme for *Vibrio cholerae*. Curr Microbiol 28:175–178
4. Klena JD, Parker CT, Knibb K, Ibbitt JC, Devane PM, Horn ST, Miller WG, Konkel ME (2004) Differentiation of *Campylobacter coli*, *Campylobacter jejuni*, *Campylobacter lari*, and *Campylobacter upsaliensis* by a multiplex PCR developed from the nucleotide sequence of the lipid A gene *lpxA*. J Clin Microbiol 42:5549–5557

5. Dotson GD, Kaltashov IA, Cotter RJ, Raetz CR (1998) Expression cloning of a *Pseudomonas* gene encoding a hydroxydecanoyl-acyl carrier protein-dependent UDP-GlcNAc acyltransferase. J Bacteriol 180:330–337
6. Williamson JM, Anderson MS, Raetz CRH (1991) Acyl-acyl carrier protein specificity of UDP-GlcNAc acyltransferases from gram-negative bacteria: relationship to lipid A structure. J Bacteriol 173:3591–3596
7. Reeves PR, Hobbs M, Valvano M, Skurnik M, Whitfield C, Coplin D, Kido N, Klena J, Maskell D, Raetz C, Rick P (1996) Bacterial polysaccharide synthesis and gene nomenclature. Trends Microbiol 4:495–503
8. Hobbs M, Reeves PR (1994) The JUMPstart sequence: a 39 bp element common to several polysaccharide gene clusters. Mol Microbiol 12:855–856
9. Marolda CL, Valvano MA (1998) Promoter region of the *Escherichia coli* O7-specific lipopolysaccharide gene cluster: structural and functional characterization of an upstream untranslated mRNA sequence. J Bacteriol 180:3070–3079
10. Valvano MA (2003) Export of O-specific lipopolysaccharide. Front Biosci 8:S452–S471
11. Wang Q, Torzewska A, Ruan X, Wang X, Rozalski A, Shao Z, Guo X, Zhou H, Feng L, Wang L (2010) Molecular and genetic analyses of the putative *Proteus* O antigen gene locus. Appl Environ Microbiol 76:5471–5478
12. Duda KA, Duda KT, Beczala A, Kasperkiewicz K, Radziejewska-Lebrecht J, Skurnik M (2009) ECA-immunogenicity of *Proteus mirabilis* strains. Arch Immunol Ther Exp 57: 147–151
13. Xu DQ, Cisar JO, Ambulos N Jr, Burr DH, Kopecko DJ (2002) Molecular cloning and characterization of genes for *Shigella sonnei* form I O polysaccharide: proposed biosynthetic pathway and stable expression in a live *Salmonella* vaccine vector. Infect Immun 70: 4414–4423
14. Salerno A, Deletoile A, Lefevre M, Ciznar I, Krovacek K, Grimont P, Brisse S (2007) Recombining population structure of *Plesiomonas shigelloides* (*Enterobacteriaceae*) revealed by multilocus sequence typing. J Bacteriol 189:7808–7818
15. Rush JS, Alaimo C, Robbiani R, Wacker M, Waechter CJ (2010) A novel epimerase that converts GlcNAc-P-P-undecaprenol to GalNAc-P-P-undecaprenol in *Escherichia coli* O157. J Biol Chem 285:1671–1680
16. Hu B, Perepelov AV, Liu B, Shevelev SD, Guo D, Senchenkova SN, Shashkov AS, Feng L, Knirel YA, Wang L (2010) Structural and genetic evidence for the close relationship between *Escherichia coli* O71 and *Salmonella enterica* O28 O-antigens. FEMS Immunol Med Microbiol 59:161–169
17. Liu B, Perepelov AV, Li D, Senchenkova SN, Han Y, Shashkov AS, Feng L, Knirel YA, Wang L (2010) Structure of the O-antigen of *Salmonella* O66 and the genetic basis for similarity and differences between the closely related O-antigens of *Escherichia coli* O166 and *Salmonella* O66. Microbiology 156:1642–1649
18. Perepelov AV, Li D, Liu B, Senchenkova SN, Guo D, Shashkov AS, Feng L, Knirel YA, Wang L (2011) Structural and genetic characterization of the closely related O-antigens of *Escherichia coli* O85 and *Salmonella enterica* O17. Innate Immun. doi:10.1177/1753425910369270
19. Perepelov AV, Liu B, Senchenkova SN, Shevelev SD, Feng L, Shashkov AS, Wang L, Knirel YA (2010) The O-antigen of *Salmonella enterica* O13 and its relation to the O-antigen of *Escherichia coli* O127. Carbohydr Res 345:1808–1811
20. Perepelov AV, Liu B, Shevelev SD, Senchenkova SN, Shashkov AS, Feng L, Knirel YA, Wang L (2010) Relatedness of the O-polysaccharide structures of *Escherichia coli* O123 and *Salmonella enterica* O58, both containing 4,6-dideoxy-4-{*N*-[(*S*)-3-hydroxybutanoyl]-D-alanyl}amino-D-glucose; revision of the *E. coli* O123 O-polysaccharide structure. Carbohydr Res 345:825–829
21. Wang W, Perepelov AV, Feng L, Shevelev SD, Wang Q, Senchenkova SN, Han W, Li Y, Shashkov AS, Knirel YA, Reeves PR, Wang L (2007) A group of *Escherichia coli* and

Salmonella enterica O antigen lipopolysaccharides sharing a common backbone structure. Microbiology 153:2159–2167

22. Samuel G, Hogbin J-P, Wang L, Reeves PR (2004) The relationship of the *Escherichia coli* O157, O111 and O55 O-antigen gene clusters with those of *Salmonella enterica* or *Citrobacter freundii* that express identical O antigens. J Bacteriol 186:6536–6543
23. Reeves PR (1993) Evolution of *Salmonella* O antigen variation by interspecific gene transfer on a large scale. Trends Genet 9:17–22
24. Kingsley RA, Bäumler AJ (2000) Host adaptation and the emergence of infectious disease: the *Salmonella* paradigm. Mol Microbiol 36:1006–1014
25. Liu D, Verma NK, Romana LK, Reeves PR (1991) Relationships among the *rfb* regions of *Salmonella* serovars A, B, and D. J Bacteriol 173:4814–4819
26. Curd H, Liu D, Reeves PR (1998) Relationships among the O-antigen gene clusters of *Salmonella enterica* groups B, D1, D2, and D3. J Bacteriol 180:1002–1007
27. Xiang SH, Hobbs M, Reeves PR (1994) Molecular analysis of the *rfb* gene cluster of a group D2 *Salmonella enterica* strain: evidence for its origin from an insertion sequence-mediated recombination event between group E and D1 strains. J Bacteriol 176:4357–4365
28. Reeves PR, Pacinelli E, Wang L (2003) O antigen gene clusters of *Yersinia pseudotuberculosis*. In: Skurnik M, Bengoechea JA, Granfors K (eds) The genus *Yersinia*: entering the functional genomics era. Kluwer, Plenum, New York, pp 199–206
29. Cunneen MM, Reeves PR (2007) The *Yersinia kristensenii* O11 O antigen gene cluster was acquired by lateral gene transfer from an *Escherichia coli*-like ancestor and incorporated at a novel chromosomal locus. Mol Biol Evol 24:1355–1365
30. De Castro C, Kenyon JJ, Cunneen MM, Reeves PR, Molinaro A, Holst O, Skurnik M (2011) Genetic characterisation and structural analysis of the O-specific polysaccharide of *Yersinia pseudotuberculosis* serotype O:1c. Innate Immun. doi:10.1177/1753425910364425
31. Pacinelli E, Wang L, Reeves PR (2002) Relationship of *Yersinia pseudotuberculosis* O antigens IA IIA and IVB: the IIA gene cluster was derived from that of IVB. Infect Immun 70:3271–3276
32. Liu B, Knirel YA, Feng L, Perepelov AV, Senchenkova SN, Wang Q, Reeves PR, Wang L (2008) Structure and genetics of *Shigella* O antigens. FEMS Microbiol Rev 32:627–653
33. Cheng J, Wang Q, Wang W, Wang Y, Wang L, Feng L (2006) Characterization of *E. coli* O24 and O56 O antigen gene clusters reveals a complex evolutionary history of the O24 gene cluster. Curr Microbiol 53:470–476
34. Feng L, Perepelov AV, Zhao G, Shevelev SD, Wang Q, Senchenkova SN, Shashkov AS, Geng Y, Reeves PR, Knirel YA, Wang L (2007) Structural and genetic evidence that the *Escherichia coli* O148 O antigen is the precursor of the *Shigella dysenteriae* type 1 O antigen and identification of a glucosyltransferase gene. Microbiology 153:139–147
35. King JD, Kocincova D, Westman EL, Lam JS (2009) Lipopolysaccharide biosynthesis in *Pseudomonas aeruginosa*. Innate Immun 15:261–312
36. Amor K, Heinrichs DE, Frirdich E, Ziebell K, Johnson RP, Whitfield C (2000) Distribution of core oligosaccharide types in lipopolysaccharides from *Escherichia coli*. Infect Immun 68:1116–1124
37. Heinrichs DE, Yethon JA, Whitfield C (1998) Molecular basis for structural diversity in the core regions of the lipopolysaccharides of *Escherichia coli* and *Salmonella enterica*. Mol Microbiol 30:221–232
38. Keenleyside WJ, Perry MB, MacLean LL, Poppe C, Whitfield C (1994) A plasmid-encoded *rfb*O:54 gene cluster is required for biosynthesis of the O:54 antigen in *Salmonella enterica* serovar Borreze. Mol Microbiol 11:437–448
39. Keenleyside WJ, Whitefield C (1996) A novel pathway for O-polysaccharide biosynthesis in *Salmonella enterica* serovar Borreze. J Biol Chem 271:28581–28592
40. Feng L, Senchenkova SN, Yang Y, Shashkov AS, Tao J, Guo H, Cheng J, Ren Y, Knirel YA, Reeves PR, Wang L (2004) Synthesis of the heteropolysaccharide O antigen of *Escherichia*

coli O52 requires an ABC transporter: structural and genetic evidence. J Bacteriol 186: 4510–4519
41. Perepelov AV, Li D, Liu B, Senchenkova SN, Guo D, Shevelev SD, Shashkov AS, Guo X, Feng L, Knirel YA, Wang L (2009) Structural and genetic characterization of *Escherichia coli* O99 antigen. FEMS Immunol Med Microbiol 57:80–87
42. Kelly RF, Whitfield C (1996) Clonally diverse *rfb* gene clusters are involved in expression of a family of related D-galactan O antigens in *Klebsiella* species. J Bacteriol 178:5205–5214
43. Shu HY, Fung CP, Liu YM, Wu KM, Chen YT, Li LH, Liu TT, Kirby R, Tsai SF (2009) Genetic diversity of capsular polysaccharide biosynthesis in *Klebsiella pneumoniae* clinical isolates. Microbiology 155:4170–4183
44. Whitfield C (2006) Biosynthesis and assembly of capsular polysaccharides in *Escherichia coli*. Annu Rev Biochem 75:39–68
45. Sugiyama T, Kido N, Kato Y, Koide N, Yoshida T, Yokochi T (1998) Generation of *Escherichia coli* O9a serotype, a subtype of *E. coli* O9, by transfer of the *wb** gene cluster of *Klebsiella* O3 into *E. coli* via recombination. J Bacteriol 180:2775–2778
46. Rahn A, Drummelsmith J, Whitfield C (1999) Conserved organization in the cps gene clusters for expression of *Escherichia coli* group 1 K antigens: relationship to the colanic acid biosynthesis locus and the *cps* genes from *Klebsiella pneumoniae*. J Bacteriol 181:2307–2313
47. Skurnik M, Zhang L (1996) Molecular genetics and biochemistry of *Yersinia* lipopolysaccharide. APMIS 104:849–872
48. Davies RL (1990) O-serotyping of *Yersinia ruckeri* with special emphasis on European isolates. Vet Microbiol 22:299–307
49. Beynon LM, Richards JC, Perry MB (1994) The structure of the lipopolysaccharide O antigen from *Yersinia ruckeri* serotype O1. Carbohydr Res 256:303–317
50. Abeyrathne PD, Daniels C, Poon KK, Matewish MJ, Lam JS (2005) Functional characterization of WaaL, a ligase associated with linking O-antigen polysaccharide to the core of *Pseudomonas aeruginosa* lipopolysaccharide. J Bacteriol 187:3002–3012
51. Rocchetta HL, Burrows LL, Lam JS (1999) Genetics of O-antigen biosynthesis in *Pseudomonas aeruginosa*. Microbiol Mol Biol Rev 63:523–553
52. Stover CK, Pham XQ, Erwin AL, Mizoguchi SD, Warrener P, Hickey MJ, Brinkman FS, Hufnagle WO, Kowalik DJ, Lagrou M, Garber RL, Goltry L, Tolentino E, Westbrock-Wadman S, Yuan Y, Brody LL, Coulter SN, Folger KR, Kas A, Larbig K, Lim R, Smith K, Spencer D, Wong GK, Wu Z, Paulsen IT, Reizer J, Saier MH, Hancock RE, Lory S, Olson MV (2000) Complete genome sequence of *Pseudomonas aeruginosa* PAO1, an opportunistic pathogen. Nature 406:959–964
53. Mullane N, O'Gaora P, Nally JE, Iversen C, Whyte P, Wall PG, Fanning S (2008) Molecular analysis of the *Enterobacter sakazakii* O-antigen gene locus. Appl Environ Microbiol 74: 3783–3794
54. Allison GE, Verma NK (2000) Serotype-converting bacteriophages and O-antigen modification in *Shigella flexneri*. Trends Microbiol 8:17–23
55. Stagg RM, Tang SS, Carlin NI, Talukder KA, Cam PD, Verma NK (2009) A novel glucosyltransferase involved in O-antigen modification of *Shigella flexneri* serotype 1c. J Bacteriol 191:6612–6617
56. Perepelov AV, Shevelev SD, Liu B, Senchenkova SN, Shashkov AS, Feng L, Knirel YA, Wang L (2010) Structures of the O-antigens of *Escherichia coli* O13, O129, and O135 related to the O-antigens of *Shigella flexneri*. Carbohydr Res 345:1594–1599
57. Pupo GM, Lan R, Reeves PR (2000) Multiple independent origins of *Shigella* clones of *Escherichia coli* and convergent evolution of many of their characteristics. Proc Natl Acad Sci USA 97:10567–10572
58. Lindberg AA (1973) Bacteriophage receptors. Annu Rev Microbiol 27:205–241
59. Liu D, Haase AM, Lindqvist L, Lindberg AA, Reeves PR (1993) Glycosyl transferases of O-antigen biosynthesis in *Salmonella enterica*: identification and characterization of transferase genes of groups B, C2, and E1. J Bacteriol 175:3408–3413

60. Caugant DA, Levin BR, Selander RK (1984) Distribution of multilocus genotypes of *Escherichia coli* within and between host families. J Hyg 92:377–384
61. Ochman H, Whittam TS, Caugant DA, Selander RK (1983) Enzyme polymorphism and genetic population structure in *Escherichia coli* and *Shigella*. J Gen Microbiol 129:2715–2726
62. Ochman H, Selander RK (1984) Evidence for clonal population structure in *Escherichia coli*. Proc Natl Acad Sci USA 81:198–201
63. Whittam TS, Ochman H, Selander RK (1983) Multilocus genetic structure in natural populations of *Escherichia coli*. Proc Natl Acad Sci USA 80:1751–1755
64. Selander RK, Smith NH (1990) Molecular population genetics of *Salmonella*. Rev Med Microbiol 1:219–228
65. Milkman R (1997) Recombination and population structure in *Escherichia coli*. Genetics 146: 745–750
66. Tenaillon O, Skurnik D, Picard B, Denamur E (2010) The population genetics of commensal *Escherichia coli*. Nat Rev Microbiol 8:207–217
67. Nelson K, Selander RK (1994) Intergenic transfer and recombination of the 6-phosphogluconate dehydrogenase gene (*gnd*) in enteric bacteria. Proc Natl Acad Sci USA 91:10227–10231
68. Mandel M, Higa A (1970) Calcium dependent bacteriophage DNA infection. J Mol Biol 53: 159–162
69. Milkman R, Jaeger E, McBride RD (2003) Molecular evolution of the *Escherichia coli* chromosome. Genetics 163:475–483
70. Touchon M, Hoede C, Tenaillon O, Barbe V, Baeriswyl S, Bidet P, Bingen E, Bonacorsi S, Bouchier C, Bouvet O, Calteau A, Chiapello H, Clermont O, Cruveiller S, Danchin A, Diard M, Dossat C, Karoui ME, Frapy E, Garry L, Ghigo JM, Gilles AM, Johnson J, Le Bouguenec C, Lescat M, Mangenot S, Martinez-Jehanne V, Matic I, Nassif X, Oztas S, Petit MA, Pichon C, Rouy Z, Ruf CS, Schneider D, Tourret J, Vacherie B, Vallenet D, Medigue C, Rocha EP, Denamur E (2009) Organised genome dynamics in the *Escherichia coli* species results in highly diverse adaptive paths. PLoS Genet 5:e1000344
71. McQuiston JR, Herrera-Leon S, Wertheim BC, Doyle J, Fields PI, Tauxe RV, Logsdon JM Jr (2008) Molecular phylogeny of the *Salmonellae*: relationships among *Salmonella* species and subspecies determined from four housekeeping genes and evidence of lateral gene transfer events. J Bacteriol 190:7060–7067
72. Thampapillai G, Lan R, Reeves PR (1994) Molecular evolution in the *gnd* locus of *Salmonella enterica*. Mol Biol Evol 11:813–828
73. Li Q, Reeves PR (2000) Genetic variation of dTDP-L-rhamnose pathway genes in *Salmonella enterica*. Microbiology 146:2291–2307
74. Li Q, Hobbs M, Reeves PR (2003) The variation of dTDP-L-rhamnose pathway genes in *Vibrio cholerae*. Microbiology 149:2463–2474
75. Whittam TS, Wolfe ML, Wachsmuth IK, Ørskov F, Ørskov I, Wilson RA (1993) Clonal relationships among *Escherichia coli* strains that cause hemorrhagic colitis and infantile diarrhea. Infect Immun 61:1619–1629
76. Hall RH, Khambaty FM, Kothary MH, Keasler SP, Tall BD (1994) *Vibrio cholerae* non-O1 serogroup associated with cholera gravis genetically and physiologically resembles O1 E1 Tor cholera strains. Infect Immun 62:3859–3863
77. Zhou Z, Li X, Liu B, Beutin L, Xu J, Ren Y, Feng L, Lan R, Reeves PR, Wang L (2010) Derivation of *Escherichia coli* O157:H7 from its O55:H7 Precursor. PLoS ONE 5:e8700
78. Albert MJ (1996) Epidemiology and molecular biology of *Vibrio cholerae* O139 Bengal. Indian J Med Res 104:14–27
79. Sueoka N (1992) Directional mutation pressure, selective constraints, and genetic equilibria. J Mol Evol 34:95–114
80. Sueoka N (1988) Directional mutation pressure and neutral molecular evolution. Proc Natl Acad Sci USA 85:2653–2657
81. Reeves PR, Wang L (2002) Genomic organization of LPS-specific loci. Curr Top Microbiol Immunol 264:109–135

82. Daniels C, Vindurampulle C, Morona R (1998) Overexpression and topology of the *Shigella flexneri* O-antigen polymerase (Rfc/Wzy). Mol Microbiol 28:1211–1222
83. Kogan G, Haraguchi G, Hull SI, Hull RA, Shashkov AS, Jann B, Jann K (1993) Structural analysis of O4-reactive polysaccharides from recombinant *Escherichia coli:* changes in the O-specific polysaccharide induced by cloning of the *rfb* genes. Eur J Biochem 214:259–265
84. Stevenson G, Neal B, Liu D, Hobbs M, Packer NH, Batley M, Redmond JW, Lindquist L, Reeves PR (1994) Structure of the O-antigen of *E. coli* K-12 and the sequence of its *rfb* gene cluster. J Bacteriol 176:4144–4156
85. D'Souza JM, Samuel G, Reeves PR (2005) Evolutionary origins and sequence of the *Escherichia coli* O4 O-antigen gene cluster. FEMS Microbiol Lett 244:27–32
86. Wang L, Huskic S, Cisterne A, Rothemund D, Reeves PR (2002) The O antigen gene cluster of *Escherichia coli* O55:H7 and identification of a new UDP-GlcNAc C4 epimerase gene. J Bacteriol 184:2620–2625
87. Stevenson G, Diekelmann M, Reeves PR (2008) Determination of glycosyltransferase specificities for the *Escherichia coli* O111 O antigen by a generic approach. Appl Environ Microbiol 74:1294–1298
88. Haraguchi GE, Zähringer U, Jann B, Jann K, Hull RA, Hull SI (1991) Genetic characterisation of the O4 polysaccharide gene cluster from *Escherichia coli*. Microb Pathogen 10:351–361
89. Skurnik M, Peippo A, Ervelä E (2000) Characterization of the O-antigen gene clusters of *Yersinia pseudotuberculosis* and the cryptic O-antigen gene cluster of *Yersinia pestis* shows that the plague bacillus is most closely related to and has evolved from *Y. pseudotuberculosis* serotype O:1b. Mol Microbiol 37:316–330

The Molecular Basis of Lipid A and Toll-Like Receptor 4 Interactions 12

Georgina L. Hold and Clare E. Bryant

12.1 Introduction

In 1989 Charles Janeway proposed the concept of 'Immune recognition'. He suggested that '*a critical issue for future study is the analysis of microbial signals that induce second signalling capacity in antigen-presenting cells, and the receptors on antigen presenting cells that detect these microbial signals. ... I term these receptors pattern recognition receptors (PRRs)*' [1]. From the early 1990s genetic studies in *Drosophila* and vertebrates led to the identification of the membrane associated Toll and Toll-like receptors (TLRs), the canonical PRRs predicted by Janeway. This was followed by the identification of different families of cytosolic PRRs including Retinoic acid-Inducible Gene-Like Receptors, Nucleotide Oligomerisation Domain-like receptors and Absent in melanoma-like receptors all of which play a role in pathogen recognition.

There are ten TLRs encoded in the human genome, which bind directly to conserved structures associated with pathogenic microorganisms. These molecules are sometimes termed pathogen associated molecular patterns (PAMPs). TLR PAMPs can be divided broadly into two groups, microbial lipids such as lipopolysaccharide (LPS) and non-self nucleic acids from bacteria, viruses and other pathogenic microorganisms. Microbial lipids are recognized by TLRs 1, 2, 4 and 6, bacterial flagellin by TLR5, RNA by TLRs 3, 7 and 8 and DNA by TLR9. These PAMPs bind and activate TLRs by promoting the dimerization of two receptor

G.L. Hold
Division of Applied Medicine, University of Aberdeen, Institute of Medical Sciences, Foresterhill, Aberdeen, UK AB25 2ZD
e-mail: g.l.hold@abdn.ac.uk

C.E. Bryant (✉)
Department of Veterinary Medicine, University of Cambridge, Madingley Road, Cambridge, UK CB3 0ES
e-mail: ceb27@cam.ac.uk

Y.A. Knirel and M.A. Valvano (eds.), *Bacterial Lipopolysaccharides*,
DOI 10.1007/978-3-7091-0733-1_12, © Springer-Verlag/Wien 2011

ectodomains causing the cytosolic Toll/IL1 domains (TIR) to associate, creating a signal induced scaffold for the assembly of a post receptor complex [2].

Lipid A was identified in the 1960s as the hydrophobic moiety of lipopolysaccharide (LPS), but its bioactivity was not determined for another 20 years [3]. How the host detects lipid A remained unknown until the discovery of TLRs, specifically TLR4, in the late 1990s [4, 5]. Most Gram-negative bacteria synthesize lipid A molecules resembling those made by *Escherichia coli* [6]. The characteristic structural features of *E. coli* lipid A are a 1,4'-bisphosphorylated β-(1 → 6)-linked D-glucosamine disaccharide backbone that is hexaacylated with acyl chains of length C_{12}–C_{14} which are distributed asymmetrically [7]. This lipid A structure – often referred to as canonical lipid A structure – is required to trigger full TLR4 activation in human cells. More details on lipid A structure, biosynthesis and genetics are discussed in Chaps. 1 and 6.

The molecular basis for how *E. coli* lipid A interacts with TLR4 has been resolved by the solving of high-resolution ligand-bound crystal structures. The molecular basis for how other lipid A structures drive the formation of active TLR4 signalling complexes is less clear. Various natural lipid A forms and analogues may be useful therapeutic compounds (see Chap. 13) and therefore a clear understanding of the molecular basis for ligand-receptor interaction is required. In this chapter we review the molecular requirements for lipid recognition by TLR4, how these differ when other lipid A structures interact with TLR4 and how single nucleotide polymorphisms (SNPs) in TLR4 may influence lipid A signalling.

12.2 The Essential Protein Components of the Lipid A Receptor Complex

There are various proteins involved in lipid recognition by mammals. Lipid A is extracted from blood and solubilised by a serum protein, LPS binding protein (LBP) [8]. LBP then transfers the lipid A to a lymphocyte extrinsic membrane protein, CD14 [9]. The major role for CD14 is to enhance the sensitivity of cells to lipid A, reducing the binding affinity to picomolar concentrations [10]. Mice without CD14, despite expressing the other proteins required for lipid A recognition, are resistant to endotoxic shock [11]. The identity of the PRR that signals in response to lipid A, TLR4 was not established until 1999. TLR4 was one of the first TLRs to be identified [4] and mapping studies in the LPS-resistant mouse strains, C3H/HeJ and C57BL/10ScCr, identified the *Tlr4* gene as the LPS receptor [5, 12]. In C3H/HeJ mice the *Tlr4* gene has a single A to C point mutation, resulting in a pro712his substitution in the TIR of TLR4 [5, 12] conferring dominant-negative activity on TLR4 [13]. This role for TLR4 in LPS signalling was confirmed when $TLR4^{-/-}$ mice were shown to be hyporesponsive to LPS [14]. Genetic and biochemical studies showed that expression of TLR4 alone does not confer responsiveness of cells to LPS and that an additional co-receptor protein, MD-2, is required [15].

Like the TLR4 knockouts, mice lacking MD-2 do not respond to LPS [16] and are resistant to endotoxic shock.

12.3 The Molecular Basis for Lipid Recognition by TLR4/MD-2

The primary sequence of TLR4 has the characteristic features of a class 1 transmembrane receptor, with an extracellular domain, a single membrane spanning helix and a globular cytoplasmic domain, the TIR. The extracellular domain contains a number of leucine-rich repeat (LRRs) motifs and an associated capping structure [17]. The LRR framework of TLRs provides binding specificity for a wide range of biological molecules for example lipid A interacts with the LRRs of TLR4 [18]. As with other class 1 receptors TLR4 signal transduction is expected to require stimulus-induced dimerization of two receptor molecules [19].

An important step forward in our understanding of lipid A recognition by the TLR4/MD-2 heterodimer came with the discovery that MD-2 belongs to a small family of lipid binding proteins [20]. These proteins fold into a β-sandwich structure similar to that formed by the immunoglobulin domains of antibody molecules. Modelling studies suggested that binding to LPS is mediated by the intercalation of the lipid A acyl chains into the hydrophobic core of the β-sandwich [21, 22]. This model was confirmed by structural analyses of MD-2 alone bound to lipid IV_A and a TLR4/MD-2 heterodimer in complex with the antagonist Eritoran [23, 24]. The four acyl chains of Eritoran and lipid IV_A are fully accommodated within the MD-2 structure and occupy approximately 90% of the solvent-accessible volume of the pocket (Fig. 12.1). Two of the acyl chains are in the fully extended conformation within the binding pocket, but two of them are bent in the middle. The di-glucosamine backbones are fully exposed to solvent [25]. In both the lipid IV_A and Eritoran MD-2 structures the ligand does not induce a conformational change in the protein, but this is an expected result as these molecules are antagonists. The TLR4 ectodomain forms a rigid curved solenoid with the MD-2 bound at two conserved sites in the N-terminal part of TLR4. The entrance to the LPS binding pocket is on the opposite side of MD-2 that is exposed to solvent.

These structures did not show how hexaacyl lipid A induced the dimerization of the TLR4/MD-2 heterodimer structure to initiate signal transduction. Functional studies indicated that mutation of the MD-2 residues phenylalanine-126 and histidine-155 abolished the ability of the TLR4/MD-2 to form the activated heterotetramer, suggesting that these residues form part of the dimerization interface [24]. A study that investigated why lipid IV_A is an agonist for horse TLR4 but an antagonist in human [26] showed that the species differences arise due to sequence variations in both MD-2 and in TLR4. A short region in the horse MD-2 (residues 57–107) is sufficient, when transplanted into a human MD-2 framework, to confer responsiveness to lipid IV_A. Equally, a region in the C-terminus of the horse TLR4, between LRRs 14 and 18, is essential for signalling activity in response to lipid IV_A. Strikingly, a horse TLR4 mutant with a single change of arginine-285 to glycine (the residue found at the equivalent position in the human protein) lost the ability to

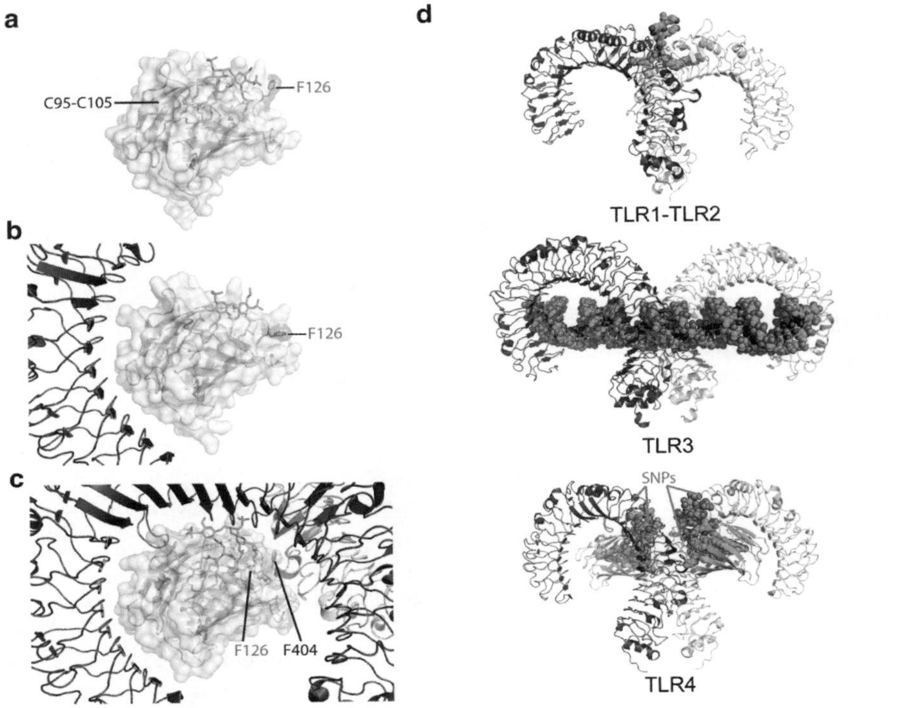

Fig. 12.1 MD-2 and TLR crystal structures. (**a**) MD-2 with lipid IV_A (PDF accession code 2E59). MD-2 is shown in pale cyan as a semi transparent molecular surface and lipid IV_A in stick representation (atoms of *carbon in green*, *oxygen in red*, *phosphate in orange* and *nitrogen in blue*). The two phosphorylated glucosamine head groups of lipid IV_A are solvent exposed. The acyl chains are buried in the hydrophobic cavity of MD-2. (**b**) The Eritoran (E5564) TLR4/MD-2 complex (PDB accession code 2Z65) shows a similar ligand binding mode compared to lipid IV_A. The position of phenylalanine-126 (F126) is indicated. (**c**) A close-up view of the active TLR4/MD-2/LPS complex (PDB accession code 3FXI) reveals a different binding mode for LPS that involves MD-2 and two TLR4 molecules. MD-2 proteins are in the same orientation in (**a**), (**b**) and (**c**). The positions of phenylalanine-126 (F126) and phenylalanine 404 (F404) are indicated. (**d**) Active TLR complexes. Top panel: TLR1/TLR2 in complex with a triacylated lipopeptide (PDB accession code 2Z7X). TLR1 in purple, TLR2 in yellow cartoon and the ligand in sphere representation. Middle panel: TLR3 in complex with double stranded RNA complex (PDB accession code 3CIY). TLR3 ectodomains are in yellow and purple cartoons. Double stranded RNA ligand is shown in sphere representation. Bottom panel: TLR4/MD-2/LPS complex. TLR4 ectodomains are in yellow and purple, MD-2 in pale cyan and pale blue and LPS in sphere representation. (Figure reproduced from Bryant et al.: Nat. Rev. Microbiol. **8**: 8–14 (2010))

signal in response to lipid IV_A. On the basis of these results a structural model for the activated, heterotetrameric complex of TLR4 and MD-2 was generated by protein-protein docking methods. This indicated that there are two regions of contact between the TLR4/MD-2 heterodimers. The first interface involves the MD-2 residue phenylalanine-126 and a hydrophobic region of the TLR4 ectodomain at leucine-444. The second site forms on the lateral surfaces of the two ectodomain molecules, centred on LRR 16, the region identified as important for signalling in the mutagenesis study. This arrangement of the two TLR4/MD-2 heterodimers brings the C-terminal, juxtamembrane sequences of the ectodomains into close proximity and has a similar 'M' shaped conformation to that of TLR1 and TLR2 complexed by triacylated lipid and the TLR3 ectodomain bound to double stranded RNA (Fig. 12.1) [27, 28].

This model of activation has now been confirmed by the elucidation of a high-resolution structure for TLR4/MD-2 bound to hexaacyl lipid A and mutagenesis studies of the predicted interface residues (Fig. 12.1C) [29, 30]. When accommodating lipid A with more than four acyl chains there is no conformational change in MD-2 and this causes the acyl chain at the 2 position to become exposed on the surface of the MD-2 structure. Together with MD-2 F126, this creates a hydrophobic patch that forms the dimerization interface with TLR4, an interaction involving leucine-444 and in addition the nearby residues phenylalanine-440 and phenylalanine-463. This forces the glucosamine backbone upwards, repositioning the phosphate groups to contact positively charged residues of both TLR4 subunits. The second dimerization interface is also as predicted by the model with the lateral surfaces of two ectodomains creating an extensive area of protein-protein interaction centred on LRR16 [31, 32]. A key point identified by Lee and colleagues is that their structure only identifies how *E. coli* lipid A binds to TLR4/MD-2. What is not known is whether similar active complexes are formed when lipid A structures from other bacterial species bind TLR4/MD-2. The interaction of mouse TLR4/MD-2 with lipid IV_A highlights lysine-367 and arginine-434 as being important for inducing agonist activity [33] emphasising that ligand-specific pharmacological studies will be critical in developing novel LPS derivatives for therapeutic use.

12.4 Host Recognition of Lipid A and LPS Structures from Bacterial Species Other than *E. coli*

Variations in lipid A composition such as altering the degree of phosphorylation, changing the length or nature of the acyl chains or changes in the number of phosphate substituents results in alteration of the biological effects observed (see Chaps. 1 and 6). These chemical changes impact on the lipid A 3D structure whereby conical lipid A shapes (increased hydrophobic region compared to hydrophilic) can form cylindrical shapes (equal hydrophobic and hydrophilic regions) [34]. These variations are unique to bacterial strains but can be regulated by environmental factors. This often results in the presence of heterogeneous lipid A species that can facilitate bacterial colonization through antimicrobial resistance in the form of cationic antimicrobial peptides (CAMPs) as well as sub-optimal TLR4 endotoxin recognition.

Gram-negative bacteria have evolved mechanisms to modify the structure of lipid A in different environments. Different lipid A structures have different binding affinities to TLR4 complex constituents and this can lead to altered host recognition. For example, lipid A of *Helicobacter pylori* in comparison to that of

E. coli, contains a major monophosphorylated tetraacylated lipid A species which is not thought to bind efficiently to the TLR4/MD-2 receptor complex due to the lack of a phosphate group in the lipid A anchor. This loss of one or both phosphate groups from the lipid A anchor is seen in a number of pathogenic organisms including *Porphyromonas gingivalis, Bacteroides fragilis, Francisella tularensis* and also *Leptospira interogans* although it was first demonstrated in the nitrogen-fixing endosymbiont *Rhizobium leguminosarum*. Certain *H. pylori* strains (especially clinical isolates) and also *L. interrogans* strains also possess a minor bisphosphorylated hexaacylated lipid A form that binds to the TLR4/MD-2 receptor complex. In the case of *H. pylori* this hexaacylated species binds with around 20-fold less efficiency compared to *E. coli* lipid A. *H. pylori* and *P. gingivalis* LPS are 10–100 fold less efficient at binding to LBP compared to canonical LPS [35]. This reduced ability is also seen with LPS-LBP mediated transfer to CD14 and also subsequent presentation to the TLR4/MD-2 complex [36].

Lipid A modifications were first identified in *Salmonella enterica* serovar Typhimurium (*S.* Typhimurium) but have subsequently been identified in other human pathogens as well as commensals. There is evidence of environmentally regulated enzymes which have been acquired through horizontal gene transfer suggesting a positive selection for altered lipid A structures, although other data supports these enzymes being ancient mechanisms of lipid A regulation [37]. *S.* Typhimurium regulates its lipid A structure through a two-component system in order to promote intracellular survival and resistance to CAMPs. It does this through the environmental sensor-kinase transcriptional regulatory system PhoP-PhoQ which is activated following phagocytosis by macrophages [38]. PhoP-PhoQ regulated lipid A is poorly recognized by human TLR4 and less stimulatory than non-regulated lipid A which allows the bacterium to evade immune detection and thus replicate more effectively within macrophages [39].

During the natural transmission cycle of *Yersinia pestis*, the causative agent of plague, the bacterium has to survive at varying temperatures. The shifts in temperature (20–25°C within the flea host and up to 37°C within the mammalian host) has been shown to modulate the degree of acylation of lipid A and ultimately its immunostimulatory activity making it a weaker inducer of TLR4-mediated innate immunity responses at 37°C [40]. Research has shown that at ambient temperature *Y. pestis* lipid A is a heterogeneous mix of tri-, tetra-, penta- and hexa-acylated species however at 37°C, the hexaacylated lipid A along with most of the pentaacylated lipid A is absent. Interestingly there is also a host specific difference in biologic activity with human macrophages showing a stronger difference in biological activity compared to murine macrophages [41]. In the mammalian host a less acylated and less glycosylated LPS is produced compared with that produced by the bacteria in the arthropod host. This supports the hypothesis that *Y. pestis* LPS is altered in order to down-regulate the host immune response during the early stages of infection in order to allow the organism to establish itself [42]. Changes in lipid A have also been induced in response to antibiotic treatment, which also has strong implication for the pathogenesis of infection [43]. In contrast another

Yersinia species, *Y. tuberculosis*, which is transmitted via the faecal-oral route synthesizes more biologically active lipid A at 37°C.

Pseudomonas aeruginosa is an example of a bacterium that can alter its lipid A in response to long-term colonisation of a host. *P. aeruginosa* is an opportunistic pathogen which is a major cause of infection in individuals with host-defense defects, including cystic fibrosis (CF) [44]. *P. aeruginosa* can change the isoelectric properties of its lipid A which makes it less immunogenic. It does this by adding new acyl chains and lengthening existing chains through the course of CF disease. Analysis of lipid A from *P. aeruginosa* strains from CF patients with mild lung disease shows a hexaacylated lipid A species which is similar to environmental isolates although the acyl chain lengths have been extended compared to environmental strains. Analysis of *P. aeruginosa* strains from CF patients with severe lung disease indicates ~48% of isolates are now heptaacylated (seven acyl chains) and this form is associated with β-lactam antibiotic resistance [45].

MD-2 plays an important role in discriminating different lipid A structures. Underacylated LPS, for example tetra- and penta-acylated lipid A species show decreased MD-2 binding and subsequent biological activity compared to the more favoured hexaacylated lipid A forms. These findings are, however, host specific in that lipid A species in one host can act as an agonist but as an antagonist in another. For example, lipid IV_A is an antagonist in human cells but an agonist in horse, murine and hamster cells. Similarly, LPS and lipid A from *Rhodobacter sphaeroides* are antagonists in human and murine cells, while the lipid A is an agonist in hamster and horse [46]. The range of lipid A structures and the ability of MD-2 to bind these structures has potentially important clinical relevance in terms of infection risk (both acute and chronic) as well as more fundamental health relevance relating to gut homeostasis and immune surveillance.

Various ligands other than lipid A have also been identified as TLR4 agonists. These include endogenous ligands, other pathogen-derived ligands (such as *Streptococcus pneumoniae* pneumolysin, *Chlamydia pneumoniae* HSP60, mouse mammary tumour virus envelope proteins, and respiratory syncytial virus fusion protein), the house dust mite protein Derp2 [47, 48] and plant ligands (paclitaxel) [49]. A complication when considering protein ligands as TLR4 activators is the potential for contamination of recombinant protein with LPS leading to false annotation of a protein as a ligand for this receptor complex. These technical issues are now largely resolved and a number of these proteins remain candidate agonists for TLR4/MD-2.

12.5 The Signalling Pathways Activated by LPS Stimulation of TLR4/MD-2

Binding of *E. coli* lipid A, a full agonist at TLR4/MD-2, induces the formation of the activated TLR4-MD-2-LPS complex. This activated receptor complex undergoes conformation changes, with lipid A interacting predominantly with the large hydrophobic MD-2 pocket [30]. Activation of TLR4 induced by ligand binding involves dimerization or oligomerization of receptor chains [50]. This in turn results in protein conformational changes in the receptor and homodimerization of two receptor TIR domains [19]. Fluorescence resonance energy transfer microscopy showed that the TLR9 TIR domains undergo a large positional change on ligand binding [51] therefore it is likely this also occurs with other TLRs on dimerization. The association of the receptor TIR domains would provide a scaffold to recruit the specific adaptor proteins to form a post-receptor signalling complex.

This leads to recruitment of adaptor proteins to the TIR domains (Fig. 12.2) and activates downstream signalling pathways (Fig. 12.3). The four primary adaptor proteins are myeloid differentiation primary-response protein 88 (MyD88), MyD88-adapter-like protein (Mal), TIR-domain-containing adaptor protein-inducing IFN-β (TRIF) and TRIF-related adaptor molecule (TRAM). A fifth adaptor, SARM, acts as a negative regulator of TRAM/TRIF signalling [52]. Mal and TRAM are thought to engage directly with the receptor and to act as 'bridging adaptors' for the recruitment of MyD88 and TRIF respectively.

Mutagenesis and molecular modelling studies suggest that ligand-induced dimerization of the TLR4 extracellular domains leads to concerted protein conformational changes that in turn lead to self-association or rearrangement of the receptor TIR domains thereby creating a new molecular surface for the recruitment of signalling adaptor proteins (see Fig. 12.2) [53]. This model predicts that Mal and TRAM bind to the same region in the TLR4 dimer interface thus explaining why

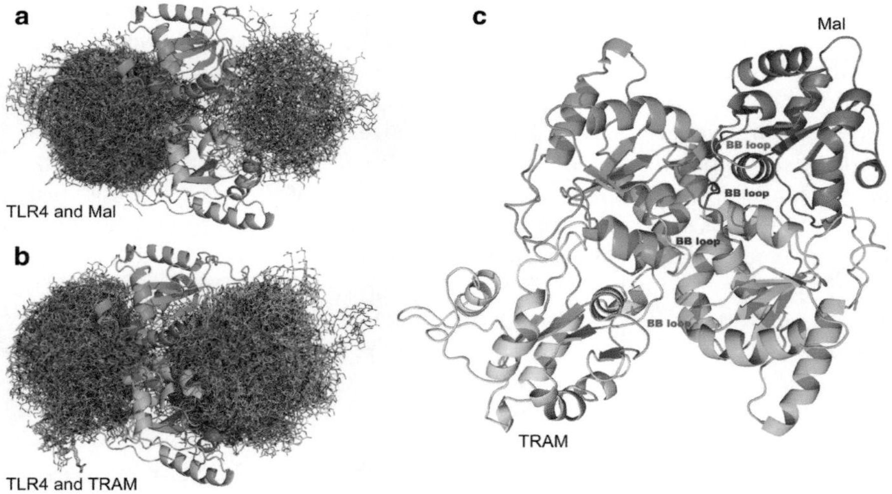

Fig. 12.2 Docking model of Mal and TRAM binding at the TLR4 homodimer interface. The TLR4 protomers, represented as ribbon diagrams are in green and cyan. Docked Mal and TRAM are represented as stick models and the 50 best docking solutions generated by GRAMM for either Mal (**a**) or TRAM (**b**) have been superimposed upon one another. (**c**) High resolution complex of TLR4 dimer (*green and cyan*), Mal (*pink*) and TRAM (*yellow*). The position of each BB loop is labeled. (Figure reproduced under Open Access License from Nunez Miguel et al. [53])

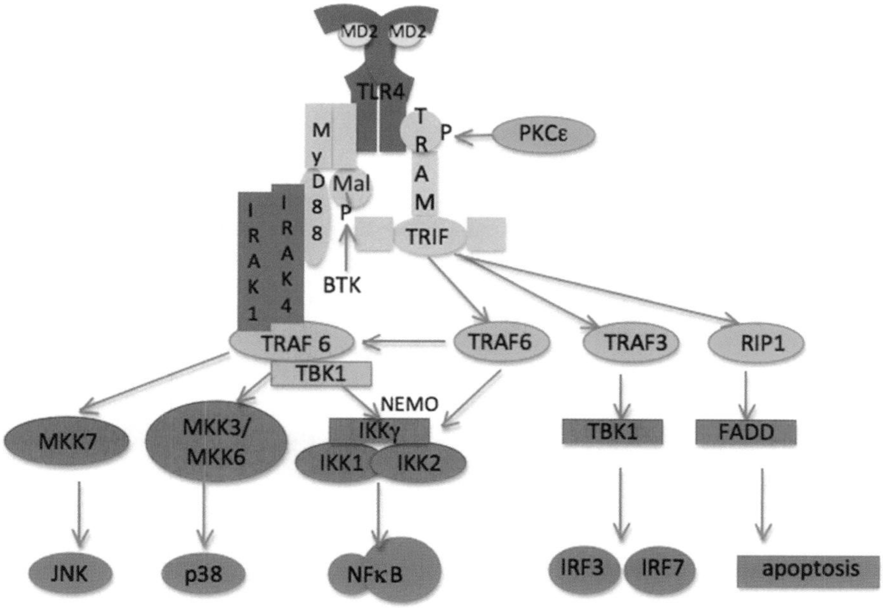

Fig. 12.3 Signalling pathways activated by lipid A binding to TLR4/MD-2. Once activated by lipid A TLR4 recruits the specific TIR adapters MyD88, Mal, TRIF and TRAM resulting in the recruitment and activation of the IRAKs and TRAF6. This leads to the activation of NEMO and the subsequent phosphorylation and degradation of IκB the inhibitor of NFκB, rendering NF-κB free to translocate from the cytosol to the nucleus and activate κB-dependent genes. The recruitment of TRAM and TRIF to TLR4 activates the non-canonical IKKs, TBK1 and IKKε, resulting in the dimerization and activation of IRF3 and the transcription of IFNβ and IFN-inducible genes

cell permeable blocking peptides compete out both Mal and TRAM directed responses simultaneously [54]. The model does not, however, resolve the question of whether a single activated receptor dimer can stimulate both the Mal and TRAM directed pathways simultaneously or whether adaptor engagement is mutually exclusive (something that would require positive cooperativity). Each activated receptor will have two symmetry related adaptor binding sites so in principle either hypothesis is feasible.

MyD88 dependent signalling activates IKK (inhibitory κB kinase) and mitogen-activated protein kinase (MAPK) pathways. The IKK pathway, through rapid activation of the transcription factor nuclear factor κB (NF-κB), controls expression of proinflammatory cytokines and other immune related genes including phenotypic activation of antigen presenting cells to enhance T cell priming [55]. Stimulation of MAPK signalling activates another transcription factor AP-1 that also plays a role in proinflammatory cytokine expression [49, 56]. NF-κB and MAPK activation also occurs through the TRAM/TRIF signalling pathway however the kinetics of activation are delayed [57]. TRAM/TRIF signalling also activates the transcription factor interferon regulatory factor 3 (IRF3) and induces the expression Type I interferon [58, 59].

It was thought initially that both Mal/MyD88 and TRAM/TRIF signalling would be fully and equally activated by ligand engagement at TLR4/MD-2, but it now appears that the balance between these two pathways in not fixed. It is now thought that lipid heterogeneity, probably along with host genetic variation in the LPS receptor proteins, may influence the balance of signalling between the two pathways. For example, *Neisseria meningitidis* LPS will induce activation of both Mal/MyD88 and TRAM/TRIF signalling pathways. *Vibrio cholera,* in contrast, selectively activates MyD88 directed signalling whereas monophosphoryl lipid A, a low-toxicity derivative of *E. coli* LPS with useful adjuvant properties, attenuates MyD88 signalling resulting in an overall bias towards TRAM/TRIF signalling [60, 61].

12.6 Polymorphisms in TLR4/MD-2 and CD14

Many mutations in the genes forming the human LPS receptor-signalling complex have been identified, but their functional and pathological significance are only now emerging. Early work identified mutations in the *Tlr4* gene, corresponding to aspartic acid-299-glycine (Asp299Gly) and threonine-399-isoleucine (Thr399Ile), which were shown to associate with hypo-responsiveness to inhaled LPS [31]. Expression of these mutants *in vitro* show reduced activation in response to LPS [62]. The TLR4 single nucleotide polymorphisms (SNPs; Asp299Gly and Thr399Ile) that reduce LPS responsiveness are located far away from the *N*-terminal TLR4 binding site for MD-2 and the TLR4 dimerization interfaces [31, 32]. The molecular mechanism underlying LPS hyporesponsiveness of these SNPs therefore remains unclear but these mutations may alter cell surface expression of TLR4 [63], affect some part of the lipid A binding process or alter the conformational changes that occur during ligand-induced signal transduction. Formation of the active TLR4/MD-2 complex changes the curvature of the TLR4 solenoid [30] and any mutations that increase rigidity could have a large effect on the kinetics of receptor activation.

TLR4 may be associated with a number of diseases and SNPs in the receptor may play an important role in disease susceptibility. Researchers have studied whether the Asp299Gly and Thr399Ile polymorphisms are associated with infectious diseases, but much of the data are conflicting [64, 65]. This may be because most of the studies consider either the Asp299Gly or the Thr399Ile polymorphism, but neglect the fact that these polymorphisms also exist in a cosegregated (Asp299Gly/Thr399Ile) way which implies that there are 4 haplotypes, namely wild type/wild type, Asp299Gly/wild type, Thr399Ile/wild type, and Asp299Gly/Thr399Ile [65]. Recent data suggests only the Asp299Gly haplotype differs in phenotype from wild type TLR4, with LPS-stimulated blood samples from this population of people showing an increased, rather than a blunted, TNF – α response [66].

Studies in knockout mice have implicated a role for TLR4 in protection against endotoxaemia [67], but suggest an increased susceptibility of TLR4 mutant mice to systemic Gram-negative infections, such as *S.* Typhimurium [68, 69]. This is because activation of TLR4 is required for protective immunity against infections, but also mediates the hyper-inflammatory effects of systemic endotoxin and sepsis. Many papers have shown a role for TLR4 in mouse infection models with several Gram-negative pathogens including *Neisseria meningitidis*, *E. coli*, *Haemophilus influenzae*, *Klebsiella pneumoniae* and *Brucella abortus* [70]. Mouse models have also shown that TLR4 is important for infection with other pathogens devoid of lipid A including *Streptococcus pneumonaie* and *Mycobacterium tuberculosis* [70]. TLR4 has also been linked to several viral infections including Respiratory Syncytial Virus [71], the murine retroviruses Mouse Mammary Tumour virus and Murine Leukaemia virus [72] as well as the picornavirus Coxsackievirus B4 [73].

The role of TLR4 in human infectious disease is emerging. There is now a vast literature on polymorphisms in TLR4 and their association with many infectious diseases including sepsis, Gram negative infections, other bacterial diseases including tuberculosis, malaria, Respiratory Syncytial virus infection and *Candida* infections [65]. Much of the data is conflicting probably because of the different populations of people studied and the variety of haplotypes involved. The strongest association of TLR4 polymorphisms with an infectious disease is with Respiratory Syncytial virus infection where high risk infants heterozygous for Asp299Gly and Thr399Ile polymorphisms showed an increased susceptibility to infection [74]. There is also increased risk of severe malaria in Ghanian children with the Asp299Gly and Thr399Ile variants [75] although there is no association between the Asp299Gly and tuberculosis in a Gambian population [76]. An association of the Asp299Gly haplotype was found only in the group of patients with septic shock, whereas the Asp299Gly/Thr399Ile haplotype was found equally in both patients and controls although patients with this genotype had a higher prevalence of Gram-negative infections [77].

TLR4 and TLR4 receptor polymorphisms have been implicated in a number of non-infectious diseases. There is genetic data emerging to support the association of TLR4 with several of these diseases. This is perhaps unsurprising given the range of endogenous ligands identified for TLR4 and the number of diseases (cancer, atherosclerosis and autoimmune conditions) that are now believed to have an inflammatory aetiology. The Asp299Gly SNP is implicated in gastric cancer, atherosclerosis, sepsis, asthma and a G11481C mutation has been linked to prostate cancer [78]. A number of studies also suggest a possible role for TLR4 in cardiovascular disease [79, 80], inflammatory bowel disease [81], Alzheimer's disease [82], rheumatoid arthritis [83], renal disease [84], obesity and both type I and type II diabetes [85], but whether the genetic evidence will support the disease tissue and model observations remains to be proven. In mouse models, for example, inhibition of TLR4 is beneficial in rheumatoid arthritis (RA) [86] and patients with the disease carrying the Asp299Gly have altered macrophage responses to LPS [87], but there is no clear genetic link between TLR4 and RA as yet. Whether or not endogenous ligands or the involvement of infectious disease is the underlying cause of the involvement of TLR4 in the susceptibility to these diseases remains to be clarified.

A mechanistic link between TLR4 and allergic asthma has recently been shown. Derp2 is a key allergen from the house dust mite that it is structurally similar to

MD-2. It increases the sensitivity of TLR4/MD-2 signalling to TLR4 and may therefore deliver LPS to TLR4 in airways to provoke inflammation. This might be a common allergic mechanism since several airborne allergens are lipid-binding proteins that might act analogously [48]. A TLR4 antagonist blocked the induction of asthma in a mouse model by house dust mite extract [47]. This suggests TLR4 may be a good therapeutic target for allergic airway disease.

There have been far fewer studies in the MD-2 knock out mice compared to the TLR4 knockout mice. Three human polymorphisms have been described, Thr35Ala [88], Gly56Arg mutation [89] and a C1625G in the MD-2 promoter [90]. The promoter polymorphism may be linked to increased susceptibility to complications such as organ dysfunction and sepsis after major trauma [90] whereas the other polymorphisms show no disease association as yet.

A SNP in the 5′ genomic region of CD14 at position 159 [91] is associated with infectious diseases, asthma and allergy [92]. Other diseases have also been linked to this polymorphism from cardiovascular disease to autoimmunity and from infections to malignancies [91]. Presumably diseases linked to CD14 will overlap those linked to TLR4 suggesting that therapeutic intervention with either CD14 or TLR4 should benefit patients who have genetic susceptibilities in either of these genes.

Conclusions

Major advances in our understanding of how lipid A is recognised by, and stimulates inflammation in, the host have occurred in the last 10 years. The solving of the lipid A-TLR4/MD-2 receptor complex will now allow the design of a range of potentially useful therapeutic compounds. Antagonists are already in well advanced clinical trials for the treatment of sepsis and the low activity agonist monophosphoryl lipid A is now being used as a vaccine adjuvant (see Chap. 13 for a full discussion). The increasing body of evidence suggesting a link between TLR4 and many diseases suggests that developing therapeutic compounds that target this receptor may generate some tremendously important drugs. It will be critical, though, to understand precisely how these compounds behave pharmacologically for the development of safe and effective drugs.

Acknowledgements Research by the authors was funded by the Horserace Betting Levy Board, the Wellcome Trust, and The Medical Research Council.

References

1. Janeway CA Jr (1989) Approaching the asymptote? Evolution and revolution in immunology. Cold Spring Harb Symp Quant Biol 54(Pt 1):1–13
2. O'Neill LA, Bryant CE, Doyle SL (2009) Therapeutic targeting of Toll-like receptors for infectious and inflammatory diseases and cancer. Pharmacol Rev 61:177–197
3. Loppnow H, Durrbaum I, Brade H, Dinarello CA, Kusumoto S, Rietschel ET, Flad HD (1990) Lipid A, the immunostimulatory principle of lipopolysaccharides? Adv Exp Med Biol 256:561–566

4. Medzhitov R, Preston-Hurlburt P, Janeway CA Jr (1997) A human homologue of the *Drosophila* Toll protein signals activation of adaptive immunity. Nature 388:394–397
5. Poltorak A, He X, Smirnova I, Liu MY, Van Huffel C, Du X, Birdwell D, Alejos E, Silva M, Galanos C, Freudenberg M, Ricciardi-Castagnoli P, Layton B, Beutler B (1998) Defective LPS signaling in C3H/HeJ and C57BL/10ScCr mice: mutations in Tlr4 gene. Science 282:2085–2088
6. Raetz CR, Reynolds CM, Trent MS, Bishop RE (2007) Lipid A modification systems in Gram-negative bacteria. Annu Rev Biochem 76:295–329
7. Erwin AL, Munford RS (1990) Deacylation of structurally diverse lipopolysaccharides by human acyloxyacyl hydrolase. J Biol Chem 265:16444–16449
8. Schumann RR, Leong SR, Flaggs GW, Gray PW, Wright SD, Mathison JC, Tobias PS, Ulevitch RJ (1990) Structure and function of lipopolysaccharide binding protein. Science 249:1429–1431
9. Wright SD, Ramos RA, Tobias PS, Ulevitch RJ, Mathison JC (1990) CD14, a receptor for complexes of lipopolysaccharide (LPS) and LPS binding protein. Science 249:1431–1433
10. Gioannini TL, Teghanemt A, Zhang D, Coussens NP, Dockstader W, Ramaswamy S, Weiss JP (2004) Isolation of an endotoxin-MD-2 complex that produces Toll-like receptor 4-dependent cell activation at picomolar concentrations. Proc Natl Acad Sci USA 101:4186–4191
11. Haziot A, Ferrero E, Kontgen F, Hijiya N, Yamamoto S, Silver J, Stewart CL, Goyert SM (1996) Resistance to endotoxin shock and reduced dissemination of Gram-negative bacteria in CD14-deficient mice. Immunity 4:407–414
12. Qureshi ST, Lariviere L, Leveque G, Clermont S, Moore KJ, Gros P, Malo D (1999) Endotoxin-tolerant mice have mutations in Toll-like receptor 4 (Tlr4). J Exp Med 189:615–625
13. Vogel SN, Johnson D, Perera PY, Medvedev A, Lariviere L, Qureshi ST, Malo D (1999) Cutting edge: functional characterization of the effect of the C3H/HeJ defect in mice that lack an Lps^n gene: in vivo evidence for a dominant negative mutation. J Immunol 162:5666–5670
14. Akira S, Uematsu S, Takeuchi O (2006) Pathogen recognition and innate immunity. Cell 124:783–801
15. Shimazu R, Akashi S, Ogata H, Nagai Y, Fukudome K, Miyake K, Kimoto M (1999) MD-2, a molecule that confers lipopolysaccharide responsiveness on Toll-like receptor 4. J Exp Med 189:1777–1782
16. Nagai Y, Akashi S, Nagafuku M, Ogata M, Iwakura Y, Akira S, Kitamura T, Kosugi A, Kimoto M, Miyake K (2002) Essential role of MD-2 in LPS responsiveness and TLR4 distribution. Nat Immunol 3:667–672
17. Gay NJ, Gangloff M (2008) Structure of toll-like receptors. Handb Exp Pharmacol 183:181–200
18. Alder MN, Rogozin IB, Iyer LM, Glazko GV, Cooper MD, Pancer Z (2005) Diversity and function of adaptive immune receptors in a jawless vertebrate. Science 310:1970–1973
19. Gay NJ, Gangloff M, Weber AN (2006) Toll-like receptors as molecular switches. Nat Rev Immunol 6:693–698
20. Inohara N, Nunez G (2002) ML – a conserved domain involved in innate immunity and lipid metabolism. Trends Biochem Sci 27:219–221
21. Gangloff M, Gay NJ (2004) MD-2: the Toll 'gatekeeper' in endotoxin signalling. Trends Biochem Sci 29:294–300
22. Gruber A, Mancek M, Wagner H, Kirschning CJ, Jerala R (2004) Structural model of MD-2 and functional role of its basic amino acid clusters involved in cellular lipopolysaccharide recognition. J Biol Chem 279:28475–28482
23. Ohto U, Fukase K, Miyake K, Satow Y (2007) Crystal structures of human MD-2 and its complex with antiendotoxic lipid IVa. Science 316:1632–1634
24. Kim HM, Park BS, Kim JI, Kim SE, Lee J, Oh SC, Enkhbayar P, Matsushima N, Lee H, Yoo OJ, Lee JO (2007) Crystal structure of the TLR4-MD-2 complex with bound endotoxin antagonist eritoran. Cell 130:906–917

25. Jin MS, Lee JO (2008) Structures of the toll-like receptor family and its ligand complexes. Immunity 29:182–191
26. Walsh C, Gangloff M, Monie T, Smyth T, Wei B, McKinley TJ, Maskell D, Gay N, Bryant C (2008) Elucidation of the MD-2/TLR4 interface required for signaling by lipid IVa. J Immunol 181:1245–1254
27. Liu L, Botos I, Wang Y, Leonard JN, Shiloach J, Segal DM, Davies DR (2008) Structural basis of toll-like receptor 3 signaling with double-stranded RNA. Science 320:379–381
28. Jin MS, Kim SE, Heo JY, Lee ME, Kim HM, Paik SG, Lee H, Lee JO (2007) Crystal structure of the TLR1-TLR2 heterodimer induced by binding of a tri-acylated lipopeptide. Cell 130:1071–1082
29. Nu AR, Va LJ, Oblak A, Pristov-Ek P, Gioannini TL, Weiss JP, Jerala R (2009) Essential roles of hydrophobic residues in both MD-2 and toll-like receptor 4 in activation by endotoxin. J Biol Chem 284:15052–15060
30. Park BS, Song D-H, Kim H, Choi B-S, Lee H, Lee J-O (2009) The structural basis of lipopolysaccharide recognition by the TLR4-MD-2 complex. Nature 458:1191–1195
31. Arbour NC, Lorenz E, Schutte BC, Zabner J, Kline JN, Jones M, Frees K, Watt JL, Schwartz DA (2000) TLR4 mutations are associated with endotoxin hyporesponsiveness in humans. Nat Genet 25:187–191
32. Rallabhandi P, Bell J, Boukhvalova MS, Medvedev A, Lorenz E, Arditi M, Hemming VG, Blanco JC, Segal DM, Vogel SN (2006) Analysis of TLR4 polymorphic variants: new insights into TLR4/MD-2/CD14 stoichiometry, structure, and signaling. J Immunol 177:322–332
33. Meng J, Lien E, Golenbock DT (2010) MD-2-mediated ionic interactions between lipid A and TLR4 are essential for receptor activation. J Biol Chem 285:8695–8702
34. Schromm AB, Brandenburg K, Loppnow H, Moran AP, Koch MH, Rietschel ET, Seydel U (2000) Biological activities of lipopolysaccharides are determined by the shape of their lipid A portion. Eur J Biochem 267:2008–2013
35. Cunningham MD, Seachord C, Ratcliffe K, Bainbridge B, Aruffo A, Darveau RP (1996) *Helicobacter pylori* and *Porphyromonas gingivalis* lipopolysaccharides are poorly transferred to recombinant soluble CD14. Infect Immun 64:3601–3608
36. Delude RL, Savedra R Jr, Zhao H, Thieringer R, Yamamoto S, Fenton MJ, Golenbock DT (1995) CD14 enhances cellular responses to endotoxin without imparting ligand-specific recognition. Proc Natl Acad Sci USA 92:9288–9292
37. Hyytiainen H, Sjoblom S, Palomaki T, Tuikkala A, Tapio Palva E (2003) The PmrA-PmrB two-component system responding to acidic pH and iron controls virulence in the plant pathogen *Erwinia carotovora* ssp. *carotovora*. Mol Microbiol 50:795–807
38. Ernst RK, Guina T, Miller SI (2001) *Salmonella typhimurium* outer membrane remodeling: role in resistance to host innate immunity. Microbes Infect 3:1327–1334
39. Kawasaki K, Ernst RK, Miller SI (2004) 3-O-Deacylation of lipid A by PagL, a PhoP/PhoQ-regulated deacylase of *Salmonella typhimurium*, modulates signaling through Toll-like receptor 4. J Biol Chem 279:20044–20048
40. Telepnev MV, Klimpel GR, Haithcoat J, Knirel YA, Anisimov AP, Motin VL (2009) Tetraacylated lipopolysaccharide of *Yersinia pestis* can inhibit multiple Toll-like receptor-mediated signaling pathways in human dendritic cells. J Infect Dis 200:1694–1702
41. Kawahara K, Tsukano H, Watanabe H, Lindner B, Matsuura M (2002) Modification of the structure and activity of lipid A in *Yersinia pestis* lipopolysaccharide by growth temperature. Infect Immun 70:4092–4098
42. Brubaker RR (2003) Interleukin-10 and inhibition of innate immunity to Yersiniae: roles of Yops and LcrV (V antigen). Infect Immun 71:3673–3681
43. Knirel YA, Lindner B, Vinogradov EV, Kocharova NA, Senchenkova SN, Shaikhutdinova RZ, Dentovskaya SV, Fursova NK, Bakhteeva IV, Titareva GM, Balakhonov SV, Holst O, Gremyakova TA, Pier GB, Anisimov AP (2005) Temperature-dependent variations and intraspecies diversity of the structure of the lipopolysaccharide of *Yersinia pestis*. Biochemistry 44:1731–1743

44. Lyczak JB, Cannon CL, Pier GB (2000) Establishment of *Pseudomonas aeruginosa* infection: lessons from a versatile opportunist. Microbes Infect 2:1051–1060
45. Ernst RK, Moskowitz SM, Emerson JC, Kraig GM, Adams KN, Harvey MD, Ramsey B, Speert DP, Burns JL, Miller SI (2007) Unique lipid a modifications in *Pseudomonas aeruginosa* isolated from the airways of patients with cystic fibrosis. J Infect Dis 196:1088–1092
46. Lohmann KL, Vandenplas ML, Barton MH, Bryant CE, Moore JN (2007) The equine TLR4/MD-2 complex mediates recognition of lipopolysaccharide from *Rhodobacter sphaeroides* as an agonist. J Endotoxin Res 13:235–242
47. Hammad H, Chieppa M, Perros F, Willart MA, Germain RN, Lambrecht BN (2009) House dust mite allergen induces asthma via Toll-like receptor 4 triggering of airway structural cells. Nat Med 15:410–416
48. Trompette A, Divanovic S, Visintin A, Blanchard C, Hegde RS, Madan R, Thorne PS, Wills-Karp M, Gioannini TL, Weiss JP, Karp CL (2009) Allergenicity resulting from functional mimicry of a Toll-like receptor complex protein. Nature 457:585–588
49. Gay NJ, Gangloff M (2007) Structure and function of Toll receptors and their ligands. Annu Rev Biochem 76:141–165
50. Saitoh S, Akashi S, Yamada T, Tanimura N, Kobayashi M, Konno K, Matsumoto F, Fukase K, Kusumoto S, Nagai Y, Kusumoto Y, Kosugi A, Miyake K (2004) Lipid A antagonist, lipid IVa, is distinct from lipid A in interaction with Toll-like receptor 4 (TLR4)-MD-2 and ligand-induced TLR4 oligomerization. Int Immunol 16:961–969
51. Latz E, Verma A, Visintin A, Gong M, Sirois CM, Klein DC, Monks BG, McKnight CJ, Lamphier MS, Duprex WP, Espevik T, Golenbock DT (2007) Ligand-induced conformational changes allosterically activate Toll-like receptor 9. Nat Immunol 8:772–779
52. O'Neill LA, Bowie AG (2007) The family of five: TIR-domain-containing adaptors in Toll-like receptor signalling. Nat Rev Immunol 7:353–364
53. Nunez Miguel R, Wong J, Westoll JF, Brooks HJ, O'Neill LA, Gay NJ, Bryant CE, Monie TP (2007) A dimer of the Toll-like receptor 4 cytoplasmic domain provides a specific scaffold for the recruitment of signalling adaptor proteins. PLoS ONE 2:e788
54. Toshchakov VY, Vogel SN (2007) Cell-penetrating TIR BB loop decoy peptides a novel class of TLR signaling inhibitors and a tool to study topology of TIR-TIR interactions. Expert Opin Biol Ther 7:1035–1050
55. Iwasaki A, Medzhitov R (2004) Toll-like receptor control of the adaptive immune responses. Nat Immunol 5:987–995
56. McAleer JP, Rossi RJ, Vella AT (2009) Lipopolysaccharide potentiates effector T cell accumulation into nonlymphoid tissues through TRIF. J Immunol 182:5322–5330
57. Kawai T, Adachi O, Ogawa T, Takeda K, Akira S (1999) Unresponsiveness of MyD88-deficient mice to endotoxin. Immunity 11:115–122
58. Yamamoto M, Sato S, Hemmi H, Hoshino K, Kaisho T, Sanjo H, Takeuchi O, Sugiyama M, Okabe M, Takeda K, Akira S (2003) Role of adaptor TRIF in the MyD88-independent toll-like receptor signaling pathway. Science 301:640–643
59. Covert MW, Leung TH, Gaston JE, Baltimore D (2005) Achieving stability of lipopoly-saccharide-induced NF-κB activation. Science 309:1854–1857
60. Mata-Haro V, Cekic C, Martin M, Chilton PM, Casella CR, Mitchell TC (2007) The vaccine adjuvant monophosphoryl lipid A as a TRIF-biased agonist of TLR4. Science 316:1628–1632
61. Miyake K (2007) Innate immune sensing of pathogens and danger signals by cell surface Toll-like receptors. Semin Immunol 19:3–10
62. Rallabhandi P, Awomoyi A, Thomas KE, Phalipon A, Fujimoto Y, Fukase K, Kusumoto S, Qureshi N, Sztein MB, Vogel SN (2008) Differential activation of human TLR4 by *Escherichia coli* and *Shigella flexneri* 2a lipopolysaccharide: combined effects of lipid A acylation state and TLR4 polymorphisms on signaling. J Immunol 180:1139–1147
63. Prohinar P, Rallabhandi P, Weiss JP, Gioannini TL (2010) Expression of functional D299G. T399I polymorphic variant of TLR4 depends more on coexpression of MD-2 than does wild-type TLR4. J Immunol 184:4362–4367

64. Schroder NW, Schumann RR (2005) Single nucleotide polymorphisms of Toll-like receptors and susceptibility to infectious disease. Lancet Infect Dis 5:156–164
65. Ferwerda B, McCall MB, Verheijen K, Kullberg BJ, van der Ven AJ, van der Meer JW, Netea MG (2008) Functional consequences of toll-like receptor 4 polymorphisms. Mol Med 14:346–352
66. Ferwerda B, McCall MB, Alonso S, Giamarellos-Bourboulis EJ, Mouktaroudi M, Izagirre N, Syafruddin D, Kibiki G, Cristea T, Hijmans A, Hamann L, Israel S, ElGhazali G, Troye-Blomberg M, Kumpf O, Maiga B, Dolo A, Doumbo O, Hermsen CC, Stalenhoef AF, van Crevel R, Brunner HG, Oh DY, Schumann RR, de la Rua C, Sauerwein R, Kullberg BJ, van der Ven AJ, van der Meer JW, Netea MG (2007) TLR4 polymorphisms, infectious diseases, and evolutionary pressure during migration of modern humans. Proc Natl Acad Sci USA 104:16645–16650
67. Hoshino K, Takeuchi O, Kawai T, Sanjo H, Ogawa T, Takeda Y, Takeda K, Akira S (1999) Cutting edge: Toll-like receptor 4 (TLR4)-deficient mice are hyporesponsive to lipopolysaccharide: evidence for TLR4 as the *Lps* gene product. J Immunol 162:3749–3752
68. O'Brien AD, Rosenstreich DL, Scher I, Campbell GH, MacDermott RP, Formal SB (1980) Genetic control of susceptibility to *Salmonella typhimurium* in mice: role of the LPS gene. J Immunol 124:20–24
69. Weiss DS, Raupach B, Takeda K, Akira S, Zychlinsky A (2004) Toll-like receptors are temporally involved in host defense. J Immunol 172:4463–4469
70. Schnare M, Rollinghoff M, Qureshi S (2006) Toll-like receptors: sentinels of host defence against bacterial infection. Int Arch Allergy Immunol 139:75–85
71. Kurt-Jones EA, Popova L, Kwinn L, Haynes LM, Jones LP, Tripp RA, Walsh EE, Freeman MW, Golenbock DT, Anderson LJ, Finberg RW (2000) Pattern recognition receptors TLR4 and CD14 mediate response to respiratory syncytial virus. Nat Immunol 1:398–401
72. Rassa JC, Meyers JL, Zhang Y, Kudaravalli R, Ross SR (2002) Murine retroviruses activate B cells via interaction with toll-like receptor 4. Proc Natl Acad Sci USA 99:2281–2286
73. Triantafilou K, Triantafilou M (2004) Coxsackievirus B4-induced cytokine production in pancreatic cells is mediated through toll-like receptor 4. J Virol 78:11313–11320
74. Awomoyi AA, Rallabhandi P, Pollin TI, Lorenz E, Sztein MB, Boukhvalova MS, Hemming VG, Blanco JC, Vogel SN (2007) Association of TLR4 polymorphisms with symptomatic respiratory syncytial virus infection in high-risk infants and young children. J Immunol 179:3171–3177
75. Mockenhaupt FP, Cramer JP, Hamann L, Stegemann MS, Eckert J, Oh NR, Otchwemah RN, Dietz E, Ehrhardt S, Schroder NW, Bienzle U, Schumann RR (2006) Toll-like receptor (TLR) polymorphisms in African children: common TLR-4 variants predispose to severe malaria. J Commun Dis 38:230–245
76. Newport MJ, Allen A, Awomoyi AA, Dunstan SJ, McKinney E, Marchant A, Sirugo G (2004) The toll-like receptor 4 Asp299Gly variant: no influence on LPS responsiveness or susceptibility to pulmonary tuberculosis in The Gambia. Tuberculosis (Edinb) 84:347–352
77. Lorenz E, Mira JP, Frees KL, Schwartz DA (2002) Relevance of mutations in the TLR4 receptor in patients with gram-negative septic shock. Arch Intern Med 162:1028–1032
78. El-Omar EM, Ng MT, Hold GL (2008) Polymorphisms in Toll-like receptor genes and risk of cancer. Oncogene 27:244–252
79. Satoh M, Ishikawa Y, Minami Y, Takahashi Y, Nakamura M (2008) Role of Toll like receptor signaling pathway in ischemic coronary artery disease. Front Biosci 13:6708–6715
80. Frantz S, Ertl G, Bauersachs J (2007) Mechanisms of disease: Toll-like receptors in cardiovascular disease. Nat Clin Pract Cardiovasc Med 4:444–454
81. Fukata M, Abreu MT (2007) TLR4 signalling in the intestine in health and disease. Biochem Soc Trans 35:1473–1478
82. Balistreri CR, Grimaldi MP, Chiappelli M, Licastro F, Castiglia L, Listi F, Vasto S, Lio D, Caruso C, Candore G (2008) Association between the polymorphisms of TLR4 and CD14 genes and Alzheimer's disease. Curr Pharm Des 14:2672–2677
83. van den Berg WB, van Lent PL, Joosten LA, Abdollahi-Roodsaz S, Koenders MI (2007) Amplifying elements of arthritis and joint destruction. Ann Rheum Dis 66(Suppl 3):iii45–iii48

84. Anders HJ, Banas B, Schlondorff D (2004) Signaling danger: toll-like receptors and their potential roles in kidney disease. J Am Soc Nephrol 15:854–867
85. Kim JK (2006) Fat uses a TOLL-road to connect inflammation and diabetes. Cell Metab 4:417–419
86. Eder W, Klimecki W, Yu L, von Mutius E, Riedler J, Braun-Fahrlander C, Nowak D, Martinez FD (2004) Toll-like receptor 2 as a major gene for asthma in children of European farmers. J Allergy Clin Immunol 113:482–488
87. Roelofs MF, Wenink MH, Toonen EJ, Coenen MJ, Joosten LA, van den Berg WB, van Riel PL, Radstake TR (2008) The functional variant (Asp299Gly) of toll-like receptor 4 (TLR4) influences TLR4-mediated cytokine production in rheumatoid arthritis. J Rheumatol 35:558–561
88. Hamann L, Kumpf O, Muller M, Visintin A, Eckert J, Schlag PM, Schumann RR (2004) A coding mutation within the first exon of the human MD-2 gene results in decreased lipopolysaccharide-induced signaling. Genes Immun 5:283–288
89. Vasl J, Prohinar P, Gioannini TL, Weiss JP, Jerala R (2008) Functional activity of MD-2 polymorphic variant is significantly different in soluble and TLR4-bound forms: decreased endotoxin binding by G56R MD-2 and its rescue by TLR4 ectodomain. J Immunol 180:6107–6115
90. Gu W, Shan YA, Zhou J, Jiang DP, Zhang L, Du DY, Wang ZG, Jiang JX (2007) Functional significance of gene polymorphisms in the promoter of myeloid differentiation-2. Ann Surg 246:151–158
91. Martinez FD (2007) CD14, endotoxin, and asthma risk: actions and interactions. Proc Am Thorac Soc 4:221–225
92. Wiertsema SP, Khoo SK, Baynam G, Veenhoven RH, Laing IA, Zielhuis GA, Rijkers GT, Goldblatt J, Lesouef PN, Sanders EA (2006) Association of CD14 promoter polymorphism with otitis media and pneumococcal vaccine responses. Clin Vaccine Immunol 13:892–897

Modulation of Lipopolysaccharide Signalling Through TLR4 Agonists and Antagonists

13

Francesco Peri, Matteo Piazza, Valentina Calabrese, and Roberto Cighetti

13.1 Introduction

Innate immunity is the first line of defense against invading pathogens in mammals. Innate immune cells evolved the Toll family of receptors (Toll-like receptors, [TLRs]) that detect microbial components with high sensitivity and selectivity [1]. Ten functional TLRs in humans comprise two distinct subpopulations with regard to subcellular distribution and ligand specificity. Indeed, current knowledge supports the view, perhaps oversimplified, that cell surface TLRs evolved as receptors recognizing external molecules of bacterial, fungal and protozoan pathogens, while internally expressed TLRs detect virus-derived nucleic acids in intracellular compartments. TLRs tend to form noncovalent dimers in the absence of ligand. TLR2 preferentially forms heterodimers with TLR1 or TLR6, while other TLRs associate as homodimers. Cell surface TLR dimers including TLR2-TLR1, TLR2-TLR6, as well as the $[TLR4\text{-}MD\text{-}2]_2$ tetramer, recognize microbial membrane lipids and lipopeptides, while TLR5 dimers recognize bacterial flagellin protein. In contrast, TLR3, 7, 8, 9 reside in intracellular organelles (endosomes/lysosomes) and recognize nucleic acids [2, 3].

Another unique role for cell surface TLRs is sensing tissue damage by responding to endogenous ligands released from broken tissues or necrotic cells [1]. For example, in the central nervous system, TLRs are expressed predominantly on microglia cells [4] and trigger an immune response not only to infectious agents, but also to neuronal

F. Peri (✉) • M. Piazza • V. Calabrese • R. Cighetti
Dipartimento di Biotecnologie e Bioscienze, Università di Milano-Bicocca, Piazza della Scienza 2, 20126 Milan, Italy
e-mail: francesco.peri@unimib.it; matteo.piazza1@unimib.it; valentina.calabrese@unimib.it; cighetti.roberto@hotmail.it

Y.A. Knirel and M.A. Valvano (eds.), *Bacterial Lipopolysaccharides*,
DOI 10.1007/978-3-7091-0733-1_13, © Springer-Verlag/Wien 2011

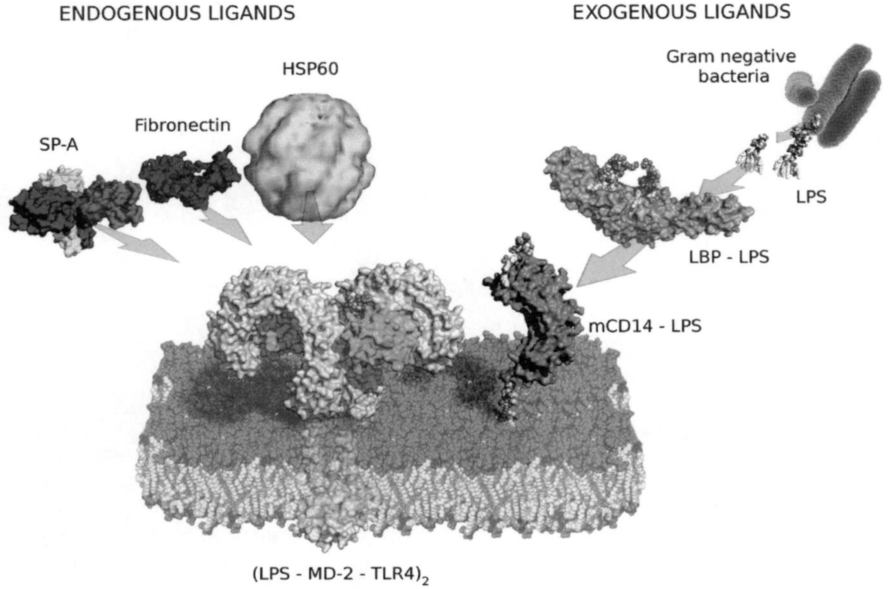

Fig. 13.1 TLR4 is a receptor for endogenous and exogenous ligands

injury [5, 6]. Increasing evidence suggests that release of endogenous pain mediators activates TLRs, resulting in tissue-injury associated inflammation and nerve-injury-associated neuropathic pain, which are both debilitating pathologies with no pharmacological treatment [7, 8].

This chapter focuses on TLR4 activation and signalling (Fig. 13.1) with special emphasis on molecules that interfere with these processes. Natural and synthetic compounds with activity as TLR4 agonists or antagonists will be reviewed and structure-function relationships will be analyzed for different classes of compounds.

13.2 Role of TLR4 and Therapeutic Application of TLR4-Active Compounds

Among TLRs, TLR4 selectively recognizes bacterial lipopolysaccharide (LPS) or endotoxin [9–11], which results in a rapid elicitation of pro-inflammatory processes. LPS induces inflammatory responses by the coordinate and sequential action of four principal LPS-binding proteins: the LPS binding protein (LBP), the cluster differentiation antigen 14 (CD14), the myeloid differentiation protein-2 (MD-2) and the Toll-like receptor 4 (TLR4). This recognition process starts with the binding of LBP to LPS aggregates, in the form of micelles or membrane blebs, and ends up with the formation of the activated TLR4-MD-2-LPS complex that has a pivotal role in initiating the inflammatory cascade (Fig. 13.1). LPS recognition by TLR4 mediates rapid cytokine production and the recruitment of inflammatory cells to the site of infection [12]. TLR4 activation also regulates adaptive immune responses [13].

TLR4-active compounds with pharmacological potential can be grouped depending on the molecular target: LBP, CD14, MD-2 and/or MD-2-TLR4 complex, and TLR4 itself. There are also TLR4 modulators whose exact molecular target still remains unidentified, which will be discussed in the final section of this chapter. The most obvious and important use of TLR4 antagonists is to inhibit LPS-triggered TLR4 activation, which plays a central role in Gram-negative bacterial sepsis and septic shock [14, 15]. Molecules with endotoxin antagonistic activity that can inhibit TLR4 activation are potential lead compounds for antisepsis drug development. Also, vaccine adjuvants played a central role in the clinical development of TLR4 agonists [16]. Another important clinical application of TLR4 (and TLR9) agonists is in the development of agents against allergic rhinitis and asthma [17–19].

Recent research uncovered a previously unappreciated role for TLRs and in particular TLR4 in some types of inflammatory disorders that are not caused by viruses and microbes ("sterile" inflammation) but rather due to tissue injury [8, 16]. Indeed, TLR4 also responds to endogenous ligands such as heat shock proteins, extracellular matrix degradation products, high-mobility group box 1 protein (HMGB-1), β-defensin, surfactant protein A, and minimally modified low-density lipoprotein (LDL) (Fig. 13.1) [1]. Recent work demonstrates that in the absence of LPS or an exogenous pathogen, TLR4 is a key microglial receptor for the initiation of nerve injury-induced behavioural hypersensitivity [20]. TLR4 might therefore be a key contributor in microglial activation connecting innate immunity with the initiation of neuropathic pain [21]. The contribution of CD14 in TLR4-dependent neuropathic pain has also been described [22]. Therefore, TLR4 antagonists can be used for the treatment of neuropathic pain and chronic pain [23].

13.3 Lipopolysaccharide Recognition

LPS, found at the surface of Gram-negative bacteria, is the primary exogenous ligand recognized by TLR4. LPS is the major structural component of the outer leaflet of the bacterial outer membrane, and its structure, biogenesis and function has been reviewed in other chapters of this book. From the three components of the LPS molecule, O-antigen, core oligosaccharide (OS), and lipid A, the latter is responsible for TLR4-dependent proinflammatory activity, and it is also known as endotoxin [24]. Lipid A consists of β-(1' → 6)-GlcNAc disaccharide bisphosphorylated at 1 and 4′ positions and with linear or branched acyl chains attached through amide or ester bonds, respectively to 2, 2′, 3 and 3′ positions (Fig. 13.2) [25] (see also Chaps. 1 and 6).

The relationship between LPS structure and biological function has been investigated for decades. The three-dimensional structure of the lipid A moiety

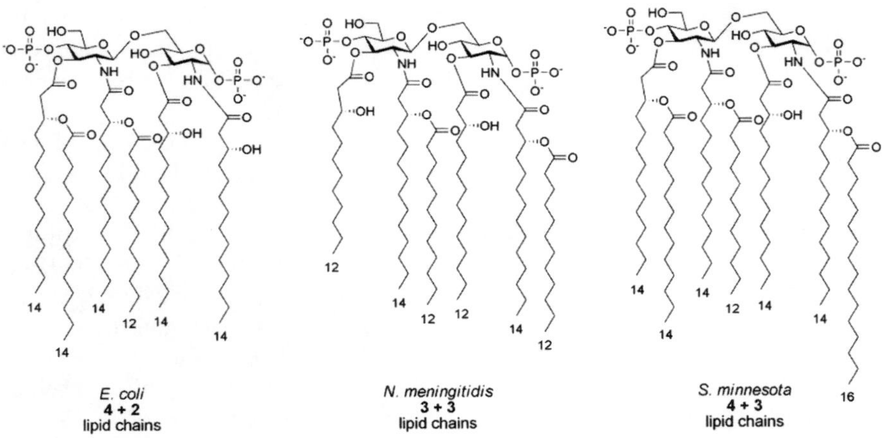

Fig. 13.2 Structures of lipid As of *E. coli*, *S. minnesota*, *N. meningitidis*

directly influences TLR4-mediated host responses. Earlier studies suggest that the D-*gluco* configuration of hexosamines, the two phosphates, and the six fatty acids define the optimal endotoxic conformation. Also, alterations of the hydrophobic region in terms of the nature, number, and distribution of acyl chains have dramatic effects on endotoxic activity. For example, the tetraacylated lipid A biosynthetic precursor (lipid IV_A) not only lacks endotoxic activity but also has antagonistic (anti-endotoxic) activity [26]. Since the high-resolution structures of hexaacylated *Escherichia coli* lipid A [27] and tetraacylated lipid IV_A [28] bound to MD-2-TLR4 complex have been recently solved (see Chap. 12 for a detailed review), the different activities of lipid A variants can now be explained by the different disposition of the lipid A forms in the hydrophobic binding cavity of MD-2 (see next section).

The chemical structure of the oligosaccharide core OS and O-antigen can also influence host immune and inflammatory responses. LPS containing O-antigen (smooth or S-LPS) activates a narrower spectrum of TLR4-MD-2-expressing cells than LPS devoid of O-antigen (rough or R-LPS), and shows less potency in vitro and in vivo. These differences could be due to a differential requirement for CD14 in the activation of cells by the two types of LPS. While S-LPS requires CD14, R-LPS can activate cells irrespective of the presence or absence of this CD14 [29].

LPS is an amphiphilic molecule that upon extraction from the outer membrane forms micelles in aqueous media. LPS has very low critical micelle concentration values [30]. Thus, stable aggregates predominate in the concentration range relevant for biological responses. Chemical changes in primary structures of natural lipid A variants lead to corresponding changes of the aggregate structure. Hexaacylated lipid A molecules adopt a conical shape that facilitates the formation of micellar structures in solution, while underacylated lipid A variants adopt cylindrical shapes forming lamellar structures [31, 32]. Lipid A molecules with strong preference to form lamellar structures (naturally occurring in isolates of *Rhodobacter capsulatus* and *R. viridis*) are endotoxically inactive and therefore lack cytokine-inducing capacity. In contrast, lipid A species with a strong tendency to form non-lamellar inverted structures (lipid A from *E. coli* and *Salmonella* strains) exhibit full endotoxicity in vitro and in vivo [33]. Therefore, it has been proposed to extend the term "endotoxic conformation," which is used to describe the conformation of a single lipid A molecule that is required for optimal triggering of biological effects, to "endotoxic supramolecular conformation" which denotes the particular organization of lipid A aggregates in physiological fluids causing biological activity [31, 34].

13.4 The TLR4 Complex

The induction of inflammatory responses by endotoxin results from the coordinate and sequential engagement of the four LPS-binding proteins LBP, CD14, MD-2 and TLR4 [35]. LBP interacts with endotoxin-rich bacterial membranes and purified endotoxin aggregates [36], catalyzing extraction and transfer of LPS monomers to CD14 [37], which that in turn transfers LPS monomers to MD-2 [38] and to TLR4-MD-2 heterodimers (Fig. 13.3) [27, 39]. Receptor dimerization leads to the recruitment of adaptor proteins to the intracellular domain of TLR4, initiating the intracellular signal cascade that culminates in translocation of transcription factors to the nucleus and the biosynthesis of cytokines.

CD14 is expressed on the surface of myelomonocytic cells as a glycosylphosphatidylinositol-linked glycoprotein or in soluble form (sCD14) in the serum [40]. Monomeric LPS-CD14 and LPS-MD-2 complexes are the proximal vehicles for LPS activation of MD-2-TLR4 and TLR4, respectively. However, prior interactions with other host LPS-binding proteins can either promote or preclude transfer of LPS to CD14 and play key roles in modulating the potency of LPS-mediated TLR4 activation [41]. Also, variability in the aggregation state and three-dimensional forms of endotoxin aggregates may directly influence the kinetics and potency of TLR4 activation and signalling [34].

MD-2 binds to the ectodomain of TLR4 and is essential for LPS signalling [42]. The TLR4-MD-2 heterodimer has complex ligand specificity and can be activated by structurally diverse molecules (see below). Indeed, minor chemical changes in synthetic derivatives can dramatically alter their activity and provoke a switch between agonist and antagonist actions. The majority of synthetic TLR4 agonists and antagonists are MD-2 ligands, so that MD-2 is considered the principal target for the pharmaceutical intervention on innate response to LPS [43].

MD-2 folds into a β-cup structure composed of two antiparallel β-sheets that form a large hydrophobic pocket for ligand binding [28, 44] (Fig. 13.4). This cavity has a volume of 1,720 cubic Å with approximate dimensions of 15 Å by 8 Å by 10 Å. The crystal structure of the dimeric TLR4-MD-2-LPS complex [27], together with crystallographic data of MD-2 bound to TLR4 antagonists such as lipid IV_A [28] and Eritoran (E5564) [44], has uncovered fundamental structural aspects of

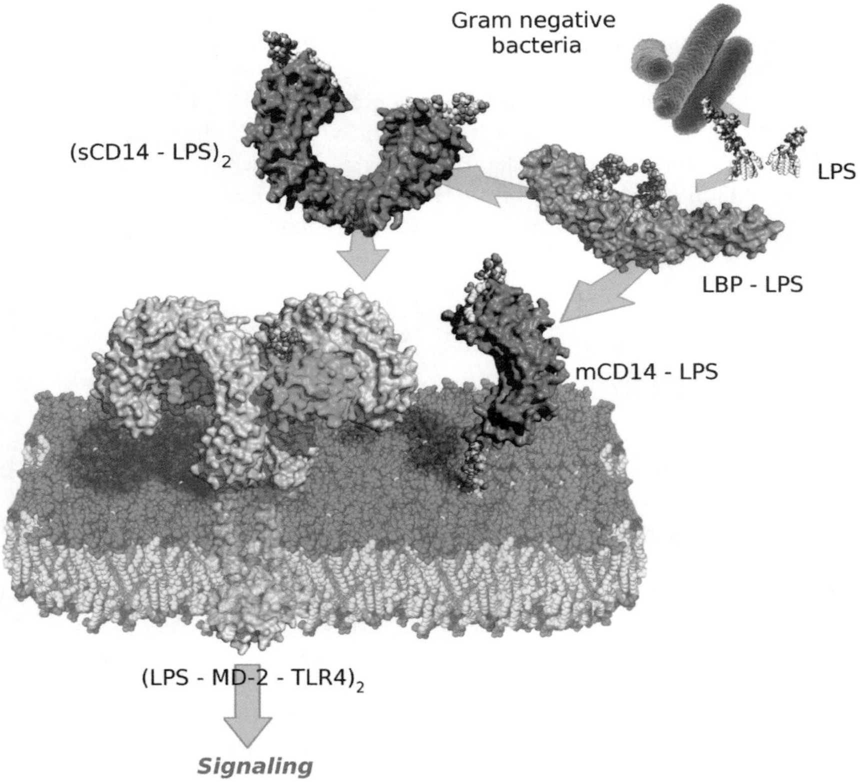

Fig. 13.3 The LPS transport chain and signal amplification: from LPS_{agg} to the dimeric TLR4-MD-2-LPS activated complex

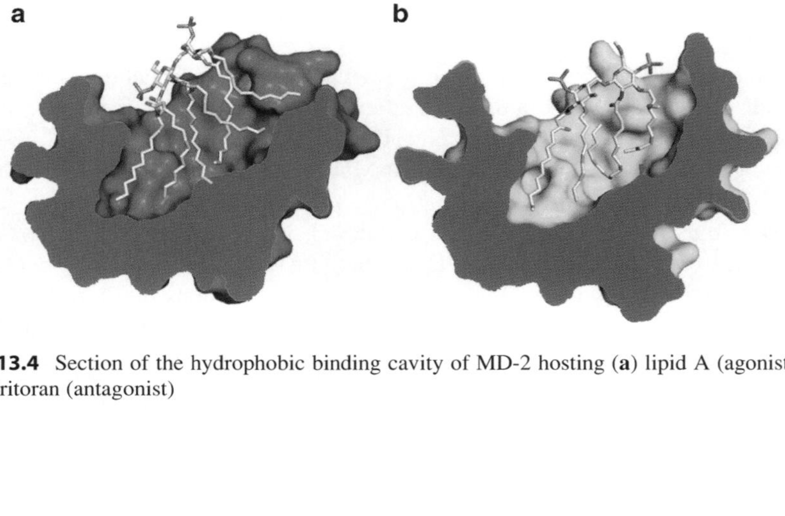

Fig. 13.4 Section of the hydrophobic binding cavity of MD-2 hosting (**a**) lipid A (agonist) and (**b**) Eritoran (antagonist)

the TLR4 dimerization process and the molecular basis of TLR4 agonism and antagonism.

In the crystal structures of MD-2 bound to lipid IV_A and Eritoran, the four acyl chains of the antagonists completely fill all the available space of the MD-2 binding cavity (Fig. 13.4b). *E. coli* LPS has two additional lipid acyl chains, so the question is how the MD-2 structure can accommodate these extra acyl chains (Fig. 13.4a) [28, 44]. The crystal structure of the dimeric TLR4-MD-2-LPS complex shows that the size of the MD-2 cavity remains unchanged and that additional space for acyl chains is generated, at least in the case of hexaacylated *E. coli* LPS, by displacing the phosphorylated glucosamine disaccharide moiety upward by about 5 Å (Fig. 13.4) [27]. This shift of the diglucosamine backbone repositions the phosphate groups such that they can interact with positively charged residues of the two TLR4 molecules, thus promoting dimerization and formation of the activated dimeric TLR4-MD-2-LPS complex. Interestingly, the glucosamine backbones of the antagonists are not only translocated but also rotated by 180° respect to the LPS agonist, thus interchanging the two phosphate groups (Fig. 13.4) [27].

LPS binding and dimerization do not disturb the overall folding of TLR4 and MD-2. In the dimeric TLR4-MD-2-LPS structure the acyl chains of lipid A are buried in the MD-2 cavity, but the chain on N-2 is partially exposed on the MD-2 surface generating an hydrophobic interface for the interaction with the second TLR4 of the complex. The ester and amide groups connecting the lipids to the glucosamine backbone or to other lipid chains are exposed on the surface of MD-2. They interact with hydrophilic side chains on the MD-2 surface and on the surface of the two TLR4 molecules. The two phosphate groups of the lipid A bind to the TLR4-MD-2 complex by interacting with positively charged residues in the two TLR4 and MD-2 and establishing hydrogen bonding to the serine-118 of MD-2 [27].

13.5 TLR4 Pathway Agonists and Antagonists

A wide definition of TLR4 agonist or antagonist will be adopted here because compounds able to modulate TLR4 activity can, in principle, act at different levels of the sequential LPS transfer by targeting LBP, CD14, MD-2 and TLR4 (Fig. 13.3). The synthetic or natural TLR4 modulators discussed below will be therefore grouped as: (1) compounds binding and sequestering LPS; (2) compounds targeting LBP and CD14/LPS interaction; (3) compounds targeting LPS/MD-2 or LPS/MD-2/TLR4 interaction; (4) compounds directly targeting TLR4; and finally (5) molecules modulating TLR4 activity whose exact target(s) are not yet elucidated.

13.5.1 LPS Sequestrants

A possible approach to develop TLR4 antagonists as antisepsis agents is to target LPS itself by the use of an agent that would bind and sequester it, thus abrogating its toxicity. LPS-sequestering agents have been reviewed exhaustively [45]. The anionic amphiphilic nature of lipid A enables it to interact with a variety of cationic hydrophobic ligands [46, 47].

Various proteins such as LBP, bactericidal/permeability-increasing protein (BPI), and *Limulus* anti-LPS factor (LALF) have strong affinity for LPS and also display antimicrobial activity against Gram-negative bacteria. LBP, BPI and LALF carry domains that share a common LPS-binding motif, which is also functionally independent of effects on LPS transport or neutralization [48]. Synthetic peptides based on the putative LPS-binding domains of LBP, BPI, LALF proteins, were especially designed and synthesized to target Gram-negative bacteria [49–51]. These peptides consist of a conserved motif of long β-strands with alternating basic and bulky hydrophobic amino acids. The up-and-down arrangement of side chains in an antiparallel β-strand produces a topological amphipatic motif that pairs the basic amino acids on one face with the hydrophobic amino acids of the opposite face [52].

Cationic antimicrobial peptides with diverse structures bind LPS and suppress its ability to stimulate TLR4-dependent cytokine production [53–57]. Polymyxin B is a membrane-active peptide antibiotic that binds to LPS and inhibits its toxicity in vitro and in animal models of endotoxemia [58]. The pronounced oto- and nephrotoxicity of polymyxin B precludes its systemic use, but has not prevented topical applications as well as the development of an extracorporeal hemoperfusion cartridge based on polymyxin B covalently immobilized on polystyrene-based fibres [59, 60]. Approved for clinical use in Japan in late 2000, polymyxin B provides a clinically validated proof-of-concept of the therapeutic potential of sequestering circulating LPS. A major goal over the past decade was to develop small-molecule analogues of polymyxin B that would sequester LPS with similar potency and be non-toxic and safe, so that it can be used parenterally for the prophylaxis or therapy of Gram-negative sepsis. To this end, various classes of synthetic cationic amphiphiles were developed including acyl [61] and sulfonamido [62] homospermines. Structure-activity relationship studies on these synthetic compounds have established that the pharmacophore necessary for optimal recognition and neutralization of lipid A requires two positively charged groups (protonatable amines) at the same distance than the two anionic phosphates in lipid A (about 14 Å) [45].

13.5.2 Compounds Targeting LBP and CD14/LPS Interactions

Serum LBP enhances binding of LPS to CD14 [36]. Human LBP is a 58–60 kDa serum glycoprotein, with 44% sequence identity to human BPI [63]. The crystal structure of murine BPI has been determined, which allows the construction of a reliable tertiary structural model of LBP [64, 65]. The disaggregation of LPS micelles requires an ordered interaction with LBP and sCD14 [66]. LBP has a concentration-dependent dual role: low concentrations of LBP enhance the LPS-induced activation [67, 68], while the acute-phase rise in LBP concentration inhibits LPS-induced cellular stimulation [50, 69, 70].

The participation of CD14 is essential for the activation of TLR4-MD-2 complex when LPS concentration is low and when LPS chemotype corresponds to S-LPS, but R-LPS and highly concentrated LPS can activate TLR4 in the absence of CD14 [29]. That TLR4-MD-2 complex can be activated by R-LPS in the absence of CD14, but not by the S-LPS, reflects a different intracellular activation level and signalling, because R-LPS binding to TLR4-MD-2 activates MyD88-dependent pathway while S-LPS, by binding to CD14, activates an MyD88-independent pathway [71]. Moreover, CD14 regulates the life cycle of dendritic cells after LPS exposure through a signal pathway based on activation of the nuclear factor of activated T-cells (NFAT), which is independent from the TLR4-activated intracellular pathway [72]. The monomeric subunit of CD14 has an horseshoe-shaped structure with a concave surface formed by a large β-sheet with the repetition of leucine-rich repeats motifs [73] (Fig. 13.5). The concave surface contains both helices and loops, in no regular pattern. As a result, it is rough rather than smooth and contains several grooves and pockets that are crucial for ligand binding. The most characteristic of the CD14 structure is the N-terminal pocket. The pocket is located on the side of the horseshoe near the N-terminus, and it is completely hydrophobic, except for the rim (Fig. 13.5). The binding sites for LPS in CD14 have been identified within the 65 N-terminal residues, clustering around the hydrophobic pocket [74], which is the only lipophilic cleft large enough to accommodate the acyl chains of lipid A (Fig. 13.5) [73]. Nuclear magnetic resonance (NMR) data from binding studies with CD14 and Kdo_2-lipid A show the interaction of terminal methyl and methylene groups of Kdo_2-lipid A lipid chains with CD14 [75, 76]. Interestingly, and in analogy with the LPS-MD-2 complex [27], the hydrophilic part

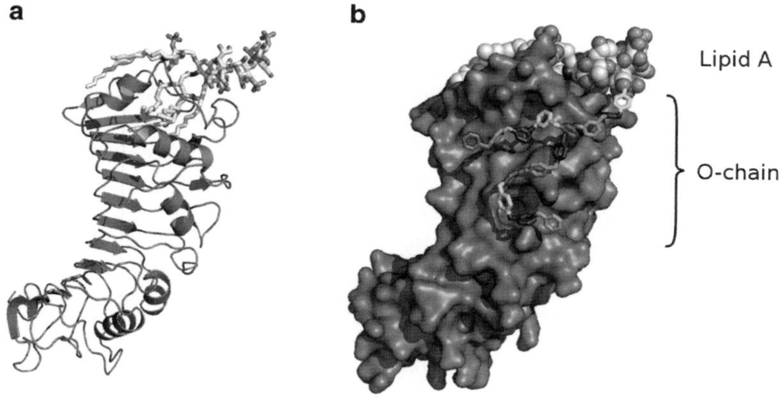

Fig. 13.5 (a) Ribbon and (b) surface CD14 structures with bound LPS

of lipid A composed by the phosphorylated GlcNAc disaccharide also interacts with CD14 residues that localize on the rim of its hydrophobic binding pocket (Fig. 13.5b). The long polysaccharide O-chain is hydrophilic and negatively charged and must have its own binding site (Fig. 13.5b), as previous research has shown that enzymatically delipidated LPS retains some affinity for CD14 [77]. It is unlikely that binding of LPS induces a global structural change in CD14, and it has been reported that binding of LPS induces only minor changes in the tryptophan fluorescence and circular dichroism spectra of CD14 [78]. Although LPS is the most studied and characterized ligand, CD14 can interact also with other molecules.

Both LBP and CD14 can bind lipoteichoic acid derived from the Gram-positive bacterium *Bacillus subtilis* [79]. In addition, whole bacteria are recognized by CD14 in an LBP-dependent reaction but only after preincubation with serum. Electron microscopy suggests that the serum pretreatment results in the opening up of the bacterial cell wall. When myelomonocytic cells expressing CD14 bind serum-pretreated bacteria, the bacteria may be engulfed. The CD14-LBP system may thus play a role in counteracting Gram-positive bacterial infections [79]. Soluble peptidoglycans (PGN) are recognized by TLR2 and CD14 is important to enhance their recognition [80]. In vitro studies provide evidence that the PGN binding interface on CD14 can overlap with the binding site of LPS polysaccharide O-chain, although the biological relevance of PGN binding to CD14-TLR2 is under debate. The hydrophobic N-terminal pocket of CD14 is unlikely to be involved in PGN binding, since PGN is a completely hydrophilic molecule. The LPS- and PGN- binding sites must overlap, at least in part, because PGN competes with LPS for binding to CD14 [63, 81, 82].

A glycoconjugate preparation from spirochetes inhibits the interaction of TLR4 (lipid A, LPS, taxol) and TLR2 (PGN) ligands with LBP and CD14, acting as an antagonist of the corresponding TLR4- or TLR2-dependent pathways [83]. This preparation is chemically heterogeneous for a detailed mechanistic analysis, but these experiments show that TLR4 and TLR2 pharmacological inhibition by interfering with LBP and/or CD14 is possible [84]. With the aim to develop new TLR4-active compounds, our research group synthesized glycosylamino- and benzylammonium-lipids (Fig. 13.6), which inhibited LPS- and lipid A-promoted cytokine production in macrophages and dendritic cells [85]. Compounds **1–4** (Fig. 13.6) inhibit LPS-induced TLR4 activation in HEK293 cells stably transfected with TLR4, MD-2 and CD14 genes and containing a secreted alkaline phosphatase reporter gene, and they also efficiently inhibit LPS-induced septic shock in mice [85]. Structure-activity studies suggest the pharmacophore consists of glucose or a phenyl ring linked to two C_{14} ether lipid chains and a basic nitrogen. Compounds **1–4** containing the complete pharmacophore are active in blocking TLR4-mediated cytokine production in innate immunity cells, while very similar compounds lacking the positively charged group (such as molecules **5** and **6**) are inactive [86]. Compound **1** can also reduce in vivo neuropathic pain by reversing mechanical allodynia and thermal hyperalgesia in mice [23].

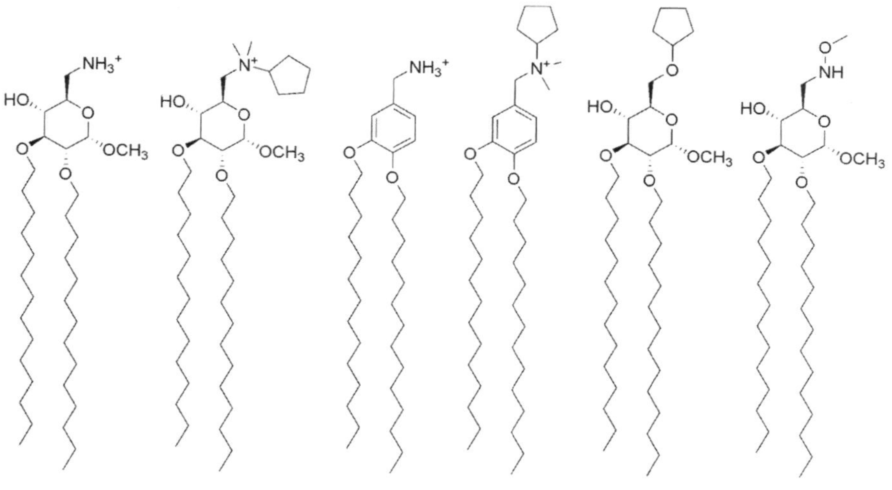

Fig. 13.6 Compounds **1–4** are active in inhibiting TLR4 signalling by targeting CD14-LPS interaction; compounds **5** and **6**, while having chemical structures very similar to 1–4, are inactive

The mechanism of action for molecules **1–4** has been investigated by analyzing all possible interactions of LPS with LBP, CD14, and MD-2 (free and TLR4-bound) [87]. Using tritiated lipooligosaccharide we tested compounds **1–4** for their ability to inhibit in vitro the formation of the activated dimeric complex TLR4-MD-2-LPS [87]. We observed that the formation of the activated complex was reproducible in the absence of synthetic compounds, but was inhibited in a dose-dependent manner by compounds **1–4** and not by compounds **5** and **6**. Inhibition was associated with the ability of the compounds to selectively block the interaction of LPS with CD14, as demonstrated by NMR Saturation Transfer Difference experiments [85]. Work is in progress to determine the exact structure and stoichiometry of the CD14-glycolipid complex.

13.5.3 MD-2-TLR4 Agonists

13.5.3.1 Monophosphoryl Lipid A Disaccharides

Previously, it was shown that the toxic effects of heptaacylated *Salmonella enterica* serovar Minnesota (*S*. Minnesota) R595 lipid A (compound **7**, Fig. 13.7) could be ameliorated by selective hydrolysis of the 1-phosphate and by removing the acyl chain in position 3 [88]. The resulting chemically modified lipid A product, monophosphoryl lipid A (MPL) is a mixture of the two differently acylated variants **8** and **9** (Fig. 13.7), and is an effective adjuvant in prophylactic and therapeutic vaccines. Comparison of the biological activities of compounds **8** and **9** reveals that both have similar adjuvant activity, but compound **8** is up to 20-times more active

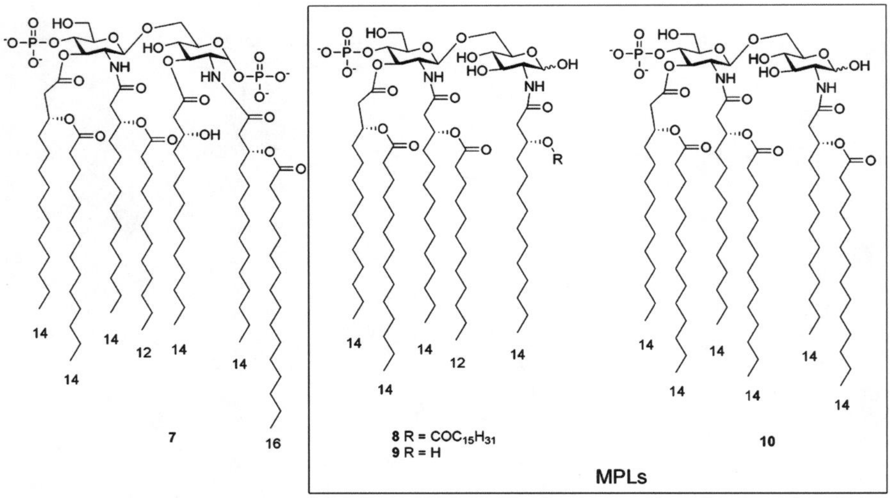

Fig. 13.7 *S. minnesota* lipid A (heptaacylated, compound **7**) and MPL TLR4 agonists' molecular components: hexaacyl (**8**) and pentaacyl (**9**) monophosphates. Synthetic hexaacyl monophosphate **10** with six C_{14} lipid chains

than **9** in inducing nitric oxide synthase in murine macrophages [88]. These observations parallel studies on lipid A variants showing that hexaacylation is a prerequisite to the full expression of endotoxic activities, and that underacylated lipid A displays reduced activity and may even possess antagonist activity. MPL agonists with reduced toxicity but increased potency meet the stringent safety criteria required for prophylactic vaccines, and MPL was the first TLR4 agonist approved for use in a human vaccine for hepatitis B virus. Currently, two approved hepatitis B virus vaccines [89] and an almost approved papilloma virus vaccine [90] use MPL as an adjuvant. Preclinical studies suggest that MPL and Ribi.529 TLR4 agonists have the potential to enhance therapeutic vaccination for cancer and chronic viral infections, including human immunodeficiency virus and hepatitis B virus [91, 92].

Synthetic compound **10** is hexaacylated and contains all lipid chains (primary and secondary) with the same length (C_{14}) and the total number of carbons in the lipid part is conserved with respect to MPL component **8** [93]. Compound **10** shows very similar activity to **8** in terms of lethal toxicity in mice, induction of cytokine production and adjuvant activity. The effect of fatty acid structure on endotoxic (agonist) activity in the MPL series was investigated systematically by synthesizing a series of chain length homologs, by varying the length of secondary lipid chains from C_4 to C_{12} [93]. Secondary fatty chains lengths have unpredictable and profound effects on the MPL agonist activity. Short chains derivatives (C_4 or C_6) are inactive in stimulating TLR4-dependent cytokine production. Whereas the C_{10} homolog exhibits the highest level of cytokine induction and the greatest pyrogenicity, the C_{14} derivative has intermediate levels of activity and toxicity [94].

Different threshold chain lengths were observed for activity in murine and human innate immunity cell models, which indicates strict but slightly different structural requirements for the two bioactivities. This would be attributable to the known species differences in the structure of the MD-2 proteins.

Together, these data suggest that the MPL orientation in the hydrophobic binding pocket of MD-2 is sensitive to both the degree of acylation and the length of lipid chains. Changing these structural parameters of the lipid part, the disposition of the phosphodisaccharide moiety of MPL with respect to the binding residues situated in the rim of MD-2 cavity also changes. This would generate different binding interfaces and binding affinities between the two TLR4 molecules of the activated dimer, thus causing profound effects in the nature and the intensity of the downstream signalling. Another detoxified lipid A product closely related to MPL and containing an heptaacyl derivative as major component is being employed as an adjuvant in therapeutic cancer vaccines against non-small-cells lung and prostate cancer [95].

13.5.3.2 Aminoalkyl Glucosaminide 4-Phosphates (AGPS)

Numerous lipid A mimetics have been prepared by removing some subunits or substituting lipid A parts with bioisoster groups. These derivatives comprise aminoalkyl glucosaminide 4-phosphates (AGPs) in which the reducing glucosamine residue has been replaced by an acylated amino acid or another acylated function. The AGPs lipid A analogues (Fig. 13.8) have been developed by Johnson et al. [93, 94]. These compounds retain significant activity as TLR4 agonists or antagonists but have a simplified structure with a reduced number of stereogenic carbons, and can be therefore obtained by simpler syntheses than corresponding lipid As.

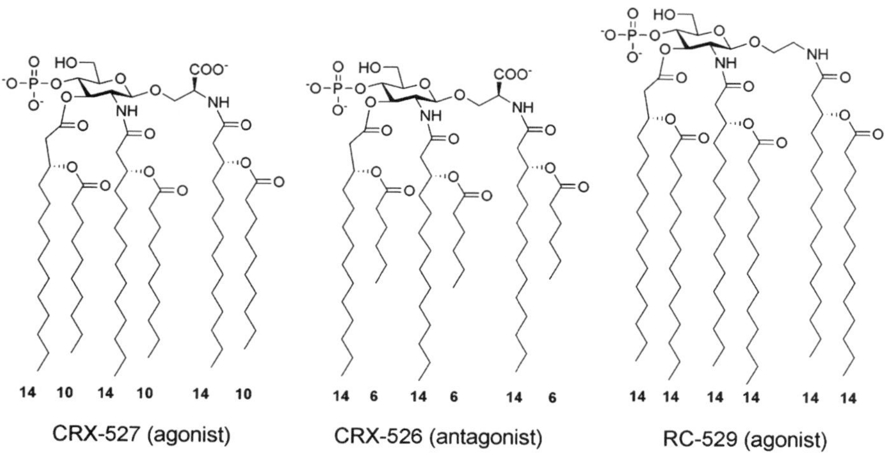

Fig. 13.8 Seryl aminoalkyl glucosaminide 4-phosphates CRX-526, CRX-527, RC-529

AGPs (Fig. 13.8) are synthetic analogues of MPL component **8** (Fig. 13.7) in which the reducing sugar has been replaced by a conformationally flexible *N*-acyl aglycon unit. The known immunostimulant activity of naturally occurring *N*-(3-acyloxyacyl)amino acids such as ornithine- or serine-containing lipids [96, 97], which were shown to require MD-2-TLR4, made the evaluation of seryl β-O-glycosides of particular interest. Pro-inflammatory responses induced by AGP show striking chain-length dependence. Compound CRX-527, with C_{14} primary and C_{10} secondary lipid chains, has adjuvant activity as demonstrated by improving humoral and cell-mediated immune responses to several different antigens in mice. CRX-527 was also evaluated in preclinical model for a respiratory syncytial virus prophylactic vaccine [98]. The action of CRX-527 was strictly dependent on TLR4 and MD-2 but not CD14, although CD14 increases the potency of this synthetic TLR4 agonist [94].

In sharp contrast to CRX-527, CRX-526 with C_{14} primary and C_6 secondary fatty acid chains has potent antagonistic activity and can block the induction of proinflammatory cytokines by LPS both in vitro and in vivo [99]. AGP derivative RC-529 has an acyl-ethanolamine instead of serine as aglycon and C_{14} primary and secondary lipid chains. This compound is less potent than CRX-527 in terms of proinflammatory activity, but preserves a potent adjuvant action. The ability of this compound with low endotoxicity to enhance adaptive response may relate, in part, to the efficiency of the interaction with CD14, which activation is required for MyD88-independent LPS signalling [71].

13.5.3.3 Monosaccharides (GLA-60)

Synthetic analogues of either the reducing or non-reducing glucosamine moieties of lipid A containing up to five fatty acids and lacking an aglycon unit (that is present in AGPs) generally present reduced TLR4 agonist activity [94]. However, monosaccharide GLA-60 with three C_{14} fatty acid groups (Fig. 13.9) showed the strongest B-cell activation and adjuvant activities among various non-reducing subunit analogues that have been examined [100]. Interestingly, the addition of a C_{14} fatty acid chain to form the tetracyl derivative GLA-47, which corresponds to the left hand side of *E. coli* and *S.* Minnesota lipid A (Fig. 13.9), abolishes TNF-α and IL-6 induction in human U937 cells and peripheral blood mononuclear cells [101].

13.5.3.4 Lipid A Mimetics with Linear Scaffold

Compound ER112022 [102], ER803022 and other similar compounds were developed at the Eisai Research Center in Boston (Fig. 13.10), as acyclic lipid A analogues. All these compounds contain six symmetrical lipid chains and two phosphate groups attached to linear linkers with different chemical structures [103]. Compound ER112022 activates NF-κB in HEK293 cells transfected with TLR4-MD-2, and although CD14 is not essential for activity it enhances the sensitivity of the response [102]. TLR4 antibodies block TNF-α release from primary human monocytes exposed to ER112022, while E5564 and E5531 TLR4 antagonists (see below) block ER112022-induced stimulation of a series of innate immunity cells.

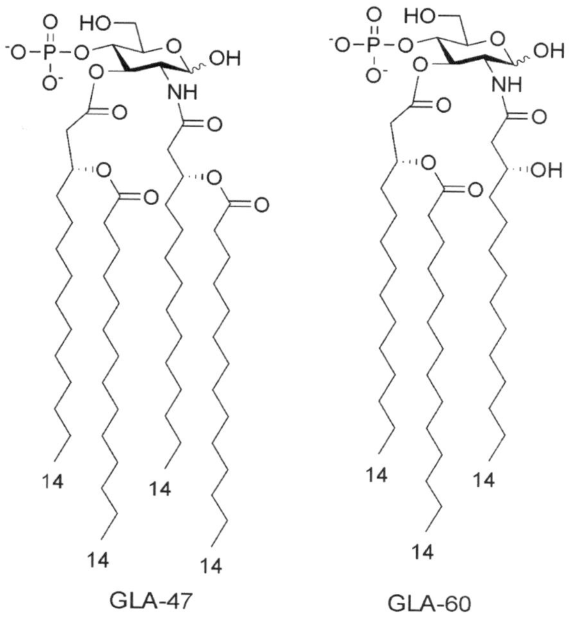

Fig. 13.9 Monosaccharide lipid A mimetics GLA-60 (triacylated) and GLA-47 (tetraacylated)

13.5.4 MD-2-TLR4 Antagonists

13.5.4.1 E5531 and E5564

LPS and lipid A fractions obtained from non-pathogenic bacteria such as *R. capsulatus* and *R. sphaeroides* are potent LPS antagonists in vitro (Fig. 13.11) [104]. Compound E5531 (Fig. 13.11), an analogue of *R. capsulatus* lipid A, was first developed in Eisai Laboratories as an LPS antagonist [105]. Although E5531 demonstrates potent inhibition of LPS toxicity when added to blood in vitro and in vivo, activity decreases as a function of time. This loss of activity is due to the interaction of E5531 with plasma lipoproteins [106]. A second-generation LPS antagonist Eritoran (E5564) (Fig. 13.11) derives from the structure of the weakly agonistic LPS of *R. sphaeroides* [107]. E5564 is a potent in vitro antagonist of endotoxin that directly binds to the hydrophobic pocket of MD-2, competitively inhibiting lipid A binding and thereby preventing dimerization of TLR4 and intracellular signalling [44, 107, 108]. E5564 is significantly protective in animal models of sepsis [108] and, in healthy volunteers, blocks the symptoms of endotoxemia in a dose-dependent manner [109]. Recently completed phase II clinical trials demonstrated that the administration of E5564 is well tolerated by patients, supporting a further phase III clinical study to assess its benefit in patients with high-risk mortality due to sepsis [110].

Fig. 13.10 Eisai's acyclic lipid A analogues

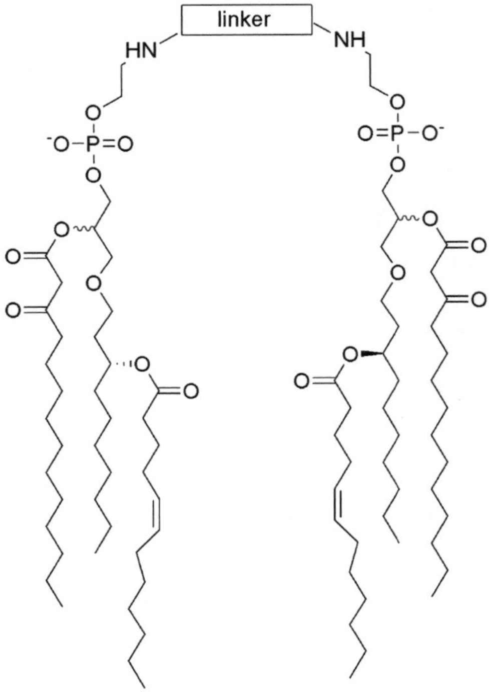

Eisai's linear TLR4 agonists

13.5.4.2 MD-2-TLR4 Antagonists with Structure Unrelated to Lipid A

Synthetic and natural compounds with chemical structures unrelated to that of lipid A are active on TLR4-LPS signalling by binding MD-2. Because their profound structural differences with lipid A, these molecules act as antagonists rather than agonists, and their binding to MD-2 does not result in the formation of the activated TLR4-MD-2 heterocomplex required for signalling. The fluorescent probe bis-aminonaphtyl sulfate (bis-ANS), which has a hydrophobic binaphtyl core, two anionic sulfates, and two positively charged anilines (Fig. 13.12), binds functional MD-2 in vitro with high affinity [111]. A free cysteine residue (Cys-133) lies inside the hydrophobic binding pocket of MD-2 and it is exposed to solvent. Compounds comprising hydrophobic MD-2 binding moieties and sulphur-reactive electrophiles could, in principle, irreversibly target MD-2 by formation of a covalent bond with Cys-133 and effectively block the LPS-TLR4 signalling. Therefore, a series of compounds such as iodoacetylaminonaphtyl sulfate (IAANS), auranofin, JTT705, *N*-pyrenemaleimide, containing a thiol-reactive functionality, were assayed as TLR4 antagonists. JTT705 and auranofin (Fig. 13.12) were used in clinical trials as antihypercholesterolemic and anti-inflammatory compounds, respectively [111].

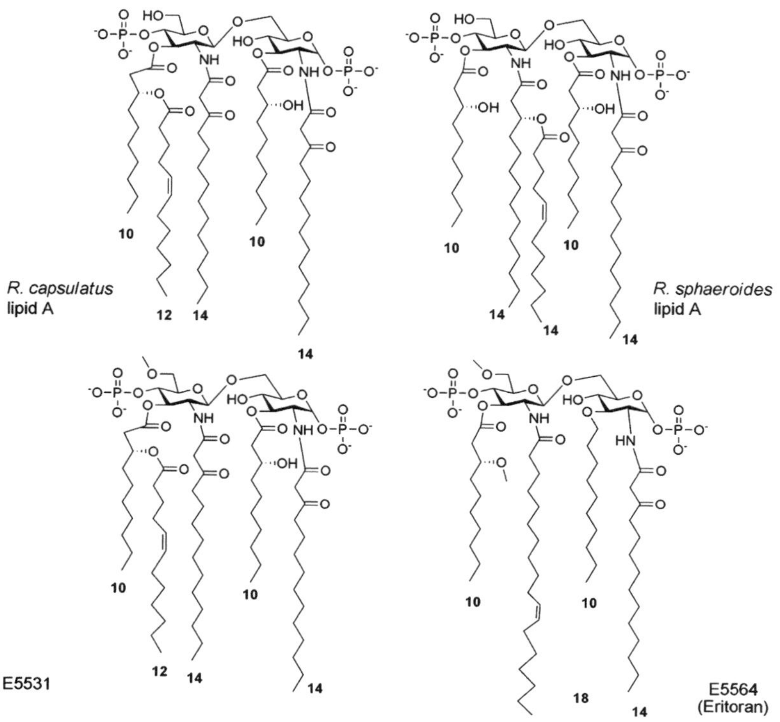

Fig. 13.11 *R. capsulatus* and *R. spheroides* lipid A structures inspired the design of synthetic lipid A analogues E5531 and E5564

While they both bind MD-2 and inhibit LPS signalling in vitro, compound JTT705 cannot prevent LPS-induced septic shock mortality in mice [111].

The taxane paclitaxel (Fig. 13.12), a well-known antitumor drug targeting tubulin, also binds to murine MD-2 and acts as agonist activating proinflammatory cascade [112]. However, paclitaxel inhibits TLR4 signalling in humans [113]. This difference parallels other observations reporting different responses in mice and humans to tetraacylated lipid A, or lipid A analogues such as MPLs and AGPs, which activate murine MD-2-TLR4 while inhibit human TLR4 activation.

Curcumin (Fig. 13.12), the main constituent of the spice turmeric used in foods and in traditional medicine, especially in India, has anti-inflammatory properties. Curcumin inhibits MyD88-dependent and -independent pathways by preventing the dimerization of TLR4 in the murine cell line BaF3 [114]. A more detailed analysis of the molecular mechanism of action revealed that curcumin very likely binds to MD-2 thus competing with LPS [115]. Curcumin behaves as a TLR4 antagonist in vitro and inhibits LPS-TLR4 signalling in HEK-TLR4 cells [115]. Curcumin is a Michael acceptor as it contains an α,β-unsaturated ketone group, and it binds MD-2 in proximity of free Cys-133. However, the curcumin-MD-2 complex is non-covalent and no covalent Michael adduct is formed upon binding [115].

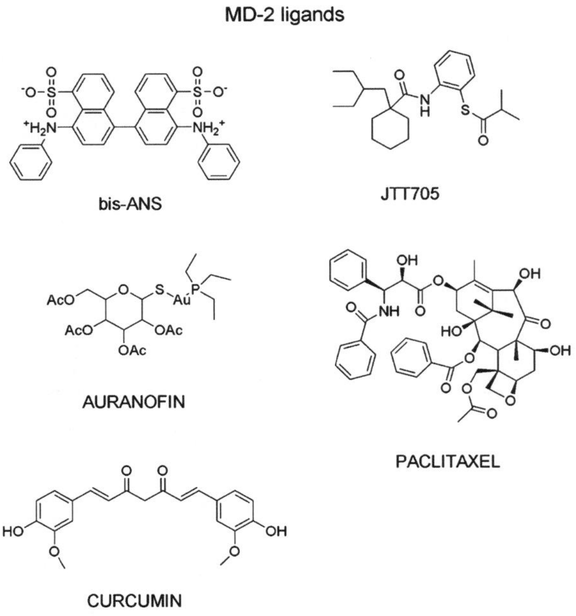

Fig. 13.12 Non-lipid A MD-2 ligands: bis-ANS, auranofin, JTT705, paclitaxel, curcumin

13.5.5 Compounds that Bind Directly to TLR4

A further possibility, yet underexploited, to target the LPS signalling is to develop molecules that directly bind to TLR4. In particular, as the TLR4 intracellular signal is highly dependent from the intracellular domain of TLR4, it would be interesting to target the hydrophobic TLR4 peptide spanning the cellular membrane. Takeda Pharmaceutical Company developed TAK-242, a cyclohexene derivative that selectively inhibits TLR4 signalling [116] and efficiently protects mice against LPS-induced lethality [117]. TAK-242 binds directly to the cysteine-747 in the intracellular domain of TLR4 [118, 119]. As TAK-242 is a Michael acceptor, it was proposed but not experimentally proven, that it forms a covalent adduct with cysteine-747. It is unclear how TAK-242 inhibits TLR4 signalling after TLR4 binding. Upon binding to TLR4, TAK-242 could inhibit myristoylation and phosphorylation of the intracellular TRAM protein, which are covalent modifications essential for TLR4 signalling [119]. Because its high potency as an antisepsis agent in animal models [117], TAK-242 was tested in Phase-I and -II clinical trials. However, the compound failed to suppress cytokine levels in patients with severe sepsis and septic shock or respiratory failure and further development was recently discontinued.

A peptide corresponding to the minimal TLR4-binding region on MD-2 blocks the TLR4-MD-2 association and TLR4 signal [120]. This seven-residue peptide was synthesized to reproduce the TLR4-binding region of the MD-2 protein that contains all the critical interacting residues. This peptide was effective in blocking the LPS-induced TLR4 activation in HEK-TLR4 cells and macrophages [120].

13.5.6 Other TLR4-Active Compounds

13.5.6.1 Heme

Hemin (Fig. 13) is a ubiquitous molecule present in organisms as prosthetic group in a large number of proteins that are essential for life and have a pivotal role in the processes of oxygen transport, storage and electron shuttling. Several pathologic situations, some but not all induced by infection, can lead to increased hemolysis and very high levels of free heme. Recently, it has been hypothesized that the activation of the TLR4 pathway is one of the ways by which the "danger signal" represented by free heme is detected and amplified. A recent investigation of the molecular mechanism whereby hemin (iron(III) heme) activates mouse macrophages shows that hemin induces the secretion of TNF-α in TLR4-CD14- and MyD88-dependent manner [121]. However, whereas the TLR4 antagonist E5564 and anti-TLR4-MD-2 antibody inhibited TNF-α secretion induced by LPS, these compounds did not inhibit cell activation by hemin. In contrast, biologically inactive heme variants such as iron-free protoporphyrin IX inhibited TLR4-dependent TNF-α secretion induced by heme but not that induced by LPS. Therefore, it may be possible that hemin interacts with MD-2-TLR4 in a CD14-dependent manner but at a site or sites that are distinct from those where LPS acts [121].

We investigated the molecular mechanism underlying the modulation of the TLR4 pathway by hemin and its metabolically oxidised derivative coprohemin (iron(III)-coproporphyrin I, Fig. 13.13) [87]. Highly concentrated hemin triggered TLR4-mediated IL-8 production in human HEK-TLR4 cell line in the absence of the co-receptors CD14 and MD-2. The observation that hemin and endotoxin have mild but reproducible additive effects when co-administrated to HEK-TLR4 cells, suggests that hemin and endotoxin interact with TLR4 through different mechanisms and probably have distinct binding sites. Coproheme, in contrast to heme, is unable to trigger TLR4-mediated interleukin production in the same HEK cells, but is active in inhibiting in a dose-dependent way endotoxin-stimulated interleukin production. This antagonistic activity of coprohemin is accompanied by reduced delivery of endotoxin to MD-2 (free or TLR4-bound) that is necessary for activation of TLR4 by endotoxin. Despite their similar chemical structure (Fig. 13.13), hemin and coprohemin have very different effects on the TLR4 pathway, the former acting as TLR4 mild agonist, the latter as an antagonist selectively targeting the endotoxin-MD-2 interaction [87].

13.5.6.2 Thalidomide

N-Phtalimidoglutarimide (thalidomide) (Fig. 13.13) is known as antiangiogenic, antitumor and antiproliferative agent and used in the treatment of some immunological disorders and cancer. Thalidomide significantly inhibits LPS-induced TNF-

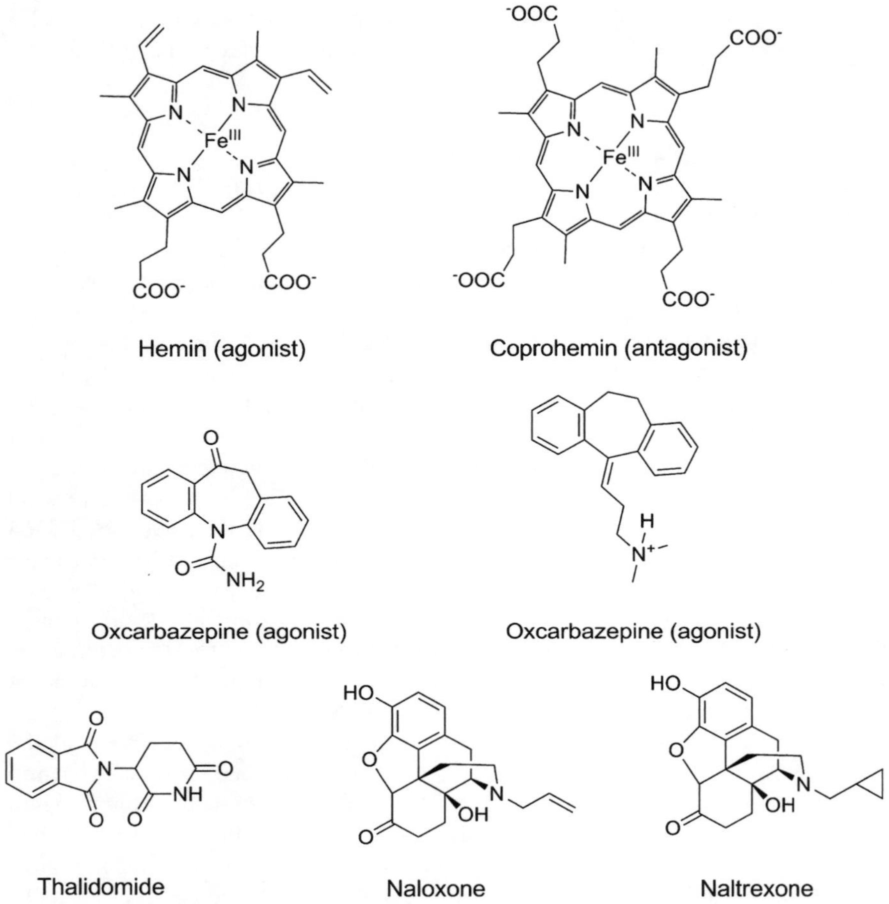

Fig. 13.13 Hemin, coprohemin, thalidomide, opioid analgesics oxcarbazepine (agonist) and amitryptiline (antagonist), naloxone, naltrexone

α production in murine macrophages [122]. It has been proposed that thalidomide effect is mainly due to the down-regulation of MyD88 expression.

13.5.6.3 Opioids

A broad-range of clinically relevant opioids (morphine, methadone, meperidine, fentanyl, oxycodone) activates TLR4 [123]. Opioid-induced glial activation suppresses acute opioid-induced analgesia, enhances the development of analgesic tolerance, dependence, and contributes to negative side effects such as respiratory depression. Importantly, opioids exert such effects via TLR4 [123] and TLR2 activation. Tricyclic compounds with opioid activity (Fig. 13.13) were tested for effects on TLR4 signalling because members of this class have been used for neuropathic pain treatment [124]. Eight tricyclics (Fig. 13.13) were tested for effects on HEK293 cells expressing human TLR4 when administered alone or together with LPS [123]. Five of them exhibited mild (desipramine), moderate (mianserin, cyclobenzaprine, imiprine) or strong (amitryptiline) TLR4 antagonist activity. In contrast, carbamazepine and oxcarbazepine (Fig. 13.13) exhibited mild and strong TLR4 activation, respectively, and no TLR4 inhibition [123]. *In silico* docking simulations of interaction between tricyclics with MD-2, suggested that these compounds could exert their action on TLR4 pathway by binding MD-2. Both enantiomers of opioid antagonist naloxone (Fig. 13.13) were reported to inhibit LPS-induced microglial production of superoxide, nitric oxide, and TNF-α [123]. Naloxone and naltrexone (Fig. 13.13) inhibited LPS-induced secreted alkaline phosphatase expression in HEK-hTLR4 cells and the inhibition was non-stereoselective in the sense that (+) and (−) enantiomers showed very similar activities. Naloxone and naltrexone reversed neuropathic pain in animals by blocking TLR4 activation and signalling [123].

13.6 Future Perspectives

Investigations on compounds that can modulate the TLR4 pathway not only offer novel pharmacological targets but they contribute to the clarification of basic structural and mechanistic aspects of TLR4 signalling, including the role of LBP, CD14 and MD-2 co-receptors. This method of investigation of biological signal pathways through the use of small molecule ligands is the so-called "chemical genetics" approach [125], which is complementary to the classical forward (mutagenesis) and reverse (gene knockout) genetic approaches. Some compounds presented in this chapter have been rationally designed to target MD-2 or MD-2-TLR4 complexes. These compounds are mainly lipid A analogues with agonist or antagonist activity on the LPS-TLR4 signal pathway, such as MPLs and AGPs that mimic the entire lipid A or part of its structure. The recent determination of crystal structures of the dimeric TLR4-MD-2 complex with bound lipid A (agonist) or lipid A antagonists (Eritoran, lipid IV_A) clarified important aspects of the structure-activity relationship in natural lipid As or synthetic lipid A analogues. Accordingly, it is possible today to use rational rules for the design of TLR4 agonists and antagonists with a lipid A-derived structure.

Other TLR4-active compounds reviewed here are natural compounds or synthetic molecules whose chemical structure is not related to that of lipid A or LPS. The activity on TLR4 has generally been discovered serendipitously for these compounds, often as pharmacological side effects or off-target activity. This has been the case of taxanes, thalidomide, synthetic opioids, and the CD14-targeting glycolipids discovered by our group. The current knowledge on the structural biology of the TLR4 pathway still not allows the rational, *ex-novo* design of chemical entities unrelated to lipid A specifically binding CD14, or MD-2 or TLR4 receptors.

Some non-lipid A compounds are competitive inhibitors of LPS binding to CD14 and MD-2 receptors. For instance, synthetic glycolipids compete with LPS for CD14 binding and curcumin, taxanes, compete with LPS for MD-2 binding. Other non-lipid A compounds such as TAK-242 are allosteric TLR4 inhibitors and bind to different sites than LPS.

A critical path for the near future would be to determine and characterize allosteric sites on TLR4 so that specific ligand can be designed to modulate TLR4 activity through non-classical CD14 and MD-2-mediated ligand presentation. This would also allow to specifically targeting sterile inflammations or autoimmune diseases not caused by the presence of a pathogen. Several research groups, included ours, are involved in the development of non classical TLR4 antagonists as lead compounds for the development of innovative and selective drugs against chronic pain, neuropathic pain and other syndromes caused by microglial TLR4 activation. On the other hand, the development of non-lipid A TLR4 mild agonists would provide innovative compounds as non-toxic vaccine adjuvant and immunotherapeutics.

The selective CD14 targeting by molecules that bind to CD14 and not to MD-2 or MD-2-TLR4 is also an innovative way to inhibit the whole TLR4 pathway and to elude bacterial resistance in the development of new generation of antisepsis agents as well as agents to target TLR4-mediated non infectious inflammatory conditions such as certain forms of neuropathic pain.

References

1. Miyake K (2007) Innate immune sensing of pathogens and danger signals by cell surface Toll-like receptors. Semin Immunol 19:3–10
2. Akira S, Takeda K (2004) Toll-like receptor signalling. Nat Rev Immunol 4:499–511
3. Akira S, Uematsu S, Takeuchi O (2006) Pathogen recognition and innate immunity. Cell 124:783–801
4. Bsibsi M, Ravid R, Gveric D, van Noort JM (2002) Broad expression of Toll-like receptors in the human central nervous system. J Neuropathol Exp Neurol 61:1013–1021
5. Kielian T (2006) Toll-like receptors in central nervous system glial inflammation and homeostasis. J Neurosci Res 83:711–730
6. Konat GW, Kielian T, Marriott I (2006) The role of Toll-like receptors in CNS response to microbial challenge. J Neurochem 99:1–12
7. Guo LH, Schluesener HJ (2007) The innate immunity of the central nervous system in chronic pain: the role of Toll-like receptors. Cell Mol Life Sci 64:1128–1136
8. Okun E, Griffioen KJ, Lathia JD, Tang SC, Mattson MP, Arumugam TV (2009) Toll-like receptors in neurodegeneration. Brain Res Rev 59:278–292
9. Beutler B (2002) TLR4 as the mammalian endotoxin sensor. Curr Top Microbiol Immunol 270:109–120
10. Beutler B, Du X, Poltorak A (2001) Identification of Toll-like receptor 4 (Tlr4) as the sole conduit for LPS signal transduction: genetic and evolutionary studies. J Endotoxin Res 7:277–280
11. Poltorak A, He X, Smirnova I, Liu MY, Van Huffel C, Du X, Birdwell D, Alejos E, Silva M, Galanos C, Freudenberg M, Ricciardi-Castagnoli P, Layton B, Beutler B (1998) Defective LPS signaling in C3H/HeJ and C57BL/10ScCr mice: mutations in Tlr4 gene. Science 282:2085–2088

12. Hayashi F, Means TK, Luster AD (2003) Toll-like receptors stimulate human neutrophil function. Blood 102:2660–2669
13. Iwasaki A, Medzhitov R (2004) Toll-like receptor control of the adaptive immune responses. Nat Immunol 5:987–995
14. Cribbs SK, Martin GS (2007) Expanding the global epidemiology of sepsis. Crit Care Med 35:2646–2648
15. Martin GS, Mannino DM, Eaton S, Moss M (2003) The epidemiology of sepsis in the United States from 1979 through 2000. N Engl J Med 348:1546–1554
16. Kanzler H, Barrat FJ, Hessel EM, Coffman RL (2007) Therapeutic targeting of innate immunity with Toll-like receptor agonists and antagonists. Nat Med 13:552–559
17. Horner AA, Redecke V, Raz E (2004) Toll-like receptor ligands: hygiene, atopy and therapeutic implications. Curr Opin Allergy Clin Immunol 4:555–561
18. Racila DM, Kline JN (2005) Perspectives in asthma: molecular use of microbial products in asthma prevention and treatment. J Allergy Clin Immunol 116:1202–1205
19. Smit LA, Siroux V, Bouzigon E, Oryszczyn MP, Lathrop M, Demenais F, Kauffmann F (2009) CD14 and toll-like receptor gene polymorphisms, country living, and asthma in adults. Am J Respir Crit Care Med 179:363–368
20. De Leo JA, Tawfik VL, LaCroix-Fralish ML (2006) The tetrapartite synapse: path to CNS sensitization and chronic pain. Pain 122:17–21
21. Tanga FY, Nutile-McMenemy N, DeLeo JA (2005) The CNS role of Toll-like receptor 4 in innate neuroimmunity and painful neuropathy. Proc Natl Acad Sci USA 102:5856–5861
22. Cao L, Tanga FY, Deleo JA (2009) The contributing role of CD14 in toll-like receptor 4 dependent neuropathic pain. Neuroscience 158:896–903
23. Bettoni I, Comelli F, Rossini C, Granucci F, Giagnoni G, Peri F, Costa B (2008) Glial TLR4 receptor as new target to treat neuropathic pain: efficacy of a new receptor antagonist in a model of peripheral nerve injury in mice. Glia 56:1312–1319
24. Rietschel ET, Wollenweber HW, Zähringer U, Lüderitz O (1982) Lipid A, the lipid component of bacterial lipopolysaccharides: relation of chemical structure to biological activity. Klin Wochenschr 60:705–709
25. Raetz CRH, Whitfield C (2002) Lipopolysaccharide endotoxins. Annu Rev Biochem 71: 635–700
26. Rietschel ET, Kirikae T, Schade FU, Mamat U, Schmidt G, Loppnow H, Ulmer AJ, Zähringer U, Seydel U, Di Padova F, Schreier M, Brade H (1994) Bacterial endotoxin: molecular relationships of structure to activity and function. FASEB J 8:217–225
27. Park BS, Song DH, Kim HM, Choi BS, Lee H, Lee JO (2009) The structural basis of lipopolysaccharide recognition by the TLR4-MD-2 complex. Nature 458:1191–1195
28. Ohto U, Fukase K, Miyake K, Satow Y (2007) Crystal structures of human MD-2 and its complex with antiendotoxic lipid IVa. Science 316:1632–1634
29. Huber M, Kalis C, Keck S, Jiang Z, Georgel P, Du X, Shamel L, Sovath S, Mudd S, Beutler B, Galanos C, Freudenberg MA (2006) R-form LPS, the master key to the activation ofTLR4/MD-2-positive cells. Eur J Immunol 36:701–711
30. Takayama K, Din ZZ, Mukerjee P, Cooke PH, Kirkland TN (1990) Physicochemical properties of the lipopolysaccharide unit that activates B lymphocytes. J Biol Chem 265:14023–14029
31. Schromm AB, Howe J, Ulmer AJ, Wiesmuller KH, Seyberth T, Jung G, Rossle M, Koch MH, Gutsmann T, Brandenburg K (2007) Physicochemical and biological analysis of synthetic bacterial lipopeptides: validity of the concept of endotoxic conformation. J Biol Chem 282:11030–11037
32. Seydel U, Labischinski H, Kastowsky M, Brandenburg K (1993) Phase behavior, supramolecular structure, and molecular conformation of lipopolysaccharide. Immunobiology 187:191–211
33. Brandenburg K, Mayer H, Koch MH, Weckesser J, Rietschel ET, Seydel U (1993) Influence of the supramolecular structure of free lipid A on its biological activity. Eur J Biochem 218:555–563
34. Gutsmann T, Schromm AB, Brandenburg K (2007) The physicochemistry of endotoxins in relation to bioactivity. Int J Med Microbiol 297:341–352

35. Jerala R (2007) Structural biology of the LPS recognition. Int J Med Microbiol 297:353–363
36. Schumann RR, Leong SR, Flaggs GW, Gray PW, Wright SD, Mathison JC, Tobias PS, Ulevitch RJ (1990) Structure and function of lipopolysaccharide binding protein. Science 249:1429–1431
37. Wright SD, Ramos RA, Tobias PS, Ulevitch RJ, Mathison JC (1990) CD14, a receptor for complexes of lipopolysaccharide (LPS) and LPS binding protein. Science 249:1431–1433
38. Gioannini TL, Teghanemt A, Zhang D, Coussens NP, Dockstader W, Ramaswamy S, Weiss JP (2004) Isolation of an endotoxin-MD-2 complex that produces Toll-like receptor 4-dependent cell activation at picomolar concentrations. Proc Natl Acad Sci USA 101:4186–4191
39. Prohinar P, Re F, Widstrom R, Zhang D, Teghanemt A, Weiss JP, Gioannini TL (2007) Specific high affinity interactions of monomeric endotoxin.protein complexes with Toll-like receptor 4 ectodomain. J Biol Chem 282:1010–1017
40. Gegner JA, Ulevitch RJ, Tobias PS (1995) Lipopolysaccharide (LPS) signal transduction and clearance. Dual roles for LPS binding protein and membrane CD14. J Biol Chem 270: 5320–5325
41. Weiss J (2003) Bactericidal/permeability-increasing protein (BPI) and lipopolysaccharide-binding protein (LBP): structure, function and regulation in host defence against gram-negative bacteria. Biochem Soc Trans 31:785–790
42. Shimazu R, Akashi S, Ogata H, Nagai Y, Fukudome K, Miyake K, Kimoto M (1999) MD-2, a molecule that confers lipopolysaccharide responsiveness on Toll-like receptor 4. J Exp Med 189:1777–1782
43. Visintin A, Halmen KA, Latz E, Monks BG, Golenbock DT (2005) Pharmacological inhibition of endotoxin responses is achieved by targeting the TLR4 coreceptor, MD-2. J Immunol 175:6465–6472
44. Kim HM, Park BS, Kim JI, Kim SE, Lee J, Oh SC, Enkhbayar P, Matsushima N, Lee H, Yoo OJ, Lee JO (2007) Crystal structure of the TLR4-MD-2 complex with bound endotoxin antagonist Eritoran. Cell 130:906–917
45. David SA (2001) Towards a rational development of anti-endotoxin agents: novel approaches to sequestration of bacterial endotoxins with small molecules. J Mol Recognit 14:370–387
46. Peterson AA, Hancock RE, McGroarty EJ (1985) Binding of polycationic antibiotics and polyamines to lipopolysaccharides of *Pseudomonas aeruginosa*. J Bacteriol 164:1256–1261
47. Vaara M, Vaara T (1983) Polycations sensitize enteric bacteria to antibiotics. Antimicrob Agents Chemother 24:107–113
48. Schumann RR, Lamping N, Hoess A (1997) Interchangeable endotoxin-binding domains in proteins with opposite lipopolysaccharide-dependent activities. J Immunol 159:5599–5605
49. Weersink AJ, van Kessel KP, van den Tol ME, van Strijp JA, Torensma R, Verhoef J, Elsbach P, Weiss J (1993) Human granulocytes express a 55-kDa lipopolysaccharide-binding protein on the cell surface that is identical to the bactericidal/permeability-increasing protein. J Immunol 150:253–263
50. Lamping N, Dettmer R, Schroder NW, Pfeil D, Hallatschek W, Burger R, Schumann RR (1998) LPS-binding protein protects mice from septic shock caused by LPS or gram-negative bacteria. J Clin Invest 101:2065–2071
51. Abrahamson SL, Wu HM, Williams RE, Der K, Ottah N, Little R, Gazzano-Santoro H, Theofan G, Bauer R, Leigh S, Orme A, Horwitz AH, Carroll SF, Dedrick RL (1997) Biochemical characterization of recombinant fusions of lipopolysaccharide binding protein and bactericidal/permeability-increasing protein. Implications in biological activity. J Biol Chem 272:2149–2155
52. Muhle SA, Tam JP (2001) Design of gram-negative selective antimicrobial peptides. Biochemistry 40:5777–5785
53. Larrick JW, Hirata M, Balint RF, Lee J, Zhong J, Wright SC (1995) Human CAP18: a novel antimicrobial lipopolysaccharide-binding protein. Infect Immun 63:1291–1297

54. Levy O, Ooi CE, Elsbach P, Doerfler ME, Lehrer RI, Weiss J (1995) Antibacterial proteins of granulocytes differ in interaction with endotoxin. Comparison of bactericidal/permeability-increasing protein, p15s, and defensins. J Immunol 154:5403–5410
55. Gough M, Hancock RE, Kelly NM (1996) Antiendotoxin activity of cationic peptide antimicrobial agents. Infect Immun 64:4922–4927
56. Scott MG, Rosenberger CM, Gold MR, Finlay BB, Hancock RE (2000) An α-helical cationic antimicrobial peptide selectively modulates macrophage responses to lipopolysaccharide and directly alters macrophage gene expression. J Immunol 165:3358–3365
57. Scott MG, Vreugdenhil AC, Buurman WA, Hancock RE, Gold MR (2000) Cutting edge: cationic antimicrobial peptides block the binding of lipopolysaccharide (LPS) to LPS binding protein. J Immunol 164:549–553
58. Morrison DC, Jacobs DM (1976) Binding of polymyxin B to the lipid A portion of bacterial lipopolysaccharides. Immunochemistry 13:813–818
59. Vincent JL, Laterre PF, Cohen J, Burchardi H, Bruining H, Lerma FA, Wittebole X, De Backer D, Brett S, Marzo D, Nakamura H, John S (2005) A pilot-controlled study of a polymyxin B-immobilized hemoperfusion cartridge in patients with severe sepsis secondary to intra-abdominal infection. Shock 23:400–405
60. Fiore B, Soncini M, Vesentini S, Penati A, Visconti G, Redaelli A (2006) Multi-scale analysis of the toraymyxin adsorption cartridge. Part II: computational fluid-dynamic study. Int J Artif Organs 29:251–260
61. Miller KA, Suresh Kumar EV, Wood SJ, Cromer JR, Datta A, David SA (2005) Lipopolysaccharide sequestrants: structural correlates of activity and toxicity in novel acylhomospermines. J Med Chem 48:2589–2599
62. Burns MR, Jenkins SA, Kimbrell MR, Balakrishna R, Nguyen TB, Abbo BG, David SA (2007) Polycationic sulfonamides for the sequestration of endotoxin. J Med Chem 50:877–888
63. Dziarski R, Tapping RI, Tobias PS (1998) Binding of bacterial peptidoglycan to CD14. J Biol Chem 273:8680–8690
64. Beamer LJ, Carroll SF, Eisenberg D (1997) Crystal structure of human BPI and two bound phospholipids at 2.4 angstrom resolution. Science 276:1861–1864
65. Beamer LJ, Carroll SF, Eisenberg D (1998) The BPI/LBP family of proteins: a structural analysis of conserved regions. Protein Sci 7:906–914
66. Gioannini TL, Zhang D, Teghanemt A, Weiss JP (2002) An essential role for albumin in the interaction of endotoxin with lipopolysaccharide-binding protein and sCD14 and resultant cell activation. J Biol Chem 277:47818–47825
67. Corradin SB, Mauel J, Gallay P, Heumann D, Ulevitch RJ, Tobias PS (1992) Enhancement of murine macrophage binding of and response to bacterial lipopolysaccharide (LPS) by LPS-binding protein. J Leukoc Biol 52:363–368
68. Dentener MA, Von Asmuth EJ, Francot GJ, Marra MN, Buurman WA (1993) Antagonistic effects of lipopolysaccharide binding protein and bactericidal/permeability-increasing protein on lipopolysaccharide-induced cytokine release by mononuclear phagocytes. Competition for binding to lipopolysaccharide. J Immunol 151:4258–4265
69. Zweigner J, Gramm HJ, Singer OC, Wegscheider K, Schumann RR (2001) High concentrations of lipopolysaccharide-binding protein in serum of patients with severe sepsis or septic shock inhibit the lipopolysaccharide response in human monocytes. Blood 98:3800–3808
70. Hamann L, Alexander C, Stamme C, Zähringer U, Schumann RR (2005) Acute-phase concentrations of lipopolysaccharide (LPS)-binding protein inhibit innate immune cell activation by different LPS chemotypes via different mechanisms. Infect Immun 73:193–200
71. Jiang Z, Georgel P, Du X, Shamel L, Sovath S, Mudd S, Huber M, Kalis C, Keck S, Galanos C, Freudenberg M, Beutler B (2005) CD14 is required for MyD88-independent LPS signaling. Nat Immunol 6:565–570
72. Zanoni I, Ostuni R, Capuano G, Collini M, Caccia M, Ronchi AE, Rocchetti M, Mingozzi F, Foti M, Chirico G, Costa B, Zaza A, Ricciardi-Castagnoli P, Granucci F (2009) CD14 regulates the dendritic cell life cycle after LPS exposure through NFAT activation. Nature 460:264–268

73. Kim JI, Lee CJ, Jin MS, Lee CH, Paik SG, Lee H, Lee JO (2005) Crystal structure of CD14 and its implications for lipopolysaccharide signaling. J Biol Chem 280:11347–11351
74. Cunningham MD, Shapiro RA, Seachord C, Ratcliffe K, Cassiano L, Darveau RP (2000) CD14 employs hydrophilic regions to "capture" lipopolysaccharides. J Immunol 164: 3255–3263
75. Albright S, Chen B, Holbrook K, Jain NU (2008) Solution NMR studies provide structural basis for endotoxin pattern recognition by the innate immune receptor CD14. Biochem Biophys Res Commun 368:231–237
76. Albright S, Agrawal P, Jain NU (2009) NMR spectral mapping of lipid A molecular patterns affected by interaction with the innate immune receptor CD14. Biochem Biophys Res Commun 378:721–726
77. Kitchens RL, Munford RS (1995) Enzymatically deacylated lipopolysaccharide (LPS) can antagonize LPS at multiple sites in the LPS recognition pathway. J Biol Chem 270: 9904–9910
78. Kobe B, Kajava AV (2001) The leucine-rich repeat as a protein recognition motif. Curr Opin Struct Biol 11:725–732
79. Fan X, Stelter F, Menzel R, Jack R, Spreitzer I, Hartung T, Schutt C (1999) Structures in *Bacillus subtilis* are recognized by CD14 in a lipopolysaccharide binding protein-dependent reaction. Infect Immun 67:2964–2968
80. Mitsuzawa H, Wada I, Sano H, Iwaki D, Murakami S, Himi T, Matsushima N, Kuroki Y (2001) Extracellular Toll-like receptor 2 region containing Ser40-Ile64 but not Cys30-Ser39 is critical for the recognition of *Staphylococcus aureus* peptidoglycan. J Biol Chem 276: 41350–41356
81. Rietschel ET, Schletter J, Weidemann B, El-Samalouti V, Mattern T, Zähringer U, Seydel U, Brade H, Flad H, Kusumoto S, Gupta D, Dziarski R, Ulmer AJ (1998) Lipopolysaccharide and peptidoglycan: CD14-dependent bacterial inducers of inflammation. Microb Drug Resist 4:37–44
82. Gupta D, Wang Q, Vinson C, Dziarski R (1999) Bacterial peptidoglycan induces CD14-dependent activation of transcription factors CREB/ATF and AP-1. J Biol Chem 274: 14012–14020
83. Asai Y, Hashimoto M, Ogawa T (2003) Treponemal glycoconjugate inhibits Toll-like receptor ligand-induced cell activation by blocking LPS-binding protein and CD14 functions. Eur J Immunol 33:3196–3204
84. Fujita M, Into T, Yasuda M, Okusawa T, Hamahira S, Kuroki Y, Eto A, Nisizawa T, Morita M, Shibata K (2003) Involvement of leucine residues at positions 107, 112, and 115 in a leucine-rich repeat motif of human Toll-like receptor 2 in the recognition of diacylated lipoproteins and lipopeptides and *Staphylococcus aureus* peptidoglycans. J Immunol 171: 3675–3683
85. Piazza M, Rossini C, Della Fiorentina S, Pozzi C, Comelli F, Bettoni I, Fusi P, Costa B, Peri F (2009) Glycolipids and benzylammonium lipids as novel antisepsis agents: synthesis and biological characterization. J Med Chem 52:1209–1213
86. Piazza M, Yu L, Teghanemt A, Gioannini T, Weiss J, Peri F (2009) Evidence of a specific interaction between new synthetic antisepsis agents and CD14. Biochemistry 48: 12337–12344
87. Piazza M, Damore G, Costa B, Gioannini TL, Weiss JP, Peri F (2011) Hemin and a metabolic derivative coprohemin modulate the TLR4 pathway differently through different molecular targets. Innate Immun. doi:10.1177/1753425910369020
88. Ulrich JT, Myers KR (1995) Monophosphoryl lipid A as an adjuvant. Past experiences and new directions. Pharm Biotechnol 6:495–524
89. Dupont J, Altclas J, Lepetic A, Lombardo M, Vazquez V, Salgueira C, Seigelchifer M, Arndtz N, Antunez E, von Eschen K, Janowicz Z (2006) A controlled clinical trial comparing the safety and immunogenicity of a new adjuvanted hepatitis B vaccine with a standard hepatitis B vaccine. Vaccine 24:7167–7174

90. Harper DM, Franco EL, Wheeler CM, Moscicki AB, Romanowski B, Roteli-Martins CM, Jenkins D, Schuind A, Costa Clemens SA, Dubin G (2006) Sustained efficacy up to 4.5 years of a bivalent L1 virus-like particle vaccine against human papillomavirus types 16 and 18: follow-up from a randomised control trial. Lancet 367:1247–1255
91. Evans JT, Cluff CW, Johnson DA, Lacy MJ, Persing DH, Baldridge JR (2003) Enhancement of antigen-specific immunity via the TLR4 ligands MPL adjuvant and Ribi.529. Expert Rev Vac 2:219–229
92. Butts C, Murray N, Maksymiuk A, Goss G, Marshall E, Soulieres D, Cormier Y, Ellis P, Price A, Sawhney R, Davis M, Mansi J, Smith C, Vergidis D, MacNeil M, Palmer M (2005) Randomized phase IIB trial of BLP25 liposome vaccine in stage IIIB and IV non-small-cell lung cancer. J Clin Oncol 23:6674–6681
93. Johnson DA, Keegan DS, Sowell CG, Livesay MT, Johnson CL, Taubner LM, Harris A, Myers KR, Thompson JD, Gustafson GL, Rhodes MJ, Ulrich JT, Ward JR, Yorgensen YM, Cantrell JL, Brookshire VG (1999) 3-O-Desacyl monophosphoryl lipid A derivatives: synthesis and immunostimulant activities. J Med Chem 42:4640–4649
94. Johnson DA (2008) Synthetic TLR4-active glycolipids as vaccine adjuvants and stand-alone immunotherapeutics. Curr Top Med Chem 8:64–79
95. North S, Butts C (2005) Vaccination with BLP25 liposome vaccine to treat non-small cell lung and prostate cancers. Expert Rev Vac 4:249–257
96. Kawai Y, Akagawa K (1989) Macrophage activation by an ornithine-containing lipid or a serine-containing lipid. Infect Immun 57:2086–2091
97. Kawai Y, Takasuka N, Inoue K, Akagawa K, Nishijima M (2000) Ornithine-containing lipids stimulate CD14-dependent TNF-α production from murine macrophage-like J774.1 and RAW 264.7 cells. FEMS Immunol Med Microbiol 28:197–203
98. Baldridge JR, McGowan P, Evans JT, Cluff C, Mossman S, Johnson D, Persing D (2004) Taking a Toll on human disease: Toll-like receptor 4 agonists as vaccine adjuvants and monotherapeutic agents. Expert Opin Biol Ther 4:1129–1138
99. Fort MM, Mozaffarian A, Stover AG, Correia Jda S, Johnson DA, Crane RT, Ulevitch RJ, Persing DH, Bielefeldt-Ohmann H, Probst P, Jeffery E, Fling SP, Hershberg RM (2005) A synthetic TLR4 antagonist has anti-inflammatory effects in two murine models of inflammatory bowel disease. J Immunol 174:6416–6423
100. Matsuura M, Kiso M, Hasegawa A (1999) Activity of monosaccharide lipid A analogues in human monocytic cells as agonists or antagonists of bacterial lipopolysaccharide. Infect Immun 67:6286–6292
101. Tamai R, Asai Y, Hashimoto M, Fukase K, Kusumoto S, Ishida H, Kiso M, Ogawa T (2003) Cell activation by monosaccharide lipid A analogues utilizing Toll-like receptor 4. Immunology 110:66–72
102. Lien E, Chow JC, Hawkins LD, McGuinness PD, Miyake K, Espevik T, Gusovsky F, Golenbock DT (2001) A novel synthetic acyclic lipid A-like agonist activates cells via the lipopolysaccharide/toll-like receptor 4 signaling pathway. J Biol Chem 276:1873–1880
103. Brandenburg K, Hawkins L, Garidel P, Andra J, Muller M, Heine H, Koch MH, Seydel U (2004) Structural polymorphism and endotoxic activity of synthetic phospholipid-like amphiphiles. Biochemistry 43:4039–4046
104. Loppnow H, Libby P, Freudenberg M, Krauss JH, Weckesser J, Mayer H (1990) Cytokine induction by lipopolysaccharide (LPS) corresponds to lethal toxicity and is inhibited by nontoxic *Rhodobacter capsulatus* LPS. Infect Immun 58:3743–3750
105. Christ WJ, Asano O, Robidoux AL, Perez M, Wang Y, Dubuc GR, Gavin WE, Hawkins LD, McGuinness PD, Mullarkey MA, Lewis MD, Kishi Y, Kawata T, Bristol JR, Rose JR, Rossignol DP, Kobayashi S, Hishinuma L, Kimura A, Asakawa N, Katayama K, Yamatsu I (1995) E5531, a pure endotoxin antagonist of high potency. Science 268:80–83
106. Wasan KM, Strobel FW, Parrott SC, Lynn M, Christ WJ, Hawkins LD, Rossignol DP (1999) Lipoprotein distribution of a novel endotoxin antagonist, E5531, in plasma from human subjects with various lipid levels. Antimicrob Agents Chemother 43:2562–2564

107. Rossignol DP, Lynn M (2002) Antagonism of in vivo and *ex vivo* response to endotoxin by E5564, a synthetic lipid A analogue. J Endotoxin Res 8:483–488
108. Mullarkey M, Rose JR, Bristol J, Kawata T, Kimura A, Kobayashi S, Przetak M, Chow J, Gusovsky F, Christ WJ, Rossignol DP (2003) Inhibition of endotoxin response by e5564, a novel Toll-like receptor 4-directed endotoxin antagonist. J Pharmacol Exp Ther 304: 1093–1102
109. Lynn M, Rossignol DP, Wheeler JL, Kao RJ, Perdomo CA, Noveck R, Vargas R, D'Angelo T, Gotzkowsky S, McMahon FG (2003) Blocking of responses to endotoxin by E5564 in healthy volunteers with experimental endotoxemia. J Infect Dis 187:631–639
110. Tidswell M, Tillis W, Larosa SP, Lynn M, Wittek AE, Kao R, Wheeler J, Gogate J, Opal SM (2010) Phase 2 trial of eritoran tetrasodium (E5564), a Toll-like receptor 4 antagonist, in patients with severe sepsis. Crit Care Med 38:72–83
111. Mancek-Keber M, Jerala R (2006) Structural similarity between the hydrophobic fluorescent probe and lipid A as a ligand of MD-2. FASEB J 20:1836–1842
112. Fitzpatrick FA, Wheeler R (2003) The immunopharmacology of paclitaxel (Taxol), docetaxel (Taxotere), and related agents. Int Immunopharmacol 3:1699–1714
113. Resman N, Gradisar H, Vasl J, Keber MM, Pristovsek P, Jerala R (2008) Taxanes inhibit human TLR4 signaling by binding to MD-2. FEBS Lett 582:3929–3934
114. Youn HS, Saitoh SI, Miyake K, Hwang DH (2006) Inhibition of homodimerization of Toll-like receptor 4 by curcumin. Biochem Pharmacol 72:62–69
115. Gradisar H, Keber MM, Pristovsek P, Jerala R (2007) MD-2 as the target of curcumin in the inhibition of response to LPS. J Leukoc Biol 82:968–974
116. Yamada M, Ichikawa T, Ii M, Sunamoto M, Itoh K, Tamura N, Kitazaki T (2005) Discovery of novel and potent small-molecule inhibitors of NO and cytokine production as antisepsis agents: synthesis and biological activity of alkyl 6-(N-substituted sulfamoyl)cyclohex-1-ene-1-carboxylate. J Med Chem 48:7457–7467
117. Sha T, Sunamoto M, Kitazaki T, Sato J, Ii M, Iizawa Y (2007) Therapeutic effects of TAK-242, a novel selective Toll-like receptor 4 signal transduction inhibitor, in mouse endotoxin shock model. Eur J Pharmacol 571:231–239
118. Kawamoto T, Ii M, Kitazaki T, Iizawa Y, Kimura H (2008) TAK-242 selectively suppresses Toll-like receptor 4-signaling mediated by the intracellular domain. Eur J Pharmacol 584: 40–48
119. Takashima K, Matsunaga N, Yoshimatsu M, Hazeki K, Kaisho T, Uekata M, Hazeki O, Akira S, Iizawa Y, Ii M (2009) Analysis of binding site for the novel small-molecule TLR4 signal transduction inhibitor TAK-242 and its therapeutic effect on mouse sepsis model. Br J Pharmacol 157:1250–1262
120. Slivka PF, Shridhar M, Lee GI, Sammond DW, Hutchinson MR, Martinko AJ, Buchanan MM, Sholar PW, Kearney JJ, Harrison JA, Watkins LR, Yin H (2009) A peptide antagonist of the TLR4-MD-2 interaction. Chembiochem 10:645–649
121. Figueiredo RT, Fernandez PL, Mourao-Sa DS, Porto BN, Dutra FF, Alves LS, Oliveira MF, Oliveira PL, Graca-Souza AV, Bozza MT (2007) Characterization of heme as activator of Toll-like receptor 4. J Biol Chem 282:20221–20229
122. Noman ASM, Koide N, Hassan F, I.-E-Khuda I, Dagvadorj J, Tumurkhuu G, Islam S, Naiki Y, Yoshida T, Yokochi T (2009) Thalidomide inhibits lipopolysaccharide-induced tumor necrosis factor-α production via down-regulation of MyD88 expression. Innate Immun 15:33–41
123. Hutchinson MR, Zhang Y, Shridhar M, Evans JH, Buchanan MM, Zhao TX, Slivka PF, Coats BD, Rezvani N, Wieseler J, Hughes TS, Landgraf KE, Chan S, Fong S, Phipps S, Falke JJ, Leinwand LA, Maier SF, Yin H, Rice KC, Watkins LR (2010) Evidence that opioids may have toll like receptor 4 and MD-2 effects. Brain Behav Immun 24:83–95
124. Verdu B, Decosterd I, Buclin T, Stiefel F, Berney A (2008) Antidepressants for the treatment of chronic pain. Drugs 68:2611–2632
125. Stockwell BR (2000) Frontiers in chemical genetics. Trends Biotechnol 18:449–455

Lipopolysaccharide and Its Interactions with Plants

14

Gitte Erbs and Mari-Anne Newman

14.1 Introduction

In an environment that is rich in potentially pathogenic microorganisms, the survival of higher eukaryotic organisms depends on efficient pathogen sensing and rapidly mounted defence responses. Such protective mechanisms are found in all multicellular organisms and are collectively referred to as innate immunity. Innate immunity is the first line of defence against invading microorganisms in vertebrates and the only line of defence in invertebrates and plants [1]. Plants interact with a variety of microorganisms, and like insects and mammals, they respond to a broad range of microbial molecules. The recognition of non-self induces plant defence responses such as the oxidative burst, nitric oxide (NO) generation, extracellular pH increase, cell wall strengthening and pathogenesis-related (PR) protein accumulation, leading to basal resistance or innate immunity. Recognition of non-self, such as an invading pathogen, is crucial for an effective defence response.

Plants perceive several general elicitors from both host and non-host pathogens. These elicitors are essential structures for pathogen survival and are for that reason conserved among pathogens. These conserved microbe-specific molecules, also referred to as microbe- or pathogen-associated molecular patterns (MAMPs or PAMPs), are recognised by the plant innate immune systems pattern recognition receptors (PRRs). MAMPs are invading evolutionarily conserved microbe-derived molecules that distinguish hosts from pathogens [2, 3]. The term MAMP, which we will use here, was coined to reflect that these elicitor molecules are not

G. Erbs • M.-A. Newman (✉)
Department of Plant Biology and Biotechnology, University of Copenhagen, Thorvaldsensvej 40, 1871 Frederiksberg, Denmark
e-mail: ger@life.ku.dk; mari@life.ku.dk

Y.A. Knirel and M.A. Valvano (eds.), *Bacterial Lipopolysaccharides*,
DOI 10.1007/978-3-7091-0733-1_14, © Springer-Verlag/Wien 2011

restricted to pathogens, but can also be found in non-pathogenic and saprophytic organisms. MAMPs from bacteria include for example: (1) lipopolysaccharide (LPS), the major component of the outer membrane of Gram-negative bacteria [4–7]; (2) peptidoglycan, which provides rigidity and structure to the cell envelopes of both Gram-negative and Gram-positive bacteria [8]; (3) flagellin, the main component of the bacterial motility organelles [9]; and (4) the elongation factor Tu (EF-Tu), which is essential for protein translation and is the most abundant bacterial protein [10]. These examples illustrate common features of MAMPs: they are usually indispensable for microbial fitness and relatively invariant in structure.

In this chapter, we will review the current knowledge of the role of LPS as a MAMP in plant innate immunity. We will give an overview of a range of responses induced by LPS, the substructures within LPS that are recognised by plants and variations within the LPS structure that can alter its activity as a MAMP. We will also discuss new work that suggests a role for the plasma-membrane plant protein encoded by the *PENETRATION1* locus (syntaxin PEN1) in the transduction of the LPS signal.

14.2 LPS Induces Basal Plant Defence Responses

As the main surface component of the bacterial cell envelope LPS is thought to contribute to the restrictive Gram-negative outer membrane permeability, allowing bacterial growth in unfavourable environments, such as those that may be encountered within or on plants. The exclusion of antimicrobial substances of plant origin probably contributes to the ability of pathogenic bacteria to parasitize plants. LPS-defective mutants show increased in vitro sensitivity to antibiotics and antimicrobial peptides and the numbers of viable bacteria often decline very rapidly upon introduction into plants. LPS may also promote bacterial adherence to plant surfaces [11].

In contrast to this role in promoting plant disease, e.g. protection and barrier function against host compounds, there are various reports detailing the effects of LPS on the induction of basal plant defences, consistent with its designation as a MAMP [11]. LPS preparations from a number of bacteria induced NO synthesis in suspension cultures and in leaves of *Arabidopsis thaliana* [12]. This common effect of LPS from diverse bacteria suggested the involvement of a shared molecular determinant, the lipid A moiety, and indeed isolated lipid A was also active. LPS can induce the production of reactive oxygen species (ROS), although this has not always been observed [5, 11]. For example, although LPS from the plant pathogen *Xanthomonas campestris* pv. *campestris* (*Xcc*) induces an oxidative burst in culture cells of tobacco, no effects are seen with LPS from the enteric bacterium *Salmonella enterica* serovar Typhimurium [13]. Furthermore, *Xcc* LPS did not elicit generation of ROS in cultured soybean cells [5]. Recent studies revealed that LPS from various pathogenic and nonpathogenic bacteria induce the generation of ROS and defence-related gene expression in rice, indicating that the machinery recognising LPS is evolutionary conserved in monocots and dicots [14].

LPS also has effects on cell wall alterations such as callose deposition [15] and on PR gene induction [7, 12]. In some cases, specific effects of a particular LPS on plant gene induction are observed. LPS from the crucifer pathogen *Xcc* induce expression of a gene encoding a defence-related β-(1–3) glucanase when applied to turnip leaves at 1 μg mL^{-1}. In contrast, LPS from *Escherichia coli* and *S. enterica* are ineffective at concentrations up to 50 μg mL^{-1} [6].

Several attempts have been made to identify plant components involved in LPS recognition and perception. Interestingly, Livaja et al. [16] found that in *Arabidopsis* cells, *Burkholderia cepacia* LPS induced a leucine-rich repeat receptor-like kinase At5g45840 by nearly 17-fold after 30 min. Furthermore, in a proteomic analysis of the changes following perception of LPS from an endophytic strain of *B. cepacia* in *Nicotiana tabacum* BY-2 cells, 88 LPS induced/regulated proteins and phosphoproteins were identified, many of which were found to be involved in metabolism- and energy-related processes. Moreover, proteins were found that are known to be involved in protein synthesis, protein folding, vesicle trafficking and secretion [17, 18].

Livaja et al. [16] performed a transcription profiling of *A. thaliana* cells treated with 100 μg mL^{-1} LPS from *B. cepacia* or 50 μg mL^{-1} harpin from *Pseudomonas syringae*. The transcriptional changes in the treated and non-treated cells were monitored at 0.5, 1, 2, 4, 8 and 24 h after elicitor treatment. Focusing on changes induced by *B. cepacia* LPS, the authors surprisingly did not find any genes involved in callose synthesis. Furthermore, genes involved in ROS production were found to be upregulated at a very low level by *B. cepacia* LPS, except after 8 h where a superoxide dismutase (SOD) and a ferritin one precursor gene were strongly induced. In addition, Livaja et al. [16] found that *B. cepacia* LPS only induced the pathogenesis-related (PR) genes *PR3* and *PR4*, whereas studies in *B. cepacia* LPS treated *Arabidopsis* leaves revealed induction of several PR genes [12]. Other LPS preparations, from *Pseudomonas aeruginosa* and *E. coli* respectively, induce *PR1* and *PR5* in *Arabidopsis* leaves [19]. The conflict in results both reflects the different plant systems (*A. thaliana* cell cultures contra the whole plant) and the origin of the LPS. All the above very specific effects show the ability of particular plants to recognise structural features within LPS that are not necessarily widely conserved.

14.2.1 LPS as Primer and Modulator of Plant Defence Responses

In addition to direct effects on plant tissue, treatment with LPS affects the pattern of accumulation of gene expression and accumulation of certain phenolics in plants in response to subsequent inoculation with virulent or avirulent bacteria [20]. LPS pre-treatment of pepper leaves altered patterns of gene expression induced by subsequent challenge with bacteria. *S. enterica* serovar Minnesota LPS alone did not induce genes encoding the PR proteins such as P6, and acidic and basic β-1,3-glucanases, while *Xcc* LPS alone gave a weak, transient expression [21]. However, pretreatment of pepper leaves with LPS affected the pattern of expression and accumulation of the above-mentioned genes following a subsequent challenge with *Xcc* (avirulent) and *Xc* pv. *vesicatoria* (virulent) [21]. Newman et al. [22] also examined the effects of LPS pretreatment on the accumulation of salicylic acid (SA) and the synthesis of the phenolic conjugates coumaroyl tyramine (CT) and feruloyl tyramine (FT). The hypersensitive response (HR, see below) in pepper is associated with increased levels of SA [23]. CT and FT are suggested to have two possible roles in plant defence, both as direct antimicrobial agents and in cell-wall reinforcement [23, 24]. LPS had apparently little effect on the timing of accumulation of SA, while the timing of accumulation of FT and CT was considerably altered. LPS pretreatment caused these two compounds to accumulate much more rapidly upon inoculation with *Xcc*. Yet LPS alone did not induce SA, CT or FT synthesis. As part of their virulence strategy, many phytopathogenic bacteria inject a suite of effector proteins directly into the host cell through a type III secretion system [25]. These effector molecules contribute to bacterial virulence in susceptible plants by interfering with or subverting host cell processes, including the triggering of innate immunity [26–28]. However in some plants, effectors can be recognised to trigger the HR, a programmed cell death associated with plant disease resistance [29]. This recognition involves the protein products of plant resistance (R) genes and the effectors that are recognised have been called avirulence (Avr) proteins, although this term does not reflect their role in promoting virulence in susceptible hosts. Pretreatment of leaves of *Arabidopsis* with purified *Xcc* LPS prevented the HR caused by subsequently inoculated avirulent strains of *P. syringae* pv. *tomato* DC3000 carrying genes expressing different effectors (AvrRpm1 or AvrRps4). The one or more mechanisms by which LPS prevents the HR are as yet unknown, but this phenomenon is associated with an enhanced resistance of the plant tissue to bacteria, which is presumed to occur through LPS-dependent induction or priming for enhanced plant defence responses [7, 22].

The effects of LPS on preventing HR (which is associated with plant resistance) while also inducing basal defences appear to present a conundrum. However, measurement of bacterial growth in LPS-treated leaves indicates that prevention of HR does not increase the susceptibility of the plant tissue. This is consistent with the notion that LPS perception allows the plant to express resistance (through enhanced expression of basal defences) without the catastrophic collapse of the HR. The underlying mechanisms are still unknown. Intriguingly, although LPS has never been shown to elicit HR in dicots [5], LPS from various bacteria induces programmed cell death in rice cells [14].

14.2.2 Substructures of LPS Recognised by Plants

LPS from plant-associated and plant pathogenic bacteria possess the same tripartite structure comprising lipid A, core oligosaccharide (OS) and an O-polysaccharide or O-antigen seen in LPS from other bacteria [30, 31]. The lipid A and the core OS are linked in the majority of cases by the sugar 3-*deoxy*-D-*manno*-2-octulosonate (Kdo). LPS molecules that lack an O-antigen are called lipooligosaccharides (LOS).

Several laboratories have investigated the contribution of the different moieties within LPS to the MAMP elicitor activity. Silipo et al. [7] determined the complete structure of purified LOS from *Xcc*, the lipid A and core OS derived from it by mild acid hydrolysis and in parallel examined the activity of these (structurally-defined) components in defence gene induction in *Arabidopsis*. *Xcc* LOS was found to be a unique molecule with a high negative charge density and a phosphoramide group, which had never been found previously as a component of LPS [7]. *Xcc* LOS induced the defence-related *PR1* and *PR2* genes in *Arabidopsis* leaves in two temporal phases; the core OS induced only the early phase and the lipid A moiety only the later phase. These findings suggest that although both *Xcc* lipid A and the *Xcc* core OS are active in defence gene induction, they may be recognised by different plant receptors [7]. This elicitor activity of *Xcc* lipid A correlates with earlier studies by Zeidler et al. [12] who showed that lipid A preparations from various bacteria induced a rapid burst of NO production that was associated with the induction of defence-related genes in *Arabidopsis*.

Interestingly, the core OS from *E. coli* and *Ralstonia solanacearum* does not prevent HR or induce defence-related genes [32], indicating that the effect of the *Xcc* core OS could be due to the unique phosphoramide group in that particular LPS molecule [7]. In contrast, in tobacco cells *Xcc* lipid A could not induce the oxidative burst, but rather it was the inner core OS part of the LPS molecule that was responsible [33]. This disparity in outcomes might be a reflection of the use of different plants, the difference in the age of the plants used (plant cell cultures versus seedlings versus fully developed plants) and the different defence responses measured after treatment with LPS and its derivatives.

Evidence for a role of the O-antigen in eliciting defence responses stems from the different ability of LPS derived from wild type *Pseudomonas fluorescens* and a mutant lacking the O-antigen in induction of induced systemic resistance (ISR) [34, 35]. More recently the role of the O-antigen has been directly examined by studies of the biological activity of synthetic O-antigen polysaccharides. Structural studies of LPS from many phytopathogenic bacteria have revealed that the O-antigen comprises a rhamnan with the trisaccharide repeating unit [α-L-Rha-(1–3)-α-L-Rha-(1–2)-α-LRha-(1–3)] [36]. This trisaccharide was synthesised and the trimer oligomerised to generate a set of OSs of increasing chain length. The tri-, hexa- and nona-saccharide synthetic O-antigens were found to suppress the HR and induce *PR1* and *PR2* transcript accumulation in *Arabidopsis*. Interestingly, the efficiency of HR suppression and PR gene induction improved with increasing chain length [4]. Moreover, this increasing chain length was associated with the formation of a coiled structure, suggesting a role for this structure as a MAMP. By extension, these findings suggest a role for the O-antigen from many phytopathogenic bacteria in triggering plant innate immunity [4].

14.2.3 Structural Variations in LPS Influence its Activity in Plants

LPS is recognised by mammalian cells through the lipid A moiety and this recognition governs the interactions with the innate immune system [37]. *E. coli* lipid A, which is an effective agonistic structure of immune responses in mammalian cells, consists of a bisphosphorylated hexaacylated disaccharide backbone with an asymmetric distribution of the acyl residues. Modifications of the lipid A structure influence the biological activity of the molecule in mammals [38]. Schromm et al. [39, 40] showed that the molecular conformation of the lipid A correlated with its biological activity. The molecular shape of lipid A is influenced by the net negative charge usually associated with the degree of phosphorylation [41]. Molecules with several negative charges adopt a conical shape and have endotoxin activity. Molecules with very few or no negative charges adopt a cylindrical shape, are less potent as endotoxins and can even have antagonistic activity.

Structural differences on the lipid A skeleton such as the level of acylation can affect its agonist/antagonist activity [42]. For example, LPS from *Shigella flexneri* elicits a weaker TLR4-mediated response in mammalian cells than *E. coli* LPS due to differences in the acylation status of their lipid A moieties [43]. Alteration of the lipid A structure also influences the biological activity in plants. Dephosphorylation of *Xcc* LOS leaves only one negative charge on the Kdo residue. The resultant molecule is unable to prevent HR in *Arabidopsis* leaves, suggesting that the charged groups present in LOS play a key role in inducing defence responses in plants [7]. Other findings suggest that plants are sensitive to the same structures of lipid A that determine biological activity in humans [44]. The unusual and very low degree of acylation in the lipid A from *Halomonas magadiensis*, a Gram-negative extremophilic and alkaliphilic bacteria isolated from a soda lake in an East African Rift Valley, suggests that the molecule may act as a LPS antagonist in human cells [45]. Indeed, *H. magadiensis* lipid A inhibited *E. coli* lipid A-induced immune responses in human cells [46], and Erbs et al. [44] found that *H. magadiensis* lipid A antagonise the action of *E. coli* lipid A when inducing *PR1* gene expression in *A. thaliana*. Even though the mode of perception of LPS in plants is far less understood than in mammals and insects, these results indicate that *A. thaliana* is sensitive to the same structures of lipid A that determine biological activity in humans.

Characterisation of the complete structure of LOS from a nonpathogenic *Xcc* mutant strain 8530, which carries a *Tn5* insertion in a gene of unknown function [47], revealed that this mutant had a truncated core OS region and also lipid A modifications in the level of acylation (pentaacetylated rather than hexaacetylated) and phosphorylation (a phosphoethanolamine substitution). These changes influence the ability of *Xcc* 8530 LPS to trigger innate immune responses in *Arabidopsis* [48]. The core OS sugars provide protection against antimicrobial compounds and attenuate the endotoxic properties of lipid A, similar to lipid A modifications seen in mammalian pathogens [38]. Therefore, *Xcc* may have the capacity to modify its structure of lipid A and thus reduce its activity as a MAMP in plants [48]. It is not known whether these (or other) modifications to lipid A occur when bacteria are within plants.

14.2.4 LPS and Systemic Effects in Plants

In addition to all the effects described above, which are induced locally, LPS can elicit systemic resistance responses in plants. Two such systemic resistance responses have been described: systemic acquired resistance (SAR) and ISR. SAR involves systemic activation of defence-related responses such as PR gene expression upon infection with a locally applied necrotising pathogen. SAR is accompanied by a systemic increase in SA, and SA is required for SAR signalling [49, 50]. In contrast, ISR is induced via the root system; it is associated with jasmonic acid and ethylene rather than SA as signals and no PR gene expression [51]. Plants exhibiting SAR show enhanced expression of defence-related genes in distant leaves in the absence of any pathogen attack on those leaves. This is not seen in plants exhibiting ISR, in which defence responses are only activated after pathogen challenge.

Early studies showed that LPS from the root colonising *P. fluorescens* induced ISR in carnation and radish, whereas mutant bacteria, lacking the O-antigen side chain could not induce ISR [34, 35]. Treatment of *Arabidopsis* with *P. aeruginosa* LPS, flagellin or bacteria triggering necrosis was shown to be associated with accumulation of SA, expression of the PR genes and expression of the SAR marker gene *Flavindependent monooxygenase 1* in treated as well as in distant leaves [19, 52]. These studies suggest that recognition of the MAMPs, LPS or Flg, rather than the necrotic lesion formation contributes to the bacterial induction of SAR in *Arabidopsis*.

The body of work outlined above demonstrates conclusively that LPS from diverse bacteria can act as a MAMP to either directly induce a range of plant defence responses or prime induction of those responses in both a local and systemic fashion.

14.3 LPS Perception in Plants

Although plant receptors for flagellin and EF-Tu have been described, the mechanisms by which plants perceive LPS is not understood. The mammalian innate immunity system perceives invading pathogens through Toll-like receptors (TLRs), an interleukin 1 receptor (IL-1R) [53], that resembles the Toll receptor found in *Drosophila* [54, 55]. TLRs, one class of PRRs, comprise a family of transmembrane receptors that have an extracellular leucine rich repeat (LRR) domain, by which pathogen components are recognised, and a cytoplasmic Toll/IL-1R (TIR) domain, through which the signal is transduced. The Toll-like receptor TLR4 is responsible for LPS perception in mammals (Fig. 14.1). Once the TLRs are activated by MAMP recognition, adaptor molecules are recruited to initiate

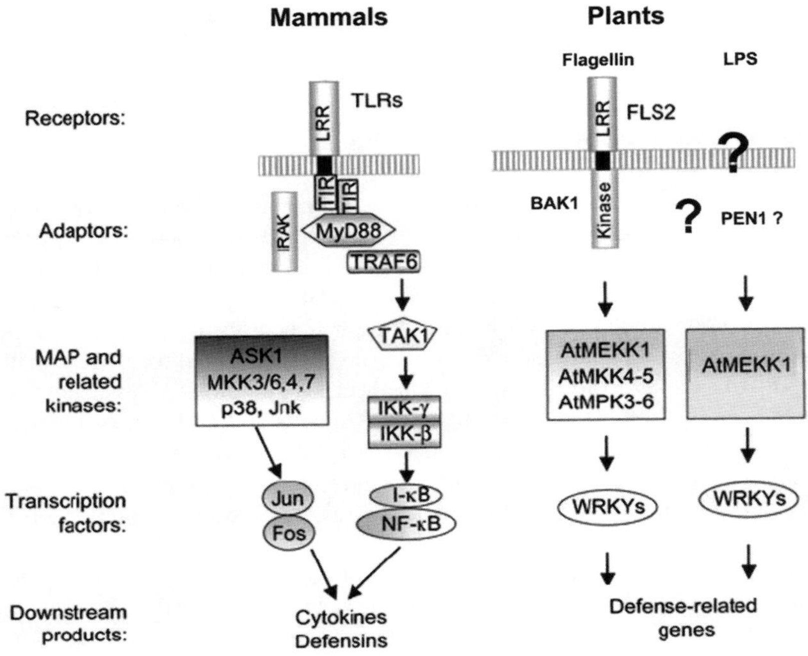

Fig. 14.1 Perception and signaling pathways downstream of PRRs in mammals and plants

downstream signalling, which involves activation of transcription factors and MAP kinases [56]. In addition to the surface localised TLR4, a second type of LPS receptor, the intracellular Nod proteins have been described in animal cells [57].

Intriguingly proteins with structural similarities to TLRs and Nods are found in plants [11]. The Toll-like receptor TLR5 is responsible for flagellin perception in mammals. The FLS2 flagellin receptor of *A. thaliana* also has extracytoplasmic LRRs, but a cytoplasmic serine/threonine kinase domain replaces the TIR domain found in mammalian TLR5. FLS2 belongs to a large family of plant receptor-like kinases containing extracytoplasmic LRRs, other members of which are responsible for perception of the MAMP EF-Tu, certain bacterial effector proteins, as well as plant signalling molecules and hormones. Some intracellular plant receptors for type III-secreted effectors contain TIR domains and additionally have the nucleotide-binding/apoptotic ATPase (NBS) domain and LRRs, which are also found in the mammalian Nod proteins [11]. On the basis of these structural similarities between plant and animal receptors for flagellin, it is tempting to speculate that perception of LPS by plants could involve surface-localised leucine-rich repeat receptor-like kinases and/or intracellular TIR-NBS-LRR proteins. Unfortunately this does not substantially narrow the search for an LPS receptor, since many such putative receptor proteins are encoded by plant genomes. In *A. thaliana*, there are at least 135 proteins with a TIR domain, 82 of which have the TIR-NBS-LRR domain organisation. In addition, *A. thaliana* encodes 600 receptor-like kinases, many of which like FLS2 have extracellular LRRs.

Despite apparent similarities between innate defence systems in plants, mammals and insects, some differences do occur [58] (Fig. 14.1). Most plant defence responses thus far described require LPS application at the 5–50 μg mL^{-1} level, whereas TLR4-mediated perception of LPS is extremely sensitive and is activated by the ligand at concentrations in the pg to ng mL^{-1} range [59]. These considerations have led to suggestions that plants possess only low affinity systems to detect LPS [12]. It is also plausible that high affinity recognition-response systems in plants do not act to directly trigger plant defences, but prime the plant so that in response to further pathogen-derived signals, such responses are mounted more rapidly or to a greater extent [11]. This suggestion is open to experimental testing with structurally characterised LPS and pathogen-derived elicitors such as Flg22. Available evidence suggests that plants recognise similar structures in lipid A as do mammals but also raise the possibility of different plant receptors for the core OS and the lipid A moieties.

14.3.1 LPS Signal Transduction

In mammals, LPS activates macrophages, a white blood cell type that plays a key role in immune responses. This recognition of LPS induces a rapid synthesis and secretion of proinflammatory cytokines such as tumour necrosis factor-α (TNF-α) starting the TIR signalling pathway [60]. A study of intracellular trafficking of TNF-α in LPS activated macrophages revealed that initially synthesised 26-kDa TNF-α, a type II membrane precursor, was cleaved by TNF-α converting enzyme to a soluble mature 17-kDa secreted form [61], with a transient appearance at the cell surface. The TNF-α precursor was found to accumulate in the Golgi-complex [62, 63] suggesting a role in the secretory pathway, but little is known about the molecules involved in regulating the vesicle budding from the Golgi-complex to and fusion with the target membrane. However, soluble *N*-ethylmaleimide-sensitive factor adaptor proteins (SNAPs) receptors (SNAREs) are required for docking and fusion of intracellular transport vesicles with acceptor/target membranes. The fusion of vesicles in the secretory pathway involves target-SNAREs (t-SNAREs) on the target membrane and vesicle-SNAREs (v-SNAREs) on vesicle membranes that recognise each other and assemble into trans-SNARE complexes [64].

SNARE proteins play a role in mediating effects of LPS on mammalian cells. Studies by Pagan et al. [65] showed that a subset of t-SNAREs syntaxin 4/SNAP23/Munc18c, known to control regulated exocytosis in other mammalian cell types than macrophages [66], was up-regulated in response to LPS whereas the level of those involved in endocytosis was decreased or unaffected. This t-SNARE complex is required for TNF-α delivery and function at the plasma membrane. Furthermore, an intracellular Q-SNARE complex located on similar Golgi derived vesicles as TNF-α was upregulated during TNF-α secretion; these SNAREs are involved in the post-Golgi secretory trafficking of TNF-α to the cell surface [67]. The regulation of SNAREs involved in vesicle docking and fusion in LPS stimulated macrophages, revealed a SNARE complex necessary for LPS induced exocytosis of TNF-α, indicating that individual SNAREs are regulated to perform specialised functions in the cell [67]. Furthermore, in LPS-induced macrophages surface delivery of TNF-α was found to be dependent on relocation of syntaxin 4 into cholesterol-dependent lipid rafts [68].

Arabidopsis possesses a relatively large number of SNAREs, and at least twice as many syntaxins compared to worms and flies [69]. This high number has been suggested to have relation to the rise of multicellularity in plants [70] and furthermore, be indicative of conceivably unique vesicle transport systems in plants. Specific SNARE proteins have roles in defence triggering in plants. Silencing of the *Nicotiana benthamiana* SNARE, NbSYP132, an ortholog of an *Arabidopsis* plasma membrane-resident syntaxin AtSYP132, revealed that NbSYP132 contributes to R-gene mediated resistance, basal resistance and SA-associated defence, and is involved in mediating secretion of PR1 into the extracellular space. In contrast to this, PR1 secretion and R-gene mediated responses were not affected by silencing NbSYP121, an ortholog of the *Arabidopsis* plasma membrane-resident PEN1 syntaxin (AtSYP121) [71]. Nevertheless, the syntaxin PEN1 (AtSYP121) is known to be a component of the vesicle-targeting machinery involved in non-host penetration resistance in *Arabidopsis* against the barley powdery mildew fungus *Blumeria graminis* f. sp. *hordei* [72]. A model for SNARE mediated vesicle trafficking in basal immunity has been proposed [73–75]. PEN1 and AtSNAP33, a synaptosomal-associated protein of 33 kDa (SNAP33), form a binary t-SNARE complex involved in marking the target membrane for vesicle trafficking. The cognate v-SNAREs are then trafficked to the targeted membrane and together with the t-SNARE complex form a ternary complex resulting in vesicle fusion and antimicrobial compounds secretion [74, 75].

14.3.2 Is PEN1 Required for LOS Signalling in *Arabidopsis*?

Recognition of LPS/LOS in mammals is rather complex; how complex this recognition is in plants is still not known. Our earlier studies have suggested that different LPS fragments (the core OS and the lipid A moiety) are recognised by different plant receptors [7]. However, the mechanism of this recognition and consequent transduction steps remain obscure. The work establishing a role for PEN1 in pathogen resistance in *Arabidopsis* prompted us to test the role of this syntaxin in induction of defence responses by LOS from the plant pathogen *Xcc*. The effect of infiltration of *Xcc* LOS on *PR1* gene expression, production of ROS and callose deposition in leaves of *A. thaliana* (cv. Columbia) wild type, *pen1-1* and *pen1-2* mutants were studied (authors' unpublished data). For comparative purposes, parallel experiments with a second MAMP, the flg22 peptide derived from flagellin, were performed. Flg22 had a marked effect on *PR1* gene transcription observed both in wild type and *pen1* mutant *Arabidopsis* plants. In contrast, *Xcc* LOS induced a 340-fold increase of *PR1* transcripts in wild-type *A. thaliana*, while only a low (2.5 fold) transient accumulation was seen in the *pen1* mutant. Flg22 was active in triggering ROS production, with *pen1* and wild-type showing a similar response. Although wild-type *A. thaliana* responded rapidly to *Xcc* LOS with an oxidative burst, a delayed and substantially reduced response was observed in the *pen1* mutant. Flg22 induced abundant callose deposition in both *pen1* and wild-type *A. thaliana*. In contrast, *Xcc* LOS induced a much lower formation of callose in the *pen1* mutant than in the wild type (authors' unpublished data). Together, these results suggest that PEN1 has a role in triggering of the immune responses in *Arabidopsis* in response to *Xcc* LOS but not in response to flg22.

A possible function for PEN1 is to provide the correct localisation at the plant cell plasma membrane of the putative receptor(s) for LOS. Alternatively, PEN1 may be required for endocytosis of an LOS complex, which may allow signalling to cytoplasmic proteins to trigger defence responses. Gross et al. [76] found that, in tobacco cells, *Xcc* LPS was internalised 2 h after its introduction to the cell suspension, where it co-localised with Ara6, a plant homolog of Rab5 which is known to regulate early endosomal functions in mammals. It was speculated that this endocytosis in tobacco cells was, in correlation with the mammalian system, part of a down regulation of defence responses [76]. In a recent study by Zeidler et al. [77] localisation and mobilisation of fluorescein-labeled *S.* minnesota LPS was studied in *Arabidopsis*. Leaves of *A. thaliana* were pressure infiltrated with 100 μg mL^{-1} of fluorescein-labeled *S.* minnesota LPS and the mobility of LPS was studied over time by fluorescence microscopy. After 1 h a fluorescent signal was observed in the intercellular space of the infiltrated leaf. The labeled LPS were visible in the midrib of the leaves after 4 h, whereas this fluorescence had spread to the smaller leaf veins near the midrib after 6 h. After 24 h it was detectable in the lateral veins. Moreover, cross-sections of the midrib 3 h after supplementation with fluorescein-labeled LPS revealed a fluorescent signal in the xylem. Using capillary zone electrophoresis they found a distribution of fluorescein-labeled *S.* minnesota LPS in treated as well as in systemic leaves [77]. In contrast to the results by Gross et al. [76], no intracellular accumulation of the labeled LPS was observed in *Arabidopsis*. The conflict in results could reflect the different LPS and plant systems used.

In mammalian macrophage cells, the LPS receptor complex is engulfed and appears on endosome-like structures. Furthermore, an inhibition of the endosomal pathway increased LPS-induced NF-κB activation [78]. Interestingly, it has also been shown that, upon stimulation with flg22, the cell membrane resident flagellin receptor FLS2 is transferred into intracellular mobile vesicles and targeted for degradation [79].

If *Arabidopsis*, PEN1 is a component of the endosomal complex responsible for endocytosis of the LPS/LOS-receptor complex, similar to endocytosis of the LPS-complex in mammals and the observed endocytosis of LPS in tobacco cells discussed above. An increased induction of innate immune responses would have presumably been observed in LOS-stimulated *Arabidopsis pen1* mutants compared to the wild-type. Although we favour a model in which PEN1 is involved in exocytosis required for *Xcc* LOS triggered immunity in *Arabidopsis*, we cannot discount an alternate role in endocytosis. Importantly, the findings indicate that PEN1 may have roles in plant disease resistance (e.g. those associated with LPS perception/signalling) that have not been appreciated thus far. The involvement of SNAREs in contributing to fusion specificity is still debated [73], and our understanding of the regulatory role of PEN1 in fusion of intracellular transport vesicles with target membranes is still limited. Only further experimental work will establish the exact role of PEN1 in secretory pathways acting in LOS triggered immunity.

14.4 Concluding Remarks

The effect of MAMPs such as LPS on the induction of basal plant defences raises the issue of how bacteria can ever cause disease in plants. Successful pathogens have evolved mechanisms to subvert or suppress MAMP-triggered immunity. Many type III secreted effectors act to block induction of basal defences, thus promoting disease [26–28]. Other bacterial products such as extracellular cyclic glucans and extracellular polysaccharides have also been shown to suppress defences [80–82]. Extracellular polysaccharides may exert their suppressive effect through sequestration of Ca^{2+} ions, thus preventing influx from the extracellular apoplastic pool [80, 82]. Ca^{2+} influx occurs as an early local response to pathogen attack and is thought to act as a signal and to activate callose synthetase. The mechanistic basis for suppression of defences by cyclic glucan is unknown.

Although plant receptors for the bacterial proteinaceous MAMPs flagellin and EF-Tu elongation factor have been identified, those involved in perception of LPS remain obscure. The cloning and characterisation of these genes remain a major goal. The development of a range of molecular genetic tools for model plants such as *A. thaliana* affords more opportunities for success. Thus far, LPS preparations used for the analysis of plant responses and for structural studies have been derived from bacteria grown in culture. We know almost nothing about the alterations in LPS that occur when bacteria are within plants, although this may be highly relevant to signalling. Changes could occur in both the size distribution of LPS (alteration in the ratio of LOS to LPS) and/or in decoration of LPS with saccharide, fatty acid, phosphate or other constituents. Increases in the sensitivity of mass spectrometric methodologies may allow development of micro-methods to analyse such changes in bacteria isolated from plants. Transcriptome or proteome profiling of bacteria isolated from plants may also give clues as to possible LPS modifications. In conclusion we expect that we in the next few years will see a substantial increase in our understanding of the processes of LPS perception and signal transduction in plants through the deployment of cross-disciplinary approaches and ever-expanding range of molecular experimental tools. A greater understanding of the mechanisms by which LPS elicits defence responses may have considerable impact on the improvement of plant health and disease resistance.

Acknowledgements Authors acknowledge funding by The Danish Council for Independent Research, Technology and Production Sciences (FTP).

References

1. van Baarlen P, van Belkum A, Thomma PHJ (2007) Disease induction by human microbial pathogens in plant-model systems: potential, problems and prospects. Drug Discov Today 12:167–173
2. Ausubel F (2005) Are innate immune signalling pathways in plants and animals conserved? Nat Immunol 6:973–979
3. Janeway CA (1992) The immune-system evolved to discriminate infectious nonself from non-infectious self. Immunol Today 13:11–16
4. Bedini E, De Castro C, Erbs G, Mangoni L, Dow JM, Newman MA, Parrilli M, Unverzagt C (2005) Structure-dependent modulation of a pathogen response in plants by synthetic O-antigen polysaccharides. J Am Chem Soc 127:2414–2416
5. Dow M, Newman MA, von Roepenack E (2000) The induction and modulation of plant defence responses by bacterial lipopolysaccharides. Annu Rev Phytopathol 38:241–261
6. Newman MA, Daniels MJ, Dow JM (1995) Lipopolysaccharide from *Xanthomonas campestris* induces defence-related gene expression in *Brassica campestris*. Mol Plant Microb Interact 8:778–780
7. Silipo A, Molinaro A, Sturiale L, Dow JM, Erbs G, Lanzetta R, Newman MA, Parrilli M (2005) The elicitation of plant innate immunity by lipooligosaccharide of *Xanthomonas campestris*. J Biol Chem 280:33660–33668
8. Erbs G, Silipo A, Aslam S, De Castro C, Liparoti V, Flagiello A, Pucci P, Lanzetta R, Parrilli M, Molinaro A, Newman MA, Cooper RM (2008) Peptidoglycan and muropeptides from pathogens *Agrobacterium* and *Xanthomanas* elicit innate immunity: structure and activity. Chem Biol 15:438–448
9. Gómez-Gómez L, Boller T (2000) FLS2: an LRR receptor-like kinase involved in the perception of the bacterial elicitor flagellin in *Arabidopsis*. Mol Cell 5:1003–1011
10. Zipfel C, Kunze G, Chinchilla D, Caniard A, Jones JDG, Boller T, Felix G (2006) Perception of the bacterial PAMP EF-Tu by the receptor EFR restricts *Agrobacterium*-mediated transformation. Cell 125:749–760
11. Newman MA, Dow JM, Molinaro A, Parrilli M (2007) Priming, induction and modulation of plant defence responses by bacterial lipopolysaccharides. J Endotoxin Res 13:69–84
12. Zeidler D, Zähringer U, Gerber I, Dubery I, Hertung T, Bors W, Hutzler P, Durner J (2004) Innate immunity in *Arabidopsis thaliana* lipopolysaccharides activate nitric oxide synthase NOS and induce defense genes. Proc Natl Acad Sci USA 101:15811–15816
13. Meyer A, Pühler A, Niehaus K (2001) The lipopolysaccharides of the phytopathogen *Xanthomonas campestris pv. campestris* induce an oxidative burst reaction in cell cultures of *Nicotiana tabacum*. Planta 213:214–222
14. Desaki Y, Miya A, Venkatesh B, Tsuyumu S, Yamane H, Kaku H, Minami E, Shibuya N (2006) Bacterial lipopolysaccharides induce defense responses associated with programmed cell death in rice cells. Plant Cell Physiol 47:1530–1540
15. Keshavarzi M, Soylu S, Brown I, Bonas U, Nicole M, Rossiter J, Mansfield J (2004) Basal defenses induced in pepper by lipopolysaccharides are suppressed by *Xanthomonas campestris pv. vesicatoria*. Mol Plant Microb Interact 17:805–815

16. Livaja M, Zeidler D, von Rad U, Durner J (2008) Transcriptional responses of *Arabidopsis thaliana* to the bacteria-derived PAMPs harpin and lipopolysaccharide. Immunobiology 213:161–171
17. Gerber IB, Laukens K, De Vijlder T, Witters E, Dubery IA (2008) Proteomic profiling of cellular targets of lipopolysaccharide-induced signaling in *Nicotiana tabacum* BY-2 cells. Biochim Biophys Acta 1784:1750–1762
18. Gerber IB, Laukens K, Witters E, Dubery IA (2006) Lipopolysaccharide-responsive phosphoproteins in *Nicotiana tabacum* cells. Plant Physiol Biochem 44:369–379
19. Mishina TE, Zeier J (2007) Pathogen-associated molecular pattern recognition rather than development of tissue necrosis contributes to bacterial induction of systemic acquired resistance in *Arabidopsis*. Plant J 50:500–513
20. Conrath U, Beckers GJ, Flors V, Garcia-Agustin P, Jakab G, Mauch F, Newman MA, Pieterse CM, Poinssot B, Pozo MJ, Pugin A, Schaffrath U, Ton J, Wendehenne D, Zimmerli L, Mauch-Mani B (2006) Priming: getting ready for battle. Mol Plant Microb Interact 19:1062–1071
21. Newman MA, von Roepenack EC, Daniels MJ, Dow JM (2000) Lipopolysaccharides and plant responses to phytopathogenic bacteria. Mol Plant Pathol 1:25–31
22. Newman MA, von Roepenack-Lahaye E, Parr A, Daniels MJ, Dow JM (2002) Prior exposure to lipopolysaccharide potentiates expression of plant defenses in response to bacteria. Plant J 29:485–497
23. Newman MA, von Roepenack-Lahaye E, Parr A, Daniels MJ, Dow JM (2001) Induction of hydroxycinnamoyl-tyramine conjugates in pepper by *Xanthomonas campestris*, a plant defense response activated by hrp gene-dependent and hrp gene-independent mechanisms. Mol Plant Microb Interact 14:785–792
24. Keller H, Hohlfeld H, Wray V, Hahlbrock K, Scheel D, Strack D (1996) Changes in the accumulation of soluble and cell wall-bound phenolics in elicitor-treated cell suspension cultures and fungus infected leaves of *Solanum tuberosum*. Phytochemistry 42:389–396
25. Alfano JR, Collmer A (2004) Type III secretion system effector proteins. Double agents in bacterial disease and plant defense. Annu Rev Phytopathol 42:385–414
26. He P, Shan L, Lin NC, Martin GB, Kemmerling B, Nürnberger T, Sheen J (2006) Specific bacterial suppressors of MAMP signalling upstream of MAPKKK in *Arabidopsis* innate immunity. Cell 125:563–575
27. Jamir Y, Guo M, Oh HS, Petnicki-Ocwieja T, Chen S, Tang X, Dickman MB, Collmer A, Alfano JR (2004) Identification of *Pseudomonas syringae* type III effectors that can suppress programmed cell death in plants and yeast. Plant J 37:554–565
28. Nomura K, DebRoy S, Lee YH, Pumplin N, Jones J, He SY (2006) A bacterial virulence protein suppresses host innate immunity to cause plant disease. Science 313:220–223
29. Jones JDG, Dangl JL (2006) The plant immune system. Nature 444:323–329
30. Holst H, Molinaro A (2009) Core region and lipid A components of lipopolysaccharides. In: Moran A, Brennan P, Holst O, von Itszstein M (eds) Microbial glycobiology: structures, relevance and applications. Elsevier, Oxford, pp 803–820
31. Raetz CRH, Whitfield C (2002) Lipopolysaccharide endotoxins. Annu Rev Biochem 71:635–700
32. Newman MA, Daniels MJ, Dow JM (1997) The activity of lipid A and core components of bacterial lipopolysaccharides in the prevention of the hypersensitive response in pepper. Mol Plant Microb Interact 10:926–928
33. Braun SG, Meyer A, Holst O, Pühler A, Niehaus K (2005) Characterization of the *Xanthomonas campestris* pv. *campestris* lipopolysaccharide substructures essential for elicitation of an oxidative burst in tobacco cells. Mol Plant Microb Interact 18:674–681
34. Leeman M, Vanpelt JA, Denouden FM, Heinsbroek M, Pahm B, Schippers B (1995) Induction of systemic resistance against fusarium-wilt of radish by lipopolysaccharides of *Pseudomonas fluorescens*. Phytopathology 85:1021–1027
35. van Loon LC, Bakker PA, Pieterse CM (1998) Systemic resistance induced by rhizosphere bacteria. Annu Rev Phytopathol 36:453–483

36. Bedini E, Parrilli M, Unverzagt C (2002) Oligomerization of a rhamnanic trisaccharide repeating unit of O-chain polysaccharides from phytopathogenic bacteria. Tetrahedron Lett 43:8879–8882
37. Loppnow H, Brade H, Durrbaum I, Dinarello CA, Kusumoto S, Rietschel ET, Flad HD (1989) IL-1 induction-capacity of defined lipopolysaccharide partial structures. J Immunol 142:3229–3238
38. Raetz CRH, Reynolds CM, Trent MS, Bishop RE (2007) Lipid A modification systems in gram-negative bacteria. Annu Rev Biochem 76:295–329
39. Schromm AB, Brandenburg K, Loppnow H, Moran AP, Koch MHJ, Rietschel ET, Seydel U (2000) Biological activities of lipopolysaccharides are determined by the shape of their lipid A portion. Eur J Biochem 267:2008–2013
40. Schromm AB, Brandenburg K, Loppnow H, Zähringer U, Rietschel ET, Carroll SF, Koch MHJ, Kusumoto S, Seydel U (1998) The charge of endotoxin molecules influences their conformation and IL-6-inducing capacity. J Immunol 161:5464–5471
41. Gutsmann T, Schromm AB, Brandenburg K (2007) The physicochemistry of endotoxins in relation to bioactivity. Int J Med Microbiol 297:341–352
42. Munford RS, Varley AW (2006) Shield as signal lipopolysaccharide and the evolution of immunity to gram-negative bacteria. PLoS Pathog 2:467–471
43. Rallabhandi P, Awomoyi A, Thomas KE, Phalipon A, Fujimoto Y, Fukase K, Kusumoto S, Qureshi N, Sztein MB, Vogel SN (2008) Differential activation of human TLR4 by *Escherichia coli* and *Shigella flexneri* 2a lipopolysaccharide. Combined effects of lipid A acylation state and TLR4 polymorphisms on signaling. J Immunol 180:1139–1147
44. Erbs G, Jensen TT, Silipo A, Grant W, Dow JM, Molinaro A, Parrilli M, Newman MA (2008) An antagonist of lipid A action in mammals has complex effects on lipid A induction of defence responses in the model plant *Arabidopsis thaliana*. Microb Infect 10:571–574
45. Silipo A, Sturiale L, Garozzo D, de Castro C, Lanzetta R, Parrilli M, Grant WD, Molinaro A (2004) Structure elucidation of the highly heterogeneous lipid A from the lipopolysaccharide of the gram-negative extremophile bacterium *Halomonas magadiensis* strain 21 M1. Eur J Org Chem 2263–2271
46. Ialenti A, Di Meglio P, Grassia G, Maffia P, Di Rosa M, Lanzetta R, Molinaro A, Silipo A, Grant W, Ianaro A (2006) Novel lipid A from *Halomonas magadiensis* inhibits enteric LPS-induced human monocyte activation. Eur J Immunol 36:354–360
47. Dow JM, Osbourn AE, Wilson TJG, Daniels MJ (1995) A locus determining pathogenicity of *Xanthomonas campestris* is involved in lipopolysaccharide biosynthesis. Mol Plant Microb Interact 8:768–777
48. Silipo A, Sturiale L, Garozzo D, Erbs G, Jensen TT, Lanzetta R, Dow JM, Parrilli M, Newman MA, Molinaro A (2008) The acylation and phosphorylation pattern of lipid A from *Xanthomonas campestris* strongly influence its ability to trigger the innate immune response in *Arabidopsis*. Chembiochem 9:896–904
49. Ryals JA, Neuenschwander UH, Willits MG, Molina A, Steiner HY, Hunt MD (1996) Systemic acquired resistance. Plant Cell 8:1809–1819
50. Schneider M, Schweizer P, Meuwly P, Métraux JP (1996) Systemic acquired resistance in plants. Int Rev Cytol 168:303–340
51. Pieterse CMJ, van Wees SCM, van Pelt JA, Knoester M, Laan R, Gerrits N, Weisbeek PJ, van Loon LC (1998) A novel signaling pathway controlling induced systemic resistance in *Arabidopsis*. Plant Cell 10:1571–1580
52. Mishina TE, Zeier J (2006) The *Arabidopsis* flavin-dependent monooxygenase FMO1 is an essential component of biologically induced systemic acquired resistance. Plant Physiol 141:1666–1675
53. Medzhitov R, Preston-Hurlburt P, Janeway CA (1997) A human homologue of the *Drosophila* Toll protein signals activation of adaptive immunity. Nature 388:394–397
54. Hashimoto C, Hudson KL, Anderson KV (1988) The Toll gene of *Drosophila*, required for dorsal-ventral embryonic polarity, appears to encode a transmembrane protein. Cell 52:269–279

55. Lemaitre B, Nicolas E, Michaut L, Reichhart JM, Hoffmann JA (1996) The dorsoventral regulatory gene cassette spätzle/Toll/cactus controls the potent antifungal response in *Drosophila* adults. Cell 86:973–983
56. Carpenter S, O'Neill LAJ (2007) How important are Toll-like receptors for antimicrobial responses? Cell Microbiol 9:1891–1901
57. Inohara N, Ogura Y, Chen FF, Muto A, Nuñez G (2001) Human Nod1 confers responsiveness to bacterial lipopolysaccharides. J Biol Chem 276:2551–2554
58. Nürnberger T, Brunner F (2002) Innate immunity in plants and animals emerging parallels between the recognition of general elicitors and pathogen-associated molecular patterns. Curr Opin Plant Biol 5:1–7
59. Miyake K (2004) Innate recognition of lipopolysaccharide by Toll-like receptor 4-MD-2. Trends Microbiol 12:186–192
60. Akira S, Takeda K (2004) Toll-like receptor signalling. Nat Immunol 4:499–511
61. Black RA, Rauch CT, Kozlosky CJ, Peschon JJ, Slack JL, Wolfson MF, Castner BJ, Stocking KL, Reddy P, Srinivasan S, Nelson N, Boiani N, Schooley KA, Gerhart M, Davis R, Fitzner JN, Johnson RS, Paxton RJ, March CJ, Cerretti DP (1997) A metalloproteinase disintegrin that releases tumour-necrosis factor-α from cells. Nature 385:729–733
62. Shurety W, Merino-Trigo A, Brown D, Hume DA, Stow JL (2000) Localization and post-Golgi trafficking of tumor necrosis factor-α in macrophages. J Interferon Cytokine Res 20:427–438
63. Shurety W, Pagan JK, Prins JB, Stow JL (2001) Endocytosis of uncleaved tumor necrosis factor-α in macrophages. Lab Invest 81:107–117
64. Söllner T, Whiteheart SW, Brunner M, Erdjument-Bromage H, Geromanos S, Tempst P, Rothman JE (1993) SNAP receptors implicated in vesicle targeting and fusion. Nature 362:318–324
65. Pagan JK, Wylie FG, Joseph S, Widberg C, Bryant NJ, James DE, Stow JL (2003) The t-SNARE syntaxin 4 is regulated during macrophage activation to function in membrane traffic and cytokine secretion. Curr Biol 13:156–160
66. Bryant NJ, Govers R, David E, James DE (2002) Regulated transport of the glucose transporter GLUT4. Nature 3:267–277
67. Murray RZ, Wylie FG, Khromykh T, Hume DA, Stow JL (2005) Syntaxin 6 and Vti1b form a novel SNARE complex which is up-regulated in activated macrophages to facilitate exocytosis of tumour necrosis factor-α. J Biol Chem 280:10478–10483
68. Kay JG, Murray RZ, Pagan JK, Stow JL (2006) Cytokine secretion via cholesterol-rich lipid raft-associated SNAREs at the phagocytic cup. J Biol Chem 281:11949–11954
69. Sanderfoot AA, Assaad FF, Raikhel NV (2000) The *Arabidopsis* genome. An abundance of soluble N-ethylmaleimide-sensitive factor adaptor protein receptors. Plant Physiol 124:1558–1569
70. Sanderfoot A (2007) Increases in the number of SNARE genes parallels the rise of multicellularity among the green plants. Plant Physiol 144:6–17
71. Kalde M, Nühse TS, Findlay K, Peck SC (2007) The syntaxin SYP132 contributes to plant resistance against bacteria and secretion of pathogenesis-related protein 1. Proc Natl Acad Sci USA 104:11850–11855
72. Collins NC, Thordal-Christensen H, Lipka V, Bau S, Kombrink E, Qiu JL, Hückelhoven R, Stein M, Freialdenhoven A, Somerville SC, Schulze-Lefert P (2003) SNARE-protein-mediated disease resistance at the plant cell wall. Nature 425:973–977
73. Kwon C, Bednarek P, Schulze-Lefert P (2008) Secretory pathways in plant immune responses. Plant Physiol 147:1575–1583
74. Kwon C, Neu C, Pajonk S, Yun HS, Lipka U, Humphry M, Bau S, Straus M, Kwaaitaal M, Rampelt H, El Kasmi F, Jürgens G, Parker J, Panstruga R, Lipka V, Schulze-Lefert P (2008) Co-option of a default secretory pathway for plant immune responses. Nature 451:835–840
75. Robatzek S, Bittel P, Chinchilla D, Kîchner P, Felix G, Shiu SH, Boller T (2007) Molecular identification and characterization of the tomato flagellin LeFLS2, an orthologue of

Arabidopsis FLS2 exhibiting characteristically different perception specificities. Plant Mol Biol 64:539–547

76. Gross A, Kapp D, Nielsen T, Niehaus K (2005) Endocytosis of *Xathomonas campestris* pathovar *campestris* lipopolysaccharides in non-host plant cells of *N. benthamiana*. New Phytol 165:215–226
77. Zeidler D, Dubery IA, Schmitt-Kopplin P, Von Rad U, Durner J (2010) Lipopolysaccharide mobility in leaf tissue of *Arabidopsis thaliana*. Mol Plant Pathol 11:747–755
78. Husebye H, Halaas Ø, Stenmark H, Tunheim G, Sandanger Ø, Bogen B, Brech A, Latz E, Espevik T (2006) Endocytic pathways regulate Toll-like receptor 4 signaling and link innate and adaptive immunity. EMBO J 25:683–692
79. Robatzek S, Chinchilla D, Boller T (2006) Ligand-induced endocytosis of the pattern recognition receptor FLS2 in *Arabidopsis*. Genes Dev 20:537–542
80. Aslam SN, Newman MA, Erbs G, Morrissey KL, Chinchilla D, Boller T, Jensen TT, De Castro C, Ierano T, Molinaro A, Jackson RW, Knight MR, Cooper RM (2008) Bacterial polysaccharides suppress induced innate immunity by calcium chelation. Curr Biol 18:1078–1083
81. Rigano LA, Payette C, Brouillard G, Marano MR, Abramowicz L, Torres PS, Yun M, Castagnaro AP, El Oirdi M, Dufour V, Malamud F, Dow JM, Bouarab K, Vojnov A (2007) A bacterial cyclic β-1,2 glucan acts in systemic suppression of plant immune responses. Plant Cell 19:2077–2089
82. Yun MH, Torres PS, El Oirdi M, Rigano LA, Gonzalez-Lamothe R, Marano MR, Castagnaro AP, Dankert MA, Bouarab K, Vojnov AA (2006) Xanthan induces plant susceptibility by suppressing callose deposition. Plant Physiol 141:178–187

Index

A
ABC transport, 292
ABC transporter
  Wzk, 284
  Wzm, 292, 293, 352–355, 361
  Wzt, 292, 352–355, 361
Abequose, 43, 46, 50, 343, 344, 346
Acidithiobacillus ferrooxidans, 166, 243
Acinetobacter, 9, 26, 35, 71–73
  A. baumanni, 5, 31, 71–73, 244, 249, 329
  A. haemolyticus, 31, 71, 72, 214, 244, 249
  A. lwoffii, 31, 32, 241
Actinobacillus pleuropneumoniae, 69, 70, 245, 256
Acyl
  amide-linked, 4, 9, 10
  ester-linked, 4, 9, 14
Acyloxyacyl, 10, 11, 242
Aeromonadaceae, 66
Aeromonas, 5, 33, 66, 67, 153, 213, 214, 250
  A. hydrophila, 33, 66, 67, 250
  A. salmonicida, 5, 33, 66, 153, 213, 214, 250
Aggregatibacter actinomycetemcomitans, 69, 201, 202, 205, 295
Agrobacterium, 9, 15, 81, 241
Alanine, 283, 285
Alcaligenaceae, 85
Aminoalkyl glucosaminide 4-phosphates, 401–402
3-Amino-3-deoxy-D-fucose, 43
4-Amino-4-deoxy-L-arabinose, 2, 145, 171, 173, 340
4-Amino-4,6-dideoxy-D-glucose, 202
2-Aminoethanol, 45
Antibacterial, 217, 329–331
Aquifex, 6, 168, 174, 243
  A. aeolicus, 168, 243, 249
  A. pyrophilus, 5–7, 243
Arabidopsis thaliana, 204, 418, 419, 422, 424–428
Avirulence, 420
Azospirillum, 82

B
Blochmannia floridanus, 323
Bordetella, 9, 85, 173, 174, 178, 181, 241, 244
  B. bronchiseptica, 6, 7, 10, 85, 166, 179, 241, 245
  B. parapertussis, 5, 85, 216, 241
  B. pertussis, 5–7, 85, 166, 168, 174, 210, 215, 216, 241, 249–251, 253
Brucella, 81
Burkholderia, 9, 11, 12, 15, 33, 82–84, 142, 145–148
  B. caryophylli, 33, 83
  B. cepacia, 5, 12, 55, 56, 83, 201, 241, 250, 419
  B. mallei, 82, 83
  B. pseudomallei, 82, 83
Burkholderiaceae, 26, 33–34, 82–84

C
Callose, 419, 426–428
Campylobacter, 250, 340
  C. fetus, 86, 205
  C. jejuni, 6, 7, 9, 86, 166, 169, 174, 175, 184, 210, 212, 251, 253, 282, 286, 291, 294
Campylobacteraceae, 86
Caryophillose, 44
Caryose, 44, 83
Cationic antimicrobial peptide, 10, 145, 170, 198, 329, 375, 396
CD14, 372, 376, 380–382, 390–393, 395–399, 402, 407, 409, 410
  polymorphism, 380–382
Chlamydia, 5, 31, 141–145, 166–168, 241, 377
  C. psitacci, 168

Y.A. Knirel and M.A. Valvano (eds.), *Bacterial Lipopolysaccharides*,
DOI 10.1007/978-3-7091-0733-1, © Springer-Verlag/Wien 2011

Chlamydia (cont.)
C. trachomatis, 5, 141, 166–169
Chlamydiaceae, 141, 241
Chlamydophila pneumoniae, 141, 241, 249
Choline, 45, 59
Chronobacter, 58
Citrobacter, 23, 49–51, 53, 78, 342
Colitose, 43, 53, 66, 74, 206, 347
Core
biosynthesis, 23, 168, 196, 209, 237, 248, 250, 251, 253–254
chemical synthesis, 131–157
inner, 2, 31, 36, 131, 146, 149, 150, 217, 218, 238, 240, 241, 244, 245, 247–251, 253–256, 317, 350–352, 355, 421
outer, 2, 23, 153–157, 197, 199, 209–211, 238–240, 245, 251–256, 350, 351, 354, 355
regulation, 254–257, 341
structure, 23, 25–35, 131, 141–145, 152, 153, 217, 218, 238, 240, 241, 248, 250, 252, 256, 257
types, 23, 238–241, 251, 252, 256, 257, 351

D
4-Deoxy-D-*arabino*-hexose, 43
6-Deoxy-D-*manno*-heptose, 43, 53
3-Deoxy-D-*manno*-oct-2-ulosonic acid
chemical synthesis, 138
chemoenzymatic synthesis, 139
6-Deoxy-L-talose, 201
D-Glyceric acid, 45
2,3-Diamino-2,3-dideoxy-L-rhamnose, 43
5,7-Diamino-3,5,7,9-tetradeoxynon-2-ulosonic acid, 70
3,6-Dideoxyhexose, 46, 61, 62, 196, 219
Dolichyl, 276, 277, 296

E
Edwardsiella, 50–52
Endotoxin, 2, 21, 117, 118, 121, 125, 128, 375, 381, 390, 391, 393, 403, 407, 422
Enterobacter, 55–59
Enterobacteriaceae, 21, 26–27, 45–66, 205, 213, 217, 323, 342, 343, 351
Enterobacterial common antigen, 21, 45, 213, 277
Envelope, 1, 36, 163, 164, 176, 311–315, 319, 320, 323, 326, 329, 377, 418
Erwinia, 65
Erwiniose, 44, 66
Escherichia, 3, 23, 41, 52–56, 118, 153–154, 163, 197, 238, 279, 311, 339, 372, 392, 419
E. alberti, 53, 353
E. coli, 3, 4, 9–11, 14, 23, 26, 36, 41, 49, 50, 52–56, 59, 62, 118–126, 128, 153–154, 163–165, 167–169, 171–178, 182, 184, 197, 279, 311, 372, 392, 419
E. fergusonii, 353
Ethanolamine, 45, 59, 164, 174, 238, 257

F
Fatty acid
distribution, 4, 9, 13, 14, 118
3-hydroxylated, 3, 9
long-chain, 9, 22
primary, 3, 14
secondary, 4, 9, 10, 14
Flavobacteria, 87–88
Flippase, 284, 285, 291, 294, 316, 321, 328
Formyl, 45, 49, 70
Francisella, 5, 7, 13, 78, 171, 173, 174, 181, 184, 249, 323, 376
F. novicida, 171, 172, 174, 249
F. tularensis, 6, 7, 9, 78, 174, 184, 376
Fusobacterium necrophorum, 88

G
Glucosylation, 49, 53, 56, 66, 357, 358
Glycerol, 45, 53, 56, 59, 164
Glycoforms, 26, 30, 31, 155, 220, 240
Glycosyltransferase
galactosyltransferase, 282
glucosyltransferase, 251, 253, 283
heptosyltransferase, 218, 248–250, 256
mannosyltransferase, 250

H
Haemophilus, 27, 29, 69, 148–153, 240, 244
H. ducreyi, 251, 253
H. influenzae, 11, 27–29, 69, 148, 150, 168, 208, 282, 381
Hafnia alvei, 5, 9, 50, 52, 56, 57
Halomonas, 79
Helicobacter, 87, 181, 238, 244
Helicobacteraceae, 86–87
Helicobacter pylori, 9, 86, 87, 123, 124, 140, 169, 171, 174, 175, 181–184, 205, 206, 208, 212, 238, 243, 249, 250, 293–295, 375, 376

Heme, 407
Heptose
  D-*glycero*-D-*manno*, 26, 43, 62, 218, 245, 250
  L-*glycero*-D-*manno*, 26, 43, 131, 148–153, 196, 218, 250
Heteropolysaccharide, 50, 69
Hexuronic acid, 42, 43, 46, 53, 56, 59, 66, 70, 71, 75, 81, 85
  amide, 42, 46, 66
Histophilus somnus, 29
Homopolysaccharide, 46, 50, 55, 85, 202
3-Hydroxybutanoyl, 45, 49, 50, 56, 58, 62, 70–73, 75, 82, 85, 87

I
Idiomarinaceae, 66–68
Immune system, 3, 4, 12, 164, 170, 171, 178, 179, 182, 202, 257, 417, 422
Infection, 2, 3, 7, 10, 12, 46, 50, 52, 54, 55, 58, 59, 62, 69–71, 77, 85, 86, 88, 126, 164, 170, 172, 179, 180, 183, 256, 275, 314, 326, 329, 358, 360, 376, 377, 380–382, 391, 398, 400, 407, 423
Innate immunity, 170–171, 389, 391, 398, 401, 417, 418, 420, 423

K
Kdo
  biosynthesis, 217, 323
  CMP-Kdo, 141, 217–219, 242, 244, 248
  transferases, 141, 248–249
Klebsiella pneumonia, 25, 50, 52, 54–56, 59, 69, 199, 202, 203, 248, 249, 253, 256, 257, 285, 292, 293, 329, 353, 361, 381
Ko, 23, 26, 31, 33, 35, 131, 142, 146, 147, 238, 241–244

L
Lactic acid ether, 45
Legionaminic acid, 44, 49, 59, 61, 67, 75, 78
Legionella, 6, 71, 178, 179, 181, 241, 249
Legionella pneumophila, 78, 179
Leptospira interogans, 376
Leucine-rich repeat, 373, 397, 419, 424
L-Fucose, 43, 66, 79, 204, 205
Limulus, 396
Lipid A
  analogue, 117–128, 401, 404, 405, 409
  bioactivity, 3, 7, 14, 372
  biosynthesis
    FtsH, 170
    LpxA, 165–168, 340
    LpxC, 167, 170, 351
    LpxH, 165, 167, 168, 340
    LpxL, 168–170, 240, 320, 328
    LpxM, 168, 169, 242, 329
    LpxP, 169, 329
    WaaA, 168, 169, 242, 248, 249, 329, 340
  chemical synthesis, 117–128
  conformation, 14, 15, 126, 128, 373, 375, 377, 380, 392, 393, 422
  mimetics, 401–403
  modification
    ArnT, 174, 182
    EptA, 175, 176, 182
    EptB, 182
    LmtA, 175, 176
    LpxE, 171, 172, 176, 323
    LpxF, 171, 172, 174, 176
    LpxO, 11, 173, 181
    LpxR, 10, 173, 176, 178, 180, 181, 184
    LpxT, 173, 175, 176, 184
    PagP, 10, 176–182, 184
    PhoP, 10, 11, 172, 176, 180–182, 376
    PhoQ, 10, 11, 171, 172, 176, 180–182, 376
    PmrA, 171, 172, 174, 176, 182
    PmrB, 171, 172, 174, 176, 182
  monophosphryl, 4, 14, 173, 376, 380, 382, 399–401
  precursor, 180, 184, 249, 321
  structure, 1–16, 118, 170, 171, 174, 178, 179, 184, 253, 340, 372, 375–377, 405, 422
  transport
    Lpt, 319, 320, 322, 324
    MsbA, 164, 171, 172, 176, 291, 316, 319–322, 328
Lipid binding protein, 373
Lipid IV_A, 4, 10, 14, 15, 36, 118, 168, 242, 250, 319, 320, 328, 329, 373–375, 377, 392, 393, 395, 409
Lipooligosaccharide (LOS), 15, 21, 44, 148, 154–155, 399, 421
Lipopolysaccharide (LPS)
  evolution, 339–365

Lipopolysaccharide (LPS) (*cont.*)
export, 311–331
recognition, 391–393
rough, 2, 22, 23, 34, 314, 317, 392, 397
smooth, 2, 21, 238, 392, 397
Lipoprotein, 1, 312–318, 324, 325, 391
transport, 316, 317, 325
Listonella anguillarum, 74
L-Lysine, 42, 45, 59, 67
L-Quinovose, 43
L-Rhamnose, 43, 75, 79, 82, 83, 87, 197, 199, 200, 240

M
Mannheimia haemolytica, 69
Membarane
asymmetry, 178, 314, 315
complex, 289, 290, 327
cytoplasmic, 1, 248, 275, 287, 291, 312, 329
diffusion barrier, 313–315
integrity, 278, 315
outer, 1, 21, 41, 164, 170, 171, 176–178, 180, 181, 195, 237, 238, 243, 247, 248, 256, 275, 277, 289, 290, 298, 311–331, 391, 392, 418
Mimicry, 35, 86, 87
molecular, 86, 87
Morganella morganii, 61
Mutant, 10, 11, 15, 21–23, 30, 36, 80, 82, 118, 124, 140, 167, 171, 172, 174, 178–182, 197, 204, 210, 213, 215, 216, 247, 248, 251, 281, 283, 287, 290, 294, 314, 315, 317–325, 328, 329, 331, 346, 373, 380, 418, 421–423, 426–428
MyD88, 378–380, 397, 402, 405, 407, 408
Myristoyl, 3, 4, 22, 168, 170, 406

N
N-Acetylfucosamine, 279
N-Acetylgalactosamine, 279
N-Acetylglucosamine, 196, 238, 295
N-Acetylneuraminic acid, 29, 155
N-Acetylquinovosamine, 279
Neisseria, 7, 11, 34–35, 86, 148–155, 166, 175, 238, 282, 311, 324, 380, 381
N. gonorrhoeae, 154, 155
N. meningitidis, 11, 33, 149, 150, 151, 154, 155, 169, 174, 238, 242, 243, 248, 250, 252, 253, 254, 295, 313, 316, 321, 322, 324, 392
N-Glycosylation, 275, 278, 286, 290, 291, 294
Nucleotide, 195–220, 245, 248, 250–252, 277–279, 285, 291, 294, 319, 321–322, 372, 380
diphosphate, 195–220
monophosphate, 195, 219

O
O-Acetylation, 42, 46, 49, 53–56, 64, 66, 70, 81, 82, 156, 240, 357, 358
O-Antigen
assembly, 275–298
biosynthesis, 23, 210, 216, 275–298, 352
chain length regulation, 287–289
composition, 42, 50, 56
evolution, 277, 342, 355
gene clusters
evolution, 342–350
G+C content, 341, 356, 361, 362
origin, 343, 360, 362, 365
genes, 294, 343, 351, 362, 363
initiation, 277, 278, 283, 287, 343.
ligation, 277, 283, 289, 294–296, 319, 323
polymerization, 4, 285, 287–289, 294
structure, 71, 207, 341, 343, 357, 362
translocase, 285, 286, 290
O-Chain, 2, 36, 46, 50, 85, 238, 398
Oligosaccharide (OS), 2, 21, 42, 148, 163, 164, 168, 174, 195, 198, 237–257, 275, 316, 339, 391, 392, 420
Opioids, 408, 409
O-Unit, 42, 44–46, 53, 56, 58–62, 64, 66, 67, 69–72, 74–80, 82, 83, 85, 87–89, 215, 276, 278, 283–288, 290, 343, 346, 348, 349, 357, 363

P
Paclitaxel, 377, 405, 406
Palmitoyl, 10, 176
Pantoea, 56–59
Paratose, 42, 43, 46, 61, 343–346
Pasteurellaceae, 27–29, 69–70
Pasteurella haemolytica, 69, 241
Pathogen-associated molecular patterns (PAMPs), 371, 417
Pectinatus, 88, 145, 201
Pectobacterium, 65, 66
Peptidoglycan, 1, 59, 176, 199, 207, 277, 278, 290, 296, 312, 313, 316, 318, 398, 418

Phosphate
2-aminoethyl, 2, 24, 131, 149–152, 240
substitutions, 4, 247, 248
Phosphoethanolamine, 69, 88, 171, 183, 238, 240, 340, 422
Phospholipid, 1, 164, 170, 178, 312–317, 320, 321, 325
transport, 316, 317, 321
Photoaffinity, 331
Phytopathogenic, 83, 420, 421
Plant
defence response, 417–423
perception, 419, 420, 428
signal transduction, 428
Plesiomonas shigelloides, 62, 210, 214, 248
Polyisoprenol
polyisoprenyl-phosphate hexose–1-phosphate transferases, 281–283
polyisoprenyl-phosphate N-acetylaminosugar–1-phosphate transferases, 278–281
Polysaccharide
capsular, 21, 45, 86, 131, 198, 242, 244, 281, 313
O-specific, 2, 21, 33, 36, 41, 275, 284, 288, 290, 291
Porphyromonas, 5, 88, 376
Proteobacteria, 46–87
Proteus, 59–63, 145–147, 198, 207, 211, 214, 342, 352, 356
Providencia, 26, 27, 59–62, 145
Pseudaminic acid, 44, 59
Pseudoalteromonadaceae, 66
Pseudoalteromonas, 5, 9, 66, 68, 87, 206, 211
Pseudomonadaceae, 29, 70, 71
Pseudomonas, 9, 11, 71, 72, 174, 178, 180, 181, 240, 244, 355, 356, 419
P. aeruginosa, 13, 29, 30, 45, 70, 155–157, 166, 197, 279, 342, 377, 419
P. fluorescens, 201, 214, 421
P. syringae, 31, 201, 205
Pyruvic acid acetal, 45, 71

R
Rahnella aquatilis, 63
Ralstonia solanacearum, 33, 34, 83, 421
Raoultella, 54, 55
RfaH, 254, 255
Rhizobiaceae, 9, 79–81
Rhizobium, 7–9, 79, 123, 169, 171–174, 182, 198, 201, 205, 209, 238, 240, 241, 244, 248, 376
R. etli, 7, 8, 172, 209
R. leguminosarum, 9, 169, 171, 198, 248, 376
Rhodobacter, 9, 14, 126, 127, 377, 393
R. capsulatus, 393
R. sphaeroides, 9, 126, 127
Ribitol, 45, 49, 53, 59

S
Salicylic acid, 420
Salmonella, 4, 22, 44, 46–50, 145, 168, 198, 238, 287, 314, 339, 376, 393
Salmonella enterica, 9, 44, 314, 339, 376, 399
Serratia marcescens, 55, 146, 241
Shewanella, 5, 61, 67, 68, 74
Shewanellose, 44, 61
Shigella, 52–54, 62, 178, 202, 207, 211, 214, 218, 339, 341–343, 349, 350, 357–359, 364, 422
Short chain dehydrogenase/reductase (SDR), 198
Sialic acid, 69, 324
Signalling pathways, 315, 377–390, 425
Soluble *N*-ethylmaleimide sensitive factor adapto receptors (SNAREs), 425, 426, 428
Stenotrophomonas maltophilia, 77, 201
Sugar nucleotides
ADP-L-*glycero*-D-*manno*-heptose, 218–219, 250
CMP-3-deoxy-D-*manno*-oct-2-ulosonic acid, 217–218
dTDP-4-acetamido-4,6-dideoxy-D-glucose, 201, 202
dTDP-4-amino-4,6-dideoxy-D-glucose, 202
dTDP-D-fucose, 201–202
dTDP-L-pneumose, 201
dTDP-L-rhamnose, 197, 199–201
GDP-4-acetamido-4,6-dideoxy-D-mannose (GDP-D-Rha4NAc), 204, 206, 207
GDP-colitose, 204, 206, 219
GDP-D-mannose, 202–205
GDP-D-pneumose, 204–205
GDP-D-rhamnose, 204–205
GDP-L-fucose, 204, 205
UDP-2-acetamido-3-acetimidoylamino-2,3-dideoxy-D-mannuronic acid, 215, 216
UDP-2-acetamido-2-deoxy-D-galactose, 199, 213–215
UDP-2-acetamido-2-deoxy-D-galacturonic acid2, 213–215

Sugar nucleotides (*cont.*)
UDP-2-acetamido-2-deoxy-D-glucose, 199, 207, 208
UDP-2-acetamido-2-deoxy-D-mannose, 213
UDP-2-acetamido-2-deoxy-D-mannuronic acid, 213
UDP-2-acetamido-2,6-dideoxy-D-galactose, 209–211
UDP-2-acetamido-2,6-dideoxy-D-glucose, 209–211
UDP-2-acetamido-2,6-dideoxy-D-xylo-hexos-4-ulose, 209–211
UDP-2-acetamido-2,6-dideoxy-L-galactose, 211–213
UDP-2-acetamido-2,6-dideoxy-L-glucose, 211–213
UDP-2-acetimidoylamino-2,6-dideoxy-L-galactose, 211
UDP-D-galactose, 197, 198
UDP-D-galacturonic acid, 197, 198
UDP-D-glucose, 197
UDP-D-glucuronic acid, 197, 198
UDP-2,3-diacetamido-2,3-dideoxy-D-glucoronic acid, 215–216
UDP-2,3-diacetamido-2,3-dideoxy-D-mannuronic acid, 215–216
Synthase, 140, 167, 201, 207, 208, 217, 218, 242, 243, 284, 293, 296, 352, 400

T
Taylorella, 85
Teichoic acid, 49, 56, 62, 88, 277
Thalidomide, 407–409
Toll-like receptor
polymorphism, 372, 380–382
TLR4
agonist, 373, 375, 377, 382, 389–410, 422
antagonist, 15, 373, 377, 382, 389–410, 422
complex, 15, 164, 372–378, 380, 382, 390, 392–395, 397, 405, 409
TLR4-MD2, 15, 164, 170, 178, 373–377, 379, 380, 382, 389, 390, 392–395, 397, 402, 404, 406, 407, 409
Tyvelose, 43, 46, 56, 343, 344, 346, 347

U
Undecaprenyl
monophosphate, 173
pyrophosphate
recycling, 176, 184, 296–298
synthesis, 176, 184, 244, 296

V
Vibrio, 75, 153, 244
Vibrio cholera, 23, 32, 66, 72, 74, 168, 181, 198, 199, 206, 207, 211, 212, 245, 247, 248, 295, 321, 360, 361, 380
Vibrionaceae, 32–33, 72–76

W
WaaL, 254, 288, 294–297, 319, 320, 323, 350, 356
WbaP, 281–283, 342, 343
WbbF, 293
WecA, 215, 278–281, 287, 290, 292, 293, 342, 343
Wzx, 284–292, 294, 340, 345, 346, 352–357, 362, 365
Wzy, 283–292, 340, 345, 347, 349, 352–357, 362, 364, 365
Wzz, 283, 285, 288–290, 292, 354, 357

X
Xanthomonas campestris, 5, 76–78, 205, 250, 418
Xylulose, 42, 43, 53, 62

Y
Yersinia, 11, 26, 36, 53, 61–62, 64, 83, 145, 174, 178, 179, 182, 342, 352–356, 361, 377
Y. enterocolitica, 61, 62, 81, 179, 181, 205, 210, 211, 251, 255, 256, 354, 355
Y. pseudotuberculosis, 10, 61–63, 179, 206, 339, 341, 343, 347–349, 351, 352, 354, 356, 365
Yersiniose, 44, 62–64, 83, 84
Yokenella, 238
Yokenella regensburgei, 62, 238